Nanotechnology in the Life Sciences

Series Editor

Ram Prasad
Department of Botany
Mahatma Gandhi Central University
Motihari, Bihar, India

Nano and biotechnology are two of the 21st century's most promising technologies. Nanotechnology is demarcated as the design, development, and application of materials and devices whose least functional make up is on a nanometer scale (1 to 100 nm). Meanwhile, biotechnology deals with metabolic and other physiological developments of biological subjects including microorganisms. These microbial processes have opened up new opportunities to explore novel applications, for example, the biosynthesis of metal nanomaterials, with the implication that these two technologies (i.e., thus nanobiotechnology) can play a vital role in developing and executing many valuable tools in the study of life. Nanotechnology is very diverse, ranging from extensions of conventional device physics to completely new approaches based upon molecular self-assembly, from developing new materials with dimensions on the nanoscale, to investigating whether we can directly control matters on/in the atomic scale level. This idea entails its application to diverse fields of science such as plant biology, organic chemistry, agriculture, the food industry, and more.

Nanobiotechnology offers a wide range of uses in medicine, agriculture, and the environment. Many diseases that do not have cures today may be cured by nanotechnology in the future. Use of nanotechnology in medical therapeutics needs adequate evaluation of its risk and safety factors. Scientists who are against the use of nanotechnology also agree that advancement in nanotechnology should continue because this field promises great benefits, but testing should be carried out to ensure its safety in people. It is possible that nanomedicine in the future will play a crucial role in the treatment of human and plant diseases, and also in the enhancement of normal human physiology and plant systems, respectively. If everything proceeds as expected, nanobiotechnology will, one day, become an inevitable part of our everyday life and will help save many lives.

More information about this series at http://www.springer.com/series/15921

Inamuddin • Abdullah M. Asiri
Editors

Sustainable Green Chemical Processes and their Allied Applications

Springer

Editors
Inamuddin
Chemistry Department
King Abdulaziz University
Jeddah, Saudi Arabia

Department of Applied Chemistry
Aligarh Muslim University
Aligarh, India

Abdullah M. Asiri
Chemistry Department
King Abdulaziz University
Jeddah, Saudi Arabia

ISSN 2523-8027 ISSN 2523-8035 (electronic)
Nanotechnology in the Life Sciences
ISBN 978-3-030-42286-8 ISBN 978-3-030-42284-4 (eBook)
https://doi.org/10.1007/978-3-030-42284-4

This Springer imprint is published by the registered company Springer Nature Switzerland AG
The registered company address is: Gewerbestrasse 11, 6330 Cham, Switzerland

Preface

Green chemistry and sustainability principles are the basis to develop alternative strategies for a rational use of renewable resources within a closed-looped system and with limited generation of waste, aiming to overcome the current challenges related with industrial progress, impact on environment and economic growth. These strategies are well-aligned with the circular economy concept that includes the reduced waste; atom economy; low toxicity; safe water-based processes; use of mild conditions leading to energy efficiency; renewable feedstock; reduced catalysis; reuse, recycle, real-time analysis; and inherently safer processes. In addition to being cost-effective and environmentally friendly, usually these strategies are considered sustainable as they are based on the use of natural resources at rates that do not excessively deplete supplies in the long term, and residues are generated at lower rates than the ones at which they can be readily assimilated by the environment. It has been discovered that the production of cost-effective renewable energy, especially from sustainable material, is the only dependable solution that could resolve climate change, transportation sector, and energy security. Consequently, the industry needs to measure impacts at all levels to be able to manage and adapt sustainable approaches which should be evolutionary in nature and not just be concerned by the yield and financial aspects. Therefore, awareness and knowledge about sustainable chemical processes and their allied applications with conceptual understanding are essential for advanced research community.

The book *Sustainable Green Chemical Process and their Allied Applications* provides an overview of interdisciplinary green chemical processes and their renewable applications. Topics include green chemical processes such as wastewater treatments, textile processing, dairy by-products, pharmaceutical contamination removal, residues, new adsorbents, polyelectrolytes, sonochemical processes, and livestock manure. The book also highlights sustainable technologies that include advanced generation of biofuels, biodiesel, advanced oxidation technologies, hydrogen production, industrial-scale water purification, energy conversion, and storage technologies. The book bridges together the experts in interdisciplinary science of green chemistry and renewable energy technologies. These are faculty, researchers, postgraduates, and industry professionals involved with green chemistry, environmental science, renewable energy, and sustainable technologies. Based on thematic topics, the book edition contains the following 21 chapters:

Chapter 1 details various experimental and theoretical progresses of sonochemistry for the production of hydrogen. Factors influencing the sonochemical production of hydrogen are discussed in detail. The progress reported herein allows researchers to determine some interesting tendencies and perspectives for this research, but also highlights some needs for innovation.

Chapter 2 discusses the recent trends on the production of biodiesel as a safe, ecofriendly, economical, and sustainable biofuel. Moreover, the application of biotechnology such as genetic engineering, strain improvement, recent advancement in fermentation, optimization, and production are discussed in detail. Some recent strategies used in structural biology that could lead to an increase in the yield of biodiesel are discussed in detail.

Chapter 3 details the basic composition and operation of Li-O_2 and Zn-air batteries. The role of metal and oxide electrocatalysts are discussed in detail. The focus is to communicate trends in cathode electrode composition and architecture, as well as the performance that is achieved in both systems.

Chapter 4 summarizes the most important green chemistry and green engineering metrics. Additionally, a simplification of concepts and principles and the standardization of the green metrics are presented. The focus is given to simplify their use to elaborate a diagnosis, quantify the greenness, and guide the decision-making.

Chapter 5 reviews available and emergent biotechnological environmentally friendly approaches to generate valuable products from residues under the scope of circular economy. Recent developments in omics, gene editing, systems and synthetic biology, bioreactors, and downstream processes are discussed. The potential of many existing residues from different sources is addressed.

Chapter 6 articulates the structure and classification, structure–property relationship, and applications of gels. In particular, the proper tuning of components in gel facilitates the formation of the desired supramolecular structure, which enables them to find applications in the field of lubricants, cosmetics, food industries, pharmaceuticals, and medicine.

Chapter 7 focuses on the hydrothermal carbonization of livestock manure. The effects of process parameters, including reaction temperature, catalyst, reaction time, heating mode, solid–liquid ratio, and feedstock, are analyzed. Next, the transport and conversion behaviors of heavy metals are explored. Lastly, the application and treatment of process water are discussed.

Chapter 8 discusses the removal of pharmaceutical compounds with adsorption processes. Accordingly, the influence of main parameters on the pharmaceuticals removal as well as the chief kinetic and isotherm models are investigated. The recently utilized adsorbents for pharmaceuticals elimination are also discussed.

Chapter 9 emphasizes dairy by-products or waste matter issues in the environment generated from different dairy industries and also discusses the treatment approach needed for value-added products (including bioenergy) synthesis via degradation of milk waste organic compounds as well as reduction in chemical oxygen demand or biological oxygen demand in various aquatic systems.

Chapter 10 discusses the chemistry of polyelectrolytes including their properties, viz. geometrical and chemical properties, charge density, and crystallinity. Different kinds of polyelectrolytes are discussed. The role of polyelectrolytes in the removal of suspension from water and wastewater is discussed in detail. The flocculation mechanism of polyelectrolytes as kinetic aspects of flocculation is also explained in detail.

Chapter 11 discusses various treatment mechanisms for diary waste management. Several treatment methods such as physico-chemical, mechanical, green technologies, and various advanced techniques are discussed in detail. Furthermore, the major focus is given to communicate the advantages, drawbacks, and future feasibility of emerging dairy-based biorefinery route reported in the literature.

Chapter 12 presents the use of purple okra waste as natural coagulant to intensify the textile wastewater treatment. This coagulant removed both organic matter and suspended and dissolved solids.

Chapter 13 delivers practical implementation guidelines addressing computational fluid dynamics (CFD) analysis of gasification processes using the ANSYS Fluent framework. The CFD workflow is clarified and broken down into manageable pieces aiding the reader throughout the complete implementation process while clarifying many aspects of performing a suitable waste-to-energy analysis in CFD.

Chapter 14 focuses on the use of lignocellulosic adsorbent materials for the purification of industrial water, as they are able to bind both organic (dyes, pesticides, petroleum and derivatives, pharmaceuticals, and cosmetic products) and inorganic (heavy metal oxides and salts, nitrogen and phosphorus anions) pollutants. The report critically assesses their advantages and factors influencing the adsorption mechanisms and presents some future directions of development.

Chapter 15 discusses the experimental and modeling approaches of sonochemistry in a green process perspective. Firstly, the attributes "sonochemical," "green," and "benign by design" are deeply questioned in link with ultrasound-assisted processes. Secondly, sonochemistry modeling steps are incrementally examined toward upscaling. Finally, several sonochemical designs are reviewed for features and limitations.

Chapter 16 discusses the catalytic utilization of waste red mud for the removal of environmental pollutants. Various preparation and characterization techniques of red mud-based catalysts used so far for the oxidation and advanced oxidation process are discussed. The process mechanism followed for different catalysts are also mentioned in this study.

Chapter 17 discusses recent discovery about the bioremediation and decolorization prospective of biosurfactant especially for ecorestoration of heavily polluted with textile waste, industrial effluents. Furthermore, this chapter highlights recent biotechnological techniques that could be applied for the identification, screening, and characterization of beneficial microorganisms that possess a very high capability to produce a high amount of biosurfactant. Structural elucidation techniques are also mentioned.

Chapter 18 deals with water treatments by Advanced Oxidation Processes (AOPs) based on chemical and photochemical reactions. The most used processes are described, and for each of them, the main reaction mechanisms, principles, advantages, drawbacks, performances and formation of by-products, coupled with technologies and their applications to waters and wastewater depollution have been analysed, supporting the main results of studies published in the pertinent literature.

Chapter 19 describes that finding innovative microalgal wastewater treatments that are cost-effective and efficient is a challenge. Recently developed microalgal cultivation methods maximize pollutant elimination and enhance biofilm growth. Understanding of the key factors that influence nutrient elimination and biomass generation promise significant improvements in the efficiency of wastewater treatments.

Chapter 20 describes details of conventional textile processing which include dyeing, finishing, bleaching, and printing. This chapter also summarizes detailed ideas of ecofriendly textile processing which include plant-based materials explored for textile processing, chemical-free irradiation treatment for different value addition of textile, and nano-based low chemical-loaded processing of textile materials.

Chapter 21 discusses that high potential yields of lipids, rapid growth using industrial wastes, and their small production-plant footprints make it desirable to exploit microalgae for biofuels production. However, technological improvements in cultivation, harvesting, extraction, conversion, and blending are required to make large-scale microalgae to biofuel plants commercially competitive.

Jeddah, Saudi Arabia

Inamuddin
Abdullah M. Asiri

The original version of this book was revised. The correction is available at https://doi.org/10.1007/978-3-030-42284-4_22

Contents

Chapter 1
The Sonochemical Approach for Hydrogen Production

Slimane Merouani and Oualid Hamdaoui

Contents

S. Merouani
Faculty of Process Engineering, Department of Chemical Engineering, Salah Boubnider Constantine 3 University, Constantine, Algeria

O. Hamdaoui (✉)
Chemical Engineering Department, College of Engineering, King Saud University, Riyadh, Saudi Arabia

Inamuddin, A. M. Asiri (eds.), *Sustainable Green Chemical Processes and their Allied Applications*, Nanotechnology in the Life Sciences,
https://doi.org/10.1007/978-3-030-42284-4_1

1.1 Introduction

Hydrogen is the perfect fuel of the future, its thermal energy is very high, and products resulting from its combustion are very ecological, since they do not include CO_2 or other non-environmentally friendly substances (Haryanto et al. 2005). Currently, fossil fuels are the main source of hydrogen production through several methods such as steam reforming, gasification, and partial oxidation (Haryanto et al. 2005; Dincer 2012; Dincer and Acar 2015). Alternative ways, i.e., water electrolysis, biological photosynthesis, and photocatalysis, were developed as clean and renewable technologies for hydrogen production (Ni et al. 2007; Das and Veziroglu 2008; Chakik et al. 2017).

The high-energy phenomena provoked by power ultrasound (20–1000 kHz) in water, i.e., water sonolysis, has attracted attention to involve ultrasound as an alternative technique for the production of hydrogen (Islam et al. 2019). Several researches were conducted at variable experimental conditions and allowed to conclude that sonolysis generates hydrogen at a rate higher than that produced by photocatalysis by factor of 200 (Gentili et al. 2009).

Organic sonochemistry, sonocatalysis, and sonochemical preparation of nanomaterials are the most common applications of sonication in these last two decades, even though other uses in extraction, water treatment, biomass valorization, and polymer chemistry were also explored (Chatel 2019). All these sonochemical applications of ultrasound can simultaneously be accompanied by hydrogen production. However, it is surprisingly viewed that there exist only very limited studies focusing on hydrogen generation by ultrasound, despite the fact that this technique offers clean hydrogen with higher yield than photocatalysis (Gentili et al. 2009), which is recently classified as an alternative technique for hydrogen production. Sonolysis does not need to add any additives or catalyst, and it is simply to handling at ambient temperature and pressure.

The scope of this chapter is focused to efforts made in the area of hydrogen production by ultrasound, with one part on the literature available data and the other part on the main factors influencing the process and the principal characteristics of acoustic cavitation for hydrogen generation. Besides, some reported intensification techniques of the process have been reviewed and discussed in detail.

1.2 Sonochemistry and Sonoreactors

The sonochemical effect originates from the acoustic cavitation phenomenon, that is, the nucleation, growth, and violent collapse of billions of microbubbles (filled with water vapor and dissolved gases) in a liquid subjected to ultrasound (Fig. 1.1a). The quick collapse of these bubbles is nearly adiabatic, yielding temperature of several thousands of Kelvin and pressure of several hundreds of atmospheres therein (Fig. 1.1b). These conditions are making each cavity as a micro-reactor in which

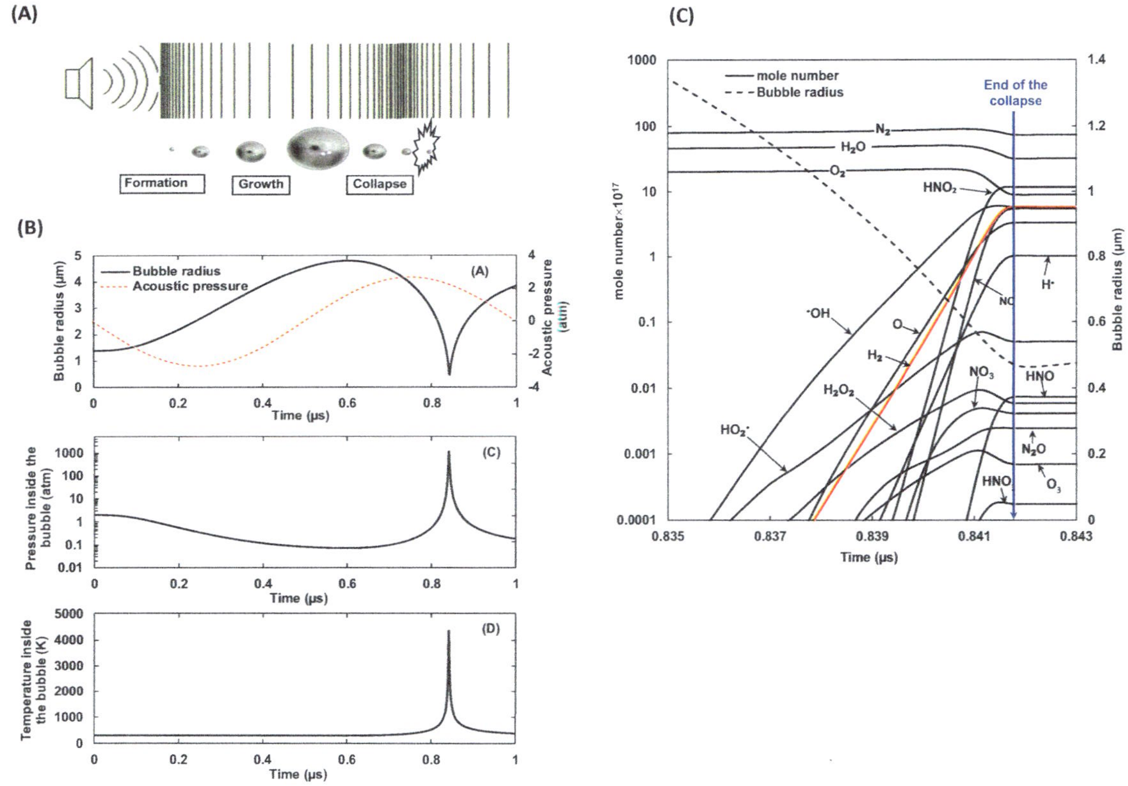

Fig. 1.1 Acoustic cavitation phenomenon (**a**), typical evolution of the bubble radius, and the bubble temperature and pressure (**b**) as well as reactions evolution inside the bubble at the end of the bubble collapse (**c**). The numerical calculations were made for an air bubble trapped in an ultrasound field of 1 MHz and 2.5 W/cm^2. Modified from Merouani and Hamdaoui (2017)

Table 1.1 Sample of the most existing studies on hydrogen production by ultrasound

Entry	Water matrix	Environmental conditions	Significant results/other remarks	Ref.
1	Deionized water (DI) and DI-formate mixtures	f = 800 kHz (bath type B) In = 1.6 ergs/cm^2 V = 4 mL, T = 20 ± 1 pH 0.6–14 Gas: Argon $[\text{formate}]_0$ = 0–0.05 M	– Sonolysis of pure water under the mentioned conditions yield H_2 with a rate of 25 μM/min – Hydrogen yield in neutral solutions of sodium formate is independent of the concentration of the solute (25 μM/min of yield) – Hydrogen yield was found to be diminished in formate (0.05 M) acid solution, but strong acceleration was observed in strong basic medium (28 μM/min) – Hydrogen yield under air atmosphere is 9.4 μM/min against 25 μM/min for argon atmosphere	Anbar and Pecht (1964)
2	Deionized water (DI) and DI-formate solutions	f = 300 kHz (bath type B) In = 3.5 W/cm^2 V = 55 mL, T = 20 °C pH 10, gas: Argon	– H_2 was formed at a rate of 26 μM/min in pure water under argon saturation, and the presence of formate has practically no effect on this accumulation rate	Hart and Henglein (1985)
3	Deionized water (DI)	f = 300 kHz (bath type B) In = 3.5 W/cm^2 V = 20 mL, T = 20 °C Gas: Argon, CO_2, Ar-CO_2, Ar-N_2O, and Ar-methane mixtures	– H_2 was produced at a rate of 46.6 μM/min under pure argon, but this rate decreased quickly as %CO_2 increased in the gas matrix until no H_2 was detected for %CO_2 > 7% – The yield of H_2, calculated after 15 min of sonication, decreased monotonically from 360 to 30 μM when N_2O was added to the argon matrix at only 4% – The H_2 yield strongly increases and decreases again after having reached a maximum at %CH_4 in argon equal to 8%	Henglein (1985)
4	Deionized water (DI) and DI-formate mixtures	f = 300 kHz (bath type B) In = 1.6 ergs/cm^2 V = 55 mL, T = NI pH 10 Gas: Argon $[\text{formate}]_0$ = 0–0.1 M	– Hydrogen was formed at a rate of ~25 μM/min, and no effect of formate concentration on the rate of hydrogen production was observed	Hart and Henglein (1985)

(continued)

Table 1.1 (continued)

Entry	Water matrix	Environmental conditions	Significant results/other remarks	Ref.
5	Deionized water (DI) and DI-acetate mixtures	f = 300 kHz (bath type B) P = 16 W, V = 55 mL T = NI, pH: NI Gas: Argon	– 500 μM of hydrogen was accumulated in the sonicated argon-saturated water after 30 min – H_2 concentration increased to 600 and 700 μM when acetate was added at 0.1 and 0.2 M, respectively. – Above 0.2 M of acetate, no further enhancement in the production of H_2 was observed	Guitièrrres et al. (1986)
6	Deionized water (DI)	f = 300 kHz (bath type B) P = 12 W, V = 37.5 mL T = NI, pH: NI, gas: Argon, nitrogen (N_2), and Ar-N_2 atmospheres	– H_2 was calculated at 14 μM/min for argon saturation – The yield of H_2 formation continuously decreased as the concentration of N_2 increases in the solution – The lowest formation rate of H_2 (~4 μM/min) was obtained under nitrogen atmosphere	Hart et al. (1986)
7	Deionized water (DI)	f = 300 kHz (bath type B) P = 12 W, V = 51.5 mL, T = NI, pH: NI, gas: Argon, deuterium (D_2), and Ar-D_2 atmospheres	– In argon-saturated water, the yield of H_2 was 10 μM/min – The yields of H_2 increase with increasing D_2 content of the gas mixture, reach the maxima (40 μM/min) at 35% D_2, and fall off at higher D_2 concentrations – In D_2-saturated water, the yield of H_2 was null	Fischer et al. (1986)
8	Deionized water (DI)	f = 300 kHz (bath type B) In = 2 W/cm^2 V = 50 mL, T = NI Gas: Argon and Ar-hydrocarbon mixtures	– The production rate of hydrogen increased monotonically with the addition of methane to argon; 10% (v/v) of CH_4 in argon provide the maximum production rate of 100 μM/min. Above 10% CH_4, the production rate falls rapidly to near zero for %CH_4 > 80% – Similar observation has been obtained for the case of ethane, but the maximum production rate of hydrogen at 10% C_2H_6 was 150 μM/min, and the limit above which no hydrogen was observed shifts to 40% C_2H_6 in argon rather than 80% for the case of CH_4-Ar mixture	Hart et al. (1990b)

(continued)

Table 1.1 (continued)

Entry	Water matrix	Environmental conditions	Significant results/other remarks	Ref.
9	Deionized water (DI)	f = 1 MHz (bath type B) In = 2 W/cm^2 V = 37.5 mL, T = NI Gas: Argon and Ar-acetylene atmospheres	– During the sonication of water containing 0.16 mM acetylene, the production of H_2 increased monotonically up to 2000 μM after 30 min of irradiation – The production rate of H_2 increased gradually with increasing acetylene concentration up to 0.6 mM where H_2 was produced at a rate of about 120 μM/min – For higher concentration of acetylene (>0.6 mM), the yield of H_2 decreased until no H_2 formation occurred at 10 mM acetylene	Hart et al. (1990a)
10	Deionized water (DI) and DI-methanol mixtures	f = 1 MHz (bath type B) In = 2 W/cm^2 V = 40 mL, T = NI Gas: Argon atmosphere	– In pure water, hydrogen is formed at a rate of 22 μM/min – The hydrogen yield strongly increases with increasing methanol concentration up to a maximum at 10 vol % – At higher methanolconcentrations, the yield decreases gradually until almost no chemistry occurs above about 80 vol %	Buettner et al. (1991)
11	Deionized water (DI) and N_3^- aqueous solutions	f = 1 MHz (bath type B) In = 2 W/cm^2 V = 25 mL, T = NI Gas: Argon atmosphere	– H_2 was formed at a rate of 4 μM/min by sonolysis of argon-saturated water – The yield of H_2 formation decreased with increasing N_3^- concentration up to 2 M, i.e., the yield at this point was at 1 μM/min	Gutierrez and Henglein (1991)
12	Methanol-water mixtures	f = 724 MHz (bath type B) In = 50 W, V = 40 mL T = −30 to 58 °C Gas: Argon atmosphere	– Changing the bulk temperature dramatically influenced the rate of the sonolytic formation of hydrogen – With increasing temperature, the rate of H_2 formation initially grows, reaches a maximum, and then falls – Increasing the concentration of methanol in the mixture decreases the temperature at which the maximum H_2 formation rate is reached	Rassokhin et al. (1995)

(continued)

Table 1.1 (continued)

Entry	Water matrix	Environmental conditions	Significant results/other remarks	Ref.
13	Deionized water (DI)	$f = 20$ kHz (probe) In = 0.3–4.7 W/cm^2 $V = 100$ mL $T = 22 \pm 2$ °C Gas: Argon atmosphere	– The production of H_2 increased linearly with time, i.e., the rate was 9.8 μM/min at 4.7 W/cm^2 – The production rate of H_2 increased linearly with increasing acoustic intensity in the range of 0.3–4.7 W/cm^2 – For all intensities, the production rate of H_2 is 1.25 of that of H_2O_2	Venault (1997)
14	Deionized water (DI)	$f = 200$ kHz, P = 200 W $V = 50$ mL, $T = 25$ °C pH: NI Gas: Argon atmosphere	– Hydrogen is one of the main products of the sonolytic decomposition of CO_2 inside the cavities	Harada (1998)
15	Deionized water (DI) and DI-tert-butyl alcohol (TBA) solution	$f = 321$ kHz (bath type B) $P_d = 170$ W/kg $V = 8$ mL T = NI, pH: NI Gas: Argon atmosphere	– Sonicating argon-saturated water containing trace of TBA increase the G-yield of H_2 with more than ten-fold, i.e. from 10×10^{-10} mol/J to ~150×10^{-10} mol/J – Increasing [TBA] > ~0.01 M rapidly reduced the yield of H_2	Tauber et al. (1999)
16	Deionized water (DI)	$f = 200$ kHz (bath type B) $P = 200$ W, $V = 40$ mL T = NI, pH: NI Gas: Argon atmosphere	– Hydrogen was produced at a flow rate of 3.16 mL/h	Harada (2001)
17	$NaHCO_3$ aqueous solution (0.41 M)	$f = 200$ kHz, $P = 200$ W $V = 50$ mL, $T = 25$ °C pH: NI Gas: Argon atmosphere	– Hydrogen accumulated at a rate of 0.16 μM/min	Harada and Sogawa (2004)
18	Deionized water (DI)	$f = 22.5$ kHz (bath type B with two transducers) $P = 50$ W for each transducer (total 100 W) $V = 100$ mL $T = 22 \pm 2$ °C Gas: NI	– Hydrogenproduction increased linearly with time with a rate of 0.045 μM/min – The production rate of hydrogen decreased with increasing external static pressure from 1 to 2.5 atm	Cotana et al. (2007)

(continued)

Table 1.1 (continued)

Entry	Water matrix	Environmental conditions	Significant results/other remarks	Ref.
19	Deionized water (DI) and DI-methanol in the presence of TiO_2 or Au/TiO_2 catalysts	f = 40 kHz, P = 50 W V = 150 mL T = 21–25 °C Gas: Argon atmosphere [catalyst] = 0.5 and 1 g/L	– Au/TiO_2 catalyst significantly increased the yields of hydrogen in the sonolysis of water – The production rate of H_2 increased by factors of 54 and 16 when sonolysis was assisted with Au/TiO_2 and TiO_2, respectively – The sono-generation yields were more pronounced in the presence of 4% of methanol	Wang et al. (2010)
20	Deionized water (DI) and DI-formic acid	f = 20 kHz (probe), 200 and 607 kHz (bath type B) P_{ac} = 13 W at 20 kHz, 45 W at 200 kHz and 65 W at 607 kHz V = 50 mL for 20 kHz and 250 mL for 200 and 607 kHz, T = 20 °C Gas: Argon atmosphere	– The chemical yield for hydrogen generation increased with increasing frequency, i.e., 0.12 µmol/kJ at 20 kHz, 0.28 µmol/kJ at 200 kHz, and 0.49 µmol/kJ at 607 kHz	Navarro et al. (2011)
21	Deionized water (DI) and DI-alcohol (methanol, ethanol, propyl alcohol)	f = 40 kHz (probe) P = 500 W V = 60 mL T = 10–30 °C Gas: Argon atmosphere	– Methanol, ethanol, and propyl alcohol showed a rate enhancement on hydrogen yield as compared to pure water – Best enhancement was observed at 5–10% (v/v) of each alcohol in water, and methanol provided the best yield	Xuemin et al. (2012)
22	Deionized water (DI) and DI-ethanol mixtures	f = 20 kHz (probe) P = 50 W V = 300 mL T = 25 °C Gas: Argon atmosphere	– The production rate of H_2 in pure water was 80 µM/h, that in pure ethanol was 5.5 µM/h, and that in ethanol/water mixture (20% v/v of ethanol) was 112 µM/min	Penconi et al. (2015)
23	Deionized water (DI)-t-butanol (t-BuOH) solutions	f = 20 kHz (probe) and 359 kHz (bath type B) P_{ac} = 25 W at 20 kHz and 43 W at 359 kHz V = 250 mL, T = ~20 °C Gas: Argon atmosphere	– Production rates of hydrogen were higher at higher frequency and *t*-butanol concentration, i.e., values of 14 and 19 µM/min were obtained at 359 kHz with 1 and 20 mM t-BuOH and 5 and 7 µM/min were registered at 20 kHz with 0.125 and 0.5 M t-BuOH.	Pflieger et al. (2015)

(continued)

Table 1.1 (continued)

Entry	Water matrix	Environmental conditions	Significant results/other remarks	Ref.
24	Deionized water (DI)-KI solution (0.1 M KI)	f = 2.4 MHz (bath type B) P = 15 W, V = 20 mL T = 25 °C Gas: Argon and Ar-CO_2 atmospheres	– The production of hydrogen after 180 min of irradiation decreased from 17 μmol under pure argon atmosphere to 5.5 μmol when the gas matrix contained only 2% CO_2	Harada and Ono (2015)

Abbreviations: *DI* deionized water, *NI* not indicated, *V* solution volume, *T* operating temperature, *f* frequency of ultrasound, *P* delivered electric power, P_{ac} acoustic power, P_d power density, I_n acoustic intensity

1.4 Mechanism of the Sonochemical Production of Hydrogen

The sono-production mechanism of hydrogen is until nowadays under debate. Some researches (Anbar and Pecht 1964; Venault 1997) postulated that H_2 is predominately produced in the gas phase of the cavities whereas other ones (Henglein 1985, 1987; Hart et al. 1986; Henglein et al. 1993) reported that H_2 is formed at the bubble-solution interface, i.e., the liquid shell, via the recombination of $H^•$ radicals diffused from the bubble interior ($H^• + H^• \rightleftharpoons H_2$). However, at present, there is not any experimental evidence confirming the exact mechanism for the acoustical production of hydrogen. A theoretical tentative has been made by Merouani et al. (2015b) for clarifying the mechanism by which hydrogen was produced during the ultrasonic irradiation of water. The authors have developed a model for single bubble sonochemistry combining the oscillation dynamic of single bubble with a complex kinetics model based on 25 reversible chemical reactions that my take place inside the bubble during the strong collapse in argon-saturated water. The material balance for hydrogen in both the liquid and gas phases gives the following overall production rate of hydrogen:

$$r_{H_2} = N \times n_{H_2} + k'\left[{}^{\bullet}H\right]^2 \tag{1.1}$$

where N (in L^{-1} s^{-1}) is the number of collapsing bubbles per unit time per unit volume, n_{H_2} (in mol) is the quantity of H_2 released by one bubble (from the gas phase), and k' is the rate constant of the reaction $H^• + H^• \rightleftharpoons H_2$. The first term of Eq. (1.1), i.e., $N \times n_{H_2} n_{H_2}$, is the global rate of hydrogen production from the gas-phase reaction, and the second part, i.e., k' $[H^•]^2$, is the production rate of H_2 from the reaction $H^• + H^• \rightleftharpoons H_2$ at the liquid shell. Steady-state conditions for $H^•$ gives:

$$k'\left[{}^{\bullet}\mathrm{H}\right]^2 = \frac{1}{2} N \times n_{H^{\bullet}} \tag{1.2}$$

where $n_{H\bullet}$ is the quantity of $H^\bullet$ produced by a single cavity. Equation (1.1) will then be:

$$r_{H_2} = N \times n_{H_2} + \frac{1}{2} N \times n_{H^\bullet} \tag{1.3}$$

The ratio (R) between H_2 production from the gas phase and the liquid phase is then given by:

$$R = \frac{r_g}{r_l} = 2 \times \frac{n_{H_2}}{n_{H^\bullet}} \tag{1.4}$$

In this equation, both n_{H_2} and $n_{H^\bullet}$ are predicted from the early single bubble sonochemistry model (Merouani et al. 2015b). The value of R was determined under different simulation conditions of frequency, saturation gases, and liquid temperature (Fig. 1.3). This figure shows that whatever the simulation conditions, R was superior than 4. For higher frequencies, much higher values of R were recorded. Additionally, R values were in general much higher when the saturation gas is nitrogen and air than argon and oxygen. Besides, higher liquid temperatures induced higher values of R. Consequently, on the one hand, these high values of R confirmed that, whatever the sonication conditions, the principal source of H_2 released during water sonolysis is the gas-phase chemistry whereas the contribution of the reaction $H^\bullet + H^\bullet \rightleftharpoons H_2$ at the liquid shell of the bubble is inappreciable.

In another hand, by analyzing the gas-phase reaction kinetics, it is observed that >99% of H_2 was due to the reaction $H^\bullet + {}^\bullet OH \rightleftharpoons H_2 + O$. This finding was practically unique whatever the simulation conditions. Therefore, the recombination of $H^\bullet$ and ${}^\bullet OH$ inside the bubble is the main pathway of H_2 production during water sonication.

1.5 Factors Influencing the Sonochemical Production of Hydrogen

Sonochemical activity depends on several parameters including the liquid properties (vapor pressure, viscosity, etc.), gas atmosphere, temperature of sonicated solutions, applied pressure, and ultrasonic intensity and frequency. All these parameters are necessary in optimizing the sonochemical production for hydrogen. However, in global, the effective influence of these factors is as yet poorly understood, due to the scarcity of data in this research area, i.e., hydrogen production by sonochemistry. However, based on some published data and numerical simulation results, the effect of these operational conditions can be discussed.

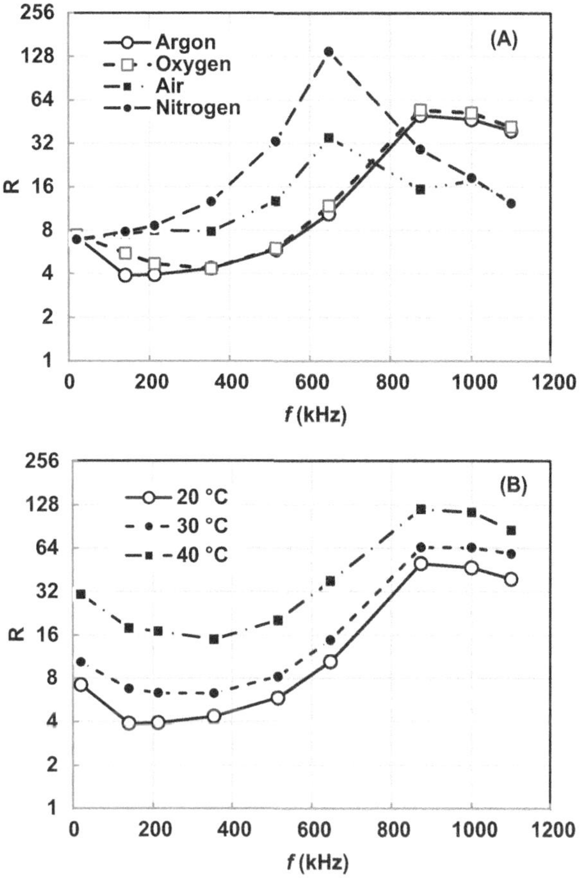

Fig. 1.3 Calculated values of R for various saturation gases (**a**) and liquid temperatures (**b**) under different frequencies

1.5.1 Frequency

Available data indicated that using high frequencies enhanced the production of hydrogen. Navarro et al. (2011) have reported that the chemical yield of hydrogen production during water sonolysis under argon atmosphere at 20 °C varies as 0.12 μmol/kJ at 20 kHz, 0.28 μmol/kJ at 200 kHz, and 0.49 μmol/kJ at 607 kHz. Similarly, Pflieger et al. (2015) have reported higher yields of hydrogen at 359 kHz than 20 kHz. Thus, it seems that the production rate of hydrogen increased with frequency rise and always higher frequencies (>100 kHz) favor much production rates than lower frequency, i.e., generally 20 kHz. In general, an optimum frequency in the range of 200–600 kHz has been reported for the sonolytic production of H_2O_2 (Pétrier and Francony 1997; Beckett and Hua 2001; Jiang et al. 2006; Torres et al. 2008). However, facing the scarcity of data about the sonolytic production of hydrogen, optimum frequency was not reported. Numerical simulations of single bubble yield showed that the bubble temperature, the bubble collapse duration, and the chemical bubble yield on H_2 decreased with frequency increase between 20 kHz and 1 MHz (Merouani et al. 2016a). But, simultaneously, the number of collapsing bubbles (*N*) showed the reverse trend, i.e., monotonic increase of *N* with frequency increase (Merouani et al. 2014a). Therefore, confronting the experimental results of Navarro et al. (2011) with the predicted effect of frequency on the single bubble yield and the number of bubbles, i.e., Merouani et al. (2014a, 2016a), it seems that the number of active cavities is the dominant factor governing the frequency impact on the sono-production of hydrogen.

1.5.2 Intensity

The 20 kHz sonolysis of argon-saturated water at 0.6 W cm^{-2} of acoustic intensity produced hydrogen at 0.8 μM/min (Venault 1997). When the acoustic intensity was increased to 1.1 and 2.5 W cm^{-2}, hydrogen was produced at 2.1 and 5 μM/min, respectively (Venault 1997). This observation is in great accordance with that reported for the sonolytic production rate of H_2O_2 (Merouani et al. 2010; Guzman-Duque et al. 2011). In general, the acoustic intensity should be operated between two limits, the Black threshold below which no sonochemical effect could be observed and a high intensity value above which the sound wave is attenuated (Henglein 1987, 1995; Henglein et al. 1993). Increasing acoustic intensity in this interval should result in a great sonochemical effect during water sonolysis. The bubble temperature and pressure, the bubble collapse time, and the single bubble yield on hydrogen were all increased with acoustic intensity increase due to the augmentation of acoustic amplitude with intensity (Ferkous et al. 2015; Merouani et al. 2016a). The number of active cavitation bubbles may also increase with an increase in acoustic intensity between the Black and the attenuated thresholds (Sunartio et al. 2007; Kanthale et al. 2008; Ferkous et al. 2015). Consequently, an increase in both the single bubble yield and the number of bubbles at high acoustic intensities explains the high performance of hydrogen production at higher applied intensities.

1.5.3 Static Pressure

Cotana et al. (2007) have experienced the sonochemical production of hydrogen in sonochemical reactor equipped with two transducers operating at 20 kHz and 100 W (global energy). They found that hydrogen accumulation after 3 h of irradiation decreased from 7.32 μM at 1 atm to 6.82 at 1.5 atm and 3.65 μM at 2.5 atm. These findings where in accordance with the results of the sonochemical oxidation of KI and sonoluminescence where atmospheric pressure (1 atm) was found to be the best operating pressure for sonochemistry and sonoluminescence (Weissler 1953; Henglein et al. 1993; Iersel et al. 2007). In consistent with these results, the numerical simulation results showed that the single bubble yield exhibited a maximum at about atmospheric pressure (Merouani et al. 2015a). In general, it has been reported that increasing the static pressure above 1 atm acts as decreasing the acoustic intensity (Tuziuti et al. 2002; Merouani et al. 2015a). Therefore, lower bubble temperature and pressure would be achieved at higher static pressures. Besides, the very low bubble temperature was predicted at a pressure inferior than 0.5 atm. Between these two limits, an optimum static pressure at about 1 atm has been reported for the bubble temperature as well as the chemical bubble yield (Merouani et al. 2015a).

1.5.4 Liquid Temperature

Rassokhin et al. (1995) have studied the effect of methanol-water mixtures at various methanol concentrations (from pure methanol to pure water) on the sonochemical (724 kHz and 50 W) production of hydrogen at different liquid temperatures (−30 to 50 °C). For pure water, the formation rate of hydrogen increased with temperature increase in the interval 5–50 °C, i.e., the maximum rate was 14 μM/min at 50 °C. In the presence of alcohol, the rates of H_2 formation were several magnitudes higher than those of pure water and being maximum at certain temperatures (Fig. 1.4). Formation rates of hydrogen as well as the observed optimum temperatures depend on the methanol dosage in the sonicated solution (Fig. 1.4). All curves, including the optimum points, shift toward higher temperatures as the concentration of methanol decreased and makes the temperature dependencies less sharp. The maximum formation rates of 100 μM/min at 0.18% CH_2O and 35 °C, 200 μM/min at 1.8% CH_2O and 25 °C, 170 μM/min at 11% CH_2O and 25 °C, and 50 μM/min at 25% CH_2O and 20 °C were recorded. However, lower production rates were observed at higher alcohol contents (>60%). Xuemin et al. (2012) have also reported similar trends of H_2 production with respect to methanol and temperature during sonolysis at 40 kHz.

Numerical simulations of liquid temperature effect on the single bubble yield in pure water have been conducted by Merouani et al. (2016a). On the one hand, the results exhibited an optimum bulk temperature between 20 and 30 °C for the generation of hydrogen. On the other hand, the number of active bubbles was increased substantially when the liquid temperature increased in the range of 25–55 °C (Merouani et al. 2016b; Chadi et al. 2018b). Thus, the overall impact of liquid temperature on the sonolytic production of hydrogen must be controlled by the number of bubbles, as a linear increase of the production rate of H_2 with the liquid temperature rise has been reported in pure water (Rassokhin et al. 1995). However, the presence of volatile substrate such as methanol could change the impact of the bulk temperature on the global generation rate of hydrogen. Alcohols were known to reduce the bubble temperature at the collapse, and the higher the dosage of the alcohol in the liquid, the lower was the bubble temperature (Ashokkumar and Grieser 2005; Ciawi et al. 2006a, b; Ashokkumar 2011). Therefore, the presence of

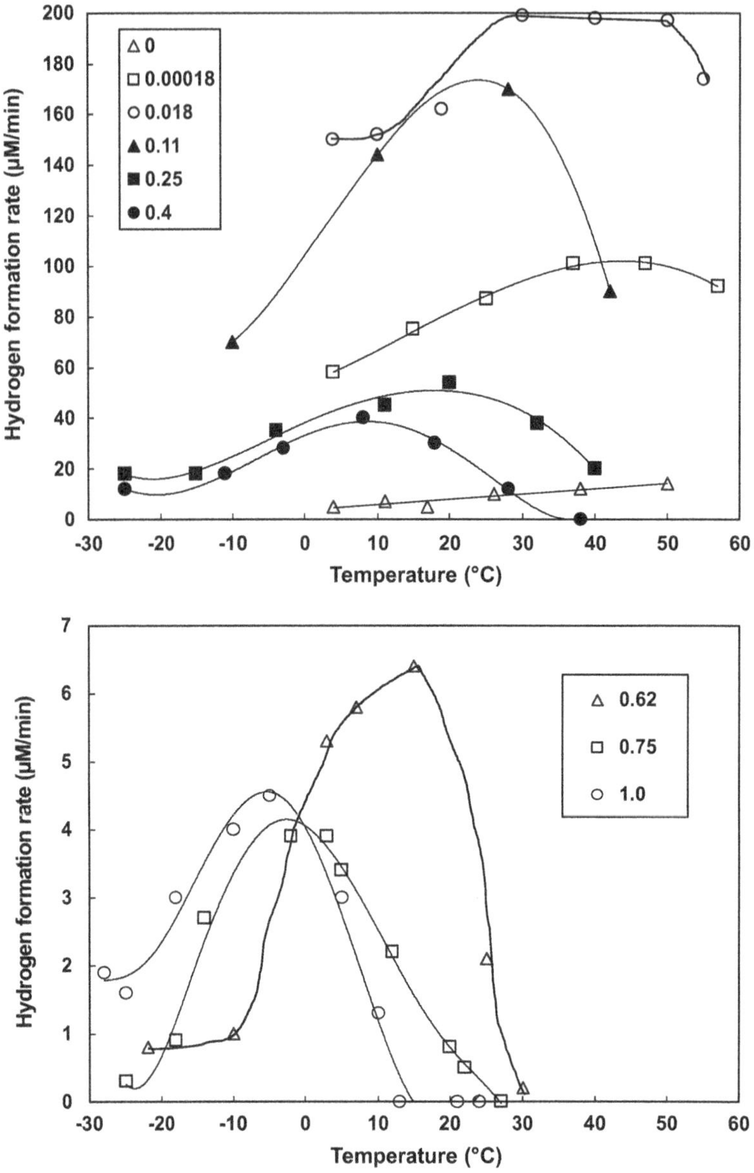

Fig. 1.4 The temperature dependencies of H_2 formation rate in the sonolysis of methanol/water mixtures with various mole fractions of methanol (legend). Modified from Rassokhin et al. (1995)

methanol at low concentration would reduce slowly the bubble temperature, but the methanol combustion inside the bubble produces an excess of hydrogen, which compensates the loss in H_2 yield provoked by the loss in the bubble temperature. However, higher concentration of alcohols in the bubble acts negatively on hydrogen yield via two ways: (1) huge reduction in the bubble temperature and (2) scavenging of H• and •OH radicals in the bubble (i.e., precursors of H_2 during sonolysis of pure water (Merouani et al. 2015b). Consequently, an optimum alcohol concentration and liquid temperature may effectively be obtained when sonolysis is proceeded in the presence of methanol.

1.5.5 Saturation Gas

Many studies have examined the impact of saturation gases on the sonochemical production of hydrogen. Anbar and Pecht (1963) found that the yield of hydrogen at 800 kHz and 1.6 ergs/cm^2 is 25 μM/min under argon saturation face 9.4 μM/min upon oxygen saturation (decrease of 2.6-fold). At 300 kHz, Hart et al. (1986) have reported that the yield of H_2 was more pronounced for argon than N_2 saturation atmosphere (14 μM/min for Ar against 4 μM/min for N_2). Henglein (1985) have studied the effect of argon-CO_2 and argon-N_2O on the production rate of hydrogen at 300 kHz and 3.5 W/cm^2. The rate was found to decrease quickly with the addition of CO_2 or N_2O in trace in the argon matrix, i.e., the hydrogen yield was null under pure CO_2 and N_2O saturations. Similar observation has been obtained by Harada and Ono (2015) during the sonication of KI solution under various Ar-CO_2 gas proportions.

Thanks to its higher water solubility, higher polytropic index (γ), and lower thermal conductivity (λ), argon could provide higher bubble temperature and thus efficient pyrolytic reaction inside the bubble than other polyatomic gases (Mead et al. 1976). Thus, it would provide the best sonochemical effect than diatomic and polyatomic gases. This statement has been proved experimentally and theoretically (Mead et al. 1976; Merouani et al. 2015c). The single bubble yield calculations showed that the order of saturation gases on the chemical bubble yield for different species could follow: Ar > O_2 > air > N_2 (Merouani et al. 2015c). However, this order may be affected by the operational conditions, i.e., frequency and power. In fact, it has been predicted that for frequencies inferior than 200 kHz, argon and air have the same impact on H_2 generation from a single bubble (Merouani et al. 2016a). Also, the single bubble calculations showed that the concentration of the pyrolytic products inside the bubble decreased as the N_2-bubble content increased (Merouani et al. 2014b). Thus, between O_2, N_2, and air (which have practically similar solubility, polytropic index, and thermal conductivity (Merouani et al. 2014b)), the order on the overall hydrogen yield could be controlled by the concentration of N_2 as O_2 > air > N_2 (Merouani et al. 2015c). Concerning CO_2, its effect was attributed to the too high solubility of CO_2 in water (46-fold than air), which provokes a suppression of the inertial cavitation responsible for the chemical effects of ultrasound (Chadi et al. 2018a). In fact, bigger visible bubbles have been observed when CO_2 is used as saturation gas (Brotchie et al. 2010; Rooze et al. 2011).

1.5.6 pH

Anbar and Pecht (1963) have investigated the production of hydrogen at 800 kHz in formate aqueous solution. The yield of H_2 production in pure water and formate solution is the same (25 μM/min), and the authors have concluded that the presence of formate could not affect the acoustic generation of H_2 and assume that H_2 was not formed by the self-recombination of $H^•$ radicals as formate is known as a strong $H^•$ scavenger. After this conclusion, the researchers found that the production rate of H_2 at pH 0.6 (16.2 μM/min) is lower than that obtained for pH 6–12 (25 μM/min). Xuemin et al. (2012) have showed that neutral conditions provide the best production rate of hydrogen from methanol sonolysis in dilute aqueous solution than acidic conditions. Therefore, the yield of hydrogen production via sonolysis increased with pH increase. Ashokkumar et al. (1999) have reported that the sonoluminescence (SL) intensity increases with increasing pH up to 10 and slowdowns afterward. Given that sonoluminescence is an effective probe of the peak temperature inside the bubble (Ashokkumar 2011), higher pH values favor higher bubble temperature, and therefore, higher concentration of hydrogen would be obtained with increasing pH from acidic to neutral conditions.

1.6 Active Bubble Sizes for the Sono-Production of Hydrogen

Active bubbles are those able to produce sonochemistry and sonoluminescence (SL). The measure of these parameters is difficult due to the chaotic nature of acoustic cavitation. Because of this, only limited works have investigated the size of active bubbles, in which the majority of studies were conducted at one frequency ultrasound (Burdin et al. 1999; Tsochatzidis et al. 2001; Labouret and Frohly 2002; Thiemann et al. 2011). Brotchie et al. (2009) have used sonoluminescence as probe and determine the range of sonoluminescing bubbles at different frequency and power. However, experimental evidence showed that the range of sonoluminescing bubbles differs from that capable to produce sonochemistry, and thus, the SL results could not reflect the sonochemical ones (Lee et al. 2005). More recently, Merouani and Hamdaoui (2016) have provided good simulation results on the size of active bubble for the sonolytic production of hydrogen. Excepting this paper, there was no any work addressing this issue neither experimentally nor theoretically. Their results have been compared with the limited data available in the literature, and a good concordance between them has been established. The simulation findings indicated that the size of active bubbles is an interval rather narrow, i.e., several micrometers, which comprises an optimum ambient radius at which the generation rate of hydrogen is maximal. The most significant findings of the study of Merouani and Hamdaoui are summarized in Table 1.2. It is rapidly observed that the range of active bubble sizes and the optimum radius are becoming lesser with increasing frequency. Moreover, the size of active bubbles, i.e., the range and optimum radius, increased with augmenting intensity and diminished with increasing the bulk temperature of the liquid.

With increasing acoustic intensity, the higher temperature and amount of vapor coexist inside the bubble at the collapse. These conditions favor a strong implosion of bubbles that have large sizes, i.e., inactive bubble at lower intensity, which then become active for hydrogen production (Merouani et al. 2016a). Note that increasing frequency acts exactly as decreasing the acoustic intensity and higher frequencies engender lower values of bubble temperature and collapse duration (Merouani et al. 2014b). As a consequence, the range of active bubbles becomes less extended at elevated frequencies. Besides, lower bubble core temperatures were recorded at higher liquid temperatures due to the large evaporation of water vapor within the bubble that resulted from the liquid heating (Merouani et al. 2015a, 2016a). This reduces the production of hydrogen through decreasing the yield of free radical generation (Merouani et al. 2015a, 2016a). Therefore, the consequence of increasing the liquid temperature could be the deactivation of active bubbles having low internal temperatures at 20 °C, as indicated in Table 1.2.

Table 1.2 The size (the range and the optimum ambient radius) of active bubbles for the production of hydrogen in the acoustic cavitation field

(A)

	Range of active ambient bubble radius (μm)		Optimum ambient bubble radius (μm)	
Frequency (kHz)	0.75 W/cm²	1 W/cm²	0.75 W/cm²	1 W/cm²
140	1.11–25.56	0.69–27.4	11.1	13.25
213	1.2–14.7	0.73–17.9	6.1	8.15
355	1.26–9.19	0.8–8.58	3.6	4.8
515	1.15–6.1	0.87–7.1	2.5	3
647	1.11–5.41	0.88–5.75	2.1	2.35
875	0.99–4.68	0.81–4.45	1.6	1.86
1000	0.96–4.11	0.79–3.92	1.5	1.6
1100	0.95–3.72	0.83–3.73	1.4	1.5

(B)

	Range of active ambient bubble radius (μm)		Optimum ambient bubble radius (μm)	
Frequency (kHz)	20 °C	50 °C	20 °C	50 °C
140	0.69–27.4	0.72–19.5	13.25	8.9
213	0.73–17.9	0.74–12.77	8.15	5.48
355	0.80–8.58	0.80–7.81	4.8	3.2
515	0.87–7.1	0.78–5.81	3	2.63
647	0.88–5.75	0.8–4.72	2.35	1.8
875	0.81–4.45	0.77–3.69	1.86	1.5
1000	0.80–3.92	0.69–3.32	1.6	1.4
1100	0.79–3.73	0.67–3.06	1.5	1.4

(A) Bubble size dependence of frequency-acoustic intensity and (B) bubble size dependence of frequency-liquid temperature. Adapted from Merouani and Hamdaoui (2016)

1.7 Relationship Between the Bubble Temperature and Pressure and the Production Rate of Hydrogen

The key parameters controlling the formation of oxidants and hydrogen are the maximum temperature (T_{max}) and pressure (p_{max}) achieved within the bubble at the strong implosion. Many researchers have correlated the single bubble yield to T_{max} and p_{max} for several cases. An optimum bubble temperature of ~5500 K has been firstly reported by Yasui et al. (2004) for the production of the oxidants inside an air bubble. Authors have attributed this tendency to the strong consumption of radicals inside the bubble by oxidizing nitrogen at higher bubble temperatures (>5500 K). For an oxygen bubble, Merouani et al. (2014c) have predicted an optimal temperature and pressure of 5200 K and 250 atm for the production of hydroxyl radical. Above these values, the ${}^{\bullet}$OH yield decreased, allowing to conclude that there exist a strong competition between consumption and production rates of ${}^{\bullet}$OH with respect to bubble temperatures and pressures (Merouani et al. 2014c). Lately, an optimum temperature of ~6000 K has also been predicted for the production of H_2 inside an oxygen bubble (Kerboua and Hamdaoui 2018).

More recently, Merouani and Hamdaoui (2018) have provided more detailed results on the correlation between the single bubble yield of hydrogen and the T_{max} and p_{max} for an argon bubble (Fig. 1.5). These correlations have been obtained for more than 800 combinations between diverse cavitation parameters including frequency, ambient bubble radius, acoustic intensity, and liquid temperature. As clearly seen, the production rate of hydrogen increased linearly with increasing T_{max} and p_{max} up to reaching the steady-state levels, which start at 3500 ± 200 K and

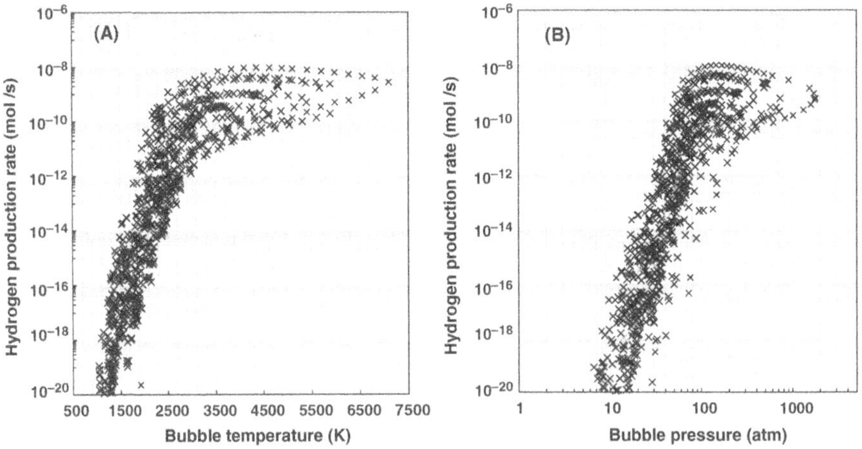

Fig. 1.5 Strong correlation between the hydrogen production rate from a single bubble and the maximum bubble temperature (**a**) and pressure (**b**). Vertical axes are in logarithmic scale. This correlation was obtained from 835 combinations of various parameters, i.e., frequency, 140–1100 kHz; acoustic intensity, 0.5–1.0 W cm^{-2}; liquid temperature, 20–50 °C; and ambient bubble radius. Modified from Merouani and Hamdaoui (2018)

100 ± 10 atm. The authors have analyzed the simultaneous progresses of H_2 precursors ($H^{\bullet}$ and $^{\bullet}OH$, because H_2 was produced mainly from $H^{\bullet} + {}^{\bullet}OH \rightleftharpoons H_2 + O$). They found similar progresses for $H^{\bullet}$ and $^{\bullet}OH$ as that of H_2. The authors have then concluded that the production of H_2 was mainly controlled by the production and consumption of $H^{\bullet}$ and $^{\bullet}OH$ inside the bubble. When T_{max} and p_{max} are inferior than 3500 K and 100 atm, reactions producing free radicals are more important than their consuming ones. Contrarily, once T_{max} and p_{max} surpass 3500 K and 100 atm, the consumption reactions overlap those of the production.

1.8 Dependence of the Sonochemical Production of Hydrogen to Liquid Depth/Height: A Scale-Up Approach

The efficiency of the sonochemical reaction was sensitively related to the height/depth of the liquid in the sonoreactors, i.e., solution volume at fixed diameter for cylindrical reactors. This parameter is one of the missing links between lab-scale sono-processes and industrial-scale sono-processes. Several studies have tried to investigate the impact of this parameter taking the oxidation of KI solution, H_2O_2 quantification, or sonoluminescence as probes. In general, it was noted that the liquid height in sonoreactors affects the cavitation process, and small changes in this parameter produces perceptible variation in sonochemical efficiency (Renaudin et al. 1994; Kojima et al. 1998; Asakura et al. 2008; Tuziuti et al. 2008; Merouani et al. 2010).

However, for the case of hydrogen sono-production, there was no available data investigating the impact of liquid height. However, Kerabchi et al. (2019) have recently published a new theoretical approach on the impact of depth on the acoustical production of hydrogen. In their single bubble sonochemistry model, the bubble depth (z) as well as the sound wave attenuation due to absorption has been integrated. Additionally, the effect of depth was investigated for up to 8 m, giving an engineering view for large-scale operations. The effect of liquid depth on the single bubble yield, i.e., including H_2, monotonically decreased with increasing z (Fig. 1.6a), and this effect become more pronounced with the elevation of the frequency of ultrasound or the decrease in liquid temperature and acoustic intensity, i.e., Fig. 1.6b illustrates the effect of frequency.

Depth affects the intensity of bubble collapse via (1) elevating the absolute pressure, i.e., $p_0 = p_\infty + \rho g z$, and (2) diminishing the acoustic intensity of the ultrasound wave through attenuation phenomena (Fig. 1.7). For both cases, numerical simulations showed a diminution in both the bubble temperature and the single bubble yields (Yasui et al. 2007; Merouani et al. 2015a; Kerabchi et al. 2018a, b). Therefore, the combined effects of depth on both the sound wave attenuation and the absolute pressure could provoke a significant inhibition of the acoustic generation of hydrogen from a single bubble. Kerabchi et al. (2019) have concluded that that attenuation phenomenon, which has no effective impact when sonolysis is carried out at lab-scale sonoreactors, should imperatively be considered when designing large-scale ultrasonic reactors.

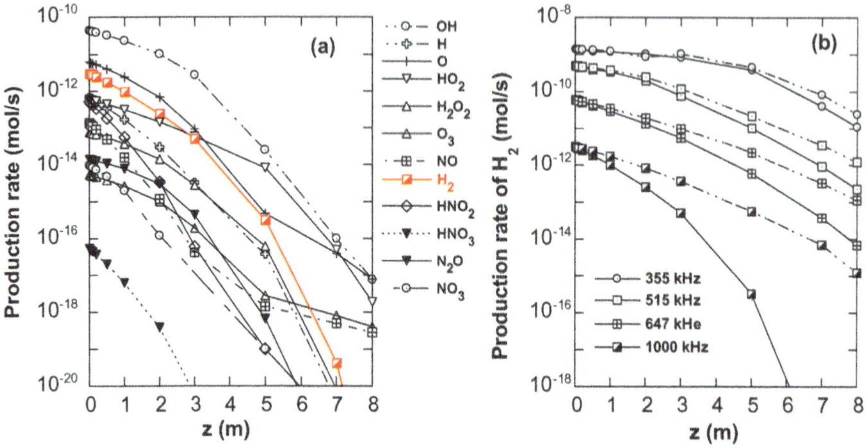

Fig. 1.6 Dependence of the single bubble yield, including that of hydrogen (red line) to depth (**a**) and the impact of frequency on the effect of depth toward the acoustic generation of hydrogen (**b**). The results of figure (**a**) were obtained for an air bubble trapped in water irradiated by an acoustic wave of 1 MHz and 2 W/cm². Modified from Kerabchi et al. (2019)

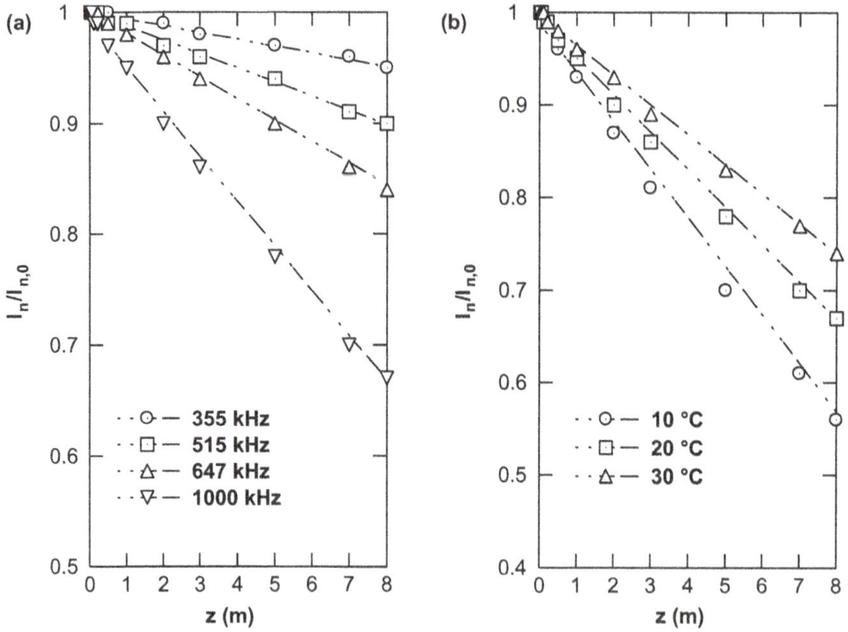

Fig. 1.7 Acoustic intensity attenuation in pure water as function of depth (z) for various frequencies (**a**) and diverse liquid temperatures (**b**) (conditions: (**a**) temperature 20 °C and pH 7, (**b**) frequency 1000 kHz and pH 7)

1.9 Intensification Techniques for Hydrogen Production by Ultrasound

Sonolysis of hydrocarbons in water is the first way for intensifying the production rate of hydrogen by ultrasound. Hart et al. (1990b) have investigated the sonochemistry of water at 300 kHz under different proportion of argon-hydrocarbon gases. Firstly, the production rate of hydrogen increased monotonically with the addition of methane to argon; 10% (v/v) of CH_4 in argon provide the maximum production rate of 100 μM/min. Above 10% of CH_4, the production rate falls rapidly to near zero for $\%CH_4 > 80\%$. Similar observation has been obtained for the case of ethane, but the maximum production rate of hydrogen at 10% C_2H_6 was 150 μM/min, and the limit above which no hydrogen was observed shifts to 40% C_2H_6 in argon rather than 80% for the case of CH_4–Ar mixture. In another paper, Hart et al. (1990a) have shown that during the sonication of water at 1 MHz under argon-acetylene mixture, the production rate of hydrogen increased from 40 μM/min in pure water to 90 and 120 μM/min in the presence of C_2H_2 at 1.5 and 6 mM, respectively. However, with the continuous rise of C_2H_2 concentration above 6 mM, the yield of H_2 was slowdowned until no production was observed at 10 mM of C_2H_2. The intensifying effect of hydrocarbons was not limited to methane, ethane, and acetylene, but also other hydrocarbons such as ethylene, propane, and butane accelerated the sono-generation rate of hydrogen (Hart et al. 1990b).

Another intensification technique of H_2 sono-generation is focused on the sonolysis of dilute aqueous solution of alcohols. Buettner et al. (1991) have investigated the sonolysis of water-methanol mixtures at 1 MHz under argon saturation. In pure water, H_2 was produced at a rate of 22 μM/min. When methanol is added to the sonicated solution, H_2production rate increased significantly to 100 μM/min at 5% (v/v) of methanol and to 150 μM/min at 10% (v/v) of methanol. However, at higher methanol concentrations, the yield decreases gradually until almost no H_2 was detected above about 80% (v/v) of methanol. Penconi et al. (2015) have registered an increase in H_2 production rate by a factor of 1.4 when applying ultrasound at 38 kHz in water and water/ethanol (20% v/v) under argon saturation. Tauber et al. (1999) have reported that the sonication of water containing a trace of tert-butyl alcohol (TBA) increased monotonically the yield of hydrogen generation from 10×10^{-10} mol/J to $\sim 150 \times 10^{-10}$ mol/J, but lower yields were recorded at high concentration of TBA. More interestingly, the intensifying effect of alcohol was liquid temperature-dependent. Rassokhin et al. (1995) have reported that relatively the higher liquid temperature and the lower percentage of methanol yield drastic acceleration of hydrogen than in pure water, i.e., with about tenfold increase.

Sonocatalysis, a new research area of applied sonochemistry, has proved this potential for accelerating the sonochemical production rate of hydrogen. Several new catalysts have been developed for this issue. Wang et al. (2010) have used Au/TiO_2 as an efficient catalyst to produce H_2. Au/TiO_2 increased the yield of H_2 in the sonolysis of water at 40 kHz and 50 W by 54-fold, as compared with sonolysis alone, and by 16.6-fold, as compared with US/TiO_2 system. Moreover, the addition of methanol in the presence of Au/TiO_2 has resulted in a 13-fold increase in the rate of H_2 evolution, as compared to the Au/TiO_2 sonolytic system (i.e., 282 μM/min against 21.6 μM/min). Similar observations have been reported by other researches using different forms of catalyst (Gentili et al. 2009; Rossi et al. 2009; Penconi et al. 2015).

Finally, other methods such as coupling sonochemistry with photocatalysis have also been recently proposed for ameliorating the yield of H_2 sono-production. Even though a synergy was sometimes observed (Gentili et al. 2009; Rossi et al. 2009), yields of H_2 from these techniques remain much lower than those observed with the three first intensification techniques, mentioned above.

1.10 Conclusion and Future Directions

This chapter deals with the sonochemical production of hydrogen, one of the applied sonochemistry branches. With this technique, clean hydrogen is produced upon water sonolysis by high-frequency ultrasonic waves (20–1000 kHz). The combustion chemistry occurring inside the bubble is the main source of the yielded hydrogen, and the operational conditions that affect the acoustic cavitation event could directly affect the sonochemical yield of hydrogen. Higher frequency, acoustic intensity, and liquid temperature could yield a higher generation rate of hydrogen. The presence of alcohols, i.e., methanol, hydrocarbon, and methane, and solid catalysts intensifies the sonochemical yield of hydrogen. However, even though this technique generates hydrogen at much rate than photocatalysis, experimental studies focusing on this topic are very limited, which is perhaps due to the limitation in designing sophistical sonochemical reactors. Therefore, the following points need to be investigated in depth for innovation:

1. Reactor design including reactor shape, material of construction, and transducers and their distribution around the reactor wall.
2. Development of continuous-flow sonochemical reactors.
3. Extrapolating the sonochemical process to large scale (industrial application).

References

Adewuyi YG (2001) Sonochemistry : environmental science and engineering applications. Ind Eng Chem Res 40:4681–4715. https://doi.org/10.1021/ie010096l

Anbar M, Pecht I (1963) On the sonochemical formation of hydrogen peroxide in. J Phys Chem 1077:1959–1962. https://doi.org/10.1021/j100784a025

Anbar M, Pecht I (1964) The sonolytic decomposition of organic solutes in dilute aqueous solutions. I. Hydrogen abstraction from Sodium formate. J Phys Chem 68:1460–1462

Asakura Y, Nishida T, Matsuoka T, Koda S (2008) Effects of ultrasonic frequency and liquid height on sonochemical efficiency of large-scale sonochemical reactors. Ultrason Sonochem 15:244–250. https://doi.org/10.1016/j.ultsonch.2007.03.012

Ashokkumar M (2011) The characterization of acoustic cavitation bubbles—an overview. Ultrason Sonochem 18:864–872. https://doi.org/10.1016/j.ultsonch.2010.11.016

Ashokkumar M, Grieser F (2005) A comparison between multibubble sonoluminescence intensity and the temperature within cavitation bubbles. J Am Chem Soc 127:5326–5327. https://doi.org/10.1021/ja050804k

Ashokkumar M, Mulvaney P, Grieser F (1999) The effect of pH on multibubble sonoluminescence from aqueous solutions containing simple organic weak acids and bases. J Am Chem Soc 121(32):7355–7359

Beckett MA, Hua I (2001) Impact of ultrasonic frequency on aqueous sonoluminescence and sonochemistry. J Phys Chem A 105:3796–3802. https://doi.org/10.1021/jp003226x

Bhangu SK, Ashokkumar M (2016) Theory of sonochemistry. Top Curr Chem 374:56. https://doi.org/10.1007/s41061-016-0054-y

Brotchie A, Grieser F, Ashokkumar M (2009) Effect of power and frequency on bubble-size distributions in acoustic cavitation. Phys Rev Lett 102:1–4. https://doi.org/10.1103/PhysRevLett.102.084302

Brotchie A, Statham T, Zhou M, Dharmarathne L, Grieser F, Ashokkumar M (2010) Acoustic bubble sizes, coalescence, and sonochemical activity in aqueous electrolyte solutions saturated with different gases. Langmuir 26:12690–12695. https://doi.org/10.1021/la1017104

Buettner J, Gutierrez M, Henglein a. (1991) Sonolysis of water-methanol mixtures. J Phys Chem 95:1528–1530. https://doi.org/10.1021/j100157a004

Burdin F, Tsochatzidis NA, Guiraud P, Wilhelm AM, Delmas H (1999) Characterisation of the acoustic cavitation cloud by two laser techniques. Ultrason Sonochem 6:43–51. https://doi.org/10.1016/S1350-4177(98)00035-2

Chadi NE, Merouani S, Hamdaoui O (2018a) Characterization and application of a 1700 kHz-acoustic cavitation field for water decontamination : a case study with toluidine blue. Appl Water Sci 8(160):1–11. https://doi.org/10.1007/s13201-018-0809-4

Chadi NE, Merouani S, Hamdaoui O, Bouhelassa M (2018b) New aspect of the effect of liquid temperature on sonochemical degradation of nonvolatile organic pollutants in aqueous media. Sep Purif Technol 200:68–74. https://doi.org/10.1016/j.seppur.2018.01.047

Chakik FE, Kaddami M, Mikou M (2017) Effect of operating parameters on hydrogen production by electrolysis of water. Int J Hydrog Energy 42:2–9. https://doi.org/10.1016/j.ijhydene.2017.07.015

Chatel G (2019) Sonochemistry in nanocatalysis: the use of ultrasound from the catalyst synthesis to the catalytic reaction. Curr Opin Green Sustain Chem 15:1–6. https://doi.org/10.1016/j.cogsc.2018.07.004

Ciawi E, Rae J, Ashokkumar M, Grieser F (2006a) Determination of temperatures within acoustically generated bubbles in aqueous solutions at different ultrasound frequencies. J Phys Chem B 110:13656–13660. https://doi.org/10.1021/jp061441t

Ciawi E, Ashokkumar M, Grieser F (2006b) Limitations of the methyl radical recombination method for acoustic cavitation bubble temperature measurements in aqueous solutions. J Phys Chem B 110:9779–9781. https://doi.org/10.1021/jp0618734

Cotana F, Rossi F, Urbani M (2007) Study of water photosonolyis for hydrogen production. In: 3rd International green energy conference, pp 3–8

Das D, Veziroglu TN (2008) Advances in biological hydrogen production processes. Int J Hydrog Energy 33:6046–6057. https://doi.org/10.1016/j.ijhydene.2008.07.098

Dincer I (2012) Green methods for hydrogen production. Int J Hydrog Energy 37:1954–1971. https://doi.org/10.1016/j.ijhydene.2011.03.173

Dincer I, Acar C (2015) Review and evaluation of hydrogen production methods for better sustainability. Int J Hydrog Energy 40:11094–11111. https://doi.org/10.1016/j.ijhydene.2014.12.035

Entezari MH, Pétrier C, Devidal P (2003) Sonochemical degradation of phenol in water: a comparison of classical equipment with a new cylindrical reactor. Ultrason Sonochem 10:103–108. https://doi.org/10.1016/S1350-4177(02)00136-0

Ferkous H, Merouani S, Hamdaoui O, Rezgui Y, Guemini M (2015) Comprehensive experimental and numerical investigations of the effect of frequency and acoustic intensity on the sonolytic degradation of naphthol blue black in water. Ultrason Sonochem 26:30–39. https://doi.org/10.1016/j.ultsonch.2015.02.004

Fischer C-H, Hart EJ, Henglein A (1986) H/D isotope exchange in the D_2-H_2O system under the influence of ultrasound. J Phys Chem 90:222–224

Gentili PL, Penconi M, Ortica F, Cotana F, Rossi F, Elisei F (2009) Synergistic effects in hydrogen production through water sonophotolysis catalyzed by new $La_{2x}Ga_{2y}In_{2(1-x-y)}O_3$ solid solutions. Int J Hydrog Energy 34:9042–9049. https://doi.org/10.1016/j.ijhydene.2009.09.027

Guitièrrres M, Heinglein A, Fischer C-H (1986) Hot spot kinetics of the sonolysis aqueuos acetate solutions. Int J Radiat Biol Relat Stud Phys Chem Med 50:313–321

Gutierrez M, Henglein A (1991) Radical scavenging in the sonolysis of aqueous solutions of I^-, Br^-, and N_3^-. J Phys Chem 95:6044–6047

Guzman-Duque F, Pétrier C, Pulgarin C, Penuuela G, Torres-Palma RA, Peñuela G, Torres-Palma RA, Penuuela G, Torres-Palma RA (2011) Effects of sonochemical parameters and inorganic ions during the sonochemical degradation of crystal violet in water. Ultrason Sonochem 18:440–446. https://doi.org/10.1016/j.ultsonch.2010.07.019

Harada H (1998) Sonochemical reduction of carbon dioxide. Ultrason Sonochem 5:73–77

Harada H (2001) Sonophotocatalytic decomposition of water using TiO_2 photocatalyst. Ultrason Sonochem 8:55–58. https://doi.org/10.1016/S1350-4177(99)00050-4

Harada H, Ono Y (2015) Improvement of the rate of sono-oxidation in the presence of CO_2. Jpn J Appl Phys 54:52–55. https://doi.org/10.7567/JJAP.54.07HE10

Harada H, Sogawa R (2004) Sonolysis of sodium hydrogen carbonate solution. Jpn J Appl Phys Part 1 Regul Pap Short Notes Rev Pap 43:6484–6487. https://doi.org/10.1143/JJAP.43.6484

Hart EJ, Henglein A (1985) Free radical and free atom reactions in the sonolysis of aqueous iodide and formate solutions. J Phys Chem 89:4342–4347. https://doi.org/10.1021/j100266a038

Hart EJ, Fischer C-H, Henglein A (1986) Isotopic exchange in the sonolysis of aqueous solutions containing nitrogen-14 and nitrogen-15 molecules. J Phys Chem 90:5989–5991

Hart EJ, Fischer C, Henglein A (1990a) Pyrolysis of acetylene in sonolytic cavitation bubbles in aqueous solution. J Phys Chem 94:284–290

Hart EJ, Fischer CH, Henglein A (1990b) Sonolysis of hydrocarbons in aqueous solution. Int J Radiat Appl Instrum Part 36:511–516. https://doi.org/10.1016/1359-0197(90)90198-Q

Haryanto A, Fernando S, Murali N, Adhikari S (2005) Current status of hydrogen production techniques by steam reforming of ethanol : a review. Energy Fuels 19(5):2098–2106. https://doi.org/10.1021/ef0500538

Henglein A (1985) Sonolysis of carbon dioxide, nitrous oxide and methane in aqueous solution. Z Naturforsc B 40:100–107

Henglein A (1987) Sonochemistry: historical developments and modern aspects. Ultrasonics 25:6–16. https://doi.org/10.1016/0041-624X(87)90003-5

Henglein A (1995) Chemical effects of continuous and pulsed ultrasound in aqueous solutions. Ultrason Sonochem 2:115–121. https://doi.org/10.1016/1350-4177(95)00022-X

Henglein A, Gutierrez A, Gutierrez M (1993) Sonochemistry and sonoluminescence: effects of external pressure. J Phys Chem 97:158–162. https://doi.org/10.1021/j100103a027

Islam MH, Burheim OS, Pollet BG, Islam H, Burheim OS, Pollet BG, Islam MH, Burheim OS, Pollet BG (2019) Sonochemical and sonoelectrochemical production of hydrogen. Ultrason Sonochem 51:533–555. https://doi.org/10.1016/j.ultsonch.2018.08.024

Jiang Y, Petrier C, Waite TD (2006) Sonolysis of 4-chlorophenol in aqueous solution: effects of substrate concentration, aqueous temperature and ultrasonic frequency. Ultrason Sonochem 13:415–422. https://doi.org/10.1016/j.ultsonch.2005.07.003

Kanthale P, Ashokkumar M, Grieser F (2008) Sonoluminescence, sonochemistry (H_2O_2 yield) and bubble dynamics: frequency and power effects. Ultrason Sonochem 15:143–150. https://doi.org/10.1016/j.ultsonch.2007.03.003

Keil FJ, Swamy KM (1999) Reactors for sonochemical engineering-present status. Rev Chem Eng 15:85–155. https://doi.org/10.1515/REVCE.1999.15.2.85

Kerabchi N, Merouani S, Hamdaoui O (2018a) Depth effect on the inertial collapse of cavitation bubble under ultrasound: special emphasis on the role of the wave attenuation. Ultrason Sonochem 48:136. https://doi.org/10.1016/j.ultsonch.2018.05.004

Kerabchi N, Merouani S, Hamdaoui O (2018b) Liquid depth effect on the acoustic generation of hydroxyl radical for large scale sonochemical reactors. Sep Purif Technol 206:118–130. https://doi.org/10.1016/j.seppur.2018.05.039

Kerabchi N, Merouani S, Hamdaoui O (2019) Relationship between liquid depth and the acoustic generation of hydrogen: design aspect for large cavitational reactors with special focus on the role of the wave attenuation. Int J Green Energy 16:1–12. https://doi.org/10.1080/15435075.2019.1577741

Kerboua K, Hamdaoui O (2018) Numerical estimation of ultrasonic production of hydrogen: effect of ideal and real gas based models. Ultrason Sonochem 40:194. https://doi.org/10.1016/j.ultsonch.2017.07.005

Kojima Y, Koda S, Nomura H (1998) Effects of sample volume and frequency on ultrasonic power in solutions on sonication. Jpn J Appl Phys 37:2992–2995. https://doi.org/10.1143/JJAP.37.2992

Labouret S, Frohly J (2002) Bubble size distribution estimation via void rate dissipation in gas saturated liquid. Application to ultrasonic cavitation bubble field. Eur Phys J Appl Phys 19:39–54. https://doi.org/10.1051/epjap:2002047

Lee J, Ashokkumar M, Kentish S, Grieser F (2005) Determination of the size distribution of sonoluminescence bubbles in a pulsed acoustic field. J Am Chem Soc 127:16810–16811. https://doi.org/10.1021/ja0566432

Leong T, Ashokkumar M, Sandra K (2011) The fundamentals of power ultrasound—a review. Acoust Aust 39:54–63

Mead EL, Sutherland RG, Verrall RE (1976) The effect of ultrasound on water in the presence of dissolved gases. Can J Chem 54:1114–1120. https://doi.org/10.1139/v76-159

Merouani S, Hamdaoui O (2016) The size of active bubbles for the production of hydrogen in sonochemical reaction field. Ultrason Sonochem 32:320–327. https://doi.org/10.1016/j.ultsonch.2016.03.026

Merouani S, Hamdaoui O (2017) Computational and experimental sonochemistry. Process Eng J 1:10–18

Merouani S, Hamdaoui O (2018) Correlations between the sonochemical production rate of hydrogen and the maximum temperature and pressure reached in acoustic bubbles. Arab J Sci Eng 43:6109–6117. https://doi.org/10.1007/s13369-018-3266-3

Merouani S, Hamdaoui O, Saoudi F, Chiha M (2010) Influence of experimental parameters on sonochemistry dosimetries: KI oxidation, Fricke reaction and H_2O_2 production. J Hazard Mater 178:1007–1014. https://doi.org/10.1016/j.jhazmat.2010.02.039

Merouani S, Ferkous H, Hamdaoui O, Rezgui Y, Guemini M (2014a) A method for predicting the number of active bubbles in sonochemical reactors. Ultrason Sonochem 22:51–58. https://doi.org/10.1016/j.ultsonch.2014.07.015

Merouani S, Hamdaoui O, Rezgui Y, Guemini M (2014b) Sensitivity of free radicals production in acoustically driven bubble to the ultrasonic frequency and nature of dissolved gases. Ultrason Sonochem 22:41–50. https://doi.org/10.1016/j.ultsonch.2014.07.011

Merouani S, Hamdaoui O, Rezgui Y, Guemini M (2014c) Theoretical estimation of the temperature and pressure within collapsing acoustical bubbles. Ultrason Sonochem 21:53–59. https://doi.org/10.1016/j.ultsonch.2013.05.008

Merouani S, Hamdaoui O, Rezgui Y, Guemini M (2015a) Computer simulation of chemical reactions occurring in collapsing acoustical bubble: dependence of free radicals production on operational conditions. Res Chem Intermed 41:881–897. https://doi.org/10.1007/s11164-013-1240-y

Merouani S, Hamdaoui O, Rezgui Y, Guemini M (2015b) Mechanism of the sonochemical production of hydrogen. Int J Hydrog Energy 40:4056–4064. https://doi.org/10.1016/j.ijhydene.2015.01.150

Merouani S, Ferkous H, Hamdaoui O, Rezgui Y, Guemini M (2015c) New interpretation of the effects of argon-saturating gas toward sonochemical reactions. Ultrason Sonochem 23:37–45. https://doi.org/10.1016/j.ultsonch.2014.09.009

Merouani S, Hamdaoui O, Rezgui Y, Guemini M (2016a) Computational engineering study of hydrogen production via ultrasonic cavitation in water. Int J Hydrog Energy 41:832–844. https://doi.org/10.1016/j.ijhydene.2015.11.058

Merouani S, Hamdaoui O, Boutamine Z, Rezgui Y, Guemini M (2016b) Experimental and numerical investigation of the effect of liquid temperature on the sonolytic degradation of some organic dyes in water. Ultrason Sonochem 28:382–392. https://doi.org/10.1016/j.ultsonch.2015.08.015

Navarro NM, Chave T, Pochon P, Bisel I, Nikitenko SI (2011) Effect of ultrasonic frequency on the mechanism of formic acid sonolysis. J Phys Chem B 115:2024–2029. https://doi.org/10.1021/jp109444h

Ni M, Leung MKH, Leung DYC, Sumathy K (2007) A review and recent developments in photocatalytic water-splitting using TiO_2 for hydrogen production. Renew Sust Energ Rev 11:401–425. https://doi.org/10.1016/j.rser.2005.01.009

Penconi M, Rossi F, Ortica F, Elisei F, Gentili PL (2015) Hydrogen production from water by photolysis, sonolysis and sonophotolysis with solid solutions of rare earth, gallium and indium oxides as heterogeneous catalysts. Sustainability 7:9310–9325. https://doi.org/10.3390/su7079310

Pétrier C, Francony A (1997) Ultrasonic waste-water treatment: incidence of ultrasonic frequency on the rate of phenol and carbon tetrachloride degradation. Ultrason Sonochem 4:295–300. https://doi.org/10.1016/S1350-4177(97)00036-9

Pflieger R, Ndiaye AA, Chave T, Nikitenko SI (2015) Influence of ultrasonic frequency on Swan band sonoluminescence and sonochemical activity in aqueous tert-butyl alcohol solutions. J Phys Chem B 119:284–290. https://doi.org/10.1021/jp509898p

Rashwan SS, Dincer I, Mohany A, Pollet BG (2019) The sono-hydro-gen process (ultrasound induced hydrogen production): challenges and opportunities. Int J Hydrog Energy 44:14500–14526. https://doi.org/10.1016/j.ijhydene.2019.04.115

Rassokhin DN, Kovalev GV, Bugaenko LT (1995) Temperature effect on the sonolysis of methanol/water mixtures. J Am Chem Soc 117:344–347. https://doi.org/10.1021/ja00106a037

Renaudin V, Gondrexon N, Boldo P, Pétrier C, Bernis A, Gonthier Y (1994) Method for determining the chemically active zones in a high-frequency ultrasonic reactor. Ultrason Sonochem 1:3–7. https://doi.org/10.1016/1350-4177(94)90002-7

Rooze J, Rebrov EV, Schouten JC, Keurentjes JTF (2011) Effect of resonance frequency, power input, and saturation gas type on the oxidation efficiency of an ultrasound horn. Ultrason Sonochem 18:209–215. https://doi.org/10.1016/j.ultsonch.2010.05.007

Rossi F, Nicolini A, Filipponi M, Corsi N (2009) Study of catalysts for water photosonolysis to increase the hydrogen production photosonolysis process. In: Hypothesis VIII Lisbon, Portugal, pp 1–6

Sunartio D, Ashokkumar M, Grieser F (2007) Study of the coalescence of acoustic bubbles as a function of frequency, power, and water-soluble additives. J Am Chem Soc 129:6031–6036. https://doi.org/10.1021/ja068980w

Tauber A, Mark G, Schuchmann H, Sonntag CV (1999) Sonolysis of tert-butyl alcohol in aqueous solution. J Chem Soc Perkin Trans 2:1129–1135

Thiemann A, Nowak T, Mettin R, Holsteyns F, Lippert A (2011) Characterization of an acoustic cavitation bubble structure at 230 kHz. Ultrason Sonochem 18:595–600. https://doi.org/10.1016/j.ultsonch.2010.10.004

Torres RA, Pétrier C, Combet E, Carrier M, Pulgarin C (2008) Ultrasonic cavitation applied to the treatment of bisphenol A. effect of sonochemical parameters and analysis of BPA by-products. Ultrason Sonochem 15:605–611. https://doi.org/10.1016/j.ultsonch.2007.07.003

Tsochatzidis NA, Guiraud P, Wilhelm AM, Delmas H (2001) Determination of velocity, size and concentration of ultrasonic cavitation bubbles by the phase-Doppler technique. Chem Eng Sci 56:1831–1840. https://doi.org/10.1016/S0009-2509(00)00460-7

Tuziuti T, Hatanaka S, Yasui K, Kozuka T, Mitome H (2002) Effect of ambient-pressure reduction on multibubble sonochemiluminescence. J Chem Phys 116:6221. https://doi.org/10.1063/1.1461357

Tuziuti T, Yasui K, Lee J, Kozuka T, Towata A, Iida Y (2008) Mechanism of enhancement of sonochemical-reaction efficiency by pulsed ultrasound. J Phys Chem A 112:4875–4878. https://doi.org/10.1021/jp802640x

Van Iersel MM, van den Manacker J-PAJ, Benes NE, Keurentjes JTF (2007) Pressure-induced reduction of shielding for improving sonochemical activity. J Phys Chem B 111:3081–3084

Venault L (1997) De l'influence des ultrasons sur la réactivité de l'uranium (U(IV)/U(VI)) et du plutonium (Pu(III)/Pu(IV)) en solution aqueuse nitrique. Université Paris Sud, Orsay

Wang Y, Zhao D, Ji H, Liu G, Chen C, Ma W, Zhu H, Zhao J (2010) Sonochemical hydrogen production efficiently catalyzed by Au/TiO_2. J Phys Chem C 114:17728–17733. https://doi.org/10.1021/jp105691v

Weissler A (1953) Sononochemistry: the production of chemical changes with sound waves. J Acoust Soc Am 24:651–657

Xuemin D, Zherui Z, Zheng D, Befeng B (2012) Hydrogen production form sonolysis of alcohol solution. CIESC 63:924–928. https://doi.org/10.3969/j.issn.0438-1157.2012.035

Yasui K, Tuziuti T, Iida Y (2004) Optimum bubble temperature for the sonochemical production of oxidants. Ultrasonics 42:579–584. https://doi.org/10.1016/j.ultras.2003.12.005

Yasui K, Tuziuti T, Kozuka T, Towata A, Iida Y (2007) Relationship between the bubble temperature and main oxidant created inside an air bubble under ultrasound. J Chem Phys 127:154502. https://doi.org/10.1063/1.2790420

Chapter 2
Production of Next-Generation Biodiesel from High Yielding Strains of Microorganisms: Recent Advances

Charles Oluwaseun Adetunji, Olugbemi Tope Olaniyan, and Nonso Evaristus Okeke

Contents

Abbreviations

ACP	Acyl-acyl carrier protein
ATP	Adenosine triphosphate
CMC	Carboxymethyl cellulose
CRISPRs	Clustered regularly interspaced short palindromic repeats

C. O. Adetunji (✉)
Applied Microbiology, Biotechnology and Nanotechnology Laboratory, Department of Microbiology, Edo University Iyamho, Iyamho, Edo State, Nigeria
e-mail: adetunji.charles@edouniversity.edu.ng

O. T. Olaniyan
Laboratory for Reproductive Biology and Developmental Programming, Department of Physiology, Edo University Iyamho, Iyahmo, Edo State, Nigeria

N. E. Okeke
Department of Chemical Engineering, Edo University Iyamho, Iyamho, Edo State, Nigeria

Inamuddin, A. M. Asiri (eds.), *Sustainable Green Chemical Processes and their Allied Applications*, Nanotechnology in the Life Sciences,
https://doi.org/10.1007/978-3-030-42284-4_2

FAME	Fatty acid methyl ester
GC/MS	Gas chromatography-mass spectroscopy
NAD(P)H	Nicotinamide adenine dinucleotide phosphate
NaOH	Sodium hydroxide
RNA	Ribonucleic acid
SCO	Single cell oil
TALENs	Transcription activator-like effector nucleases

2.1 Introduction

The global fuel consumption was estimated to be 87 million barrels per day, and this value has been forecasted to rise to 122 million barrels of liquid fuel per day in 2040 (US Energy Information Administration 2014). This has been a long time global challenges because of the high rate of reduction in fossil fuels reserve. Therefore, there is of a necessity to search for a substitute and sustainable biofuel to meet the request of ever-increasing population. Biodiesel has been discovered as a sustainable fuel that could form a permanent replacement to all the adverse effect of petroleum diesel including lower emissions generation and presence of high sulfur content (Shahir et al. 2015). The International Energy Agency has stated that the fabrication of biofuels is the greatest demand of our time because of the following reasons which include prevention of greenhouse effect, sustainability, diversity, and energy security as well as increase in price of crude oil, world decrease of fossil fuels, and increase in the rate of environmental pollution (Kent Hoekman 2009).

Biofuel has been discovered as the next reliable solution that could meet the global demand for fuel utilized for transportation by 2050 (Volk et al. 2000). It has been discovered that the process of transesterification reaction could be catalyzed enzymatically or chemically (Akhil et al. 2010) during biodiesel production, but the utilization of engineered microorganisms has been shown as an important and a greater capability that could increase the high yield of biodiesel at industrial scale (Rainer et al. 2006).

The fabrication of biodiesel is mainly performed through the process of transesterification that happens between a short chain alcohol which might be ethanol or methanol and an acylglycerol obtained from fats of animal or oils obtained from vegetable through the influence of a catalyst. The history of biodiesel started from the utilization of some vegetable oil derived from coconut oil, soybean, palm oil, sunflower, and rapeseed. The biodiesel derived from this is referred to as the first-generation biodiesel. The main disadvantage is that the cost of feedstock may account for about 80% out of the overall cost of producing biofuel (Lam et al. 2010). The second-generation biodiesel are normally obtained from nonedible vegetable oil such as pongamia, rubber seed, jatropha, tobacco, castor, neem, karanja, and babassu. It has advantage over the first-generation biodiesel because the oil used for the production is not used as edible cooking oil (Dorado 2008). The third-generation biodiesel fabrication is generated mainly from the oils derived from microorganisms such as fungi, bacteria, microalgae, and yeast (Azócar et al. 2010; Revellame et al. 2012; Raimondi et al. 2014; Hidalgo et al. 2014; Zhang et al. 2014; Mattanovich et al. 2014).

Moreover, many of such organisms that produce oil used in the production of biodiesel are referred to as oleaginous. Triacylglycerol is the type of lipid deposited in the oleaginous microorganisms which form an alternative to the oil derived from animal and agricultural wastes (Qiang et al. 2008; Hull 2010; Xin et al. 2009). A typical example of biodiesel produced using microbial lipids is referred to as single cell oils (Joseph 2006). The high rate of oil buildup in these microorganisms which improve high-quality microbial diesel has been recognized to be economically inexpensive (Xin et al. 2009). It has been discovered that the biomass of some beneficial microorganism consists of lipid which ranges between 40 and 70% mainly from *Yarrowia*, *Rhodosporidium*, *Lipomyces*, and *Cryptococcus*, respectively (Li et al. 2007). Moreover, some fungi most especially from zygomycetes like *Cunninghamella* and *Mortierella* pose some capability to accumulate lipid (Papanikolaou et al. 2007). Examples of typical fatty acids available in the single cell oil are stearic acid, oleic acid, palmitic acid, and linoleic acid (Chen and Liu 1997; Peng and Chen 2008).

This chapter intends to provide the current trends about the application of beneficial microorganism that could produce biodiesel as a permanent replacement to convention fuel.

2.2 Application of Metabolomics Engineering for Production of Biofuel

The microorganism host's metabolism has been shown to have thousands of chemical reactions that regulates energy and carbon most especially NAD(P)H and ATP. In order to produce high quantity and quality biofuel, it requires numerous enzymatic steps that will catalyze several biosynthetic pathways. There are several molecular biology techniques that have been discovered to induce changes in the level of enzymes flux during biofuel synthesis. Some of these novel strategies includes synthetic scaffolds, knockout/knockdown of competitive pathways, types of plasmids, enhancement of ribosome binding sites, promoter engineering, modification of key enzymes, directed evolution, and codon optimization (Dueber et al. 2009; Carneiro et al. 2013; Nowroozi et al. 2014). Moreover, several biotechnological intervention which leverage on genetic techniques have been documented to show new potential that could help in the edition of microbial metabolisms such as TALENs, CRISPRs, and RNA interference, respectively (Jiang et al. 2013; Sun and Zhao 2013). There are significant strategies that have been identified that could enhance the carbon fluidity of the finished products whenever metabolomics engineering is applied. They include "push-pull-block," utilized for improving the flux movement during the biofuel production pathway (Atsumi et al. 2010; Kind et al. 2013). Moreover, the utilization of synthetic biology led to the development of genetic circuits which can accurately control the expression of gene whether there is presence of environmental or chemical inputs (Khalil and Collins 2010). Some of these synthetic biological gadgets which have relevant application in this regards include genetic oscillators, a toggle switch, and trigger-memory system (Way et al. 2014). Moreover, the metabolomics engineers have started utilizing synthetic biology for the manipulation of fluxes toward regulation of biosynthetic pathways available at several stages during fermentation. In this regard, Soma et al. (2014) performed an experiment on toggle switch into *E. coli* that could switch off the process of tricarboxylic acid cycle and change the direction of flux toward isopropanol. Also, some other synthetic biology tools like biosensor-regulator systems have been established to regulate several host metabolisms depending on the changes in the environment (Neupert et al. 2008; Topp et al. 2010; Gorochowski et al. 2013). This will eventually lead to increase in the microbial host's metabolic rate (Zhang et al. 2012).

2.3 Production of Biodiesel from Different Oleaginous Microorganisms

Shafiq and Ali (2017) isolated and identified oleaginous fungi from numerous cereals (sesame, fenugreek, corn black grain, mustard) available in various locations at Baghdad City. The result showed that the species and genera vary from the frequency and percentage of occurrence. The strain of *Penicillium* was shown to exhibit 44.4%, 50%, and 42.8% from the following samples of cereals sesame, cereals, and corn black grain. The several strains of the fungal isolates show an enhanced extracellular lipase activity when assayed on the solid agar which varies from nonproducer, producer, and weak producer. The following seven fungal strains exhibited the capability to produce lipase activity when tested in liquid media containing yeast extract sucrose. They include *Aspergillus fumigatus, Trichoderma harzianum, Aspergillus niger, Fusarium graminearum, Aspergillus terreus, Penicillium* spp., and *Aspergillus flavus*. It was later discovered that *Penicillium* spp. was shown to exhibit the highest biomass value of 20.15 g/L while the total amount of the lipid content showed 20% more than their dry weight with great exception of *A. niger*. The structural elucidation performed by gas chromatography-mass spectroscopy (GC/MS) has showed the occurrence of stearic acid, palmitic, linoleic acid, and oleic acids in all the isolated isolates which showed similar properties when compared with standard biodiesel. One of the major problems faced in the commercialization of biodiesel is the exorbitant in the total price of their production. Therefore, there is a need to search for sustainable biodiesel. The utilization of microbial lipids has been shown as a sustainable replacement to some other biodiesel.

Neema and Kumari (2013) utilized novel strains of fungi and yeast strains obtained from secondary sludge obtained from wastewater treatments soil and plants. They were selected based on the fact that they could utilize glycerol as their carbon source. The staining of the fugal isolates with Sudan Black B showed the availability of lipid inclusion bodies inside their cells when cultivated in nitrogen-limiting environments. The structural elucidation performed by GC-MS showed that the most predominant lipids present in these microorganisms were C12:0-C20:2 while the main fatty acids available in the lipids of the yeast cell were linoleic acid (C18:2) (28%) followed by oleic acid (C18:1) (18%) and palmitic acid (C16:0) (19%) when cultured under nitrogen-limited environment. The result obtained demonstrates that the fatty acid composition obtained from the lipids from these oleaginous yeast and fungi is comparable to that of the vegetable oils. The authors suggested that more research should be channeled to the utilization of cheap carbon source obtained from wastewater obtained from sewage sludge and food industry for improved lipid production which will enhance the rate of biodiesel production because of the availability of cheap substrate for mass production.

Cea et al. (2015) evaluated sewage sludge as a cheap source of microbial oils feedstock for high fabrication of biodiesel. The authors obtained several samples of wastewater generated from treatment plants mainly from La Araucanía region in Southern Chile. It was revealed that the total lipids content ranges between 7.7 and 12.6%. It was observed that 50% of the overall removed lipids obtained from Vilcún region exhibited the maximum transesterifiable lipids content. Also, it was established that the following bacterial samples containing *Bacillus*, *Acinetobacter*, and *Pseudomonas* were isolated from sludge samples. Moreover, it was observed that *Bacillus* sp. of strain V10 demonstrated the highest lipid content of 7.4%. The culturing of *Bacillus* sp. V10 utilizing urban wastewater augmented with glucose was used to obtained nitrogen-exhausted medium, while a low-cost carbon source was obtained from milk processing wastewater. The structural analysis showed that the lipid profile of *Bacillus* sp. V10 reveals the presence of unsaturated long-chain fatty acids which are of low degree like C18:1 constitute 50% of the lipids' content that showed it could be exploited as a crude material for fabrication of biodiesel.

Kitchaa and Cheirsilp (2011, b) isolated 889 yeast strains from wastes of palm oil mill, soils, and biodiesel plant available in southern region of Thailand. The cultural condition utilized glycerol and glucose as carbon source with 0.0001% chloramphenicol pH of 6.0 and acidic condition with pH of 4.0. The analysis performed using Sudan Black B tests reveals the 23 strains that are oleaginous yeast or prospective lipid producer. The amount of lipid value available in these 23 strains was associated to crude glycerol-based medium. It was revealed that strain BY4–523 of *Kodamaea ohmeri* accrued maximum lipid content of 53.28%, while strain JU4–57 of *Trichosporonoides spathulata* showed a lipid content of 41.50%. The blend of yeast extract and peptone (1:1) serving as nitrogen sources showed maximum biomass of 11.1 g/L for *K. ohmeri* and 17.05 g/L for *T. spathulata* while maximum lipid value of 4.53 g/L for *K. ohmeri* and 10.43 g/L for *T. spathulata.* Moreover, it was observed that ammonium sulfate was the best inorganic nitrogen source that exhibited the greatest activity with 10.45 g/L for *K. ohmeri* and 9.17 g/L for *T. spathulata* while the highest lipid production was 3.17 g/L for *K. ohmeri* and (3.85 g/L for *T. spathulata.* Furthermore, high lipid content was observed from the new isolate of yeast which could utilize crude glycerol-containing medium augmented with only ammonium sulfate. Also, 0.5% ammonium sulfate and 10% crude glycerol were the best optimum conditions that showed the best activity with lipid production of 4.45 g/L and highest biomass of 10.40 g/L from *T. spathulata*, while strain *K. ohmeri* showed lipid production of 3.22 g/L and maximum biomass of 10.50 g/L.

Thancharoen et al. (2017) screened the biodiesel potential of oleaginous yeasts and isolated the best strain that stored the maximum amount of lipid for fabrication of lipid when glucose was used as a carbon source. It was discovered that 10 oleaginous yeasts showed the highest potential to produce biodiesel among the 24 yeast microorganisms from the wastewater obtained from canteen screened when subjected to Sudan Black B. The application of Sudan Black B staining has shown the availability of lipid inclusion bodies present in their cells when cultivated in nitrogen-limiting environments. Moreover, it was observed that the total oleaginous strains stored amounts of lipid greater than 20% of their biomass when glucose was used as a source of carbon. Also, it was observed that strain CTWY07 accumulated up to 71.59% of its biomass which consists of 6.85 g/L biomass and 4.904 g/L lipid under a non-static condition. Their study shows that yeast isolates could be utilized as a possible candidate for the fabrication of biodiesel.

Nawabi et al. (2011) reported the utilization of bacterial fatty acid methyltransferase that catalyzes the development of 3-hydroxyl fatty acid methyl esters from *S*-adenosylmethionine and from particular free acids. The results revealed that fatty acid methyltransferase showed enhanced specificity toward 3-hydroxy free fatty acids when compared to free fatty acids leading to the formation of 3-hydroxy fatty acid methyl esters in vitro assay. The study also showed a member of bacterial that poses fatty acyl-acyl carrier protein and thioesterase enzyme that poses specific acyl chain. It was observed that the fatty acyl-acyl carrier protein (ACP) thioesterase showed enhanced specificity on 3-hydroxyacyl-ACP leading to 3-hydroxy free fatty acids which can be utilized by fatty acid methyltransferase to generate 3-hydroxyl fatty acid methyl esters while the co-expression of PhaG using fatty acid methyltransferase led to the storage of 3-hydroxy free fatty acids and fatty acid methyltransferase. Also, it was revealed that the occurrence of fatty acid methyltransferase constitutes a substantial factor that controls the rate of methyl esters synthesis by the bacterial cell. Their result showed that elimination of *metJ* which is a global methionine regulator as well as the overexpression of methionine adenosyltransferase led to enhanced methyl ester synthesis.

Wahlen et al. (2013) described the performance of biodiesel produced from numerous oleaginous microbes including bacteria, microalgae, and yeast. The stains utilized for the engine with a two-cylinder diesel engine performance include *Rhodococcus opacus*, *Chaetoceros gracilis*, and *Cryptococcus curvatus*. The biodiesel obtained from these strains was compared to commercial biodiesel from soybeans and petroleum diesel. Some of the physicochemical properties of the microbial-derived biodiesel evaluated include cetane index, heating value, density, and viscosity. The result of their analysis of the exhaust emissions showed that all the biofuel tested shows minimum hydrocarbon and CO when compared to the petroleum diesel. Moreover, it was observed that microalgae biodiesel yield the smallest level of NOx when compared to the petroleum diesel. Their study showed that the composition of the fatty acids obtained from the microbial-derived biodiesel could form a permanent replacement to petroleum diesel.

de Moura et al. (2015) isolate fungi that are capable of producing biodiesel from swine waste stabilization lake using standard procedures. The various microbiology media used include standard count, potato dextrose agar, and nutrient agar. The serial dilution and the direct counting of plates for the amount of colony-forming units for this strain were determined after 9 and 5 days after incubation in the dark at 28 °C. The total amount of oil yield was determined using solvent extraction technique. The results obtain indicated that 13 fungal isolates pose lipid content that was above 25% of their total biomass which substantiate them as oleaginous microorganisms. The following isolates exhibit maximum potential for the production of biodiesel containing 33 (*Rhizopus* sp.), 25 (*Mucor* sp.), and 31 (*Rhizopus* sp.)

Chioke et al. (2018) wrote a comprehensive report on the application of enzyme-catalyzed transesterification for the production biodiesel as a permanent replacement to solve the problem facing mankind in the area of environmental science. It was stated that biodiesel which is a methyl ester could be produced using enzymatic catalysis or chemical reaction. The process of producing biodiesels using enzyme method has been discovered to be very costly most especially during the process of production and purification of the catalyst as well as their low activity most especially during the biological conversion of the process. Their review work stated some solution that could reduce the production cost implication of lipase-catalyzed transesterification, recent techniques, and merits of enzymatic transesterification when compared to synthetic chemical-catalyzed transesterification. Also, some significant factors that regulate the process of catalyzed transesterification are highlighted in detail, while several possible solutions that could help in the mitigation of challenges encountered during enzymatic transesterification are well elucidated.

The production of microbial oil from microorganisms has been recognized as a sustainable alternative to bases of triglycerides for the biofabrication of biodiesel. However, it has been discovered that the process of producing biodiesel using microorganisms has been reorganized to be very costly. The application of cheaper nutrient sources for the culturing and enforcement of cheaper production processes could decrease the cost of production which will make the utilization of microbial lipids inexpensive when compared to commodity-type oils utilized for the fabrication of biodiesel. Moreover, it has been discovered that the fabrication of biodiesel using oleaginous seeds is based on a multistep process. The application of in situ transesterification happens in the biomass without passing through the process of lipid extraction. Martinez-Silveira et al. (2019) evaluate the application of in situ transesterification from the biofabrication of biodiesel using oleaginous yeasts. The best condition that favors the stages involved in the biodiesel production was obtained through in situ transesterification using oleaginous yeast (*Rhodotorula graminis* S1/S2) which was performed through response surface methodology. The result obtained using response surface methodology established that the highest optimum conditions that led to the highest production of biodiesel include methanol-to-biomass ratio of 60:1, incubation at 70 °C for 3 h, and 0.4 M H_2SO_4. It was also observed that the process of optimized in situ process led to the production that was

123% higher when compared to the two-step techniques in which fatty acids derived from saponifiable lipids were removed before the process of esterification with methanol. The results obtained from the composition of fatty acid ester showed that *R. graminis* S1/S2 using in situ transesterification conform to the Uruguayan standards for biodiesel. Their study shows that microbial oil could be utilized as a potential replacement to the synthetic biodiesels most especially from yeast when produced at industrial scale.

Winayanuwattikun et al. (2011) evaluated the lipase producing ability of microorganisms and assess their possibility to biofabricate biodiesel utilizing palm oil as substrate. The authors isolated and screened 360 strains of yeast, fungi, and bacteria from wastewater and oil-contaminated soil. They observed that strains with high lipase-generation include *Fusarium solani, Candida rugosa, and Staphylococcus warneri*. The lipase producing capability of these strain was performed using palm oil as inducer, while the lipase activity was compared to synthetic and hydrolytic catalysis. Moreover, the best strain that produced the highest capability for catalyzing biodiesel production was later immobilized using different hydrophobic support after the process of purification. Their study shows that utilization of enzymatic transesterification feedback for the production of biodiesel using vegetable oils as a substrate is a very promising biotechnological technique.

Kuan et al. (2018) revealed that lipid produced from oleaginous microbes has magnitude higher than those of energy crops and can serve as alternative feedstock for biodiesel production, but the cost of production, organic solvents used, and energy could impose a great barrier to the biotransformation of the lipids to biodiesel. These authors decrease the amount of organic solvents and energy in the presence of sodium hydroxide or sulfuric acid as a catalyst; they reported that *Rhodotorula glutinis* (oleaginous yeast) was directly transesterified without passing through the stage of lipid pre-extraction given rise to biodiesel. The results obtained by different methods confirmed that 1 g of dry *R. glutinis* biomass of the fatty acid methyl ester (FAME) yielded 111%; 1 g/L NaOH of the FAME yielded 102%; and the acid-catalyzed method displayed a superior moisture tolerance. They concluded by alluding to the fact that direct transesterification simplifies, reduces the reaction time, and improves the FAME yield associated to orthodox transesterification which necessitates lipid pre-extraction.

Nithya and Velayutham (2011) revealed that in the near future biofuel will soon become the world's source of renewable and sustainable energy due to the decline in the conventional energy sources and its accompanied environmental and health challenges. These authors confirmed that during metabolic processes, microorganisms such as fungi, bacteria, and algae could accrued over 70% intracellular lipid, especially fungi lipid which is less studied as compared to others.

El-haj et al. (2015) revealed that single-cell oil (SCO) oleaginous microorganisms such as yeast, fungi, and bacteria when subjected to certain culture conditions are capable of accumulating intracellular lipids. These authors evaluated the biodiversity of oleaginous microorganism in Lebanese using 39 isolates SCO producers selected from different habitats identified as bacteria (*Arthrobacter* sp., *Escherichia coli*, *Agrobacterium* sp., *Pantoea* sp., and *Chryseobacterium* sp.), yeasts, and filamentous fungi (*Cryptococcus* sp., *Candida* sp., *Lipomyces* sp., *Fusarium oxysporum*, *Yarrowia* sp., *Mucor hiemalis*, *Aspergillus tamari*, *Penicillium citrinum*, and *Aspergillus niger*). The results revealed varied biodiversities of isolated microbes as 22 (56.4%) in farmland, 8 (20.5%) in sand, 5 (12.8%) in lawn, and 4 (10.3%) in wetland. When suggested to carbon utilization test, the results indicated that for maximum lipid production, most of the filamentous fungi can use carboxymethyl cellulose (CMC) and xylose as a sole carbon source considered as a promising SCO producers using agro-industrial waste materials.

Shruthi et al. (2014) screened 22 isolated bacteria for lipid content for biodiesel production using Sudan Black B staining method; they discovered that three of the isolates are rich in lipid—*F. oryzihabitans*, *Morococcus* sp., and *P. aeruginosa*—taken for lipid extraction by Bligh and dyer method and thin-layer chromatography for the determination of fatty acid. The results showed that *Morococcus* sp. has the highest amount of lipid of lauric acid (C12). The optimization showed carbon (0.70 mL/100 mL) of palmitic acid (C18), nitrogen (0.65 mL/100 mL) of arachidonic acid (C20), and pH 7.0 (1.96 mL/100 mL) of oleic acid (C18) and in strain enhancement through mutagenesis of *F. oryzihabitans* got enhanced and accrued high lipid at 10 min (1.50 mL/100 mL) of linoleic acid (C16).

Abu-Elreesh and Abd-El-Haleem (2014) investigated numerous fungal isolates obtained from various Egyptian ecosystems for their capability to investigate intracellular lipid buildup using Nile-red viable colony staining assay. The various four strains which were coded F1, F2, F3, and F4 were selected out of all the isolated strains as potential lipid producers. These potential isolates were cultured in a basal-based medium having yeast and glucose as its source of carbon and nitrogen. The process of lipid production was performed using the standard procedure. The potential lipid-producing strains were characterized using molecular and morphological techniques as F1, F2, F3, and F4 as *Drechslera* sp., *Fusarium* sp., *Aspergillus fumigates*, and lipid content of 50, 49, 46, and 71%, respectively. Also subjecting the fatty acids through GC/MS, the fatty acid profiles revealed the presence of octadecanoic acid, hexadecanoic acid, and 11-octadecenoic acid in all isolates, while 9,12-octadecadienoic acid was detected only in the lipids of strain F4 with the percentage value of 40%. They concluded that utilization of beneficial fungal isolates in biodiesel fabrication will have energy balance and decrease the emissions of greenhouse gases.

2.4 Synopsis of Oil Deposition in Microalgae

Microalgae bioproduction of fat molecules especially the "family of green algal" is an interconnected complex of plethora metabolic pathway. The production starts in the microalgal cells by chloroplast photosynthesizing mechanism exploiting the carbon from the atmosphere to generate starch and then catabolized by glycolytic pathway leading to the production of the backbone of fatty acids and triacylglycerol. Combining all the enzymes through acetyl-CoA carboxylase to generate malonyl-CoA induces fatty acid bioproduction from acetyl-CoA. Malonyl-CoA biotransformation into malonyl-ACP signifies the commencement of elongation stage of fatty acid bioproduction through a prokaryotic type II fatty acid synthase catalyzing the process which is confined in the cell stroma (Gresham et al. 2008). This process is disturbed occasionally by fatty-ACP thioesterases leading to the newly formed fatty acid leak from complex acyl-ACP (Wenger et al. 2011). Free fatty acid molecules' form is integrated and manufacturing several cellular fats. The production of lipids is linked to the reactions in Kennedy pathway, involving the combination of fatty acids to form glycerol backbone synthesizing triacylglycerol. According to Kennedy pathway, acyl-CoA or acyl-ACP is used as acyl donor, whereas microalgae alternatively can move in as pathway for obtaining acyl groups expending phospholipids by way of acyl donation for triglyceride production (Gresham et al. 2008). Rapid plant manipulation for oil production and development provides evidences aimed at microalgal bioengineering in lipid metabolic pathways, providing the alterations in lipid metabolism improvement in plants and microalgae are understood.

The identification of important genes of microalgae triglyceride production sets the basis for plant's lipid metabolic processes which has enabled the prediction of fat metabolic genes. In *Chlamydomonas* genome, some of the anticipated genes involved in fatty acid generation are shown as a single copy, demonstrating their encoded enzymes in both mitochondria and chloroplast, but different in higher plants where compartment-specific enzymes rise the complexity of fatty acid generation. Different duplicates of genes encoding acyl-CoA: diacylglycerol acyltransferases were predictable in *Chlamydomonas* gene, linked with the fewer copies noticed in developed plants, showing an important role of triacylglycerol microalgal cell physiology (Goddard et al. 2005). *Chlamydomonas* is a microalgae model which accrues starch as a major source of energy, but due to stress, it changes to triacylglycerol, suggestive of adaptation changes necessary for preserving membrane integrity. Additionally, triacylglycerol is broken down releasing fatty acids for the membrane synthesis when stress is reversed. Moreover, triacylglycerol fragments actions as reservoir for directing surplus energy and equivalents, which may then predispose cellular breakdowns (Gray and Goddard, 2012). Additional evidence is the existence of distinctive triglycerides "betaines," which provides a benefit to microalgae in adapting to nutrient-limiting environments.

2.5 Conclusion, Future Direction, and Recommendation

This chapter has discussed in detail all the recent biotechnological technologies and trends used in the process of transforming substrates into sustainable, economical, and eco-friendly biodiesel. Moreover, it has been observed that despite the several breakthroughs that have been recorded, there is a need to develop more biotechnological solution that could lead to increase in the scale of production majorly through the biotransformation of the fatty acid methyl esters present in oleaginous microbes.

Moreover, there is a need to search for more novel microorganisms that can tolerate any adverse environmental factor produced during the fermentation process without affecting the quality and quantity of biodiesel produced. Also, there is a need for mutual and effective collaboration between the government, researchers, and policy makers to work together. This will lead to the development of sustainable biodiesel as an alternate to fossil fuels through the development application of recent biotechnological techniques that could ensure energy balance and decrease the emissions of greenhouse gases and reduce the cost of production there making biodiesel as the next-generation biofuel to meet the problem of transportation and prevent climate as well as environmental challenges through the maintenance of healthy planet.

References

Abu-Elreesh G, Abd-El-Haleem D (2014) Promising oleaginous filamentous fungi as biodiesel feed stocks: screening and identification. Eur J Exp Biol 4(1):576–582

Akhil B, Purva L, Prabhat N, Jha R, Narain M (2010) Biodiesel production through lipase catalyzed transesterification. J Mol Catal B Enzym 59(9):21

Atsumi S, Wu TY, Eckl EM, Hawkins S, Buelter T, Liao J (2010) Engineering the isobutanol-biosynthetic pathway in *Escherichia coli* by comparison of three aldehyde reductase/alcohol dehydrogenase genes. Appl Microbiol Biotechnol 85:651–657. https://doi.org/10.1007/s00253-009-2085-6

Azócar L, Giudad G, Heipieper H, Navia R (2010) Biotechnological processes for biodiesel production using alternative oils. Appl Microbiol Biotechnol Appl Microbiol Biotechnol 88:621–636

Carneiro S, Ferreira EC, Rocha I (2013) Metabolic responses to recombinant bioprocesses in *Escherichia coli*. J Biotechnol J Biotechnol 164:396–408. https://doi.org/10.1016/j.jbiotec.2012.08.026

Cea M, Sangaletti-Gerhard N, Acuña P, Fuentes I, Jorquera M, Godoy K, Osses F, Naviaa R (2015) Screening transesterifiable lipid accumulating bacteria from sewage sludge for biodiesel production. Biotechnol Rep 8:116–123

Chen HC, Liu TM (1997) Inoculum effects on the production of gammalinolenic acid by the shake culture of *Cunninghamella echinualata* CCRC 31840. Enzyme Microb Technol 21(137):142

Chioke OJ, Ogbonna CN, Onwusi C, Ogbonna JC (2018) Lipase in biodiesel production. Afr J Biochem Res 12(8):73–85. https://doi.org/10.5897/AJBR2018.0999

Dorado MP (2008) In: Nag A (ed) Raw materials to produce low cost biodiesel. Biofuels refining and performance. McGraw-Hill, New York, pp 107–147

Dueber JE, Wu GC, Malmirchegini GR, Moon TS, Petzold CJ, Ullal AV, Prather KL, Keasling JD (2009) Synthetic protein scaffolds provide modular control over metabolic flux. Nat Biotechnol 27:753–759. https://doi.org/10.1038/nbt.1557

El-haj M, Olama Z, Holail H (2015) Biodiversity of oleaginous microorganisms in the Lebanese environment. Int J Curr Microbiol Appl Sci 4(5):950–961

Goddard MR, Godfray HC, Burt A (2005) Sex increases the efficacy of natural selection in experimental yeast populations. Nature 434:636–640

Gorochowski TE, van den Berg E, Kerkman R, Roubos JA, Bovenberg RAL (2013) Using synthetic biological part sand microbioreactors to explore the protein expression characteristics of *Escherichia coli*. ACS Synth Biol 3:129–139. https://doi.org/10.1021/sb4001245

Gray JC, Goddard MR (2012) Sex enhances adaptation by unlinking beneficial from detrimental mutations in experimental yeast populations. BMC Evol Biol 12:43

Gresham D, Desai MM, Tucker CM, Jenq HT, Pai DA, Ward A, DeSevo CG, Botstein D, Dunham MJ (2008) The repertoire and dynamics of evolutionary adaptations to controlled nutrient-limited environments in yeast. PLoS Genet 4(12):e1000303. https://doi.org/10.1371/journal.pgen.1000303. Epub 2008 Dec 12

Hidalgo P, Toro C, Ciudad G, Schober S, Mittelbach M, Navia R (2014) Evaluation of different operational strategies for biodiesel production by direct transesterification of microalgal biomass. Energy Fuel 28:3814–3820

Hull (2010) Patterns of accumulation. J Mol Catal B Enzym 62(9):14

Jiang W, Bikard D, Cox D, Zhang F, Marraffini LA (2013) RNA-guided editing of bacterial genomes using CRISPR-Cassystems. Nat Biotechnol 31:233–239. https://doi.org/10.1038/nbt.2508

Joseph Gonsalves B (2006) An assessment of the biofuels industry in India. United Nations conference on trade and development. UNCTAD/DITC/TED/2006/6

Kent Hoekman S (2009) Biodiesel in US-challenges and opportunities. Renew Energy 34(14):22

Khalil AS, Collins JJ (2010) Synthetic biology: applications come of age. Nat Rev Genet 11:367–379. https://doi.org/10.1038/nrg2775

Kind S, Becker J, Wittmann C (2013) Increased lysine production by flux coupling of the tricarboxylic acid cycle and the lysine biosynthetic pathway—metabolic engineering of the availability of succinyl-CoA in *Corynebacterium glutamicum*. Metab Eng 15:184–195. https://doi.org/10.1016/j.ymben.2012.07.005

Kitchaa S, Cheirsilp B (2011) Screening of oleaginous yeasts and optimization for lipid production using crude glycerol as a carbon source. Energy Procedia 9(2011):274–282. https://doi.org/10.1016/j.egypro.2011.09.029

Kuan IC, Kao WC, Chen CL, Yu CY (2018) Microbial biodiesel production by direct transesterification of *Rhodotorula glutinis* biomass. Energies 2018(11):1036. https://doi.org/10.3390/en11051036

Lam MK, Lee KT, Mohamed AR (2010) Homogeneous, heterogeneous and enzymatic catalysis for transesterification of high free fatty acid oil (waste cooking oil) to biodiesel: a review. Biotechnol Adv 28:500–518

Li Y, Zhao ZK, Bai F (2007) High density cultivation of oleaginous yeast *Rhodosporidium toruloides Y4* in fed batch culture. Enzyme Microb Technol 41(312):317

Martinez-Silveira A, Villarreal R, Garmendia G, Rufo C, Vero S (2019) Process conditions for a rapid in situ transesterification for biodiesel production from oleaginous yeasts. Electron J Biotechnol 38:1–9

Mattanovich D, Sauer M, Gasser B (2014) Yeast biotechnology: teaching the old dog new tricks. Microb Cell Factories 13:34

de Moura JB, Moreira RM, Jakoby ICMC, de Souza Castro CF, Soares MA, Andrade RDA, Souchie EL (2015) Isolation of microorganisms from a swine waste stabilization Lake for biodiesel production American-Eurasian. J Agric Environ Sci 15(8):1630–1636. https://doi.org/10.5829/idosi.aejaes.2015.15.8.12756

Nawabi P, Bauer S, Kyrpides N, Lykidis A (2011) Engineering *Escherichia coli* for biodiesel production utilizing a bacterial fatty acid methyltransferase. Appl Environ Microbiol 77(22):0099–2240. https://doi.org/10.1128/AEM.05046-11

Neema PM, Kumari A (2013) Isolation of lipid producing yeast and fungi from secondary sewage sludge and soil. Aust J Basic Appl Sci 7(9):283–288

Neupert J, Karcher D, Bock R (2008) Design of simple synthetic RNA thermometers for temperature-controlled gene expression in *Escherichia coli*. Nucleic Acids Res 36:e124. https://doi.org/10.1093/nar/gkn545

Nithya DM, Velayutham P (2011) Biodiesel production from Fungi. Indian J Nat Sci 1(5):275

Nowroozi F, Baidoo EK, Ermakov S, Redding-Johanson AM, Batth TS, Petzold CJ, Keasling JD (2014) Metabolic pathway optimization using ribosome binding site variants and combinatorial gene assembly. Appl Microbiol Biotechnol 98:1567–1581. https://doi.org/10.1007/s00253-013-5361-4

Papanikolaou S, Galiotou-Panayotou M, Fakas S, Aggelis G (2007) Lipid production by o*leaginous Mucorales* cultivated on renewable carbon sources. Eur J Lipid Sci Technol 109(1060):1070

Peng X, Chen H (2008) Rapid estimation of single cell oil content of solid-state fermented mass using near-infrared spectroscopy. Bioresour Technol 299(8869):8872

Qiang L, Wei D, Dehua L (2008) Perspectives of microbial oils for biodiesel production. Appl Microbial Biotechnol 80(749):756

Raimondi S, Rossi M, Leonardi A, Bianchi M, Rinaldi T, Amaretti A (2014) Getting lipids from glycerol: new perspectives on biotechnological exploitation of *Candida freyschussii*. Microb Cell Factories 13:83

Rainer K, Torsten S, Alexander S (2006) Microdiesel: *Escherichia coli* engineered for fuel production. Microbiologica 152(2529):2536

Revellame E, Hernandez R, French W, Holmes W, Benson T, Pham J, Forks A, Callahan R (2012) Lipid storage compounds in raw activated sludge microorganisms for biofuels and oleochemicals production. RSC Adv 2:2015–2031

Shafiq SA, Ali RH (2017) Myco-diesel production by oleaginous fungi. Res J Pharm Biol Chem Sci 8(2):1252

Shahir VK, Jawahar CP, Suresh PR (2015) Comparative study of diesel and biodiesel on CI engine with emphasis to emissions—a review. Renew Sustain Energy Rev 45:686–697

Shruthi P, Rajeshwari T, Mrunalini BR, Girish V, Girisha ST (2014) Original research evaluation of oleaginous Bacteria for potential biofuel. Int J Curr Microbiol Appl Sci 3(9):47–57

Soma Y, Tsuruno K, Wada M, Yokota A, Hanai T (2014) Metabolic flux redirection from a central metabolic pathway toward a synthetic pathway using a metabolic toggle switch. Metab Eng 23:175–184. https://doi.org/10.1016/j.ymben.2014.02.008

Sun N, Zhao H (2013) Transcription activator-like effect or nucleases (TALENs): a highly efficient and versatile tool for genome editing. Biotechnol Bioeng 110:1811–1821. https://doi.org/10.1002/bit.24890

Thancharoen K, Malasri A, Leamsingkorn W, Boonyalit P (2017) Selection of oleaginous yeasts with lipid accumulation by the measurement of Sudan black B for benefits of biodiesel. J Pharm Med Biol Sci 6(2):53–37

Topp S, Reynoso CMK, Seeliger JC, Goldlust IS, Desai SK, Murat D, Shen A, Puri AW, Komeili A, Bertozzi CR, Scott JR, Gallivan JP (2010) Synthetic riboswitches that induce gene expression in diverse bacterial species. Appl Environ Microbiol 76:7881–7884. https://doi.org/10.1128/AEM.01537-10

U.S. Energy Information Administration (EIA) (2014) International Energy Outlook (DOE/EIA-0484/2014). www.eia.gov/forecasts/ieo/pdf/0484(2014).pdf. Accessed January 2015

Volk TA, Abrahamson LP, White EH, Neuhauser E, Gray E, Demeter C, Linsey C, Jarnefeld J, Anerhansley DJ, Pellerin R, Edick S (2000) Developing a willow biomass crop enterprises for bioenergy and bioproducts in the United States. https://bioenergy.ornl.gov/papers/bioen00/volk.html

Wahlen BD, Morgan MR, McCurdy AT, Willis RM, Morgan MD, Dye MJ, Bugbee B, Wood BD, Seefeldt LC (2013) Biodiesel from microalgae, yeast, and bacteria: engine performance and exhaust emissions. Energy Fuels 27:220–228. https://doi.org/10.1021/ef3012382

Way JC, Collins JJ, Keasling JD, Silver PA (2014) Integrating biological redesign: where synthetic biology came from and where it needs to go. Cell 157:151–161. https://doi.org/10.1016/j.cell.2014.02.039

Wenger JW, Piotrowski J, Nagarajan S, Chiotti K, Sherlock G, Rosenzweig F (2011) Hunger artists: yeast adapted to carbon limitation show trade-offs under carbon sufficiency. PLoS Genet 7(8):e1002202. https://doi.org/10.1371/journal.pgen.1002202. Epub 2011 Aug 4

Winayanuwattikun P, Kaewpiboon C, Piriyakananon K, Chulalaksananukul W, Yongvanich T, Svasti J (2011) Afr J Biotechnol 10(9):1666–1673. https://doi.org/10.5897/AJB10.1802

Xin M, Jianming Y, Xin X, Lie Z, Qingjuan N, Mo X (2009) Biodiesel production from oleaginous microorganisms. Renew Energy 34(1):5

Zhang FZ, Carothers JM, Keasling JD (2012) Design of a dynamic sensor-regulator system for production of chemicals and fuels derived from fatty acids. Nat Biotechnol 30:354–359. https://doi.org/10.1038/nbt.2149

Zhang X, Yan S, Tyagi RD, Surampalli RY, Valéro JR (2014) Wastewater sludge as raw material for microbial oils production. Appl Energy 135:192–201

Chapter 3
New Approaches for Renewable Energy Using Metal Electrocatalysts for Lithium-O_2 and Zinc-Air Batteries

Josiel Martins Costa and Ambrósio Florêncio de Almeida Neto

Contents

3.1 Introduction

Global warming and energy demand from fossil fuels characterize two issues pressing the traditional vehicle industry. Unlike gasoline-powered vehicles, electric vehicles, which use rechargeable batteries, not only reduce fossil fuel consumption but also eradicate the pollution generated in the exhaust, giving credit to the auto industry (Yang et al. 2011). Indicatives in Fig. 3.1 demonstrate the rise of 2011 to 2017 of plug-in electric vehicles, which are hybrid vehicles composed of higher capacity batteries.

J. M. Costa (✉) · A. F. de Almeida Neto
Laboratory of Electrochemical Processes and Anticorrosion, Department of Products and Processes Design, School of Chemical Engineering, University of Campinas (UNICAMP), Avenida Albert Einstein, Campinas, SP, Brazil

Inamuddin, A. M. Asiri (eds.), *Sustainable Green Chemical Processes and their Allied Applications*, Nanotechnology in the Life Sciences,
https://doi.org/10.1007/978-3-030-42284-4_3

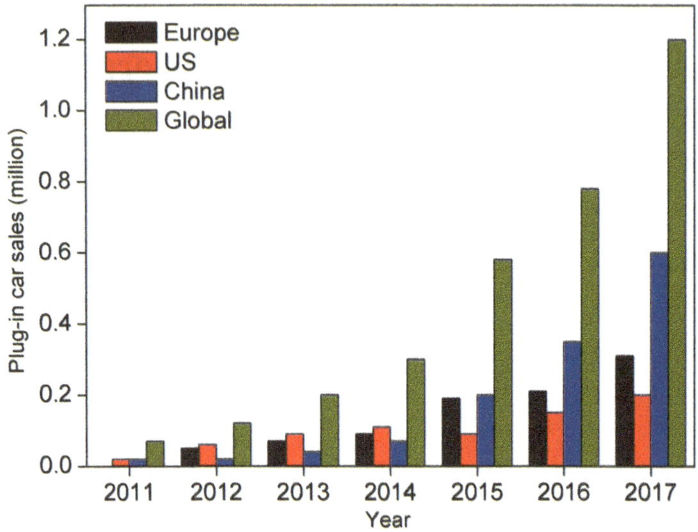

Fig. 3.1 Plug-in vehicles sold annually, including global number. Modified after (Pelegov and Pontes 2018) Copyright 2018, Batteries. There is an increase in the number of sales of plug-in vehicles, as a result of concern for the environment and the development of new technologies. With the advancement of battery technology, the tendency is to decrease the sale of combustion engine-powered cars

Among the technologies available, lead-acid batteries have a shorter useful life cycle of 3–4 years; however, it is still used as a source of renewable energy. Nickel-cadmium batteries have a longer life cycle than lead-acid; however, the cost of this technology is expensive. Concerning lithium-ion batteries, the energy density limitation occurs due to the chemical intercalation of the electrode material. Thus, this type of battery does not present relevant applications in electric vehicles (Akhtar and Akhtar 2015). The specific energy and power density of the lithium-ion battery, even when applied to electric vehicles as a power source presented deficiencies, and ongoing research and the development of rechargeable batteries have led to the fundamentals of Li-air battery.

Scientific advances are fundamental to understanding the reactions that occur in rechargeable batteries, as well as the development of new materials to overcome the challenges related to energy storage (Bruce et al. 2011; Christensen et al. 2011; Imanishi and Yamamoto 2014). In this way, it is necessary to develop more efficient and safe rechargeable batteries from economically more accessible technologies.

Specifically, metal-air cells, which electrochemical reactions occur between electrodes that consume air or O_2, have been studied (Girishkumar et al. 2010; Neburchilov et al. 2010). In this sense, this chapter aims to discuss the main challenges and current trends in the architecture of cathode electrodes of lithium-O_2 and zinc-air batteries, with a focus on electrocatalysts that hold up the oxygen reduction and evolution reactions.

3.2 Li-O_2 Batteries

In 2009, the Battery 500 Project was initiated by scientists from the International Business Machines Almaden Research Center which aimed to develop a Li-air battery that had a capacity of 500 miles per charge. For Li-O_2 systems, in the cathode of nanostructured carbon or other metallic materials (nickel foam, steel mesh), oxygen is extracted and is where reactions of lithium ions with electrons occur to generate electricity. When connected to the car, oxygen is released into the atmospheric air. It is intended to transform this technology into a complete product by 2030 (Wilcke and Kim 2016). The authors reported that in an ideal cell, the amount of oxygen consumed during the flush should be exactly equal to the amount released in the refill. The separation of Li_2O_2 in the recharge was attacking the cell's electrolyte during the polls, which caused it to be replaced by another. There is no solvent that is stable for commercially available Li-O_2 cells; however, the authors stated that a combination of solvents worked suitably.

In recent years, two types of reversible Li-O_2 batteries classified as aqueous and nonaqueous systems have been proposed (Bruce et al. 2011; Zhang et al. 2011). Equation (3.1) shows the reaction for nonaqueous systems:

$$2\mathrm{Li} + \mathrm{O_2} \rightarrow \mathrm{Li_2O_2} \tag{3.1}$$

For aqueous systems, Eq. (3.2) describes the redox reaction of the air cathode that occurs with the presence of water molecules:

$$4\mathrm{Li} + 6\mathrm{H_2O} + \mathrm{O_2} \rightarrow 4\left(\mathrm{LiOH{\cdot}H_2O}\right) \tag{3.2}$$

The voltage of the cell is influenced by the OH^-concentration; meanwhile, it was observed that the OH^- concentration increases during the discharge process. The first studies on Li-air rechargeable cells used an anode composed of Li metallic, a polymeric electrolyte, and a porous carbon-based electrode as a catalyst (Abraham and Jiang 1996; Li and Dai 2014; Read 2006).

The electrolyte serves to promote the transport of lithium ions, the dissolution of gaseous oxygen, and the transfer to the reaction sites and also to protect the lithium anode. The water present in the aqueous electrolytes also participates in the electrochemical reactions, acting as reagent (Tan et al. 2017). An ideal nonaqueous electrolyte for the Li-O_2 system must (1) be stable with the lithium anode and lithium salts, (2) have superior oxidation potential, and (3) present a lower vapor pressure and a superior boiling point. The formulation of the electrolyte influences the ability to charge and discharge. As examples of electrolytes, there may be mentioned gel-type polymer of $LiPF_4$, carbonates, ethers, ionic liquids at room temperature, sulfoxides, and nitrites (Abraham and Jiang 1996; Read 2006).

In the electrode is the place where the oxygen evolution and reduction reaction occur. The objective is to achieve superior energy density and capacity, as well as reversibility. The electrochemical processes involved in the electrode, composed of the interaction between gaseous and ionic species, should remain in contact long enough to result in high capacity.

In general, structures composed of pores are more effective to promote the oxygen transport in the electrodes, in addition to a large surface to react and facilitate electrical conductivity. The product of the lithium reaction with oxygen gives rise to Li_2O_2, which is in the solid phase. The solid present inside the porous cathode impairs oxygen transport and lowers battery capacity in nonaqueous systems. Even in aqueous systems, such as hybrid batteries, in which Li_2O_2 can be dissolved, the electrochemical reactions can be affected, directly impacting the capacity. Therefore, the structural shape of the electrode also is a major problem for porous cathodes (Tan et al. 2017). Examples of materials used as electrodes are mesoporous carbon, carbon nanotubes, graphene, and macroporous foams (Ding et al. 2014). However, carbon instability has led to the search of non-carbon cathodes capable of being used in nonaqueous Li-air batteries, for example, a nickel foam, which with potentials above 3.5 V promoted the degradation of carbonate electrolytes (Veith and Dudney 2011).

The kinetics of oxygen reduction limits the performance of LiO_2 systems. The formation of Li_2O_2 and/or Li_2O depends on the reduction of O_2, according to Eqs. (3.3), (3.4), and (3.5).

$$Li^+ + e^- + O_2 \rightarrow LiO_2 \quad U = 3.00 \text{ V} \tag{3.3}$$

$$2Li^+ + 2e^- + O_2 \rightarrow Li_2O_2 \quad U = 2.96 \text{ V} \tag{3.4}$$

$$4Li^+ + 4e^- + O_2 \rightarrow 2Li_2O \quad U = 2.91 \text{ V} \tag{3.5}$$

Some authors suggest that Eq. (3.4) does not occur directly (Bruce et al. 2011) and is divided into three stages, according to Eqs. (3.6), (3.7), and (3.8).

$$O_2 + e^- \rightarrow O_2^- \tag{3.6}$$

$$O_2^- + Li^+ \rightarrow LiO_2 \tag{3.7}$$

$$2LiO_2 \rightarrow Li_2O_2 + O_2 \tag{3.8}$$

A voltage range for charge-discharge potentials can be obtained through these different oxygen reduction reactions. From the formation of LiO_2, the mechanism becomes dependent on the material type of the electrode. Materials in which O_2 adsorb weakly, for example, carbon, reaction with LiO_2 rapidly generates Li_2O_2. However, for surfaces that have strong adsorption of oxygen, such as in platinum, the reaction can trigger the formation of Li_2O, being similar to the process that occurs in an aqueous medium (Lu et al. 2010).

3.2.1 *Li-O_2 Batteries Challenge*

Several factors contribute to decreasing energy storage in the practice of Li-O_2 batteries. A porous conductive matrix, generally composed of carbon, is used as a cathode, in which, with the discharge, electrode mass and volume are increased due to the accumulation of product generated. Larger quantities of the Li metal are added than necessary for the stoichiometric ratio to suppose the regeneration inefficiency. Besides, energy storage can also be reduced with the use of electric current collectors and, in particular for Li-O_2 batteries, gas diffusion channels (Bruce et al. 2011).

There are other factors such as the following: the instability of organic electrolytes; the behavior of the electrode and Li metal anode when in contact with reduced oxygen species, specifically the superoxide radical anion; and overpotentials during oxygen reduction/evolution reactions, which characterize a technological challenge to be overcome (Ottakam Thotiyl et al. 2013; Zou et al. 2016).

Regarding the use of Li metallic anode, the impregnation of nonhomogeneous ions leads to the formation of Li dendrites during the cycle, promoting the neglect of Li metal. The dendrites pierce the separator giving rise to short circuits, as well as causing the contact with the electrode to be lost. The battery cycle duration is reduced as the dendrites grow and the solid electrolytic interface consumes electrolyte. The distribution of uneven current can cause this phenomenon (Bai et al. 2016). Seeking to increase anode stability, materials have been developed that involve homogeneous artificial layers of protection, highly conductive Li-ion, to reduce the formation of dendrites during the cycles. For example, polymer electrolytes and thin ceramic films present a promising direction to minimize problems of formation of such lithium dendrites (Girishkumar et al. 2010).

Another relevant issue in the batteries is the high polarization of the electrode, which results in the loss of energy efficiency, during the discharge of the cell, approximately 2.7 V, which is significantly lower than the standard potential (U_o). This potential difference is the discharge overpotential (η_{des}). The same analysis is valid for recharging, where the voltage increases up to 4.0 V, which results in the charge overpotential (η_{car}). The deviation from the standard potential is greater than the current flowing through the system and is known as electrode polarization. In this condition, the electrode is polarized (Girishkumar et al. 2010).

Three types of polarization are reported: concentration polarization or mass transport when it is the result of the depletion of the reactant gas at the electrode/electrolyte interface, which emphasizes the need for electrodes with high surface area and porosity that allow easy access to the reagents gaseous; the activation polarization, caused by the high concentration of reagent that forms a barrier to electron transfer; and the polarization by ohmic fall, which is the result of the electrical resistance inside the cell, being the process conditioned by Ohm's law, which distances the potential of balance electrode (Zheng et al. 2016; Zouhri and Lee 2016).

The stability of the electrolyte is also an aggravating factor since the nonaqueous electrolyte must be conductive, have electrochemical stability, and have considerable oxygen solubility (Balaish et al. 2014). Even with ongoing research, the instability of the solvent and lithium salts characterizes issues that require studies. To improve this instability, additives have been used to increase cell performance (Tan et al. 2017).

Electrolytes based on ionic liquids are currently a strategy aimed at improving the cycling of the Li-O_2 battery with high discharge rate. Monaco et al. (2013) analyzed the O_2 reduction reaction on a mesoporous carbon electrode using an ionic liquid as an electrolyte. The study demonstrated that the response of the oxygen electrode under high currents was governed by mass transport of O_2. The current density used was 0.2 mA cm^{-2}, and a value of 600 mAh g^{-1} of specific capacity was obtained.

3.3 Zn-Air Batteries

Known for expressive energy density, such as 1086 Wh kg^{-1}, Zn-air batteries also stand out for the low cost of production, lower than 10$ kW^{-1} h^{-1}, which is approximately two times lower compared to lithium-ion systems. Zn-air batteries become an alternative to lithium-air batteries, which may come to fruition in the future due to the higher available energy density when compared to any rechargeable system. However, as well as other metal-air cells, problems related to the metallic Zn electrode and the catalyst impair the performance of this system.

Among the applications of Zn-air batteries, in the medical and telecommunication sector, it has already been tested. Even with superior energy densities, Zn-air batteries have lower power, less than 10 mW for devices applied to the hearing aid, due to the inefficiency of the available catalysts. Furthermore, dissolution of zinc and inadequate bifunctional catalysts have also hindered the life cycle (Li and Dai 2014).

A negative zinc electrode, an air electrode (generally porous), and a separation membrane both in contact through an alkaline electrolyte make up a Zn-air battery. During the discharge, oxidation of the zinc occurs, solubilizing the Zn in the electrolyte, giving rise to ions $Zn(OH)_4^{-2}$ (Lee et al. 2011). This reaction occurs until the electrolyte is supersaturated. After supersaturation, zinc ions are presented as insoluble zinc oxide, described by the equations (Meng et al. 2018):

$$\text{Negative electrode} : Zn + 4OH \rightarrow Zn(OH)_4^{2-} + 2e^- \qquad (3.9)$$

$$Zn(OH)_4^{2-} \rightarrow ZnO + H_2O + 2OH \quad (3.10)$$

$$\text{Positive electrode}: O_2 + 4e^- + 2H_2O \rightarrow 4OH \quad (3.11)$$

$$\text{General reaction}: 2Zn + O_2 \rightarrow 2ZnO \quad (3.12)$$

$$\text{Parasitic reaction}: Zn + 2H_2O \rightarrow Zn(OH)_2 + H_2 \quad (3.13)$$

Thermodynamically, the evolution and oxidation reactions occur spontaneously and have a voltage value of 1.65 V. Meanwhile, during the charge and discharge cycles, they are kinetically slow, requiring catalysts to accelerate the process (Park et al. 2017).

During the heterogeneous catalysis process, the reactions occur on the surface of the catalyst. In this way, the catalytic activity depends on the mass transport that proceeds between the catalyst surface and the bulk phase. Both oxygen evolution and oxygen reduction reactions are composed of electron transfer and gas/liquid diffusion (Tang et al. 2018). To enhance the reactions, gas diffusion layers are used as the air electrode. However, the material must have high electrical conductivity, be hydrophobic, and have considerable porosity and surface area (Danner et al. 2016). These diffusers are usually composed of thin porous films manufactured of carbon materials, meshes, or metal foam (Chen et al. 2018).

The gas diffusion layer must allow uniform contact between the air and the catalyst and, simultaneously, serve as a barrier for electrolyte penetration, which justifies hydrophobicity and the formation of interpenetrating subsystems capable of facilitating the diffusion of O_2 to the catalyst site, where a three-phase electrolyte-gas-catalyst interface is created (Fu et al. 2017; Su et al. 2017).

The morphology of the electrocatalyst must be designed to expose the active sites to electrolyte and sources of oxygen. Vertically aligned macropores and matrices grown on carbon nanotubes, for example, increase ionic transport and facilitate the diffusion of oxygen. Metal nanoparticles provide regions for reactions to occur on the surface (Jia et al. 2017).

3.3.1 Zn-Air Batteries Challenge

The Zn electrodes have performance impaired due to several factors, such as dendrite growth, mass change, passivation, and hydrogen evolution. Battery performance is decreased when one or more of these phenomena occur simultaneously. Retention of charge and discharge cycles, a considerable utilization rate of Zn, and high efficiency with minimum hydrogen evolution are expected on an ideal Zn electrode. To obtain these characteristics, researches have been developed by varying the design of the Zn electrode structure, the air-electrode, and the electrolyte composition. The incorporation of heavy metals in the Zn electrodes, for example, can suppress these three phenomena, just as the electrolyte optimization can reduce the effect of the shape change, the dendrite growth, and the hydrogen evolution (Chen et al. 2018). However, the incorporation of heavy metal in the Zn electrodes are not ideal due to their environmental toxicity.

Due to the slow kinetics, the electrocatalysts are applied to reduce the overpotentials and facilitate the occurrence of oxygen reduction and evolution reactions. The most efficient and usual electrocatalysts are the base of precious metals and compounds, usually Pt, IrO_2, and RuO_2 (Chen et al. 2019). High efficiencies in these two reactions depend on the different reactions associated with the electronic structure and adsorption energy, which consequently has electrolyte pH dependence (Jiao et al. 2015; Vij et al. 2017). Moreover, unavailability, high cost, and the poisonous effect caused by precious metals the hindered practical applications. It is important to seek alternative multifunctional electrocatalysts than conventional (Chen et al. 2019).

As the reactions have different demands for the active sites, the bifunctional catalysts need to be adequate to regulate the intrinsic activity, in addition to the enlarged surface area, the diffusion, and the electron transport being accelerated, expounding the active sites in the catalytic process (Kuang and Zheng 2016).

Zinc-air batteries work in an alkaline environment, such as Na and K hydroxides, for a better activity for the zinc electrode and cathode, since Zn can react quickly in acid solution and generate anodic corrosion (Lee et al. 2016). KOH is generally more appropriate over NaOH due to its low viscosity, high O_2 diffusion coefficient, and better ionic effect. However, since these are open systems, the loss of water from the liquid electrolyte is a relevant cause of performance reduction. Research has been carried out aiming at electrolyte gelling to minimize loss and improve battery performance and cycle life (Othman et al. 2001). The function of the gelling agent is to immobilize and store the KOH electrolyte with a retention capacity of 20–100 times its weight. Besides, the use of gel can improve the specific capacity, according to the study of the author (Mohamad 2006).

3.4 Electrocatalysts for Li-O_2 and Zn-Air Batteries

In rechargeable systems, the heterogeneous electrocatalysis has the function of reducing the overpotential generated during the oxygen reduction and evolution reaction, since this overpotential decreases the energy storage (McCloskey et al. 2011). In order to accelerate the reactions that occur in the discharge and charge steps, catalysts are incorporated, as shown in Fig. 3.2.

In aqueous electrolytes, developing bifunctional catalysts with significant catalytic activity compared to noble metals is still a challenge. For nonaqueous Li-O_2 systems, understanding the catalyst function concerning the formed and decomposed solid product is indispensable for the development of effective catalysts. As the discharge process in nonaqueous electrolytes is followed by the generation of solid Li_2O_2, which can cover the catalyst surface over time, and in order to shorten

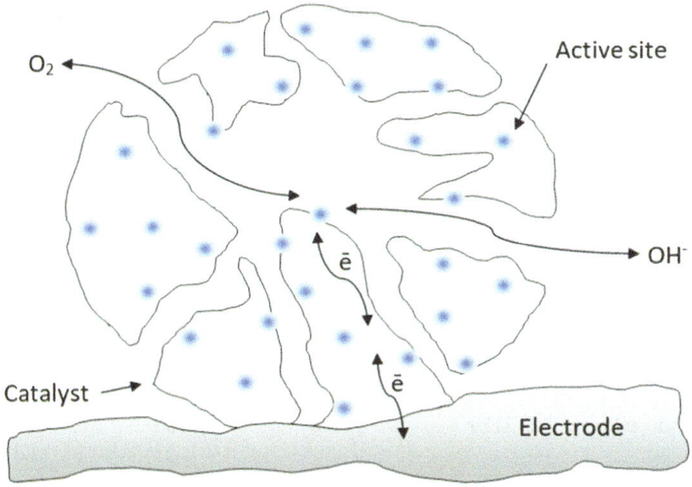

Fig. 3.2 Catalytic process diagram in air electrodes. Modified after (Tang et al. 2018) Copyright 2018, American Chemical Society. The active site is the region where occur the reactions involved in the charge-discharge process of the batteries. More porous materials favor the catalysis of the reactions

this time and improve the discharge, electrocatalysts that are repeatedly adsorbed against the increasing formation of Li_2O_2 are required (Tan et al. 2017).

Song et al. (2017) developed Co–W–C nanocomposites containing encapsulated functional species doped with nitrogen. The nanoparticles of cobalt and tungsten carbide were encapsulated using induced evaporation and heat treatment. The oxygen reduction reaction was facilitated, showing higher yield and good transferability of electrons with the material obtained.

Cho et al. (2017) studied Ag-Co catalysts with varied stoichiometric ratios. The alloys were obtained by electrodeposition with Triton X-100 (surfactant), Ag and Co, and KNO_3 in an aqueous solution. The alloy composed of Ag_1Co_7 electrodeposited for 200 s, with 14% Co, revealed the superior catalytic activity to the oxygen reduction reaction, with a similar or better result to platinum.

Yu et al. (2016) manufactured nanotubes of cobalt oxide (Co_3O_4) by electrospinning and subsequent air calcination. Cobalt oxide nanotubes were reduced to cobalt by calcination under H_2 and argon atmosphere and served as a model to react with the silver precursor (silver nitrate), through galvanic substitution, giving rise to the AgCo nanotubes. Several AgCo nanotubes were prepared under different conditions, varying the reduction reaction time and the galvanic substitution. The reduction time of 1 h and galvanic substitution of 3 h were shown to be better for the discharge, demonstrated by the higher current density and the number of electrons that were transferred during the reaction. In addition, the discharge performance using AgCo nanotubes was better than cobalt, bare platinum, and silver nanowires.

Table 3.1 Main bifunctional electrocatalysts used in rechargeable Li-O_2 systems

Electrocatalyst	Discharge capacity (mAh g_c^{-1})	Charge capacity (mAh g_c^{-1})	Current density (mA g_c^{-1})	Reference
Fe_3O_4 at CoO	5100	5100	40	Shang et al. (2018)
Co@CoO carbon nanofibers	4427	4427	100	Cao et al. (2019)
Ru/N carbon nanotubes	7100	7100	100	Zhang et al. (2019b)
δ-MnO_2 at multi-walled carbon nanotubes	~28,400	–	100	Ma et al. (2019)
W_2C carbon layers	11,000	11,000	100	Gao et al. (2017)
Fe/Fe_3C N-doped graphene	7150	~7000	100	Lai et al. (2016)
V_2O_5/Al_2O_3	2875	2875	100	Lim et al. (2013)
$LaMnO_3$	6852	6852	100	Li et al. (2018)
Ag/NiO-Fe_2O_3	5138	5138	200	Lu et al. (2019a)
Ta_2O_5 nanoparticles	~15,000	~15,000	200	Zhang et al. (2018)
Pd-Cu	12,000	–	200	Choi et al. (2014)
RuO_2 at $NiCo_2O_4$	17,633	~16,500	200	Zou et al. (2018)

Reduced graphene and AgCo oxidized electrocatalyst was synthesized by an electrochemical method. The synthesis of the AgCo alloy in the reduced graphene catalyst was performed using polyethylene glycol as an additive. The electrocatalyst presented the superior activity of electrochemical surface area and mass, resulting from well-dispersed nano-linked nanoparticles that provided transfer rate during the oxygen reduction reaction. The AgCo in reduced graphene electrocatalyst showed higher current density in the oxygen reduction reaction, in addition to being stable and selective when compared to the other catalysts tested (Joo et al. 2017).

As shown in Table 3.1, there is a trend of carbon-based electrocatalysts and their respective discharge and charge capacities in the last 6 years. The values of the capacities vary according to the method used for the synthesis of nanotubes, the use or not of heat treatment, and the metallic elements that make up the cathode. As an example, the study of Gao et al. (2017) used a layer of carbon doped with nitrogen and encapsulated with hybrid W_2C. The authors obtained high capacity in both O_2 reactions, with an initial capacity close to 11,000 mAh g^{-1}, smaller overpotential, and greater life cycle, which was assigned to the synergic effect between the nanoparticles of W_2C and the layers of carbon doped with nitrogen. The specific cathode architecture allowed the decomposition of the undesired $LiCO_3$ product and the effective conduction of electrons and Li^+.

Cobalt nanoparticles immobilized internally in CoO and moored in nitrogen-doped porous carbon nanofibers were produced using a coaxial electrospinning process and thereafter a heat treatment (Cao et al. 2019). The cathode Co showed a porous hierarchical structure of a high surface area, composed by a large amount of active sites that showed synergistic interactions between the two chemical species, Co and CoO. The petal-shaped Li_2O_2 discharge products from the Co and CoO cathode were easily decomposed when compared to the film-type Li_2O_2. The results provided a model for the design of Co in CoO cathodes with charge and discharge capacity of 4427.3 mAh g^{-1}.

Ultrafine bimetallic Fe-Co nanoparticles fixed on nitrogen-doped porous carbon were used in Zn-air batteries (Zhong et al. 2019). The catalyst exhibited superior performance for the discharge with a half-wave potential = 0.84 V and limiting current density = −5.3 mA cm^{-2}. A high potential circuit = 1.50 V and high specific capacity = 726.2 mAh g^{-1} with stability, even higher than the Pt in C catalyst, were also observed.

The Fe-Co-Ni ternary alloy electrodeposited by electrophoresis was used in mesoporous carbon nanofibers to catalyze the charge reaction (Li et al. 2019b). After carbonization, the morphology of fibrous appearance with a diameter of 250 nm was maintained. In the Zn-air battery, the alloy presented a differential charge voltage of 0.88 V at 20 mA cm^{-2}, considerable stability, and superior power density of 73 mW cm^{-2} at a current density of 80 mA cm^{-2}, having surpassed the results of electrodes containing Pt in C.

To further enhance the discharge reaction activities, Cu was incorporated in carbon nanotubes containing Co (Li et al. 2019c). The synergistic effect observed between the Cu-N_x and Co-N_x species increased the O_2 reduction reaction activities, being applied as a cathode of Zn-air cell. The specific capacity achieved was 760 mAh g^{-1}, which equates to an energy density of 821 Wh kg^{-1}, with a discharge rate of 100 mA cm^{-1}. The superior result of the bimetallic catalyst proposed a viable option for the production of energy conversion cells.

In the electrodes of Zn-air systems, the electrode architecture is also extremely relevant, according to Table 3.2, which shows the trends of the electrocatalysts of the last 2 years. A strategy developed with carbon-supported nitrogen-doped Zn-Co demonstrated the ease of O–O bond cleavage, presenting an overpotential of 0.335 V for the discharge process (Lu et al. 2019b). As shown in Table 3.2, the cathodic catalyst showed a power density of 230 mW cm^{-2}. Furthermore, the authors concluded that the catalytic activity can be sustained even in the acid medium and that the Zn-air battery presented a greater power density and stability.

In addition to alloys and metal oxides, the composites are also applicable to the Zn-air electrode. According to Table 3.2, Co_9S_8/S and nitrogen dual-doped graphene composite showed enhanced electrochemical kinetics for O_2 reactions, with better activity parameters such as lower Tafel slope of 47.7 mV dec^{-1} for reduction reaction and 69.2 mV dec^{-1} for evolution reaction than traditional Pt in C and RuO_2 catalysts (Shao et al. 2019). To extend the practical application, the Zn-air battery

Table 3.2 Main bifunctional electrocatalysts used in rechargeable Zn-air systems

Electrocatalyst	Oxygen reduction reaction activity	Oxygen evolution reaction activity	Battery performance	Electrolyte	Reference
(Ni, Co)S_2 nanosheets	Onset potential: 0.82 V	Onset potential: 1.47 V	Power density: 153.5 mW cm^{-2}	6 M KOH	Zhang et al. (2019a)
B-N co-doped porous carbon	Onset potential: 0.894 V	Onset potential: 1.38 V	Cycling ability: 600 cycles (100 h) at 2 mA cm^{-2}	6 M KOH	Qian et al. (2017)
Co_9S_8/S and N dual doped graphene composite	Onset potential: 0.92 V Tafel slope: 47.7 mV dec^{-1}	Tafel slope: 69.2 mV dec^{-1}	Power density: 36.2 mW cm^{-2}	6 M KOH	Shao et al. (2019)
(Ni, Co)Se_2	Onset potential: 0.70 V	Onset potential ~1.58 V	Power density: 110 mW cm^{-2}	6 M KOH with 0.2 M zinc acetate	Sun et al. (2019)
Fe/Co-N-C nanofibers with embedding Fe-Co alloy nanoparticles	Tafel slope: 80 mV dec^{-1}	Tafel slope: 78 mV dec^{-1}	Power density: 115 mW cm^{-2}	6 M KOH with 0.2 M zinc acetate	Li et al. (2019a)
Co-Ni nitrogen-doped graphitic carbon	Tafel slope: 51 mV dec^{-1}	Tafel slope: 78 mV dec^{-1}	Power density: 130.5 mW cm^{-2}	6 M KOH with 0.2 M ZnO	Yang et al. (2019)
FeCoNi alloy	–	Tafel slope: 78 mV dec^{-1}	Power density: 130.5 mW cm^{-2}	6 M KOH with 0.2 M ZnO	Li et al. (2019b)
Zn/CoN-C	Onset potential: 1.004 V	–	Power density: 230 mW cm^{-2}	6 M KOH	Lu et al. (2019b)
Iron-imidazolate	Onset potential: 0.90 V	Onset potential: 0.56 V	Cycling ability: 600 cycles (160 h) at 2 mA cm^{-2}	6 M KOH with 0.2 M ZnO	Lee et al. (2019)
$NiFe_2O_4$	Onset potential: 0.83 V	Onset potential: 1.56 V	Power density: 211 mW cm^{-2}	KOH with 0.2 M $ZnCl_2$	Naik and Sampath (2018)
$Co_2P_2O_7$ particles	Onset potential: 0.90 V	Onset potential: 1.45 V	Power density: 138 mW cm^{-2}	6 M KOH	Ren et al. (2018)

demonstrated excellent charge-discharge results and cycle stability and was fully prepared in the solid state.

3.5 Conclusions

Significant research has been developed on materials and structural designs to obtain better results of electrochemically rechargeable Li-O_2 and Zn-air batteries in the last 10 years. Despite the significant effort, the technology of rechargeable Li and Zn batteries is still not finalized enough for widespread applications. It is suggested the use of computational simulation and screening on the high yield, aiming to accelerate the development of electrocatalysts with high catalytic activity. Active sites should be more exposed, reducing the impediment of mass transfer; for this, the geometric structures of the electrodes must be controlled. The temperature must be considered since for practical applications the batteries will be used in a temperature range, not at a fixed temperature.

Even with several efficient bifunctional metal catalysts, carbon materials are essential, due to applications such as substrates, catalysts, or additives. However, the high oxidation potential generates carbon corrosion during charging, reducing performance and stability. An alternative to this issue has been to use graphitized carbon, foam, or mesh metal material to restrict the effect of corrosion. Therefore, the control of morphology is necessary, due to the influence on the properties of the material, such as the specific surface area, which determines the catalytic activity. We hope that these approaches in this chapter will be used for researchers and motivate the innovation on catalysts and that in the future occurs the commercialization of rechargeable Li-O_2 and Zn-air batteries.

Acknowledgment This work was financed in part by the Coordenação de Aperfeiçoamento de Pessoal de Nível Superior—Brasil (CAPES) (Finance Code 001).

References

Abraham KM, Jiang Z (1996) A polymer electrolyte-based rechargeable lithium/oxygen battery. J Electrochem Soc 143:1–5. https://doi.org/10.1149/1.1836378

Akhtar N, Akhtar W (2015) Prospects, challenges, and latest developments in lithium-air batteries. Int J Energy Res 39:303–316. https://doi.org/10.1002/er.3230

Bai P, Li J, Brushett FR, Bazant MZ (2016) Transition of lithium growth mechanisms in liquid electrolytes. Energ Environ Sci 9:3221–3229. https://doi.org/10.1039/C6EE01674J

Balaish M, Kraytsberg A, Ein-Eli Y (2014) A critical review on lithium-air battery electrolytes. Phys Chem Chem Phys 16:2801–2822. https://doi.org/10.1039/C3CP54165G

Bruce PG, Freunberger SA, Hardwick LJ, Tarascon JM (2011) Li-O_2 and Li-S batteries with high energy storage. Nat Mater 11:19–29. https://doi.org/10.1038/nmat3191

Cao Y et al (2019) Synthesis of Ag/Co@CoO NPs anchored within N-doped hierarchical porous hollow carbon nanofibers as a superior free-standing cathode for LiO_2 batteries. Carbon 144:280–288. https://doi.org/10.1016/j.carbon.2018.12.048

Chen X, Zhou Z, Karahan HE, Shao Q, Wei L, Chen Y (2018) Recent advances in materials and design of electrochemically rechargeable zinc-air batteries. Small 14:1801929. https://doi.org/10.1002/smll.201801929

Chen J et al (2019) Chitin-derived porous carbon loaded with Co, N and S with enhanced performance towards electrocatalytic oxygen reduction, oxygen evolution, and hydrogen evolution reactions. Electrochim Acta 304:350–359. https://doi.org/10.1016/j.electacta.2019.03.028

Cho Y-B, Moon S, Lee C, Lee Y (2017) One-pot electrodeposition of cobalt flower-decorated silver nanotrees for oxygen reduction reaction. Appl Surf Sci 394:267–274. https://doi.org/10.1016/j.apsusc.2016.10.068

Choi R et al (2014) Ultra-low overpotential and high rate capability in Li-O_2 batteries through surface atom arrangement of PdCu nanocatalysts. Energ Environ Sci 7:1362–1368. https://doi.org/10.1039/C3EE43437K

Christensen J et al (2011) A critical review of Li/air batteries. J Electrochem Soc 159:R1–R30. https://doi.org/10.1149/2.086202jes

Danner T, Eswara S, Schulz VP, Latz A (2016) Characterization of gas diffusion electrodes for metal-air batteries. J Power Sources 324:646–656. https://doi.org/10.1016/j.jpowsour.2016.05.108

Ding N, Chien SW, Hor TSA, Lum R, Zong Y, Liu Z (2014) Influence of carbon pore size on the discharge capacity of Li-O_2 batteries. J Mater Chem A 2:12433–12441. https://doi.org/10.1039/C4TA01745E

Fu J, Cano ZP, Park MG, Yu A, Fowler M, Chen Z (2017) Electrically rechargeable zinc-air batteries: progress, challenges, and perspectives. Adv Mater 29:1604685. https://doi.org/10.1002/adma.201604685

Gao R, Zhou Y, Liu X, Wang J (2017) N-doped defective carbon layer encapsulated W_2C as a multifunctional cathode catalyst for high performance Li-O_2 battery. Electrochim Acta 245:430–437. https://doi.org/10.1016/j.electacta.2017.05.177

Girishkumar G, McCloskey B, Luntz AC, Swanson S, Wilcke W (2010) Lithium-air battery: promise and challenges. J Phys Chem Lett 1:2193–2203. https://doi.org/10.1021/jz1005384

Imanishi N, Yamamoto O (2014) Rechargeable lithium–air batteries: characteristics and prospects. Mater Today 17:24–30. https://doi.org/10.1016/j.mattod.2013.12.004

Jia G, Zhang W, Fan G, Li Z (2017) Three-dimensional hierarchical architectures derived from surface-mounted metal-organic framework membranes for enhanced electrocatalysis. Angew Chem Int Ed Engl 56:13781–13785. https://doi.org/10.1002/anie.201708385

Jiao Y, Zheng Y, Jaroniec M, Qiao SZ (2015) Design of electrocatalysts for oxygen- and hydrogen-involving energy conversion reactions. Chem Soc Rev 44:2060–2086. https://doi.org/10.1039/C4CS00470A

Joo Y, Ahmed MS, Han HS, Jeon S (2017) Preparation of electrochemically reduced graphene oxide-based silver-cobalt alloy nanocatalysts for efficient oxygen reduction reaction. Int J Hydrogen Energ 42:21751–21761. https://doi.org/10.1016/j.ijhydene.2017.07.123

Kuang M, Zheng G (2016) Nanostructured bifunctional redox electrocatalysts. Small 12:5656–5675. https://doi.org/10.1002/smll.201600977

Lai Y, Chen W, Zhang Z, Qu Y, Gan Y, Li J (2016) Fe/Fe_3C decorated 3-D porous nitrogen-doped graphene as a cathode material for rechargeable Li-O_2 batteries. Electrochim Acta 191:733–742. https://doi.org/10.1016/j.electacta.2016.01.134

Lee J-S, Tai Kim S, Cao R, Choi N-S, Liu M, Lee KT, Cho J (2011) Metal-air batteries with high energy density: Li-air versus Zn-air. Adv Energy Mater 1:34–50. https://doi.org/10.1002/aenm.201000010

Lee J, Hwang B, Park M-S, Kim K (2016) Improved reversibility of Zn anodes for rechargeable Zn-air batteries by using alkoxide and acetate ions. Electrochim Acta 199:164–171. https://doi.org/10.1016/j.electacta.2016.03.148

Lee J-SM, Sarawutanukul S, Sawangphruk M, Horike S (2019) Porous Fe-N-C catalysts for rechargeable zinc-air batteries from an iron-imidazolate coordination polymer. ACS Sustain Chem Eng 7:4030–4036. https://doi.org/10.1021/acssuschemeng.8b05403

Li Y, Dai H (2014) Recent advances in zinc–air batteries. Chem Soc Rev 43:5257–5275. https://doi.org/10.1039/C4CS00015C

Li C, Yu Z, Liu H, Chen K (2018) High surface area $LaMnO_3$ nanoparticles enhancing electrochemical catalytic activity for rechargeable lithium-air batteries. J Phys Chem Solids 113:151–156. https://doi.org/10.1016/j.jpcs.2017.10.039

Li C, Wu M, Liu R (2019a) High-performance bifunctional oxygen electrocatalysts for zinc-air batteries over mesoporous Fe/Co-N-C nanofibers with embedding FeCo alloy nanoparticles. Appl Catal B Environ 244:150–158. https://doi.org/10.1016/j.apcatb.2018.11.039

Li C, Zhang Z, Wu M, Liu R (2019b) FeCoNi ternary alloy embedded mesoporous carbon nanofiber: an efficient oxygen evolution catalyst for rechargeable zinc-air battery. Mater Lett 238:138–142. https://doi.org/10.1016/j.matlet.2018.11.160

Li Z et al (2019c) Structural modulation of Co catalyzed carbon nanotubes with Cu-Co bimetal active center to inspire oxygen reduction reaction. ACS Appl Mater Interfaces 11:3937–3945. https://doi.org/10.1021/acsami.8b18496

Lim SH, Kim DH, Byun JY, Kim BK, Yoon WY (2013) Electrochemical and catalytic properties of V_2O_5/Al_2O_3 in rechargeable Li-O_2 batteries. Electrochim Acta 107:681–685. https://doi.org/10.1016/j.electacta.2013.06.045

Lu Y-C, Gasteiger HA, Crumlin E, McGuire R, Shao-Horn Y (2010) Electrocatalytic activity studies of select metal surfaces and implications in Li-Air batteries. J Electrochem Soc 157:A1016–A1025. https://doi.org/10.1149/1.3462981

Lu X et al (2019a) 3D Ag/NiO-Fe_2O_3/Ag nanomembranes as carbon-free cathode materials for Li-O_2 batteries. Energy Storage Mater 16:155–162. https://doi.org/10.1016/j.ensm.2018.05.002

Lu Z et al (2019b) An isolated zinc-cobalt atomic pair for highly active and durable oxygen reduction. Angew Chem Int Ed 58:2622–2626. https://doi.org/10.1002/anie.201810175

Ma L, Meng N, Zhang Y, Lian F (2019) Improved electrocatalytic activity of δ-MnO_2@MWCNTs by inducing the oriented growth of oxygen reduction products in Li-O_2 batteries. Nano Energy 58:508–516. https://doi.org/10.1016/j.nanoen.2019.01.089

McCloskey BD, Scheffler R, Speidel A, Bethune DS, Shelby RM, Luntz AC (2011) On the efficacy of electrocatalysis in nonaqueous Li-O_2 batteries. J Am Chem Soc 133:18038–18041. https://doi.org/10.1021/ja207229n

Meng F-L, Liu K-H, Zhang Y, Shi M-M, Zhang X-B, Yan J-M, Jiang Q (2018) Recent advances toward the rational design of efficient bifunctional air electrodes for rechargeable Zn–air batteries. Small 14:1703843. https://doi.org/10.1002/smll.201703843

Mohamad AA (2006) Zn/gelled 6M KOH/O_2 zinc–air battery. J Power Sources 159:752–757. https://doi.org/10.1016/j.jpowsour.2005.10.110

Monaco S, Soavi F, Mastragostino M (2013) Role of oxygen mass transport in rechargeable Li/O_2 batteries operating with ionic liquids. J Phys Chem Lett 4:1379–1382. https://doi.org/10.1021/jz4006256

Naik KM, Sampath S (2018) Two-step oxygen reduction on spinel $NiFe_2O_4$ catalyst: rechargeable, aqueous solution- and gel-based, Zn-air batteries. Electrochim Acta 292:268–275. https://doi.org/10.1016/j.electacta.2018.08.138

Neburchilov V, Wang H, Martin JJ, Qu W (2010) A review on air cathodes for zinc–air fuel cells. J Power Sources 195:1271–1291. https://doi.org/10.1016/j.jpowsour.2009.08.100

Othman R, Basirun WJ, Yahaya AH, Arof AK (2001) Hydroponics gel as a new electrolyte gelling agent for alkaline zinc–air cells. J Power Sources 103:34–41. https://doi.org/10.1016/S0378-7753(01)00823-0

Ottakam Thotiyl MM, Freunberger SA, Peng Z, Bruce PG (2013) The carbon electrode in nonaqueous Li-O_2 cells. J Am Chem Soc 135:494–500. https://doi.org/10.1021/ja310258x

Park J, Park M, Nam G, Kim MG, Cho J (2017) Unveiling the catalytic origin of nanocrystalline yttrium ruthenate pyrochlore as a bifunctional electrocatalyst for Zn-air batteries. Nano Lett 17:3974–3981. https://doi.org/10.1021/acs.nanolett.7b01812

Pelegov VD, Pontes J (2018) Main drivers of battery industry changes: electric vehicles—a market overview batteries. Batteries 4(4):65. https://doi.org/10.3390/batteries4040065

Qian Y et al (2017) A metal-free ORR/OER bifunctional electrocatalyst derived from metal-organic frameworks for rechargeable Zn-air batteries. Carbon 111:641–650. https://doi.org/10.1016/j.carbon.2016.10.046

Read J (2006) Ether-based electrolytes for the lithium/oxygen organic electrolyte battery. J Electrochem Soc 153:A96–A100. https://doi.org/10.1149/1.2131827

Ren J-T, Yuan G-G, Chen L, Weng C-C, Yuan Z-Y (2018) Rational dispersion of $Co_2P_2O_7$ fine particles on N,P-codoped reduced graphene oxide aerogels leading to enhanced reversible oxygen reduction ability for Zn-air batteries. ACS Sustain Chem Eng 6:9793–9803. https://doi.org/10.1021/acssuschemeng.8b00873

Shang C et al (2018) Fe_3O_4@CoO mesospheres with core-shell nanostructure as catalyst for Li-O_2 batteries. Appl Surf Sci 457:804–808. https://doi.org/10.1016/j.apsusc.2018.07.026

Shao Q, Liu J, Wu Q, Li Q, Wang H-G, Li Y, Duan Q (2019) In situ coupling strategy for anchoring monodisperse Co_9S_8 nanoparticles on S and N dual-doped graphene as a bifunctional electrocatalyst for rechargeable Zn-air battery. Nano-Micro Lett 11:4. https://doi.org/10.1007/s40820-018-0231-3

Song L et al (2017) Functional species encapsulated in nitrogen-doped porous carbon as a highly efficient catalyst for the oxygen reduction reaction chemistry. Eur J 23:3398–3405. https://doi.org/10.1002/chem.201605026

Su H, Xu Q, Chong J, Li H, Sita C, Pasupathi S (2017) Eliminating micro-porous layer from gas diffusion electrode for use in high temperature polymer electrolyte membrane fuel cell. J Power Sources 341:302–308. https://doi.org/10.1016/j.jpowsour.2016.12.029

Sun C, Guo X, Zhang J, Han G, Gao D, Gao X (2019) Rechargeable Zn-air batteries initiated by nickel-cobalt bimetallic selenide. J Energy Chem 38:34–40. https://doi.org/10.1016/j.jechem.2019.01.001

Tan P et al (2017) Advances and challenges in lithium-air batteries. Appl Energy 204:780–806. https://doi.org/10.1016/j.apenergy.2017.07.054

Tang C, Wang H-F, Zhang Q (2018) Multiscale principles to boost reactivity in gas-involving energy electrocatalysis accounts of chemical research. Acc Chem Res 51:881–889. https://doi.org/10.1021/acs.accounts.7b00616

Veith GM, Dudney NJ (2011) Current collectors for rechargeable Li-air batteries. J Electrochem Soc 158:A658–A663. https://doi.org/10.1149/1.3569750

Vij V et al (2017) Nickel-based Electrocatalysts for energy-related applications: oxygen reduction, oxygen evolution, and hydrogen evolution reactions. ACS Catal 7:7196–7225. https://doi.org/10.1021/acscatal.7b01800

Wilcke WW, Kim H (2016) The 800-km battery lithium-ion batteries are played out. Next up: lithium-air. IEEE Spectrum 53:42–62. https://doi.org/10.1109/MSPEC.2016.7420398

Yang Z, Zhang J, Kintner-Meyer MC, Lu X, Choi D, Lemmon JP, Liu J (2011) Electrochemical energy storage for green grid. Chem Rev 111:3577–3613. https://doi.org/10.1021/cr100290v

Yang L, Wang D, Lv Y, Cao D (2019) Nitrogen-doped graphitic carbons with encapsulated CoNi bimetallic nanoparticles as bifunctional electrocatalysts for rechargeable Zn-Air batteries. Carbon 144:8–14. https://doi.org/10.1016/j.carbon.2018.12.008

Yu A, Lee C, Lee N-S, Kim MH, Lee Y (2016) Highly efficient silver-cobalt composite nanotube electrocatalysts for favorable oxygen reduction reaction. ACS Appl Mater Interfaces 8:32833–32841. https://doi.org/10.1021/acsami.6b11073

Zhang T, Imanishi N, Takeda Y, Yamamoto O (2011) Aqueous lithium/air rechargeable batteries. Chem Lett 40:668–673. https://doi.org/10.1246/cl.2011.668

Zhang RH, Zhao TS, Wu MC, Jiang HR, Zeng L (2018) Mesoporous ultrafine Ta_2O_5 nanoparticle with abundant oxygen vacancies as a novel and efficient catalyst for non-aqueous Li-O_2 batteries. Electrochim Acta 271:232–241. https://doi.org/10.1016/j.electacta.2018.03.164

Zhang J et al (2019a) Bimetallic nickel cobalt sulfide as efficient electrocatalyst for Zn-air battery and water splitting. Nano-Micro Lett 11:2. https://doi.org/10.1007/s40820-018-0232-2

Zhang P-F et al (2019b) High-performance rechargeable Li-CO_2/O_2 battery with Ru/N-doped CNT catalyst. Chem Eng J 363:224–233. https://doi.org/10.1016/j.cej.2019.01.048

Zheng Q, Xing F, Li X, Ning G, Zhang H (2016) Flow field design and optimization based on the mass transport polarization regulation in a flow-through type vanadium flow battery. J Power Sources 324:402–411. https://doi.org/10.1016/j.jpowsour.2016.05.110

Zhong B, Zhang L, Yu J, Fan K (2019) Ultrafine iron-cobalt nanoparticles embedded in nitrogen-doped porous carbon matrix for oxygen reduction reaction and zinc-air batteries. J Colloid Interface Sci 546:113–121. https://doi.org/10.1016/j.jcis.2019.03.038

Zou L, Jiang Y, Cheng J, Gong Y, Chi B, Pu J, Jian L (2016) Dandelion-like $NiCo_2O_4$ hollow microspheres as enhanced cathode catalyst for Li-oxygen batteries in ambient air. Electrochim Acta 216:120–129. https://doi.org/10.1016/j.electacta.2016.08.151

Zou L, Jiang Y, Cheng J, Chen Y, Chi B, Pu J, Jian L (2018) Bifunctional catalyst of well-dispersed RuO_2 on $NiCo_2O_4$ nanosheets as enhanced cathode for lithium-oxygen batteries. Electrochim Acta 262:97–106. https://doi.org/10.1016/j.electacta.2018.01.005

Zouhri K, Lee S-Y (2016) Evaluation and optimization of the alkaline water electrolysis ohmic polarization: exergy study. Int J Hydrogen Energ 41:7253–7263. https://doi.org/10.1016/j.ijhydene.2016.03.119

Chapter 4
Green Chemistry Metrics for Environmental Friendly Processes: Application to Biodiesel Production Using Cooking Oil

Nawel Outili and Abdeslam Hassen Meniai

Contents

N. Outili (✉) · A. H. Meniai
Process Engineering Faculty, Laboratory of Environmental Processes Engineering (LIPE), Constantine3 University, Constantine, Algeria
e-mail: nawel.outili@univ-constantine3.dz

Inamuddin, A. M. Asiri (eds.), *Sustainable Green Chemical Processes and their Allied Applications*, Nanotechnology in the Life Sciences,
https://doi.org/10.1007/978-3-030-42284-4_4

Abbreviations

AE	Atom economy
C	MSDS indication for corrosive
CE	Carbon economy
cEF	Complete environmental factor (kg/kg)
C_k	The nondiscounted yearly cash flows assuming that they follow at the end of each year *k*
CMR	Mutagen carcinogen and reprotoxic
E	MSDS indication for explosive
E_Factor	Environmental factor (kg/kg)
EATOS	Environmental Assessment Tool for Organic Syntheses
EE	Energy efficiency (kg/kJ)
EM	Effective mass yield (kg/kg)
EPE	Energetic process expenditure (kJ/kg)
ERP	Energy recovery parameter (kJ/kJ)
F and F+	MSDS indication for flammable and very flammable, respectively
FFA	Free fatty acids
GAL	The Green Aspiration Level method
Hazard	Hazard parameter (kg/kg)
Hazard_coeff	Hazard coefficient
iGAL	The innovation Green Aspiration Level method
IRR	Internal rate of return (%)
LCA	Life-cycle assessment
MRP	Material recovery parameter (kg/kg)
MSDS	Material safety data sheet
MW	Molecular weight (g/mol)
N	MSDS indication for dangerous material toward the environment
NC	Number of carbons
NPV	Net present value ($)
PMI	Process mass intensity (kg/kg)
Productivity	The productivity in (kg/h)
r	The discount rate
REI	Renewable energy index (kJ/kg)
REP	Renewable energy percentage (%)
RI	Renewable intensity (kg/kg)
Risk	Hazard metric taking into account exposure time
RME	Reaction mass efficiency (kg/kg)
ROI	Return on investment (%)
RP	Renewable percentage (%)
RY	Reaction yield (kg/kg)
sEF	Simple environmental factor (kg/kg)
SF	Stoichiometric factor (kg/kg)

SI	Solvent intensity (kg/kg)
SRE	Solvent recovery energy (kJ/kg)
STY	Space time yield (kg/m^3/h)
Syn_Ideality	Synthesis ideality
T and T^+	MSDS indication for toxic and very toxic, respectively
Toxicity	Toxicity parameter (kg/kg)
Toxicity_coeff	Toxicity coefficient
WI	Water intensity (kg/kg)
WTE	Waste treatment (kJ/kg)
X_i	MSDS indication for irritant
X_n	MSDS indication for harmful

4.1 Introduction

Human beings are continuously seeking the improvement of their standards of living, lifestyle, and comfort through the development of new products, services, facilities, and technologies by tapping into nature. Manufactured products are invading our everyday life (Lapkin 2018) with an ever-increasing demand, requiring an important industrial development which generates waste and pollution in most cases, mainly at the expense of the environment, the human health, and the earth resources.

Since the early 1980s, pollution had caused considerable damage to the main components of the environment—air, water, and soil—hence the great number of major incidents involving highly hazardous materials that have occurred around the world. One can cite the historical worst air pollution disaster that occurred in December 1952, characterized by vast and lethal smog and caused by trapped combustion products at ground level (Kelly and Fussell 2015). Also the diseases associated with the widespread soil contamination by radionuclides from the Chernobyl disaster in 1986 are still in the mind for many people (Rodriguez-Eugenio et al. 2018). The Songhua River toxic spill which occurred in November 2005 following an explosion that took place at Jilin Chemical Industrial plant is a concrete example of water pollution disaster (Wang et al. 2011).

Consequently and in recent years, the extent of the damages caused by human urged him to think about new production initiatives with an increasing concern for sustainability issues (Phan et al. 2015). There is a trend for reducing waste, saving energy, and replacing hazardous substances (Albini and Protti 2016). The chemists and process engineers have therefore focused in the way of reducing environment impacts by reinventing chemical reactions, production routes, and chemical processes. This has introduced a new concept for chemistry, namely, "green chemistry."

Green chemistry is the conception of chemical products and processes that uses and generates less hazardous substances, reduces pollution and waste, encourages the use of renewable feedstock, leads to safer designs, optimizes materials and

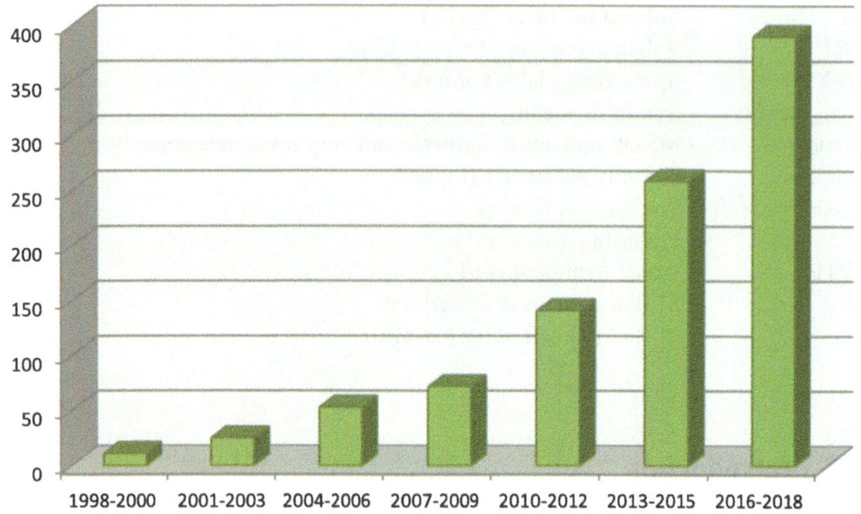

Fig. 4.1 Number of publication with "green chemistry metrics" for each period of 3 years from 1998 to 2018 (obtained using Google Scholar citation data)

energy usage, and minimizes costs (Andraos and Sayed 2007; Albini and Protti 2016). This definition marks a major deviation from the way environmental issues were regarded in the processes' up-front design. Paul Anastas and John C. Warner elaborated 12 principles of green chemistry, which are considered as a practical illustration of the concept (Tundo and Arico 2007).

To be truly green, chemistry must completely satisfy the 12 principles. However, it is approaching green if most of these principles are satisfied, but it is clear that it will be necessary to bring precise indicators to assess the "greenness" of the chemical processes (Augé and Scherrmann 2017). Those metrics allow gauging process improvement and performance comparison (Washington national academy of engineering 1999). Several works were devoted to the green chemistry metrics over the past two decades as shown in Fig. 4.1 by the increasing number of publications every year.

The most common metric used by chemists to assess reaction efficiency is yield which is useful but failed to account for the reagents quantities, solvents and catalysts, recycling, resources, energy consuming, waste generation, hazardous, and toxicity (Clark 2005). With the publishing of the 12 principles of green chemistry, other metrics are necessarily needed to take into account the possible maximum of the principles simultaneously (Mcelroy et al. 2015).

The main objective of this review is to summarize the most used and the easiest to implement green metrics in the literature. Indeed, over the past 20 years, much had been written about the green chemistry metrics (Fig. 4.1) and what constituted a good metric (Dichiarante et al. 2010; Constable et al. 2002). A large number of parameters considering different principles of green chemistry had been proposed (Jiménez-González 2012; Sheldon 2017; Albini and Protti 2016; Roschangar et al.

2015). In order to be implemented in the "greenness" quantification of the processes, parameters must be clearly defined, simple, measurable, and objective, rather than subjective, and must ultimately drive the desired behavior (Mcelroy et al. 2015; Constable et al. 2002).

As a good example of the application of the principles of green chemistry metrics, one can cite biodiesel production from cooking oil. This process comprises renewable feedstock, design for degradation (Henrie 2015), waste generation, use of reagent in excess, and energy consumption. Practically all the principles of green chemistry can be easily applied and the corresponding metrics handily calculated.

4.2 History

Since 1940, the relationship between environmental problems and the growth of industrial activities began to appear (De Marco et al. 2019). Public concern increased following the publication of the scientific book "Silent Spring" by Rachel Carson in 1962 (Carson 2002; Linthorst 2010). The book outlined the damage caused by chemicals on ecosystems and was considered as a warning signal to the entire scientific, industrial, and even political communities. Increased pollution and resource depletion caused by industrialization had motivated several conferences on resources use, conservation of biosphere resources, pollution reduction, and environment protection that marked the 1980s (De Marco et al. 2019). A real awareness of environmental degradation has created a set of important decision-making and regulatory restrictions (Wynne 1992) leading to the birth of the green chemistry concept.

The history of green chemistry had been marked by important events as presented below; the most important steps are summarized in Fig. 4.2.

- In 1969, the US National Environmental Policy Act (NEPA) was enacted (Haklay 2017), and the aim of the law was to "create and preserve conditions under which man and nature can coexist in harmony" (Benson and Garmestani 2011).
- In 1987, the World Commission on Environment and Development published the "Brundtland Report" on world development and environment, which described the notion of sustainable development for the first time (Gendron and Reveret 2000).
- The Pollution Prevention Act of 1990 (PPA) created a national policy based on the principle of prevention or reduction of pollution at the source and the idea of green chemistry emerged (Anastats and Beach 2009; Linthorst 2010).
- In 1990s, the first journals on green chemistry were published out, and one can cite the Journal of Clean Processes and Products by Springer-Verlag and Green Chemistry by the Royal Society of Chemistry (Karagölge and Bahri 2016). Nowadays, a great number of journals, books, and conferences are currently available.

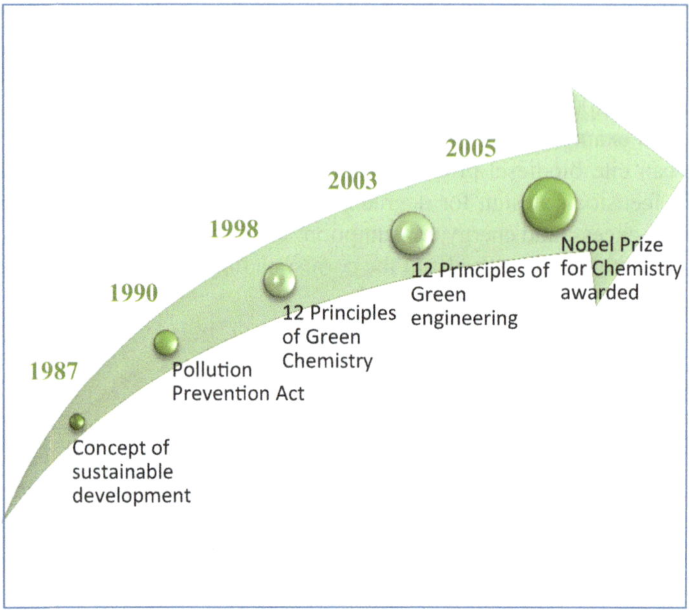

Fig. 4.2 The most important steps in green chemistry history from the first time that the concept of sustainable development was defined to the Nobel Prize for chemistry awarded in 2005

- In 1997, the foundation of the Green Chemistry Institute (GCI) in the USA is a significant step in green chemistry growth (Tundo and Arico 2007; Linthorst 2010).
- In 1998, the publication of the 12 Principles of Green Chemistry by Paul T. Anastas and John C. Warner (Anastas and Warner 1998) providing a clear set of rules for better implementation (Anastats and Beach 2009).
- In 2003, the proposal of the 12 principles of green engineering by Anastas et al. (Anastas and Zimmerman 2003). But before that, in 1998, Malone published his scientific paper "Introduction to green engineering" where he defined green engineering as "Designing with concern for the natural environment" (Malone 1998). Also, Anastas et al. published in 2000 a book entitled: "Green engineering" (Anastas et al. 2000).
- One of the most important steps in the history of green chemistry marking the importance of this concept was the citation for the 2005 Nobel Prize for Chemistry awarded to Chauvin, Grubbs, and Schrock, for the development of a new greener technique in organic synthesis (Zhao and Tuhlar 2009).

The successful story of green chemistry continues to inspire a lot of researchers to develop and innovate greener synthesis, materials, energies, solvent, and sustainable processes in fields as varied as chemistry, pharmacy, energy, building, agronomy, and everything related to life and survival. As examples, one can cite a green synthesis of new nanostructured components for different pharmaceutical and biomedical uses that was proposed by Kumar et al. (2019):

The development of a safe and green method to synthesize graphene oxide that is highly required for electronic field, catalysis, energy storage, separation membranes, biomedicine, and composites (Pei et al. 2018).
The production of renewable energy using waste resources such as the production of liquid fuel from plastic waste using pyrolysis process (Sharuddin et al. 2018) or using municipal solid waste as a feedstock for bioenergy (Pour et al. 2018).
The development of green methods for organic synthesis in medicinal field (Zhang and Cue 2018).
The use of supercritical fluids and ionic liquids as green solvents (Sokmen et al. 2018; Elgharbawy et al. 2019).
The utilization of biomass as renewable feedstock to acquire high-tech bioproducts (Popa 2018) and so on.

4.3 Green Chemistry Principles

Green chemistry implies the application of the 12 principles developed by Anastas and Warner in 1998 (Anastas and Warner 1998) and which are considered as guidelines to carry out a greener action (Albini and Protti 2016) such as: waste reduction, use of more sustainable feedstock's, conservation of energy, reduction of hazardous components, and insurance of a safer process. Table 4.1 regroups those principles which are reported in all publications dealing with green chemistry as (Albini and Protti 2016; Augé and Scherrmann 2017).

Tang et al. have proposed a mnemonics to give a very pleasant manner of communicating and learning the green chemistry values (Tang et al. 2005). They produced a simple acronym "PRODUCTIVELY" that states each of the 12 principles of green chemistry in a short way (see Table 4.2).

According to the foundation of green chemistry as claimed by its first principle "Prevention," all the elements influencing a process are to be considered from the design. The 12 principles of green chemistry focused on three main elements of the reaction or process: the materials and waste, the energy, and the process safety (Fig. 4.3).

The application of green chemistry principles leads to an ideal chemical industry (Augé and Scherrmann 2017): low polluting, efficient in mass and energy uses, and safer process.

Several works have been performed by academicians and industrials on green chemistry principles application (Dirgha and Nisha 2019) considering in details published laboratory experiments and proven case studies (Dicks 2016). Also, a great interest on green chemistry was noticed in the education field as well as in the undergraduate and postgraduate levels to facilitate the use and integration of green chemistry principles and metrics (Andraos 2018a, b; Fennie and Roth 2016; Morra and Dicks 2016).

Table 4.1 The 12 principles of green chemistry and their definition

N°	Principle	Definition
1	Prevention	Preventing the formation of waste is better than treating or cleaning waste after it has been produced
2	Atom economy	Maximize the inclusion in the final products of all components used in the reaction
3	Less hazardous chemical syntheses	Use and produce less toxic and hazardous substances and any material that can constitute a danger for human health and environment
4	Designing safer chemicals	Chemical products should be designed to maintain their desired function while minimizing their toxicity
5	Safer solvents and auxiliaries	The use of auxiliary substances (e.g., solvents, separation agents, etc.) should be minimized wherever possible and safer when used
6	Design for energy efficiency	Processes energy demands should be minimized for their environmental and economic effects
7	Use of renewable feedstock	A raw material should be renewable and not be depleted wherever technically and economically feasible
8	Reduce derivatives	If feasible, unnecessary generation of derivatives should be minimized or prevented, as such steps require extra reagents and may result in waste
9	Catalysis	Catalytic reagents (as selective as possible) are preferred to those stoichiometric
10	Design for degradation	Chemical products should be designed to break down into harmless degradation products at the end of their lives
11	Real-time analysis for pollution prevention	To enable real-time, in-process surveillance and control before the creation of hazardous substances, analytical methodologies need to be further developed
12	Inherently safer chemistry for accident prevention	To minimize the potential for chemical accidents, including discharge, explosion, and fire, substances used in a chemical process should be selected

Table 4.2 Condensed principles of green chemistry following the acronym PRODUCTIVELY

P	Prevention of wastes
R	Renewable materials
O	Omit derivatization steps
D	Degradable chemical products
U	Use safe synthetic methods
C	Catalytic reagents
T	Temperature, pressure ambient
I	In-process monitoring
V	Very few auxiliary substances
E	E-factor, maximize feed in product
L	Low toxicity in chemical products
Y	Yes, it is safe

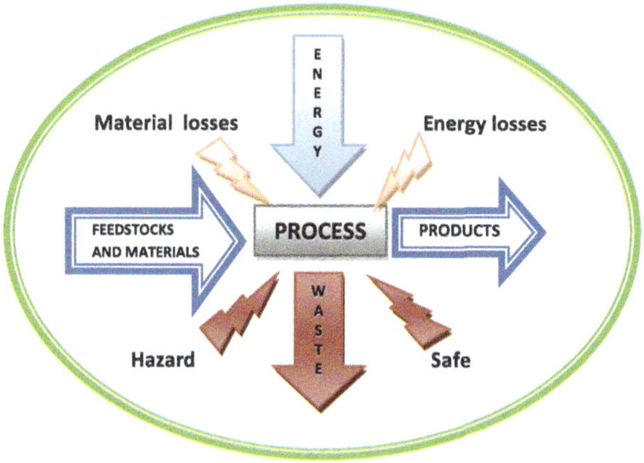

Fig. 4.3 Main elements of the process considered by the 12 green chemistry principles: materials, waste, energy, and safety

4.3.1 Materials Use and Waste Prevention Principles

The raw material and all the components used by the process and generated as waste are considered by the first, second, seventh, eighth, and 11th principles. The main objective is preventing waste instead of remediation (Sheldon 2016a) by optimizing the efficiency of resource utilization (Sheldon 2016b) and using renewable materials (Sperry and García-Álvarez 2016) to preserve the environment.

Also, the use of catalysis following the ninth principle enables an efficient use of resources, improves the reaction rate, and reduces the number of process steps and waste generation (Delidovich and Palkovits 2016). Also, combining principle 9 and 7 leads to the use of biocatalysts that allows performing reactions under moderate temperature and pressure circumstances (Dunn et al. 2010).

Large amounts of solvents are used for the isolation, separation, and purification of components (Poliakoff and Licence 2007), and if in addition they are toxic, the process will be less respectful of the environment. To reduce their negative impact, recently research has focused on replacing traditional organic reaction media by green solvents (Sperry and García-Álvarez 2016) respecting the green chemistry principles 1, 3, 5, and 7.

Also, the use of real-time analysis outlined by the 11th principle prevents pollution by providing essential feedback to ensure that processes are working properly, prevent material losses, and evaluate quality (Glasgow et al. 2004).

4.3.2 Energy Principles

Energy efficiency is considered in principles 6 and 7 in order to reduce the energy consumption of the process, increasing its efficiency and promoting the use of renewable energy. The ideal situation should be to conduct syntheses at ambient temperature and pressure. Also, the 11th principle allows controlling the energy losses.

Renewable energies (solar, biofuels, hydro, wind, geothermal, wave, tidal, and ocean) are essential for energy security and for sustainability (Twidell and Weir 2015). Intensification of the process and related synergy can enhance the effectiveness of resource use (Gude 2015).

4.3.3 Hazard and Safe Principles

The green chemistry principle of inherently safer chemistry has an important role to play in accident prevention (Salah and Koller 2018). Indeed, principles 3, 4, 5, 10, 11, and 12 recommend the minimization or elimination of hazardous materials, the production of degradable products, pollution prevention, and detection of problems before major emissions or accidents to occur.

The choice of safe, nonhazardous, renewable, and low-cost solvents is a key objective of organic green synthesis (Sperry and Garcia-Álvarez 2016).

4.4 Sustainable and Green Chemistry

As mentioned in the history section, the idea of sustainable development was defined for the first time in 1987. Sustainable development was described as "Meeting current generation requirements without compromising future generation's capacity to satisfy their own requirements" (Robert et al. 2005). This concept reflects the necessity to find a balance between financial growth, impact on the environmental, and societal equity, which are represented by three indicators: profit, planet, and people (Sheldon 2017). The general objective of sustainable development is a long-term economic and environmental stability (Emas 2015).

Sustainability could be said to be the objective and green chemistry is the means to accomplish it (Sheldon et al. 2007). However, other concepts can also allow achieving sustainability at different scales. Indeed, sustainability concepts have to face global-scale environmental hazards such as depletion of ozone, climate change, and loss of biodiversity or nitrogen cycle modifications (Geissdoerfer et al. 2017). Otherwise, green chemistry can be applied to change the synthesis routes at the laboratory scale, while green engineering intervenes at the process scale (Fig. 4.4). For a large scale that takes into account the future of the products after their use: "cradle-to-grave", a life cycle assessment (LCA) is a method for extending the environmental impact assessment beyond the usual manufacturing plant boundaries, a territorial scale (Loiseau et al. 2018). Life cycle assessment considers the resource consumption, pollutants emitted, energy consumption, and efficiency and includes extraction, manufacture, transport, sales, distribution, use, and final destiny (Augé and Scherrmann 2017). A green chemist must be conscious that the system boundaries go beyond the current process, so that the requirements of the life cycle also need to be taken into account (Jiménez-González et al. 2012).

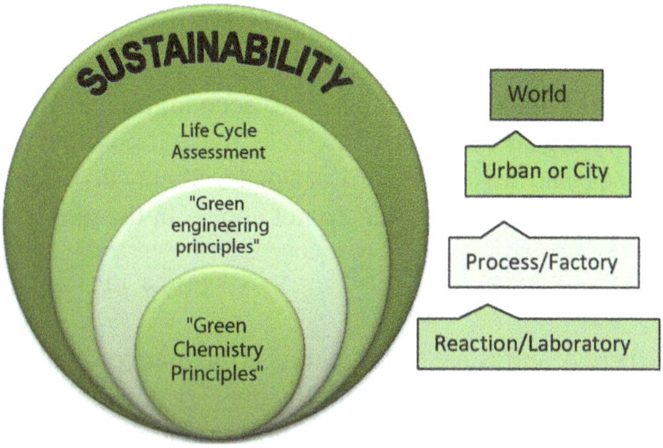

Fig. 4.4 Concepts used to achieve sustainability and their scale of action: laboratory scale for green chemistry, process scale for green engineering, city scale for life cycle assessment, and global scale for sustainability

4.5 Green Chemistry Metrics

While useful, Green Chemistry's 12 principles are qualitative and cannot be quantified. However, what cannot be measured cannot be managed (Tickner and Becker 2016). Indeed, measuring the process "greenness" may indicate how far the process respects the principles of green chemistry. Therefore, defining clearly metrics for researchers and decision-makers is crucial for the application of the green chemistry principles (Roschangar et al. 2015).

Based on the 12 principles of green chemistry, various simple metrics were developed (Phan et al. 2015) for quantification purposes, allowing a more objective comparison of alternatives (Eissen et al. 2004) and a decision-making approach which is based on numerical data (Andraos and Dicks 2012). Green chemistry metrics have also played a role to make chemists aware of the necessity to innovate new chemical syntheses, products, and processes so that they are less pollutant (Constable et al. 2002).

Depending on the 12 principles, the proposed green metrics were classified by researchers around three aspects—economic, environmental, and social for sustainable development (Washington National Academy of engineering 1999). The green metrics cover, among others, different areas of materials, resources, energy, equipment, manufacturing, cleaning, life cycle, and renewability (Jiménez-González et al. 2012). Several types of metrics (Ribeiro and Machado 2013) emerged such as follows: mass metrics (Song and He 2018), energy metrics (Calvo-Flores 2009), environmental metrics (Jiménez-González et al. 2012; Ribeiro and Machado 2013), social metrics (Onsrud and Simon 2013), economic metrics (Lima-Ramos et al. 2014), and safety-hazard metrics (Andraos et al. 2015; Andraos 2016; Naidu et al. 2008).

Based on the classifications cited above and considerations taken into account by the principles of green chemistry reported earlier (Fig. 4.1), four categories are proposed: material-based metrics, energy-based metrics, hazardous and safety metrics, and economic metrics.

4.5.1 Material-Based Metrics

Material-based metrics take into account the quantities of all the materials used by the synthesis or the process. They make it possible to measure mass yields, excess, losses, and waste. The process material may be reactants, solvents, water, catalysts, and other auxiliaries that do not appear in the global reaction scheme:

$$a\mathrm{A} + b\mathrm{B} \rightarrow p\mathrm{P} + q\mathrm{Q}$$

where A and B are the reagents, P the main product, and Q a coproduct. The corresponding stoichiometric coefficients are noted *a*, *b*, *p*, and *q*, respectively.

Reaction yield (RY): This early metric is the most used by chemists to quantify the limiting reactant being transformed to the desired product. Its value is between 0 and 1 (Albini and Protti 2016):

$$\mathrm{RY} = \frac{\text{Real Mass of product}\,(\mathrm{kg})}{\text{Theoritical Mass of product}\,(\mathrm{kg})} \tag{4.1}$$

Atom economy (AE): In 1991, Trost introduced the concept of atomic economy (Augé and Scherrmann 2017) that *related* the reaction efficiency to the degree of integration of atoms of the reagents in the desired product. It is considered as the most commonly used measure of green chemistry efficiency (Jiménez-González et al. 2012):

$$\mathrm{AE} = \frac{\mathrm{MW\ product}\left(\dfrac{\mathrm{g}}{\mathrm{mol}}\right)}{\mathrm{MW\ reagents}\left(\dfrac{\mathrm{g}}{\mathrm{mol}}\right)} \tag{4.2}$$

where MW is the molecular weight (g/mol).

Maximum value for this metric (AE equal to 1) leads to design synthetic methods that incorporate all the used reactants into the final product. Indeed, reaction mechanisms have a great influence on this parameter, so that addition mechanisms will have an atom economy equal to 1, while substitution or elimination will always have an atom economy less than 1 (Louaer and Outili 2017).

Eissen et al. have suggested a more elaborate expression of AE for synthesis with several steps which takes into account the reactants and atom economy of each intermediate reaction of the global mechanism (Eissen et al. 2004):

$$\mathrm{AE}_{\mathrm{Multi}} = \frac{\mathrm{MW\ product}}{\dfrac{\mathrm{MW\ reagents}_1}{\mathrm{AE}_1} + \dfrac{\mathrm{MW\ reagents}_2}{\mathrm{AE}_2} + \ldots} \tag{4.3}$$

where MW is the molecular weight (g/mol) and AE_i is the atom economy of step i.

Carbon economy (CE): This metric was proposed by Curzons et al. in 2001 (Curzons et al. 2001) and quantifies the number of carbons (N_C) in the reagents that is integrated in the main product, according to Eq. (4.4) (Albini and Protti 2016):

$$\mathrm{CE} = \frac{N_\mathrm{C}\ \text{in product}}{N_\mathrm{C}\ \text{in reagents}} \tag{4.4}$$

According to this definition, a reaction with a coproduct containing no carbon will have a CE of 1 and an AE less than 1. Indeed, the AE reports the loss of atoms and the CE the loss of carbons.

Reaction mass efficiency (RME): All the factors cited above does not account for solvents, catalysts, and other reactants; thus, the efficiency of a real system is not fully represented. The concept of reaction mass efficiency was proposed to tackle this problem (Jiménez-González et al. 2012):

$$\mathrm{RME} = \frac{\text{Mass of product}\left(\text{kg}\right)}{\text{Mass of reactants}\left(\text{kg}\right)} \tag{4.5}$$

A reactant designates any organic or inorganic material that participates to the elaboration of the main product or any intermediates (Jiménez-González et al. 2012).

Stoichiometric factor (SF): As mentioned above, reagent excess is considered as reaction waste and the stoichiometric factor quantifies that excess to account for reactions that run under non-stoichiometric conditions (Andraos 2005):

$$\mathrm{SF} = 1 + \frac{\sum \text{Mass excess reagents}\left(\text{kg}\right)}{\sum \text{Mass stoichiometric reagents}\left(\text{kg}\right)} \tag{4.6}$$

Material recovery parameter (MRP): This factor includes all the recycled materials (reagents in excess, solvents, and catalysts) (Andraos 2005):

$$\mathrm{MRP} = \frac{\text{Mass of all materials recovered}\left(\text{kg}\right)}{\text{Mass of product}\left(\text{kg}\right)} \tag{4.7}$$

The reaction mass efficiency (RME) given by Eq. (4.5) is controlled by four independent factors, each varying in the interval [0–1]: reaction yield (RY), reciprocal of stoichiometric factor (1/SF), atom economy (AE), and material recovery parameter (MRP) (Andraos 2006):

$$\mathrm{RME} = \left(\mathrm{RY}\right)\left(\frac{1}{\mathrm{SF}}\right)\left(\mathrm{AE}\right)\left(\mathrm{MRP}\right) \tag{4.8}$$

Process mass intensity (PMI): Another strategy takes advantage of the process mass intensity (PMI), which evaluates effectiveness by taking into account masses of all materials and water used in a process for the production of the unit mass of the desired product (Roschangar et al. 2015; Jiménez-González et al. 2012):

$$\mathrm{PMI} = \frac{\text{Mass of all materials}\left(\text{kg}\right)}{\text{Mass of product}\left(\text{kg}\right)} \tag{4.9}$$

In some works, water was excluded from the total mass materials in the PMI calculation (Sheldon 2017; Albini and Protti 2016).

Solvent and water intensity (SI) and (WI): Solvent and water losses are a significant cause of waste in chemical procedures, especially in the pharmaceutical sector (Sheldon 2017; Roschangar et al. 2015). Also, the water shortage in many parts of the world justifies its preservation and its importance as a green chemistry parameter. Thus, two mass-based parameters are used to quantify those losses to respect the principle of preventing waste (Albini and Protti 2016; Jiménez-González et al. 2012):

$$\mathrm{SI} = \frac{\text{Mass of all solvent used excluding water}\left(\text{kg}\right)}{\text{Mass of product}\left(\text{kg}\right)} \tag{4.10}$$

$$\mathrm{WI} = \frac{\text{Mass of all water used}\left(\text{kg}\right)}{\text{Mass of product}\left(\text{kg}\right)} \tag{4.11}$$

Environmental factor (E_Factor): The first principle, known as the prevention principle, can be quantified by the *E_Factor* (or environmental factor of Sheldon), defined as early as 1992 as follows (Augé and Scherrmann 2017):

$$\mathrm{E\_Factor} = \frac{\text{Mass of waste}\left(\text{kg}\right)}{\text{Mass of product}\left(\text{kg}\right)} \tag{4.12}$$

Therefore, any output materials, excluding recyclable ones, from the reaction other than the main product, are considered waste. Its ideal value is zero corresponding to a production without generating any waste (Albini and Protti 2016).

Typical E_factors for various industries indicated that the pharmaceutical industry is the one that generates the most waste (see Table 4.3) (Calvo-Flores 2009, Phan et al. 2015). This is primary owing to the high molecular complexity of the products requiring a high number of steps for their synthesis (Roschangar et al. 2015).

Table 4.3 Typical E_Factors for various industries

Sector	Product tonnage	E_Factor
Oil refining	10^6–10^8	<0.1
Bulk chemicals	10^4–10^6	<1–5
Fine chemicals	10^2–10^4	5 to >50
Pharmaceuticals	10–10^2	25 to >100

More lately, depending on the process development stage, Roschangar proposed the use of simple environmental factor sEF and complete environmental factor cEF (Roschangar et al. 2015). The sEF takes no account of water and solvents and is better suitable for early path planning, while the cEF takes all process products including solvents and water into consideration, and it is better suitable for complete waste stream assessment (Sheldon 2017):

$$\text{sEF} = \frac{\sum \text{Mass}_{\text{raw materials}} + \sum \text{Mass}_{\text{reagents}} - \text{Mass of product}\left(\text{kg}\right)}{\text{Mass of product}\left(\text{kg}\right)} \tag{4.13}$$

$$\text{cEF} = \frac{\sum \text{Mass}_{\text{raw materials}} + \sum \text{Mass}_{\text{reagents}} + \sum \text{Mass}_{\text{solvents}} + \sum \text{Mass}_{\text{water}} - \text{Mass of product}\left(\text{kg}\right)}{\text{Mass of product}\left(\text{kg}\right)} \tag{4.14}$$

Typical E_factors for various industries indicate that the pharmaceutical industry is the one that generates the most waste (Table 4.3) (Augé and Scherrmann 2017). This is because of the high molecular complexity of the products requiring high number of steps for their synthesis (Roschangar et al. 2015).

Synthesis ideality (Syn_Ideality): This factor account for the reaction steps and was proposed by Gaich and Baran (2010) as shown in Eq. (4.15):

$$\text{Syn\_Ideality} = \frac{\text{No.of construction reactions} + \text{No.of strategic redox reactions}}{\text{No.of reactions}} \tag{4.15}$$

Construction reactions are described as chemical modifications forming bonds of carbon-carbon or carbon-heteroatom. Strategic redox reactions are a sort of building reactions that immediately set the right functions found in the main product and include asymmetric reductions or oxidations.

This parameter allows quantifying the sixth principle of green chemistry by reducing chemical transformations required and the comparison of different routes for achieving the final complex structure (Albini and Protti 2016; Gaich and Baran 2010). In 1997, the pharmaceutical group BHC developed greener synthetic pathways for ibuprofen production process reducing the number of steps from six to three (Roschangar and Colberg 2018).

Renewable intensity RI: The use of renewable raw materials is one of the principles of green chemistry, and the straightforward metric to evaluate how renewable a reaction is would be that of renewable intensity (Mcelroy et al. 2015):

$$\text{RI} = \frac{\text{Mass of all renewably derived materials used}\left(\text{kg}\right)}{\text{Mass of product}\left(\text{kg}\right)} \tag{4.16}$$

The RI to PMI comparison allows the calculation of a renewable percentage (RP):

$$\text{RP} = \frac{\text{RI}}{\text{PMI}} \times 100 \tag{4.17}$$

$$\mathrm{RP} = \frac{\text{Mass of all renewably derived materials used}\,(\mathrm{kg})}{\text{Mass of all materials}\,(\mathrm{kg})} \tag{4.18}$$

4.5.2 Toxicity and Safety Metrics

According to green chemistry principles of less hazardous chemicals and safer processes, those metrics must take into account the quality of process chemicals and their environmental and human health risks. There are several tools to identify the intrinsic danger of a compound and to evaluate the risk associated with its use (Jiménez-González et al. 2012) such as the Inherently Safer Design Tool proposed by Eljack (2019)) or the software elaborated by Eissen and Metzger (2002) EATOS (Environmental Assessment Tool for Organic Syntheses).

However, the simpler method is the use of material safety data sheet MSDS to obtain information on the chemical's intrinsic health and safety risks. Indeed, the format of the MSDS contains information such as toxicity, flammability, and exposure hazards (Hodson et al. 2019).

Effective mass yield (EMY): A simple green metric to account of safety was performed by the term effective mass yield defined as the fraction of main product mass to the all non-benign materials mass used in its synthesis (Albini and Protti 2016), according to Eq. (4.18):

$$\mathrm{EMY} = \frac{\text{Mass of products}\,(\mathrm{kg})}{\text{Mass of non}\quad\text{benign materials}\,(\mathrm{kg})} \tag{4.19}$$

This parameter accounts for the product fraction produced by toxic materials (Albini and Protti 2016). There are several compound characteristics that fulfill the green requirements: renewable resources, safer, nonflammable, nontoxic, environmentally not dangerous, and biodegradable (Manahan 2006).

Hazard: Using the material safety data sheet MSDS, this parameter allows to measure the hazard of all the materials used in the reaction (Kerras et al. 2019):

$$\mathrm{Hazard} = 1 - \frac{\sum_i \left(\mathrm{Hazard\_coeff} \times \mathrm{Mass}_{\mathrm{material}_i}\right)}{\text{Mass all materials}} \tag{4.20}$$

The hazard coefficient is calculated as:

$$\mathrm{Hazard\_coeff} = \frac{X_n + X_i + C + (5 \times E) + F + (5 \times F^{+})}{14} \tag{4.21}$$

The parameters of Eq. (4.21) are obtained from each material MSDS and refer to several dangers: harmful for X_n, irritant for X_i, corrosive for C, explosive for E, flammable for F, and very flammable for F^+.

If the danger is on the MSDS of a given material, its corresponding parameter will have the value 1; else, its value is 0.

Toxicity: The same way as hazard, the toxicity is measured by Eq. (4.22) (Kerras et al. 2019):

$$\text{Toxicity} = 1 - \frac{\sum_i \left(\text{Toxicity\_coeff} \times \text{Mass}_{\text{material}_i}\right)}{\text{Mass all materials}} \tag{4.22}$$

The toxicity coefficient is calculated using Eq. (4.23):

$$\text{Toxicity\_Coeff} = \frac{T + 5 \times T^{+} + 2 \times N}{8} \tag{4.23}$$

The toxicity parameters are also obtained from MSDS of each material, with toxic and very toxic for T and T^+, respectively, and N refers to the dangerous material toward the environment.

Mutagen carcinogen and reprotoxic (CMR): The three types of hazards are calculated using the CMR parameter which is calculated as the same way as for the two previous. It is calculated by (Kerras et al. 2019):

$$\text{CMR} = 1 - \frac{\sum_i \left(\text{CMR\_Coeff} \times \text{Mass}_{\text{material}_i}\right)}{\text{Mass all materials}} \tag{4.24}$$

with:

$$\text{CMR\_Coeff} = \frac{\left(\left(R40R45R49\right) + \left(R46\right) + \left(R60R61R62R63\right)\right)}{3} \tag{4.25}$$

R40R45R49, *R46*, and *R60R61R62R63* are risk R-phrases corresponding to carcinogenic effect, heritable genetic damage, and possible risk of impaired fertility and during pregnancy respectively (Uhlman and Saling 2010).

Risk: The previous hazard metrics take into account the quantities of materials but neglect a major aspect of risk that is the exposure time. Therefore, another parameter is defined as risk and estimated using Eq. (4.26):

$$\text{Risk} = F\left\{\text{hazard} \times \text{exposure}\right\} \tag{4.26}$$

This relationship demonstrates that risk can be lowered by hazard decrease, exposure decrease, and different combinations of both (Manahan 2006).

4.5.3 Energy-Based Metrics

The perfect scenario would be a quick reaction or process at ambient temperature and atmospheric pressure. However, most of the time, temperature and/or pressure circumstances need to be adjusted to enhance the response affecting energy consumption (Phan et al. 2015).

As power costs increase, higher attempts are made to reduce consumption and regulate energy-related emissions. This stimulates the use of renewable energy resources and other heating modes such as the use of ultrasonic and microwave reactors. Both are based on the use of extensive direct radiation, which can result in very brief synthesis times or enhanced product reaction yields and selectivity (Clark 2005; Razzaq and Kappe 2008).

Chemical processes have been implemented with different energy metrics, most of which are comparable to mass metrics. The most used energy metrics are the following: energy efficiency, waste treatment energy, and solvent recovery energy (Calvo-Flores 2009).

Energy efficiency (EE): The simpler metric to be used is the energy efficiency (EE), which is the proportion between the quantity of the main product and the energy consumed to obtain it (Albini and Protti 2016):

$$\mathrm{EE} = \frac{\text{Mass of product}\,(\text{kg})}{\text{Energy consumption}\,(\text{kJ})} \tag{4.27}$$

Energetic process expenditure (EPE): The reciprocal EE is described as the energetic process expenditure. Energy consumption is the sum of the energy consumed during the reaction and during workup, but often the latter contribution is larger (Albini and Protti 2016):

$$\mathrm{EPE} = \frac{\text{Reaction energy consumption}\,(\text{kJ}) + \text{Workup energy consumption}\,(\text{kJ})}{\text{Mass of product}\,(\text{kg})} \tag{4.28}$$

Renewable energy index (REI): According to green chemistry principles, it is crucial to use energy from renewable resources rather than from fossil ones (Sheldon 2017). Renewable energy index is a metric proposed to quantify this principle, and it can be written as (Jiménez-González 2012):

$$\mathrm{REI} = \frac{\text{Renewable energy}\,(\text{kJ})}{\text{Mass of product}\,(\text{kg})} \tag{4.29}$$

This parameter can be written as a percentage by similarity to the mass metric renewable percentage to get the renewable energy percentage (REP) as:

$$\mathrm{REP} = \frac{\text{Renewable energy index}}{\mathrm{EPE}} \tag{4.30}$$

$$\mathrm{REP} = \frac{\text{Renewable energy}\left(\text{kJ}\right)}{\text{Energy consumption}\left(\text{kJ}\right)} \tag{4.31}$$

Waste treatment and solvent recovery energy (WTE and SRE): As waste treatment and solvent recovery consume a lot of energy, two metrics are usually used to quantify the percentage of these consumptions (Calvo-Flores 2009):

$$\mathrm{WTE} = \frac{\text{Waste treatment energy requirement}\left(\text{kJ}\right)}{\text{Mass of product}\left(\text{kg}\right)} \tag{4.32}$$

$$\mathrm{SRE} = \frac{\text{Solvent recovery energy requirement}\left(\text{kJ}\right)}{\text{Mass of product}\left(\text{kg}\right)} \tag{4.33}$$

The two metrics can be expressed as percentages by diving them by the *EPE*.

Energy recovery parameter (ERP): It is important to take into account in the process conception step the optimization of power recovery by matching cold and warm streams to limit energy losses (Bakhtiari et al. 2015). To quantify this principle, one can use energy metric similar to the material recovery parameter expressed as:

$$\mathrm{ERP} = \frac{\text{Enegy recovered}\left(\text{kJ}\right)}{\text{Energy consumption}\left(\text{kJ}\right)} \tag{4.34}$$

4.5.4 Economic-Based Metrics

In many cases, there is a compromise between economically and environmentally optimal solutions (Lapkin 2018) because waste treatment may have significant cost implications (Constable et al. 2002). In a number of cases, the benefit of using a more environmentally sustainable method may be leaved due to energy costs (Augé and Scherrmann 2017). Both economic and environmental considerations have motivated the emergence of new greener technologies to ensure the survival of chemical industry in increasingly demanding market. (Tundo et al. 2000).

Furthermore, since the early 1970s, the notion of circular economy has been commonly used, and it is seen as a sustainability situation (Geissdoerfer et al. 2017). Indeed, the prevention concept is economically reinforced by the reuse and recycling concept of materials, and circular economy plays a part in all the three characters of sustainable development (Korhonen et al. 2018), economic, environmental, and social (Fig. 4.5).

Using the atom economy metric, it has been shown that low atom economy affected the cost of synthesizing a new chemical (Constable et al. 2002). Also, Jolliffe and Gerogiorgis have concluded from their techno-economical evaluations of pharmaceutical manufacturing that reducing E-factor improved the process costs (Jolliffe and Gerogiorgis 2016). Furthermore, using renewable feedstocks, reducing

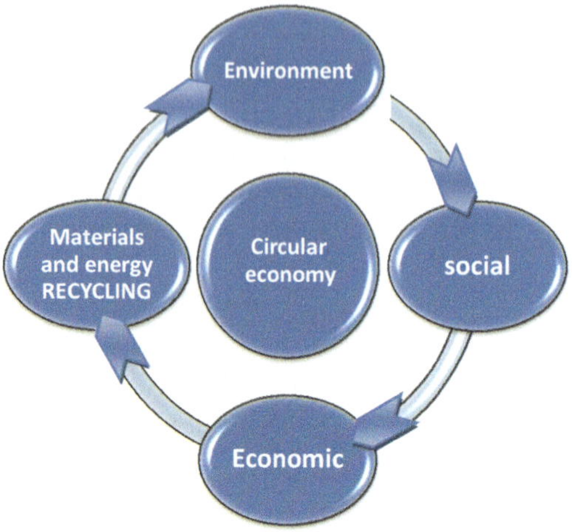

Fig. 4.5 Circular economy as a sustainable concept and its relation with the environment, social, and economic dimensions of sustainability

reaction steps, designing safer, minimizing energy requirements, and activating reactions using catalysis and real-time analysis are all principles of green chemistry in favor of the economic profit of the process. Thus, economic metrics are crucial in the decision of a greener alternative reaction or process.

Techno-economic analysis is usually performed for designing chemical processes or optimizing economic, energy, and environmental impacts (Novak Pintarič and Kravanja 2017). A number of parameters have been examined for such studies (Dunn et al. 2010; Novak Pintarič and Kravanja 2017; Jiang and Bhattacharyya 2016), and the most used ones are the following: total cost, profit, discounted payback period (the time required to recover the investment in fixed capital), internal rate of return (IRR), return on investment (ROI), and the net present value (NPV) (additive cash flows). It has been shown that selecting the option with the greatest possible NPV is a preferred strategy (Jiang and Bhattacharyya 2016; Jolliffe and Gerogiorgis 2016) which is calculated as:

$$\mathrm{NPV} = \sum_{i=0}^{k} \frac{C_k}{(1+r)^k} \tag{4.35}$$

where C_k is the nondiscounted yearly cash flows assuming that they follow at the end of each year k and r is the discount rate.

The NPV increases the value of a company over the long term and gives an appropriate trade-offs between the profitability, process efficiency, and environmental completion (Novak Pintarič and Kravanja 2017).

Regarding the significant cost implications of waste treatment and energy, relevant economic metrics must be calculated, and similarly to the previous ones, they can be defined as:

- Waste treatment cost percentage expressing the total cost of the waste treatment over the total process cost
- Utilities cost percentage defined as total cost of the process utilities over the total process cost

4.6 Principles of Green Engineering

The 12 principles of green engineering were proposed in 2003 by Anastas and Zimmerman (2003) with the objective to achieve process efficiency in addition to the chemical reaction efficiency. Indeed, plant operations and operating conditions optimization are needed to achieve process efficiency (Jiménez-González et al. 2012). The objective is to extend the use of the green chemistry principles and the sustainable concept to engineers beside to chemists to reach the maximum of decision-makers (Augé and Scherrmann 2017). The 12 principles are summarized in Table 4.4 (Anastats and Zimmerman 2003).

As for green chemistry principles, Calvo-Flores has proposed a mnemonics with a simple acronym "IMPROVEMENTS," which summarizes the spirit of the 12 green engineering principles (Table 4.5) (Calvo-Flores 2009).

The comparison of the principles presented in Tables 4.1 and 4.4 shows a great similarity, and the major principle of "prevention at the source" is always present.

Table 4.4 The 12 principles of green engineering and their definition

N°	Principle	Definition
1	Inherent rather than circumstantial	Designers must strive to guarantee that all inputs and outputs of material and power are as inherently nonhazardous as possible
2	Prevention instead of treatment	Preventing waste is better than treating or cleaning waste after it has been created
3	Design for separation	To minimize energy consumption and use of materials, separation and purification activities should be intended
4	Maximize mass, energy, space, and time efficiency	To maximize mass, energy, space, and time effectiveness, products, procedures, and systems should be intended
5	Output-pulled versus input-pushed	Products, processes, and systems should be "output-pulled" rather than "input-pushed" through the use of energy and materials
6	Conserve complexity	When making design decisions on recycling, reuse, or useful disposition, embedded entropy and complexity must be regarded as an investment
7	Durability rather than immortality	Targeted durability should be a design objective, not immortality
8	Meet need, minimize excess	Designing unnecessary ability or capability alternatives (e.g., "one-size-fits-all") should be regarded a design flaw
9	Minimize material diversity	To encourage disassembly and value preservation, material diversity in multicomponent goods should be minimized
10	Integrate local material and energy flows	Design of products, processes, and systems must include integration and interconnectivity with available energy and materials flows
11	Design for commercial "afterlife"	Products, processes, and systems should be designed for performance in a commercial "afterlife"
12	Renewable rather than depleting	Material and energy inputs should be renewable rather than depleting

Table 4.5 Condensed principles of green engineering following the acronym "PRODUCTIVELY"

I	Inherently nonhazardous and safe
M	Minimize material diversity
P	Prevention instead of treatment
R	Renewable materials and energy
O	Output-led design
V	Very simple
E	Efficient use of mass, energy, space, and time
M	Meet the need
E	Easy to separate by design
N	Networks for exchange local mass and energy
T	Test the life cycle of the design
S	Sustainability throughout product life cycle

The basic principles of green chemistry are generally taken up by those of green engineering with a greater attention on the process and reaction technology aspects that are outlined using specific terms such as follows: design, optimization, capacity, flows, process, and separation. Also, concepts referring to sustainability and life cycle reflect the scale of the engineering principles that must consider the green assessment outside the boundaries of the plant (Fig. 4.4).

Consequently, the green chemistry metrics may be used to quantify those of green engineering requiring a focus not only on a reaction but also on the technology involved by this reaction (Curzons et al. 2001). Indeed, an industrial plant must be a sequence of units for physical and/or chemical transformation of materials, with a sequence of input-output streams with different composition and operating conditions. Therefore, the metrics may be applied for each unit operation. In addition, the operation time and equipment capacity must be taken into consideration using other metrics such as productivity and space time yield (*STY*) (Albini and Protti 2016).

$$\text{Productivity} = \frac{\text{Mass of product}\left(\text{kg}\right)}{\text{time unit}\left(\text{h}\right)} \tag{4.36}$$

$$\text{STY} = \frac{\text{Mass of product}\left(\text{kg}\right)}{\text{time unit}\left(\text{h}\right)\times\text{unit volume}\left(\text{m}^3\right)} \tag{4.37}$$

Analogically to the global metric for multistage synthesis suggested by Eissen et al., the global greenness of the multunit plant can be estimated taking into account the greenness of each unit (Eissen et al. 2004).

4.7 Evaluation of the Process Greenness

Using the green chemistry metrics allows the quantification of the 12 principles and the estimation of the "greenness" of chemical processes. This quantification iessential to enable analyses and numerical diagnoses of processes (Calvo-Flores, 2009) in order to support a decision-making. However, there is no single metric that allows quantifying the overall process efficiency (Jiménez-González 2012).

According to its 12 principles, the impacts of green chemistry are multidimensional (De Marco et al. 2019), and each dimension has several metrics for its quantification. In addition, all the metrics are not in the same scale (Jiang and Bhattacharyya 2016) which makes it difficult to determine which principles are the most influential.

As presented above, even for the same principle, several metrics are used to take it into account, and often the same principle is considered up by several parameters (Henderson et al. 2010). Instead of considering the principles as separated parameters to be individually optimized, they can be viewed as a united structure with mutually strengthening components (Beach et al. 2009). So, the evaluation of the process greenness needs a standardization step before the estimation of a *green chemistry balance* as an average of the standard metrics used to allow diagnosis analysis.

4.7.1 Metrics Standardization

Normalizing the metrics representing each green principle can greatly enhance the comparison and the choice of the greener alternative whether for a reaction or a process. Researcher's efforts have also been made to unify the used metrics and to propose methods and tools that make the implementation of green chemistry principles easier. As an example, the Green Aspiration Level (GAL) method was proposed as a new process efficiency measure to quantify the influence of preparing a particular pharmaceutical compound on the environment while taking into consideration the complexity of the reaction path of synthesizing the desired molecule (Roschangar et al. 2015). This was later improved to give rise to "innovation Green Aspiration Level" (iGAL) methodology (Roschangar et al. 2018). Another metric tool named "GREEN MOTION" was designed to enable the enhancement of the health, security, and environmental effects of cosmetic manufacturing on a scale ranging from 0 to 100 (Phan et al. 2015), using the Environmental Assessment Tool for Organic Syntheses (EATOS) which is a software elaborated by Eissen and Metzger in 2004 (Albini and Protti 2016). For this kind of tools with normalized metrics, the degree of involvement of each factor in the estimation of the greenness is often taken into consideration through penalty points (Phan et al. 2015; Roschangar and Colberg 2018). Other tool kits have been developed by academicians for purely educational reasons and are based in most cases on excel sheets for an easy use (Andraos and Sayed 2007).

From the metrics section presented in this chapter, most of them are in the scale (0–1 or 100) with 0 assigned to the worst and 1 (or 100) to the best values. If this is not the case, these values can be standardized and adimensionalized as presented in Table 4.6. Once normalized, the metrics are used in the diagnosis concerning materials, energy, and safety. The economic metrics can be used to decide between several green solutions alternatives (see Fig. 4.6).

Table 4.6 Summary of metrics of the chapter in the standardized form in the scale 0–1 assigned to the worst and best values, respectively, and also the standardized ones

Metrics	Range value	Standardized form 0–1	Observation
Material-based metrics			
Reaction yield (RY)	0–1		
Atom economy (AE)	0–1		
Carbon economy (CE)	0–1		
Stoichiometric factor (SF)	1–∞	1/SF	Reciprocal SF
Material recovery parameter (MRP)	0–1		
Reaction mass efficiency (RME)	0–1		
Process mass intensity (PMI)	1–∞	1/PMI	Reciprocal PMI
Solvent intensity (SI)	0–∞	1 − (SI/PMI)	Inversed solvent percentage
Water intensity (WI)	0–∞	1 − (WI/PMI)	Inversed solvent percentage
E_Factor	0–∞	1 − (F_Factor/PMI)	Inversed waste percentage
Syn_Ideality	1–∞	*1 − Syn_Ideality*	Inversed Syn_Ideality
Renewable intensity (RI)	0–∞	RI/PMI	Renewable percentage
Hazard and safe metrics			
Effective mass yield (EMY)	0–∞	1 − (PMI/EMY)	Inversed non-benign
Hazard	0–1		
Toxicity	0–1		
CMR	0–1		
Energy-based metrics			
Reaction energy consumption (REC)	0–∞	REC × energy efficiency	REC percentage
Workup energy consumption (WEC)	0–∞	(1 − WEC × energy efficiency)	Inversed REC percentage
Renewable energy percentage (REP)	0–1		
Waste treatment energy (WTE)	0–∞	(1 − WTE × energy efficiency)	Inversed WTE percentage
Solvent recovery energy (SRE)	0–∞	(1 − SRE × energy efficiency)	Inversed SRE percentage
Energy recovery parameter (ERP)	0–1		

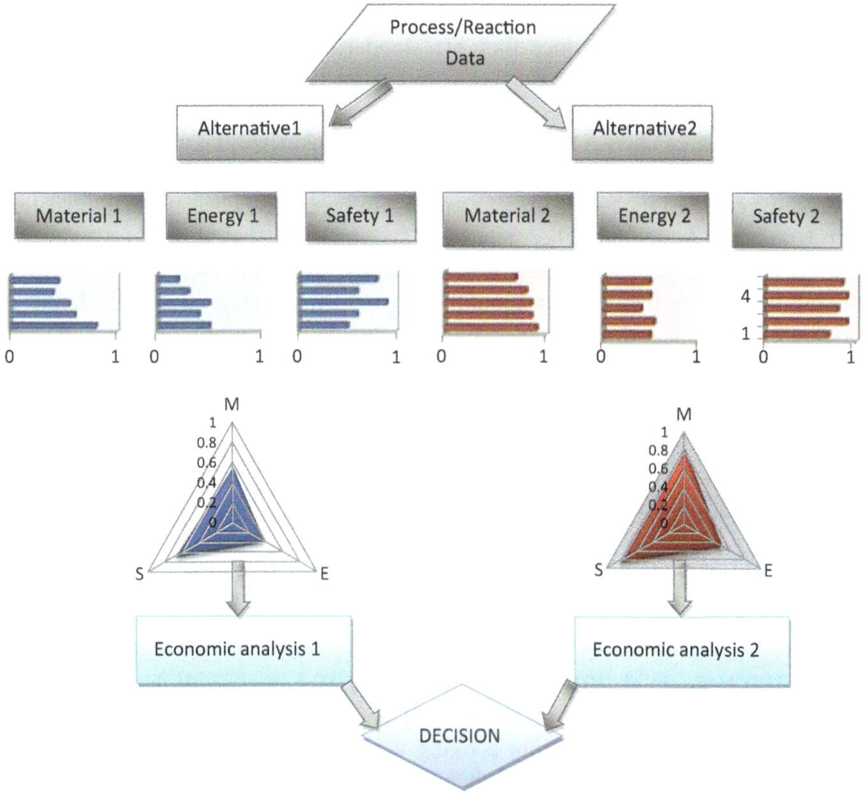

Fig. 4.6 Flowchart of the steps of alternatives study using the green chemistry metrics and methods for decision with graphic representations. Average material standard metrics noted M, average energy standard metrics noted E, and average safety standard metrics noted S

4.7.2 Graphic Representation of "Greenness"

The standardization of metrics facilitates its graphical representation. The most current representation was to construct radar, which is a polygon with a number of lines equal to the number of metrics considered by the greenness evaluation of the process. Each corner of a polygon was with a length in the interval [0–1] on which the degree of accomplishment of the corresponding metric is represented. The more the surface of the radar is full, the more the process is considered green (Ribeiro et al. 2010; Clavos-Flors 2009; Augé and Scherrmann 2017). Other researchers used the histogram representations (Andraos and Sayed 2007) to classify the metrics effects. The two systems allow making a diagnosis by visualizing which metrics are to improve and facilitate the decision-making of the action to be implemented according to a specific category of principles (see Fig. 4.6).

However, the green chemistry principles of the reaction or the process are classified into several categories – materials, energy, hazard, and economics – and each category have its own metrics. One metric can be chosen to represent a category or to average some of them once they are standardized. The most important is to respect the coherence in the case of comparison between several processes and alternatives.

The global greenness representation must also visualize the degree of involvement of each category. Multi-criteria graphical representations were developed to give an overview of the greenness of a product or process, such as the method called Eco-Compass (Augé and Scherrmann 2017), which was based on six axes representing the energy intensity, the material intensity, the safety and the environment, the preservation of the resources, and the reuse of waste, respectively. Each axis is graduated from 0 to 5. In this way, graphs are obtained in the form of radars, making it possible to compare various options. The shape of the radar depends on the number of axes representing the selected categories.

4.8 Application to Biodiesel Production

The creation of a cost-effective and greener alternative fuel such as biodiesel is a major focus of science today (Henrie 2015). The preparation of biodiesel from waste vegetable oils has been mainly studied over the past few years (Mahgoub et al. 2015; Keera et al. 2018; Uddin et al. 2013). However, in all these works, the reaction yield and conversion were the only parameters considered in the production optimization. Biodiesel from cooking vegetable oils is usually obtained by a transesterification reaction using methanol in excess and base catalyzed. It also generated glycerol by-product as presented by the reaction:

$$\text{Triglyceride}(\text{oil}) + 3\,\text{methanol} \leftrightarrow \text{fatty acid methyl esters}(\text{biodiesel}) + \text{glycerol}$$

The production of biodiesel meets the criteria of green chemistry, including the use of renewable feedstock and design for biodegradation (Henrie 2015), so it is a suitable example for the application of the green chemistry principles.

However, for the implementation of green chemistry metrics in biodiesel synthesis, little prior experimental information was reported (Martinez-Guerra and Gude 2017). One can cite the work of Martinez-Guerra and Gude that studied the application of green chemistry parameters to microwave and ultrasound synergistic effects when converting waste frying oil into biodiesel. The considered green metrics were as follows: environmental factor, atom economy, process mass intensity, reaction mass efficiency, and stoichiometry factor (Martinez-Guerra and Gude 2017). Frang and Guo investigated a "green" strategy to biodiesel manufacturing by using sustainable strong acid catalysts as substitutes for liquid acid catalysts in order to avoid the use of damaging substances and waste generation (Fang and Guo 2014). The research was conducted without estimating green metrics.

In the work of Outili et al., several green metrics were applied to the reaction of production of biodiesel from cooking oil. The protocol of the reaction of transesteri-

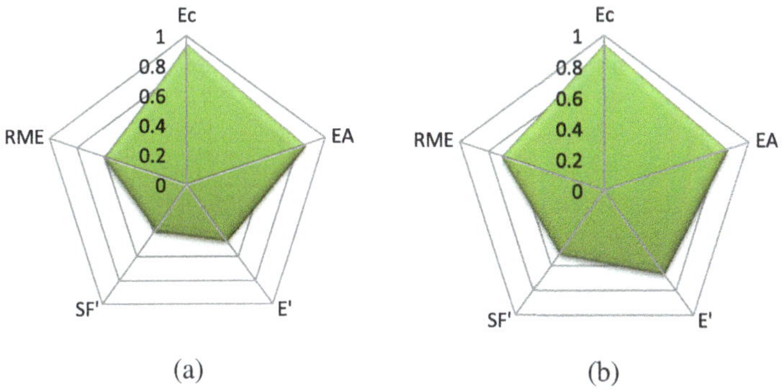

Fig. 4.7 Material metrics radars of the transesterification reaction with (**a**) operating conditions (temperature = 55 °C, Ratio methanol/oil = 8, catalyst% = 1.25) and (**b**) operating conditions (temperature = 45 °C, ratio methanol/oil = 5, catalyst% = 2). With *Ec* and *EA* carbon and atom economy, respectively, *RME* the reaction mass efficiency, *SF′* and *E′*, standardized stoichiometric and environmental factors

fication was reported in (Outili et al. 2019). The green chemistry parameter was used as a response of a design of experiments that was used to optimize the operating conditions of the transesterification reaction of biodiesel production. In Fig. 4.7, a comparison of the green chemistry radars of two different reaction conditions is presented. The reaction greenness is evaluated using five standard material metrics as shown in Fig. 4.7, and the green chemistry balance was estimated as an average value of the used parameters.

The results show the importance of using standardized metrics to detect the worst parameters. Indeed, the "greenness" of the reaction was improved by using less methanol excess which led to the improvement of the E_factor (in its standard form), the stoichiometric factor, and the reaction mass efficiency. The "greenness" of the production estimated using a green chemistry balance had gone from 65 to 74% corresponding to operating conditions of cases (a) and (b), respectively (Outili et al. 2019).

In the second example, the green chemistry metrics are used to compare the greenness of two processes without reaction, which were applied as pretreatments of the frying oil to be used for biodiesel production (Miroud and Koliai 2019). Indeed, as the presence of free fatty acids (FFA) in the waste oils reduces the transesterification yield (Susilowati et al. 2019), two physical pretreatments methods were applied to reduce the FFA from the collected cooking oil before performing the reaction of biodiesel production, namely, microwave pretreatment and charcoal adsorption. The results using materials and hazard's metrics are shown in Fig. 4.8.

The results show the application of green chemistry material and hazard metrics to the case of a separation process without reaction. Indeed, the reaction metrics, such as atom economy (AE) and carbon economy (CE) were eliminated, and the reaction mass efficiency and E_factor were adapted to the separation process materials and waste. The greenness of the pretreatment using microwave was improved,

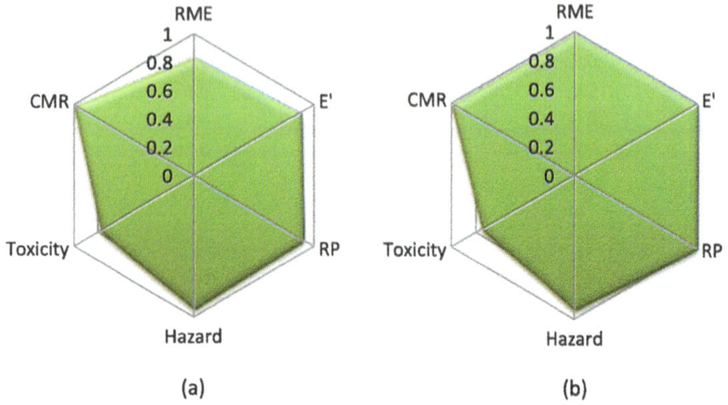

Fig. 4.8 Material and safety metrics radars of the pretreatments applied to reduce free fatty acids of the cooking oil with (**a**) charcoal adsorption and (**b**) microwave. RME, reaction mass efficiency adapted to a pretreatment process; *E* the standardized environmental factor, *RP* the renewable percentage; hazard and toxicity parameters, *CMR* the mutagen carcinogen and reprotoxic parameter

and the green chemistry balance had gone from 83.05 to 90.69%. In addition, a better efficiency of elimination of FFA was obtained with microwave pretreatment: 67% against 46% in the case of adsorption on activated carbon. Also, for this case, the energy consumption was a crucial metric, and it was calculated using the equipment power (Watt) × pretreatment time (s), and the results showed that the microwave pretreatment consumed 16 kJ where the adsorption consumed 900 kJ (Miroud and Koliai 2019).

From these results, it is clear that the application of green chemistry and green engineering principles improve the greenness character of a process, especially when combining several categories of metrics: material and waste, hazardous, and energy metrics. However, an economic study is recommended to complete and strengthen the selection, especially by investigating several alternatives.

4.9 Conclusion

Chemists and engineers face the challenge of creating and designing new products by proceeding differently. They must implement new processes with social, environmental, and economic advantages that improve sustainability. Metrics are very useful to achieve this goal, allowing comparison of reaction routes and processes alternatives.

Researchers are getting increasingly interested in green chemistry, proposing more parameters. This raises the selection of the most suitable parameter to be adopted since for a same principle there may be several candidates. However, the parameters must be relevant, easy to implement, and standardized, to provide an effective tool for industrial and also educational people to let the future decision-makers be aware about environmental issues during their formation.

Also, the study of a process cannot be based on a principle at the expense of others. Indeed, for a reliable study and a fairly decision between several alternatives, all the main elements must be taken into consideration, namely, materials, waste, toxicity, safety, energy, and economy.

Several efforts have been made by researchers and industrials for greener and more sustainable products and processes, but the actual damages to the environment, the human health, and the ecosystem are such that much remains to be done.

References

Albini A, Protti S (2016) Paradigms in green chemistry and technology. Springer, Heidelberg

Anastas PT, Beach ES (2009) Green Chemistry Education. Changing the Course of Chemistry. ACS Symposium Series; American Chemical Society: Washington, DC

Anastas PT, Warner JC (1998) Green chemistry: theory and practice. Oxford University Press, Oxford

Anastas PT, Zimmerman JB (2003) Design through the 12 principles of green engineering. Sustainability requires objectives at the molecular, product, process, and system levels. Environ Sci Technol 37(5):94A–101A. https://doi.org/10.1021/es032373g

Anastas PT, Heine LG, Williamson TC (2000) Green engineering. American Chemical Society, Washington, DC

Andraos J (2005) Unification of reaction metrics for green chemistry: applications to reaction analysis. Org Process Res Dev 9(2):149–163. https://doi.org/10.1021/op049803n

Andraos J (2006) On using tree analysis to quantify the material, input energy, and cost throughput efficiencies of simple and complex synthesis plans and networks: towards a blueprint for quantitative total synthesis and green chemistry. Org Process Res Dev 10(2):212–240

Andraos J (2016) Critical evaluation of published algorithms for determining material efficiency green metrics of chemical reactions and synthesis plans. ACS Sustain Chem Eng 4(4):1917–1933

Andraos J (2018a) Reaction green metrics: problems, exercises, and solutions. CRC Press, Boca Raton. https://doi.org/10.1201/9780429424243

Andraos J (2018b) Synthesis green metrics: problems, exercises, and solutions, vol 2018. CRC Press, Boca Raton. https://doi.org/10.1201/9780429400681

Andraos J, Dicks AP (2012) Green chemistry teaching in higher education: a review of effective practices. Chem Educ Res Pract 13(2):69–79

Andraos J, Sayed M (2007) On the use of "green" metrics in the undergraduate organic chemistry lecture and lab to assess the mass efficiency of organic reactions. J Chem Educ 84(6):1004

Andraos J, Ballerini E, Vaccaro L (2015) A comparative approach to the most sustainable protocol for the β-azidation of α, β-unsaturated ketones and acids. Green Chem 17(2):913–925

Augé, J, Scherrmann M-C (2017) Chimie verte: concepts et applications. EDP sciences

Bakhtiari B, Pylkkanen V, Retsina T (2015) Pinch analysis-an essential tool for energy optimization of pulp and paper mills. O PAPEL 76(4):51–54

Beach ES, Cui Z, Anastas PT (2009) Green chemistry: a design framework for sustainability. Energy Environ Sci 2(10):1038–1049

Benson MH, Garmestani AS (2011) Embracing panarchy, building resilience and integrating adaptive management through a rebirth of the National Environmental Policy Act. J Environ Manag 92(5):1420–1427. https://doi.org/10.1016/j.jenvman.2010.10.011

Calvo-Flores FG (2009) Sustainable chemistry metrics. ChemSusChem: Chem Sustain Energy Mater 2(10):905–919. https://doi.org/10.1002/cssc.200900128
Carson R (2002) Silent spring. Houghton Mifflin Harcourt, Boston
Clark JH (2005) Green chemistry and environmentally friendly technologies. In: Separation processes: fundamentals and applications, pp 1–18
Constable DJC, Curzons AD, Cunningham VL (2002) Metrics to 'green'chemistry—which are the best? Green Chem 4(6):521–527
Curzons AD, Constable David JC, Mortimer DN et al (2001) So you think your process is green, how do you know?—using principles of sustainability to determine what is green–a corporate perspective. Green Chem 3(1):1–6. https://doi.org/10.1039/B007871
De Marco BA, Rechelo BS, Tótoli EG et al (2019) Evolution of green chemistry and its multidimensional impacts: a review. Saudi Pharm J 27(1):1–8
Delidovich I, Palkovits R (2016) Catalytic versus stoichiometric reagents as a key concept for Green Chemistry. Green Chem 18(3):590–593. https://doi.org/10.1039/C5GC90070K
Dichiarante V, Ravelli D, Albini A (2010) Green chemistry: state of the art through an analysis of the literature. Green Chem Lett Rev 3(2):105–113. https://doi.org/10.1080/17518250903583698
Dicks AP (2016) Green organic chemistry in lecture and laboratory, CRC Press, Boca Raton. https://doi.org/10.1201/b11236
Dirgha RJ, Nisha A (2019) Green chemistry: beginning, recent progress, and future challenges. World J Pharm Pharm Sci 8(7):280–293. https://doi.org/10.20959/wjpps20197-14208
Dunn PJ, Wells AS, Williams MT (2010) Future trends for green chemistry in the pharmaceutical industry. In: Green chemistry in the pharmaceutical industry. Wiley-VCH, pp 333–355
Eissen M, Metzger JO (2002) Environmental assessment tool for organic syntheses. EATOS. Online verfügbar unter http://www.metzger.chemie.uni-oldenburg.de/eatos/eatosanleitung.pdf, zuletzt aktualisiert am, vol 5, p 2002
Eissen M, Mazur R, Quebbemann H-G et al (2004) Atom economy and yield of synthesis sequences. Helv Chim Acta 87(2):524–535
Elgharbawy AA, Muniruzzaman M, Salleh HM et al (2019) Ionic liquids as a green solvent for lipase-catalyzed reactions. Ind Appl Green Solvents Vol I 50:21–60
Eljack F, Kazi M-K, Kazantzi V (2019) Inherently safer design tool (i-SDT): A property-based risk quantification metric for inherently safer design during the early stage of process synthesis. J Loss Prev Process Ind 57:280–290. https://doi.org/10.1016/j.jlp.2018.12.004
Emas R (2015) The concept of sustainable development: definition and defining principles. Brief for GSDR 2015
Fang S, Guo Y (2014) Advancements in solid acid catalysts for biodiesel production. Green Chem 16(6):2934–2957. https://doi.org/10.1039/c3gc42333f
Fennie MW, Roth JM (2016) Comparing amide-forming reactions using green chemistry metrics in an undergraduate organic laboratory. J Chem Educ 93(10):1788–1793. https://doi.org/10.1021/acs.jchemed.6b00090
Gaich T, Baran PS (2010) Aiming for the ideal synthesis. J Org Chem 75(14):4657–4673
Geissdoerfer M, Savaget P, Bocken NMP et al (2017) The circular economy—a new sustainability paradigm? J Cleaner Prod 143:757–768. https://doi.org/10.1016/j.jclepro.2016.12.048
Gendron C, Revéret J-P (2000) Le développement durable. Écon Soc 37(91):111–124
Glasgow HB, Burkholder JAM, Reed RE et al (2004) Real-time remote monitoring of water quality: a review of current applications, and advancements in sensor, telemetry, and computing technologies. J Exp Mar Biol Ecol 300(1–2):409–448. https://doi.org/10.1016/j.jembe.2004.02.022
Gude VG (2015) Synergism of microwaves and ultrasound for advanced biorefineries. Resour Eff Technol 1(2):116–125. https://doi.org/10.1016/j.reffit.2015.10.001
Haklay MM (2017) The three eras of environmental information: the roles of experts and the public. In: Participatory sensing, opinions and collective awareness. Springer, Cham, pp 163–179
Henderson RK, Constable DJC, Jiménez-González (2010) Concepción. Green chemistry metrics. In: Green chemistry in the pharmaceutical industry, pp 21–48
Henrie SA (2015) Green chemistry laboratory manual for general chemistry. CRC Press, Boca Raton

Hodson L, Eastlake A, Herbers R (2019) An evaluation of engineered nanomaterial safety data sheets for safety and health information post implementation of the revised hazard communication standard. J Chem Health Saf 26(2):12–18. https://doi.org/10.1016/j.jchas.2018.10.002

Jiang Y, Bhattacharyya D (2016) Sustainable engineering economic and profitability analysis. In: Sustainability in the design, synthesis and analysis of chemical engineering processes. Butterworth-Heinemann, pp 141–167

Jiménez-González C, Constable David JC, Ponder CS (2012) Evaluating the "greenness" of chemical processes and products in the pharmaceutical industry—a green metrics primer. Chem Soc Rev 41(4):1485–1498

Jolliffe HG, Gerogiorgis DI (2016) Plantwide design and economic evaluation of two continuous pharmaceutical manufacturing (CPM) cases: ibuprofen and artemisinin. Comput Chem Eng 91:269–288. https://doi.org/10.1016/j.compchemeng.04.005

Karagölge ZR, Bahri GÜR (2016) Sustainable chemistry: green chemistry. Iğdır Üniversitesi Fen Bilimleri Enstitüsü Dergisi 6(2):89–96

Keera ST, El Sabagh SM, Taman AR (2018) Castor oil biodiesel production and optimization. Egyptian J Petroleum 27(4):979–984. https://doi.org/10.1016/j.ejpe.2018.02.007

Kelly FJ, Fussell JC (2015) Air pollution and public health: emerging hazards and improved understanding of risk. Environ Geochem Health 37(4):631–649. https://doi.org/10.1007/s10653-015-9720-1

Kerras H, Merouani R, Nekkab C, Outili N, Meniai AH (2019) Green chemistry metrics application to biodiesel production from waste vegetable oil. In: 10th international renewable energy congress (IREC) IEEE. https://doi.org/10.1109/IREC.2019.8754649

Korhonen J, Honkasalo A, Seppälä J (2018) Circular economy: the concept and its limitations. Ecol Econ 143:37–46. https://doi.org/10.1016/j.ecolecon.2017.06.041

Kumar I, Mondal M, Sakthivel N (2019) Green synthesis of phytogenic nanoparticles. In: Green synthesis, characterization and applications of nanoparticles. Elsevier, pp 37–73. https://doi.org/10.1016/B978-0-08-102579-6.00003-4

Lapkin AA (2018) Chemical engineering science and green chemistry the challenge of sustainability. In: Handbook of green chemistry, pp 1–16

Lima-Ramos J, Tufvesson P, Woodley JM (2014) Application of environmental and economic metrics to guide the development of biocatalytic processes. Green Process Synthesis 3(3):195–213

Linthorst JA (2010) An overview: origins and development of green chemistry. Found Chem 12(1):55–68. https://doi.org/10.1007/s10698-009-9079-4

Loiseau E, Aissani L, Le Féon S et al (2018) Territorial life cycle assessment (LCA): what exactly is it about? A proposal towards using a common terminology and a research agenda. J Cleaner Prod 176:474–485. https://doi.org/10.1016/j.jclepro.2017.12.169

Louaer M, Outili N (2017) Reaction mechanism choice using green chemistry principles. Algerian J Eng Res (1):16–20

Mahgoub HA, Nimir AS, Amani AM (2015) Suitable condition of biodiesel production from waste cooking oil–Al-Baha City–KSA. Int J Multidiscip Curr Res 3:447–451

Malone PM (1998) Introduction to green engineering. IA: J Soc Ind Archeol 24(1):5–8

Manahan SE (2006) Green chemistry and the ten commandments of sustainability. ChemChar Research, Inc Publishers Columbia, Missouri U.S.A.

Martinez-Guerra E, Gude V (2017) Assessment of sustainability indicators for biodiesel production. Appl Sci 7(9):869

Mcelroy CR, Constantinou A, Jones LC et al (2015) Towards a holistic approach to metrics for the 21st century pharmaceutical industry. Green Chem 17(5):3111–3121

Miroud I, Koliai R (2019) Experimental study of used oil pretreatment methods and their effect on biodiesel production. Unpublished master's thesis, Process Engineering Faculty. Constantine3 University. Algeria

Morra B, Dicks AP (2016). Recent progress in green undergraduate organic laboratory design. In: Green chemistry experiments in Undergraduate Laboratories. American Chemical Society, pp 7–32. https://doi.org/10.1021/bk-2016-1233.ch002

Naidu S, Sawhney R, Li X (2008) A methodology for evaluation and selection of nanoparticle manufacturing processes based on sustainability metrics. Environ Sci Technol 42(17):6697–6702. https://doi.org/10.1021/es703030r

Novak Pintarič Z, Kravanja Z (2017) The importance of using discounted cash flow methodology in technoeconomic analyses of energy and chemical production plants. J Sustain Dev Energy Water Environ Syst 5(2):163–176. https://doi.org/10.13044/j.sdewes.d5.0140

Onsrud H, Simon R (2013) The social, business, and policy environment for green manufacturing. In: Green manufacturing. Springer, Boston, pp 25–47

Outili N, KerVras H, Nekkab C, Merouani R, Meniai AH (2019) Biodiesel production optimization from waste cooking oil using green chemistry metrics. Renew Energy 145:2575. https://doi.org/10.1016/j.renene.2019.07.152

Pei S, Wei Q, Huang K et al (2018) Green synthesis of graphene oxide by seconds timescale water electrolytic oxidation. Nat Commun 9(1):145

Phan T, Gallardo C, Mane J (2015) GREEN MOTION: a new and easy to use green chemistry metric from laboratories to industry. Green Chem 17(5):2846–2852

Poliakoff M, Licence P (2007) Sustainable technology: green chemistry. Nature 450(7171):810

Popa VI (2018) Biomass for fuels and biomaterials. In: Biomass as renewable raw material to obtain bioproducts of high-tech value, pp 1–37

Pour N, Webley PA, Cook PJ (2018) Potential for using municipal solid waste as a resource for bioenergy with carbon capture and storage (BECCS). Int J Greenhouse Gas Control 68:1–15. https://doi.org/10.1016/j.ijggc.2017.11.007

Razzaq T, Kappe CO (2008) On the energy efficiency of microwave-assisted organic reactions. ChemSusChem Chem Sustain Energy Mater 1(1–2):123–132. https://doi.org/10.1002/cssc.200700036

Ribeiro MGTC, Machado AASC (2013) Greenness of chemical reactions—limitations of mass metrics. Green Chem Lett Rev 6(1):1–18. https://doi.org/10.1080/17518253.2012.669798

Ribeiro M, Gabriela TC, Dominique AC, Machado AASC (2010) "Green star": a holistic green chemistry metric for evaluation of teaching laboratory experiments. Green Chem Lett Rev 3(2):149–159

Robert KW, Parris TM, Leiserowitz AA (2005) What is sustainable development? Goals, indicators, values, and practice. Environ Sci Policy Sustain Dev 47(3):8–21

Rodríguez-Eugenio N, Mclaughlin M, Pennock D (2018) Soil pollution: a hidden reality. FAO

Roschangar Frank, Colberg Juan. (2018). Green chemistry metrics. Green Techniques for Organic Synthesis and Medicinal Chemistry, pp 1–19. https://doi.org/10.1002/9781119288152.ch1

Roschangar F, Sheldon RA, Senanayake CH (2015) Overcoming barriers to green chemistry in the pharmaceutical industry–the green aspiration level™ concept. Green Chem 17(2):752–768

Roschangar F, Zhou Y, Constable David JC et al (2018) Inspiring process innovation via an improved green manufacturing metric: iGAL. Green Chem 20(10):2206–2211

Saleh HE-DM, Koller M (2018) Introductory chapter: principles of green chemistry. In: Green chemistry. IntechOpen

Sharuddin SDA, Abnisa F, Daud WMAW, et al. (2018). Pyrolysis of plastic waste for liquid fuel production as prospective energy resource. In: IOP conference series: materials science and engineering. IOP Publishing, p 012001. https://doi.org/10.1088/1757-899X/334/1/012001

Sheldon RA (2016a) Biocatalysis and green chemistry. In: Green biocatalysis, pp 1–15. https://doi.org/10.1002/9781118828083.ch1

Sheldon RA (2016b) Green chemistry and resource efficiency: towards a green economy. Green Chem 18(11):3180–3183. https://doi.org/10.1039/C6GC90040B

Sheldon RA (2017) Metrics of green chemistry and sustainability: past, present, and future. ACS Sustain Chem Eng 6(1):32–48. https://doi.org/10.1021/acssuschemeng.7b03505

Sheldon RA, Arends I, Hanefeld U (2007) Green chemistry and catalysis. Wiley, New York

Sökmen M, Demir E, Alomar SY (2018) Optimization of sequential supercritical fluid extraction (SFE) of caffeine and catechins from green tea. J Supercrit Fluids 133:171–176. https://doi.org/10.1016/j.supflu.2017.09.027

Song Q-W, Liang-Nian HE (2018) Atom economy. In: Encyclopedia of sustainability science and technology, pp 1–21

Sperry J, García-Álvarez J (2016) Organic reactions in green solvents. https://doi.org/10.3390/molecules21111527
Susilowati E, Hasan A, Syarif A (2019) Free fatty acid reduction in a waste cooking oil as a raw material for biodiesel with activated coal ash adsorbent. J Phys Conf Ser. IOP Publishing, p 012035
Tang SLY, Smith RL, Poliakoff M (2005) Principles of green chemistry: Productively. Green Chem 7(11):761–762
Tickner JA, Becker M (2016) Mainstreaming green chemistry: the need for metrics. Curr Opin Green Sustain Chem 1:1–4
Tundo P, Arico F (2007) Thoughts on the short history of the field. Chem Int 29(5):4–7
Tundo P, Anastas P, Black David SC et al (2000) Synthetic pathways and processes in green chemistry. Introductory overview. Pure Appl Chem 72(7):1207–1228
Twidell J, Weir T (2015) Renewable energy resources. Routledge. https://doi.org/10.4324/9781315766416
Uddin MR et al (2013) Synthesis of biodiesel from waste cooking oil. Chem Eng Sci 1(2):22–26. https://doi.org/10.12691/ces-1-2-2
Uhlman BW, Saling P (2010) Measuring and communicating sustainability through eco-efficiency analysis. Chem Eng Prog 106(12):17–26
Wang XY, Zhu GS, Wan YQ, et al (2011) Water pollution emergencies in China: actualities, prevention and response. In: Advanced materials research. Trans Tech Publications, pp 589–594
Washington National Academy of Engineering (1999) Industrial environmental performance metrics: challenges and opportunities. National Academy Press
Wynne B (1992) Uncertainty and environmental learning: reconceiving science and policy in the preventive paradigm. Glob Environ Chang 2(2):111–127. https://doi.org/10.1016/0959-3780(92)90017-2
Zhang W, Cue BW (eds) (2018) Green techniques for organic synthesis and medicinal chemistry. John Wiley & Sons, New York
Zhao Y, Truhlar DG (2009) Benchmark energetic data in a model system for Grubbs II metathesis catalysis and their use for the development, assessment, and validation of electronic structure methods. J Chem Theory Comput 5(2):324–333

Chapter 5
Biotech Green Approaches to Unravel the Potential of Residues into Valuable Products

Eduardo J. Gudiña, Cláudia Amorim, Adelaide Braga, Ângela Costa, Joana L. Rodrigues, Sara Silvério, and Lígia R. Rodrigues

Contents

E. J. Gudiña · C. Amorim · A. Braga · Â. Costa · J. L. Rodrigues
S. Silvério · L. R. Rodrigues (✉)
CEB-Centre of Biological Engineering, Universidade do Minho, Braga, Portugal
e-mail: egudina@deb.uminho.pt; claudia.oliveira.amorim@ceb.uminho.pt; abraga@deb.uminho.pt; angela.costa@ceb.uminho.pt; joana.joanalucia@deb.uminho.pt; sarasilverio@deb.uminho.pt; lrmr@deb.uminho.pt

Inamuddin, A. M. Asiri (eds.), *Sustainable Green Chemical Processes and their Allied Applications*, Nanotechnology in the Life Sciences,
https://doi.org/10.1007/978-3-030-42284-4_5

Abbreviations

ATPS	Aqueous two-phase systems
CCR	Carbon catabolite repression
cmc	Critical micelle concentration
COS	Chitooligosaccharides
CRISPR-Cas9	Clustered regularly interspaced short palindromic repeats-associated caspase 9 endonuclease
CSL	Corn step liquor
DNA	Deoxyribonucleic acid
E24	Emulsifying index
epPCR	Error-prone polymerase chain reaction
FOS	Fructooligosaccharides
GOS	Galactooligosaccharides
IMO	Isomaltooligosaccharides
NGS	New-generation sequencing
OMW	Olive mill wastewater
PCR	Polymerase chain reaction
PEG	Polyethylene glycol
PUF	Polyurethane foam
RL	Rhamnolipid
SmF	Submerged fermentation
SSF	Solid state fermentation
ST	Surface tension
STR	Stirred tank reactor
XOS	Xylooligosaccharides

5.1 Introduction

Green chemistry and sustainability principles are the basis to develop alternative strategies for a rational use of resources (preferably renewable) within a closed-looped system and with limited generation of waste, aiming to overcome the current challenges related with industrial progress, impact on environment, and economic growth (Winans et al. 2017; Sheldon and Woodley 2018). These strategies are well aligned with the circular economy concept that includes the well-known 3Rs (reduce, reuse, recycle) (Vea et al. 2018). Bioeconomy involves the use of biotech processes to sustainably produce and convert renewable biological resources and waste streams into valuable products across several economic sectors, including food, feed, bio-based products, and bioenergy. Merging the circular economy and bioeconomy concepts is a priority (Aguilar et al. 2018). A circular economy without added-value and wealth generation as well as a noncircular bioeconomy will not be widely adopted and accepted by society.

Biotech processes are a good fit to the green chemistry principles, namely, in what concerns reduced waste, atom economy, low toxicity, safe water-based processes, use of mild conditions leading to energy efficiency, renewable feedstock (e.g., enzymes), reduced derivatization, catalysis, real-time analysis, and inherently safer processes (Sheldon and Woodley 2018). In addition to being cost-effective and environmentally friendly, usually these strategies are considered sustainable as they are based on the use of natural resources at rates that do not excessively deplete supplies on the long term, and residues are generated at lower rates than the ones at which they can be readily assimilated by the environment (Graedel 2002). Remarkable advances in biotechnology, bioinformatics, metabolic engineering and molecular, systems, and synthetic biology made this possible. High-throughput DNA sequencing unraveled hundreds of microbial genome sequences which are currently available, opening countless opportunities for the identification of a target gene through bioinformatics analysis (data mining). In addition, DNA synthesis became affordable and feasible; thus with the sequence retrieved from databases it is easy to get the gene synthesized, cloned into an adequate chassis (host microorganism), and further produced at industrial scales in a cost-efficient manner. Two decades ago, the low number of enzymes available dictated the process conditions which often had to be adjusted as there were no other alternatives. Using in vitro directed evolution it is nowadays possible to create libraries of mutant enzymes to be further screened for enhanced properties (Denard et al. 2015) and engineered to specifically match predefined process requirements (e.g., activity, specificity, stability), leading to bioprocesses that are truly sustainable by design (Bommarius et al. 2011; Illanes et al. 2012). Actually, through protein engineering not only the optimization of existing enzymes but also the assembly of entirely new pathways and reactions using systems and synthetic biology principles has been possible. Site-directed mutagenesis enables the creation of point mutations by replacing an amino acid at a predetermined site in the protein by the other canonical amino acids. Hence, it can be used to rationally design an enzyme given that its three-dimensional structure and mechanism is known. Alternatively, if that structural information is missing, random mutagenesis can be used. This can be achieved using the error-prone polymerase chain reaction (epPCR) which has been extensively used to generate mutant enzymes (Chen and Arnold 1993; Moore and Arnold 1996). DNA shuffling that enables the in vitro homologous recombination of pools of selected mutant genes by random fragmentation and PCR reassembly (Stemmer 1994) led to several advances in this topic. Indeed, this procedure of random mutagenesis, recombination, and screening can be repeated until the desired mutant phenotypes are generated, for instance, the ability of an enzyme to catalyze a non-natural substrate under severe operational conditions (Renata et al. 2015; Sun et al. 2016). The exciting new gene editing tool CRISPR-Cas9 (clustered regularly interspaced short palindromic repeats-associated caspase 9 endonuclease) and its further developments are broadening even more the protein engineering possibilities (Doudna and Charpentier 2014). For instance, tools like EvolvR that constantly vary all nucleotides within a tunable window length at user-defined loci will allow advancing genetic strategies in organisms/systems that do not easily and effectively integrate homology-directed oligonucleotides. Additionally, these tools will likely simplify the fast development of valuable phenotypes through accelerated and parallelized mutagenesis and selection cycles. Using engineered DNA polymerases targeted to loci via CRISPR-guided nickases, EvolvR directly generates mutations (Halperin et al. 2018).

Additionally, the diversity of unexplored microorganisms and communities that inhabit the planet makes possible the continuous screening and discovery of promising biocatalysts. Metagenomics has emerged as an innovative and strategic approach to explore these unculturable microorganisms through the analysis of DNA extracted from environmental samples (Batista-García et al. 2016). Functional metagenomics involving the direct cloning and heterologous expression of environmental DNA without prior knowledge of its sequence is a useful technique to discover novel and promising enzymes. The total genomic DNA is extracted from the microbial communities present in a given sample and further used to construct metagenomics libraries. A functional screening is then performed to retrieve the most robust catalysts for different industrial applications. After identification of the genes involved in the biosynthesis of the enzymes, the most promising biocatalysts can be characterized and further engineered for a desired application. Successful examples of this approach have been reported (Ko et al. 2013; Rabausch et al. 2013; Wang et al. 2016; Zafra et al. 2016).

Besides the potential use of protein engineering strategies, there are other optimizations at the level of the substrate, culture medium, and reactor design that can greatly enhance cost-effectiveness and efficiency, thus contributing to the processes sustainability. In addition, emergent downstream solutions are expected to lead not only to purer added-value products but also to lower amounts of waste and effluent streams, as well as to a lower use of hazardous solvents, hence with a less negative impact on the environment. All these topics will be further detailed in the next subsections.

In the latest years, the European Union has been strongly committed in promoting a sustainable growth based on bioeconomy, through a change toward the use of renewable resources and environmentally friendly processes to produce energy. Mandatory regulatory rules to effectively accomplish greenhouse gas reductions and to diminish risks related to zones of carbon stock and high biodiversity value have been established. Overall, this urged the search for cleaner and more resource-efficient alternatives that simultaneously produce less waste and generate economic value. The biorefinery concept is well positioned to overcome such needs and to provide interesting solutions (Sheldon and Woodley 2018). A biorefinery is grounded on the use of renewable raw materials to produce several bio-based products, chemicals, and biofuels (Elbersen et al. 2012; Gullon et al. 2012). Industrial and agro-food residues and wastes are considered renewable feedstock that are rich in valuable compounds (Devesa-Rey et al. 2011; Lin et al. 2014) that can be reused and recycled to produce energy or other products (e.g., enzymes, bulk chemical, prebiotics, biosurfactants, biopolymers), thus setting the foundations for the circular bioeconomy (Wysokinska 2016).

In order to effectively base the European growth on a sustainable and rational use of resources, it is essential that all those existing resources suitable for a biorefinery processing are properly identified and quantified. For instance, the amount of food waste currently produced is a global concern and the development of sustainable green technologies is crucial (Dahiya et al. 2018). Bioprocesses like acidogenesis, fermentation, methanogenesis, solventogenesis, photosynthesis, oleaginous process, and bioelectrogenesis, among others, can be used for food waste valorization either alone or integrated (Lin et al. 2014). On the other hand, many residues have already a commercial use, as, for example, sawdust that is generated in the wood production process and is used to make fibreboard or some crop residues that are used for animal feed or in some agricultural activities. Consequently, only a portion of those wastes (around 225 million tons/year) can indeed be used in biorefineries. These wastes include crop residues (around 122 million tons/year), municipal waste (gardens, food, paper, and wood) (around 63 million tons/year), and forestry residues (around 40 million tons/year) (Searle and Malins 2013). Mapping of resources available and other potential uses from a biorefinery perspective is an unmet need (Scoma et al. 2016). A significant percentage of available residues are comprised by agriculture and forestry residues that could potentially contribute to satisfy global energy needs (De Wit and Faaij 2010; Scarlet et al. 2010). Nevertheless, agriculture residues are heterogeneous and are produced seasonally. Therefore, residues generated in industrial processes (e.g., citrus pulp, spent coffee grounds, winery residues, olive mill wastewater (OMW), corn steep liquor (CSL), seafood waste, and brewer spent grain, among others) comprise an alternative renewable source and an opportunity for process economy improvement, new market opportunity development, and innovation.

5.2 Metagenomics Approaches to Unravel Novel Biocatalysts and Microorganisms

Microbial ecosystems are viewed as enormous reservoirs of genetic diversity, representing the largest proportion of biomass on Earth, although only 0.1–1% microorganisms are cultivable from any given niche (Handelsman et al. 1998). The discovery and characterization of this vast microbial and metabolic diversity has been possible due to culture-independent method advances. To unlock the genetic diversity contained within microbiomes, analysis of nucleic acids, proteins, and lipids using molecular approaches, such as genetic fingerprinting, metagenomics, and a combination study of the meta-omics techniques, revealed to be very useful (Bragg and Tyson 2014; Panigrahi et al. 2019). Currently, metagenomics is accepted as the most promising methodology for identifying enzymes with novel (bio)catalytic activities (so-called biocatalysts), by exploring the genetic material of whole microbial communities (the metagenome), as if it were a single large genome in an environmental sample. In addition, the rapidly expanding field of omics-mediated survey (e.g., genomics, transcriptomics, proteomics, and metabolomics, among others) coupled with an enhanced repertoire of bioinformatics tools has revolutionized the ability to analyze microbial communities, providing access to the diversity of taxonomically and phylogenetically relevant genes, catabolic genes, and whole operons and correlation of genomes with particular functions in the environment (Béjà et al. 2000; Aguiar-Pulido et al. 2016; Ameen and Raza 2017). This knowledge has a wide range of potential applications under the areas of systems biology and biodiversity (Allen and Banfield 2005; DeLong 2005; Maruthamuthu 2017) and to produce novel natural products for several biotechnological and therapeutic applications (Montella et al. 2016; Maruthamuthu 2017).

5.2.1 Metagenomics on the Road for Green Chemistry

Metagenomics is the most successful tool for uncovering novel biocatalysts with great applicability and potential in green sustainable technologies for the industrial production of relevant chemicals (Castilla et al. 2018). Ahmad et al. (2019) reported that between 2014 and 2017, 332 metagenome-sourced enzymes were identified and characterized. These consist of one β-agarase, nitrilase, exonuclease, and acylase; two epoxides; four transferases; six proteases, amylases, chitinases, and phosphatases/phytases; 19 oxidoreductases, 118 cellulases/hemicellulases; and 161 esterases/lipases. Various environmental substrates can be explored using metagenomics, including marine waters and sediment, soil, compost, activated sludge, and animal rumen, among others (Montella et al. 2016; Berini et al. 2017; Ahmad et al. 2019; Castilla et al. 2018). For instance, enzymes naturally adapted to the constraints of certain industrial processes have been identified in extreme environments, namely, halotolerant esterases and glycoside hydrolases, thermostable lipases, or even psychrophilic DNA polymerases (Ufarte et al. 2015; DeCastro et al. 2016). With the application of these enzymes and others yet to be discovered, the future of green chemistry looks very promising.

The greatest advantage of metagenomics is that no prior knowledge of the DNA sequence is required and several different screening strategies may be applied. The critical step is the extraction of the environmental DNA. Any protocol needs to be optimized for each environmental sample and no protocol can provide an accurate determination of species distribution. Hence, different protocols for genomic DNA extraction can be employed and the DNA pools can be mixed to maximize the number of different species represented, thus increasing the final level of metagenomics diversity (Delmont et al. 2011). Moreover, some adjustments may be included in the protocol, like enrichment of the sample in the target microorganisms to improve the specificity of the sample's DNA (Jiao et al. 2006), although most of studies use the non-enrichment approach to preserve the natural diversity and relative abundance of microbial communities (Montella et al. 2016).

Metagenomics screening can be function-based or sequence-based. Function-based screening is a straightforward way to identify novel biocatalysts and biochemical mechanisms that have desired industrial functions. This technique is based on the principle of discriminating biochemical activities of the reactions catalyzed by enzymes and the detection of novel functional genetic elements with different functions from the known enzymes (Ahmad et al. 2019). Numerous novel enzymes and bioactive compounds that have potential commercial applications were identified through functional screening, such as hydrolases, oxidoreductases, transferases, or catabolic enzymes with environmental cleanup purposes (Sjöling and Cowan 2008; DeCastro et al. 2016; Madhavan et al. 2017). This strategy involves the creation of metagenomics libraries where genomic DNA extracted from environmental samples is directly cloned in a suitable vector like phage, plasmid, cosmid, fosmid, or bacterial artificial chromosome. Subsequently, the ligated product is transferred into a suitable host to evaluate the expression of the target enzyme. Cosmid and fosmid libraries are the most commonly used because of their stability, their capacity to accept large fragments, and their feasibility of transduction (Montella et al. 2016) and also because they enable recovering of full-length genes, gathering information on their genetic context in natural sources, and unveiling cocktails of synergistic activities that degrade complex substrates (Berini et al. 2017). The common function-based screening strategy is assaying enzyme activities by harvesting a metagenomics library on agar plates incorporated with dyes and substrates of target enzymes. This technique provides a simple and straightforward approach to identify novel biocatalysts that function under diverse conditions (Simon and Daniel 2011) and even though the low throughput, a large number of unique enzymes from various environments have been successfully isolated (Ngara and Zhang 2018). Despite the great potentialities of functional screening, there are constraints limiting its success; the hit rate (probability) of identifying a particular gene depends on multiple elements that are inextricably linked to each other including the host-vector system, size of the target gene, the efficiency of heterologous gene expression, and screening conditions employed. Nevertheless, several strategies to overcome these drawbacks are being evaluated (Batista-García et al. 2016).

The sequence-based screening can be achieved by either the homology-based screening of libraries or the direct large-scale sequencing of extracted DNA, using engineered degenerate primers from highly conserved target regions in PCR or probes for hybridization (Batista-García et al. 2016). Most metagenomics studies involving high-throughput sequencing adopt either an amplicon-based or a shotgun-based approach using next-generation sequencing (NGS) technology. NGS technologies are enabling the sequencing of thousands of genomes in parallel (Montella et al. 2016). A number of sequencing platforms are currently available, offering varying levels of sequence coverage and depth (Goodwin et al. 2016). The collected data offer an insight into the potential function of the population while also aiding direct gene mining for biotechnological applications (Aguiar-Pulido et al. 2016). These techniques already proved to be useful to find enzymes with high activity and efficiency leading to the identification of several novel enzymes such as bacterial laccases, dioxygenases, nitrite reductases, hydrogenases, hydrazine oxidoreductases, or chitinases from various ecosystems (DeCastro et al. 2016; Berini et al. 2017; Madhavan et al. 2017; Ahmad et al. 2019). The continuous development of bioinformatics tools suggests that the shotgun strategy will be more suitable than the function-based screening or amplicon sequencing for the discovery of novel target genes from environmental DNA (Montella et al. 2016).

5.2.2 Novel Metagenomics Biocatalysts on the Market

The whole process of finding novel biocatalysts and also the optimization of the efficiency of the processes is very expensive and time-consuming. Also, independently of the novelty of the sequence encoding an enzyme, the key is its application. Novel enzymes such as oxygenases, esterases, and laccases with very diverse origins have been highlighted for their competence to degrade pollutants such as nitriles, lindane, styrene, naphthalene, aliphatic and aromatic carbohydrates, organophosphorus, or plastic materials (Ufarte et al. 2015). Nevertheless, very few cases of new environmental biocatalyst products have been yet patented and translated to market (Berini et al. 2017). Some examples of patented enzymes obtained through metagenomics for application on green chemistry include nitrile hydratases (EP2369009A3), laccases (GB01P006EP), cellulases (EP04015680.4), an esterase able to degrade terephthalate esters (component of bioplastics) (WO 2007017181), and cow rumen-derived esterases (EP04015920.4) (Ngara and Zhang 2018).

5.3 Synthetic Biology Approaches to Construct Microbial Cell Factories Able to Degrade Agro-industrial Wastes

Society still depends on fossil fuels and other petrochemicals that are nonrenewable and contribute to the global climate change. Besides, the reliance on the extraction from plants of essential compounds such as drugs, foods, and cosmetics, among others, is also not environmentally friendly and demands valuable resources such as water and high area of arable land. In addition, the slow and seasonal growth of the plants, in combination to the small amounts produced in some cases, can cause a serious supply depletion (Jullesson et al. 2015). Taking all this into account, new sustainable and greener processes ought to be developed. Synthetic biology allows to design, construct, and engineer parts, devices, systems, and organisms with useful, predictable, and novel functions by applying engineering principles (Weber and Fussenegger 2012). Nowadays, the design and construction of biological systems to produce drugs (Westfall et al. 2012; Paddon and Keasling 2014; Rodrigues et al. 2015a, b, c; Couto et al. 2017; Rodrigues et al. 2017a; Kung et al. 2018; Abdallah et al. 2019), biofuels (Lee et al. 2008; Majidian et al. 2018), biosensors (Rodrigues et al. 2014; Rodrigues et al. 2017a; Rodrigues and Rodrigues 2017b; Kim et al. 2018), and added-value products such as natural fibers (Bowen et al. 2018) became a routine (Rodrigues and Rodrigues 2017a; Rodrigues et al. 2017b). However, the success of industrial synthetic biology strategies depends not only on the ability of getting products in high concentrations but also in an economically viable manner of using renewable and inexpensive feedstock (Protzko et al. 2018). Sustainable feedstock cannot compete directly or indirectly with food production and should include waste lignocellulose derived from agriculture and forestry residues and by-products from fruit and vegetable processing processes (Fig. 5.1).

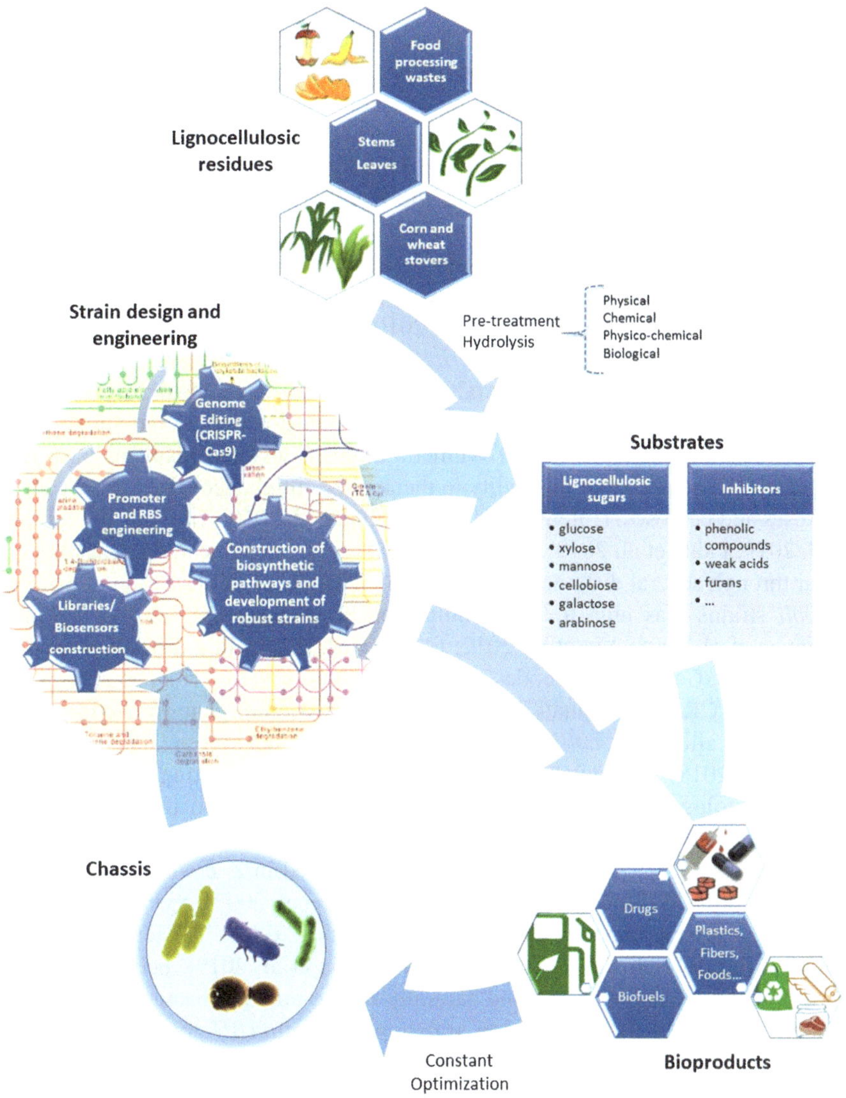

Fig. 5.1 Overview of synthetic biology approaches to produce added-value products from lignocellulosic residues. A selected chassis is engineered using parts retrieved from databases and literature. After the parts/pathway(s) assembly, the engineered strain is used to degrade the lignocellulosic residues or to directly convert the lignocellulosic sugars into the envisaged (bio)product. The chassis can be continuously optimized until the productivity of the target added-value product meets the expected level

As previously mentioned, lignocellulosic residues are the most abundant renewable source of sugars. Lignocellulose is mostly composed of cellulose, hemicellulose, lignin, and pectin. Its hydrolysis results in a mixture of sugars that includes hexoses (glucose, mannose, and galactose), pentoses (xylose and arabinose), and uronic acids (galacturonic acid). Glucose is the hexose present in higher amount and the only monomer present in cellulose. Hemicellulose contains different sugar monomers with xylose usually being the one present in higher concentration. Simultaneous co-fermentation of these carbon sources is required for the efficient production of compounds derived from lignocellulosic residues. Many microbes (e.g., *Saccharomyces cerevisiae* and *Zymomonas mobilis*) cannot naturally metabolize xylose or other pentoses (Kim et al. 2010). Even microorganisms that are able to metabolize xylose, such as *Escherichia coli*, experience repression from glucose – carbon catabolite repression (CCR). Therefore, diauxic growth is observed – pentose consumption only occurs after glucose exhaustion (Görke and Stülke 2008). Using metabolic engineering and synthetic biology, several organisms have been engineered to confer them the ability to metabolize simultaneously different sugars (hexoses and pentoses) (Eiteman et al. 2008; Kim et al. 2010; Xia et al. 2012; Flores et al. 2019; Paula et al. 2019). For example, *E. coli* was engineered to use only one sugar through several different gene deletions. The use of a consortium of different *E. coli* strains was able to use simultaneously glucose, xylose, and arabinose (Eiteman et al. 2008; Xia et al. 2012; Flores et al. 2019). In a different study, the deletion of *ptsG* gene that encodes a subunit of glucose transporter and plays a central role in CCR, in combination to the (over)expression of genes related to xylose degradation, allowed *E. coli* to co-utilize both glucose and xylose (Jung et al. 2015; Chae et al. 2018). The deletion of key enzymes in the competing pathways using synthetic biology tools such as CRISPR-Cas9 also allows to increase the titers (Wang et al. 2018; Abdelaal et al. 2019;). In addition, *E. coli* metabolic pathway for xylose and/or arabinose consumption was constructed in *Z. mobilis* and *S. cerevisiae* (Lawford and Rousseau 2002; Hahn-Hägerdal et al. 2007). Xylose assimilation in these organisms has been achieved by inserting xylose isomerase or xylose reductase-xylitol dehydrogenase pathways (Cunha et al. 2019), being these pathways found in anaerobic fungi and naturally xylose-utilizing bacteria, respectively (Hahn-Hägerdal et al. 2007). In the arabinose case, *araABD* genes from different sources were tested throughout the years and it was concluded that they were effective in arabinose consumption but only in combination with adaptive evolution (Caballero and Ramos 2017). Since these organisms do not have specific transporters for xylose or arabinose (Kim et al. 2010), transporters were also engineered (Li et al. 2016a, b; Caballero and Ramos 2017; Wang et al. 2019). More recently, Protzko et al. (2018) engineered *S. cerevisiae* to use both glucose and galacturonic acid from citrus peel waste (pectin-rich biomass) as carbon sources and to produce meso-galactaric acid, a building block used in the production of numerous compounds (nylon, plastic monomers). The expression of the heterologous transporter, GatA, from *Aspergillus niger* was essential.

In addition to introduce pathways to consume the sugars from lignocellulosic residues, other issues need to be considered, such as the presence of inhibitory by-products (fermentation inhibitors). These by-products that include phenolic compounds, furans, and acids are produced during the lignocellulose pretreatment and reduce the microorganism growth. Therefore, the engineering of stress response regulators can be of utmost importance to increase tolerance to these by-products (Paula et al. 2019). Transcription factors are transcriptional regulatory proteins that allow the organism to use the available nutrients without wasting energy. Several studies proved that the overexpression or inactivation of specific transcription factors can increase *S. cerevisiae* resistance to acids and ethanol (Li et al. 2017; Balderas-Hernández et al. 2018). Synthetic biology also provides new tools, such as promoter engineering, to improve the expression in biological systems. Synthetic promoters are designed to optimize expression levels. For example, Hector and Mertens (2017) designed a synthetic hybrid promoter in *S. cerevisiae* controlled by xylose presence/absence. For that purpose, they constructed a promoter that combined *Ashbya gossypii* TEF promoter and the operator sequence (*xylO*) for binding the XylR repressor. This new hybrid promoter was repressed in the absence of xylose. The generation of promoter libraries is also a synthetic biology approach that can be very useful to find new promoters that allow an appropriate expression level (Rodrigues et al. 2014; Rodrigues and Rodrigues 2017a; Rodrigues and Rodrigues 2017b). For example, Redden and Alper (2015) used this strategy to construct short synthetic yeast promoters that maintain a strong function. For that purpose, the authors identified minimal core elements and minimal upstream activating sequences and combined them to obtain short and strong constitutive and inducible promoters.

Synthetic biology can also have a major role in the use of lignocellulose wastes to produce added-value compounds by designing trees with customized lignin. Lignin, the complex polymer that confers mechanical support to the plant, acts as major impediment to industrial processing due to the high energy demand needed to break it down. Yang et al. (2013) used synthetic biology tools in *Arabidopsis thaliana* to replace some promoters of key lignin genes in order to decrease the lignin content without the collapsing of the vessels. In a different study (Tsuji et al. 2015), the introduction of LigD enzyme from bacterium *Sphingobium* sp. in *A. thaliana* allowed to alter the lignin structure according to the design due to the enzyme ability to oxidize α-hydroxyl groups in β-O-4 units in lignin to α-keto analogues that are chemically more labile. More recently, the curcumin biosynthetic pathway was introduced in *A. thaliana* and curcumin and other phenylpropanoids produced were incorporated into the cell wall (Oyarce et al. 2019). This allowed to significantly increase the saccharification efficiency after chemical treatment of the heterologous *A. thaliana* as compared to the wild type. All these strategies can be used in the future to enhance the lignocellulosic biomass processing efficiency.

5.4 Bioprocess Development: Trends in Bioreactor Design and New Engineering Approaches

Another important aspect to be considered in the development of a sustainable development evolution of a bio-industry is the process engineering. In bioreactor design two main issues should be addressed, namely, the optimization of volumetric productivity and the efficient use of the starting material and/or catalyst (Andri et al. 2010). Additionally, the control of some parameters and variables addition, it is also essential to control some parameters and variables in order to ensure the optimal conditions for the reaction system, being the most relevant the dissolved oxygen concentration, medium pH, temperature and mixing (Zhong 2010; Caramihai and Severin 2013; Sofía et al. 2018).

In the last decades, research efforts have been concentrated on bioreactor design and engineering to overcome the main problems with mixing, mass, and heat transfer (Mitchell et al. 2000; Ferreira et al. 2017). Different bioreactor configurations have been developed, being the most common the stirred tank reactor (STR). In 2017, Guneser et al. (2017) studied the production of D-limonene, a flavor compound, by *Rhizopus oryzae* and *Candida tropicalis* using OMW as raw material in a 5-L STR. A maximum D-limonene concentration of 87.73 mg/kg was attained. Although the STR is the most commonly used bioreactor at an industrial scale, nowadays a variety of bioreactor type and configurations have been explored and developed such as airlift bioreactors and bubble columns (Chisti and Moo-Young 1987; Werner et al. 2018). The production of β-carotene and single cell oil with *Rhodotorula glutinis* from cellulose hydrolysate was studied by Yen and Chang (2015). The authors compared the cultivation of this oleaginous yeast in an airlift and STR bioreactors and observed that the highest lipid content (34.3%) was obtained in the airlift, as well as the highest biomass concentration. However, the highest β-carotene content was achieved in the STR (1.2 mg/g). Currently, fermentation processes can be classified as submerged fermentation (SmF) or solid state fermentation (SSF). In SmF, microorganisms are cultivated in a liquid nutrient medium (Pandey 2003; Singhania et al. 2009). Instead, SSF comprises the microbial growth in a solid substrate in the absence of free water (Pandey et al. 2000; Singhania et al. 2009).

In fact, SSF approach allows the valorization of unexploited biomass and offers the possibility of using residues and wastes as substrates, helping in solving some environmental issues (Soccol et al. 2017). SSF has many advantages comparing to traditional SmF such as higher yields and product titers, easier downstream process, reactors with smaller volumes, and lower energy and cost requirements (Singhania et al. 2009; Manan and Webb 2017). One of the key factors in the design of a SSF process is the selection of the microbial host. Filamentous fungi are the dominant microorganisms used in SSF since this technique simulates their natural environment, as well as their mycelia growth (Farinas 2015). However, the utilization of yeasts (e.g., *Pichia pastoris*, *Kluyveromyces marxianus*, *S. cerevisiae*) and some bacterial cultures (e.g., *Bacillus subtilis*, *B. thuringiensis*, and *Clostridium thermo-*

cellum) have also been described in SSF processes (Singhania et al. 2009; López-Pérez and Viniegra-González 2015). Another important aspect in this process is the substrate selection. The most common residues applied in SSF include starchy substrates as rice, oats, cassava, and wheat bran, among others, that are rich in carbohydrates (Soccol et al. 1994; El-Shishtawy et al. 2014); substrates with protein such as oil cakes (e.g., soybean, coconut, sesame, sunflower, and olive oil cakes) (Ali et al. 2012; Paithankar and Rewatkar 2014); cellulosic or lignocellulosic substrates that include sugarcane bagasse, corncob, barley husk, barley straw, and wood (Elegbede and Lateef 2017; Martínez-Avila et al. 2018b; Sadh et al. 2018); and substrates rich in soluble sugars that are mostly produced in fruit industries such as sugar beets and sugar beet pulp, molasses, and grape pomace (Zeng et al. 2018; Teles et al. 2019). Palm oil mill and OMW, organic wastes and textile wastes, and tea, coffee, and cassava wastes have also been used as substrates in SSF as an alternative mean of waste valorization integrated in the biorefinery concept (Lopes et al. 2016; Hu et al. 2018; Mejias et al. 2018). Usually these residues and wastes provide the nutrients and the carbon necessary for microbial growth, in addition to be a solid support for microbial growth and nutrient adsorption. Nevertheless, sometimes it is also necessary to additionally supplement the medium with macro- and micronutrient essentials for an optimum microbial growth (Pandey 2003; Farinas 2015).

In the last decade, SSF has been applied in the production of a broad range of metabolites, such as organic acids, enzymes, secondary metabolites, biofuels, flavors, pigments, and phenolic compounds (Kumar et al. 2003; Cavalcanti et al. 2005; Gadhe et al. 2011; Try et al. 2017; Martínez-Avila et al. 2018a; Mejias et al. 2018). Table 5.1 presents a summary of the most recent studies (last 4 years) regarding the production of some added-value molecules by SSF.

Bioreactor design for SSF is another important research area under development (Ashok et al. 2018). In SSF, the bioreactors employed can be classified according to the mixed system used and the type of aeration (Durand 2003; Spier et al. 2011). In the last years, bioreactors such as packed beds (Khanahmadi et al. 2006; Melikoglu et al. 2015), multilayer packed beds (Shojaosadati and Babaeipour 2002), rotating drum bioreactors (Saithi and Tongta 2016), column bioreactors (Linde et al. 2007; Salum et al. 2010), column-tray bioreactors (Ruiz et al. 2012), magnetic rotating biological contactors (Saha 1997), fixed beds (Castro et al. 2015), and tray systems (Khanahmadi et al. 2004), among others, have been used in SSF processes. The most commonly used bioreactors for SSF are the static bioreactors such as Erlenmeyer flasks, trays, fixed bed, or packed bed bioreactors, being their main advantage its simplicity (Durand 2003). Tray bioreactors are chambers or rooms, with air circulation and controlled humidity and temperature, in which different recipients laid with the substrate are placed. Usually, the top of the tray is open and the bottom is perforated to allow air convection (Durand 2003). The scale-up of this system is simple; however it needs large areas, since it is performed by increasing the number of trays, is intensive labor, and is a non-sterile process, which makes the contamination control difficult. Another bottleneck in these systems can occur with oxygen transfer, which is strongly affected by substrate bed height and bioreactor design. Oliveira et al. (2017b) compared the lipase production by SSF of olive pom-

Table 5.1 Overview of the production of some added-value molecules by solid state fermentation (SSF) in the recent years

Products	Strain	Substrate/support	Concentration/ activity	Bioreactor	Reference
Enzymes					
Amylase	*Bacillus amyloliquefaciens*	Orange peels and cheese whey	220 U/mL	Flask	Uygut and Tanyildizi (2018)
	Bacillus sp. *BBXS-2*	Wheat straw	6900 U/g	Flask	Qureshi et al. (2016)
	Bacillus cereus and *B. thuringiensis*	Wheat brain	14.5 U/mL	Flask	Abdullah et al. (2018)
Xylanase	*Aspergillus tubingensis* FDHN1	Sorghum straw	5177.23 U/g	Static flask	Adhyaru et al. (2016)
	Aspergillus awamori IOC-3914	Palm kernel cake and palm pressed fiber	134.2 U/g	Tray-type bioreactor	Oliveira et al. (2018)
	Aspergillus oryzae		27.2 U/g		
	Aspergillus niger		18.8 U/g		
Cellulase	*Trichoderma reesei CECT 2114*	Rice bran, rice husk, rice straw	1.317 U/g	Flask	Darabzadehet al. (2019)
Laccase	*Funalia trogii IFP0027*	Rice straw	172.74 U/g	Flask	Li et al. (2019)
Lipase	*Burkholderia cenocepacia*	Sugarcane bagasse, sunflower seed, and olive oil	72.3 U/g	Flask	Liu et al. (2016)
	Rhizomucor miehei	Babassu cake	30 U/g	Fixed-bed bioreactor	Ávila et al. (2018)
	Rhizomucor miehei	Cottonseed meal	93 U/g	Tray-type bioreactor	Aguieiras et al. (2018)
Phytase	*Aspergillus ustus*, *Aspergillus niger* van Tieghem, *Trichoderma atroviride*, *Trichoderma koningii*, *Trichoderma harzianum Rifai*, *Bacillus subtilis*, *Bacillus megaterium*, *Bacillus amyloliquefaciens*, *Aspergillus niger*	Cassava bagasse, soybean bran, wheat bran, sorghum, and maize distiller dried grains with solubles	0.33–2.03 U/ $mg_{protein}$	Flask	Barbosa et al. (2019)

Products	Strain	Substrate/support	Concentration/ activity	Bioreactor	Reference
Pigments					
Astaxanthin	*Sporidiobolus salmonicolor* (ATCC 24259), *Yamadazyma guilliermondii* (ATCC 90197), *Yarrowia lipolytica* (ATCC 24060), *Xanthophyllomyces dendrorhous* (ATCC 24202)	Wheat wastes	109.23 μg/g	Flask	Dursun and Dalgiç (2016)
β-Carotene	*Blakeslea trispora* MTCC 884	Fruit and vegetable waste (orange, carrot, and papaya peels)	0.127 mg/mL	Flask	Kaur et al. (2019)
Organic acids					
Oxalic acid	*Trichoderma asperellum* MG323528	Corn stover	27.55 mg/g	Flask	Al-Askar et al. (2018)
Citric acid	*Aspergillus niger MTCC 281*	Brewer's spent grain	0.22 %	Flasks	Pathania et al. (2018)
Fumaric acid	*Aspergillus oryzae*	Wheat bran	0.54 mg/g	Flask	Jiménez-Quero et al. (2017)
Flavor and fragrances					
2-Phenylethanol	*Kluyveromyces marxianus*	Sugarcane bagasse	16 mg/g	Flask	Martínez-Avila et al. (2018a)
γ-Decalactone	*Yarrowia lipolytica W29*	Castor oil	196 mg/L	Forced aeration mini-reactor	Try et al. (2017)
			75 mg/L	Small-headspace bottle	
6-Pentyl-α-pyrone	*Trichoderma asperellum* G7	Sugarcane bagasse	85.1μg/g	Flask	Hamrouni et al. (2019)

ace using a tray-type bioreactor and a novel pressurized bioreactor. In the latter one, the increase in the total pressure was used as a tool to enhance the oxygen solubility in the media, improving the oxygen transfer rate (Belo et al. 2003). The authors observed a twofold increase in the lipase production using the pressurized bioreactor (126 U/g), at a pressurized aeration of 200 and 400 kPa, when compared with the tray-type bioreactor (61 U/g). Another interesting bioreactor for SSF is the packed bed bioreactor. The main attractive features of this bioreactor are its easier operation and the possibility of continuous operation, besides allowing the in situ extraction of enzymes (Ganguly and Nandi 2015). Oliveira et al. (2017a) studied the lipase production by *Aspergillus ibericus* MUM 03.49 in a packed bed bioreactor. The authors used SSF of olive pomace and wheat brain, and after the process optimization an enzyme activity of 223 U/g (dry basis) was obtained after 7 days of fermentation. Another relevant concept to consider on bioreactor design for SSF is the use of continuous or intermittent agitation in order to decrease the heterogeneity of the solid medium and improve the oxygen transfer to the microorganism. The application of a rotating drum bioreactors was described by Saithi and Tongta (2016) in phytase production with *A. niger* from soybean meal. In this bioreactor, agitation is applied by a rotating system tumbling the solid medium. The authors reported a phytase activity of 580 U/g substrate dry weight.

Generally, the most common operation mode applied in SSF systems is the batch process where all nutrients are added before the cultivation starts and remain in the bioreactor until the end of fermentation. However, this approach has several issues mainly due to the problems with mass and heat transfer in the solid-liquid-gas interphases, thus making the scale-up of a SSF process very challenging (Lonsane et al. 1985). Therefore, it is necessary to develop alternative operational strategies to overcome these limitations and enhance the SSF performance. Semicontinuous operation allows the substrate supplementation with partial feeding or sequential batch of nutrients. This is an interesting approach that has been used to control the microorganism growth rates (Cerda et al. 2017; Martínez-Avila et al. 2019). This feature was explored by Martínez-Avila et al. (2019). The authors compared the 2-phenylethanol and 2-phenethyl acetate production by *K. marxianus* using fed-batch and sequential-batch strategies in a SSF process with sugarcane bagasse as substrate, supplemented with L-phenylalanine. The authors observed that in the fed-batch strategy it was possible to increase the production of both flavors in 12.5%. Moreover, in the sequential-batch the production was also increased in 2.4% compared to the batch. Cheirsilp and Kitcha (2015) compared the lipid production using different cellulolytic oleaginous fungi and compared its production in SSF using a fed-batch strategy with intermittent feeding and a repeated-batch strategy. Comparing to the batch experiments, with this approach it was possible to increase the lipid yield from 79.9 to 86.6 mg/g dry substrate. Nevertheless, the industrial applications of SSF are still limited mainly due to the necessity for continuous inoculum production that becomes expensive and time-consuming and the low productivity inherent to the repeated-batch operation. A continuous SSF process can potentially overcome these bottlenecks. Lagemaat and Pyle (2004) firstly described a SSF process operating in a mixed continuous mode without an inoculation feed.

The authors cultivated *Penicillium glabrum* on tannin-rich model substrate for tannase production, attaining concentrations that range from 50 to 140 μmol (min/gPUF).

Another important feature that has been applied in SSF is the utilization of co- or mixed-culture systems. In fact, most studies with SSF reported the application of a single strain for (bio)production or bioconversion of the desired metabolite (Qiao et al. 2018; Martínez-Avila et al. 2019). However, the substrate utilization in a monoculture processes is often limited to more simple substrates (Lin et al. 2011). On the other hand, the utilization of two or more microorganisms that are cultured together in the same medium (co-culture strategies) allow the utilization of more complex substrates, as lignocellulosic and agriculture residues, as well as mixtures of substrates with different compositions (Lin et al. 2011). Co-cultures appear to be advantageous over single microorganism ones since with this approach it will be possible to use the metabolic abilities of all involved strains in the co-culture setup (Yao and Nokes 2013). The production of cellulase and hemicellulase from various agriculture residues using single and co-cultures of *A. niger* and *Trichoderma reesei* and SSF technology was studied by Dhillon et al. (2011). They observed that the utilization of mixed cultures allowed a higher cellulase activity (3106.34 $IU/g_{dry\ substrate}$) when compared to the activities attained with single cultures. These experiments demonstrated the potential of co-cultivation as an alternative strategy to produce enzymes in a competitive way from cheaper substrates. In fact, SSF has a high potential to be a more economical industrial process than SmF. However, although SSF industrial applications are well established for certain traditional processes in oriental countries, their use in western countries is still very limited. The main drawback is that a simple and practical automated fermenter for SSF processes has not yet been developed. In fact, the main constraints for SSF scale-up are still the selection of the microbial host, the bioreactor system employed, and substrate/support used.

5.5 Green and Sustainable Downstream Processing of Biomolecules Using Aqueous Two-Phase Systems

Aqueous two-phase systems (ATPSs) have been widely described as a versatile and promising separation/recovery tool in biotechnology. These biphasic systems are formed when aqueous solutions of two different constituents are mixed above certain critical conditions, namely, concentration and temperature. Typically, the equilibrium phases are constituted mostly by water (>70%) and each one is enriched in a specific component. Depending on the type of ATPS prepared, the predominant component in each equilibrium phase may significantly differ. ATPSs formed by two polymers (such as dextran and polyethylene glycol (PEG)) or a polymer and a salt (such as PEG-Na_2SO_4) constitute the classical systems. However, other alternative biphasic systems have been proposed using surfactants (Cordisco et al. 2016), alcohols (Xu et al. 2017), organic acids (Saravanan et al. 2008), carbohydrates (Silva et al. 2007), ionic liquids (Lee et al. 2017), or deep eutectic solvents (Xu et al. 2015). The main advantages reported for the use of ATPSs in the recovery of biomolecules include the simplicity of operation, rapid separation, relative low cost and energy consumption, ease to scale up, high selectivity, ability to be coupled with different extraction techniques, biocompatibility of most of the constituents, and the high water content in both equilibrium phases (Phong et al. 2018). Since water is the main component of ATPSs, a gentle environment is generated for the efficient recovery and separation of biological compounds without compromising their integrity. Furthermore, some compounds frequently used in ATPS formation, such as polyols (sorbitol or xylitol), polymers (PEG or dextran), and surfactants (Tween or Triton) are known for their additional capacity to stabilize the structure and activity of several biological constituents (Silva et al. 2018). ATPSs offer innumerous advantages in biomolecule recovery when compared to the classical liquid-liquid extraction using organic solvents which can present several environmental issues and toxicity and frequently promote the damage of biological products. However, to increase the potential of ATPSs as a green and sustainable downstream strategy it is important to use phase-forming components with proven biocompatibility and biodegradability and preferably with the ability to be reused and recycled. In the last decade, several studies have been focused on the recyclability and environmental impact of ATPSs. Several phase-forming constituents from natural origin such as choline derivatives, salts derived from organic acids, or carbohydrates have been proposed (Li et al. 2012; Kalaivani and Regupathi 2013; Ramalakshmi et al. 2014). Also, several strategies have been adopted to recycle the phase-forming components such as the application of thermo-separating polymers (Leong et al. 2016) or the preparation of successive ATPSs to achieve an efficient back-extraction of the target product and the regeneration of the most expensive components (Cláudio et al. 2014). Some practical examples of the successful applications of ATPSs as greener and eco-friendly alternatives for the extraction and recovery of different biological compounds will be further discussed below.

5.5.1 Extraction of Natural Products from Plants or Animal Tissues

Higher yield and/or purity has been reported for the extraction of several phytochemicals using ATPSs when compared with the traditional extraction procedures. Mejía-Manzano et al. (2019) improved the extraction of the natural anticancerigen pristimerin from *Mortonia greggii* root bark previously ground using an ethanol-Na_3PO_4 ATPS. The extraction of pro-anthocyanidins (catechin, epicatechin, and procyanidin B2) from powdered grape seeds was also significantly raised using the ethanol-$(NH_4)_2SO_4$ ATPS with the ionic liquid 1-hexyl-3-methylimidazolium tetrafluoroborate ($[C_6mim]BF_4$) as adjuvant (Ran et al. 2019). Additionally, the use of ultrasonic-assisted ethanol-$(NH_4)_2SO_4$ ATPSs proved to be an effective strategy for the extraction of curcumin from powdered *Curcuma longa* rhizomes (Xu et al. 2017). In all these examples, the natural products were extracted to the ethanol-rich phase. However, less organic solvent was consumed, shorter extraction time was required, and safer methodologies were developed when compared to the traditional solvent extraction.

The extraction of enzymes directly from animal tissues was also reported using ATPSs. Tjerneld et al. (1987) studied the affinity extraction of lactate dehydrogenase from pig muscle using polymer-polymer ATPS composed of Aquaphase PPT and PEG 8000. Boland et al. (1991) performed pilot-scale trials to extract superoxide dismutase from bovine liver using PEG 1550-potassium phosphate buffer ATPS. Although these strategies were described as economically favorable and presenting efficient use of space and time, they are not commonly used. Alternatively, previous crude extracts from animal tissues and/or animal tissue processing residues can be prepared and the target biomolecules be further recovered through ATPSs (Sripokar et al. 2017).

5.5.2 Recovery and Purification of Biomolecules from Alternative Fermentation Broths and Industrial Residues

The recovery and separation of several enzymes/proteins from synthetic fermentation broths using ATPSs has been extensively studied (Iqbal et al. 2016) with recognized advantages over the conventional methodologies (Aguilar et al. 2006; Moreira et al. 2013). However, product recovery from fermentation broths composed of agro-industrial residues is generally described as a more difficult task. Fermentation media containing residues are considered a cheaper and sustainable approach to obtain some biomolecules, but these fermentation media are also more complex in composition and impurities which can make the downstream process difficult. ATPSs have demonstrated likewise potential for product recovery from these alternative fermentation broths. Some examples of enzymes recovered and partially purified from fermentation broths containing agro-industrial residues include esterase from *A. pullulans*, cellulase from *Schizophyllum commune*, laccase from *Peniophora cinerea*, or keratinolytic protease from *Serratia marcescens* which were produced in fermentation broths containing OMW (Lemes et al. 2019), wheat bran (Kumar et al. 2018), CSL (Moreira et al. 2013), and feather meal (Bach et al. 2012), respectively. Furthermore, ATPSs can also contribute for the direct valorization of industrial by-products and residues. The extraction and recovery of added-value products such as enzyme/proteins from cheese whey (Anandharamakrishnan et al. 2005), tannery wastewater (Raja and Murty 2013), salted egg white (Jiang et al. 2019), or brewery yeast waste (Léon-González et al. 2016) have been effectively achieved using different PEG-salt ATPSs.

5.5.3 Process Integration

ATPSs have enormous potential for process integration. The simultaneous mechanical cell disruption and extraction of intracellular proteins from yeasts has been successfully reported (Rito-Palomares and Lyddiatt 2002; Gurpilhares et al. 2015). Furthermore, extractive fermentation for the production and in situ product recovery can be performed in ATPSs to improve the production yield and reduce both the costs and the operation time. Pullulan production by *A. pullulans* through extractive fermentation in PEG 4000-Dextran 500 ATPS effectively integrated the upstream and downstream processes for continuous production and proved to be environmentally friendly, reliable, and reproducible (Badhwar et al. 2019). The extractive fermentation using a thermo-separating ATPS to obtain polyhydroxyalkanoates from *Cupriavidus necator* contributed to reduce the costs and environmental impact and increase the productivity and purity of the target product (Leong et al. 2019). The scale-up of the extractive fermentation process using a thermo-separating ATPS to a bench-scale bioreactor allowed the increase of the production and recovery yield of *Burkholderia cepacia* lipase (Show et al. 2012). Also, extractive bioconversions using enzymes and whole cells as biocatalysts have been performed in ATPSs. The simultaneous synthesis and recovery of cyclodextrins in PEG 3000-potassium phosphate buffer ATPS using *Bacillus cereus* cyclodextrin glycosyltransferase proved to be a cost- and time-saving technique in repetitive batch process (Lin et al. 2016). The hydrolysis of poly-ε-caprolactone by *B. cepacia* lipase in PEG 3000-potassium phosphate ATPSs was optimized and allowed 80% recovery of hydrolyzed product in the top phase (Chew et al. 2015). Also, lactose was successfully hydrolyzed by β-galactosidase from bacteria, yeast, and fungi in polymer-polymer ATPSs (Chen and Wang 1991). The whole cells of recombinant *E. coli* expressing penicillin acylase were also used as biocatalysts for penicillin G hydrolysis in PEG 6000-potassium phosphate ATPSs allowing the recovery of the target product in the top phase and the reuse of the biocatalyst in ten cycles (Liao et al. 1999).

5.5.4 Aqueous Two-Phase Flotation

This technique combines solvent sublation with aqueous two-phase extraction (Bi et al. 2009). Basically, the surface-active compounds present in an aqueous solution are adsorbed on bubble surfaces of an ascending gas stream and further collected in the top phase of the system. The mass transfer by bubble adsorption is reported as more effective than the mechanical vibration frequently used in conventional liquid-liquid extraction. Aqueous two-phase flotation has been applied for the recovery and concentration of penicillin G from *Penicillium chrysogenum* (Bi et al. 2009), puerarin from *Puerariae* (Bi et al. 2010), baicalin from *Scutellaria baicalensis* Georgi (Bi et al. 2013), proteins from wet microalgae (Phong et al. 2017), betacyanins from peel and flesh of *Hylocereus polyrhizus* (Leong et al. 2018), or lipase from *B. cepacia* (Sankaran et al. 2018a). Process integration of fermentation and liquid biphasic flotation was also reported and this approach enabled the acceleration of product formation, improved the production yield, and facilitated the downstream processing (Sankaran et al. 2018b). Compared with solvent sublation and liquid-liquid extraction, the combined strategy of aqueous two-phase flotation presented clear advantages in separation efficiency, providing high concentration coefficients and reducing the consumption of organic solvents (Bi et al. 2013).

5.6 Valorization of Agro-industrial Residues Through the Production of Added-Value Compounds: Case Studies

5.6.1 *Xylooligosaccharides (XOS)*

From an economic and environmental point of view, the use of agro-residues is an attractive approach for the production of bioactive compounds that offer potential health benefits, among other added-value (bio)products (Chapla et al. 2012; Linares-Pastén et al. 2018). As a consequence of the negative impact *of globalization*, the current food processing technologies are leading to a decrease of nutrients and functional properties of food products (Adamberg et al. 2014). Additionally, the people's lifestyles have significantly changed, including their eating behaviors, which in turn are related to their health and morbidity. The high prevalence of diet-related chronic diseases worldwide including diabetes mellitus, cardiovascular disease, stroke, and hypertension (WHO 2017, 2018) has increased the consumer awareness for the prevention of these diseases. In this sense, the consumers have shifted their attention toward healthier food options, even when these are more expensive (Liu et al. 2017b).

Among several bioactive molecules being studied, prebiotics have been the focus of many studies due to their broad range of beneficial effects on human and animal health (Ashwini et al. 2019). The prebiotic definition was recently reviewed by the International Scientific Association for Probiotics and Prebiotics to "a substrate that is selectively utilized by host microorganisms conferring a health benefit" (Gibson et al. 2017), including human and animal hosts. Several mechanisms promoted by the use of prebiotics have been linked to health benefits. The most widely reported phenomenon is the selective fermentation of prebiotics by beneficial gut bacteria, such as bifidobacteria and lactobacilli, which are known producers of short-chain fatty acids, such as acetate, propionate, and butyrate (Gibson and Roberfroid 1995; Hutkins et al. 2016). Added to the selective growth stimulation of beneficial gut microflora, the use of prebiotics has been related with cholesterol lowering, enhanced mineral absorption, immune modulation, pathogen exclusion, glucose homeostasis, anticarcinogenic properties, and antioxidant, among others (Samanta et al. 2015; Gibson et al. 2017). Prebiotics have been employed in a wide range of applications including food and beverage processing, dietary supplements, animal feed (Mano et al. 2018), and less explored cosmetics applications (Bockmühl et al. 2007). In fact, the increased demand of prebiotics in the past years has been reflected on the significant growth of global prebiotic market which is expected to reach nearly 6.0 billion $ by 2022 (Gaurav 2017). Oligosaccharides are the most widely reported and stabilized type of prebiotics. In particular, inulin, fructooligosaccharides (FOS), and galactooligosaccharides (GOS) dominate the market (Global Market Insights Inc. 2017). These compounds are classified as established prebiotics in the market (Fig. 5.2), according to the number of commercial applications and regulatory status, while isomaltooligosaccharides (IMO), chitooligosaccharides (COS), and xylooligosaccharides (XOS) are considered as emerging ones.

The development of more cost-effective, simple, and green technology for the production of prebiotics is currently a major concern. At the present, the production strategies that better meet those criteria are enzymatic hydrolysis (including isolated and immobilized enzyme biocatalysis) and the whole-cell fermentation, using abundant and cheap agro-residues as substrates, e.g., cheese whey, molasses, corncob, and orange straw, among others (Fig. 5.2). Besides the direct use as substrates for prebiotics production, these agro-residues have also been widely explored for

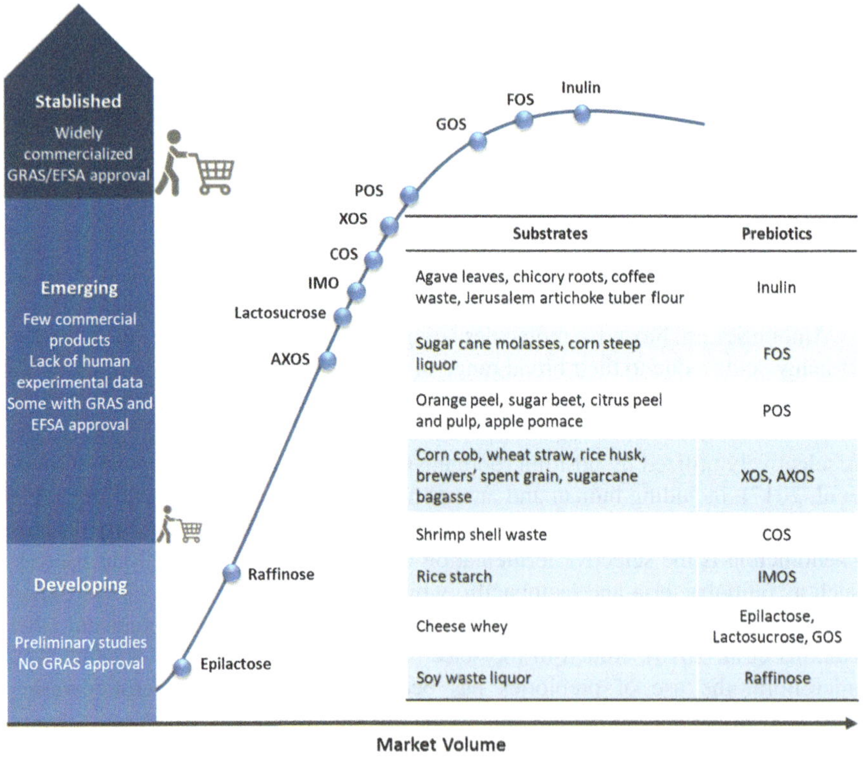

Fig. 5.2 Prebiotics positioning in the market and possible substrates for their production. *Abbreviations*: *AXOS* arabinoxylooligosaccharides, *IMO* isomaltooligosaccharides, *COS* chitooligosaccharides, *XOS* xylooligosaccharides, *POS* pectin oligosaccharides, *GOS* galactooligosaccharides, *FOS* fructooligosaccharides

enzyme production, which in turn are applied on production processes of prebiotics. For instance, sugar molasses and cassava husk, cheese whey, wheat bran, and corncob have been used for the microbial production of β-fructofuranosidase and fructosyltransferase (Babu et al. 2008; Fernandes et al. 2017), ß-galactosidase (Barbosa et al. 1985; Cardoso et al. 2017), and xylanase (Narang et al. 2001; Khanahmadi et al. 2018), which can be further applied on the production of FOS, GOS, and XOS, respectively.

XOS have raised particular interest to both industry and academy, especially because these compounds can be produced from lignocellulosic agro-residues. XOS are oligomers composed of a main chain of xylose residues linked through (β1,4)-linkages, which can present different side elements attached, including acetyl groups, arabinose, and glucuronic acids (Kumar and Satyanarayana 2011). These compounds present a remarkable potential as food ingredients being considered one of the most promising emerging prebiotics. Indeed, XOS stand out due to their price competitiveness, pH and temperature stability, organoleptic properties (Courtin et al. 2009), resistance to digestion (Amorim et al. 2019a), and a broad range of beneficial effects on human and animal health (Aachary et al. 2015) with a lower recommended dose, 2 g/day (Yang et al. 2015), when compared to other prebiotics. Due to the increasing market demand of XOS (The Market Reports 2018), the industry is currently motivated on developing alternative production approaches of XOS, such as the use of agro-residues, in order to reduce costs, while improving efficiency and sustainability. XOS can be obtained from xylan which is the major constituent of hemicelluloses present in the lignocellulosic biomass (Aragon et al. 2013; Samanta et al. 2015). XOS have been produced by different techniques, including chemical, auto-hydrolytic, enzymatic, and fermentation processes or a combination of these techniques (Carvalho et al. 2013; Kumar and Satyanarayana, 2015; Amorim et al. 2018). The XOS production by chemical or auto-hydrolytic processes may encounter several disadvantages, as the production of undesired products, in particular toxic compounds (e.g., furfural and hydroxymethylfurfural), which increases the costs of the downstream process (Bian et al. 2014). Other common limitations are the use of toxic chemicals and more expensive and robust equipment operating at more extreme conditions and low control over the degree of polymerization (DP) (Bian et al. 2014). On the contrary, the enzymatic process is considered a greener approach, being better aligned with the perspective of biosustainability. It does not require the use of harmful chemicals, operates at milder conditions, presents high specificity and efficiency and low production of undesired by-products, and allows a higher control over DP (Akpinar et al. 2007). The enzymatic hydrolysis is currently considered a more efficient and environmentally friendly (Carvalho et al. 2013) for XOS production, especially for food applications. However, the lignocellulosic biomass generally comprises a xylan-lignin matrix structure (Samanta et al. 2012), reducing the xylan accessibility to enzymes. Hence, XOS are mainly obtained by combined methods, usually in a two-step process (Table 5.2). The first step, usually known as pretreatment, includes the pretreatment of the lignocellulosic residue in order to obtain soluble xylan or to reduce the amount of lignin from the residue, followed by the second step of xylan hydrolysis by xylanases (Carvalho et al. 2013). The limitations of this approach rely on the low yields of the pretreatment step added to the cost of producing or purchasing xylanases, which may compromise the economic viability of the process (Reddy and Krishnan, 2016a).

Recent advances have been done on the development of more integrated production approaches, as the single-step fermentation of lignocellulosic residues by microorganisms and the co-production of XOS and other added-value products (Table 5.2). The single-step fermentation is a promising strategy, in the sense that, by not including the purchase or production of enzymes and by reducing the process stages, it may reduce the production cost and benefit the overall XOS production yield. This approach was successfully applied for the production of AXOS from brewers' spent grain by submerged fermentation (Amorim et al. 2018, 2019b). A mixture composed by a high amount of XOS and low concentration of monosaccharides, mainly xylose, was obtained, since the free sugars in the medium are assimilated by the microorganism before starting the step of XOS degradation. In this context, single-step fermentation also may allow to reduce downstream costs. Furthermore, the authors concluded that the use of single-step fermentation was more advantageous than the application of commercial enzymes. Nevertheless, this approach is fraught with different challenges. Besides the great influence of the type of microorganism and agro-residue used on the economic feasibility of this approach, the cross-reactivity with cellular metabolites, the suboptimal use of cofactors that may be used in other metabolic networks, and the decomposition of substrates and products through competing cellular reactions may contribute to lowering the XOS production yield (Sheldon and Woodley 2018). Reddy and Krishnan (2016b) and da Silva Menezes et al. (2017) applied single-step solid state fermentation to co-produce XOS and xylanase using different agro-residues (Table 5.2). However, the authors reported low yields of XOS, which may be explained by the diversion between the specific optimal process conditions required for XOS and xylanase production, including different optimal fermentation times (Amorim et al. 2019b).

5.6.2 *Biopolymers*

Polymers constitute a versatile group of compounds that perform essential functions in our society, being their production estimated in more than 180 million tons per year. The application of these compounds includes fields such as food, textile, paper, painting, pharmaceutics, cosmetics, and petroleum industries, among others, where they are used as emulsifiers, stabilizers, or thickening agents (Sharma et al. 2013; Mehta et al. 2014; Wu et al. 2016). For instance, in food industries, polymers enhance the rheology, texture, viscosity, flavor release, appearance, or water control of juices, fruit pulps, chocolates, jellies, desserts, margarine, yoghurts, bakery products, frozen foods, or sauces. In the petroleum industry, they are applied to enhance oil production from oil fields that exhibit low productivities by improving the water-

Table 5.2 Valorization of different agro-residues for xylooligosaccharide (XOS) production and other added-value products using greener bioprocess approaches

Agro-residue	Pretreatment (PTT)	Bioprocess	Biocatalyst	Products	Reference
Areca nut husk	Alkali PTT (two-step process)	EH	*Trichoderma viride* endo-1,4-β-xylanase M1	XOS	Singh et al. (2018)
Brewers' spent grain	No PTT	EH	Commercial xylanase (*Trichoderma longibrachiatum*)		Amorim et al. (2019b)
		SmF	*Trichoderma reesei*		
			Wild-type and recombinant *Bacillus subtilis* 3610		Amorim et al. (2018)
Corncob	Steam explosion with acidic electrolyzed water	EH	*Paenibacillus barengoltzii* (PbXyn10A) xylanase		Liu et al. (2018)
	Ground and prehydrolysis with acetic acid		Commercial cellulase (*Trichoderma reesei*)	XOS, cellobiose, glucose, and xylose	Zhang et al. (2017)
	Ultrahigh-pressure PTT		*Streptomyces thermovulgaris* TISTR1984 xylanase	XOS	Seesuriyachan et al. (2017)
Finger millet seed coat	De-starch and water extraction		Commercial xylanase (*Thermomyces lanuginosus*)		Palaniappan et al. (2017)
Mahogany	Thermal PTT with NaOH		*Clostridium* sp. BOH3 xylanase		Rajagopalan et al. (2017)
Mango					
Quinoa stalks	Alkaline extraction		*Rhodothermus marinus* RmXyn10ACM		Salas-Veizaga et al. (2017)
Reed scraps	Liquid hot water		Commercial xylanase and cellulase	XOS and glucose	Chen et al. (2019a, b)
Rice husk	No PTT	SSF	*Aspergillus brasiliensis*	Xylanase and XOS	da Silva Menezes et al. (2017)

(continued)

Table 5.2 (continued)

Agro-residue	Pretreatment (PTT)	Bioprocess	Biocatalyst	Products	Reference
Rice straw	Milling	EH	Aggregate of magnetic crosslinked xylanase (*Acinetobacter pittii* MASK 25)	XOS	Purohit et al. (2017)
Sugarcane bagasse	Ammonia PTT		β-Xylosidase-free xylanase of *Bacillus subtilis* KCX006		Reddy and Krishnan (2016a)
Wheat bran	Thermal and enzymatic PTT		1,4-β-Xylanase (*Thermomyces lanuginosus*) and *Neocallimastix patriciarum* NpXyn11A		Mathew et al. (2018)
	Sodium acetate buffer washing and alkali extraction		Recombinant *Bacillus amyloliquefaciens* xylanase A		Liu et al. (2017a)
	De-starch		Recombinants xylanase and feruloyl esterase	XOS and ferulic acid	Wu et al. (2017)
Groundnut oil cake and wheat bran	Finely ground	SSF	*Bacillus subtilis* KCX006	XOS and xylanase	Reddy and Krishnan (2016b)
Wheat straw	Alkaline extraction	EH	Recombinant *Bacillus halodurans* S7 endoxylanase	XOS	Faryar et al. (2015)

Abbreviations: *EH* enzymatic hydrolysis, *SmF* submerged fermentation, *SSF* solid state fermentation

flooding performance during tertiary oil recovery operations (Niknezhad et al. 2015; Salah et al. 2015; Couto et al. 2019). Nowadays, the global market is dominated by synthetic polymers (polyacrylamides) and biopolymers extracted from plants (gums, cellulose, pectins, starch) or algae (alginate, carrageenan, agar). Biopolymers of microbial origin exhibit better environmental compatibility and biodegradability than the synthetics, and their production is faster when compared with those obtained from plants (Assis et al. 2014; Ai et al. 2015; Antunes et al. 2017).

Microbial biopolymers display a broad variety of chemical structures, which results in different physicochemical and rheological properties. Furthermore, they usually exhibit high viscosifying activity at low concentration, high solubility, high water retention capacity, stability at extreme temperatures, pHs and salinities, resistance to shear degradation, and compatibility with many polysaccharides and salts. Consequently, they represent a promising alternative for application in several fields (Savvides et al. 2012; Gunasekar et al. 2014; Li et al. 2016a, b). Another advantage of microbial biopolymers is the possibility of producing them from renewable resources. Furthermore, several microbial biopolymers exhibit distinctive biological activities that are not present in their synthetic counterparts, including antitumor, antimicrobial, antioxidant, or prebiotic activities. Therefore, some of them can be incorporated in high-value products, including functional foods and medical devices (Antunes et al. 2015; Salah et al. 2015; Wu et al. 2016).

Despite all these advantages, the application of microbial biopolymers is limited as a result of the high costs associated to their production. In the recent years, the search for new molecules with improved rheological properties has increased due to the growing demand for new biopolymers; at the same time, cost-effective and more efficient production processes have been developed to make biopolymers more accessible to the industry. Several microbial biopolymers, such as xanthan gum, pullulan, gellan gum, scleroglucan, dextran, or levan, are already commercialized for application in medicine, food industries, cosmetics, and the petroleum industry (Sharma et al. 2013; Li et al. 2016a, b; Couto et al. 2019). Xanthan gum, synthesized by *Xanthomonas campestris* (a Gram-negative bacterium), is the microbial biopolymer with the highest production volume. Due to the extraordinary performance of xanthan gum, it is mainly used in the food industry, but also in the petroleum and textile industries. Its annual worldwide production is around 30 kilotons, corresponding to a market of 360 million €. However, its price (3500–4500 €/ton) is high when compared with the synthetic polymers (Savvides et al. 2012; Salah et al. 2015; Li et al. 2016a, b).

The high production costs of microbial biopolymers are mostly due to the price of the culture media, in particular the carbon sources. Traditional substrates, including sucrose, glucose, or fructose (400–600 €/ton), in which concentration in the culture medium can achieve 200 g/L, can represent up to one-third of the total production costs (Ai et al. 2015; Niknezhad et al. 2015; Li et al. 2016a, b; Antunes et al. 2017). Therefore, the use of alternative inexpensive substrates is crucial to reduce the price of the final product. Accordingly, several wastes and by-products generated by agricultural and industrial processes are being evaluated as low-cost carbon and nitrogen sources for the production of biopolymers by different microorganisms. Some examples are summarized in Table 5.3. Among the different alternative carbon sources stands out molasses, an inexpensive residue (85–150 €/ton) resulting from the extraction of sugar from dates, sugar beet, or sugarcane. Molasses is the residual syrup from the final crystallization stage, from which further crystallization of sugar is uneconomical. This residue is rich in carbohydrates (between 40 and 70% (w/w)), but is also an important source of micronutrients (vitamins, minerals, and organic acids) (Banik et al. 2007; Joshi et al. 2008; Al-Bhary et al. 2013; Ai et al. 2015; Gudiña et al. 2015a). Accordingly, this residue from the sugar industries is commonly used as a suitable alternative source of carbon for the biosynthesis of xanthan gum, gellan gum, and welan gum (Table 5.3). Another important alternative carbon source is raw glycerol, the main by-product from biodiesel production. Each liter of biodiesel results in the production of approximately 0.12 kg of glycerol. It is estimated that the production of this residue (around 15 million m^3 per year) will grow in the upcoming years as a result of a higher production of biodiesel. Raw glycerol contains impurities (e.g., organic compounds, salts), which avoids its use in food or pharmaceutical-related applications, being a low-cost source of carbon for the biosynthesis of added-value compounds by different microorganisms (de Faria et al. 2011; Sousa et al. 2012; Assis et al. 2014; de França et al. 2015; Cruz et al. 2018). Cheese whey is a residue generated by the dairy industry in large amounts that contains a high concentration of lactose. It represents about 85–95% of the milk's volume and its worldwide production is around 140 million tons per year. Besides lactose, it also contains soluble proteins, minerals, lactic acid, and fats.

The high oxygen demand of this residue makes necessary its treatment before being discharged into the environment (Savvides et al. 2012; Antunes et al. 2015; Niknezhad et al. 2015). CSL is a by-product generated by the corn wet milling industry. This inexpensive (40 €/ton) and abundant liquid residue contains proteins, vitamins, amino acids, and minerals and represents an excellent nitrogen source for the growth of microorganisms in many industrial processes (Sharma et al. 2013; Mehta et al. 2014; Gudiña et al. 2015a, b). The applicability of CSL as a low-cost source of nitrogen for the synthesis of microbial biopolymers has also been demonstrated (Table 5.3).

The use of some of these residues to produce added-value compounds can contribute to reduce the negative impact associated to their disposal into the environment. Some of these residues have been successfully used as substrates to synthesize biopolymers without the need of performing previous pretreatments, which reduces the costs associated to their use (Banik et al. 2007; Sharma et al. 2013; Assis et al. 2014; Antunes et al. 2015, 2017). However, in other cases, specific pretreatments are necessary in order to remove growth inhibitors or to breakdown complex carbohydrates into simple reducing sugars, which increases the overall production costs (Kalogiannis et al. 2003; Göksungur et al. 2011; Savvides et al. 2012; Gunasekar et al. 2014; Ai et al. 2015; Li et al. 2016a, b; Wu et al. 2016). In some cases, even using low-cost carbon sources, it was necessary to add expensive nitrogen sources (e.g., tryptone, yeast extract) to the culture medium, which results in high production costs (Banik et al. 2007; Göksungur et al. 2011; Gunasekar et al. 2014; Niknezhad et al. 2015; Wu et al. 2016). However, in some cases it was possible to design culture media containing residues as sole ingredients (Mehta et al. 2014; Li et al. 2016a, b). The results shown in Table 5.3 demonstrate that the replacement of expensive nutrients by low-cost residues can result in the biosynthesis of high amounts of biopolymer (40–88 g/L). Furthermore, in some cases, that production was achieved in 24 h. The variability observed among the biopolymer productions obtained (Table 5.3) can be due to the different degree of purification achieved in the different works, which is not always reported, and make the comparison among the different studies difficult. Nevertheless, according to several authors, the use of agricultural and industrial wastes as low-cost substrates to produce microbial biopolymers can reduce their production costs between 50 and 75%, making their production processes more cost-effective (Sharma et al. 2013; Mehta et al. 2014; Salah et al. 2015; Wu et al. 2016). However, this can have a negative impact on the properties of the product obtained, or it can be necessary to apply additional downstream purification steeps to obtain a compound with a similar purity to those obtained using synthetic culture media.

Table 5.3 Biopolymer production using different agricultural and industrial by-products and residues as substrates

Biopolymer	Strain	Substrate	[BP] g/L	Productivity (g/L/day)	Reference
Xanthan gum	*Xanthomonas campestris* NCIM 2954	Hydrolyzed tapioca pulp	7.1	2.4	Gunasekar et al. (2014)
	Xanthomonas campestris ATCC 13951	Sugar beet molasses	53.0	53.0	Kalogiannis et al. (2003)
	Xanthomonas campestris NRRL B-1459	Palm date juice	43.3	43.3	Salah et al. (2015)
	Xanthomonas campestris LRELP-1	Kitchen waste hydrolysate	11.7	2.8	Li et al. (2016a, b)
	Xanthomonas campestris 2103	Raw glycerol	5.6	1.1	Assis et al. (2014)
	Xanthomonas campestris ATCC 13951	Cheese whey	28.0	7.0	Savvides et al. (2012)
	Xanthomonas campestris PTCC1473	Cheese whey	16.4	8.2	Niknezhad et al. (2015)
Pullulan	*Aureobasidium pullulans* P56	Hydrolyzed potato starch	19.2	4.1	Göksungur et al. (2011)
	Aureobasidium pullulans CJ001	Hydrolyzed potato starch	36.2	9.0	Wu et al. (2016)
	Aureobasidium pullulans RBF 4A3	Corn steep liquor	88.6	22.1	Sharma et al. (2013)
	Aureobasidium pullulans RBF 4A3	Corn steep liquor + de-oiled Jatropha seed cake + jaggery	66.2	22.1	Mehta et al. (2014)
Welan gum	*Alcaligenes* sp. ATCC 31555	Sugarcane molasses	41.0	8.2	Ai et al. (2015)
Gellan gum	*Sphingomonas paucimobilis* ATCC 31461	Molasses	13.8	6.9	Banik et al. (2007)
FucoPol	*Enterobacter* A47	Cheese whey	6.4	2.0	Antunes et al. (2015)
		Out-of-specification tomato paste	8.7	2.9	Antunes et al. (2017)

[BP] biopolymer concentration

5.6.3 *Biosurfactants*

Surfactants or surface-active compounds are an outstanding class of chemicals that are present in most of the products used in our daily life. These compounds are included in the formulation of cleaning products (e.g., laundry formulations, detergents), paints, pesticides, herbicides, cosmetic products, and pharmaceutical products and are also used in agriculture, paper, textile, food, and petroleum industries, among others (Gudiña et al. 2016; Chen et al. 2018; Cruz et al. 2018). Surfactants are amphiphilic compounds containing at least one hydrophilic and one hydrophobic moiety that accumulate at interfaces between fluid phases with different degrees of polarity (e.g., oil–water, air–water interfaces). Due to their amphiphilic structure, surfactants reduce both surface and interfacial tensions and decrease the repulsive forces that exist between different phases, which makes easier the mixture of immiscible phases. Consequently, surfactants allow the formation of dispersions, emulsions, and foam, being indispensable in applications that require the stabilization of emulsions or foam, lubrication, the solubilization of immiscible compounds, or phase dispersion (Moshtagh et al. 2018; Chen et al. 2019a, b; Pérez-Armendáriz et al. 2019).

More than 15 million tons of surfactants are produced per year, and these figures are expected to increase up to 24 million tons, corresponding to a market of 37300 million € per year by 2020 (Markets and Markets 2015). Currently, the vast majority of surfactants commercialized are chemically synthesized from petrochemical resources and are only partially biodegradable. Due to the ubiquity of surfactants in daily life, questions regarding their environmental impact in the long term constitute a growing concern. Consequently, the demand for eco-friendly and sustainable surfactants has grown in recent years (Sathi-Reddy et al. 2016; Roelants et al. 2019). The market for these environmentally friendly surface-active compounds is projected to grow at a CAGR of 5.5%, from 1600 million € in 2017 to more than 2400 million € by 2024, with consumption exceeding 540 kilotons (Grand Market Insights Research 2018).

Surfactants of biological origin (i.e., biosurfactants) are attracting significant interest as a potential alternative to the synthetic ones in many fields. Among them, microbial biosurfactants (synthesized by bacteria, yeasts, and filamentous fungi) are particularly attractive to replace the chemical surfactants in a wide variety of applications, given their unique properties and the possibility of large-scale sustainable production using renewable substrates. The performance of the microbial biosurfactants is similar or even better comparing with the synthetic ones; their effectiveness is not lost at extreme salinities, temperatures, and pH values; and they exhibit higher biodegradability and lower toxicity comparing with the conventional synthetic surfactants (Cruz et al. 2018; Pérez-Armendáriz et al. 2019; Roelants et al. 2019). According to their chemical structure, microbial biosurfactants can be classified as glycolipids, lipopeptides, phospholipids, fatty acids, and neutral lipids, being glycolipids and lipopeptides the most widely studied (Gudiña et al. 2013; Roelants et al. 2019).

Regarding the lipopeptide biosurfactants, the hydrophobic domain consists in a fatty acid of variable length, whereas the hydrophilic moiety is a peptide ring comprising seven or ten amino acids. Lipopeptide biosurfactants are mainly synthesized by species belonging to the genus *Bacillus*. Among the broad spectrum of lipopeptide biosurfactants identified until now, surfactin, produced by *B. subtilis* strains, is one of the most effective. This lipoheptapeptide reduces the surface tension of water from 72 mN/m to values as low as 26 mN/m; the concentration at which the formation of micelles is initiated (critical micelle concentration, *cmc*) can be as low as 10 mg/L; furthermore, it also exhibits a high emulsifying activity (Al-Wahaibi et al. 2014; Gudiña et al. 2015b; Moshtagh et al. 2018; Chen et al. 2019a, b).

Among the glycolipid biosurfactants, rhamnolipids, synthesized mainly by *Pseudomonas aeruginosa*, are the most studied. Rhamnolipids are classified in mono-rhamnolipids and di-rhamnolipids, according to the number of rhamnose molecules that constitute their hydrophilic domain. The hydrophobic domain consists of two (or more rarely one) β-hydroxy fatty acids; the chain length of those fatty acids varies between 8 and 24 carbons, and they can be saturated or unsaturated. To the date, more than 60 different rhamnolipid congeners have been reported. Different congeners differ in their surface and emulsifying activities, solubilities, and *cmcs*. These glycolipid biosurfactants are usually synthesized as mixtures of different congeners, in which composition depends on the microorganism used and is also affected by the culture medium and other parameters such as the pH and the concentration of oxygen. Rhamnolipids reduce the surface tension to values around 28 mN/m and exhibit high emulsifying activities, and their *cmcs* are between 10 and 200 mg/L (depending on the rhamnolipid congeners) (Abdel-Mawgoud et al. 2008; Gudiña et al. 2015a; Chen et al. 2018).

Besides the properties described above, rhamnolipids and surfactin also exhibit antimicrobial, antifungal, antiviral, and antitumor activities, which are not present in their synthetic counterparts; this contributes to increase their value and their potential applications (Gudiña et al. 2013; Duarte et al. 2014). Despite their outstanding properties, the commercialization of microbial biosurfactants will be only possible if their production costs are reduced to make them competitive with the synthetic surfactants. The main obstacles to achieve this objective are the relatively low amounts of biosurfactants usually produced and the high price of the culture media used. The price of chemical surfactants such as sodium lauryl sulfate is around 1–2 €/kg, whereas the sales price of sophorolipids (a glycolipid biosurfactant), the cheapest and most widely available microbial biosurfactant, is around 30 €/kg. In the case of rhamnolipids (95% purity) and surfactin (98% purity), the current market price is in the range of thousands of euros per gram (Roelants et al. 2019). As a result, a limited number of companies commercialize microbial biosurfactants, usually incorporated in high-value products such as cosmetics. Some examples are Rhamnolipid Companies, AGAE Technologies, TOYOBO, Logos Technologies, Evonik Industries, and KANEKA.

The culture medium contributes significantly (up to 50%) to the total production costs of microbial biosurfactants. The expensive synthetic media commonly used for their production can be replaced by low-cost agricultural and industrial by-products or wastes in order to make their production economically viable and increase their competitiveness in the market. Furthermore, the valorization of these residues as substrates for the production of added-value compounds reduces the environmental impact caused by their disposal (Al-Bhary et al. 2013; Lan et al. 2015; Gudiña et al. 2016; Ozdal et al. 2017; Chen et al. 2018). As it can be seen in Tables 5.4 and 5.5, several agricultural and industrial residues and by-products have been used to produce surfactin and rhamnolipids by different microorganisms. Most of these residues are the same previously described for the production of biopolymers. In most of the cases, these residues were used as alternative carbon sources, which means that nitrogen sources (organic or inorganic), salts, micronutrients, or other supplements have to be added to the media, increasing the costs of the production process. However, in some cases it was possible to design culture media that contain as sole ingredients agricultural and industrial residues, without the addition of other nutrients (Nitschke and Pastore 2006; Joshi et al. 2008; Gudiña et al. 2015a, 2015b, 2016). In the case of rhamnolipids, a considerable number of residues associated with the extraction or the consumption of different vegetable oils have been used as alternative carbon sources (Table 5.4). These residues contain different long-chain fatty acids that stimulate the biosynthesis of rhamnolipids by *P. aeruginosa* (Gudiña et al. 2016; Samykannu and Achary 2017; Pérez-Armendáriz et al. 2019). Waste cooking oils are generated in large amounts during food preparation and their disposal represents an environmental problem. Therefore, their use as substrates to produce biosurfactants is an environmentally friendly and cost-effective strategy. However, the use of these water-immiscible substrates made difficult the subsequent recovery and purification processes. A particular residue is OMW, an effluent generated during the extraction of olive oil. In the Mediterranean countries, around 30 million m^3 of OMW are generated each year. OMW is rich in long-chain fatty acids, but also contains phenolic compounds, carbohydrates, tannins, pectins, organic acids, and minerals. OMW is a hazardous waste, being toxic for microorganisms, aquatic ecosystems, and plants. Consequently, the bioconversion of OMW into rhamnolipids is an interesting strategy from an environmental and economic point of view. Furthermore, the only cost associated to the use of this residue is the handling and transportation cost (Ramírez et al. 2015; Gudiña et al. 2016). As it can be seen in Table 5.4, high amounts of rhamnolipids (up to 13 g/L) have been produced using different agricultural and industrial residues as substrates.

However, the productivities reported in the different works are difficult to compare, since different methods can be used to quantify the rhamnolipid concentrations. For instance, the use of colorimetric methods can indicate rhamnolipid concentrations up to nine times higher than the real ones (Ramírez et al. 2015). The variability observed in the amount of surfactin produced in the different studies (Table 5.5) is due to the different purification processes applied, which results in products with different purity. The promising future of microbial biosurfactants depends on the exploitation of abundant and inexpensive agricultural and industrial residues together with the improvement of the culture conditions to increase the current production yields.

Table 5.4 Rhamnolipid production by *Pseudomonas aeruginosa* strains using different agricultural and industrial by-products and residues as substrates

Substrate	Strain	[RL] g/L	Time (h)	ST (mN/m)	E_{24} (%)	cmc (mg/L)	Reference
Molasses (1%, v/v)	2B	4.97	96	29	84	100	Aparna et al. (2012)
Sugarcane molasses (10%, w/v) + corn steep liquor (CSL) (10%, v/v)	#112	3.19	96	29	54	30	Gudiña et al. (2015a)
Sugarcane molasses (10%, w/v) + CSL (10%, v/v) + olive oil mill wastewater (25%, v/v)	#112	5.12	168	29	58	13	Gudiña et al. (2016)
Raw glycerol (2%, v/v)	P6	7.54[a]	144	–	–	–	El-Housseiny et al. (2016)
Orange peels (3%, w/v)	MTCC 2297	9.18[a]	192	31	73	–	George and Jayachandran (2009)
Sunflower oil refinery waste	LBI	7.50[a]	72	34	83	120	Benincasa and Accorsini (2008)
Soybean oil refinery waste (2%, w/v)	LBI	11.70[a]	144	26	55	51	Nitschke et al. (2008)
Coconut oil cake (1.5%, w/v) + coconut oil sludge (2%, w/v)	AMB AS7	5.53[a]	60	31	88	50	Samykannu and Achary (2017)
Mango kernel oil (1%, v/v)	DR1	2.80	72	30	73	80	Sathi-Reddy et al. (2016)
Olive oil mill wastewater (10%, w/v)	PAO1	0.19	192	33	–	–	Ramírez et al. (2015)
Waste cooking oil (40 g/L)	SWP-4	13.93[a]	60	24	59	27	Lan et al. (2015)
Waste cooking coconut oil (2%, w/v)	D	3.55[a]	168	24	71	–	George and Jayachandran (2013)
Waste cooking olive/sunflower (1:1, v/v) oil (40 g/L)	47T2	8.10[a]	96	32	90	108	Haba et al. (2003)
Waste cooking canola oil (3%, v/v)	MK307837	3.58[a]	336	–	–	–	Pérez-Armendáriz et al. (2019)
Kitchen waste oil (2%, w/v)	–	2.47[a]	120	–	58	56	Chen et al. (2018)
Waste cooking oil (5.2%, w/v) + chicken feather peptone (0.9%, w/v)	OG1	13.31[a]	168	–	80	–	Ozdal et al. (2017)

[RL] rhamnolipid concentration, *ST* surface tension, E_{24} emulsifying index, *cmc* critical micelle concentration

[a]Rhamnolipids were quantified as rhamnose equivs alents through colorimetric methods

Table 5.5 Surfactin production by *Bacillus subtilis* strains using different agricultural and industrial by-products and residues as substrates

Substrate	Strain	[Surf] g/L	Time (h)	ST (mN/m)	E_{24} (%)	cmc (mg/L)	Reference
Molasses (16%, v/v)	BS5	1.12	60	25	–	–	Abdel-Mawgoud et al. (2008)
Molasses (5%, w/w)	R1	–	72	29	–	–	Joshi et al. (2008)
Date molasses (8%, w/v)	B20	2.29	24	27	–	–	Al-Bhary et al. (2013)
Date molasses (2%, w/v)	B30	0.30	12	26	50	–	Al-Wahaibi et al. (2014)
Raw glycerol (2%, v/v)	LAMI005	0.44	72	29	–	28	Sousa et al. (2012)
Raw glycerol (2%, v/v)	ICA56	1.29	54	28	90	25	de França et al. (2015)
Raw glycerol (5%, v/v)	LSFM-05	0.93	72	29	67	72	de Faria et al. (2011)
Raw glycerol (5%, v/v)	ATCC 6633	0.79	72	35	38	1500	Cruz et al. (2018)
Potato peels (2%, w/v)	DM-03	2.04	48	34	55	140	Das and Mukherjee (2007)
Cashew apple juice (67%, v/v)	LAMI005	0.32	72	30	67	63	Oliveira et al. (2013)
Corncob hydrolysate (80%, v/v) + feather hydrolysate waste (1%, v/v)	BS-37	0.52	72	–	–	–	Chen et al. (2019a, b)
Brewery wastewater (7%, v/v)	N3-1P	0.66	72	27	63	107	Moshtagh et al. (2018)
Corn steep liquor (10%, v/v)	#573	1.31	48	30	55	30	Gudiña et al. (2015b)
Cassava wastewater	LB5a	3.00	48	26	74	33	Nitschke and Pastore (2006)

[Surf] surfactin concentration, *ST* surface tension, E_{24} emulsifying index, *cmc* critical micelle concentration

5.7 Conclusion

Nowadays, all stakeholders and society in general recognize that it is urgent for the humanity, environment, and progress the study and implementation of alternative strategies for the rational use of resources and zero-waste generation. Among these strategies, the design of new, sustainable, and green biotechnological processes encloses a great promise. The possibility of using microorganisms (single or microbial consortia) and/or enzymes (crude, pure, or mixtures) able to degrade complex substrates (such as the C5 compounds derived from lignocellulosic residues) to produce added-value products (as the ones herein discussed, functional food ingredients, biopolymers, and biosurfactants, among others) is unique to these strategies. The recent developments of the omics techniques (e.g., metagenomics), gene editing (CRISPR-Cas9), sequencing (speed, costs, and accuracy), metabolic engineering, synthetic biology, and bioinformatics make, in principle, possible to improve and/or create any organism for any given purpose, opening great opportunities under the circular bioeconomy and biorefinery concepts. In addition, new bioreactors and production techniques together with new recovery strategies are expected to greatly contribute to the development of cost-effective and productive processes. Nevertheless, in order to assure a sustainable and rational use of resources (viz., residues from different origins) by these bioprocesses, it is crucial to map and deeply characterize all existing resources, as well as to develop integrated processes that are versatile enough to deal with the resources heterogeneity and availability (related to their sazonality in some cases). Several agricultural, forestry, and industrial residues remain underused and these comprise an extremely interesting alternative to fossil resources and a great opportunity for the development of an innovative bioeconomy.

Acknowledgements C.A. and A.C. acknowledge their grants (UMINHO/BPD/4/2019 and UMINHO/BPD/37/2018, respectively) from Portuguese Foundation for Science and Technology (FCT). The study received financial support from FCT under the scope of the strategic funding of UID/BIO/04469/2013 unit, COMPETE 2020 (POCI-01-0145-FEDER-006684), and the projects FoSynBio (POCI-01-0145-FEDER-029549) and Lignozymes (POCI-01-0145-FEDER-029773). The authors also acknowledge BioTecNorte operation (NORTE-01-0145-FEDER-000004) and the project MultiBiorefinery (POCI-01-0145-FEDER-016403) funded by the European Regional Development Fund under the scope of Norte2020—Programa Operacional Regional do Norte.

References

Aachary AA, Gobinath D, Srinivasan K, Prapulla SG (2015) Protective effect of xylooligosaccharides from corncob on 1,2-dimethylhydrazine induced colon cancer in rats. Bioact Carbohydr Diet Fibre 5:146–152. https://doi.org/10.1016/j.bcdf.2015.03.004

Abdallah II, Pramastya H, van Merkerk R, Sukrasno QWJ (2019) Metabolic engineering of *Bacillus subtilis* toward taxadiene biosynthesis as the first committed step for taxol production. Front Microbiol 10:1–11. https://doi.org/10.3389/fmicb.2019.00218

Abdelaal AS, Jawed K, Yazdani SS (2019) CRISPR/Cas9-mediated engineering of *Escherichia coli* for n-butanol production from xylose in defined medium. J Ind Microbiol Biotechnol:1–11. https://doi.org/10.1007/s10295-019-02180-8

Abdel-Mawgoud AM, Aboulwafa MM, Hassouna NAH (2008) Optimization of surfactin production by *Bacillus subtilis* isolate BS5. Appl Biochem Biotechnol 150:305–325. https://doi.org/10.1007/s12010-008-8155-x

Abdullah R, Naeem N, Aftab M, Kaleem A, Iqtedar M, Iftikhar T, Naz S (2018) Enhanced production of alpha amylase by exploiting novel bacterial co-culture technique employing solid state fermentation. Iran J Sci Technol Trans A Sci 42:305–312. https://doi.org/10.1007/s40995-016-0015-x

Adamberg S, Sumeri I, Uusna R, Ambalam P, Kondepudi KK, Adamberg K, Wadström T, Ljungh A (2014) Survival and synergistic growth of mixed cultures of bifidobacteria and lactobacilli combined with prebiotic oligosaccharides in a gastrointestinal tract simulator. Microb Ecol Health Dis 25:1–9. https://doi.org/10.3402/mehd.v25.23062

Adhyaru DN, Bhatt NS, Modi HA, Divecha J (2016) Insight on xylanase from *Aspergillus tubingensis* FDHN1: production, high yielding recovery optimization through statistical approach and application. Biocatal Agric Biotechnol 6:51–57. https://doi.org/10.1016/j.bcab.2016.01.014

Aguiar-Pulido V, Huang W, Suarez-Ulloa V, Cickovski T, Mathee K, Narasimhan G (2016) Metagenomics, metatranscriptomics, and metabolomics approaches for microbiome analysis. Evol Bioinform Online 12(Suppl 1):5–16. https://doi.org/10.4137/EBO.S36436

Aguieiras ECG, De Barros DSN, Fernandez-Lafuente R, Freire DMG (2018) Production of lipases in cottonseed meal and application of the fermented solid as biocatalyst in esterification and transesterification reactions. Renew Energy 130:574–581. https://doi.org/10.1016/j.renene.2018.06.095

Aguilar O, Albiter V, Serrano-Carreón L, Rito-Palomares M (2006) Direct comparison between ion-exchange chromatography and aqueous two-phase processes for the partial purification of penicillin acylase produced by *E. coli*. J Chromatogr B 835:77–83. https://doi.org/10.1016/j.jchromb.2006.03.016

Aguilar A, Wohlgemuth R, Twardowski T (2018) Perspectives on bioeconomy. N Biotechnol 40:181–184. https://doi.org/10.1016/j.nbt.2017.06.012

Ahmad T, Singh RS, Gupta G, Sharma A, Kaur B (2019) Metagenomics in the search for industrial enzymes. In: Singhania RR, Pandey A, Larroche C (eds) Singh RS. Elsevier, Advances in enzyme technology, pp 419–451. https://doi.org/10.1016/B978-0-444-64114-4.00015-7

Ai H, Liu M, Yu P, Zhang S, Suo Y, Luo P, Li S, Wang J (2015) Improved welan gum production by *Alcaligenes* sp. ATCC31555 from pretreated cane molasses. Carbohydr Polym 129:35–43. https://doi.org/10.1016/j.carbpol.2015.04.0330

Akpinar O, Ak O, Kavas A, Bakir U, Yilmaz L (2007) Enzymatic production of xylooligosaccharides from cotton stalks. J Agric Food Chem 55:5544–5551. https://doi.org/10.1021/jf063580d

Al-Askar AA, Saber WIA, Ghoneem KM, Rashad YM (2018) Oxalic acid as the main molecule produced by *Trichoderma asperellum* MG323528 fermented on corn stover based medium. Biotechnology 17:95–103. https://doi.org/10.3923/biotech.2018.95.103

Al-Bhary SN, Al-Wahaibi YM, Elshafie AE, Al-Bemani AS, Joshi SJ, Al-Makhmari HS, Al-Sulaimani HS (2013) Biosurfactant production by *Bacillus subtilis* B20 using date molasses and its possible application in enhanced oil recovery. Int Biodeter Biodegr 81:141–146. https://doi.org/10.1016/j.ibiod.2012.01.006

Ali O, Moftah S, Grbav S (2012) Adding value to the oil cake as a waste from oil processing industry: production of lipase and protease by *Candida utilis* in solid state fermentation. Appl Biochem Biotechnol 166:348–364. https://doi.org/10.1007/s12010-011-9429-2

Allen EE, Banfield JF (2005) Community genomics in microbial ecology and evolution. Nat Rev Microbiol 3(6):489–498. https://doi.org/10.1038/nrmicro1157

Al-Wahaibi Y, Joshi S, Al-Bahry S, Elshafie A, Al-Bemani A, Shibulal B (2014) Biosurfactant production by *Bacillus subtilis* B30 and its application in enhancing oil recovery. Colloids Surf B Biointerfaces 114:324–333. https://doi.org/10.1016/j.colsurfb.2013.09.022

Ameen A, Raza S (2017) Metaproteomics approaches and techniques: a review. Int J Adv Sci Res Manag 3(5):49–51. https://doi.org/10.7439/ijasr.v3i5.4167

Amorim C, Silvério SC, Silva SP, Coelho E, Coimbra MA, Prather KLJ, Rodrigues LR (2018) Single-step production of arabino-xylooligosaccharides by recombinant *Bacillus subtilis* 3610 cultivated in brewers' spent grain. Carbohydr Polym 199:546–554. https://doi.org/10.1016/j.carbpol.2018.07.017

Amorim C, Silvério SC, Gonçales RFS, Pinheiro AC, Silva SP, Coelho E, Coimbra MA, Prather KLJ, Rodrigues LR (2019a) Downscale fermentation for xylooligosaccharides production by recombinant *Bacillus subtilis* 3610. Carbohydr Polym 205:176–183. https://doi.org/10.1016/j.carbpol.2018.09.088

Amorim C, Silvério SC, Rodrigues LR (2019b) One-step process for producing prebiotic arabinoxylooligosaccharides from brewer's spent grain employing *Trichoderma* species. Food Chem 270:86–94. https://doi.org/10.1016/j.foodchem.2018.07.080

Anandharamakrishnan C, Raghavendra SN, Barhate RS, Hanumesh U, Raghavarao KSMS (2005) Aqueous two-phase extraction for recovery of proteins from cheese whey. Food Bioprod Process 83:191–197. https://doi.org/10.1205/fbp.03403

Andri P, Meyer AS, Jensen PA, Dam-johansen K (2010) Reactor design for minimizing product inhibition during enzymatic lignocellulose hydrolysis: I Significance and mechanism of cellobiose and glucose inhibition on cellulolytic enzymes. Biotechnol Adv 28:308–324. https://doi.org/10.1016/j.biotechadv.2010.01.003

Antunes S, Freitas F, Alves VD, Grandfils C, Reis MAM (2015) Conversion of cheese whey into a fucose- and glucuronic acid-rich extracellular polysaccharide by *Enterobacter* A47. J Biotechnol 210:1–7. https://doi.org/10.1016/j.jbiotec.2015.05.013

Antunes S, Freitas F, Sevrin C, Grandfils C, Reis MAM (2017) Production of FucoPol by *Enterobacter* A47 using waste tomato paste by-product as sole carbon source. Bioresour Technol 227:66–73. https://doi.org/10.1016/j.biortech.2016.12.018

Aparna A, Srinikethan G, Smith H (2012) Production and characterization of biosurfactant produced by a novel *Pseudomonas* sp. 2B. Colloids Surf B Biointerfaces 95:23–29. https://doi.org/10.1016/j.colsurfb.2012.01.043

Aragon CC, Santos AF, Ruiz-Matute AI, Corzo N, Guisan JM, Monti R, Mateo C (2013) Continuous production of xylooligosaccharides in a packed bed reactor with immobilized–stabilized biocatalysts of xylanase from *Aspergillus versicolor*. J Mol Catal B: Enzym 98:8–14. https://doi.org/10.1016/j.molcatb.2013.09.017

Ashok A, Doriya K, Ram D, Rao M, Kumar DS (2018) Design of solid state bioreactor for industrial applications: an overview to conventional bioreactors. Biocatal Agric Biotechnol 9:11–18. https://doi.org/10.1016/j.bcab.2016.10.014

Ashwini A, Ramya HN, Ramkumar C, Reddy KR, Kulkarni RV, Abinaya V, Naveen S, Raghu AV (2019) Reactive mechanism and the applications of bioactive prebiotics for human health. review. J Microbiol Methods 159:128–137. https://doi.org/10.1016/j.mimet.2019.02.019

Assis DJ, Brandão LV, Costa LAS, Figueiredo TVB, Sousa LS, Padilha FF, Druzian JI (2014) A study of the effects of aeration and agitation on the properties and production of xanthan gum from crude glycerin derived from biodiesel using the response surface methodology. Appl Biochem Biotechnol 172:2769–2785. https://doi.org/10.1007/s12010-014-0723-7

Ávila SNS, Gutarra MLE, Fernandez-Lafuente R, Cavalcanti EDC, Freire DMG (2018) Multipurpose fixed-bed bioreactor to simplify lipase production by solid-state fermentation and application in biocatalysis. Biochem Eng J 144:1–7. https://doi.org/10.1016/j.bej.2018.12.024

Babu IS, Ramappa S, Mahesh DG, Kumari KS, Kumari KS, Subba G (2008) Optimization of medium constituents for the production of fructosyltransferase (ftase) by *Bacillus subtilis* using response surface methodology. Res J Microbiol 3:114–121. https://doi.org/10.3923/jm.2008.114.121

Bach E, Anna S', Daroit DJ, Corrêa APF, Segalin J, Brandelli A (2012) Production, one-step purification, and characterization of a keratinolytic protease from *Serratia marcescens* P3. Process Biochem 47:2455–2462. https://doi.org/10.1016/j.procbio.2012.10.007

Badhwar P, Kumar P, Dubey KK (2019) Extractive fermentation for process integration and amplified pullulan production by *A. pullulans* in aqueous two-phase systems. Sci Rep 9:32. https://doi.org/10.1038/s41598-018-37314-y

Balderas-Hernández VE, Correia K, Mahadevan R (2018) Inactivation of the transcription factor mig1 (YGL035C) in *Saccharomyces cerevisiae* improves tolerance towards monocarboxylic weak acids: acetic, formic and levulinic acid. J Ind Microbiol Biotechnol 45:735–751. https://doi.org/10.1007/s10295-018-2053-1

Banik RM, Santhiagu A, Upadhayay SN (2007) Optimization of nutrients for gellan gum production by *Sphingomonas paucimobilis* ATCC-31461 in molasses based medium using response surface methodology. Bioresour Technol 98:792–797. https://doi.org/10.1016/j.biortech.2006.03.012

Barbosa MDFS, Silva DO, Pinheiro AJR, Guimarães WV, Borges AC (1985) Production of beta-d-galactosidase from *Kluyveromyces fragilis* grown in cheese whey. J Dairy Sci 68:1618–1623. https://doi.org/10.3168/jds.s0022-0302(85)81004-3

Barbosa R, Pereira MS, Jahn L (2019) Screening of *Aspergillus*, *Bacillus* and *Trichoderma* strains and influence of substrates on auxin and phytases production through solid-state fermentation. Biocatal Agric Biotechnol 19:101165. https://doi.org/10.1016/j.bcab.2019.101165

Batista-García RA, Sánchez-Carbente MR, Talia P, Jackson SA, O'Leary ND, Dobson ADW et al (2016) From lignocellulosic metagenomes to lignocellulolytic genes: trends, challenges and future prospects. Biofuels Bioprod Biorefin 10(6):864–882. https://doi.org/10.1002/bbb.1709

Béjà O, Aravind L, Koonin EV, Suzuki MT, Hadd A, Nguyen LP et al (2000) Bacterial rhodopsin: evidence for a new type of phototrophy in the sea. Science 289(5486):1902–1906. https://doi.org/10.1126/science.289.5486.1902

Belo I, Pinheiro R, Mota M (2003) Fed-batch cultivation of *Saccharomyces cerevisiae* in a hyperbaric bioreactor. Biotechnol Prog 19:665–671. https://doi.org/10.1021/bp0257067

Benincasa M, Accorsini FR (2008) *Pseudomonas aeruginosa* LBI production as an integrated process using the wastes from sunflower-oil refining as a substrate. Bioresour Technol 99:3843–3849. https://doi.org/10.1016/j.biortech.2007.06.048

Berini F, Casciello C, Marcone GL, Marinelli F (2017) Metagenomics: novel enzymes from non-culturable microbes. FEMS Microbiol Lett 364(21). https://doi.org/10.1093/femsle/fnx211

Bi P-Y, Li D-Q, Dong L-R (2009) A novel technique for the separation and concentration of penicillin G from fermentation broth: aqueous two-phase flotation. Sep Purif Technol 69:205–209. https://doi.org/10.1016/j.seppur.2009.07.019

Bi P-Y, Dong H-R, Yuan Y-C (2010) Application of aqueous two-phase flotation in the separation and concentration of puerarin from *Puerariae* extract. Sep Purif Technol 75:402–406. https://doi.org/10.1016/j.seppur.2010.09.010

Bi P-Y, Chang L, Mu Y-L, Liu J-Y, Wu Y, Geng X, Wei Y (2013) Separation and concentration of baicalin from *Scutellaria Baicalensis Georgi* extract by aqueous two-phase flotation. Sep Purif Technol 116:454–457. https://doi.org/10.1016/j.seppur.2013.06.024

Bian J, Peng P, Peng F, Xiao X, Xu F, Sun RC (2014) Microwave-assisted acid hydrolysis to produce xylooligosaccharides from sugarcane bagasse hemicelluloses. Food Chem 156:7–13. https://doi.org/10.1016/j.foodchem.2014.01.112

Bockmühl D, Jassoy C, Nieveler S, Scholtysse R, Wadl A, Waldmann-Laue M (2007) Prebiotic cosmetics: an alternative to antibacterial products. Int J Cosmet Sci 29:63–64. https://doi.org/10.1111/j.1467-2494.2007.00355_2.x

Boland MJ, Hesselink PGM, Papamichael N, Hustedt H (1991) Extractive purification of enzymes from animal tissue using two phase systems: pilot scale studies. J Biotechnol 19:19–34. https://doi.org/10.1016/0168-1656(91)90072-4

Bommarius AS, Blum JK, Abrahamson MJ (2011) Status of protein engineering for biocatalysts: how to design an industrially useful biocatalyst. Curr Opin Chem Biol 15:194–200. https://doi.org/10.1016/j.cbpa.2010.11.011

Bowen CH, Dai B, Sargent CJ, Bai W, Ladiwala P, Feng H, Huang W, Kaplan DL, Galazka JM, Zhang F (2018) Recombinant spidroins fully replicate primary mechanical properties of natural spider silk. Biomacromolecules 19:3853–3860. https://doi.org/10.1021/acs.biomac.8b00980

Bragg L, Tyson GW (2014) Metagenomics using next-generation sequencing. Methods Mol Biol 1096:183–201. https://doi.org/10.1007/978-1-62703-712-9_15

Caballero A, Ramos JL (2017) Enhancing ethanol yields through D-xylose and L-arabinose co-fermentation after construction of a novel high efficient L-arabinose-fermenting *Saccharomyces cerevisiae* strain. Microbiology 163:442–452. https://doi.org/10.1099/mic.0.000437

Caramihai M, Severin I (2013) Biomass now—sustainable growth and use. In: IntechOpen (ed) Bioprocess modeling and control

Cardoso B, Silvério SC, Abrunhosa L, Teixeira JA, Rodrigues LR (2017) β-galactosidase from *Aspergillus lacticoffeatus*: a promising biocatalyst for the synthesis of novel prebiotics. Int J Food Microbiol 257:67–74. https://doi.org/10.1016/j.ijfoodmicro.2017.06.013

Carvalho AFA, Neto PO, Silva DF, Pastore GM (2013) Xylo-oligosaccharides from lignocellulosic materials: chemical structure, health benefits and production by chemical and enzymatic hydrolysis. Food Res Int 51:75–85. https://doi.org/10.1016/j.foodres.2012.11.021

Castilla IA, Woods DF, Reen FJ, O'Gara F (2018) Harnessing marine biocatalytic reservoirs for green chemistry applications through metagenomic technologies. Mar Drugs 16(7). https://doi.org/10.3390/md16070227

Castro AM, Castilho LR, Freire DMG (2015) Performance of a fixed-bed solid-state fermentation bioreactor with forced aeration for the production of hydrolases by *Aspergillus awamori*. Biochem Eng J 93:303–308. https://doi.org/10.1016/j.bej.2014.10.016

Cavalcanti EDAC, Gutarra MLE, Freire DMG, Castilho LDR, Sant'Anna Júnior GL (2005) Lipase production by solid-state fermentation in fixed-bed bioreactors. Brazilian Arch Biol Technol 48:79–84. https://doi.org/10.1590/S1516-89132005000400010

Cerda A, Gea T, Vargas-garcía MC, Sánchez A (2017) Towards a competitive solid state fermentation: cellulases production from coffee husk by sequential batch operation and role of microbial diversity. Sci Total Environ 589:56–65. https://doi.org/10.1016/j.scitotenv.2017.02.184

Chae TU, Choi SY, Ryu JY, Lee SY (2018) Production of ethylene glycol from xylose by metabolically engineered *Escherichia coli*. AIChE J 64:4193–4200. https://doi.org/10.1002/aic.16339

Chapla D, Pandit P, Shah A (2012) Production of xylooligosaccharides from corncob xylan by fungal xylanase and their utilization by probiotics. Bioresour Technol 115:215–221. https://doi.org/10.1016/j.biortech.2011.10.083

Cheirsilp B, Kitcha S (2015) Solid state fermentation by cellulolytic oleaginous fungi for direct conversion of lignocellulosic biomass into lipids: fed-batch and repeated-batch fermentations. Ind Crop Prod 66:73–80. https://doi.org/10.1016/j.indcrop.2014.12.035

Chen K, Arnold FH (1993) Tuning the activity of an enzyme for unusual environments: Sequential random mutagenesis of subtilisin E for catalysis in dimethylformamide. Proc Natl Acad Sci U S A 90:5618–5622. https://doi.org/10.1073/pnas.90.12.5618

Chen JYH-P, Wang C-H (1991) Lactose hydrolysis by β-galactosidase in aqueous two-phase systems. J Ferment Bioeng 71:168–175. https://doi.org/10.1016/0922-338X(91)90104-O

Chen C, Sun N, Li D, Long S, Tang X, Xiao G, Wang L (2018) Optimization and characterization of biosurfactant production from kitchen waste oil using *Pseudomonas aeruginosa*. Environ Sci Pollut Res Int 25:14934–14943. https://doi.org/10.1007/s11356-018-1691-1

Chen C, Lin J, Wang W, Huang H, Li S (2019a) Cost-effective production of surfactin from xylose-rich corncob hydrolysate using *Bacillus subtilis* BS-37. Waste Biomass Valor 10:341–347. https://doi.org/10.1007/s12649-017-0052-5

Chen M, Lu J, Cheng Y, Li Q, Wang H (2019b) Novel process for the coproduction of xylo-oligosaccharide and glucose from reed scraps of reed pulp mill. Carbohydr Polym 215:82–89. https://doi.org/10.1016/j.carbpol.2019.03.068

Chew PL, Annuar MSM, Show PL, Ling TC (2015) Extractive bioconversion of poly-ε-caprolactone by *Burkholderia cepacia* lipase in aqueous two-phase system. Biochem Eng J 101:9–17. https://doi.org/10.1016/j.bej.2015.04.015

Chisti MY, Moo-Young M (1987) Airlift reactors: characteristics, applications and design considerations. Chem Eng Commun 60:195–242. https://doi.org/10.1080/00986448708912017

Cláudio AFM, Marques CFC, Boal-Palheiros I, Freire MG, Coutinho JAP (2014) Development of back-extraction and recyclability routes for ionic-liquid-bases aqueous two-phase systems. Green Chem 16:259–268. https://doi.org/10.1039/c3gc41999a

Cordisco E, Haidar CN, Coscueta ER, Nerli BB, Malpiedi LP (2016) Integrated extraction and purification of soy isoflavones by using aqueous micellar systems. Food Chem 213:514–520. https://doi.org/10.1016/j.foodchem.2016.07.001

Courtin CM, Swennen K, Verjans P, Delcour JA (2009) Heat and pH stability of prebiotic arabinoxylooligosaccharides, xylooligosaccharides and fructooligosaccharides. Food Chem 112:831–837. https://doi.org/10.1016/j.foodchem.2008.06.039

Couto MR, Rodrigues JL, Rodrigues LR (2017) Optimization of fermentation conditions for the production of curcumin by engineered *Escherichia coli*. J R Soc Interface 14:1–8. https://doi.org/10.1098/rsif.2017.0470

Couto MR, Gudiña EJ, Ferreira D, Teixeira JA, Rodrigues LR (2019) The biopolymer produced by *Rhizobium viscosum* CECT 908 is a promising agent for application in microbial enhanced oil recovery. N Biotechnol 49:144–150. https://doi.org/10.1016/j.nbt.2018.11.002

Cruz JM, Hughes C, Quilty B, Montagnolli RN, Bidoia ED (2018) Agricultural feedstock supplemented with manganese for biosurfactant production by *Bacillus subtilis*. Waste Biomass Valor 9:613–618. https://doi.org/10.1007/s12649-017-0019-6

Cunha JT, Soares PO, Romaní A, Thevelein JM, Domingues L (2019) Xylose fermentation efficiency of industrial *Saccharomyces cerevisiae* yeast with separate or combined xylose reductase/xylitol dehydrogenase and xylose isomerase pathways. Biotechnol Biofuels 12:1–14. https://doi.org/10.1186/s13068-019-1360-8

Dahiya S, Kumar AN, Sravan JS, Chatterjee S, Sarkar O, Mohan SV (2018) Food waste biorefinery: sustainable strategy for circular bioeconomy. Bioresour Technol 248:2–12. https://doi.org/10.1016/j.biortech.2017.07.176

Darabzadeh N, Hamidi-Esfahani Z, Hejazi P (2019) Optimization of cellulase production under solid-state fermentation by a new mutant strain of *Trichoderma reesei*. Food Sci Nutr 7:572–578. https://doi.org/10.1002/fsn3.852

Das K, Mukherjee AK (2007) Comparison of lipopeptide biosurfactants production by *Bacillus subtilis* strains in submerged and solid state fermentation systems using a cheap carbon source: some industrial applications of biosurfactants. Process Biochem 42:1191–1199. https://doi.org/10.1016/j.procbio.2007.05.011

De Wit M, Faaij A (2010) European biomass resource potential and costs. Biomass Bioenergy 34(2):188–202. https://doi.org/10.1016/j.biombioe.2009.07.011

DeCastro ME, Rodriguez-Belmonte E, Gonzalez-Siso MI (2016) Metagenomics of thermophiles with a focus on discovery of novel thermozymes. Front Microbiol 7(1521). https://doi.org/10.3389/fmicb.2016.01521

Delmont TO, Robe P, Cecillon S, Clark IM, Constancias F, Simonet P et al (2011) Accessing the soil metagenome for studies of microbial diversity. Appl Environ Microbiol 77(4):1315–1324. https://doi.org/10.1128/AEM.01526-10

DeLong EF (2005) Microbial community genomics in the ocean. Nat Rev Microbiol 3(6):459–469. https://doi.org/10.1038/nrmicro1158

Denard CA, Ren H, Zhao H (2015) Improving and repurposing biocatalysts via directed evolution. Curr Opin Chem Biol 25:55–64. https://doi.org/10.1016/j.cbpa.2014.12.036

Devesa-Rey R, Vecino X, Varela-Alende JL, Barral MT, Cruz JM, Moldes AB (2011) Valorization of winery wastes vs. the costs of not recycling. Waste Manag 31(11):2327–2335. https://doi.org/10.1016/j.wasman.2011.06.001

Dhillon GS, Oberoi HS, Kaur S, Bansal S, Brar SK (2011) Value-addition of agricultural wastes for augmented cellulase and xylanase production through solid-state tray fermentation employing mixed-culture of fungi. Ind Crop Prod 34:1160–1167. https://doi.org/10.1016/j.indcrop.2011.04.001

Doudna JA, Charpentier E (2014) The new frontier of genome engineering with CRISPR-Cas9. Science 346(6213):1258096. https://doi.org/10.1126/science.1258096

Duarte C, Gudiña EJ, Lima CF, Rodrigues LR (2014) Effects of biosurfactants on the viability and proliferation of human breast cancer cells. AMB Express 4:40. https://doi.org/10.1186/s13568-014-0040-0

Durand A (2003) Bioreactor designs for solid state fermentation. Biochem Eng J 13:113–125. https://doi.org/10.1016/S1369-703X(02)00124-9

Dursun D, Dalgiç AC (2016) Optimization of astaxanthin pigment bioprocessing by four different yeast species using wheat wastes. Biocatal Agric Biotechnol 7:1–6. https://doi.org/10.1016/j.bcab.2016.04.006

Eiteman MA, Lee SA, Altman E (2008) A co-fermentation strategy to consume sugar mixtures effectively. J Biol Eng 2:1–8. https://doi.org/10.1186/1754-1611-2-3

Elbersen B, Startisky I, Hengeveld G, Schelhaas MJ, Naeff H (2012) Atlas of EU biomass potentials—deliverable 3.3: spatially detailed and quantified overview of EU biomass potential taking into account the main criteria determining biomass availability from different sources, IEE 08 653 SI2. 529 241.

Elegbede JA, Lateef A (2017) Valorization of corn-cob by fungal isolates for production of xylanase in submerged and solid state fermentation media and potential biotechnological applications. Waste and Biomass Valorization 9:1273–1287. https://doi.org/10.1007/s12649-017-9932-y

El-Housseiny GS, Aboshanab KM, Aboulwafa MM, Hassouna NA (2016) Optimization of rhamnolipid production by *Pseudomonas aeruginosa* isolate P6. J Surfactant Deterg 19:943–955. https://doi.org/10.1007/s11743-016-1845-4

El-Shishtawy RM, Mohamed SA, Asiri AM, Gomaa AM, Ibrahim IH (2014) Solid fermentation of wheat bran for hydrolytic enzymes production and saccharification content by a local isolate *Bacillus megatherium*. BMC Biotechnol 14:1–8. https://doi.org/10.1186/1472-6750-14-29

de Faria AF, Teodoro-Martínez DS, de Oliveira Barbosa GN, Vaz BG, Silva IS, García JS, Tótola MR, Eberlin MN, Grossman M, Alves OL, Durrant LR (2011) Production and structural characterization of surfactin (C14/Leu7) produced by *Bacillus subtilis* isolate LSFM-05 grown on raw glycerol from the biodiesel industry. Process Biochem 46:1951–1957. https://doi.org/10.1016/j.procbio.2011.07.001

Farinas CS (2015) Developments in solid-state fermentation for the production of biomass-degrading enzymes for the bioenergy sector. Renew Sustain Energy Rev 52:179–188. https://doi.org/10.1016/j.rser.2015.07.092

Faryar R, Linares-Pastén JA, Immerzeel P, Mamo G, Andersson M, Stålbrand H, Karlsson EN (2015) Production of prebiotic xylooligosaccharides from alkaline extracted wheat straw using the K80R-variant of a thermostable alkali-tolerant xylanase. Food Bioprod Process 93:1–10. https://doi.org/10.1016/j.fbp.2014.11.004

Fernandes MLP, Jorge JA, Henrique LHS (2017) Characterization of an extracellular β-D-fructofuranosidase produced by *Aspergillus niveus* during solid-state fermentation (SSF) of cassava husk. J Food Biochem 42:1–10. https://doi.org/10.1111/jfbc.12443

Ferreira A, Rocha F, Mota A, Teixeira JA (2017) Characterization of industrial bioreactors: mixing, heat, and mass transfer. In: Larroche C, Sanroman M, Du G, Pandey A (eds) Current developments in biotechnology and bioengineering: bioprocesses, bioreactors and controls, Elsevier B.V., pp 563–592

Flores AD, Ayla EZ, Nielsen DR, Wang X (2019) Engineering a synthetic, catabolically-orthogonal co-culture cystem for enhanced conversion of lignocellulose-derived sugars to ethanol. ACS Synth Biol 8:1089–1099. https://doi.org/10.1021/acssynbio.9b00007

de França IWL, Lima AP, Lemos JAM, Lemos CGF, Melo VMM, Santana HB, Gonçalves LRB (2015) Production of a biosurfactant by *Bacillus subtilis* ICA56 aiming bioremediation of impacted soils. Catal Today 255:10–15. https://doi.org/10.1016/j.cattod.2015.01.046

Gadhe A, Mudliar S, Pandey R, Elumalai S, Satpute D (2011) Statistical optimization of process parameters for the production of vanillic acid by solid-state fermentation of groundnut shell waste using response surface methodology. J Chem Technol Biotechnol 86:1535–1541. https://doi.org/10.1002/jctb.2671

Ganguly S, Nandi S (2015) Process optimization of lipase catalyzed synthesis of diesters in a packed bed reactor. Biochem Eng J:1–4. https://doi.org/10.1016/j.bej.2015.03.020

Gaurav, SG (2017) Prebiotics: ingredients, applications and global markets BCC research. https://www.bccresearch.com/market-research/food-and-beverage/prebiotics-ingredients-applications-and-global-markets.html. Accessed 21 Apr 2019.

George S, Jayachandran K (2009) Analysis of rhamnolipid biosurfactants produced through submerged fermentation using orange fruit peelings as sole carbon source. Appl Biochem Biotechnol 158:694–705. https://doi.org/10.1007/s12010-008-8337-6

George S, Jayachandran K (2013) Production and characterization of rhamnolipid biosurfactant from waste frying coconut oil using a novel *Pseudomonas aeruginosa* D. J Appl Microbiol 114:373–383. https://doi.org/10.1111/jam.12069

Gibson GR, Roberfroid MB (1995) Dietary modulation of the human colonic microbiota:introducing the concept of prebiotics. J Nutr 125:1401–1412. https://doi.org/10.1093/jn/125.6.1401

Gibson GR, Hutkins R, Sanders ME, Prescott SL, Reimer RA, Salminen SJ, Scott K, Stanton C, Swanson KS, Cani PD, Verbeke K, Reid G (2017) The International Scientific Association for Probiotics and Prebiotics (ISAPP) consensus statement on the definition and scope of prebiotics. Nat Rev Gastroenterol Hepatol 14:491–502. https://doi.org/10.1038/nrgastro.2017.75

Global Market Insights, Inc. (2017) Prebiotics market size by ingredient (inulin, GOS, FOS, MOS), by application (animal feed, food & beverages [dairy, cereals, baked goods, fermented meat, dry foods], dietary supplements [food, nutrition, infant formulations]), industry analysis report, regional outlook, application potential, price trends, competitive market share & forecast, 2017–2024. https://www.gminsights.com/industry-analysis/prebiotics-market. Accessed 23 June 2019

Göksungur Y, Uzunogullari P, Dagbagli S (2011) Optimization of pullulan production from hydrolysed potato starch waste by response surface methodology. Carbohydr Polym 83:1330–1337. https://doi.org/10.1016/j.carbpol.2010.09.047

Goodwin S, McPherson JD, McCombie WR (2016) Coming of age: ten years of next-generation sequencing technologies. Nat Rev Genet 17(6):333–351. https://doi.org/10.1038/nrg.2016.49

Görke B, Stülke J (2008) Carbon catabolite repression in bacteria: many ways to make the most out of nutrients. Nat Rev Microbiol 6:613–624. https://doi.org/10.1038/nrmicro1932

Graedel TE (2002) Green chemistry and sustainable development. In: Clark J, Macquarrie DJ (eds) Handbook of green chemistry and technology. Blackwell Science Ltd., Oxford, pp 56–61

Grand Market Insights Research (2018) Biosurfactants market worth over $2.7 bn by 2024. https://www.gminsights.com/pressrelease/biosurfactants-market-size. Accessed 8 June 2019.

Gudiña EJ, Rangarajan V, Sen R, Rodrigues LR (2013) Potential therapeutic applications of biosurfactants. Trends Pharmacol Sci 34:667–675. https://doi.org/10.1016/j.tips.2013.10.002

Gudiña EJ, Rodrigues AI, Alves E, Domingues MR, Teixeira JA, Rodrigues LR (2015a) Bioconversion of agro-industrial by-products in rhamnolipids toward applications in enhanced oil recovery and bioremediation. Bioresour Technol 177:87–93. https://doi.org/10.1016/j.biortech.2014.11.069

Gudiña EJ, Fernandes EC, Rodrigues AI, Teixeira JA, Rodrigues LR (2015b) Biosurfactant production by *Bacillus subtilis* using corn step liquor as culture medium. Front Microbiol 6:59. https://doi.org/10.3389/fmicb.2015.00059

Gudiña EJ, Rodrigues AI, de Freitas V, Azevedo Z, Teixeira JA, Rodrigues LR (2016) Valorization of agro-industrial wastes towards the production of rhamnolipids. Bioresour Technol 212:144–150. https://doi.org/10.1016/j.biortech.2016.04.027

Gullon P, Romani A, Vila C, Garrote G, Parajo JC (2012) Potential of hydrothermal treatments in lignocellulose biorefineries. Biofuels Bioprod Biorefin 6(2):219–232. https://doi.org/10.1002/bbb.339

Gunasekar V, Reshma KR, Treesa G, Gowdhaman D, Ponnusami V (2014) Xanthan from sulphuric acid treated tapioca pulp: influence of acid concentration on xanthan fermentation. Carbohydr Polym 102:669–673. https://doi.org/10.1016/j.carbpol.2013.11.006

Guneser O, Demirkol A, Karagul Y, Ozmen S, Isleten M, Elibol M (2017) Biotechnology and industrial microbiology production of flavor compounds from olive mill waste by *Rhizopus oryzae* and *Candida tropicalis*. Braz J Microbiol 48:275–285. https://doi.org/10.1016/j.bjm.2016.08.003

Gurpilhares DB, Pessoa A, Roberto IC (2015) Process integration for the disruption of *Candida guilliermondii* cultivated in rice straw hydrolysate and recovery of glucose-6-phosphate dehydrogenase by aqueous two-phase systems. Appl Biochem Biotechnol 176:1596–1612. https://doi.org/10.1007/s12010-015-1664-5

Haba E, Pinazo A, Jauregui O, Espuny MJ, Infante MR, Manresa A (2003) Physicochemical characterization and antimicrobial properties of rhamnolipids produced by *Pseudomonas aeruginosa* 47T2 NCBIM 40044. Biotechnol Bioeng 81:316–322. https://doi.org/10.1002/bit.10474

Hahn-Hägerdal B, Karhumaa K, Jeppsson M, Gorwa-Grauslund MF (2007) Metabolic engineering for pentose utilization in *Saccharomyces cerevisiae*. In: Olsson L (ed) Biofuels. Springer, New York, pp 147–177

Halperin SO, Tou CJ, Wong EB, Modavi C, Schaffer DV, Dueber JE (2018) CRISPR-guided DNA polymerases enable diversification of all nucleotides in a tunable window. Nature 560:248–252. https://doi.org/10.1038/s41586-018-0384-8

Hamrouni R, Molinet J, Miché L, Carboué Q, Dupuy N, Masmoudi A, Roussos S (2019) Production of coconut aroma in solid-state cultivation: screening and identification of *Trichoderma* strains for 6-pentyl-alpha-pyrone and conidia production. J Chem. https://doi.org/10.1155/2019/8562384

Handelsman J, Rondon MR, Brady SF, Clardy J, Goodman RM (1998) Molecular biological access to the chemistry of unknown soil microbes: a new frontier for natural products. Chem Biol 5(10):245–249

Hector RE, Mertens JA (2017) A synthetic hybrid promoter for xylose-regulated control of gene expression in *Saccharomyces* yeasts. Mol Biotechnol 59:24–33. https://doi.org/10.1007/s12033-016-9991-5

Hu Y, Du C, Leu S, Jing H, Li X, Sze C, Lin K (2018) Valorisation of textile waste by fungal solid state fermentation: An example of circular waste-based biorefinery. Resour Conserv Recycl 129:27–35. https://doi.org/10.1016/j.resconrec.2017.09.024

Hutkins RW, Krumbeck JA, Bindels LB, Cani PD, Fahey G, Goh YJ, Sanders ME (2016) Prebiotics: why definitions matter. Curr Opin Biotechnol 37:1–7. https://doi.org/10.1016/j.copbio.2015.09.001

Illanes A, Cauerhff A, Wilson L, Castro GR (2012) Recent trends in biocatalysis engineering. Bioresour Technol 115:48–57. https://doi.org/10.1016/j.biortech.2011.12.050

Iqbal M, Tao Y, Xie S, Zhu Y, Chen D, Wang X, Huang L, Peng D, Sattar A, Shabbir MAB, Hussain HI, Ahmed S, Yuan Z (2016) Aqueous two-phase system (ATPS): an overview and advances in its applications. Biol Proced Online 18:18. https://doi.org/10.1186/s12575-016-0048-8

Jiang B, Na J, Wang L, Li D, Liu C, Feng Z (2019) Eco-innovation in reusing food by-products: separation of ovalbumin from salted egg white aqueous two-phase system of PEG 1000/$(NH_4)_2SO_4$. Polymers 11:238. https://doi.org/10.3390/polym11020238

Jiao JY, Wang HX, Zeng Y, Shen YM (2006) Enrichment for microbes living in association with plant tissues. J Appl Microbiol 100(4):830–837. https://doi.org/10.1111/j.1365-2672.2006.02830.x

Jiménez-Quero A, Pollet E, Zhao M, Marchioni E, Averous L, Phalip V (2017) Fungal fermentation of lignocellulosic biomass for itaconic and fumaric acid production. J Microbiol Biotechnol 27:1–8. https://doi.org/10.4014/jmb.1607.07057

Joshi S, Bharucha C, Jha S, Yadav S, Nerurkar A, Desai AJ (2008) Biosurfactant production using molasses and whey under thermophilic conditions. Bioresour Technol 99:195–199. https://doi.org/10.1016/j.biortech.2006.12.010

Jullesson D, David F, Pfleger B, Nielsen J (2015) Impact of synthetic biology and metabolic engineering on industrial production of fine chemicals. Biotechnol Adv 33:1395–1402. https://doi.org/10.1016/j.biotechadv.2015.02.011

Jung I-Y, Lee J-W, Min W-K, Park Y-C, Seo J-H (2015) Simultaneous conversion of glucose and xylose to 3-hydroxypropionic acid in engineered *Escherichia coli* by modulation of sugar transport and glycerol synthesis. Bioresour Technol 198:709–716. https://doi.org/10.1016/j.biortech.2015.09.079

Kalaivani S, Regupathi I (2013) Partitioning studies of α-lactalbumin in environmental friendly poly(ethylene glycol)-citrate salt aqueous two phase systems. Bioprocess Biosyst Eng 36:1475–1483. https://doi.org/10.1007/s00449-013-0910-x

Kalogiannis S, Iakovidou G, Liakopoulou-Kyrikiades M, Kyriakidis A, Skaracis GN (2003) Optimization of xanthan gum production by *Xanthomonas campestris* grown in molasses. Process Biochem 39:249–256. https://doi.org/10.1016/S0032-9592(03)00067-0

Kaur P, Ghoshal G, Jain A (2019) Bio-utilization of fruits and vegetables waste to produce β-carotene in solid-state fermentation: characterization and antioxidant activity. Process Biochem 76:155–164. https://doi.org/10.1016/j.procbio.2018.10.007

Khanahmadi M, Roostaazad R, Safekordi A, Bozorgmehri R, Mitchell DA (2004) Investigating the use of cooling surfaces in solid-state fermentation tray bioreactors: modelling and experimentation. J Chem Technol Biotechnol 79:1228–1242. https://doi.org/10.1002/jctb.1117

Khanahmadi M, Roostaazad R, Mitchell DA, Miranzadeh M, Bozorgmehri R, Safekordi A (2006) Bed moisture estimation by monitoring of air stream temperature rise in packed-bed solid-state fermentation. Chem Eng Sci 61:5654–5663. https://doi.org/10.1016/j.ces.2006.04.039

Khanahmadi M, Arezi I, Amiri M, Miranzadeh M (2018) Bioprocessing of agro-industrial residues for optimization of xylanase production by solid- state fermentation in flask and tray bioreactor. Biocatal Agric Biotechnol 13:272–282. https://doi.org/10.1016/j.bcab.2018.01.005

Kim J-H, Block DE, Mills DA (2010) Simultaneous consumption of pentose and hexose sugars: an optimal microbial phenotype for efficient fermentation of lignocellulosic biomass. Appl Microbiol Biotechnol 88:1077–1085. https://doi.org/10.1007/s00253-010-2839-1

Kim HJ, Jeong H, Lee SJ (2018) Synthetic biology for microbial heavy metal biosensors. Anal Bioanal Chem 410:1191–1203. https://doi.org/10.1007/s00216-017-0751-6

Ko KC, Lee JH, Han Y, Choi JH, Song JJ (2013) A novel multifunctional cellulolytic enzyme screened from metagenomics resources representing ruminal bacteria. Biochem Biophys Res Commun 441:567–572. https://doi.org/10.1016/j.bbrc.2013.10.120

Kumar V, Satyanarayana T (2011) Applicability of thermo-alkali stable and cellulose free xylanase from a novel thermo-halo-alkaliphilic *Bacillus haloduransin* producing xyloligosaccharides. Biotechnol Lett 33:2279–2285. https://doi.org/10.1007/s10529-011-0698-1

Kumar V, Satyanarayana T (2015) Generation of xylooligosaccharides from microwave irradiated agroresidues using recombinant thermo-alkali-stable endoxylanase of the polyextremophilic bacterium *Bacillus halodurans* expressed in *Pichia pastoris*. Bioresour Technol 179:382–389. https://doi.org/10.1016/j.biortech.2014.12.049

Kumar D, Jain VK, Shanker G, Sri A (2003) Utilisation of fruits waste for citric acid production by solid state fermentation. Process Biochem 38:1725–1729. https://doi.org/10.1016/S0032-9592(02)00253-4

Kumar B, Bhardwaj N, Alam A, Agrawal K, Prasad H, Verma P (2018) Production, purification and characterization of an acid/alkali and thermo tolerant cellulase from *Schizophyllum commune* NAIMCC-F-03379 and its application in hydrolysis of lignocellulosic wastes. AMB Expr 8:173. https://doi.org/10.1186/s13568-018-0696-y

Kung SH, Lund S, Murarka A, McPhee D, Paddon CJ (2018) Approaches and recent developments for the commercial production of semi-synthetic artemisinin. Front Plant Sci 9:1–7. https://doi.org/10.3389/fpls.2018.00087

van de Lagemaat J, Pyle DL (2004) Solid-state fermentation: a continuous process for fungal tannase production. Biotechnol Bioeng 87:924–929. https://doi.org/10.1002/bit.20206

Lan G, Fan Q, Liu Y, Chen C, Li G, Liu Y, Yin X (2015) Rhamnolipid production from waste cooking oil using *Pseudomonas* SWP-4. Biochem Eng J 101:44–54. https://doi.org/10.1016/j.bej.2015.05.001

Lawford HG, Rousseau JD (2002) Performance testing of *Zymomonas mobilis* metabolically engineered for cofermentation of glucose, xylose, and arabinose. In: Finkelstein M, McMillan JD, Davison BH (eds) Biotechnology for fuels and chemicals. Humana Press, New York, pp 429–448

Lee SK, Chou H, Ham TS, Lee TS, Keasling JD (2008) Metabolic engineering of microorganisms for biofuels production: from bugs to synthetic biology to fuels. Curr Opin Biotechnol 19:556–563. https://doi.org/10.1016/j.copbio.2008.10.014

Lee SY, Khoiroh I, Ooi CW, Ling TC, Show PL (2017) Recent advances in protein extraction using ionic liquid-based aqueous two-phase systems. Sep Purif Rev 46:291–304. https://doi.org/10.1080/15422119.2017.1279628

Lemes CL, Silvério SC, Rodrigues S, Rodrigues LR (2019) Integrated strategy for purification of esterase from Aureobasidium pullulans. Sep Purif Technol 209:409–418. https://doi.org/10.1016/j.seppur.2018.07.062

Leong YK, Lan JC-W, Loh H-S, Ling TC, Ooi CW, Show PL (2016) Thermoseparating aqueous two-phase systems: recent trends and mechanisms. J Sep Sci 39:640–647. https://doi.org/10.1002/jssc.201500667

Leong HY, Ooi CW, Law CL, Julkifle AL, Ling TC, Show PL (2018) Application of liquid biphasic flotation for betacyanins extraction from peel and flesh of *Hylocereus polyrhizus* and antioxidant activity evaluation. Sep Purif Technol 201:156–166. https://doi.org/10.1016/j.seppur.2018.03.008

Leong YK, Show P-L, Lan JC-W, Krishnamoorthy R, Chu D-T, Nagarajan D, Yen H-W, Chang J-S (2019) Application of thermo-separating aqueous two-phase system in extractive bioconversion of polyhydroxyalkanoates by *Cupriavidus necator* H16. Bioresour Technol 287:121474. https://doi.org/10.1016/j.biortech.2019.121474

Léon-González G, González-Valdez J, Mayolo-Deloisa R-PM (2016) Intensified fractionation of brewery yeast waste for the recovery of invertase using aqueous two-phase systems. Biotechnol Appl Biochem 63:886–894. https://doi.org/10.1002/bab.1435

Li Z, Liu X, Pei Y, Wang J, He M (2012) Design of environmentally friendly ionic liquid aqueous two-phase systems for the efficient and high activity extraction of proteins. Green Chem 14:2941–2950. https://doi.org/10.1039/c2gc35890e

Li H, Schmitz O, Alper HS (2016a) Enabling glucose/xylose co-transport in yeast through the directed evolution of a sugar transporter. Appl Microbiol Biotechnol 100:10215–10223. https://doi.org/10.1007/s00253-016-7879-8

Li P, Li T, Zeng Y, Li X, Jiang X, Wang Y, Xie T, Zhang Y (2016b) Biosynthesis of xanthan gum by *Xanthomonas campestris* LRELP-1 using kitchen waste as the sole substrate. Carbohydr Polym 151:684–691. https://doi.org/10.1016/j.carbpol.2016.06.017

Li P, Fu X, Zhang L, Zhang Z, Li J, Li S (2017) The transcription factors Hsf1 and Msn2 of thermotolerant *Kluyveromyces marxianus* promote cell growth and ethanol fermentation of *Saccharomyces cerevisiae* at high temperatures. Biotechnol Biofuels 10:1–13. https://doi.org/10.1186/s13068-017-0984-9

Li G, Fu Y, Dang W, Hu R, Xue H (2019) The effects of aqueous ammonia-pretreated rice straw as solid substrate on laccase production by solid-state fermentation. Bioprocess Biosyst Eng 42:567–574. https://doi.org/10.1007/s00449-018-02060-y

Liao L-C, Ho CS, Wu W-T (1999) Bioconversion with whole cell penicillin acylase in aqueous two-phase systems. Process Biochem 34:417–420. https://doi.org/10.1016/S0032-9592(98)00099-5

Lin C, Wu C, Tran D, Shih M, Li W, Wu C (2011) Mixed culture fermentation from lignocellulosic materials using thermophilic lignocellulose-degrading anaerobes. Process Biochem 46:489–493. https://doi.org/10.1016/j.procbio.2010.09.024

Lin CSK, Koutinas AA, Stamatelatou K, Mubofu EB, Matharu AS, Kopsahelis N, Pfaltzgraff LA, Clark JH, Papanikolaou S, Kwan TH, Luque R (2014) Current and future trends in food waste valorization for the production of chemicals, materials and fuels: a global perspective. Biofuels Bioprod Biorefin 8(5):686–715. https://doi.org/10.1002/bbb.1506

Lin YK, Show PL, Yap YJ, Ariff AB, Annuar MSM, Lai OM, Tang TK, Juan JC, Ling TC (2016) Production of γ-cyclodextrin by *Bacillus cereus* cyclodextrin glycosyltransferase using extractive bioconversion in polymer-salt aqueous two-phase system. J Biosci Bioeng 121:692–696. https://doi.org/10.1016/j.jbiosc.2015.11.001

Linares-Pastén JA, Aronsson A, Karlsson EN (2018) Structural considerations on the use of endo-xylanases for the production of prebiotic xylooligosaccharides from biomass. Curr Protein Pept Sci 19:48–67. https://doi.org/10.2174/1389203717666160923155209

Linde GA, Magagnin G, Alberto J, Costa V, Bertolin TE, Colauto NB (2007) Column bioreactor use for optimization of pectinase production in solid substrate cultivation. Braz J Microbiol 38:557–562. https://doi.org/10.1590/S1517-83822007000300033

Liu Y, Li C, Meng X, Yan Y (2016) Biodiesel synthesis directly catalyzed by the fermented solid of *Burkholderia cenocepacia* via solid state fermentation. Fuel Process Technol 106:303–309. https://doi.org/10.1016/j.fuproc.2012.08.013

Liu MQ, Huo WK, Xu X, Weng XY (2017a) Recombinant *Bacillus amyloliquefaciens* xylanase A expressed in *Pichia pastoris* and generation of xylooligosaccharides from xylans and wheat bran. Int J Biol Macromol 105:656–663. https://doi.org/10.1016/j.ijbiomac.2017.07.073

Liu R, Hooker NH, Parasidis E, Simons CT (2017b) A natural experiment: using immersive technologies to study the impact of "all-natural" labeling on perceived food quality, nutritional content, and liking. J Food Sci. https://doi.org/10.1111/1750-3841.13639

Liu X, Liu Y, Jiang Z, Liu H, Yang S, Yan Q (2018) Biochemical characterization of a novel xylanase from *Paenibacillus barengoltzii* and its application in xylooligosaccharides production from corncobs. Food Chem 264:310–318. https://doi.org/10.1016/j.foodchem.2018.05.023

Lonsane BK, Ghildyal NP, Budiatman S, Ramakrishna SV (1985) Engineering aspects of solid state fermentation. Enzyme Microb Technol 7:258–265

Lopes VRO, Farias MA, Belo IMP, Coelho MAZ (2016) Nitrogen sources on tpomw valorization through solid state fermentation performed by *Yarrowia Lipolytica*. Brazilian J Chem Eng 33:261–270. https://doi.org/10.1590/0104-6632.20160332s20150146

López-Pérez M, Viniegra-González G (2015) Production of protein and metabolites by yeast grown in solid state fermentation: present status and perspectives. J Chem Technol Biotechnol 91:1224–1231. https://doi.org/10.1002/jctb.4819

Madhavan A, Sindhu R, Parameswaran B, Sukumaran RK, Pandey A (2017) Metagenome analysis: a powerful tool for enzyme bioprospecting. Appl Biochem Biotechnol 183(2):636–651. https://doi.org/10.1007/s12010-017-2568-3

Majidian P, Tabatabaei M, Zeinolabedini M, Naghshbandi MP, Chisti Y (2018) Metabolic engineering of microorganisms for biofuel production. Renew Sustain Energy Rev 82:3863–3885. https://doi.org/10.1016/j.rser.2017.10.085

Manan MA, Webb C (2017) Design aspects of solid state fermentation as applied to microbial bioprocessing. J Appl Biotechnol Bioeng 4:91. https://doi.org/10.15406/jabb.2017.04.00094

Mano MCR, Neri-Numa IA, da Silva JB, Paulino BN, Pessoa MG, Pastore GM (2018) Oligosaccharide biotechnology: an approach of prebiotic revolution on the industry. App Microbiol Biotechnol 102:17–37. https://doi.org/10.1007/s00253-017-8564-2

Markets and Markets (2015) Surfactants market worth $42,120.4 million by 2020. http://www.prnewswire.com/news-releases/surfactants-market-worth-421204-million-by-2020-506134181.html. Accessed 8 June 2019

Martínez-Avila O, Sánchez A, Font X (2018a) Bioproduction of 2-phenylethanol and 2-phenethyl acetate by *Kluyveromyces marxianus* through the solid-state fermentation of sugarcane bagasse. Appl Microbiol Biotechnol 102:4703–4716. https://doi.org/10.1007/s00253-018-8964-y

Martínez-Avila O, Sánchez A, Font X, Barrena R (2018b) Enhancing the bioproduction of value-added aroma compounds via solid-state fermentation of sugarcane bagasse and sugar beet molasses: operational strategies and scaling-up of the process. Bioresour Technol 263:136–144. https://doi.org/10.1016/j.biortech.2018.04.106

Martínez-Avila O, Sánchez A, Barrena R, Font X (2019) Fed-batch and sequential-batch approaches to enhance the bioproduction of 2-phenylethanol and 2-phenethyl acetate in solid-state fermentation residue-based systems. J Agric Food Chem 67:3389–3399. https://doi.org/10.1021/acs.jafc.9b00524

Maruthamuthu M (2017) Novel bacterial enzymes for plant biomass degradation discovered by meta-omics approach. University of Groningen

Mathew S, Aronsson A, Karlsson EN, Adlercreutz P (2018) Xylo- and arabinoxylooligosaccharides from wheat bran by endoxylanases, utilisation by probiotic bacteria, and structural studies of the enzymes. Appl Microbiol Biotechnol 102:3105–3120. https://doi.org/10.1007/s00253-018-8823-x

Mehta A, Prasad GS, Choudhury AR (2014) Cost effective production of pullulan from agri-industrial residues using response surface methodology. Int J Biol Macromol 64:252–256. https://doi.org/10.1016/j.ijbiomac.2013.12.011

Mejía-Manzano LA, Barba-Dávila BA, Vázquez-Villegas P (2019) Improved extraction of the natural anticancerigen pristimerin from *Mortonia greggii* root bark using green solvents and aqueous two-phase systems. Sep Purif Technol 211:667–672. https://doi.org/10.1016/j.seppur.2018.08.056

Mejias L, Cerda A, Barrena R, Gea T (2018) Microbial strategies for cellulase and xylanase production through solid-state fermentation of digestate from biowaste. Sustainability 10:2433. https://doi.org/10.3390/su10072433

Melikoglu M, Lin CSK, Webb C (2015) Solid state fermentation of waste bread pieces by *Aspergillus awamori*: analysing the effects of airflow rate on enzyme production in packed bed bioreactors. Food Bioprod Process 95:63–75. https://doi.org/10.1016/j.fbp.2015.03.011

Mitchell DA, Krieger N, Stuart DM, Pandey A (2000) New developments in solid-state fermentation. II: Rational approaches to the design, operation and scale-up of bioreactors. Process Biochem 35:1211–1225. https://doi.org/10.1016/S0032-9592(00)00157-6

Montella S, Amore A, Faraco V (2016) Metagenomics for the development of new biocatalysts to advance lignocellulose saccharification for bioeconomic development. Crit Rev Biotechnol 36(6):998–1009. https://doi.org/10.3109/07388551.2015.1083939

Moore JC, Arnold FH (1996) Directed evolution of a paranitrobenzyl esterase for aqueous organic solvents. Nat Biotechnol 14:458–467. https://doi.org/10.1038/nbt0496-458

Moreira S, Silvério SC, Macedo EA, Milagres AMF, Teixeira JA, Mussatto SI (2013) Recovery of *Peniophora cinerea* laccase using aqueous two-phase systems composed by ethylene oxide/propylene oxide copolymer and potassium phosphate salts. J Chromatogr A 1321:14–20. https://doi.org/10.1016/j.chroma.2013.10.056

Moshtagh B, Hawboldt K, Zhang B (2018) Optimization of biosurfactant production by *Bacillus subtilis* N3-1P using the brewery waste as the carbon source. Environ Technol 14:1–10. https://doi.org/10.1080/09593330.2018.1473502

Narang S, Sahai V, Bisaria VS (2001) Optimization of xylanase production by *Melanocarpus albomyces* IIS68 in solid state fermentation using response surface methodology. J Biosci Bioeng 91:425–427. https://doi.org/10.1016/s1389-1723(01)80164-x

Ngara TR, Zhang H (2018) Recent advances in function-based metagenomic screening. Genomics Proteomics Bioinformatics 16(6):405–415. https://doi.org/10.1016/j.gpb.2018.01.002

Niknezhad SV, Asadollahi MA, Zamani A, Biria D, Doostmohammadi M (2015) Optimization of xanthan gum production using cheese whey and response surface methodology. Food Sci Biotechnol 24:453–460. https://doi.org/10.1007/s10068-015-0060-9

Nitschke M, Pastore GM (2006) Production and properties of a surfactant obtained from *Bacillus subtilis* grown on cassava wastewater. Bioresour Technol 97:336–341. https://doi.org/10.1016/j.biortech.2005.02.044

Nitschke M, Costa SG, Haddad R, Gonçalves LAG, Alves NC, Eberlin MN, Contiero J (2008) Oil wastes as unconventional substrates for rhamnolipid biosurfactant production by *Pseudomonas aeruginosa* LBI. Biotechnol Prog 21:1562–1566. https://doi.org/10.1021/bp050198x

Oliveira DWF, Sousa JR, França IWL, Felix AKN, Martins JJL, Gonçalves LRB (2013) Kinetic study of biosurfactant production by *Bacillus subtilis* LAMI005 grown in clarified cashew apple juice. Colloids Surf B Biointerfaces 101:34–43. https://doi.org/10.1016/j.colsurfb.2012.06.011

Oliveira F, Salgado JM, Abrunhosa L, Pérez-Rodríguez N, Domínguez JM, Venâncio A, Belo I (2017a) Optimization of lipase production by solid-state fermentation of olive pomace: from flask to laboratory-scale packed-bed bioreactor. Bioprocess Biosyst Eng 40:1123–1132. https://doi.org/10.1007/s00449-017-1774-2

Oliveira F, Salgado JM, Pérez-Rodríguez N, Domínguez JM, Venâncio A, Belo I (2017b) Lipase production by solid-state fermentation of olive pomace in tray-type and pressurized bioreactors. J Chem Technol Biotechnol 93:1312–1319. https://doi.org/10.1002/jctb.5492

Oliveira AC, Amorim GM, Azevêdo JAG, Mateus G, Freire DMG (2018) Solid-state fermentation of co-products from palm oil processing: production of lipase and xylanase and effects on chemical composition. Biocatal Biotransformation 36:381–388. https://doi.org/10.1080/10242422.2018.1425400

Oyarce P, Meester B, Fonseca F, Vries L, Goeminne G, Pallidis A, Rycke R, Tsuji Y, Li Y, Van den Bosch S (2019) Introducing curcumin biosynthesis in *Arabidopsis* enhances lignocellulosic biomass processing. Nat Plants 5:225–237. https://doi.org/10.1038/s41477-018-0350-3

Ozdal M, Gurkok S, Ozdal OG (2017) Optimization of rhamnolipid production by *Pseudomonas aeruginosa* OG1 using waste frying oil and chicken feather peptone. 3 Biotech 7:117. https://doi.org/10.1007/s13205-017-0774-x

Paddon CJ, Keasling JD (2014) Semi-synthetic artemisinin: a model for the use of synthetic biology in pharmaceutical development. Nat Rev Microbiol 12:355–367. https://doi.org/10.1038/nrmicro3240

Paithankar A, Rewatkar A (2014) Oil cakes as substrate for improved lipase production in solid state fermentation. Int J Microbiol Res 3:71. https://doi.org/10.9790/3008-09413138

Palaniappan A, Balasubramaniam VG, Antony U (2017) Prebiotic potential of xylooligosaccharides derived from finger millet seed coat. Food Biotechnol 31:264–280. https://doi.org/10.1080/08905436.2017.1369433

Pandey A (2003) Solid-state fermentation. Biochem Eng J 13:81–84. https://doi.org/10.1016/S1369-703X(02)00121-3

Pandey A, Soccol CR, Mitchell D (2000) New developments in solid state fermentation: I-bioprocesses and products. Process Biochem 35:1153–1169. https://doi.org/10.1016/S0032-9592(00)00152-7

Panigrahi S, Velraj P, Rao TS (2019) Functional microbial diversity in contaminated environment and application in bioremediation. Microbial diversity in the genomic era. Academic Press, pp 359–385. https://doi.org/10.1016/b978-0-12-814849-5.00021-6

Pathania S, Sharma S, Kumari K (2018) Solid state fermentation of BSG for citric acid production. Indian J Nat Prod Resour 9:70–74

Paula RG, Antoniêto ACC, Ribeiro LFC, Srivastava N, O'Donovan A, Mishra P, Gupta VK, Silva RN (2019) Engineered microbial host selection for value-added bioproducts from lignocellulose. Biotechnol Adv 37(6):107347. https://doi.org/10.1016/j.biotechadv.2019.02.003

Pérez-Armendáriz B, Cal-y-Mayor-Luna C, El-Kassis EG, Ortega-Martínez LD (2019) Use of waste canola oil as a low-cost substrate for rhamnolipid production using *Pseudomonas aeruginosa*. AMB Express 9:61. https://doi.org/10.1186/s13568-019-0784-7

Phong WN, Show PL, Teh WH, Teh TX, Lim MY, Nazri NSB, Tan CH, Chang J-S, Ling TC (2017) Protein recovery from wet microalgae using liquid biphasic flotation (LBF). Bioresour Technol 244:1329–1336. https://doi.org/10.1016/j.biortech.2017.05.165

Phong WN, Show PL, Chow YH, Ling TC (2018) Recovery of biotechnological products using aqueous two-phase systems. J Biosci Bioeng 126:273–281. https://doi.org/10.1016/j.jbiosc.2018.03.005

Protzko RJ, Latimer LN, Martinho Z, de Reus E, Seibert T, Benz JP, Dueber JE (2018) Engineering *Saccharomyces cerevisiae* for co-utilization of D-galacturonic acid and D-glucose from citrus peel waste. Nat Commun 9:5059. https://doi.org/10.1038/s41467-018-07589-w

Purohit A, Rai SK, Chownk M, Sangwan RS, Yadav SK (2017) Xylanase from *Acinetobacter pittii* MASK 25 and developed magnetic cross-linked xylanase aggregate produce predominantly xylopentose and xylohexose from agro biomass. Bioresour Technol 244:793–799. https://doi.org/10.1016/j.biortech.2017.08.034

Qiao W, Tao J, Luo Y, Tang T (2018) Microbial oil production from solid-state fermentation by a newly isolated oleaginous fungus, *Mucor circinelloides* Q531 from mulberry branches. R Soc Open Sci 5:180551. https://doi.org/10.1098/rsos.180551

Qureshi AS, Khushk I, Ali HC, Chisti Y, Ahmad A, Majeed H (2016) Coproduction of protease and amylase by thermophilic *Bacillus* sp. BBXS-2 using open solid-state fermentation of lignocellulosic biomass. Biocatal Agric Biotechnol 8:146–151. https://doi.org/10.1016/j.bcab.2016.09.006

Rabausch U, Juergensen J, Ilmberger N, Böhnke S, Fischer S, Schubach B, Schulte M, Streit WR (2013) Functional screening of metagenome and genome libraries for detection of novel flavonoid-modifying enzymes. Appl Environ Microbiol 79:4551–4563. https://doi.org/10.1128/AEM.01077-13

Raja S, Murty VR (2013) Optimization of aqueous two-phase systems for the recovery of soluble proteins from tannery wastewater using response surface methodology. J Eng 2013:217483. https://doi.org/10.1155/2013/217483

Rajagopalan G, Shanmugavelu K, Yang KL (2017) Production of prebiotic -xylooligosaccharides from alkali pretreated mahogany and mango wood sawdust by using purified xylanase of *Clostridium* strain BOH3. Carbohydr Polym 167:158–166. https://doi.org/10.1016/j.carbpol.2017.03.021

Ramalakshmi S, Ramanan RN, Ariff AB, Ooi CW (2014) Liquid-liquid equilibrium of primary and secondary aqueous two-phase systems composed of sucrose + triton X114 + water at different temperatures. J Chem Eng Data 59:2756–2762. https://doi.org/10.1021/je500571k

Ramírez IM, Tsaousi K, Rudden M, Marchant R, Alameda EJ, Román MG, Banat IM (2015) Rhamnolipid and surfactin production from olive oil mill waste as a sole carbon source. Bioresour Technol 198:231–236. https://doi.org/10.1016/j.biortech.2015.09.012

Ran L, Yang C, Xu M, Yi Z, Ren D, Yi L (2019) Enhanced aqueous two-phase extraction of proanthocyanidins from grape seeds by using ionic liquids as adjuvants. Sep Purif Technol 226:154–161. https://doi.org/10.1016/j.seppur.2019.05.089

Redden H, Alper HS (2015) The development and characterization of synthetic minimal yeast promoters. Nat Commun 6:1–9. https://doi.org/10.1038/ncomms8810

Reddy SS, Krishnan C (2016a) Production of high-pure xylooligosaccharides from sugarcane bagasse using crude β-xylosidase-free xylanase of *Bacillus subtilis* KCX006 and their bifidogenic function. LWT—Food Sci Technol 65:237–245. https://doi.org/10.1016/j.lwt.2015.08.013

Reddy SS, Krishnan C (2016b) Production of xylooligosaccharides in SSF by *Bacillus subtilis* KCX006 producing β-xylosidase-free endo-xylanase and multiple xylan debranching enzymes. Prep Biochem Biotechnol 46:49–55. https://doi.org/10.1080/10826068.2014.970694

Renata H, Wang J, Arnold FH (2015) Expanding the enzyme universe: accessing non-natural reactions by mechanism-guided directed evolution. Angew Chem 54:3351–3367. https://doi.org/10.1002/anie.201409470

The Market Reports (2018) Global Xylo-oligosaccharide (XOS) market by manufacturers, regions, type and application, forecast to 2023. https://www.themarketreports.com/report/global-xylo-oligosaccharide-xos-market-by-manufacturers-regions-type-and-application-forecast-to-2023. Accessed 23 June 2019.

Rito-Palomares M, Lyddiatt A (2002) Process integration using aqueous two-phase partition for the recovery of intracellular proteins. Chem Eng J 87:313–319. https://doi.org/10.1016/S1385-8947(01)00241-8

Rodrigues JL, Rodrigues LR (2017a) Synthetic biology: perspectives in industrial biotechnology. In: Pandey A, Teixiera J (eds) . Foundations of biotechnology and bioengineering. Current developments in biotechnology and bioengineering, Elsevier, pp 239–269

Rodrigues JL, Rodrigues LR (2017b) Potential applications of the *Escherichia coli* heat shock response in synthetic biology. Trends Biotechnol 36:186–198. https://doi.org/10.1016/j.tibtech.2017.10.014

Rodrigues JL, Sousa M, Prather KL, Kluskens LD, Rodrigues LR (2014) Selection of *Escherichia coli* heat shock promoters towards their application as stress probes. J Biotechnol 188:61–71. https://doi.org/10.1016/j.jbiotec.2014.08.005

Rodrigues J, Araújo R, Prather K, Kluskens L, Rodrigues L (2015a) Heterologous production of caffeic acid from tyrosine in *Escherichia coli*. Enzyme Microb Technol 71:36–44. https://doi.org/10.1016/j.enzmictec.2015.01.001

Rodrigues J, Araújo R, Prather K, Kluskens L, Rodrigues L (2015b) Production of curcuminoids from tyrosine by a metabolically engineered *Escherichia coli* using caffeic acid as an intermediate. Biotechnol J 10:599–609. https://doi.org/10.1002/biot.201400637

Rodrigues JL, Prather KL, Kluskens L, Rodrigues L (2015c) Heterologous production of curcuminoids. Microbiol Mol Biol Rev 79:39–60. https://doi.org/10.1128/MMBR.00031-14

Rodrigues JL, Couto MR, Araújo RG, Prather KL, Kluskens L, Rodrigues LR (2017a) Hydroxycinnamic acids and curcumin production in engineered *Escherichia coli* using heat shock promoters. Biochem Eng J 125:41–49. https://doi.org/10.1016/j.bej.2017.05.015

Rodrigues JL, Ferreira D, Rodrigues LR (2017b) Synthetic biology strategies towards the development of new bioinspired technologies for medical applications. In: Rodrigues L, Mota M (eds) Bioinspired materials for medical applications. Woodhead Publishing, United Kingdom, pp 451–497

Roelants S, Van Renterghem L, Maes K, Everaert B, Vanlerberghe B, Demaeseneire S, Soetaert W (2019) Microbial biosurfactants: from lab to market. In: Banat IM, Thavasi R (eds) Microbial biosurfactants and their environmental and industrial applications. CRC Press, Boca Raton. https://doi.org/10.1201/b21950

Ruiz HA, Rodríguez-Jasso RM, Rodríguez R, Contreras-Esquivel JC, Aguilar CN (2012) Pectinase production from lemon peel pomace as support and carbon source in solid-state fermentation column-tray bioreactor. Biochem Eng J 65:90–95. https://doi.org/10.1016/j.bej.2012.03.007

Sadh PK, Duhan S, Duhan JS (2018) Agro-industrial wastes and their utilization using solid state fermentation: a review. Bioresour Bioprocess 5:1. https://doi.org/10.1186/s40643-017-0187-z

Saha MLAL (1997) Citric acid fermentation by magnetic rotating biological contactors using *Aspergillus niger* AJ 117173. J Ferment Bioeng 84:244–248. https://doi.org/10.1016/S0922-338X(97)82062-4

Saithi S, Tongta A (2016) Phytase production of *Aspergillus niger* on soybean meal by solid-state fermentation using a rotating drum bioreactor. Agric Agric Sci Procedia 11:25–30. https://doi.org/10.1016/j.aaspro.2016.12.005

Salah RB, Chaari K, Besbes S, Ktari N, Blecker C, Deroanne C, Attia H (2015) Optimisation of xanthan gum production by palm date (Phoenix dactylifera L.) juice by-products using response surface methodology. Food Chem 121:627–633. https://doi.org/10.1016/j.foodchem.2009.12.077

Salas-Veizaga DM, Villagomez R, Linares-Pastén JA, Carrasco C, Álvarez MT, Adlercreutz P, Karlsson EN (2017) Extraction of glucuronoarabinoxylan from quinoa stalks (*chenopodium quinoa* willd.) and evaluation of xylooligosaccharides produced by GH10 and GH11 xylanases. J Agric Food Chem 65:8663–8673. https://doi.org/10.1021/acs.jafc.7b01737

Salum TFC, Villeneuve P, Barea B, Itsuo C, Cristina L, Alexander D, Krieger N (2010) Synthesis of biodiesel in column fixed-bed bioreactor using the fermented solid produced by *Burkholderia cepacia* LTEB11. Process Biochem 45:1348–1354. https://doi.org/10.1016/j.procbio.2010.05.004

Samanta AK, Jayapal N, Kolte AP, Senani S, Sridhar M, Suresh KP, Sampath KT (2012) Enzymatic production of xylooligosaccharides from alkali solubilized xylan of natural grass (*Sehima nervosum*). Bioresour Technol 112:199–205. https://doi.org/10.1016/j.biortech.2012.02.036

Samanta AK, Jayapal N, Jayaram C, Roy S, Kolte AP, Senani S, Sridhar M (2015) Xylooligosaccharides as prebiotics from agricultural by-products: production and applications. Bioact Carbohydr Diet Fibre 5:62–71. https://doi.org/10.1016/j.bcdf.2014.12.003

Samykannu M, Achary A (2017) Utilization of agro-industry residue for rhamnolipid production by *Pseudomonas aeruginosa* AMB AS7 and its application in chromium removal. Appl Biochem Biotechnol 183:70–90. https://doi.org/10.1007/s12010-017-2431-6

Sankaran R, Show PL, Yap YJ, Tao Y, Ling TC, Tomohisa K (2018a) Green technology of liquid biphasic flotation for enzyme recobery utilizing recycling surfactant and sorbitol. Clean Technol Environ Policy 20:2001–2012. https://doi.org/10.1007/s10098-018-1523-5

Sankaran R, Show PL, Lee SY, Yap YJ, Ling TC (2018b) Integration process of fermentation and liquid flotation for lipase separation from *Burkholderia cepacia*. Bioresour Technol 250:306–316. https://doi.org/10.1016/j.biortech.2017.11.050

Saravanan S, Rao JR, Nair BU, Ramasami T (2008) Aqueous two-phase poly(ethylene glycol)-poly(acrylic acid) system for protein partitioning: influence of molecular weight, pH and temperature. Process Biochem 43:905–911. https://doi.org/10.1016/j.procbio.2008.04.011

Sathi-Reddy K, Yahya-Khan M, Archana K, Reddy G, Hameeda B (2016) Utilization of mango kernel oil for the rhamnolipid production by *Pseudomonas aeruginosa* DR1 towards its application as biocontrol agent. Bioresour Technol 221:291–299. https://doi.org/10.1016/j.biortech.2016.09.041

Savvides AL, Katsifas EA, Hatzinikolaou DG, Karagouni AD (2012) Xanthan production by *Xanthomonas campestris* using whey permeate medium. World J Microbiol Biotechnol 28:2759–2764. https://doi.org/10.1007/s11274-012-1087-1

Scarlat N, Martinov M, Dallemand JF (2010) Assessment of the availability of agricultural crop residues in the European Union: potential and limitations for bioenergy use. Waste Manag 30:1889–1897. https://doi.org/10.1016/j.wasman.2010.04.016

Scoma A, Rebecchi S, Bertin L, Fava F (2016) High impact biowastes from South European agro-industries as feedstock for second-generation biorefineries. Crit Rev Biotechnol 36:175–189. https://doi.org/10.3109/07388551.2014.947238

Searle S, Malins C (2013) Availability of cellulosic residues and wastes in the EU. International Council on Clean Transportation, Washington, DC

Seesuriyachan P, Kawee-ai A, Chaiyaso T (2017) Green and chemical-free process of enzymatic xylooligosaccharide production from corncob: enhancement of the yields using a strategy of lignocellulosic destructuration by ultra-high pressure pretreatment. Bioresour Technol 241:537–544. https://doi.org/10.1016/j.biortech.2017.05.193

Sharma N, Prasad GS, Choudhury AR (2013) Utilization of corn steep liquor for biosynthesis of pullulan, an important exopolysaccharide. Carbohydr Polym 93:95–101. https://doi.org/10.1016/j.carbpol.2012.06.059

Sheldon RA, Woodley JM (2018) Role of biocatalysis in sustainable chemistry. Chem Rev 118:801–838. https://doi.org/10.1021/acs.chemrev.7b00203

Shojaosadati SA, Babaeipour V (2002) Citric acid production from apple pomace in multi-layer packed bed solid-state bioreactor. Process Biochem 37:909–914. https://doi.org/10.1016/S0032-9592(01)00294-1

Show PL, Tan CP, Anuar MS, Ariff A, Yusof YA, Chen SK, Ling TC (2012) Extractive fermentation for improved production and recovery of lipase derived from *Burkholderia cepacian* using a thermoseparating polymer in aqueous two systems. Bioresour Technol 116:226–233. https://doi.org/10.1016/j.biortech.2011.09.131

Silva TM, Minim LA, Maffia MC, Coimbra JSR, Minim VPR, da Silva LHM (2007) Equilibrium data for poly(propylene glycol) + sucrose + water and poly(propylene glycol) + fructose + water systems from (15 to 45) °C. J Chem Eng Data 52:1649–1652. https://doi.org/10.1021/je700031d

da Silva MB, Rossi DM, Ayub MA (2017) Screening of filamentous fungi to produce xylanase and xylooligosaccharides in submerged and solid-state cultivations on rice husk, soybean hull, and spent malt as substrates. World J Microbiol Biotechnol 33:58. https://doi.org/10.1007/s11274-017-2226-5

Silva C, Martins M, Jing S, Fu J, Cavaco-Paulo A (2018) Practical insights on enzyme stabilization. Crit Ver Biotechnol 38:335–350. https://doi.org/10.1080/07388551.2017.1355294

Simon C, Daniel R (2011) Metagenomic analyses: past and future trends. Appl Environ Microbiol 77(4):1153–1161. https://doi.org/10.1128/AEM.02345-10

Singh RD, Banerjee J, Sasmal S, Muir J, Arora A (2018) High xylan recovery using two stage alkali pre-treatment process from high lignin biomass and its valorization to xylooligosaccharides of low degree of polymerisation. Bioresour Technol 256:110–117. https://doi.org/10.1016/j.biortech.2018.02.009

Singhania RR, Patel AK, Soccol CR, Pandey A (2009) Recent advances in solid-state fermentation. Biochem Eng J 44:13–18. https://doi.org/10.1016/j.bej.2008.10.019

Sjöling S, Cowan DA (2008) Metagenomics: microbial community genomes revealed. In: Margesin R, Schinner F, Marx JC, Gerday C (eds) Psychrophiles: from biodiversity to biotechnology. Springer, Berlin, pp 313–332. https://doi.org/10.1007/978-3-540-74335-4_18

Soccol CR, Iloki I, Marin B, Raimbault M (1994) Comparative production of alpha-amylase, glucoamylase and protein enrichment of raw and cooked cassava by rhizopus strains in submerged and solid state. J Food Sci Technol 31:320–323

Soccol CR, Scopel E, Alberto L, Letti J, Karp SG, Woiciechowski AL, Porto L, Vandenberghe DS (2017) Recent developments and innovations in solid state fermentation. Biotechnol Res Innov 1:52–71. https://doi.org/10.1016/j.biori.2017.01.002

Soffa M, Rodríguez-Jasso RM, Michelin M, Flores-Gallegos AC (2018) Bioreactor design for enzymatic hydrolysis of biomass under the biorefinery concept. Chem Eng J 347:119–136. https://doi.org/10.1016/j.cej.2018.04.057

Sousa M, Melo VMM, Rodrigues S, Santana HB, Gonçalves LRB (2012) Screening of biosurfactant-producing *Bacillus* strains using glycerol from the biodiesel synthesis as main carbon source. Bioprocess Biosyst Eng 35:897–906. https://doi.org/10.1007/s00449-011-0674-0

Spier MR, Porto L, Vandenberghe DS, Bianchi A, Medeiros P, Soccol CR (2011) Application of different types of bioreactors in bioprocesses. In: Bioreactors: design, properties and applications. Nova Science Publishers, Hauppauge, pp 53–87

Sripokar P, Chaijan M, Benjakul S, Yoshida A, Klomklao S (2017) Aqueous two-phase partitioning of liver proteinase from albacore tuna (*Thunnus alalunga*): application to starry triggerfish (*Abalistes stellaris*) muscle hydrolysis. Int J Food Prop 52:51600–51612. https://doi.org/10.1080/10942912.2017.1350705

Stemmer WPC (1994) Rapid evolution of a protein in vitro by DNA shuffling. Nature 370:389–391. https://doi.org/10.1038/370389a0

Sun Z, Wikmark Y, Baeckvall J-E, Reetz MT (2016) New concepts for increasing the efficiency in directed evolution of stereoselective enzymes. Chem A Eur J 22(15):5046–5054. https://doi.org/10.1002/chem.201504406

Teles ASC, Chávez DWH, Oliveira RA, Bon EPS, Terzi SC, Souza EF, Gottschalk LMF, Tonon RV (2019) Use of grape pomace for the production of hydrolytic enzymes by solid-state fermentation and recovery of its bioactive compounds. Food Res Int 120:441–448. https://doi.org/10.1016/j.foodres.2018.10.083

Tjerneld F, Johansson G, Joelsson M (1987) Affinity liquid-liquid extraction of lactate dehydrogenase on a large scale. Biotechnol Bioeng 30:809–816. https://doi.org/10.1002/bit.260300702

Try S, De-Coninck J, Voilley A, Chunhieng T, Waché Y (2017) Solid state fermentation for the production of γ-decalactones by Yarrowia lipolytica. Process Biochem 64:9–15. https://doi.org/10.1016/j.procbio.2017.10.004

Tsuji Y, Vanholme R, Tobimatsu Y, Ishikawa Y, Foster CE, Kamimura N, Hishiyama S, Hashimoto S, Shino A, Hara H (2015) Introduction of chemically labile substructures into *Arabidopsis* lignin through the use of LigD, the Cα-dehydrogenase from *Sphingobium* sp. strain SYK-6. Plant Biotechnol J 13:821–832. https://doi.org/10.1111/pbi.12316

Ufarte L, Potocki-Veronese G, Laville E (2015) Discovery of new protein families and functions: new challenges in functional metagenomics for biotechnologies and microbial ecology. Front Microbiol 6(563). https://doi.org/10.3389/fmicb.2015.00563

Uygut MA, Tanyildizi MŞ (2018) Optimization of alpha-amylase production by *Bacillus amyloliquefaciens* grown on orange peels. Iran J Sci Technol Trans A Sci 42:443–449. https://doi.org/10.1007/s40995-016-0077-9

Vea EB, Romeo D, Thomsen M (2018) Biowaste valorisation in a future circular bioeconomy. Procedia CIRP 69:591–596. https://doi.org/10.1016/j.procir.2017.11.062

Wang C, Dong D, Wang H, Müller K, Qin Y, Wang H, Wu W (2016) Metagenomic analysis of microbial consortia enriched from compost: new insights into the role of Actinobacteria in lignocellulose decomposition. Biotechnol Biofuels 9:22. https://doi.org/10.1186/s13068-016-0440-2

Wang Y, Xian M, Feng X, Liu M, Zhao G (2018) Biosynthesis of ethylene glycol from d-xylose in recombinant *Escherichia coli*. Bioengineered 9:233–241. https://doi.org/10.1080/21655979.2018.1478489

Wang X, Yang J, Yang S, Jiang Y (2019) Unraveling the genetic basis of fast L-arabinose consumption on top of recombinant xylose-fermenting *Saccharomyces cerevisiae*. Biotechnol Bioeng 116:283–293. https://doi.org/10.1002/bit.26827

Weber W, Fussenegger M (2012) Emerging biomedical applications of synthetic biology. Nat Rev Genet 13:21–35. https://doi.org/10.1038/nrg3094

Werner S, Maschke RW, Eibl D, Eibl R (2018) Bioreactor technology for sustainable production of plant cell-derived products. In: Bioprocessing of plant in vitro systems, pp 413–432

Westfall PJ, Pitera DJ, Lenihan JR, Eng D, Woolard FX, Regentin R, Horning T, Tsuruta H, Melis DJ, Owens A, Fickesa S, Diolaa D, Benjamina KR, Keasling JD, Leavella MD, McPheea DJ, Renningera NS, Newmana JD, Paddona CJ (2012) Production of amorphadiene in yeast, and its conversion to dihydroartemisinic acid, precursor to the antimalarial agent artemisinin. Proc Natl A Sci 109:E111–E118. https://doi.org/10.1073/pnas.1110740109

WHO—World Health Organization (2017) Cardiovascular diseases (CVDs) https://www.who.int/news-room/fact-sheets/detail/cardiovascular-diseases-(cvds). Accessed 23 June 2019

WHO—World Health Organization (2018) Global report on diabetes. https://apps.who.int/iris/bitstream/handle/10665/204871/9789241565257_eng.pdf.jsessionid=AB2C916A6DBA6120F68D15F7A47BF5BC?sequence=1. Accessed 23 June 2019

Winans K, Kendall A, Deng H (2017) The history and current applications of the circular economy concept. Renew Sustain Energy Rev 68:825–833. https://doi.org/10.1016/j.rser.2016.09.123

Wu S, Lu M, Chen J, Fang Y, Wu L, Xu Y, Wang S (2016) Production of pullulan from raw potato starch hydrolysates by a new strain of *Aureobasidium pullulans*. Int J Biol Macromol 82:740–743. https://doi.org/10.1016/j.ijbiomac.2015.09.075

Wu H, Li H, Xue Y, Luo G, Gan L, Liu J, Mao L, Long M (2017) High efficiency co-production of ferulic acid and xylooligosaccharides from wheat bran by recombinant xylanase and feruloyl esterase. Biochem Eng J 120:41–48. https://doi.org/10.1016/j.bej.2017.01.001

Wysokinska Z (2016) The "New" environmental policy of the European Union: a path to development of a circular economy and mitigation of the negative effects of climate change. Comp Econ Res 19:57–73. https://doi.org/10.1515/cer-2016-0013

Xia T, Eiteman MA, Altman E (2012) Simultaneous utilization of glucose, xylose and arabinose in the presence of acetate by a consortium of *Escherichia coli* strains. Microb Cell Fact 11:77. https://doi.org/10.1186/1475-2859-11-77

Xu K, Wang Y, Huang Y, Li N, Wen Q (2015) A green Deep eutectic solvent-based aqueous two-phase system for protein extracting. Anal Chim Acta 864:9–20. https://doi.org/10.1016/j.aca.2015.01.026

Xu G, Hao C, Tian S, Gao F, Sun W, Sun R (2017) A method for the preparation of curcumin by ultrasonic-assisted ammonium sulfate/ethanol aqueous two phase extraction. J Chromatogr B Analyt Technol Biomed Life Sci 1041-1041:167–174. https://doi.org/10.1016/j.jchromb.2016.12.029

Yang F, Mitra P, Zhang L, Prak L, Verhertbruggen Y, Kim JS, Sun L, Zheng K, Tang K, Auer M (2013) Engineering secondary cell wall deposition in plants. Plant Biotechnol J 11:325–335. https://doi.org/10.1111/pbi.12016

Yang J, Summanen PH, Henning SM, Hsu M, Lam H, Huang J, Tseng C-H, Dowd SE, Finegold SM, Heber D, Li Z (2015) Xylooligosaccharide supplementation alters gut bacteria in both healthy and prediabetic adults: a pilot study. Front Physiol 6:216. https://doi.org/10.3389/fphys.2015.00216

Yao W, Nokes SE (2013) The use of co-culturing in solid substrate cultivation and possible solutions to scientific challenges. Biofuels Bioprod Biorefin 7:361–372. https://doi.org/10.1002/bbb.1389

Yen H, Chang J (2015) Growth of oleaginous *Rhodotorula glutinis* in an internal-loop airlift bioreactor by using lignocellulosic biomass hydrolysate as the carbon source. J Biosci Bioeng 119:580–584. https://doi.org/10.1016/j.jbiosc.2014.10.001
Zafra G, Taylor TD, Absalón AE, Cortés-Espinosa DV (2016) Comparative metagenomic analysis of PAH degradation in soil by a mixed microbial consortium. J Hazard Mater 318:702–710. https://doi.org/10.1016/j.jhazmat.2016.07.060
Zeng X, Miao W, Zeng H, Zhao K, Zhou Y, Zhang J, Zhao Q, Tursun D, Xu D, Li F (2018) Production of natamycin by *Streptomyces gilvosporeus* Z28 through solid-state fermentation using agro-industrial residues. Bioresour Technol 273:377–385. https://doi.org/10.1016/j.biortech.2018.11.009
Zhang H, Xu Y, Yu S (2017) Co-production of functional xylooligosaccharides and fermentable sugars from corncob with effective acetic acid prehydrolysis. Bioresour Technol 234:343–349. https://doi.org/10.1016/j.biortech.2017.02.094
Zhong J (2010) Recent advances in bioreactor engineering. Korean J Chem Eng 27:1035–1041. https://doi.org/10.1007/s11814-010-0277-5

Chapter 6
State of the Art and New Perspectives in Oleogels and Applications

Vara Prasad Rebaka, Arun Kumar Rachamalla, Srishti Batra, and Nagarajan Subbiah

Contents

V. P. Rebaka · A. K. Rachamalla · S. Batra · N. Subbiah (✉)
Department of Chemistry, National Institute of Technology Warangal
(Institute of National Importance), Warangal, Telangana, India
e-mail: snagarajan@nitw.ac.in

Inamuddin, A. M. Asiri (eds.), *Sustainable Green Chemical Processes and their Allied Applications*, Nanotechnology in the Life Sciences,
https://doi.org/10.1007/978-3-030-42284-4_6

Abbreviations

ALS	Aromatic steroid linker
CNC	5α-Cholestan-3β-yl-*N*-(2-naphthyl)carbamate
DAG	Diacylglycerol
DBS	Dibenzylidene-D-sorbitol
DCC	*N*,*N*′-dicyclohexylcarbodiimide
DGI	Dodecyl glyceryl itaconate
DMP	*N*-methyl-2-2-pyrrolidone
DMSO	Dimethyl sulfoxide
FA	Fatty acid
GMS	Glycerol monostearate
HAS	Hydroxy stearic acid
HPMC	Hydroxypropyl methyl cellulose
LMW	Low molecular weight
MAG	Monoacylglycerol
PEA	Potential emulsifying agent
RBX	Rice bran wax
SMP	Sorbitan monopalmitate
SMS	Sorbitan monostearate
STS	Sorbitan tristearate
TAG	Triacylglycerol
TBAH	Tetrabutylammonium hydroxide
TBDMS	Tert-butyldimethylsilyl
THF	Tetrahydrofuran
UPyEMA	2-(((6-(6-Methyl-4[1H]pyrimidionylureido)hexyl)carbamoyl)oxy) ethyl-methacrylate

6.1 Introduction

In the last two decades, gels have attracted a wide range of interests because of their applications as soft materials. Gels are soft viscoelastic materials that hold a prophetic definition "easier to recognize than define" (Lloyd 1926). However, gels are generally referred as a non-fluid colloidal network or polymer network expanded throughout its whole volume by a fluid (Alemán et al. 2007). Such semisolid soft material is formed by a solid gelator, which undergoes spontaneous molecular self-assembly to form crosslinked fibrous network structure wherein solvent molecules were entrapped. Three-dimensional fibrous structure formed by the molecular assembly is stabilized by the various interactions such as H-bonding, π–π stacking, van der Waals forces, anionic and cationic–π interactions, CH–π interaction, and donor-acceptor type interactions. To a larger extent, these physical interactions are affected by pH, temperature, pressure and other environmental conditions (Thamizhanban and Lalitha 2019). Therefore, the properties of the physical gels can be altered, whereas gels formed by chemical bonding cannot be altered by external stimuli. Gels particularly play a vital role in food technology and processing, cosmetics, pharmaceutical industries and research and development laboratories. The widespread applications of these gels are due to their physicochemical properties that are obtained by using various formulations. Normally, gels are called as systems that have high concentrated solvents within a low concentrated structural agent. These structural agents form the skeleton of the gels (Buerkle and Rowan 2012), which enables no flow of solvent when present in steady state. Such two-component semisolid gel system contains a large amount of liquid, which is more rigid than jellies because of the existence of larger order of interactions. Generally, transparent and opaque nature of gel is decided by the molecular dispersity of the gelator (Duan et al. 2014). Gels are broadly classified into three types: hydrogel, organogel and oleogel. Oleogels are microstructured complex networks which immobilize an oil in its 3D gel matrix. Oleogel can be either derived from the low molecular weight (LMW) gelator or polymeric gelator, which displays adjustable physical properties and unique chemical properties. The salient feature of oleogel is the best alternative for saturated fatty acids and trans fat structuring agents (Singh et al. 2017). Recently, oleogels are more predominantly used in drug delivery networks because of their biocompatibility, crosslinking network structure and molecular stability of the bioagent used in the gel. Besides, oleogels derived from edible renewable resources can be potentially used in food industries and cosmetics as a structuring agent. In an effort to deliver alternatives to the existing saturated and trans fats, recently scientists have actively involved in the development of oleogels with tunable physical properties and stimuli-responsive character.

6.2 Structure and Classification of Gels

Gels are comprised of assembled small molecules or polymers that form a three-dimensional matrix through a dispersed medium of solvent. The network is made by the gelling particles interlocked within the matrix as shown in Fig. 6.1. Gels are rigid due to the existence of network structure and nature of force which interlocks these gelling particles decides the morphology of network and gel properties (Arsem 1926). This force of attraction between the gelling molecules ranges from strong primary valencies like in acidic silica gels to weak van der Waals forces (Wu and Morbidelli 2001).

Generally, gels are classified based on the nature of the colloid particles or by the nature of the medium used (Fig. 6.2). Based on the nature of the medium used, gels are further classified into hydrogel, oleogel, organogel and aerogel.

Hydrogels possess very tractable structural properties and controllable by altering its network density.

- They are very useful for biological processes as it can load bioactive molecules in its matrix and if reversal of the process is observed, then it can function as delivery systems, i.e. releasing these molecules in a sustained manner under appropriate condition.
- They are chemically and mechanically biocompatible with respect to the physiological condition.
- They can be made either into biodegradables or biocompatible.
- Very much compatible for formation and deformation at the target site.

Organogels or oleogels are highly hydrophobic in nature; i.e. they can take large number of bioactive molecules and can be used to deliver at target sites.

- These gels containing hydrophobic bioactive molecules are very useful nowadays in pharmaceutical, cosmetics and food-related applications.
- The lack of aqueous media in organogels deters microbial growth, which reduces the possibility of the future inflammation process.

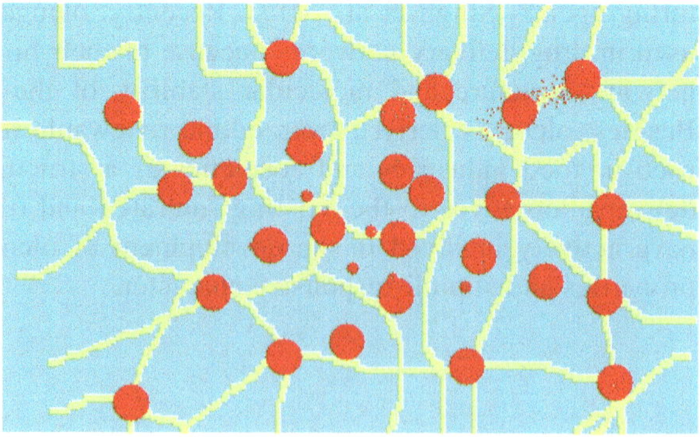

Fig. 6.1 Structure of gels; the yellow line represents the fibrous network and red dot represents the solvent molecules

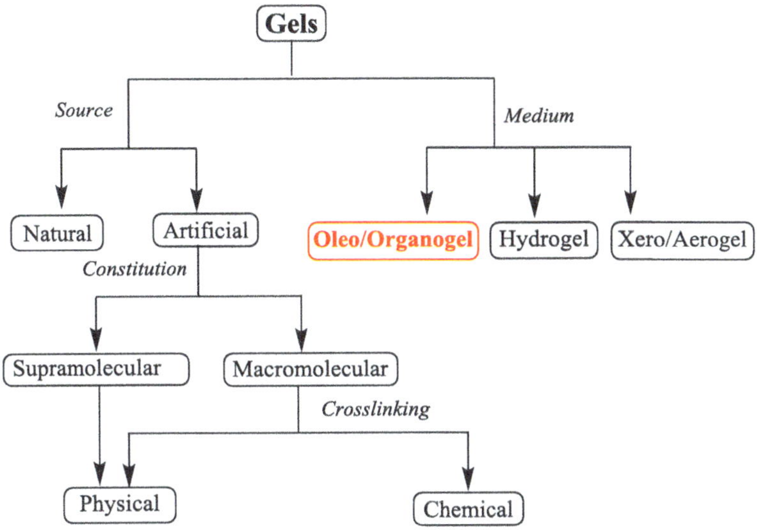

Fig. 6.2 Classification of gels

Aerogel is an ultralightweight synthetic porous material derived from gel, wherein the immobilized liquid is replaced with air or gas.

- The porous structure with lightweight property can adsorb oils and other hydrophobic substances.
- These materials replace the conventional insulation used in the construction and industrial sectors.

6.3 Oleogels

In this chapter, we discuss the structure, properties and applications of oleogels. A gel is said to be a hydrogel or organogel based upon the nature of the liquid component present as the continuous phase in it. If the liquid component is water, it is named as hydrogel and if the liquid component is an organic solvents, oil are referred as organogel or oleogel. These are semisolid system in which an organic phase liquid is immobilized in a 3D matrix made up of self-assembled, viscoelastic and interlinked network of gelators (Co and Marangoni 2018). Gelators that form these types of gels are called organogelators or oleogelators. A broad classification of oleogel based on the nature of the interaction is given in Fig. 6.3.

Oleogels have an extensive range of applications in the food industry, pharmaceutical and drug delivery systems (Singh et al. 2017). Some organic solvents have toxic impact on the body; these organogels cannot be used as drug delivery systems. The use of biocompatible oleogels derived from vegetable oils increased its applica-

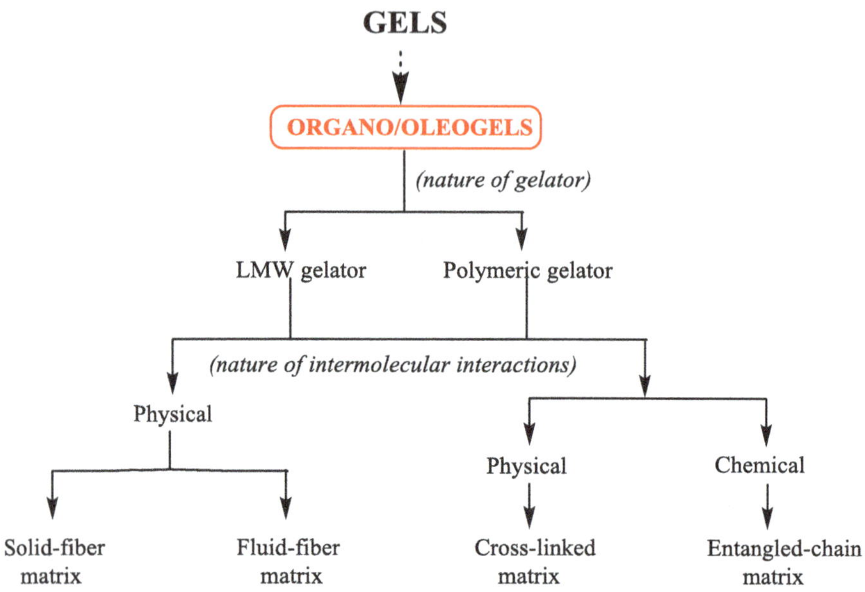

Fig. 6.3 Classification of oleogels based on the nature of intermolecular interactions

tions in the cosmetic and pharmaceutical field and drug delivery formulations. Oleogels also have antibacterial characteristics due to the absence of aquatic media (Rogers et al. 2014). They are present in a low energy state attributed to its thermodynamic stability. These gels undergo gel-to-sol transition when the temperature rises above room temperature. They are also sensitive to moisture.

Oleogelators are classified into two types such as low molecular weight oleogelators and polymeric oleogelators based on the molecular weight of the gelating molecules as shown in Fig. 6.3. Polymeric gelators form a network of entangled or crosslinked chains to immobilize the organic solvent resulted in physical and chemical gels. The stability of the physical gels is further promoted by weak interchain interactions like H-bonding, van der Waals forces and π-π stacking. Similarly, low molecular weight oleogelators depend on their physical interactions for the formation of sufficiently long aggregate to induce solvent gelation and also for overlapping. Low molecular weight oleogels are divided into solid-fibre matrix and fluid-fibre matrix according to its composition and the kinetic properties of the aggregates (Fuhrhop and Actually 1993). The structures of the solid-fibre matrix and the liquid-fibre matrix are shown below (Vintiloiu and Leroux 2008).

Solid matrix gels are very strong due to the presence of rigid solid-like networks, where the junction points are large microdomains, the circled area in Fig. 6.4. Most of the oleogels are formed from solid fibres, where the temperature is decreased below the solubility limit of the gelator (Brizard et al. 2018). These gelator molecules quickly assemble in the organic phase via various intermolecular interactions to form 3D aggregates. Fluid matrix gels have weak networks where the junction

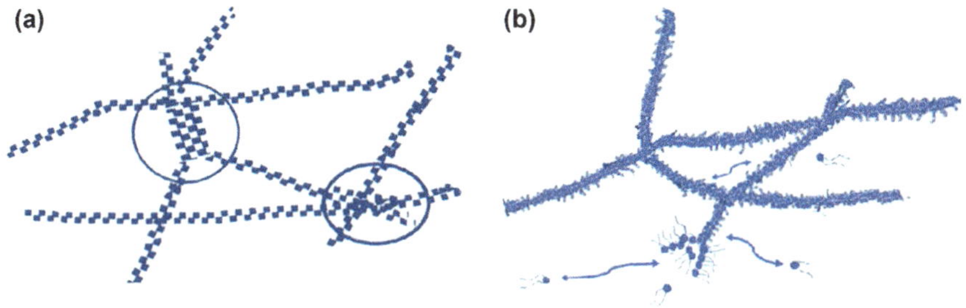

Fig. 6.4 Structures of (**a**) solid-fibre matrix and the (**b**) liquid-fibre matrix

points are the small interlinks. These are produced by the arrangement of small molecules into a single- or double-layered cylindrical networks that hold the solvent. These aggregates go through a dynamic exchange process between the gelators and the liquid media (Vintiloiu and Leroux 2008).

6.3.1 Lower Molecular Weight Oleogelators

Most of the low molecular weight compounds form highly stable and more efficient gels at low concentrations with different organic solvents. Increased solubility of these gels in organic solvents on heating and smooth gelation in low concentration are the salient features of these gelators (George and Weiss 2006). These are capable of gelling edible oils even at low concentrations ranging from 0.5 to 3% wt/v. The process of gelation with these gelators is thermoreversible and temperature sensitive. Generally, high concentration of oleogelators is required to gel oils with large amounts of impurities as the impurities interact with the gelators. These gelators have high gelation ability and the formed gel will have a long lifetime (Suzuki and Hanabusa 2010). There are different types of low molecular weight oleogelators such as waxes, fatty acids and fatty alcohols, phytosterols, ceramides, lecithin, surfactant, TAG derivatives and others.

6.3.1.1 *n*-Alkanes and Waxes

n-Alkanes are the simplest form of oleogelators obtained from the impurities of petroleum products. These gelators crystallize at low temperatures, which then convert the petroleum products into gels (Abdallah et al. 1999). Recently, Weiss and co-workers have discussed the systematic structural development in *n*-alkanes as oleogelators. In addition, a systematic discussion on oleogelation process using 12-hydroxystearic acid and its derivatives has also been included (Rogers and Weiss 2015).

Waxes are a class of organic compounds that are comprised of long alkyl chains including various functional groups like fatty acid ester, fatty alcohols, aldehydes, ketones and aromatic compounds. The main component responsible for the formation of gels is the wax ester, which constitutes 75% of the total mass of the wax. Due to the differences in the chemical compounds present in different waxes, different types of gelators are formed (Patel et al. 2015). Waxes with long alkyl chain esters produce gel more efficiently than the short alkyl chain esters. Even the small components present in the wax will have a large effect in the formation of gels and their properties. The uniqueness of wax-based oleogel is that no two gels will be identical. Gels of desired application can be achieved by changing the processing parameters such as cooling and shear rates (Mattice & Marangoni 2018). Wax-based oleogels find potential applications in cosmetics, pharmaceuticals and as structuring agent in food industries (Sagiri et al. 2018). For example, 5α-cholestan-3β-yl-*N*-(2-naphthyl)carbamate (CNC) also called as aromatic steroid linker (ALS) is an organogelator formed from n-alkane liquids (Fig. 6.5) (Huang et al. 2005).

Rice bran wax (RBX) is comprised of esters from fatty acids and fatty alcohols with carbon numbers ranging from 16 to 32 and 24 to 38, respectively (Fig. 6.6). The most abundant wax esters are comprised of 52, 54 or 56 total number of carbon atoms (Dassanayake et al. 2009).

Fig. 6.5 Structure of 5α-cholestan-3β-yl-*N*-(2-naphthyl)carbamate

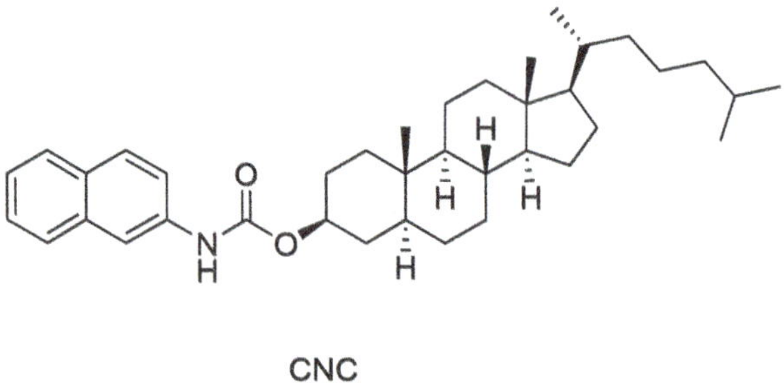

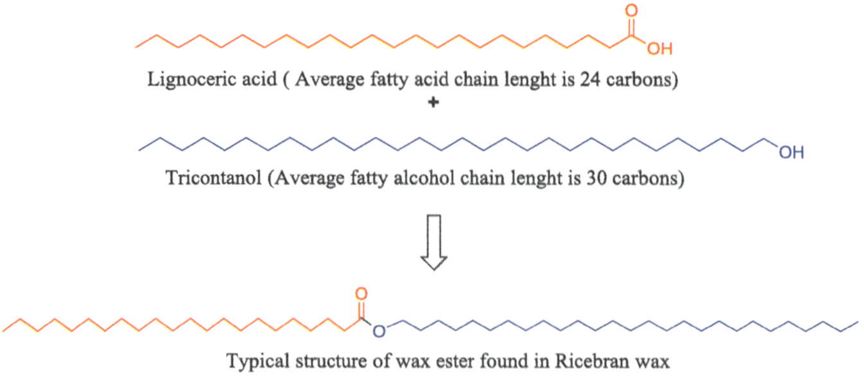

Fig. 6.6 Illustration of RBX chemical structure and their starting materials

6.3.1.2 Fatty Acids and Fatty Alcohols

The biocompatibility of the fatty acids and fatty alcohols under physiological environment enables them to behave as a good gelator to form oleogels. There is a linear relationship between the hardness of the gel and concentration of the gelators containing fatty acids and fatty alcohols (Fig. 6.7). The mechanical properties of the gel depends on the acid/alcohol ratio. The maximum mechanical properties are observed at 3:7 acid/alcohol ratio of stearic acid to stearyl alcohol mixtures. Superior gels were formed by the gelator having an equal chain length of the fatty acids and fatty alcohols, which are highly sensitive to temperature (Gandolfo et al. 2004). Recently Saboya et al. have used cation exchange resins for the esterification of fatty acids derived from castor oil and fatty alcohol, which displayed the application as a bio-lubricant in automobile industries (Saboya et al. 2017).

6.3.1.3 Phytosterols

Highly stable oleogels are formed from the mixtures of phytosterol and phytosterol esters (Matheson et al. 2018) Molecular structure of phytosterol is given in Fig. 6.8. Transparent gels are formed from phytosterol gelators because of the small size and molecular arrangement. The viscosity of the oil and polarity alters the self-assembly of the phytosterol-based gelator molecules. For example, the use of high viscous oil takes long gelling time, whereas the low viscous oil furnished the gel with good mechanical properties. The use of γ-oryzanol and β-sitosterol mixtures produces a transparent gel in sunflower oil with helical tube-like morphology (Sawalha et al. 2012).

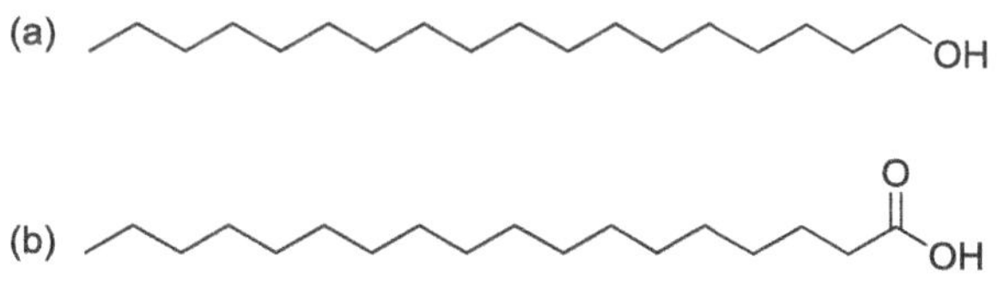

Fig. 6.7 Structures of (**a**) stearyl alcohol and (**b**) stearic acid

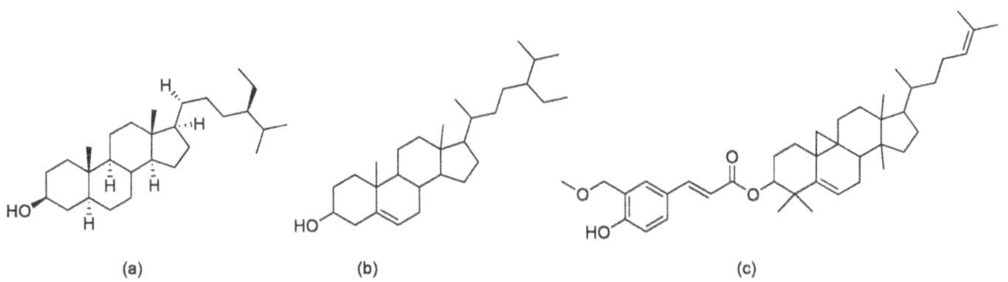

Fig. 6.8 Molecular structure of (**a**) phytosterol, which is used for the synthesis of phytosterol esters and glucosides, (**b**) β-sitosterol and (**c**) γ-oryzanol

Stearic acid and β-sitosterol form a rigid gel in sunflower oil and the rigidity increases with the increase in the quantity of stearic acid added. The phytosterol ethoxylates along with ionic liquids at room temperature displayed excellent gel formation and interfacial properties (Sakai et al. 2009). Phytosterol glucosides such as acyl steryl glycosides and steryl glucosides isolated from soybean lecithin powders displayed excellent gelation behaviour (Kang et al. 2019).

6.3.1.4 Ceramides

Ceramide-based gels are derived from sphingosine and fatty acids, which help in retaining skin moisture and protect from skin irritants and environmental pollution. A large variety of ceramide-based oleogelators are used for gelling vegetable oils and find applications in the food and cosmetic industries. Ceramides are obtained from sphingosine by amidation reaction at its amine group with a fatty acid. Molecular structure of ceramide is given in Scheme 6.1. The gelation ability of ceramides depends on the length of the fatty acid chains: short-chain fatty acids have high gel-forming ability even at lower concentrations when compared to the longer-chain fatty acids (Kang et al. 2019; Guo et al. 2019). Ceramides are expensive than phytosterols which limit its practical applications. Synthetically made ceramides are not as efficient as natural ceramides in gelling oils. Ceramides form needle-like crystals in edible oil and display nutraceutical properties, signal transduction for cell division and apoptosis. Milk sphingomyelin is an important bioactive lipid actively involved in the formation of milk fat globule membrane. The use of palmitoyl ceramide induces the gel phase formation and enhances the mechanical properties (Kang et al. 2019; Guo et al. 2019).

6.3.1.5 Sorbitan Esters and Lecithin

Sorbitan is an organic compound derived from sorbitol upon the action of dehydration. Free hydroxyl group in the sorbitan undergoes esterification with various acid generates mono, di- and tri-sorbates (Fig. 6.9), which acts as a potential emulsifying agent. Oleogelators derived from sorbitan tristearate (STS) along with sunflower lecithin can gel edible oils (Pernetti et al. 2007a, b). However, these compounds are not capable of gelling oils individually even at the higher concentrations of 20% w/w. A mixture of sorbitan with lecithin can form a gel at low concentrations at 4% w/w. The mechanical behaviour of gel depends upon the optimum ratio of lecithin and

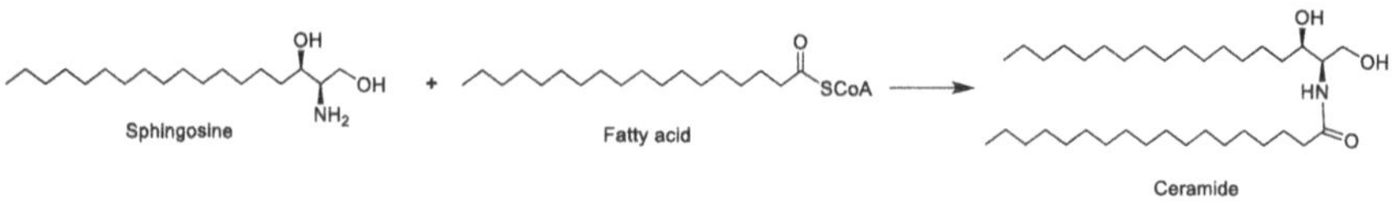

Scheme 6.1 Representation for the synthesis of ceramide

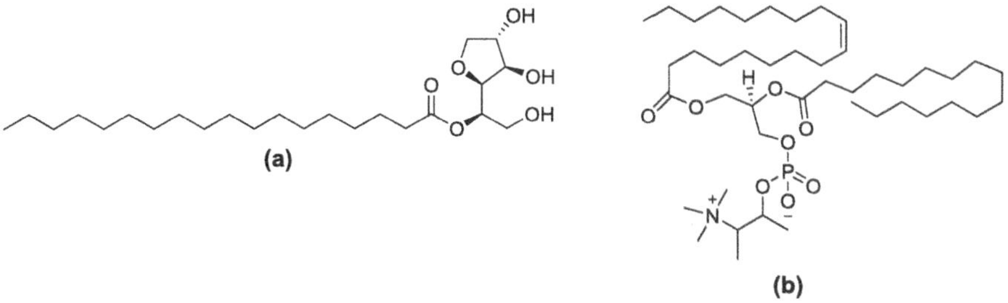

Fig. 6.9 Molecular structure of (**a**) sorbitan monostearate and (**b**) lecithin

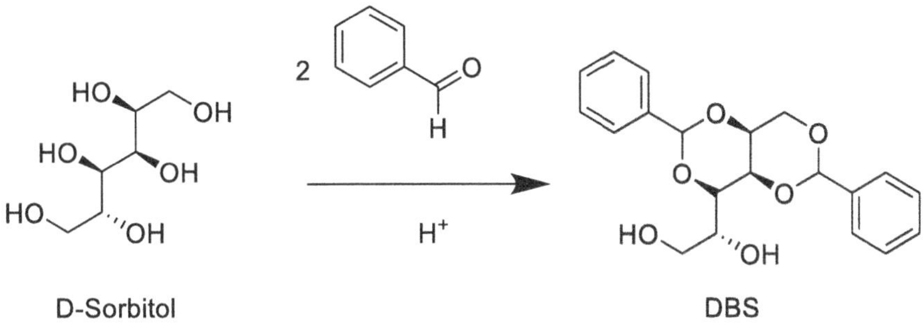

Scheme 6.2 Synthesis of 1,3:2,4-dibenzylidene-D-sorbitol organogelator

sorbitan ranging between 2:3 and 3:2 lecithin/sorbitan. These gelators form needle-like crystals of 10 μm. Gels formed by STS alone resulted in the formation of less needle-like crystals; the addition of lecithin improves the formation of needle-like crystals, thereby increasing the oil structuring efficiency. These gels are water and temperature sensitive, which start melting at the temperature of 15 °C. Oleogels using sunflower oil can be gelled by phosphatidylcholine and α-tocopherol mixtures taken in 1:1 ratio (Nikiforidis and Scholten 2014). The efficiency of the gelation can be increased by introducing polysorbate emulsifiers into the gelators. Pure lecithin alone can form gels with organic solvents at a concentration of 10% w/w. Lecithin molecules self-assemble themselves into hexagonal tube-like micelles, which in turn form oleogels. During the gelation of lecithin, an aggregate of spherical micelles are also formed in apolar solvents. Soybean lecithin can be gelled by adding water and these are largely used for drug delivery (Nikiforidis and Scholten 2014).

1,3:2,4-Dibenzylidene-D-sorbitol (DBS) is sorbitol-based organogelator obtained from D-sorbitol through an acid-catalysed condensation reaction (Scheme 6.2). The hydrophobic benzylidene part helps them to dissolve in various solvents. These gelators can gel a large amount of solvent and displayed a wide range of commercial applications, which were discussed elsewhere.

6.3.1.6 Surfactants

The compound that lowers the interfacial tension between two phases is called surfactants. The self-assembling of surfactant molecules can be brought up by many factors such as pH, ionic strength, concentration, nature of oil, type of solvent and temperature. Physical properties of the oleogels are affected by the surfactant and the type of oil (Co and Marangoni 2018). Surfactants forms lamellar structures in oils leads to crystal like platelets which forms a three-dimensional aggregate structure, thereby stabilizes the oleogels. Sorbitan monostearate, glycerol monostearate and sorbitan monopalmitate are few examples of surfactant oleogelators. Glycerol monostearate forms strong gels when compared with sorbitan monostearate (Fig. 6.10). Sorbitan monostearate is strong and thermally stable than the sorbitan monopalmitate (Murdan et al. 1999).

6.3.1.7 Monoglyceride and Triacylglycerol Derivatives

Monoglycerides are an important class of organogelators commonly used in food, agriculture and pharmaceutical industries. Generally, structured products of fat can be produced from monoglycerides in the form of emulsions (Calligaris et al. 2010). For example, hazelnut oil gelled by a monoglyceride called monostearin (Chen and Terentjev 2009) displays similar behaviour in both oil and water phase. A mixture of monoglycerides can form better oleogels than a single monoglyceride. In particular, 1:1 ratio of monostearin and monopalmitin can gel vegetable oil effectively than the individual component. Monoglyceride and phytosterol in the ratio of 8:2 can form an effective gel in sunflower oils. Here the use of phytosterols restricts the formation of large structures by monoglycerides, thus generating micro- and nano-architectures to form a harder gel.

Triacylglycerols can naturally self-assemble to form various architectures, which act as a structuring agent by trapping or immobilizing oil. The process of oleogelation was studied by mixing triacylglycerols of high melting and low melting point and found to be more effective than the individual TAG (Higaki et al. 2003). Interestingly, even low concentration of monoacylglycerol is sufficient enough to gelate oil, whereas diglycerides require high concentrations around 50–60%.

Dodecyl glyceryl itaconate (Scheme 6.3) is a low-cost and recyclable phase selective organogelator based on monoglycerides. Its high capacity of oil uptake helps in clearing marine oil spills (Wang et al. 2015a, b).

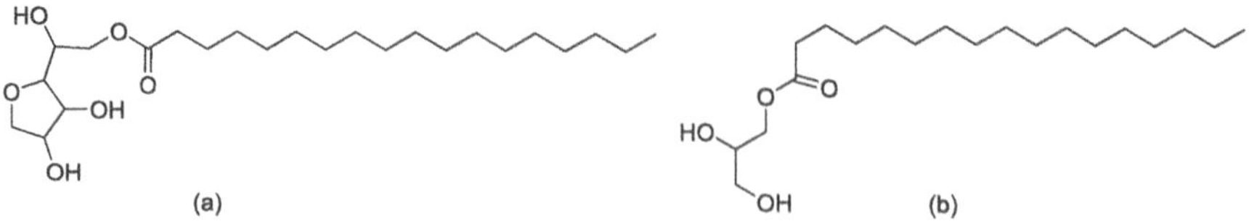

Fig. 6.10 Structures of (**a**) SMS and (**b**) GMS

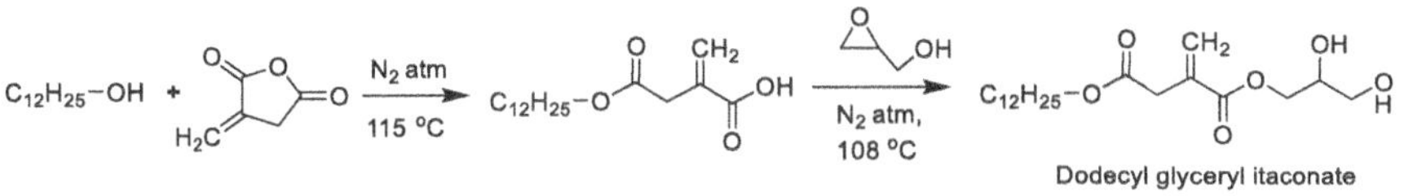

Scheme 6.3 Synthesis of dodecyl glyceryl itaconate organogelator

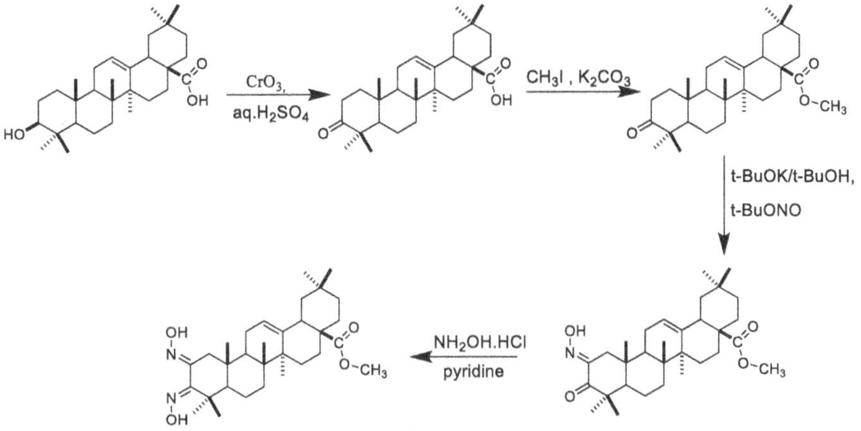

Scheme 6.4 Synthesis of 2,3-dihydroxyimino-oleanlic acid

6.3.1.8 Triterpenoid

Both synthetic and natural triterpenoids displayed various biological and medicinal properties including hepatoprotective, anti-HIV, antitumour and antiviral properties. Betulin, oleanolic acid, lupeol and betulinic acid are the representative example for the natural triterpenoid isolated from the bark of the birch tree. The gelation behaviour and pharmacological properties of triterpenes were completely discussed elsewhere. Here, 2,3-dihydroxyimino-oleanic acid has been chosen as a representative example to discuss the oleogelation (Scheme 6.4). The oleanlic acid obtained from plants as free acids act as a potent oleogelator and finds application in the drug industry and biomedical engineering because of their bioactivity, biocompatibility and low toxicity (Hu et al. 2009).

6.3.2 Polymeric Oleogelators

Mechanical processability is the major drawback of the LMW oleogels, which can easily break down during the process step itself. As LMW gelator is a small molecule, the molecules will self-assemble themselves into fibre structures; even slight tuning of environmental conditions facilitates the gel-to-sol transition (Hanabusa and

Suzuki 2014); on the other hand, polymers and oligomers are macromolecular gelators which won't undergo crystallization because of their chain entanglements and distribution of weight along the polymer (Suzuki and Hanabusa 2010). Even though polymer-based oleogel exists several limitations, their stability and easy processability argue the commercial value. Generally, in polymeric systems, oleogelation occurs via the swelling process by forming network structures through physical interactions. Oleogelation could be observed in star-shaped, linear and hyperbranched polymer. Some of the common polymeric oleogelators are the derivatives of L-lysine. The ability of these gelators to form oleogels can be varied by changing the backbone of the polymer. The gels obtained from these gelators displayed low gel-to-sol transition temperature when compared to LMW oleogelators. For example, copolymerization of 2-(((6-(6-methyl-4[1H]pyrimidionylureido)hexyl)carbamoyl)oxy) ethyl-methacrylate (UPyEMA) and styrene using reversible addition chain-fragmentation polymerization to form a poly(sty-s-UPyEMA) organogelator which uses hydrogen bonding as physical crosslinkers (Scheme 6.5) (Yu et al. 2016).

6.3.2.1 Cellulose Derivative

Ethyl cellulose, a linear polysaccharide, is the first polymer to be used as an oleogelator, which is capable of gelling the edible oil (Fig. 6.11). Ethyl cellulose-based gels were obtained by dissolving it in liquid oil above the glass transition temperature, i.e. 140 °C, and cooling down to normal room temperature. Generally, a plasticizer is added to facilitate the dissolution process. The mechanical properties of the gel are affected by the temperature because of the presence of hydrogen bonds (Davidovich-Pinhas et al. 2015) Some higher fatty alcohol like oleyl alcohol and oleic acid are added to increase the strength of the gel. Apart from ethyl cellulose, the other known polymeric gelators are methyl cellulose, α-cellulose and ethyl acetate.

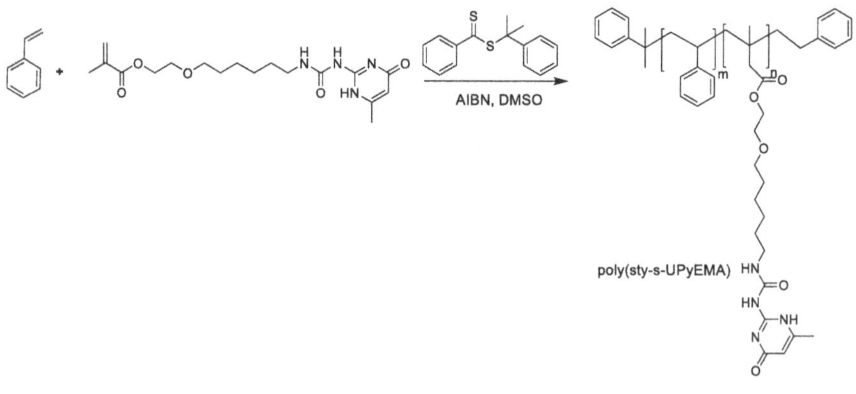

Scheme 6.5 Synthesis of poly(sty-s-UPyEMA) organogelator

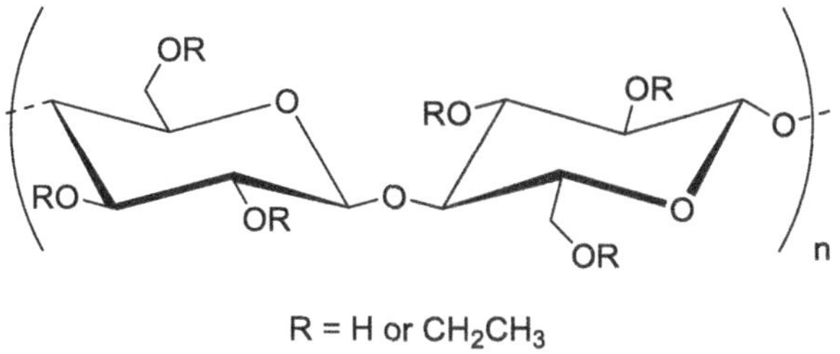

Fig. 6.11 Structure of ethyl cellulose

6.3.2.2 Proteins

Proteins are highly hydrophilic substance made of a sequence of amino acids which does not dissolve in oil well. Therefore, using proteins as oleogelators is a great challenge for researchers. The oil droplets in an oil-in-water emulsion are stabilized by a crosslinked monolayer of protein adsorbed at the surfaces. When the water is evaporated, it results in a dry foam network-like structure, where the air gaps are filled by oil and the proteins are present on the walls. The physical properties of the gel can be varied by changing the diameter of the droplets in the emulsion. Currently, whey protein isolate is used to produce oleogels through a solvent exchange system. First, gels are formed from whey protein isolate by heating them; later it undergoes solvent exchange process by a solvent intermediate (Romoscanu and Mezzenga 2006).

6.3.2.3 Water-Soluble Polysaccharides

Recently, a large variety of polysaccharides were used to form hydrogels. However, preparing oleogels from polysaccharides is less common. The common method adopted for the preparation of polysaccharide-based oleogel is as follows: oleogels are prepared by forming an aqueous foam using hydroxypropyl methylcellulose followed by freeze-drying to form a porous cryogel (Fig. 6.12). Later the added oil enters into the pores of cryogel by the action of shearing, which is having greater absorption properties (Online et al. 2013).

6.3.3 Miscellaneous Gelators

Some examples of oleogelators (Yang et al. 2019; Maffezzoni and Zanda 2008; Motulsky et al. 2005a, b; Sullivan et al. 2016; Pernetti et al. 2007a, b) along with the chemical structures (Fig. 6.13) are given below:

- Cholesterol is used for gelling liquid paraffin.
- Triazine functionalized with amino acid appendages is used in gelling haloalkanes and other aromatic solvents.

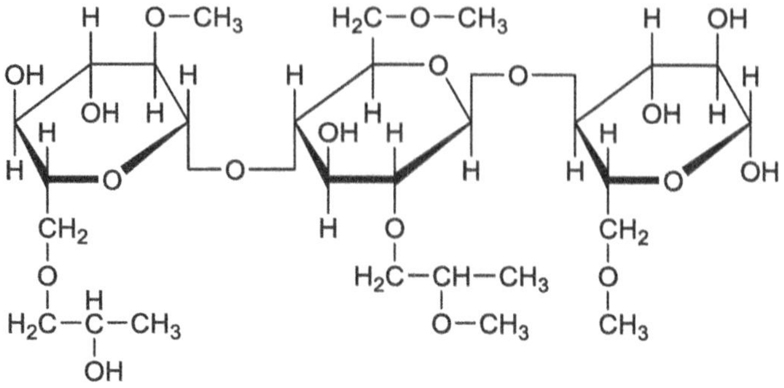

Fig. 6.12 Structure of hydroxypropyl methylcellulose

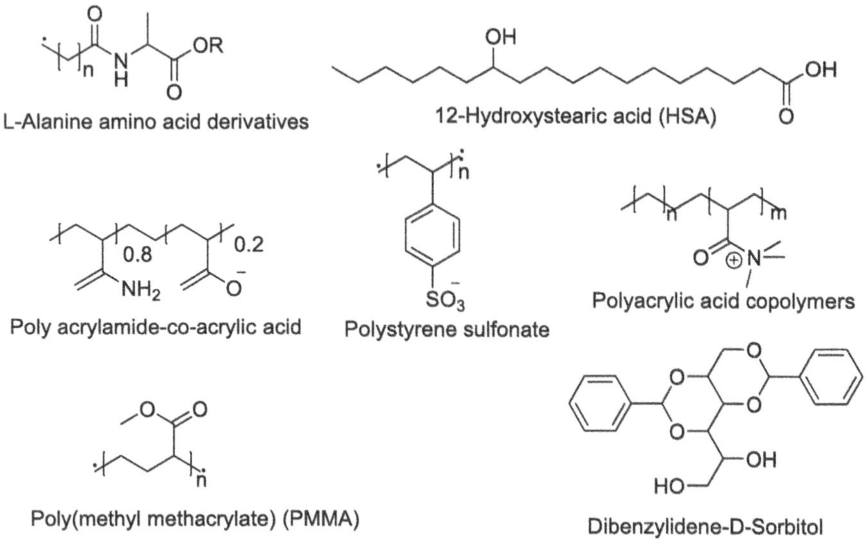

Fig. 6.13 Chemical structures of common oleogelators

- L-Alanine derivatives are used for gelling various pharmaceutical grade vegetables and synthetic oil.
- 12-Hydroxy stearic acid is used for gelling soya bean oil or caprylic triglyceride.
- Phytosterol with oryzanol, TAG, DAG, MAG, FA and waxes can gel various edible oils.
- Sorbitan monostearate with lecithin can gel sunflower oil(Co and Marangoni 2018).

6.4 Synthetic Routes of Some Important Organogelators

6.4.1 *Carbohydrate as Low Molecular Weight Gelators (Palanisamy 2017)*

Supramolecular gel derived from gluconosemicarbazide (Scheme 6.6) displayed gelation upon the action of ultra-sonification as well as heating.

Properties of sugar-based gels as follows:

- Non-toxic in nature
- Structural diversity
- Eco-friendly
- Biodegradable
- Formed from renewable plant-derived sources

6.4.2 *D-π-A Diphenyl Acrylonitrile Derivatives: Fluorescence-Enhanced Gels*

Scheme 6.7 represents the synthetic method for diphenyl acrylonitrile-based gelators, which act as an aggregation-induced emission fluorophore to impart fluorescent properties (Wang et al. 2013a, b).

- Nitro groups and alkoxy groups are important in the formation of a gel.
- This gel shows aggregation-induced emission properties due to the restriction of molecular planarity, molecular motion and J-aggregate formation upon the formation of a gel.

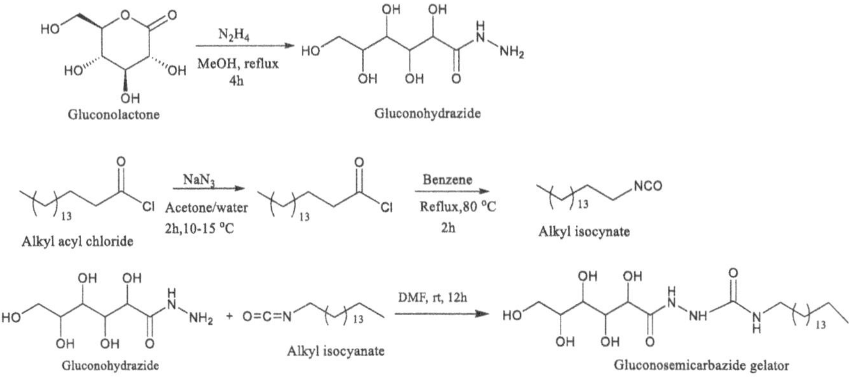

Scheme 6.6 Synthesis of gluconosemicarbazide gelator

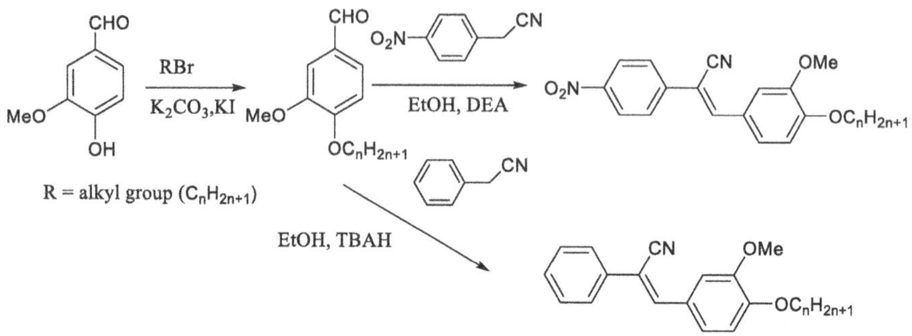

Scheme 6.7 Synthesis of diphenyl acrylonitrile gelators

6.4.3 Cholesterol-Based Salicylidene Schiff Base Gel

A cholesterol-based salicylidene Schiff base derivative shown in Scheme 6.8 is prone to self-assemble in a wide variety of organic solvents to form gels (Zang et al. 2013) Interestingly, this gel displayed photochromism, thermochromism and solvatochromism properties.

6.4.4 A Simple Tyrosine-Based Organogelator

L-tyrosine(tert-butyldimethylsilyl)-OH (L-Try(TBDMS)-OH) acts as a fluoride sensing low molecular weight organogelator because of the presence of tert-butyldimethylsilyl (TBDMS) group, which can undergo cleavage of Si–O bond in the presence of fluoride and forms L-Try-OH (Scheme 6.9). It can form gel in various solvents such as DMF, butanol, THF, and toluene even in very low concentrations of 0.1 w/v% (Aykent et al. 2019) L-Try(TBDMS)-OH can sense fluoride even at low concentrations of 0.2 ppm.

6.5 Mechanism of Formation of Oleogels

To understand the process of the formation of these gels, they are classified into three types of structures like a protein: primary structure, secondary structure and tertiary structure. The primary structures are made from the aggregation of gelator molecules in a single direction. These structures range from Å to nm scale. The morphologic structures of the aggregates such as vesicles, micelles, ribbons, fibres and sheets are called secondary structures. These structures range from nm to μm scale. These structures are affected directly by their molecular structures. The ter-

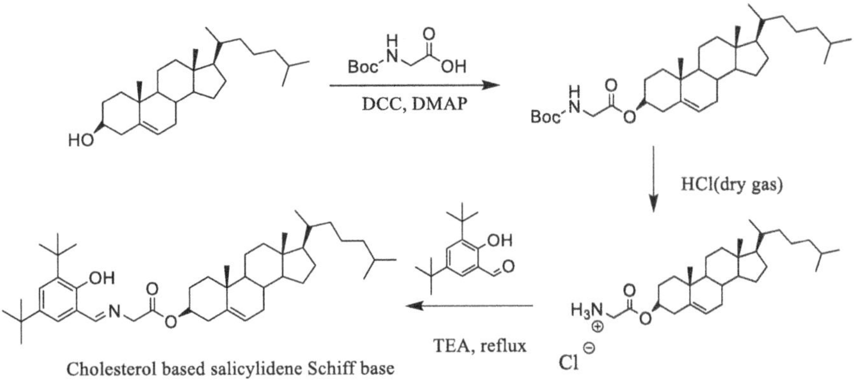

Scheme 6.8 Synthesis of cholesterol-based salicylidene Schiff base gelator

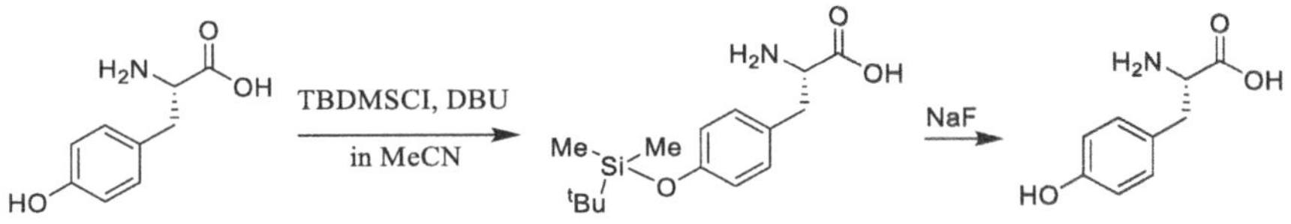

Scheme 6.9 Synthesis of L-Try(TBDMS)-OH and its reactivity with fluoride ion

tiary structures of the gels include the interconnections between the individual aggregates that form a gel network (Estroff and Hamilton 2004).

Normally, gels are obtained by mixing gelator at low concentrations around less than 15% w/v in organic solvents at high temperatures followed by cooling at room temperature. However, a low concentration of gelator up to 0.1% was used in sugar-derived supergelator. Cooling the super-heated solutions will decrease the affinity of solvent molecules and the organogelators which leads to the self-assembling of organogelators into solid aggregates via physical intermolecular interactions. Generally, upon cooling gelator assembles in three ways: (1) crystallization because of the highly ordered aggregation, (2) precipitation because of the random aggregation and (3) gelation via aggregation process (Fig. 6.14) (Sangeetha and Maitra 2005).

Oleogels are formed when a structuring agent or a gelator is added to a liquid oil which self-assembles itself to form a stable three-dimensional network which is stabilized by hydrogen bonding, π–π interactions and van der Waals forces (Balasubramanian et al. 2012) This process occurs in three steps (Fig. 6.15) (Martins et al. 2018; Patel 2017):

Step 1: Solubilization of a gelator in organic media

When a gelator is dissolved in nonpolar solvent, it forms reversible spherical micelles of 0.01 mM concentration.

Step 2: Addition of polar additive

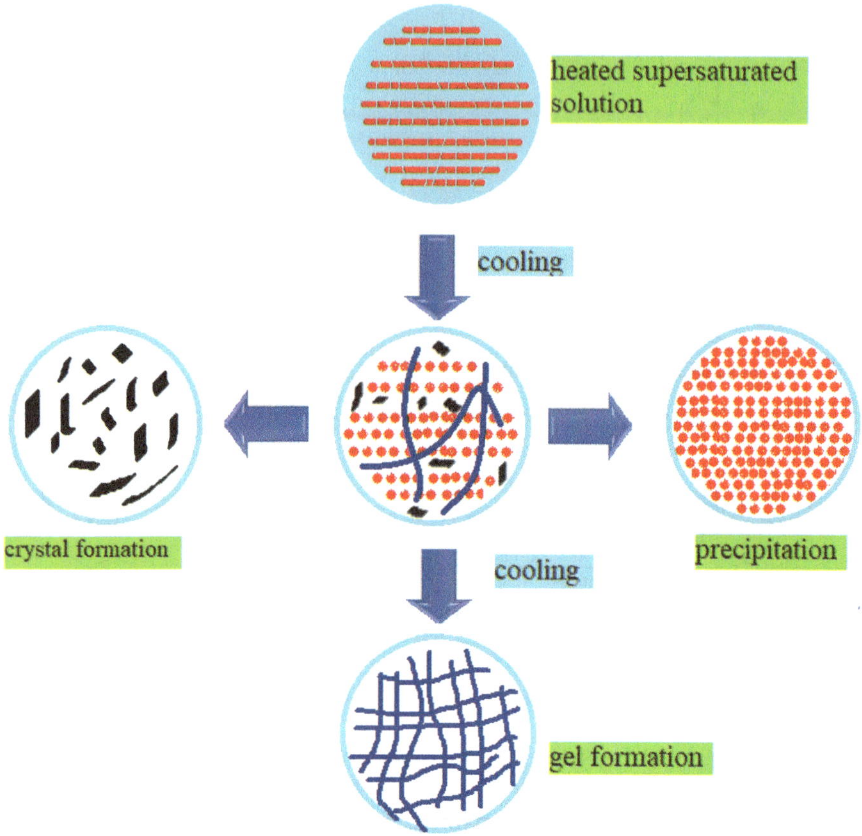

Fig. 6.14 Aggregation modes of gelator

To the gelator dispersion in the organic nonpolar liquid, a polar additive is added. This will transform the reverse spherical micelles to cylindrical micellar aggregates. The polar solvent molecule forms a bridge between the hydrophilic heads of two lecithin molecules. A linear one-dimensional network is formed with polar molecules and the lecithin molecules.

Step 3: Formation of oleogels

Increasing the amount of polar organic liquid added to the 1-D network results in the formation of long tubular micelles with a radius of 2.0–2.5 nm and the length of these micelles is ranging from hundreds to thousands of nm. After reaching the critical length of the micelles, they tend to overlap and crosslink with each other to form a transient 3D network, i.e. oleogel.

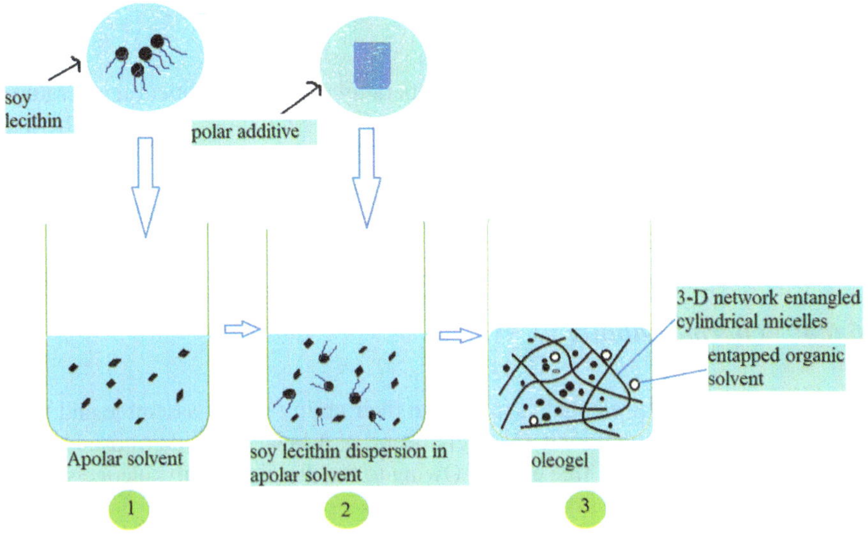

Fig. 6.15 Mechanism of oleogelation in nonpolar solvent

6.6 Properties of Oleogels

6.6.1 Viscoelasticity

A material is said to be viscoelastic if it contains both viscous and elastic properties (Sahoo et al. 2011). Oleogels display an elastic behaviour at low shear rates, whereas the increase in shear stress resulted in the weakening of the physical interactions in gel; further it tends to flow. The properties of the flow of gel can be known by observing its rheological properties in apolar medium during the gel formation. Rheological studies furnished the increase in the viscosity of the gel upon the addition of trace quantity of water in apolar solution of gelators, which is due to the formation of rigid structures by the entanglement of the fibrous structure. The viscosity of lecithin gels increased by the order of 10^4 to 10^6 upon the addition of water to lecithin in aprotic solution.

6.6.2 Non-birefringence

Generally, the morphology of oleogels displays dark matrix in the microscope, which is attributed to their isotopic nature; polarized light will not travel through the gel. This property is known as non-birefringence.

6.6.3 *Thermoreversibility*

When the oleogels are heated at high temperatures above its critical temperature, they lose their rigid solid matrix-like structure and begin to flow. The rise in thermal energy breaks the physical interactions in the oleogels which resulted in the gel-to-sol transition. However, when these organogels are cooled, the physical interactions of the gel become stronger and the gels get their solid-like matrix back.

6.6.4 *Thermostability*

Thermostability of the oleogels depends on their ability to self-assemble and form gels under suitable conditions. Oleogels occur as a low-energy thermostable network due to the release of free energy. Owing to the thermostability, oleogels act as an efficient drug delivery system.

6.6.5 *Optical Clarity*

Depending upon the composition and mode of aggregation, oleogels display a transparent and opaque character. Lecithin oleogels are transparent, whereas sorbitan monostearate oleogels are opaque.

6.6.6 *Chirality*

Generally, most of the properties are affected by the chirality in a gelator. Chirality provides various modes of molecular packing in the self-assembly process.

6.6.7 *Biocompatibility*

Biocompatibility of the oleogels depends upon the biocompatibility of the gelators chosen. The use of biocompatible oleogelators or gelators derived from renewable resources facilitates the practical applications of oleogels, especially in medicinal fields.

6.6.8 *Swelling*

Generally, polymer gels swell upon absorbing the solvent into the gel matrix; the solvent-gel interactions replace the existing gel-gel interactions and increase the volume of the gel. The extent of swelling is determined by the crosslinking network. The sol-to-gel and gel-to-sol transitions provide information about the nature of the microstructure forms during the gelation. When the elastic stress present in the gel is relaxed, the interstitial gaps for the solvent are decreased and the solvent is expelled out. This process is called syneresis.

6.7 Applications of Oleogels

Oleogels display a wide range of applications such as cosmetics, food industry, pharmaceuticals, drug delivery, plastics and coating applications (Fig. 6.16). Recent days, interest towards oleogels increased because of its diverse advantages such as the following (Esposito et al. 2017):

- They are easy to prepare.
- They are more stable when compared with other types of gels.
- Enhanced topical application of drugs.
- No first pass metabolism.
- These gels are moisture insensitive.
- Cost effective.

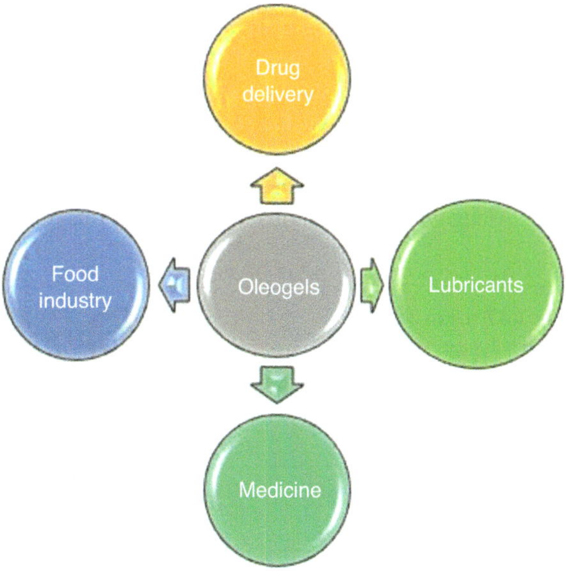

Fig. 6.16 Pictorial representation of applications of oleogels

- Thermodynamically stable.
- Drugs with short half-life are used.
- Releases the drug in a controlled way.
- Longer shelf life.
- Frequency of drug dosing is reduced.
- Easy to remove from the applied area on the skin.
- Diffusion rate of the drug is reduced.
- These gels can be incorporated with hydrophilic and lipophilic drugs.

6.7.1 Drug Delivery Systems

The applications of oleogel in drug delivery are limited because of their non-biocompatible nature. Recent research focusing on the synthesis and construction of biocompatible structural agents for oleogelation enables them suitable for medical applications (Fig. 6.17).

6.7.1.1 Topical Drug Delivery

Skin is the largest organ with the average surface area of 1.5–2 m^2; it contains large number of blood vessels which help in the transportation of the drugs. Generally, drugs are absorbed into the blood vessels by applying them topically. This mode of drug administration is very crucial for drug transportation were applied topically as ointments, creams, pastes, gels. Out of various forms of topical drug administration, gels have obtained more significance because of the simple absorption process through the layers of the skin (Patil et al. 2019) Recently extensive studies have been performed on topical drug delivery and huge progress was made in this field. Interest on these ointments is because of their increase in the absorption of water, restricting the evaporation of moisture present on the skin, high penetration and incorporation of ingredients that inhibit the skin irritation (Vintiloiu and Leroux 2008). Oleogels display high compatibility towards numerous hydrophobic drugs and generally considered as very safe to the human and release the drugs into the body in a controlled way. Along with these ointments, the other forms of semisolid dosages in the form of

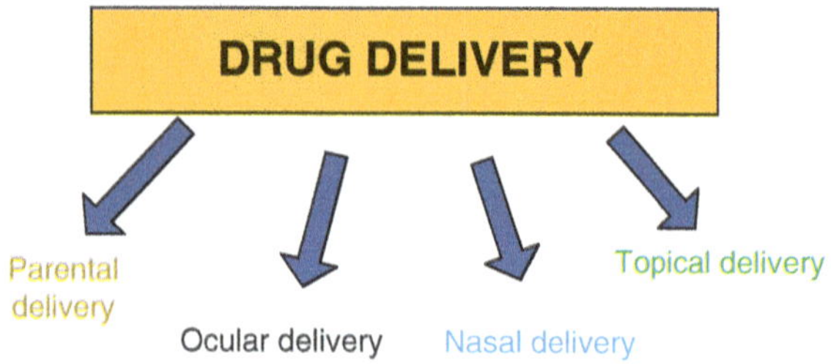

Fig. 6.17 Drug delivery applications of oleogels

emulsions also find potential in topical drug delivery (Esposito et al. 2017). Oleogels obtained from fatty acid ester and lecithins containing indomethacin display higher permeability in tissues (Fujii et al. 1996). Sorbitan monostearate and Tween-20 gelators in hexadecane can be used for carrying hydrophilic drugs (Lawrence and Rees 2012). The efficiency in drug delivering can be increased by using permeation enhancers like terpene, limonene, farnesol, cineole and linalool.

6.7.1.2 Parenteral Drug Delivery

Anhydrous oleogels are the best suitable drug delivery vehicle for parenteral administration of drugs. Sorbitan monostearate gelators along with other gelator modifiers are mainly used for this type of dosages because of their ability to release the drugs and antigens in a controlled way. For example, L-alanine-based oleogels are effective in the delivery of macromolecular biologically active agents into the body. In L-alanine derivatives, the gel-to-sol transition depends upon the nature of the solvent and concentration of the gelator. L-aniline derivatives obtained from safflower oil are used for long-term drug release up to 14–25 days of leuprolide, a luteinizing hormone for prostate cancer (Motulsky et al. 2005a, b) Overall, these gels provide a good platform for long-term drug delivery.

6.7.1.3 Ocular Drug Delivery

Generally, topical administration of a drug is preferred by many ocular treatments. There are different types of dosage forms of ocular drugs such as eye drops, ointments, suspensions and recently developed "eyelid skin delivery systems". The most commonly used dosage form is eye drops. However, the main drawback of the eye drops is that they won't stay on the eyes and get diluted by the tears which come out resulting in the delivery of a small proportion of drug at the targeted site. The use of suspensions is also not efficient because the rate at which the drug is released depends upon the dissolution of the drug in the suspension. The use of lecithin oleogels overcomes the above-mentioned limitations because of its viscoelastic character and hence cannot be removed by tears. Among the various gels reported, paraffin-based oleogels are found to be the safest of all gels (Esposito et al. 2017).

6.7.1.4 Nasal Drug Delivery

Administration of drugs through nasal route is gaining more importance recently because it is an easily accessible route, highly permeable and richly vascularized. Sorbitan monostearate-based oleogels with propranolol hydrochloride are used for trans-nasal drug delivery (Bhoyar et al. 2012).

6.7.2 Nutraceutical Applications

In the field of food chemistry and technology, research has been done to modify the physical characteristics of oils to replace saturated and trans fats. In this manner, oil-based structuring agents can bring a specific texture and rheology to the food even without changing the quality of the final product. The most suitable way of forming these materials is by incorporating specific materials like polymers, waxes and amphiphiles into oils to produce oleogels (Pehlivanoğlu et al. 2016) Ethyl cellulose has a great capacity to bind with oil at levels of 10% and below, resulting in the formation of oleogels with diverse properties. These oleogels formed from the ethyl cellulose have successfully reduced the risk of cardiovascular diseases by replacing the highly saturated beef fat in comminute meat food products (Jimenez-Colmenero et al. 2015).

6.7.3 Medicinal Applications

Oleogels and their composites can be used for the treatment of many diseases such as rheumatoid arthritis, osteoarthritis, Reiter's syndrome, dysmenorrhoea, post-operative pain, pyrexia and acute gout treatment (Knani 2019).

Lecithin organogels act as potential drug delivery vehicles for many biologically active agents that can treat ageing of cells (Raut et al. 2012) Ointments based on oleogels have many advantages when compared to conventional ointments. For example, cyclosporine encapsulated in sorbitan monooleate gelator administrated orally displayed enhanced activity. In addition, oleogel derived from Eudragit L and S finds potential in rectal drug delivery (Sahoo et al. 2011). Hydrophilic vaccines can also be delivered by microemulsion-based oleogels. They produce depot effect by slowly releasing the antigens (Vintiloiu and Leroux 2008).

6.7.4 Lubricants

A detailed investigation in the field of lubricants reveals that the decomposition temperatures of oleogels are much higher than the standard lubricating grease and other different conventional bases used for topical administration. Their chemical stability of oleogels is also higher than the lubricating greases. Oleogels made from the castor oil and ethyl cellulose, α-cellulose, methyl cellulose blends are treated as environmentally friendly lubricating greases with high thermal and mechanical stability. Oleogels composed of sorbitan and glyceryl monostearates and other various classes of vegetable oils are used as biodegradable alternatives to the lubricating greases (Garcés et al. 2011). Using low-viscosity oils like rapeseed oil and soybean oils generate oleogels with high values for the linear viscoelastic functions.

Table 6.1 Food applications of various organogelators

S. No	Type of gelator	Applications
1	12-Hydroxystearic acid	Controlled release of β-carotene, decreased syneresis in peanut butter
2	Monoglyceride and carnauba wax	Reduces saturated fat in margarine
3	Monostearin	Releases curcuminoids in a controlled manner
4	Ethyl cellulose	Reduces saturated fats in frankfurters
5	Soy lecithin and palm oil gel	Releases nutraceuticals in a controlled manner
6	Shellac	Restricts oil migration in chocolate paste
7	Lecithin	Releases nutraceuticals in a controlled manner
8	Rice bran wax	Reduces saturated fat in ice cream
9	Sunflower wax soya bean oil	Reduces saturated fat in margarine
10	Fatty acids and fatty alcohols	Meat suspensions

6.7.5 *Food Industry*

The main objective of using oleogels in the food industry is to replace solid fats which have adverse health effects (Table 6.1). These constituents of oleogels are rich in unsaturated fats which are healthier than solid fat (Botega et al. 2013) However, applications of oleogel in food industries are limited because of the unavailability of low-cost oleogels suitable for food grade products. For example, soft characteristics for a cookie can be produced using candelilla wax gelator of canola oil (Jang et al. 2015) To reduce the saturated fat content in frankfurters, γ-oryzanol and β-sitosterol-based oleogel emulsions are used in the place of pork back fat. The main advantage of ethyl cellulose oleogels is their interconnected structures with pore size ranging from 0.5 to 6 μm, where the oil is trapped. Cakes prepared with shellac oleogels have significant differences in its size, volume, stickiness and sponginess, when compared with the cakes made with shortenings (Patel et al. 2014). Shortenings are mainly used for their foamability without breaking the batter. Oleogels also show similar foamability as like traditional shortenings and can provide thermal stability for chocolates. Shellac resins are used as structuring agent in chocolates.

6.8 Conclusion and Future Perspectives

A detailed investigation in the field of gel chemistry and technology has developed many edible applications both in academic and industrial, despite their limited commercial availability (John et al. 2019; Thamizhanban and Lalitha 2019) In particular, oleogels and organogels display unique properties such as thermoreversibility, viscoelasticity, self-healing and processability (Lalitha et al. 2014; Muthusamy et al. 2016; Lalitha et al. 2015a, b, 2017; Prasad et al. 2017; Prasad et al. 2018). By considering the current potential applications of oleogels and futuristic scope, recently many research groups are showing interest in synthesizing or finding the new molecules that can act as structuring agents (Co and Marangoni 2018; Jadhav et al. 2013; Patel 2015; Silverman and John 2015). Oleogels are identified as the most efficient drug delivery vehicles because of their biocompatibility, longer shelf life, ease in preparation and thermoreversible nature (Rehman and Zulfakar 2014). In the field of drug delivery, oleogels can accommodate both hydrophilic and hydrophobic drugs and deliver in the targeted site in a controlled manner. Oleogels obtained from renewable resources and semisynthetic natural molecules play a crucial role in the food industry as a healthier version, which are capable of promoting oil gelation even with small concentrations. In the near future, oleogels may completely replace the conventional forms of drug delivery in medicine, oil structuring agents in food industries and cosmetics.

Acknowledgements Authors thank SERB, Department of Science and Technology (CRG/2018/001386), and SPARC, Ministry of Human Resource Development (SPARC/2018–2019/P263/SL), India, for financial support. S. N. thanks National Institute of Technology, Warangal, for infrastructure facilities.

References

Abdallah DJ, Lu L, Weiss RG (1999) Thermoreversible organogels from alkane gelators with one heteroatom. Chem Mater 11(10):2907–2911. https://doi.org/10.1021/cm9902826

Alemán JV, Chadwick AV, He J, Hess M, Horie K, Jones RG et al (2007) Definitions of terms relating to the structure and processing of sols, gels, networks, and inorganic-organic hybrid materials (IUPAC Recommendations 2007). Pure Appl Chem 79(10):1801–1829. https://doi.org/10.1351/pac200779101801

Arsem BYWC (1926) Gel structure. J Phys Chem C 30(3):306–311. https://doi.org/10.1021/j150261a003

Aykent G, Zeytun C, Marion A, Özçubukçu S (2019) Simple tyrosine derivatives act as low molecular weight organogelators. Sci Rep 9(1):1–8. https://doi.org/10.1038/s41598-019-41142-z

Balasubramanian R, Sughir AA, Damodar G (2012) Oleogel: a promising base for transdermal formulations. Asian J Pharm 6(1):1–9. https://doi.org/10.4103/0973-8398.100118

Bhoyar N, Giri TK, Tripathi DK, Alexander A, A. (2012) Recent advances in novel drug delivery systems through gels: review. J Pharm Allied Health Sci 2(2):21–39. https://doi.org/10.3923/jpahs.2012.21.39

Botega DCZ, Marangoni AG, Smith AK, Goff HD (2013) The potential application of rice bran wax oleogel to replace solid fat and enhance unsaturated fat content in ice cream. J Food Sci 78(9):1334–1339. https://doi.org/10.1111/1750-3841.12175

Brizard A, Oda R, Huc I (2018) Chirality effects in self-assembled fibrillar networks chirality effects in self-assembled fibrillar networks. Top Curr Chem 256:167–218. https://doi.org/10.1007/b107174

Buerkle LE, Rowan SJ (2012) Supramolecular gels formed from multi-component low molecular weight species. Chem Soc Rev 41(18):6089–6102. https://doi.org/10.1039/c2cs35106d

Calligaris S, Da S, Arrighetti G, Barba L (2010) Effect of the structure of monoglyceride–oil–water gels on aroma partition Food Research International effect of the structure of monoglycer-

ide–oil–water gels on aroma partition. Food Res Int 43(3):671–677. https://doi.org/10.1016/j.foodres.2009.10.011
Chen CH, Terentjev EM (2009) Aging and metastability of monoglycerides in hydrophobic solutions. Langmuir 25(12):6717–6724. https://doi.org/10.1021/la9002065
Co ED, Marangoni AG (2018). Oleogels: an introduction. In: Edible oleogels, pp 1–29. https://doi.org/10.1016/b978-0-12-814270-7.00001-0
Dassanayake LSK, Kodali DR, Ueno S, Sato K (2009) Physical properties of organogels made of rice bran wax and vegetable oils. In: Edible oleogels: structure and health implications, 2nd ed. https://doi.org/10.1016/B978-0-9830791-1-8.50010-3
Davidovich-pinhas M, Gravelle AJ, Barbut S, Marangoni AG (2015) Food hydrocolloids temperature effects on the gelation of ethylcellulose oleogels. Food Hydrocolloids 46:76–83. https://doi.org/10.1016/j.foodhyd.2014.12.030
Duan P, Cao H, Zhang L, Liu M (2014) Gelation induced supramolecular chirality: chirality transfer, amplification and application. Soft Matter 10(30):5428–5448. https://doi.org/10.1039/c4sm00507d
Esposito CL, Kirilov P, Roullin VG (2017) Organogels, promising drug delivery systems: an update of state- of-the-art and recent applications. J Control Release 271:1. https://doi.org/10.1016/j.jconrel.2017.12.019
Estroff LA, Hamilton AD (2004) Water gelation by small organic molecules. Chem Rev 104(3):1201–1217. https://doi.org/10.1021/cr0302049
Fuhrhop J, Actually F (1993) Fiuld and solid fibers made of lipid molecular bilayers. Chem Rev 93:1565–1582. https://doi.org/10.1021/cr00020a008
Fujii M, Shiozawa K, Henmi T, Yamanouchi S, Yamashita N, Matsumoto M (1996) Skin permeation of indomethacin from gel formed by fatty-acid ester and phospholipid. Int J Pharm 137:117–124. https://doi.org/10.1016/0378-5173(96)04510-3
Gandolfo FG, Bot A, Flöter E (2004) Structuring of edible oils by long-chain FA, fatty alcohols, and their mixtures. J Am Oil Chem Soc 81(1):1–6. https://doi.org/10.1007/s11746-004-0851-5
Garcés R, Martínez-Force E, Salas JJ (2011) Vegetable oil basestocks for lubricants. Grasas Aceites 62(1):21–28. https://doi.org/10.3989/gya.045210
George M, Weiss RG (2006) Molecular organogels, soft matter comprised of low molecular mass organic gelators and organic liquids. Account Chem Res 39(8):489. https://doi.org/10.1021/ar0500923
Guo S, Song M, He X, Yang F, Cao Y, Rogers M, Lan Y (2019) Water-induced self-assembly of mixed gelator system (ceramide and lecithin) for edible oil structuring. Food Funct 10:3923–3933. https://doi.org/10.1039/c9fo00473d
Hanabusa K, Suzuki M (2014) Development of low-molecular-weight gelators and polymer-based gelators. Polym J 46(11):776–782. https://doi.org/10.1038/pj.2014.64
Higaki K, Sasakura Y, Koyano T, Hachiya I, Sato K (2003) Physical analyses of gel-like behavior of binary mixtures of high- and low-melting fats. J Am Oil Chem Soc 80(3):263–270. https://doi.org/10.1007/s11746-003-0687-z
Hu J, Zhang M, Ju Y (2009) A simple oleanlic acid derivative as potent organogelator. Soft Matter 5(24):4971–4974. https://doi.org/10.1039/b916427h
Huang X, Terech P, Raghavan SR, Weiss RG (2005) Kinetics of 5α-cholestan-3β-yl N-(2-naphthyl) carbamate/n-alkane organogel formation and its influence on the fibrillar networks. J Am Chem Soc 127(12):4336–4344. https://doi.org/10.1021/ja0426544
Jadhav SR, Hwang H, Huang Q, John G (2013) Medium-chain sugar amphiphiles: a new family of healthy vegetable oil structuring agents. J Agric Food Chem 61(49):12005–12011. https://doi.org/10.1021/jf401987a
Jang A, Bae W, Hwang H, Gyu H, Lee S (2015) Evaluation of canola oil oleogels with candelilla wax as an alternative to shortening in baked goods. Food Chem 187:525–529. https://doi.org/10.1016/j.foodchem.2015.04.110
Jimenez-colmenero F, Salcedo-sandoval L, Bou R, Herrero AM, Ruiz-capillas C (2015) Novel applications of oil-structuring methods as a strategy to improve the fat content of meat products. Trends Food Sci Technol 44:177. https://doi.org/10.1016/j.tifs.2015.04.011

John G, Nagarajan S, Kumar P, Silverman JR, Pillai CKS (2019) Natural monomers: a mine for functional and sustainable materials—occurrence, chemical modification and polymerization. Prog Polym Sci 92:158–209. https://doi.org/10.1016/j.progpolymsci.2019.02.008

Kang J, Che Y, Yan N, Cao D (2019) Evaluation system construction and factor impact analysis of silica-gel adsorption to extract phytosterol glycosides from soybean lecithin powder. J Sci Food Agric 99(9):4287–4295. https://doi.org/10.1002/jsfa.9661

Knani D (2019) Low molecular weight hydro-and organo gelators used for medical applications. Biomed J Sci Techn Res 12(5):9553–9554. https://doi.org/10.26717/bjstr.2019.12.002315

Lalitha K, Jenifer P, Prasad YS, Muthusamy K, John G, Nagarajan S (2014) A self-assembled p-conjugated system as an anti-proliferative agent in prostate cancer cells and and a probe for intra-cellular imaging. RSC Adv 4(Scheme 1):48433–48437. https://doi.org/10.1039/C4RA07710E

Lalitha K, Prasad YS, Maheswari CU, Sridharan V, John G, Nagarajan S (2015a) Stimuli responsive hydrogels derived from a renewable resource: synthesis, self-assembly in water and application in drug delivery. J Mater Chem B 3(27):5560–5568. https://doi.org/10.1039/c5tb00864f

Lalitha K, Prasad YS, Sridharan V, Maheswari CU, John G, Nagarajan S (2015b) A renewable resource-derived thixotropic self-assembled supramolecular gel: magnetic stimuli responsive and real-time self-healing behaviour. RSC Adv 5(95):77589–77594. https://doi.org/10.1039/c5ra14744a

Lalitha K, Sridharan V, Maheswari CU, Vemula PK, Nagarajan S (2017) Morphology transition in helical tubules of a supramolecular gel driven by metal ions. Chem Commun 53(9):1538–1541. https://doi.org/10.1039/c6cc09120b

Lawrence MJ, Rees GD (2012) Microemulsion-based media as novel drug delivery systems. Adv Drug Deliv Rev 64:175–193. https://doi.org/10.1016/j.addr.2012.09.018

Lloyd J (1926) Colloid chemistry: theoretical and applied. Edited by Jerome Alexander. Vol. I. J Soc Chem Ind 1:111. https://doi.org/10.1002/jctb.5000454915

Maffezzoni R, Zanda M (2008) Stable organogels derived from triazines functionalized with chiral α-amino acid derivatives. Tetrahedron Lett 49(35):5129–5132. https://doi.org/10.1016/j.tetlet.2008.06.101

Martins AJ, Vicente AA, Cunha RL, Cerqueira MA (2018) Function edible oleogels : an opportunity for fat. Food Funct 9:758–773. https://doi.org/10.1039/c7fo01641g

Matheson A, Dalkas G, Clegg PS, Euston SR (2018) Phytosterol-based edible oleogels: a novel way of replacing saturated fat in food. Nutr Bull 43(2):189–194. https://doi.org/10.1111/nbu.12325

Mattice KD, Marangoni AG (2018) New insights into wax crystal networks in oleogels. Food Chem Funct Anal January(3):71–94. https://doi.org/10.1039/9781788010184-00069

Motulsky A, Hoarau D, Ong H, Leroux J (2005a) First report on the efficacy of l -alanine-based in situ-forming implants for the long-term parenteral delivery of drugs. J Control Release 108:433–441. https://doi.org/10.1016/j.jconrel.2005.08.016

Motulsky A, Lafleur M, Couffin-Hoarau AC, Hoarau D, Boury F, Benoit JP, Leroux JC (2005b) Characterization and biocompatibility of organogels based on L-alanine for parenteral drug delivery implants. Biomaterials 26(31):6242–6253. https://doi.org/10.1016/j.biomaterials.2005.04.004

Murdan S, Gregoriadis G, Florence AT (1999) Novel sorbitan monostearate organogels. J Pharm Sci 88(6):608–614. https://doi.org/10.1021/js980342r

Muthusamy K, Sridharan V, Maheswari CU, Nagarajan S (2016) Lipase catalyzed synthesis of fluorescent glycolipids: gelation studies and graphene incorporated self-assembled sheet formation for semiconductor applications. Green Chem 18(13):3722–3731. https://doi.org/10.1039/c6gc00347h

Nikiforidis CV, Scholten E (2014) Self-assemblies of lecithin and α-tocopherol as gelators of lipid material. RSC Adv 4(5):2466–2473. https://doi.org/10.1039/c3ra46584e

Online VA, Patel AR, Schatteman D, Dewettinck K (2013) A foam-templated approach for fabricating organogels using a water-soluble polymer. RSC Adv 3:22900–22903. https://doi.org/10.1039/c3ra44763d

Palanisamy M, Himabindu M (2017) Ultrasound- and temperature-induced gelation of gluconosemicarbazide gelator in DMSO and water mixtures. Gels 3. https://doi.org/10.3390/gels3020012

Patel AR (2015) Alternative routes to oil structuring. Springer Briefs Food Health Nutr:15–27. https://doi.org/10.1007/978-3-319-19138-6

Patel AR (2017) A colloidal gel perspective for understanding oleogelation. Curr Opin Food Sci 15:1–7. https://doi.org/10.1016/j.cofs.2017.02.013

Patel AR, Babaahmadi M, Lesaffer A, Dewettinck K (2015) Rheological profiling of organogels prepared at critical gelling concentrations of natural waxes in a triacylglycerol solvent. J Agric Food Chem 63(19):4862–4869. https://doi.org/10.1021/acs.jafc.5b01548

Patel AR, Rajarethinem PS, Grędowska A, Turhan O, Lesaffer A, De Vos WH et al (2014) Edible applications of shellac oleogels: spreads, chocolate paste and cakes. Food Funct 5(4):645–652. https://doi.org/10.1039/c4fo00034j

Patil PB, Datir SK, Saudagar RB (2019) A review on topical gels as drug delivery systems. J Drug Deliv Ther 9:989–994. https://doi.org/10.22270/jddt.v9i3-s.2930

Pehlivanoğlu H, Demirci M, Toker OS, Konar N (2016) Oleogels, a promising structured oil for decreasing saturated fatty acid concentrations: production and food-based applications. Crit Rev Food Sci Nutr 58:1–12. https://doi.org/10.1080/10408398.2016.1256866

Pernetti M, van Malssen KF, Flöter E, Bot A (2007b) Structuring of edible oils by alternatives to crystalline fat. Curr Opin Colloid Interface Sci 12:221–231. https://doi.org/10.1016/j.cocis.2007.07.002

Pernetti M, van Malssen K, Kalnin D, Flöter E (2007a) Structuring edible oil with lecithin and sorbitan tri-stearate. Food Hydrocoll 21(5–6):855–861. https://doi.org/10.1016/j.foodhyd.2006.10.023

Prasad YS, Miryala S, Lalitha K, Ranjitha K, Barbhaiwala S, Sridharan V et al (2017) Disassembly of bacterial biofilms by the self-assembled glycolipids derived from renewable resources. ACS Appl Mater Interfaces 9(46):40047–40058. https://doi.org/10.1021/acsami.7b12225

Prasad YS, Saritha B, Tamizhanban A, Lalitha K, Kabilan S, Maheswari CU et al (2018) Enzymatic synthesis and self-assembly of glycolipids: robust self-healing and wound closure performance of assembled soft materials. RSC Adv 8(65):37136–37145. https://doi.org/10.1039/C8RA07703G

Raut S, Singh S, Uplanchiwar V, Mishra V (2012) Lecithin organogel: a unique micellar system for the delivery of bioactive agents in the treatment of skin aging. Acta Pharm Sin B 2(1):8–15. https://doi.org/10.1016/j.apsb.2011.12.005

Rehman K, Zulfakar MH (2014) Recent advances in gel technologies for topical and transdermal drug delivery. Drug Dev Ind Pharm 40:433–440. https://doi.org/10.3109/03639045.2013.828219

Rogers MA, Strober T, Bot A, Toro-vazquez JF, Stortz T, Marangoni AG (2014) Edible oleogels in molecular gastronomy. Int J Gastronomy Food Sci 2(1):22–31. https://doi.org/10.1016/j.ijgfs.2014.05.001

Rogers MA, Weiss RG (2015) Systematic modifications of alkane-based molecular gelators and the consequences to the structures and properties of their gels. New J Chem 39(2):785–799. https://doi.org/10.1039/c4nj01439a

Romoscanu AI, Mezzenga R (2006) Emulsion-templated fully reversible protein-in-oil gels. Langmuir 22(18):7812–7818. https://doi.org/10.1021/la060878p

Saboya RMA, Cecilia JA, García-Sancho C, Sales AV, de Luna FMT, Rodríguez-Castellón E, Cavalcante CL (2017) Synthesis of biolubricants by the esterification of free fatty acids from castor oil with branched alcohols using cationic exchange resins as catalysts. Ind Crop Prod 104:52–61. https://doi.org/10.1016/j.indcrop.2017.04.018

Sagiri SS, Samateh M, John G (2018) Biobased molecular structuring agents. Food Chem Funct Anal 2018(January):25–52. https://doi.org/10.1039/9781788010184-00023

Sahoo S, Kumar N, Bhattacharya C, Sagiri SS, Jain K, Pal K et al (2011) Organogels: properties and applications in drug delivery. Des Monomers Polym 14:95–108. https://doi.org/10.1163/138577211X555721

Sakai H, Saitoh T, Endo T, Tsuchiya K, Sakai K, Abe M (2009) Phytosterol ethoxylates in room-temperature ionic liquids: excellent interfacial properties and gel formation. Langmuir 25(5):2601–2603. https://doi.org/10.1021/la900139d

Sangeetha NM, Maitra U (2005) Supramolecular gels: functions and uses. Chem Soc Rev 34:821–836. https://doi.org/10.1039/b417081b

Sawalha H, Venema P, Van Der Linden E, Den Adel R, Bot A, Flöter E (2012) Organogel-emulsions with mixtures of β-sitosterol and γ-oryzanol: influence of water activity and type of oil phase on gelling capability. J Agric Food Chem 60(13):3462–3470. https://doi.org/10.1021/jf300313f

Silverman JR, John G (2015) Biobased fat mimicking molecular structuring agents for medium-chain triglycerides (MCTs) and other edible oils. J Agric Food Chem 63(48):10536–10542. https://doi.org/10.1021/acs.jafc.5b04236

Singh A, Auzanneau F, Rogers MA (2017) Advances in edible oleogel technologies—a decade in review. Food Res Int 97(April):307–317. https://doi.org/10.1016/j.foodres.2017.04.022

Sullivan CMO, Barbut S, Marangoni AG (2016) Edible oleogels for the oral delivery of lipid soluble molecules: composition and structural design considerations. Trends Food Sci Technol 57:59. https://doi.org/10.1016/j.tifs.2016.08.018

Suzuki M, Hanabusa K (2010) Polymer organogelators that make supramolecular organogels through physical cross-linking and self-assembly. Chem Soc Rev 39(2):455–463. https://doi.org/10.1039/b910604a

Thamizhanban A, Lalitha K (2019) Self-assembled soft materials for energy and environmental applications. Environ Chem Sustain World 23:443–470. 10.1007%2F978-3-030-04474-9

Vintiloiu A, Leroux J (2008) Organogels and their use in drug delivery—a review. J Control Release 125:179–192. https://doi.org/10.1016/j.jconrel.2007.09.014

Wang D, Niu J, Wang Z, Jin J (2015a) Monoglyceride-based organogelator for broad-range oil uptake with high capacity. Langmuir 31(5):1670–1674. https://doi.org/10.1021/acs.langmuir.5b00053

Wang S, Xue P, Wang P, Yao B, Li K, Liu B (2015b) Fluorescence-enhanced gels of D-π-A diphenylacrylonitrile derivatives: influence of nitro group and alkoxy chain. Dyes Pigments 122:302. https://doi.org/10.1016/j.dyepig.2015.07.009

Wu H, Morbidelli M (2001) A model relating structure of colloidal gels to their elastic properties. Langmuir 17(11):1030–1036. https://doi.org/10.1021/la001121f

Yang HK, Wang XM, Liu LL, Shi HX (2019) Design and gelation behaviors of cholesterol-based derivatives as organogelators: an investigation of the correlation between molecular structures and gelation behaviors. New J Chem 43(8):3366–3373. https://doi.org/10.1039/C8NJ02899K

Yu X, Chen X, Chai Q, Ayres N (2016) Synthesis of polymer organogelators using hydrogen bonding as physical cross-links. Colloid Polym Sci 294(1):59–68. https://doi.org/10.1007/s00396-015-3797-z

Zang L, Shang H, Wei D, Jiang S (2013) A multi-stimuli-responsive organogel based on salicylidene Schiff base. Sensors Actuat B Chem 185:389–397. https://doi.org/10.1016/j.snb.2013.05.022

Chapter 7
Advances in Hydrothermal Carbonization of Livestock Manure

Chun-Huo Zhou, Hua-Jun Huang, Lin Li, Zi-Qian Pan, Xiao-Feng Xiao, and Jia-Xin Wang

Contents

Abbreviations

DTPA Diethylenetriamine pentaacetic acid
EU European Union
GHG Greenhouse gases
TCLP Toxicity characteristic leaching procedure

C.-H. Zhou · H.-J. Huang (✉) · L. Li · Z.-Q. Pan · X.-F. Xiao · J.-X. Wang
College of Land Resources and Environment, Jiangxi Agricultural University, Nanchang City, People's Republic of China

Inamuddin, A. M. Asiri (eds.), *Sustainable Green Chemical Processes and their Allied Applications*, Nanotechnology in the Life Sciences,
https://doi.org/10.1007/978-3-030-42284-4_7

7.1 Introduction

With the sustained and rapid development of social economy in the world, people's living standards are gradually rising. As a result, people's demand for animal protein continues to increase. In response to this social need, livestock farming industry shows a rapid development trend of scale, centralization, and modernization (Spielmeyer 2018). One must not overlook the fact that a lot of manure will be produced in the process of livestock breeding, which needs proper treatment. Otherwise, it will cause serious environmental problems (Shi et al. 2018). In the European Union (EU), livestock production produces about 14.0 billion tons of manure a year (Loyon 2017). In China, the amount of livestock manure produced in livestock production is about 3.2 billion tons every year (Zhang et al. 2012). In general, livestock manure will be first stored and then used in various fields of agricultural production. However, the long-term excessive direct use of livestock manure is often accompanied by a series of soil, water, and air pollution problems (Fig. 7.1) (Bidart et al. 2014; Kumar et al. 2013; Loyon 2017). The environmental pollution problems caused by livestock manure have become a limiting factor for the sustainable development of livestock breeding industry. Hence, it is particularly important to treat livestock manure properly in order to ensure that the environmental risk caused by the application of livestock manure is minimized.

At present, the treatment methods of livestock manure can be mainly divided into two categories: biochemical process (aeration or anaerobic digestion,

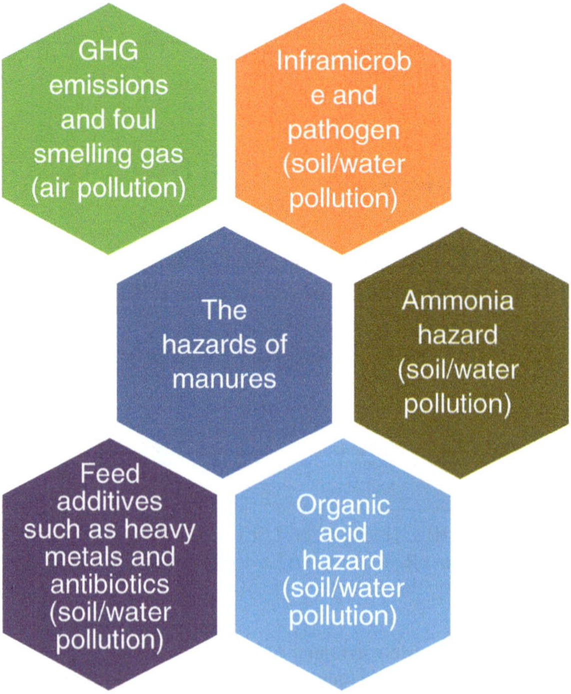

Fig. 7.1 The hazards of livestock manure (Summarized from literature, Niu and Li 2016; Kumar et al. 2013). *GHG* greenhouse gases

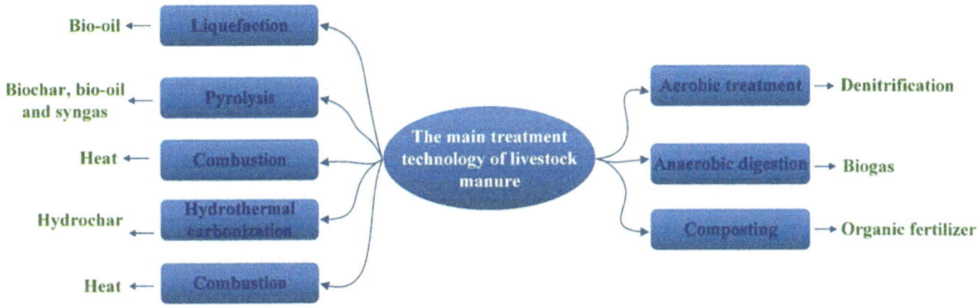

Fig. 7.2 The main treatment processes of livestock manure

composting) and thermochemical process (liquefaction, pyrolysis, combustion, gasification, and hydrothermal carbonization) (Fig. 7.2). During the process of aerobic digestion treatment, anoxic and aerobic phases can be alternated in space or time, achieving nitrification and denitrification to remove nitrogen, which can also be realized by low aeration (Béline and Martinez 2002). This process can lead to the emission of nitrogen (N_2), sometimes forming nitrous oxide under adverse conditions (Loyon et al. 2007). In the anaerobic digestion process, organic matter will be decomposed in the aid of microorganisms under anoxic conditions, producing biogas consisting mainly of CH_4 and CO_2 (Nasir et al. 2012; Vlyssides et al. 2015). Anaerobic digestion of livestock manure can produce biogas, which can be regenerated into heat or electricity in situ. Livestock manure used for anaerobic digestion would become nutrient-rich digestion, which may replace fertilizer in agriculture (Niu and Li 2016; Tambone et al. 2015). Composting is a traditional treatment method which has been used for a long time. It mainly degrades organic matter in livestock manure through the microbial aerobic process. The primary purposes of composting include three aspects: recycling nutrients defecated by animals, stabilizing organic matter before carriage and utilization, and reducing fecal pathogens (Bernal et al. 2009).

Liquefaction of livestock manure is usually carried out at medium temperatures (250–400 °C) and high pressures (5–25 MPa) with a suitable solvent (water or organic solvent) that results in bio-oil as target product (Islam and Park 2018). Pyrolysis is a useful method to get liquid (bio-oil), solid (biochar), and gas products from livestock manure, which is generally carried out at 300–700 °C in an oxygen-free environment (Erdogdu et al. 2019; Jiang et al. 2018; Zeng et al. 2018). The combustion of livestock manure is applied to recover heat production (Lundgren and Pettersson 2009; Zhang et al. 2019a). The gasification process is a partly oxidizing process, which can convert livestock manure into syngas in the presence of different gasifying agents such as steam, air, or CO_2. The gas product mainly consists of H_2, CO, CO_2, CH_4, C_2H_6, C_2H_4, C_2H_2, and other higher series of hydrocarbons (Cao et al. 2016; Hussein et al. 2017; Tańczuk et al. 2019). Hydrothermal carbonization of livestock manure is often performed at relatively low temperatures (180–280 °C) and autogenous pressures for minutes to hours, mainly producing hydrochar (Gao et al. 2018; Ghanim et al. 2017; Lang et al. 2019c).

The potential applications of hydrochar are very extensive, including soil amendment, solid energy, electrical condenser, cheaper adsorbent, and medical applications (Fang et al. 2018). Compared with other thermochemical technologies, such as pyrolysis, combustion, and gasification, hydrothermal carbonization can treat high humidity biomass, thus saving the cost of deep dehydration of biomass. In addition, the hydrothermal carbonization process is carried out at relatively low treatment temperatures, which means lower energy consumption (Nizamuddin et al. 2017; Yang et al. 2019). Thus, in recent years, the number of researches on the hydrothermal carbonization of biomass (including livestock manure) has increased rapidly. At the same time, some articles focusing on the advances in the hydrothermal carbonization of biomass have also been published (Table 7.1).

As listed in Table 7.1, Yang et al. (2019) compared the hydrothermal carbonization technology with other carbonization strategies, such as pyrolysis, ionothermal carbonization, and molten salt carbonization. Among these various workable synthetic technologies, hydrothermal carbonization is highly praised for its energy efficiency and the capacity to synthesize widely used carbon-containing materials. The influences of process factors on the hydrothermal carbonization of biomass have been systematically summarized by Nizamuddin et al. (2017), Heidari et al. (2018), Wang et al. (2018), and Zhang et al. (2019b). Reviews written by Hu et al. (2010) and Titirici and Antonietti (2010) discussed the preparation of engineering carbon materials from the high-/low-temperature hydrothermal carbonization of biomass and its application in energy, catalysis, environment, sensor, and biology. The

Table 7.1 Recent main review papers focused on the hydrothermal carbonization of biomass

Items	Main contributions	References
1	Comparing the hydrothermal carbonization technology with other carbonization strategies, such as pyrolysis, ionothermal carbonization, and molten salt carbonization	Yang et al. (2019)
2	The influence of process factors, the application of products, main challenges of hydrothermal carbonization	Heidari et al. (2018)
3	The influence of process factors	Nizamuddin et al. (2017)
4	The influence of process factors, the formation mechanism of hydrochar, the characterization of hydrochar	Wang et al. (2018)
5	Elaborating the preparation of carbonaceous material by the hydrothermal carbonization of biomass at high/low temperature and its applications	Hu et al. (2010), Titirici and Antonietti (2010)
6	Characterization and application of aqueous products	Usman et al. (2019)
7	Potential applications of hydrochar	Fang et al. (2018)
8	Hydrothermal carbonization of food waste	Zhou et al. (2018)
9	The mechanism of reaction, the properties and applications of hydrochar	Krylova and Zaitchenko (2018)
10	The influence of process factors, the properties of hydrochar, the environmental impact of hydrochar, the modification and application of hydrochar	Zhang et al. (2019a, b)

formation mechanism, properties, and application of hydrochar were systematically introduced by Heidari et al. (2018), Wang et al. (2018), Krylova and Zaitchenko (2018), Fang et al. (2018), and Zhang et al. (2019b). In the work of Heidari et al. (2018), the main challenges of hydrothermal carbonization were presented in details. The properties and application of aqueous products from the hydrothermal carbonization of biomass can consult the review paper published by Usman et al. (2019). Zhou et al. (2018) explored the possibility of treating food waste by coupling hydrothermal carbonization technology with anaerobic digestion technology and summarized the effects of process parameter on the hydrothermal carbonization of food waste. In the chapter of Zhang et al. (2019b), the environmental impact of hydrochar was systematically discussed. However, as far as we know, there are no related review papers/chapters focused on the hydrothermal carbonization of livestock manure.

This chapter aims to systematically summarize the hydrothermal carbonization of livestock manure in hopes of providing a reference for follow-up research. The specific contributions of this chapter are as follows: (1) the summary of the influences of fundamental process factors, including reaction temperature, reaction time, solid-liquid ratio, catalyst, heating mode, and feedstock; (2) the summary of the transport/conversion characteristics of heavy metals in the hydrothermal carbonization of livestock manure; and (3) the summary of the application/treatment of process water.

7.2 The Basic Process of Hydrothermal Carbonization

A schematic of the basic process of hydrothermal carbonization for livestock manure is shown in Fig. 7.3. Generally speaking, the equipment for the hydrothermal carbonization of livestock manure is made of stainless steel and is equipped with a magnetic stirrer. A sequence of structural and chemical changes will be taken place during the hydrothermal carbonization of livestock manure. In a typical hydrothermal carbonization run, livestock manure slurry is firstly loaded into the reactor, and then the air or oxygen in the reactor airspace is removed by aerating reductive or inert gases, such as H_2, N_2, and CO. Then, the reactor will be heated to the desired temperate, which will be maintained for the set reaction time. The reaction temperature of the reactor is usually controlled by a proportional-integral-derivative controller. After the reaction is finished, the reactor is cooled by circulating water or fan. When it is lowered to room temperature, the reactor will be opened and the reaction products, including hydrochar (target product), gaseous products (by-product), and process water (by-product), are separated and collected.

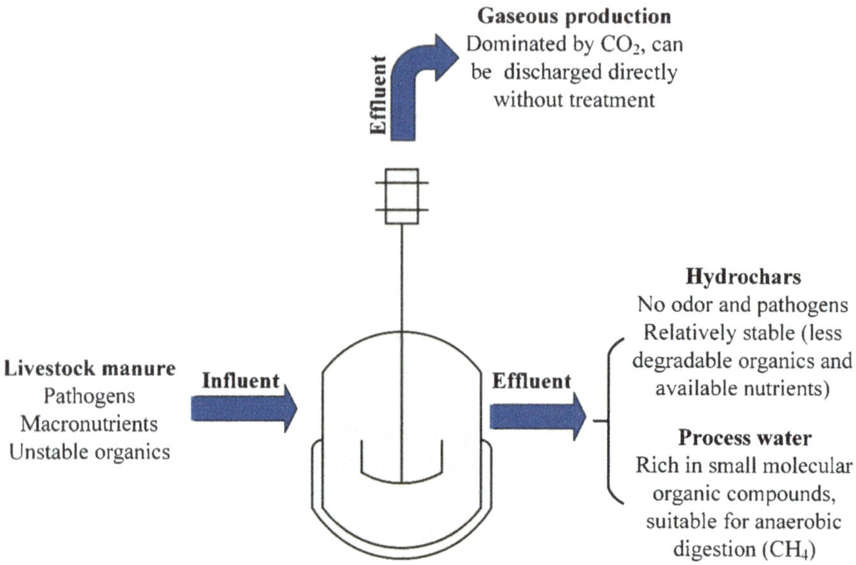

Fig. 7.3 The basic process of hydrothermal carbonization of livestock manure

7.3 The Effects of Main Process Parameters

The main process parameters affecting the hydrothermal carbonization of livestock manure can be divided into two categories: the characteristics of raw manure materials and the reaction parameters of hydrothermal carbonization (Fig. 7.4). The reaction parameters mainly include carbonization temperature, carbonization time, livestock manure-water mass or volume ratio, catalyzer, and heating modes, e.g., electrical and microwave. The next sections will further discuss the impact of each process parameter.

7.3.1 Reaction/Carbonization Temperature

Reaction temperature (*T*) is usually considered to be a key parameter in the hydrothermal carbonization process. The basic function of temperature is to provide disintegration heat for the breakdown of biomass bonds. The conversion efficiency of biomass increases with the increment of reaction temperature, which is because temperature transfers extra energy to destroy biomass bonds (Akhtar and Saidina Amin 2012). Usually, the yield of hydrochar will decrease fast firstly with the increase of reaction temperature and then level off or increase slightly. The increase of hydrochar yield at higher reaction temperature is ascribed to the recombination of free radical reactions. Figure 7.5 presents information on the outputs of

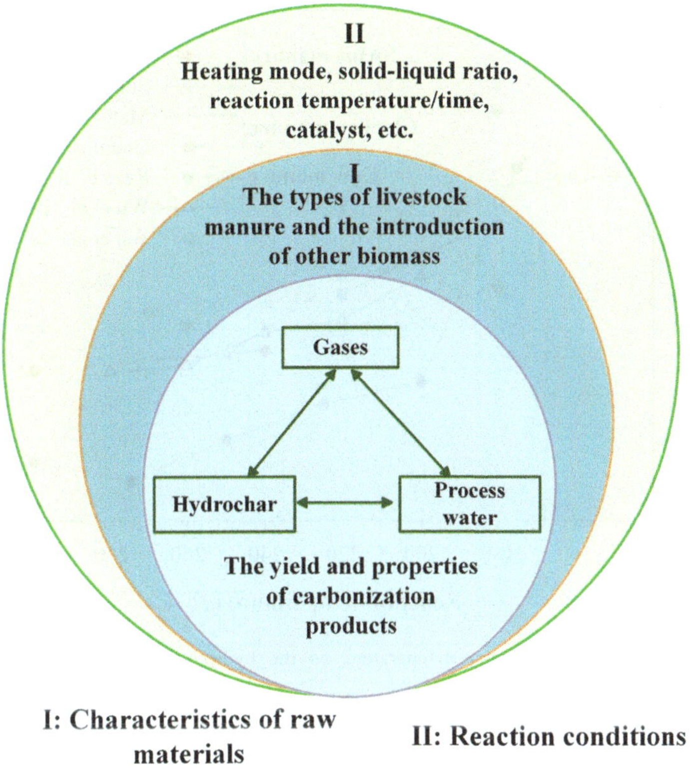

Fig. 7.4 Main influencing factors of hydrothermal carbonization for livestock manure

hydrochar produced from the hydrothermal carbonization of different livestock manures at various reaction temperatures. During the hydrothermal carbonization of swine manure (Lang et al. 2019d; Liu et al. 2017; Song et al. 2018), cow manure (Toufiq Reza et al. 2016; Wu et al. 2017, 2018), and chicken manure (Ghanim et al. 2016; Mau and Gross 2018; Mau et al. 2016), the yield of hydrochar is relatively high at temperatures below 200 °C and reduces fast with reaction temperature increasing to an intermediate range of 200–250 °C. And as the reaction temperature rises further, the yield of hydrochar levels off or slightly increases.

It should be noted that the highest temperature of hydrothermal carbonization is about 300 °C. Generally, solid products are dominant in the temperature range of 150–200 °C, liquid production will be higher in the medium temperature range of 250–350 °C, and the conversion of solid and liquid products to gas products will be strengthened above 350 °C. In the temperature range higher than hydrothermal carbonization, this process is defined as hydrothermal liquefaction. In hydrothermal liquefaction process, the liquid production is greater than solid and gaseous production. When the reaction temperature further increases, for example, in the supercritical conditions, the gas production will dominate due to the enhancement of free radical reaction. This process is called hydrothermal gasification or supercritical water gasification (Safari et al. 2018).

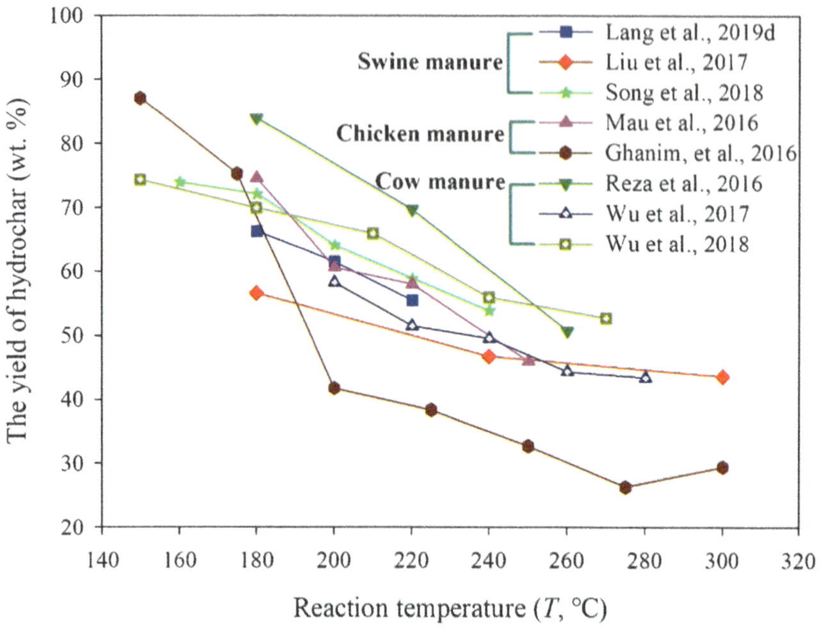

Fig. 7.5 The effects of reaction temperature on the hydrothermal carbonization for livestock manure

7.3.2 *Reaction Time*

Reaction time (*t*) is an essential parameter during the production of hydrochar as the reaction severity will be enhanced with long reaction time. The yield of hydrochar is usually high when reaction time is short, and it will decrease as reaction time increases. As the reaction time is further increased, the polyreaction of fragments dissolved in the liquid phase will be improved, which further enhances the production of secondary hydrochar with a polyaromatic structure (Wang et al. 2018). Fig. 7.6 depicts the influence of reaction time on the yield of hydrochar obtained from the hydrothermal carbonization of livestock manure (Ghanim et al. 2016; Song et al. 2018; Toufiq Reza et al. 2016). As shown in Fig. 7.6, the influence of reaction time on the yield of hydrochar is distinct at different reaction temperatures. It is suggested that the hydrothermal carbonization reaction should be carried out at high temperature with a short time or low temperature for a long time to guarantee the yield and quality of biochar at the same time. Furthermore, the hydrochars obtained after a long reaction time usually possess better physical characterizations, including porosity, pore capacity, and specific surface area (Heidari et al. 2018). In a study performed by Gao et al. (2013), the surface characteristics of hydrochar obtained from the hydrothermal carbonization of water hyacinth at 240 °C for 30 min–24 h were compared. The findings indicated that when the reaction time is

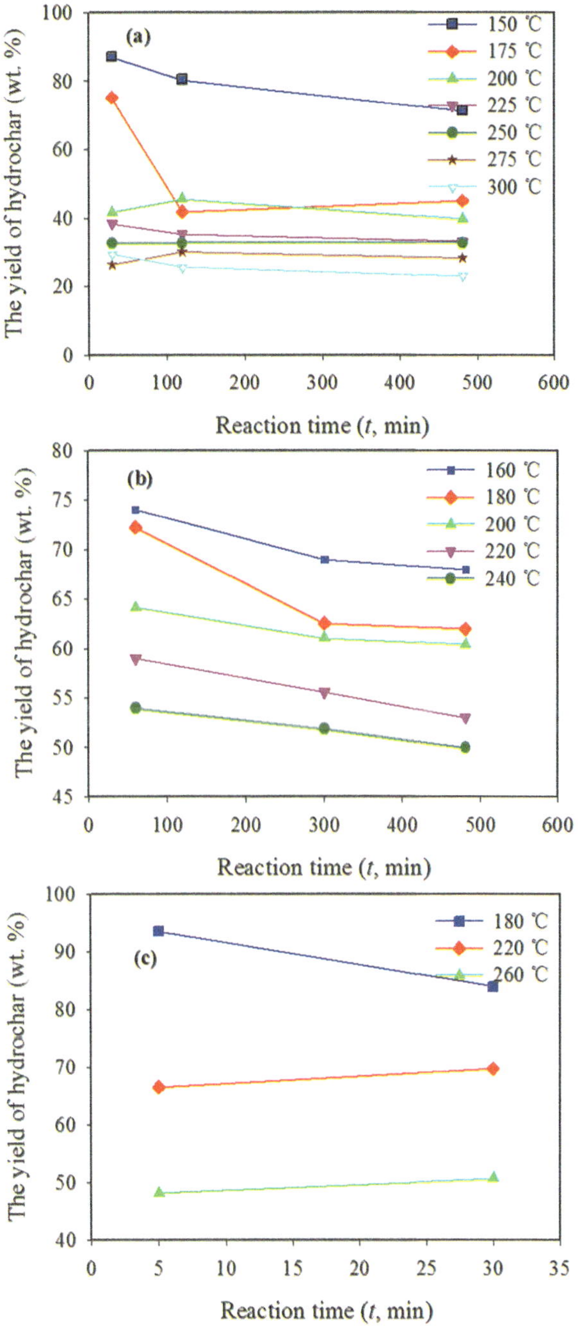

Fig. 7.6 The effects of reaction time on the hydrothermal carbonization for livestock manure. (**a**) Chicken manure (Drawn according to the literature of Ghanim et al. 2016), (**b**) swine manure (Drawn according to the literature of Song et al. 2018), (**c**) cow manure (Drawn according to the literature of Toufiq Reza et al. 2016)

short, only some cracks and trenches were present on the hydrochar surface. As the reaction time was longer than 6 h, microspheres or carbon microspheres began to form. When the reaction time is further prolonged to more than 24 h, the microspheres were aggregated on the surface of hydrochar.

Reaction time has an effect not only on the distribution of carbonization products but also on the energy expenditure intensity of hydrothermal carbonization process. Longer reaction time means higher energy consumption. Hydrothermal carbonization is considered to be a slow reaction process. For hydrothermal carbonization, the involved reaction time ranges from a few minutes to several hours (Nizamuddin et al. 2017). During the hydrothermal carbonization of livestock manure, the selection of reaction time is usually divided into two situations. Some researchers selected a longer reaction time, such as in the hydrothermal carbonization of cow manure (4–24 h) (Dai et al. 2015), swine manure (20 h) (Cao et al. 2011), and chicken manure (4 h) (Oliveira et al. 2013). And also, some researchers selected a short reaction time, such as in the hydrothermal carbonization of cow manure (5–30 min) (Toufiq Reza et al. 2016), chicken manure (15 min) (Kabadayi Catalkopru et al. 2017), and swine manure (60 min) (Zhou et al. 2019). On the whole, if you want to develop hydrochar into catalysts and adsorbents, longer reaction time is more appropriate. Hydrochar obtained with shorter reaction time can be utilized as a soil amendment or solid energy.

7.3.3 Livestock Manure-Water Mass or Volume Ratio

The mass or volume ratio of livestock manure to water (R) is usually also named as solid-to-liquid ratio, which is also an important factor affecting the hydrothermal carbonization of livestock manure. The volume of water used in hydrothermal carbonization should be sufficient to guarantee the complete diffusion of livestock manure in the reaction medium, in order to produce a more effective hydrothermal reaction. A higher solid-liquid ratio will reduce the relative interactions among the molecules of livestock manure and water, which can restrain the dissolution of manure components. In general, it is believed that the solid-liquid ratio has a certain threshold. Too high R will lead to an increase of hydrochar yield due to the incomplete hydrothermal carbonization. But attention should also be paid to the fact that at a low R, the amount of water consumed and the energy required for heating water are usually high (Heidari et al. 2018; Huang et al. 2017). Ro et al. (2017) assessed the effect of R on the hydrothermal carbonization of swine manure and chicken manure (Fig. 7.7). As clearly shown in Fig. 7.7, the yield of hydrochar increases with the increment of R at 250 °C for 4 h. In other cases, the yield of hydrochar decreases with the increase of R. However, it is sure that the energy content of hydrochar will decline with the rise of R. In addition, lower R will result in a higher content of volatile organic matter in hydrochar (Ro et al. 2017).

The amount of water in the hydrothermal carbonization process should ensure the hydrothermal reaction of the whole livestock manure sample and ensure

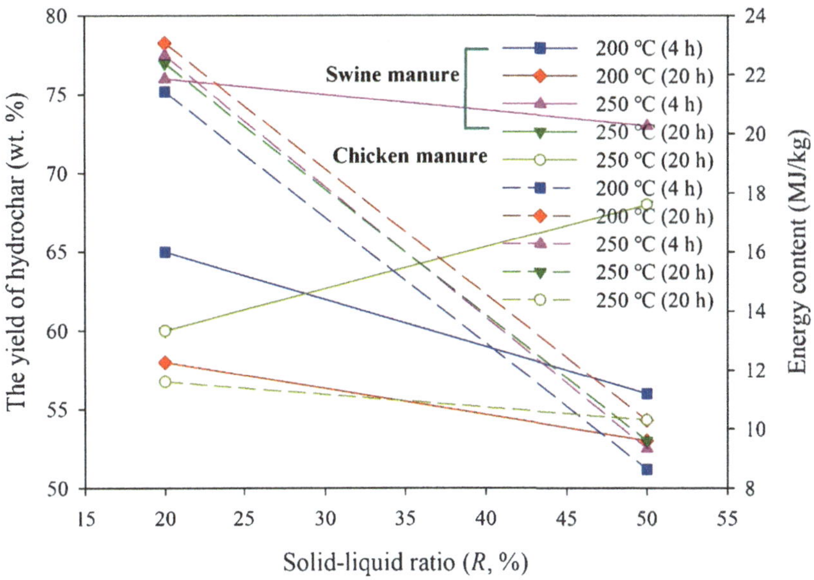

Fig. 7.7 The effects of solid-liquid ratio on the hydrothermal carbonization for livestock manure (Drawn according to the data reported by Ro et al. 2017)

adequate heat and mass transfer. In addition, some objective factors will also affect the amount of water, for example, the self-characteristics of livestock manure, such as density, structure, and hydrophobicity, and the reactor design, e.g., adding agitators. If the livestock manure possesses higher density, more water will be used to guarantee ample energy and matter transfer. The porous composition of livestock manure will make it easier for water to penetrate the pore, and less water is needed. More water for hydrothermal reactions is required for hydrophobic materials. If the design of the reactor can ensure uniform/effective heat transfer for the whole reaction system, higher *R* can be adopted, which can be achieved by adding a stirrer to the reactor (Heidari et al. 2018).

7.3.4 Catalyst

The use of catalysts in the hydrothermal carbonization process will alter the characteristics of water-biomass mixtures and leads to the expected changes in processes and products. Therefore, the choice of catalyst is decided by the end goal of the researches. Some acids and bases are usually used as a catalyst in the treatment process, resulting in an increase in the ion concentration of proton or hydroxide, then promoting the ionic strength, ultimately improving the reaction rate or adjusting the reaction path to obtain the required hydrochar. In general, acid catalysts that facilitate hydrolysis can be used to increase hydrochar formation. On the contrary, the adoption of alkaline catalyzer will usually weaken the production of hydrochar, but improve the production of liquid products (Nizamuddin et al. 2017). In addition, the nitrogen content in hydrochar can be reduced by adding corresponding catalysts, thus reducing the emission of nitrogen oxides in the incineration process. Considering that the application of CaO in acidic soils can effectively inactive the heavy metals and enhance the soil quality and productivity, Lang et al. (2019d) applied it to the hydrothermal carbonization of swine manure (Fig. 7.8a). The addition of CaO enhances the removal of nitrogen in hydrochar. Meanwhile, the yield of hydrochar is significantly improved when CaO is adopted, which is attributed to the formation of $CaCO_3$ during the hydrothermal carbonization process. In addition, the introduction of CaO significantly enhances the pH of hydrochar, which makes swine manure-derived hydrochar to be the promising soil amendment, especially for acidic soils.

Ghanim et al. (2017, 2018) investigated the effects off acidic catalysts (CH_3COOH and H_2SO_4) on the hydrothermal carbonization of chicken manure (Fig. 7.8b). Both the use of CH_3COOH and H_2SO_4 can enhance the formation of hydrochar but present different trends. In particular, the yield of hydrochar decreases with decreasing pH in the case of CH_3COOH, but, for H_2SO_4, exactly the opposite. The caloric value and C content of hydrochar can be improved, especially at lower pH conditions. In addition, the introduction of acid catalyst also has effects on the fraction/passivation of P and other nutrients in hydrochar. Acidic conditions are conducive to higher nutrient extraction and provide opportunities for nutrient recovery. In other words, the total content and solubility of nutrients in hydrochar are both decreased, which favors the utilization of hydrochar as a slow-release fertilizer.

7.3.5 Heating Mode

The mode of heating usually contains two types: conventional heating and microwave heating. These two heating modes have a distinct heat transfer way. In the traditional heating systems, the heating source is generally located in the periphery of the reactor, and energy is transferred to the reactor through convection and conduction, thus heating it up. A temperature gradient will be formed between the reactor periphery and its inner center, and energy will be transferred along the temperature gradient until it reaches a stable state. On the other hand, in the microwave system, microwave energy is transformed into internal energy by generating heat in the whole material through microwave penetration (Mubarak et al. 2016). In the past few years, microwave heating systems have attracted researchers' wide attention, which can be ascribed to the inherent advantages of microwave heating. Microwave heating belongs to molecular horizontal heating, which can make the material heated rapidly and evenly. In addition, microwave heating is a more controllable way, resulting in higher cost-effectiveness and lower energy consumption; therefore, it is thought that microwave heating can be an effective alternative to traditional heating (Nizamuddin et al. 2018).

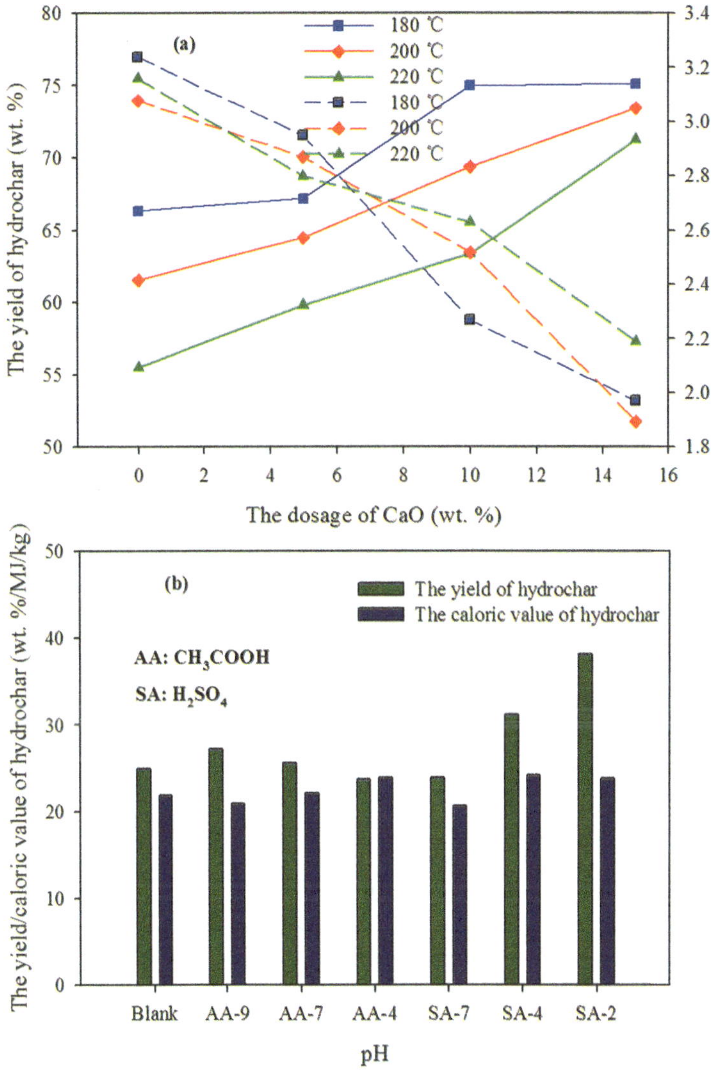

Fig. 7.8 The effects of catalysts on the hydrothermal carbonization for livestock manure. (**a**) Swine manure (Drawn according to the literature of Lang et al. 2019d), (**b**) chicken manure (Drawn according to the literature of Ghanim et al. 2017)

Microwave technology has been introduced into the hydrothermal carbonization of cow manure (Gao et al. 2018). The effects of reaction temperature and time on the yield of hydrochar are basically consistent with those described in Sects. 3.1 and 3.2 (Fig. 7.9). In addition, the energy content of hydrochar will be improved as the reaction temperature/time is increased. Compared with traditional heating, the most prominent point of microwave heating is that hydrochar has better structure. Just a

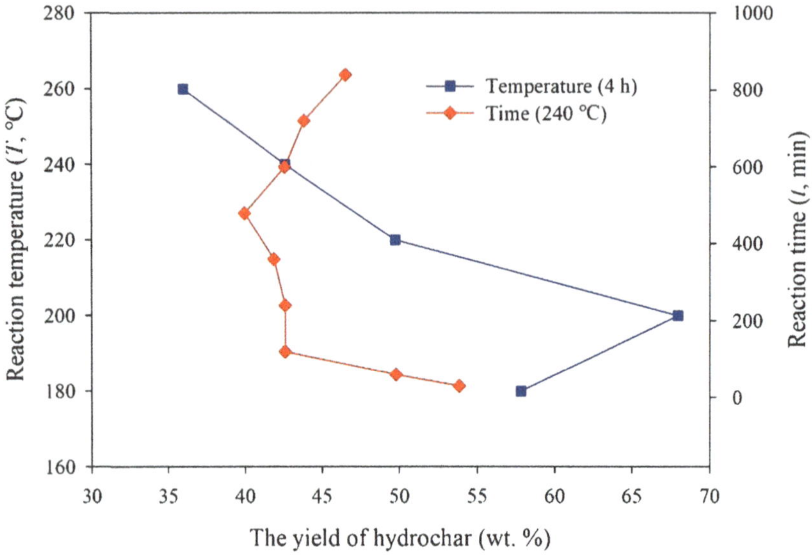

Fig. 7.9 Microwave-assisted hydrothermal carbonization for cow manure (Drawn according to the literature of Gao et al. 2018)

short reaction time can result in the formation of mesoporous structure on the surface of hydrochar, presenting spongy morphology. With 10 h of residence time, the hydrochar shows graphene-like layered structure. After 14 h, the hydrochar morphology changes into high yield microspheres. In addition, it has been found that the formation of hydrochar will be affected by the aqueous-phase products, such as saccharides, aldehydes, and carbon microspheres, which makes it feasible to realize the concentration of aqueous products and the production of carbon-based nanomaterials by microwave-assisted control of hydrothermal reaction pathway. So far, there are few studies on microwave hydrothermal carbonization of livestock manure, and more in-depth studies need to be carried out.

7.3.6 Feedstock

In addition to the reaction parameters of the hydrothermal carbonization process, the yield and properties of hydrochar also depend on the characteristics of raw livestock manure. The composition of livestock manure varies greatly with species, growth stages, and animal feed, which may lead to different carbonization behaviors (Shen et al. 2015). Zhou et al. (2019) compared the thermochemical/combustion characteristics of hydrochars produced from different types of livestock manure, including one swine manure; two chicken manures, e.g., broiler litter and layer chicken litter; and two cow manures, e.g., dairy cattle manure and beef cattle manure. The highest yield of hydrochar is obtained from swine manure, followed by cow manure and chicken manure. Cow manure-derived hydrochar has higher energy recovery rate and comprehensive combustion index than other two manure hydrochars. It is suggested that to achieve higher energy recovery rate and comprehensive combustion index, the proper reaction temperature is 210 °C for swine manure, 180 °C for cow manure, and 180–210 °C for chicken manure. In another study (Jin et al. 2019), the authors compared the physicochemical properties of hydrochar obtained from digested swine and cow manures. It is also found that the yield of hydrochar from digested swine manure is higher than that from digested cow manure, which is ascribed to the higher content of volatile matters and lower content of fixed carbon in digested cow manure. However, the hydrochar from digested swine manure possesses lower carbonization levels and lower stability in soils. Ro et al. (2019) explored the combustion behavior of hydrochars from swine/chicken manures. The activation energies of swine manure-derived hydrochars are lower than those of chicken manure-derived hydrochars, now that the swine manure-derived hydrochars possess higher energy content (Ro et al. 2017).

In recent years, lignocellulosic biomass, e.g., sawdust and corn stalk, is introduced into the hydrothermal carbonization of swine manure (Lang et al. 2018, 2019c). The original purpose consists of two parts: one is to further resolve the pollution problems of antibiotics and heavy metals, and the other is to generate synergistic effects. Compared with swine manure treated by hydrothermal carbonization alone, the introduction of lignocellulosic biomass improves the carbonization process in the following aspects: (1) to enhance the dehydration of swine manure and the aromatization of hydrochar, (2) to increase the carbon content and heating value of hydrochar, (3) to improve the energy recovery rate of hydrothermal carbonization process, and (4) to enhance the combustion performance of hydrochar by increasing ignition temperature and decreasing burnout temperature. Figure 7.10 depicts the main synergistic coefficients obtained during the co-hydrothermal carbonization of sawdust/corn stalk-swine manure. The synergistic effect in the co-hydrothermal carbonization process is mainly due to the Maillard reaction. In particular, amino acids produced by the hydrolysis of protein in swine manure react with sugars produced by hydrolysis of carbohydrate in lignocellulose (Baccile et al. 2011).

7.4 The Transport/Conversion Behaviors of Heavy Metals

In modern large-scale farms, feed additives containing copper and zinc are usually utilized to enhance the growth of animal and decrease the occurrence of diseases. But these additives have a low metabolic rate in animal, and most of them, about 90%, will be excreted with animal manure, resulting in a significant increase in heavy metal concentrations in animal manure (Wei et al. 2018). A previous study reported that the contents of copper, zinc, and chromium in livestock manure were 1726, 11,379, and 1602 mg/kg, respectively (Wang et al. 2013). When the livestock manure is used in soil, the heavy metals are nonbiodegradable and so will be

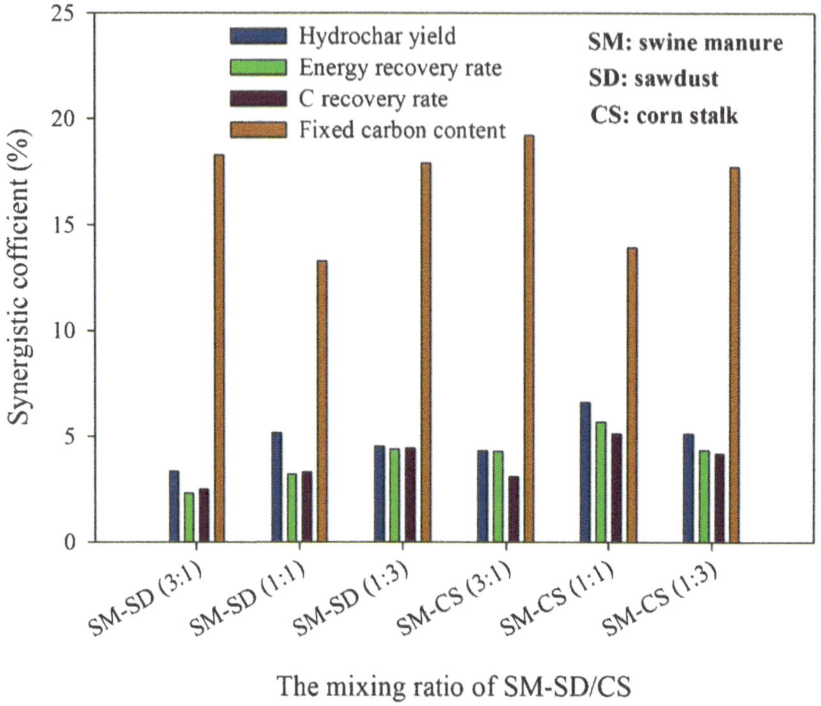

Fig. 7.10 Synergistic effects during the co-hydrothermal carbonization of swine manure-sawdust/corn stalk (Drawn according to the literature of Lang et al. 2018)

enriched along the food chain, leading to the potential environmental and human health risks. Therefore, the transport/conversion behaviors of heavy metals have become an important concern for the hydrothermal carbonization of livestock manure (Jin et al. 2019, Lang et al. 2018, 2019a).

The main research findings are summarized as follows (Fig. 7.11): (1) in the relatively low hydrothermal carbonization temperature range, the distribution of heavy metals in gas products can be neglected. The bulk of heavy metals will be concentrated in hydrochar; (2) the total of heavy metals will be obviously increased due to the decomposition and conversion of organic matters in the hydrothermal carbonization process. However, it is well known that the chemical speciation of heavy metals, rather than their total concentrations, is related to the environmental bioavailability and potential risks of heavy metals (Huang and Yuan 2015, 2016); (3) the bioavailable fraction of heavy metals will be transformed into the relatively stable fraction to a certain extent, thus making the heavy metals in hydrochars possessing low environmental risk; (4) the contents of heavy metals in TCLP (toxicity characteristic leaching procedure) and DTPA (diethylenetriamine pentaacetic acid) extractable fractions are both declined, indicating low leachability and plant bioavailability of heavy metals in hydrochar; (5) the use of proper catalyst, for example, CaO, and the introduction of lignocellulosic biomass during the hydrothermal

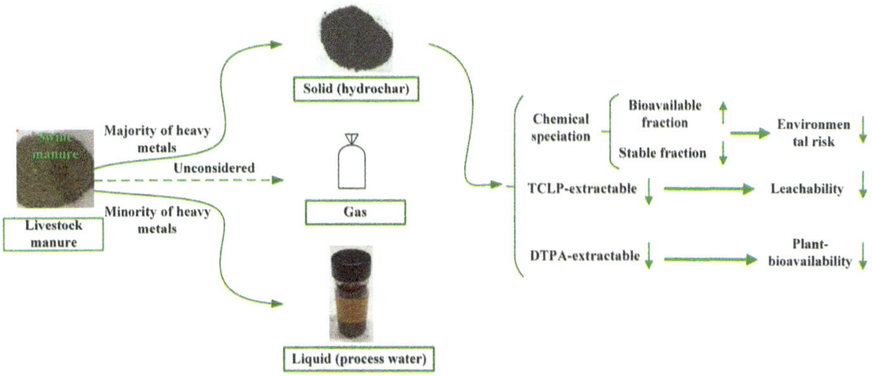

Fig. 7.11 The migration/transformation behaviors of heavy metals during the hydrothermal carbonization of livestock manure

carbonization of livestock manure could further reduce the total content of heavy metals and enhance the immobilization effects of heavy metals in hydrochar (Lang et al. 2018, 2019a).

7.5 The Application and Treatment of Process Water

As shown in Fig. 7.3, there are two by-products in the process of hydrothermal carbonization of livestock manure, i.e., process water and gaseous products. Carbon dioxide is the main component of gaseous products, so it can be directly discharged without further treatment. However, the process water contains a lot of inorganics and degradation products of livestock manure, which should be appropriately treated or be recycled/resourced. If not properly handled, they would severely contaminate the environment, and this potential resource would be wasted at the same time. Fig. 7.12 presents current feasible treatment or recycling methods of process water (Usman et al. 2019). So far, few pieces of research in the literature can be found that focuses on the treatment/reuse of process water produced from the hydrothermal carbonization of livestock manure. More methods mentioned in Fig. 7.12 can be attempted for the treatment or reuse of process water.

Reza et al. (2016) treated the process water from the hydrothermal carbonization of cow manure by wet air oxidation. They found that the total organic carbon content of treated process water was low under different oxygen loads, mainly composed of short-chain organic acids, which were not considered harmful to the environment. Kabadayi Catalkopru et al. (2017) focused on the effects of process water recirculation in the hydrothermal carbonization of chicken manure. Through recycling, the quality and energy yield of hydrochar increased. After reused three times, the mass and energy yields of hydrochar increased from 52.0 wt.% to 61.0 wt.% and from 68.4% to 76.8%, respectively. Meanwhile, it is noteworthy that

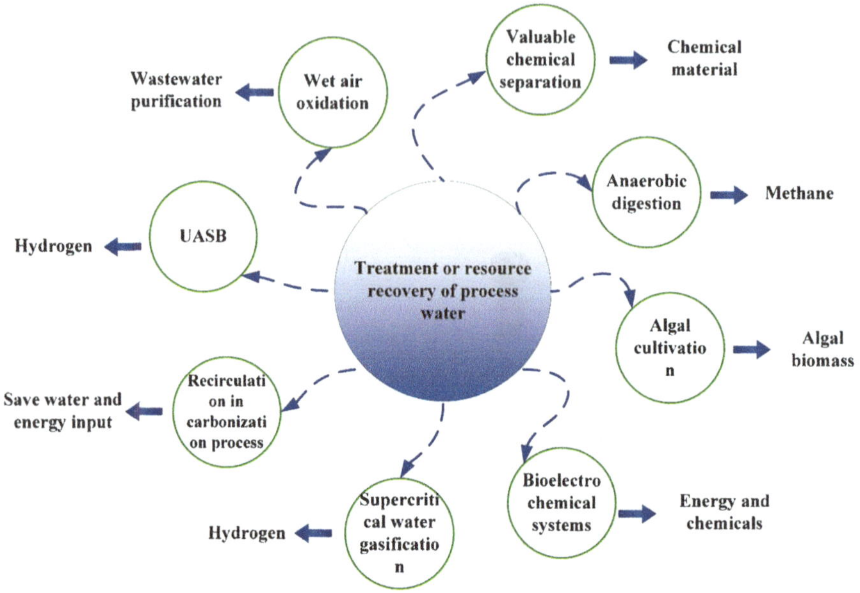

Fig. 7.12 The available treatment/reuse methods of process water (Drawn according to the literature of Usman et al. (2019))

the ignition temperature and combustion reactivity of hydrochar would decrease. The concentrations of organic and inorganic in process water increased with the number of cycles, but not as much as expected after the first cycle. In addition, the enrichment of organic and inorganic matters in process water will weaken the leachability of some inorganic matters and the dispersion of decomposed soluble fragments from chicken manure during the next round of hydrothermal carbonization process. The increase of the mass/energy yield of hydrochars is usually ascribed to two aspects of reasons: (1) through the recycling of process water, some inorganic and organic matter in process water will accumulate on the surface of hydrochar (Kabadayi Catalkopru et al. 2017); (2) compared with pure water or deionized water, the acidity of process water is higher (Kambo et al. 2018).

In another research performed by Lang et al. (2019b), sawdust was introduced in the hydrothermal carbonization of swine manure and the process water was recirculated. The authors reported some similar results. The recirculation of process water will promote the yield of hydrochar and its energy content by enhancing the dehydration and decarboxylation reactions, which was increased from 60.5 wt.% and 21.6 MJ/kg to 68.1 wt.% and 22.5 MJ/kg, respectively. The ignition temperature of hydrochar was expected to decrease, and the burnout temperature was increased. Additionally, the reuse of process water can significantly reduce the activating energy of hydrochar, especially in the first cycle of process water. Different from the research of Kabadayi Catalkopru et al. (2017), it was found that the reuse of process water improved the combustion reactivity of hydrochar, which may be due to the introduction of sawdust during the hydrothermal carbonization of swine manure.

7.6 Conclusions and Prospects

The management/treatment technologies of livestock manure should be able to effectively remove or reduce pathogens and heavy metals in order to prevent health risks to humans and other organisms. Hydrothermal carbonization is an available technology that can be applied to remove excess nutrients, eliminate pathogenic bacteria, and immobilize toxic heavy metals from livestock manure, which is then converted to multipurpose hydrochar. In the future, research on hydrothermal carbonization of livestock manure should focus on the following aspects: (1) Microwave heating has become an advanced technology of biomass processing in recent years. The application of microwave in the hydrothermal carbonization of livestock manure has only just begun and needs to be vigorously promoted. (2) At present, only a few conventional catalysts have been used for the hydrothermal carbonization of livestock manure. The development of more catalysts for the hydrothermal carbonation of livestock manure is worth further promoting. (3) The addition of lignocellulosic biomass during the hydrothermal carbonization of livestock manure has been certified to be beneficial for obtaining higher yield/quality of hydrochar. Whether the introduction of algae biomass can improve the hydrothermal carbonization of livestock manure remains to be answered. (4) As well known, in addition to heavy metals, livestock manure also contains a certain amount of organic pollutants, such as antibiotics, polybrominated diphenyl ethers, halohydrocarbon, and polycyclic aromatic hydrocarbons. The transformation effect/mechanism of these pollutants in the hydrothermal carbonization of livestock manure is also a significant subject to be solved. (5) Process water is the main by-product of hydrothermal carbonization of livestock manure. Reasonable treatment or utilization of process water will directly determine the cleanliness and economy of hydrothermal carbonization of livestock manure. Researchers need to devote more efforts to study or explore the treatment or utilization of process water produced in the hydrothermal carbonization of livestock manure.

Acknowledgments My research work on hydrothermal carbonization of livestock manure is supported by the following scientific funds: the Chinese Natural Science Research Programme (Serial number: 21707056), the Chinese Focal Study and Demonstration Project (Serial number: 2017YFD0200808), and the Scientific Research Fund of Jiangxi Provincial Education Department (Serial number: GJJ14302).

References

Akhtar J, Saidina Amin N (2012) A review on operating parameters for optimum liquid oil yield in biomass pyrolysis. Renew Sust Energ Rev 16(7):5101–5109. https://doi.org/10.1016/j.rser.2012.05.033

Baccile N, Laurent G, Coelho C, Babonneau F, Zhao L, Titirici M-M (2011) Structural insights on nitrogen-containing hydrothermal carbon using solid-state magic angle spinning ^{13}C and ^{15}N nuclear magnetic resonance. J Phys Chem C 115(18):8976–8982. https://doi.org/10.1021/jp2015512

Béline F, Martinez J (2002) Nitrogen transformations during biological aerobic treatment of pig slurry: effect of intermittent aeration on nitrous oxide emissions. Bioresour Technol 83(3):225–228. https://doi.org/10.1016/S0960-8524(01)00219-X

Bernal MP, Alburquerque JA, Moral R (2009) Composting of animal manures and chemical criteria for compost maturity assessment. A review. Bioresour Technol 100(22):5444–5453. https://doi.org/10.1016/j.biortech.2008.11.027

Bidart C, Fröhling M, Schultmann F (2014) Livestock manure and crop residue for energy generation: macro-assessment at a national scale. Renew Sust Energ Rev 38:537–550. https://doi.org/10.1016/j.rser.2014.06.005

Cao XY, Ro KS, Chappel M, Li Y, Mao JD (2011) Chemical structures of swine-manure chars produced under different carbonization conditions investigated by advanced solid-sate ^{13}C nuclear magnetic resonance (NMR) spectroscopy. Energy Fuel 25(1):388–397. https://doi.org/10.1021/ef101342v

Cao W, Cao CQ, Guo LJ, Jin H, Dargusch M, Bernhardt D, Yao XD (2016) Hydrogen production from supercritical water gasification of chicken manure. Int J Hydrogen Energy 41(48):22722–22731. https://doi.org/10.1016/j.ijhydene.2016.09.031

Dai LC, Tan FR, Wu B, He MX, Wang WG, Tang XY, Hu QC, Zhang M (2015) Immobilization of phosphorus in cow manure during hydrothermal carbonization. J Environ Manag 157:49–53. https://doi.org/10.1016/j.jenvman.2015.04.009

Erdogdu AE, Polat R, Ozbay G (2019) Pyrolysis of goat manure to produce bio-oil. Int J Eng Sci Technol 22(2):452–457. https://doi.org/10.1016/j.jestch.2018.11.002

Fang J, Zhan L, Ok YS, Gao B (2018) Minireview of potential applications of hydrochar derived from hydrothermal carbonization of biomass. J Ind Eng Chem 57:15–21. https://doi.org/10.1016/j.jiec.2017.08.026

Gao Y, Wang XH, Wang J, Li XP, Cheng JJ, Yang HP, Chen HP (2013) Effect of residence time on chemical and structural properties of hydrochar obtained by hydrothermal carbonization of water hyacinth. Energy 58:376–383. https://doi.org/10.1016/j.energy.2013.06.023

Gao Y, Liu YH, Zhu GK, Xu JY, Xu H, Yuan QX, Zhu YZ, Sarma J, Wang YF, Wang J, Ji L (2018) Microwave-assisted hydrothermal carbonization of dairy manure: chemical and structural properties of the products. Energy 165:662–672. https://doi.org/10.1016/j.energy.2018.09.185

Ghanim BM, Pandey DS, Kwapinski W, Leahy JJ (2016) Hydrothermal carbonisation of poultry litter: effects of treatment temperature and residence time on yields and chemical properties of hydrochars. Bioresour Technol 216:373–380. https://doi.org/10.1016/j.biortech.2016.05.087

Ghanim BM, Kwapinski W, Leahy JJ (2017) Hydrothermal carbonisation of poultry litter: effects of initial pH on yields and chemical properties of hydrochars. Bioresour Technol 238:78–85. https://doi.org/10.1016/j.biortech.2017.04.025

Ghanim BM, Kwapinski W, Leahy JJ (2018) Speciation of nutrients in hydrochar produced from hydrothermal carbonization of poultry litter under different treatment conditions. ACS Sustain Chem Eng 6(9):11265–11272. https://doi.org/10.1021/acssuschemeng.7b04768

Heidari M, Dutta A, Acharya B, Mahmud S (2018) A review of the current knowledge and challenges of hydrothermal carbonization for biomass conversion. J Energy Inst 92:1779. https://doi.org/10.1016/j.joei.2018.12.003

Hu B, Wang K, Wu L, Yu S-H, Antonietti M, Titirici M-M (2010) Engineering carbon materials from the hydrothermal carbonization process of biomass. Adv Mater 22(7):813–828. https://doi.org/10.1002/adma.200902812

Huang HJ, Yuan XZ (2015) Recent progress in the direct liquefaction of typical biomass. Prog Energy Combust Sci 49:59–80. https://doi.org/10.1016/j.pecs.2015.01.003

Huang HJ, Yuan XZ (2016) The migration and transformation behaviors of heavy metals during the hydrothermal treatment of sewage sludge. Bioresour Technol 200:991–998. https://doi.org/10.1016/j.biortech.2015.10.099

Huang HJ, Yuan XZ, Wu GQ (2017) Liquefaction of biomass for bio-oil products. In: Singh L, Kalia VC (eds) Waste biomass management—a holistic approach. Springer International Publishing, Cham, pp 231–250. https://doi.org/10.1007/978-3-319-49595-8_11

Hussein MS, Burra KG, Amano RS, Gupta AK (2017) Temperature and gasifying media effects on chicken manure pyrolysis and gasification. Fuel 202:36–45. https://doi.org/10.1016/j.fuel.2017.04.017

Islam MN, Park J-H (2018) A short review on hydrothermal liquefaction of livestock manure and a chance for Korea to advance swine manure to bio-oil technology. J Mater Cycles Waste 20(1):1–9. https://doi.org/10.1007/s10163-016-0566-0

Jiang B, Lin Y, Mbog JC (2018) Biochar derived from swine manure digestate and applied on the removals of heavy metals and antibiotics. Bioresour Technol 270:603–611. https://doi.org/10.1016/j.biortech.2018.08.022

Jin H, Yan D, Zhu N, Zhang S, Zheng M (2019) Immobilization of metal(loid)s in hydrochars produced from digested swine and dairy manures. Waste Manag 88:10–20. https://doi.org/10.1016/j.wasman.2019.03.027

Kabadayi Catalkopru A, Kantarli IC, Yanik J (2017) Effects of spent liquor recirculation in hydrothermal carbonization. Bioresour Technol 226:89–93. https://doi.org/10.1016/j.biortech.2016.12.015

Kambo HS, Minaret J, Dutta A (2018) Process water from the hydrothermal carbonization of biomass: a waste or a valuable product? Waste Biomass Valori 9(7):1181–1189. https://doi.org/10.1007/s12649-017-9914-0

Krylova AY, Zaitchenko VM (2018) Hydrothermal carbonization of biomass: a review. Solid Fuel Chem 52(2):91–103. https://doi.org/10.3103/S0361521918020076

Kumar RR, Park BJ, Cho JY (2013) Application and environmental risks of livestock manure. J Korean Soc Appl Bi 56(5):497–503. https://doi.org/10.1007/s13765-013-3184-8

Lang Q, Guo Y, Zheng Q, Liu Z, Gai C (2018) Co-hydrothermal carbonization of lignocellulosic biomass and swine manure: Hydrochar properties and heavy metal transformation behavior. Bioresour Technol 266:242–248. https://doi.org/10.1016/j.biortech.2018.06.084

Lang Q, Chen M, Guo Y, Liu Z, Gai C (2019a) Effect of hydrothermal carbonization on heavy metals in swine manure: speciation, bioavailability and environmental risk. J Environ Manag 234:97–103. https://doi.org/10.1016/j.jenvman.2018.12.073

Lang Q, Luo H, Li Y, Li D, Liu Z, Yang T (2019b) Thermal behavior of hydrochar from co-hydrothermal carbonization of swine manure and sawdust: effect of process water recirculation. Sustain Energ Fuels, vol 3, p 2329. https://doi.org/10.1039/C9SE00332K

Lang Q, Zhang B, Liu Z, Chen Z, Xia Y, Li D, Ma J, Gai C (2019c) Co-hydrothermal carbonization of corn stalk and swine manure: combustion behavior of hydrochar by thermogravimetric analysis. Bioresour Technol 271:75–83. https://doi.org/10.1016/j.biortech.2018.09.100

Lang Q, Zhang B, Liu Z, Jiao W, Xia Y, Chen Z, Li D, Ma J, Gai C (2019d) Properties of hydrochars derived from swine manure by CaO assisted hydrothermal carbonization. J Environ Manag 233:440–446. https://doi.org/10.1016/j.jenvman.2018.12.072

Liu Y, Yao S, Wang Y, Lu H, Brar SK, Yang S (2017) Bio- and hydrochars from rice straw and pig manure: inter-comparison. Bioresour Technol 235:332–337. https://doi.org/10.1016/j.biortech.2017.03.103

Loyon L (2017) Overview of manure treatment in France. Waste Manag 61:516–520. https://doi.org/10.1016/j.wasman.2016.11.040

Loyon L, Guiziou F, Beline F, Peu P (2007) Gaseous emissions (NH_3, N_2O, CH_4 and CO_2) from the aerobic treatment of piggery slurry-comparison with a conventional storage system. Biosyst Eng 97(4):472–480. https://doi.org/10.1016/j.biosystemseng.2007.03.030

Lundgren J, Pettersson E (2009) Combustion of horse manure for heat production. Bioresour Technol 100(12):3121–3126. https://doi.org/10.1016/j.biortech.2009.01.050

Mau V, Gross A (2018) Energy conversion and gas emissions from production and combustion of poultry-litter-derived hydrochar and biochar. Appl Energy 213:510–519. https://doi.org/10.1016/j.apenergy.2017.11.033

Mau V, Quance J, Posmanik R, Gross A (2016) Phases' characteristics of poultry litter hydrothermal carbonization under a range of process parameters. Bioresour Technol 219:632–642. https://doi.org/10.1016/j.biortech.2016.08.027

Mubarak NM, Sahu JN, Abdullah EC, Jayakumar NS (2016) Plam oil empty fruit bunch based magnetic biochar composite comparison for synthesis by microwave-assisted and conventional heating. J Anal Appl Pyrol 120:521–528. https://doi.org/10.1016/j.jaap.2016.06.026

Nasir IM, Mohd Ghazi TI, Omar R (2012) Anaerobic digestion technology in livestock manure treatment for biogas production: a review. Eng Life Sci 12(3):258–269. https://doi.org/10.1002/elsc.201100150

Niu Q, Li Y-Y (2016) Recycling of livestock manure into bioenergy. In: Karthikeyan OP, Heimann K, Muthu SS (eds) Recycling of solid waste for biofuels and bio-chemicals. Springer, Singapore, pp 165–186. https://doi.org/10.1007/978-981-10-0150-5_6

Nizamuddin S, Baloch HA, Griffin GJ, Mubarak NM, Bhutto AW, Abro R, Mazari SA, Ali BS (2017) An overview of effect of process parameters on hydrothermal carbonization of biomass. Renew Sust Energ Rev 73:1289–1299. https://doi.org/10.1016/j.rser.2016.12.122

Nizamuddin S, Baloch HA, Siddiqui MTH, Mubarak NM, Tunio MM, Bhutto AW, Jatoi AS, Griffin GJ, Srinivasan MP (2018) An overview of microwave hydrothermal carbonization and microwave pyrolysis of biomass. Rev Environ Sci Biotechnol 17(4):813–837. https://doi.org/10.1007/s11157-018-9476-z

Oliveira I, Blöhse D, Ramke H-G (2013) Hydrothermal carbonization of agricultural residues. Bioresour Technol 142:138–146. https://doi.org/10.1016/j.biortech.2013.04.125

Reza MT, Freitas A, Yang X, Coronella CJ (2016) Wet air oxidation of hydrothermal carbonization (HTC) process liquid. ACS Sustain Chem Eng 4(6):3250–3254. https://doi.org/10.1021/acssuschemeng.6b00292

Ro KS, Flora JRV, Bae S, Libra JA, Berge ND, Álvarez-Murillo A, Li L (2017) Properties of animal-manure-based hydrochars and predictions using published models. ACS Sustain Chem Eng 5(8):7317–7324. https://doi.org/10.1021/acssuschemeng.7b01569

Ro KS, Libra JA, Bae S, Berge ND, Flora JRV, Pecenka R (2019) Combustion behavior of animal-manure-based hydrochar and pyrochar. ACS Sustain Chem Eng 7(1):470–478. https://doi.org/10.1021/acssuschemeng.8b03926

Safari F, Javani N, Yumurtaci Z (2018) Hydrogen production via supercritical water gasification of almond shell over algal and agricultural hydrochars as catalysts. Int J Hydrogen Energy 43(2):1071–1080. https://doi.org/10.1016/j.ijhydene.2017.05.102

Shen X, Huang G, Yang Z, Han L (2015) Compositional characteristics and energy potential of Chinese animal manure by type and as a whole. Appl Energy 160:108–119. https://doi.org/10.1016/j.apenergy.2015.09.034

Shi L, Simplicio WS, Wu G, Hu Z, Hu H, Zhan X (2018) Nutrient recovery from digestate of anaerobic digestion of livestock manure: a review. Curr Pollut Rep 4(2):74–83. https://doi.org/10.1007/s40726-018-0082-z

Song C, Shan S, Müller K, Wu S, Niazi NK, Xu S, Shen Y, Rinklebe J, Liu D, Wang H (2018) Characterization of pig manure-derived hydrochars for their potential application as fertilizer. Environ Sci Pollut Res 25(26):25772–25779. https://doi.org/10.1007/s11356-017-0301-y

Spielmeyer A (2018) Occurrence and fate of antibiotics in manure during manure treatments: a short review. Sustain Chem Pharm 9:76–86. https://doi.org/10.1016/j.scp.2018.06.004

Tambone F, Terruzzi L, Scaglia B, Adani F (2015) Composting of the solid fraction of digestate derived from pig slurry: biological processes and compost properties. Waste Manag 35:55–61. https://doi.org/10.1016/j.wasman.2014.10.014

Tańczuk M, Junga R, Werle S, Chabiński M, Ziółkowski Ł (2019) Experimental analysis of the fixed bed gasification process of the mixtures of the chicken manure with biomass. Renew Energy 136:1055–1063. https://doi.org/10.1016/j.renene.2017.05.074

Titirici M-M, Antonietti M (2010) Chemistry and materials options of sustainable carbon materials made by hydrothermal carbonization. Chem Soc Rev 39(1):103–116. https://doi.org/10.1039/B819318P

Toufiq Reza M, Freitas A, Yang X, Hiibel S, Lin H, Coronella CJ (2016) Hydrothermal carbonization (HTC) of cow manure: carbon and nitrogen distributions in HTC products. Environ Prog Sustain 35(4):1002–1011. https://doi.org/10.1002/ep.12312

Usman M, Chen H, Chen K, Ren S, Clark JH, Fan J, Luo G, Zhang S (2019) Characterization and utilization of aqueous products from hydrothermal conversion of biomass for bio-oil and hydro-char production: a review. Green Chem 21(7):1553–1572. https://doi.org/10.1039/C8GC03957G

Vlyssides A, Mai S, Barampouti EM (2015) Energy generation potential in Greece from agricultural residues and livestock manure by anaerobic digestion technology. Waste Biomass Valori 6(5):747–757. https://doi.org/10.1007/s12649-015-9400-5

Wang H, Dong Y, Yang Y, Toor GS, Zhang X (2013) Changes in heavy metal contents in animal feeds and manures in an intensive animal production region of China. J Environ Sci 25(12):2435–2442. https://doi.org/10.1016/S1001-0742(13)60473-8

Wang T, Zhai Y, Zhu Y, Li C, Zeng G (2018) A review of the hydrothermal carbonization of biomass waste for hydrochar formation: process conditions, fundamentals, and physicochemical properties. Renew Sust Energ Rev 90:223–247. https://doi.org/10.1016/j.rser.2018.03.071

Wei X, Liu D, Li W, Liao L, Wang Z, Huang W, Huang W (2018) Biochar addition for accelerating bioleaching of heavy metals from swine manure and reserving the nutrients. Sci Total Environ 631-632:1553–1559. https://doi.org/10.1016/j.scitotenv.2018.03.140

Wu K, Gao Y, Zhu G, Zhu J, Yuan Q, Chen Y, Cai M, Feng L (2017) Characterization of dairy manure hydrochar and aqueous phase products generated by hydrothermal carbonization at different temperatures. J Anal Appl Pyrol 127:335–342. https://doi.org/10.1016/j.jaap.2017.07.017

Wu K, Zhang X, Yuan Q (2018) Effects of process parameters on the distribution characteristics of inorganic nutrients from hydrothermal carbonization of cattle manure. J Environ Manag 209:328–335. https://doi.org/10.1016/j.jenvman.2017.12.071

Yang D-P, Li Z, Liu M, Zhang X, Chen Y, Xue H, Ye E, Luque R (2019) Biomass-derived carbonaceous materials: recent progress in synthetic approaches, advantages, and applications. ACS Sustain Chem Eng 7(5):4564–4585. https://doi.org/10.1021/acssuschemeng.8b06030

Zeng X, Xiao Z, Zhang G, Wang A, Li Z, Liu Y, Wang H, Zeng Q, Lian Y, Zou D (2018) Speciation and bioavailability of heavy metals in pyrolytic biochar of swine and goat manures. J Anal Appl Pyrol 132:82–93. https://doi.org/10.1016/j.jaap.2018.03.012

Zhang F, Li Y, Yang M, Li W (2012) Content of heavy metals in animal feeds and manures from farms of different scales in Northeast China. Int J Environ Res Public Health 9(8):2658–2668. https://doi.org/10.3390/ijerph9082658

Zhang J, Liu J, Evrendilek F, Xie W, Kuo J, Zhang X, Buyukada M (2019a) Kinetics, thermodynamics, gas evolution and empirical optimization of cattle manure combustion in air and oxy-fuel atmospheres. Appl Therm Eng 149:119–131. https://doi.org/10.1016/j.applthermaleng.2018.12.010

Zhang S, Zhu X, Zhou S, Shang H, Luo J, Tsang DCW (2019b) Hydrothermal carbonization for hydrochar production and its application. In: Ok YS, Tsang DCW, Bolan N, Novak JM (eds) Biochar from biomass and waste, Elsevier, pp. 275–294. https://doi.org/10.1016/B978-0-12-811729-3.00015-7

Zhou Y, Engler N, Nelles M (2018) Symbiotic relationship between hydrothermal carbonization technology and anaerobic digestion for food waste in China. Bioresour Technol 260:404–412. https://doi.org/10.1016/j.biortech.2018.03.102

Zhou S, Liang H, Han L, Huang G, Yang Z (2019) The influence of manure feedstock, slow pyrolysis, and hydrothermal temperature on manure thermochemical and combustion properties. Waste Manag 88:85–95. https://doi.org/10.1016/j.wasman.2019.03.025

Chapter 8
A Review on Pharmaceutical Removal from Aquatic Media by Adsorption: Understanding the Influential Parameters and Novel Adsorbents

Ali Khadir, Afsaneh Mollahosseini, Ramin M. A. Tehrani, and Mehrdad Negarestani

Contents

A. Khadir
Young Researcher and Elite Club, Yadegar-e-Imam Khomeini (RAH) Shahre Rey Branch, Islamic Azad University, Tehran, Iran
e-mail: Khadir_ali@civileng.iust.ac.ir

A. Mollahosseini (✉)
Research Laboratory of Spectroscopy and Micro and Nano Extraction, Department of Chemistry, Iran University of Science and Technology, Tehran, Iran
e-mail: amollahosseini@iust.ac.ir

R. M. A. Tehrani
Department of Chemistry, Yadegar Imam Khomeini (RAH) Shahre Rey Branch, Islamic Azad University, Tehran, Iran

M. Negarestani
Department of Civil and Environmental Engineering, Iran University of Science and Technology, Tehran, Iran

Inamuddin, A. M. Asiri (eds.), *Sustainable Green Chemical Processes and their Allied Applications*, Nanotechnology in the Life Sciences,
https://doi.org/10.1007/978-3-030-42284-4_8

8.1 Introduction

There is no doubt that the Industrial Revolution has significantly ameliorated the life of people globally in every aspect including industrialization and agriculture. In fact, it is believed that the water supply is the main inevitable parameter for every society development especially once the world is facing water shortages. The situation is aggravated while water is polluted by natural sources (soil erosion, decaying of organic matter) or human activities. Water pollution may be categorized into different groups:

1. Common pollutants
2. Emerging contaminants (micropollutants)

Emerging pollutants are defined as any chemicals that are recently detected in the environment usually at low concentration and have adverse effects on ecosystem and human health (Li et al. 2019a). Among various types of micropollutants, pharmaceuticals have gained much attention over the last two decades (Khadir et al. 2020). Pharmaceuticals have been produced with the aim of human and animal treatment; however, these compounds have shown negative effects on the environment especially for their continuous consumption. Non-steroidal anti-inflammatory drugs, beta-blockers, antibiotics, neuroleptics, and hormones are the most detected compounds in the environment (Szymonik et al. 2017). Antibiotics that are used for treating infections both in people and animals are the most frequently prescribed pharmaceuticals. In Canada, hypertension drugs and in Germany and United States, beta-blockers are reported to be the most frequently prescribed drugs (Szymonik et al. 2017).

Scientists are of firm opinion that the presence of pharmaceutical compounds in the environment is attributed to the increase in the pharmaceuticals consumption. Diclofenac as a single drug, for example, its consumption increased from 877 to 940 Mg from 2007 to 2008, respectively (Zhang et al. 2008). In Germany 2001, some of the most commonly used anti-inflammatory drugs were acetylsalicylic acid, with 836 tonnes, paracetamol, 622 tonnes, and ibuprofen, with 345 tonnes (Quesada et al. 2019). It is evident that all of these drugs are not degradable in human bodies. Hence, many of them are excreted from the body to the environment in an unchanged form or biologically changed structure (Madikizela et al. 2017). An investigation showed that among 400 households in the UK, 63.2% discarded their unfinished drugs in household waste and 11.5% discarded them into the toilet or sink (Bound and Nikolaos 2005). In Germany, it was reported that amounts of up to 16,000 tonnes of pharmaceuticals were disposed of each year from human medical care (Zhang et al. 2008). Pharmaceuticals are either discarded in the household waste or excreted from human bodies; they enter the environment, especially wastewaters.

Literature declared that pharmaceuticals have been widely detected around the world in water media including influent and effluents wastewater/water treatment plants, ground water, surface water, and potable water. For instance, ciprofloxacin hydrochloride and fluoroquinolones were detected in wastewaters in the concentration of 249–405 ng/L and 0.6–2 μg/L, respectively (Dhiman and Sharma 2019). Ibuprofen showed the minimum and maximum concentration of 184 and 248 ng/L in the surface water of Mexico (Rivera-Jaimes et al. 2018). Also, caffeine was observed in the surface waters of China (Yang et al. 2018a), USA (Bean et al. 2018), and Malaysia (Praveena et al. 2018). Many researchers believe that the inability of the conventional wastewater/water treatment plants is one of the significant reasons of pharmaceutical compounds detection in waters (Greenham et al. 2019; Lee et al. 2019; Reis et al. 2019). So, it is highly necessary to implement an efficient treatment process to remove these hazardous compounds.

Since now, various treatment methods including nanofiltration, reverse osmosis, photocatalysis, ozonation, adsorption, ion exchange, and biological processes have been implemented for drug removal. This review aims to consider the pharmaceutical pollution and removal in conventional water and wastewater treatment plants around the world. In addition, the authors attempted to introduce the adsorption processes as efficient, simple, and promising treatment technique for water and wastewater purification from pharmaceutical compounds. Accordingly, some of the basic parameters which strongly influence the whole procedure including contact time, adsorbent dosage, pharmaceutical concentration, pH, and temperature were studied. Moreover, this chapter summarized the major works regarding the removal of pharmaceuticals from water and wastewater using various adsorbents.

8.2 Pharmaceutical Classification

Pharmaceutical compounds have been widely prescribed for illness treatment or prevention. There are several ways for the classification of pharmaceuticals proposed by various researchers. Generally, pharmaceuticals are categorized into antibiotics, beta-blockers and other cardiovascular drugs, cytostatic drugs, drugs affecting the nervous system, non-steroidal anti-inflammatory drugs, and natural hormones. Among them, antibiotics are the most important group that in the following section they are thoroughly discussed. For further studies about the other drugs, refer to (Teixeira-Lemos et al. 2019; Ahmed 2017a; Quesada et al. 2019; Li 2014).

8.2.1 Antibiotics (Anti-infectives)

Antibiotics, considered the most significant medical milestone of the twentieth century, are pharmaceuticals that are extensively applied to treat bacterial infections in living creatures by inhibiting or eradicating the growth of microorganism. Antibiotics own the third rank among the drugs most prescribed for human disease, and in the case of veterinary medicine, 70% of the consumed drugs are antibiotics (Van Boeckel et al. 2015; Teixeira-Lemos et al. 2019). In spite of the fact that the development of antibiotics has contributed human to have a longer life, the presence of these compounds in the environment has drawn much attention lately. Based on the literature, annual consumption of antibiotics is estimated to be between 100,000 and 200,000 tons, and of course consumption rate increased 36% between 2000 and 2010 in 71 countries (Homem and Santos 2011; Klein et al. 2018). Another study found that Brazil, Russia, India, China, and South Africa are the biggest user of antibiotics (Gelband and Duse 2011).

Furthermore, global antibiotic use enhanced by 65% between 2000 and 2015 (Klein et al. 2018). In the USA, approximately 23,000 tons antibiotics are consumed annually (Ternes and Joss 2007). In 2013 China was nominated as the global greatest manufacturer and user of 36 antibiotics (up to 92,700 tons) and released roughly 53,800 tons antibiotics into the environment (Zhang et al. 2015; Huang et al. 2019b). Cephalexin, amoxicillin, ofloxacin, tetracycline, and norfloxacin were the top five antibiotics used in China (Zhang et al. 2015). Reports have revealed than Turkish people tend to use antibiotics approximately 2–3 times higher than European countries (Aydin et al. 2019b). According to Al-Faham et al. (2011) and Bin Abdulhak et al. (2011), 87% and 77% of antibiotics are dispensed without prescription in Syria and Saudi Arabia, respectively. Owning to their extensive consumption, contamination of various matrices including water (Wei et al. 2011; Li et al. 2018; Patrolecco et al. 2018; Sabri et al. 2018; Yang et al. 2018b; Danner et al. 2019; Huang et al. 2019a; Qiu et al. 2019), soil (Song et al. 2016; Zhou et al. 2017; Chen et al. 2019a; Zeng et al. 2019; Zhao et al. 2019), sediment (Liang et al. 2018; Chen et al. 2019b; Li et al. 2019c), and even the potable water (Huo et al. 2016; Zeng et al. 2018) by antibiotics was probable and reported in previous studies. Hospitals are potential candidates for prevailing antibiotics in the environment, mainly due to the excretion via urine or feces of the patients. Kümmerer (2001) stated that of total antibiotics used for human disease, 26% are used in hospitals (Kümmerer 2001). In Europe, for instance, 20–30% of inpatients receive antibiotic treatment during their hospitalization (Ansari et al. 2009). The significance of hospital wastewaters is because they usually have higher concentration of antibiotics than other media (Pena et al. 2010; Lien et al. 2016; Wang et al. 2018). The presence of antibiotics or their metabolites in the hospital discharge has been reported in different countries including Spain (Gros et al. 2013), Italy (Verlicchi et al. 2012), Vietnam (Lien et al. 2016), Norway (Thomas et al. 2007), Romania (Szekeres et al. 2017), Iran (Shokoohi et al. 2017), Qatar (Al-Maadheed et al. 2019), Oman (Al-Riyami et al. 2018), USA (Brown et al. 2006), Taiwan (Li and Lin 2015), Sweden (Lindberg et al. 2004),

Germany (Ohlsen et al. 2003), China (Wang et al. 2018), Brazil (Santoro et al. 2015), etc. Ciprofloxacin, ofloxacin, clarithromycin, sulfamethoxazole, trimethoprim, and metronidazole are the most frequent detected antibiotics in hospital effluent. Normally, antibiotics concentrations in the hospital effluent are higher in winter as compared to summer (Verlicchi et al. 2012). Aydin et al. (2019b) investigated the concentration of antibiotics in 16 different hospital effluents. They found that ciprofloxacin, azithromycin, and clarithromycin owned the greatest concentration in the hospital discharge. Moreover, the total concentration of antibiotics in hospital effluents ranged from 21.2 to 4886 ng/L and 497 to 322,735 ng/L in summer and winter, respectively (Aydin et al. 2019b). Shokoohi et al. (2017) examined the presence of six common utilized antibiotics (amoxicillin, ciprofloxacin, erythromycin, sulfamethoxazole, imipenem, and cefixime) in wastewater effluent of two different hospitals in Iran. They found out that in one hospital three antibiotics including amoxicillin, cefixime, and imipenem with mean concentration of 5.86, 10.85, and 25.53 μg/L, respectively, were identified; however, in the second hospital none of the mentioned antibiotics was detected (Shokoohi et al. 2017). Hence, for a proper antibiotic removal, in-situ examination regarding the type and quantity of these macromolecule compounds must be conducted. The pros of antibiotics in healthcare centers are undeniable; however, the uncontrolled, immoderate, and misuse of antibiotics have engendered to propagation, prevalence, and spread of antibiotic-resistant bacteria (ARB) and antibiotic-resistant genes (ARGs) in the aquatic environment. World Health Organization characterizes antimicrobial resistance as one of the three biggest global public health crises that must be managed with the utmost urgency (Verlicchi et al. 2015; Yang et al. 2018b). It is estimated that by 2050, antibiotic resistance (AR) will account for ten million deaths and a financial burden of approximately US$100 trillion (Sanganyado and Gwenzi 2019). Unfortunately, many researchers have highlighted hospital effluent as one of the major sources of antibiotic resistance spread (Le et al. 2016; Varela et al. 2016; Timraz et al. 2017; Wang et al. 2018).

8.3 Common Wastewater Treatment Methods

There are numerous methods that have been proposed in terms of pharmaceutical removal from various water media in recent years. Generally, wastewater treatment methods are broadly divided into physical, chemical, and biological processes; however, some may employ a combination of these processes. Adsorption, ion exchange, membrane technology, ozonation, irradiation, electrochemical processes, aerobic and anaerobic microbial degradation are of the most promising and assuring techniques in terms of pollutant removal. All of these methods have their own advantages and disadvantages that sometimes may limit their applications in some cases. Table 8.1 shows the major strengths and drawbacks of different wastewater treatment methods (Ghenaatgar et al. 2019; Salleh et al. 2011; Carolin et al. 2017; Ngulube et al. 2017; Wong et al. 2019).

Table 8.1 Comparison of the pros and cons of different wastewater treatment plants

Techniques	Pros	Cons
Coagulation/ flocculation + sedimentation	– Cost-effective – Simple dewatering – Robust process	– Sludge generation that requires a further process – Utilization of chemicals is high – High toxicity and TDS content in sludge – pH-dependent system – May be expensive due to the usage of different chemicals – Recycling of chemicals is not feasible
Membrane filtration	– High separation efficiency – Lower space requirement – Low energy consumption – Enrichment of nitrifying bacteria can reduce fouling in membrane bioreactors	– Very expensive – Membrane fouling – Complex process – Concentrated sludge production – Short life span due to membrane fouling – Membranes must have excellent chemical and thermal stability to withstand high pressures and fluxes – High investment cost
Ion exchange	– High transformation of components – Produces water of high quality – Regenerable	– Operational cost is high – High cost of resins – Generates large volumes of sludge
Adsorption	– Simplicity – Easy operation – Readily available – High efficiency – Regenerable adsorbents – Less sludge production – Utilization of low-cost adsorbents	– Incomplete removal – Regeneration may be expensive – Desorption
Ozonation processes	– No chemical sludge – Residual ozone can decompose to oxygen easily – Does not increase wastewater volume – Fast and effective – Can be used in a gaseous state	– Extremely short half-life – High cost – Produces unstable toxic by-products and carcinogenic aromatic amines
Photochemical	– Can effectively eliminate pollutants – No foul odors production	– Expensive – Formation of by-products

(continued)

Table 8.1 (continued)

Techniques	Pros	Cons
Electrochemical destruction	– Low chemical sludge – No consumption of chemicals	– Initial investment is high – Need high electrical supply – High flow rates causing direct decrease in final efficiency
Biological treatment (aerobic and anaerobic processes)	– Inexpensive – High COD removal	– Need to be developed – Incomplete elimination – Forms CH_4 and H_2S – Long reaction time and inflexible

Biological treatment processes utilize a vast variety of microorganisms, mainly come from human body, to degrade pollutants under aerobic and anaerobic conditions to gain energy for their survival. Simple operation, inexpensiveness, and high organic matter removal are the significant merits of biological processes. However, these methods mostly need a large land as well as strict environmental controls (such as pH, temperature, nutrient, toxic substances, and oxygen). Also, conventional biological methods have proven to be inadequate in terms of pharmaceuticals removal (Spataro et al. 2019).

Chemical methods are often effective for a wide range of pollutants removal. Usually, chemical methods use a large number of expensive chemicals and reagents or electricity which are not always beneficial methods in view of financial supports. Moreover, advanced oxidation processes are prone to generate secondary pollutants. Beheshti et al. (2019) observed that sulfamethoxazole was completely removed from the solution by TiO_2 and WO_3 nanoparticles via photocatalytic degradation; however, secondary pollutants through oxidation were generated which required further treatment (Beheshti et al. 2019).

The common adsorption and membrane filtration have been extensively used for pollutants removal. Adsorption has gained much attention recently because of its efficiency and simplicity and less generation of sludge. Also, since the process is physical, no new pollutants are generated (Mollahosseini et al. 2019). Membrane filtration, including nanofiltration, ultrafiltration, microfiltration, and reverse osmosis is a suitable and efficient treatment method but it must be intended that the membranes are almost sensitive and may have a short life-span due to the fouling if it is not controlled preciously (Kaya et al. 2019).

8.4 Common Water and Wastewater Treatment Plants

In any country around the world, water after collecting from an aquatic media is transferred to water treatment plants (WTPs) to remove contaminants until the water meets standards for drinking. After the water is consumed in various ways,

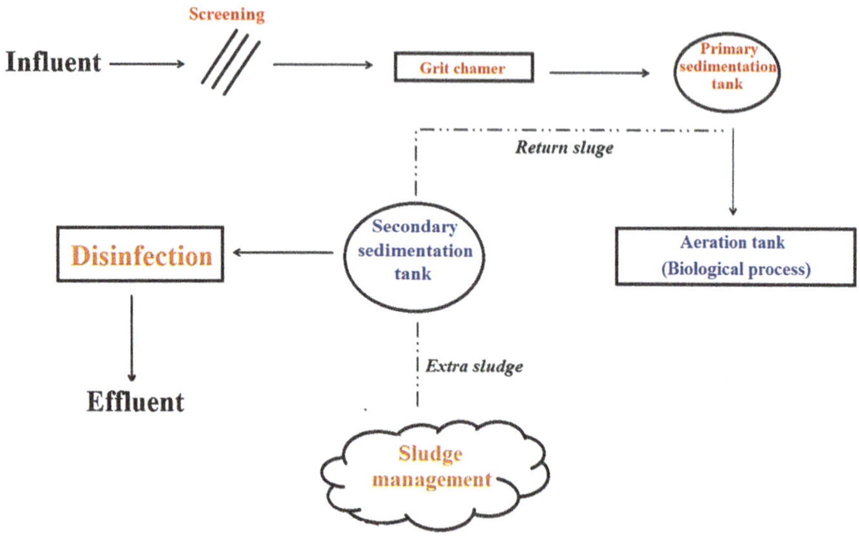

Fig. 8.1 Scheme for a conventional wastewater treatment plant having various physical, chemical, and biological units and processes

including drinking, washing, toilet, shower, etc., wastewater is generated. The generated wastewater is collected by sanitary sewer networks and sent to the wastewater treatment plants (WWTPs).

WWTPs generally employ physical, chemical, and biological methods to treat the wastewater (Fig. 8.1). Firstly, wastewater experiences physical treatment processes such as screening and grit trapping to eliminate large, rough, coarse debris, and particles. These materials are removed by sedimentation tanks allowing the suspended particles to settle out of wastewater. This process is called primary treatment. According to previous studies, primary treatment could remove lower than 10% of the pharmaceuticals and it is fair to suggest that it is inefficient at removing pharmaceuticals (Leung et al. 2012).

Then, biological processes are utilized to degrade organic materials to H_2O and CO_2 by means of flocculating activated microorganisms. Activated sludge is the most popular biological treatment process extensively utilized in most of the WWTPs. This type of treatment was developed by two English researchers, Ardern and Lockett, in 1914, and since then, it has been implemented on a global scale. Activated sludge system is comprised of a bioreactor followed by the secondary sedimentation tank. In the case of carbon, nitrogen and phosphorous removal, the bioreactor has different compartments with different operational conditions (aerobic, anaerobic, and anoxic) (Verlicchi et al. 2013). Finally, the wastewater is disinfected to kill pathogenic organisms. In fact, the role of WWTP to minimize the level of harmful contaminants in reclaimed water is highly concerned.

In terms of pharmaceutical compounds, countless studies have shown that the present WWTPs are not able to thoroughly remove these hazardous compounds and therefore they have been detected in effluent of WWTPs, surface water, ground water, and drinking water samples (Table 8.2). Cosenza et al. (2018) investigated

Table 8.2 The concentration of various pharmaceutical compounds in surface water, river, WWTP influent and effluent, urban effluent, and hospital effluent

Pharmaceuticals	Water source	Concentration (ng/L)
Atenolol	Surface water	7
	Surface water	21.7
	Surface water	31.12
Acetaminophen	Surface water	3.18
	Surface water	445
	Surface water	430
	Surface water	38.18
	Surface water	0.20
	Surface water	209
Amoxicillin	Hospital effluents	900
	WWTP influents	9.94×10^3
Ampicillin	Industrial effluents	5.8×10^3
Atenolol	River	250–600
Bezafibrate	WWTP influents	0.3–87
Caffeine	Urban effluents	23–776
	Surface water	2.9–194
	WWTP influents	2448–4865
	River	38–250
Carbamazepine	WWTP influents	33–1318
	Urban effluents	73–729
	Surface water	4.5–61
	WWTP influents	24–50
	River	56–160
	Surface water	1.17
	Surface water	412.5
	Surface water	1048
Cefaclor	WWTP influents	6.15×10^3
Cefazolin	Industrial effluents	4.2×10^3
Cefotaxime	Industrial effluents	4.2×10^3
Cephalexin	Hospital effluents	1×10^4
	WWTP influents	6.4×10^4
	Industrial effluents	3.1×10^3
Ciprofloxacin	WWTP influents	27–514
	WWTP influents	11–63
	Hospital effluents	1.5×10^4
	WWTP influents	1.1×10^3
Clofibric acid	WWTP influents	Nd-82
	Liao River	18

(continued)

Table 8.2 (continued)

Pharmaceuticals	Water source	Concentration (ng/L)
Demethyl diazepam	WWTP influents	Nd-62
	Urban effluents	8.8–127
	Surface water	1.1–6.8
	WWTP influents	9–13
	River	21–98
	Surface water	3.95
	Surface water	313
	Surface water	33.56
	Liao River	717
Dilantin	Urban effluents	8.8–181
	Surface water	1.1–8.9
Enrofloxacin	Hospital effluents	100
Erythromycin	WWTP influents	9–353
	Urban effluents	8.9–294
	Surface water	1.8–4.8
Gemfibrozil	WWTP influents	181–451
Ibuprofen	Liao River	246
	Urban effluents	10–137
	Surface water	11–38
	River	35–270
	Surface water	662.17
Ketoprofen	Surface water	29.51
Lincomycin	WWTP influents	11–629
	Hospital effluents	1.7×10^3
	Industrial effluents	1.1×10^5
Pharmaceutical	Water source	Concentration (ng/L)
Metronidazole	Industrial effluents	7.8×10^3
Nalidixic acid	Industrial effluents	6.7×10^3
Naproxen	Urban effluents	20–483
	Surface water	1.8–18
	River	81–360
Norfloxacin	Hospital effluents	200
Ofloxacin	WWTP influents	150–1081
	Industrial effluents	1.3×10^3
Oxytetracycline	Industrial effluents	1.5×10^4
Salicylic acid	WWTP influents	433–8036
	Liao River	295
Spiramycin	WWTP influents	11–129

(continued)

Table 8.2 (continued)

Pharmaceuticals	Water source	Concentration (ng/L)
Sulfadiazine	Industrial effluents	353
	WWTP influents	46–253
	Urban effluents	3.8–407
	Surface water	1.7–36
	WWTP influents	13–261
	River	9–190
	Hospital effluents	300
	WWTP influents	3×10^3
	Industrial effluents	5.8×10^3
	Surface water	9.79
	Surface water	56.19
	Surface water	299
Sulfanilamide	Industrial effluents	207
Sulfathiazole	Industrial effluents	9.6×10^3
Tetracycline	Industrial effluents	1.5×10^3
	Urban effluents	10–188
	Surface water	3.2–53.
	River	11–94
	Hospital effluents	300
	WWTP influents	4.3×10^3

the occurrence and the behavior of illicit drugs and their metabolites in a conventional WWTP located in Sicily (island of Italy). The following average removal efficiencies were reported: benzoylecgonine (77.85%), cocaine (92.34%), codeine (64.75%), and morphine (90.16%).

Al-Maadheed et al. (2019) examined the elimination of eight selected antibiotics by a domestic wastewater treatment plant built in 1990 Doha, Qatar with a flow rate of 54,000 m^3/d. The WWTP consists of several units and processes, including bar screen, grit detroiter, elevator, conventional aeration, activated sludge recirculation, sand filter, and chlorination. The removal efficiencies of clavulanic acid (65.7%), metronidazole (57.4%), amoxicillin (46.3%), ciprofloxacin (70.5%), tetracycline (19.2%), penicillin (95.6%), and erythromycin (90.9%) were obtained. More importantly, they found that concentrations of antibiotics in the influent wastewater were the same as (or even higher) their concentrations in the effluent water of the main hospital, indicating that direct disposal of drugs into the sewer system is as important as the clinics and hospital effluents as the main source of pharmaceutical compounds spread (Al-Maadheed et al. 2019). They declared that sand filtration could not remove antibiotics from wastewater.

Three WWTPs in Italy were investigated and it was found that acetaminophen and ketoprofen were both removed in spite of distinctive in-situ operations (Palli et al. 2019). However, atenolol removal ranged from medium to excellent based on the operational parameters in the WWTP. Carbamazepine, diclofenac, and amoxicillin exhibited low removal efficiency and were reluctant to biological oxidation.

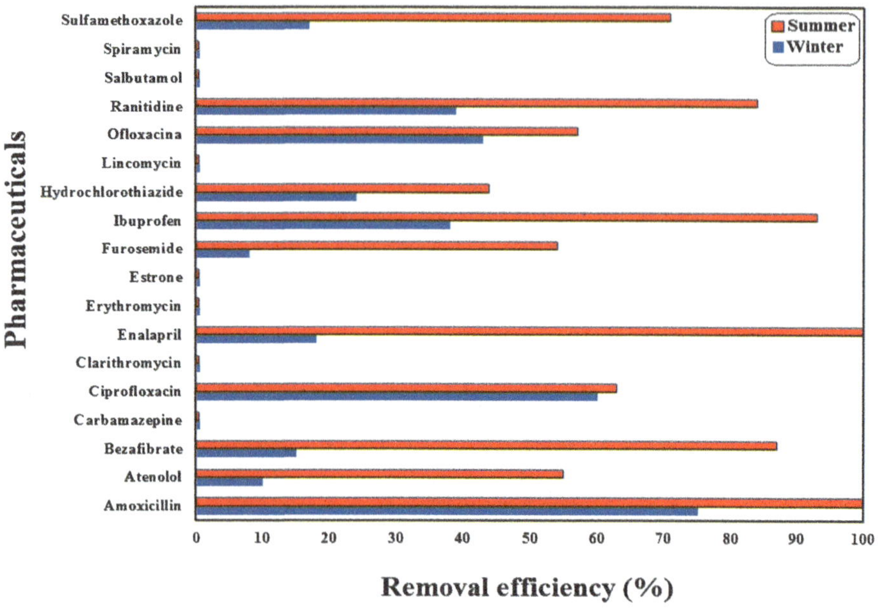

Fig. 8.2 The removal efficiency of various drugs in summer and winter

They believed that solid retention time played an important role in the pharmaceutical removal (Palli et al. 2019).

Kim et al. (2010) reported the presence of clarithromycin, metformin, atenolol, carbamazepine, and trimethoprim at high concentrations in the effluent of membrane bioreactor WWTP. Furthermore, a sewage treatment plant in Japan was named as a main source of six anti-cancer agents in its effluent (Azuma et al. 2015). The presence of various pharmaceuticals in the influent or/and effluent of WWTP in Adana, Turkey (Daglioglu et al. 2019), Nova Scotia and New Brunswick (Greenham et al. 2019), Gniewino (Kołecka et al. 2019), 11 cities of China (Yu et al. 2019), Costa Rica (Causanilles et al. 2017), Durban, South Africa (Faleye et al. 2019), Rome, Italy (Spataro et al. 2019), and India (Kurasam et al. 2018) has also been reported.

It has been reported that the removal rate of drugs is dependent on various factors, including the types and properties of the drugs (solubility, volatility, photodegradation, and biodegradability), the pattern of the drug consumption and its load during time variations, water consumption rate, geography, climate, and operating conditions in WWTPs (Tiwari et al. 2017; Kanakaraju et al. 2018). Peng et al. (2019) stated that in the activated sludge process, pharmaceutical removal occurs via adsorption, hydrolysis, oxidation, heterotrophic biodegradation, and autotrophic biodegradation. The operating conditions generally relate to parameters such as sludge retention time (SRT), hydraulic retention time (HRT), and pH of the water. Another report claimed that seasonal variations may be effective in terms of pharmaceuticals removal. In this direction, drugs may be categorized into three distinctive groups as follows (Fig. 8.2) (Castiglioni et al. 2006):

1. In summer, higher removal rate was attained including amoxicillin, atenolol, bezafibrate, and enalapril.
2. There was no difference in removal efficiency in both summer and winter including ciprofloxacin, hydrochlorothiazide, and ofloxacin.
3. The third group is of those that the removal efficiency was approximately zero in both summer and winter.

The higher removal efficiency in summer can be attributed to the influence of the temperature on the biological processes, the amount of water and drugs consumption and nitrification rate (Wang and Wang 2016). It was formerly proved that with decreasing the temperature, the nitrification rate and pharmaceutical removal both decreased (Verlicchi et al. 2013). Moreover, the variations of the drugs concentration must be intended. Kumar et al. (2019) investigated the seasonal variations of different drugs in the influent of a WWTP in New Zealand and found that some compounds only are observed in certain time of the year. Cocaine, for instance, was only detected in summer and winter. Also, the total concentrations of the studied drugs were found to be 30–40% higher during summer than winter (Kumar et al. 2019).

More interestingly, conventional WTPs have reported to be ineffective to remove pharmaceutical compounds from raw water similar to conventional WWTPs. Reis et al. (2019) conducted a research regarding occurrence, removal and seasonal variation of pharmaceuticals in six Brazilian drinking water treatment plants. Table 8.3 shows the concentration of the drugs entering in the WTPs. It can be observed from Table 8.3 that pharmaceuticals are entered into the WTPs. The authors believed that the fluctuation of pharmaceuticals among the WTPs was attributed to the area that raw water was supplied/collected. Table 8.4 summarizes the removal efficiency of the pharmaceuticals after the treatment. It is evident that some drugs were completely removed; however, the others are remained in the drinking water and probably enter the human body. These WTPs employed units and processes such as primary chlorination, coagulation, flocculation, sedimentation, sand filtration, post-chlorination, and dissolved-air flotation that were not suitable enough to remove pharmaceuticals pollution.

An interesting study was conducted regarding the surface water (including rivers, lakes, oceans, and aquifers) of the USA (Deo 2014). Figure 8.3 shows the number of drugs that were detected in the investigated waters. Briefly, of 93 pharmaceuticals that were observed, 27 antibiotics; 15 antidepressants; 9 antihypertensives; 7 analgesics; and 7 anticonvulsants owned the most groups of drugs that were detected.

8.5 Adsorption Process

Adsorption techniques have been extensively applied for the separation of both organic and inorganic pollutants from various kinds of wastewaters. Adsorption, a physical phase transfer process, utilizes a solid material as the adsorbent to eliminate pharmaceuticals from the liquid phase. The contaminant in this process is named adsorbate. In fact, the efficiency of the process is attributed to the interaction

Table 8.3 The concentration of various pharmaceutical compounds entering to six water treatment plants

Pharmaceuticals	WTP1	WTP2	WTP3	WTP4	WTP5	WTP6
Betamethasone	<MQL-6371	<MDL-4109	<MQL-11,960	<MDL-4354	<MQL-7836	<MDL-3662
Clarithromycin	–	<MDL-168	–	<MDL-199	–	–
Phenazone	–	–	–	–	<MDL-33	–
Phenylbutazone	<MDL-76	–	–	–	–	–
Fenofibrate	–	–	–	<MDL-290	<MDL-119	<MDL-1388
Fluconazole	<MQL-1413	<MQL-583	–	<MQL-763	<MQL-1294	<MQL-62
Loratadine	<MDL-123	–	<MDL-56	<MDL-214	–	<MDL-486
Prednisone	<MQL-1895	<MDL-2444	<MQL-2085	<MQL-2329	<MQL-8105	<MDL-855
Atorvastatin	<MQL-523	<MDL-455	<MDL-109	<MQL-608	<MQL-1020	–
Caffeine	–	<MDL-1385	–	–	–	–
Danofloxacin	<MDL-41	<MDL-35	–	<MDL-61	<MDL-272	<MDL-23
Enoxacin	<MDL-386	<MDL-354	–	<MDL-257	<MDL-240	–
Enrofloxacin	–	<MDL-64	–	<MDL-13	<MDL-71	–
Metformin	<MDL-203	–	–	<MDL-167	–	–
Norfloxacin	<MQL-156	<MDL-146	<MDL-89	<MDL-136	<MQL-285	–
Ketoprofen	<MQL-300	<MQL-298	–	<MQL-1020	<MQL-959	<MDL-22
Gemfibrozil	<MDL-899	<MQL-475	<MDL220	<MDL-211	<MQL-948	<MDL-12
Ibuprofen	–	–	–	<MDL-302	–	–

between adsorbent and adsorbate. Generally, adsorption is comprised of three distinctive steps:

1. Firstly, the adsorbate in the bulk solution must move to the external boundary layer around the adsorbent which is called film diffusion or external diffusion.
2. Secondly, the adsorbates may be able to enter into the pores of the adsorbent. This step is pore diffusion.
3. Finally, fixation of the adsorbate on the adsorbent sites via energetic interaction occurs, named surface reaction.

Table 8.4 The removal efficiency of different pharmaceutical compounds after treatment in six different water treatment plants

Pharmaceuticals	WTP1	WTP2	WTP3	WTP4	WTP5	WTP6
Betamethasone	84 ± 23%	70 ± 10%	70 ± 10%	61 ± 34%	75 ± 37%	91 ± 18%
Clarithromycin	Na	100%[1]	Na	100%[1]	Na	Na
Phenazone	Na	Na	Na	Na	100 ± 0%	Na
Phenylbutazone	100 ± 0%	Na	Na	Na	Na	Na
Fenofibrate	Na	Na	Na	100%[1]	100%[1]	100%[1]
Fluconazole	59 ± 21%	64 ± 32%	Na	48 ± 27%	43 ± 24%	100%[1]
Loratadine	82 ± 31%	Na	100%[1]	100 ± 0%	Na	75 ± 36%
Prednisone	69 ± 25%	60 ± 37%	32 ± 6%	56 ± 38%	45 ± 23%	99 ± 1%
Atorvastatin	57 ± 23%	84 ± 36%	100 ± 0%	55 ± 38%	61 ± 32%	Na
Caffeine	100 ± 0%	100%[1]	Na	Na	Na	Na
Danofloxacin	75 ± 24%	100 ± 0%	100 ± 0%	100 ± 0%	89 ± 22%	100%[1]
Enoxacin	72 ± 40%	100 ± 0%	Na	100 ± 0%	100 ± 0%	Na
Enrofloxacin	100 ± 0%	100%[1]	Na	100%[1]	96 ± 10%	Na
Metformin	100 ± 0%	Na	Na	100%[1]	Na	Na
Norfloxacin	86 ± 31%	83 ± 20%	100 ± 0%	100 ± 0%	50 ± 29%	Na
Gemfibrozil	91 ± 16%	96 ± 9%	100%[1]	79 ± 12%	59 ± 15%	100%[1]
Ketoprofen	82 ± 23%	100 ± 0%	Na	72 ± 32%	73 ± 38%	100%[1]
Ibuprofen	Na	Na	Na	100%[1]	Na	Na

[1]Standard deviation not available

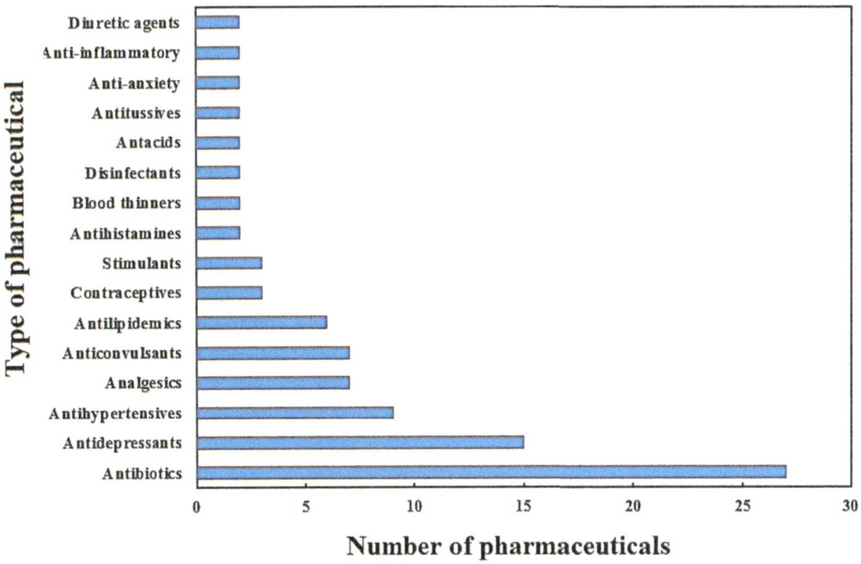

Fig. 8.3 The number of pharmaceuticals reported to occur in the surface water of the USA

Table 8.5 Comparison between some properties of physical and chemical adsorption

Physical adsorption	Chemical adsorption
Multilayer adsorption	Monolayer adsorption
Low enthalpy	High enthalpy
High entropy	Low entropy
Low activation energy	High activation energy
Reversible	Non-reversible
Low temperatures	High temperatures
Weak intermolecular forces like van der Waals forces	Strong covalent bonding involving electron exchange

Fig. 8.4 The schematic picture to describe the difference between physical and chemical adsorption

It is a generally accepted assumption that the third step is rapid. However, the external diffusion and pore diffusion determine the total rate of adsorption kinetic.

Depending on the interaction between adsorbate and adsorbent, the adsorption can be divided into two types:

1. Physisorption (physical adsorption): When no electron is shared between adsorbate and adsorbent, the removal process follows physical sorption. Weak attraction forces such as hydrophobicity, hydrogen bonding, polarity, static interactions, π-π interactions, and van der Waals forces are the major responsible of pollutant removal (Dawood and Sen 2014). Furthermore, low energy, simple desorption, and multilayer build-up of adsorbate on adsorbent is other properties of physisorption.
2. Chemisorption (chemical adsorption): Unlike physisorption, chemical adsorption involves electron sharing (or chemical bond or attraction) and has a high level of energy. Monolayer is observed in chemisorption and usually desorption is difficult.

Some of the differences between physisorption and chemisorption are shown in Table 8.5 (Ngulube et al. 2017). Figure 8.4 shows a scheme for the meaning of physical and chemical adsorption.

8.6 Influential Parameters Affecting the Adsorption of Pharmaceuticals

To implement the adsorption treatment process for the removal of drugs from aqueous media, it is highly important to figure out the effect of parameters on the process. Optimization of these parameters leads to attain the maximum efficiency of the process along with minimum cost. In the next section, some of the factors affecting the adsorption of pharmaceuticals are discussed.

8.6.1 Effect of Initial Drug Concentration and Contact Time

The removal efficiency usually depends on the initial pollutant concentration and enough contact time between adsorbate and adsorbent. As a matter of fact, if the adsorbent available sites are not completely occupied by the pollutant concentration, the removal efficiency seems to increase with drugs concentration. On the other hand, a reduction in removal efficiency can be observed once the adsorbent sites are saturated and the drugs concentration increases. Generally, removal of pharmaceuticals increases with an increase in contact time until the equilibrium is achieved. Once the equilibrium is reached, no significant further uptake of adsorbate occurs. Equilibrium state is attained once the rate of adsorption and rate of desorption becomes equal (Ahmed 2017b).

Khadir et al. (2020) investigated the removal of ibuprofen from aqueous media by using Luffa fibers/popypyrrple nanocomposite. The authors reported that after 90 min of the reaction, the removal efficiency did not fluctuate and reached a maximum value of > 60%. Tian et al. (2019) studied the effect of contact time on the adsorption of terramycin on carbon aerogel and magnetic carbon aerogel. They observed that at the first 30 min of the reaction adsorption was rapid and then the adsorption became slow and reached equilibrium after 60 min. Thus, a combination of rapid/slow trend was seen within the first 60 min (Tian et al. 2019). Duan et al. (2019) reported the removal of three fluoroquinolone antibiotics using polypyrrole functionalized *Calotropis gigantea* fiber. Duan et al. (2019) expressed that within 30 min of the treatment process, more than 75% of enrofloxacin, ciprofloxacin, and norfloxacin were removed, and the adsorption system reached its equilibrium gradually within 180 min. Similar observations were reported for adsorption of sulfonamide antibiotics by amino-functionalized biomass-derived porous carbons (Wang et al. 2019b), cephalexin antibiotic by activated carbon derived from a single-step pyrolysis of phosphoric acid-activated chitin (Khanday et al. 2019), fluoroquinolone adsorption by chitosan-derived granular hydrogel (Wang et al. 2019a) and tetracycline by magnetic graphene oxide (Miao et al. 2019).

Paunovic et al. (2019) studied naproxen removal by microwave functionalized biochar derived from novel lignocellulosic waste biomass. They noticed that by increasing the adsorbate concentration from 3.1 to 125.3 ppm, the removal efficiency decreased. Turk Sekulic et al. (2019a) investigated the adsorption capacity of multifunctional carbonous adsorbent for drugs adsorption. They observed that the adsorption capacity increases from 0.499 to 18.981 mg/g for sulfamethoxazole, from 0.500 to 21.497 mg/g for carbamazepine and from 0.505 to 19.729 mg/g for ketoprofen when the initial concentration is increased from 1 to 50 ppm (Turk Sekulic et al. 2019a). A higher driving force at elevated adsorbate concentration was the main reason for such an increase (Turk Sekulic et al. 2019a).

The initial contaminant concentration also affects the time of the equilibrium state. Wang et al. (2019a) believed that by increasing the ciprofloxacin and enrofloxacin concentrations from 10 to 250 mg/L, equilibrium was found to be 30, 60, and 120 min, respectively. Khanday et al. (2019) also reported similar observations for cephalexin antibiotic. Hence, to reach a faster equilibrium state the pharmaceuticals concentration would be at minimum. Table 8.6 presented the results of various studies about the effect of initial pharmaceuticals concentration on adsorption.

Table 8.6 Results obtained from various studies regarding the effect of the initial pharmaceutical concentrations on the adsorption capacity or the removal efficiency

Adsorbent	Pharmaceuticals	Initial concentration (ppm)	Adsorption capacity (mg/g)	Removal efficiency (%)	Sources
Polyamidoamine/silica Nanohybrid	Clofibric acid Diclofenac Ibuprofen Ketoprofen	0.05–0.2 0.05–0.2 0.05–0.2 0.05–0.2	– – – –	89–46 96–55 99–67 98–61	Lotfi et al. (2019)
Polypyrrole functionalized *Calotropis gigantea* fiber	Enrofloxacin Ciprofloxacin Norfloxacin	0–200 0–200 0–200	0–84 0–77 0–72	– – –	Duan et al. (2019)
Functionalized carbonaceous adsorbent	Sulfamethoxazole Carbamazepine Ketoprofen	1–50 1–50 1–50	0.49–18.98 0.50–21.49 0.50–19.72	– – –	Turk Sekulic et al. (2019a)
Amino-functionalized porous carbon	Sulfadiazine	0–8	0–120	–	Khanday et al. (2019)
Nanostructured pillar MOF	Amoxicillin	60-150	–	92-51	(Abazari et al. 2019)
Microwave functionalized biochar	Naproxen	3.1–125.3	7–73	97–23	Paunovic et al. (2019)
Magnetic graphene oxide	Chlortetracycline	5–50	20–133	–	Miao et al. (2019)
Activated charcoal	Propafenone Diphenhydramine Propranolol Thioridazine	0–200 0–200 0–200 0–480	0–172.5 0–126.7 0–143.4 0–338.8	– – – –	Zhao et al. (2018)
Nitrogen-doped reduced graphene oxide/Fe_3O_4	Norfloxacin Ketoprofen	10–60 10–60	91.6–160.2 85.4–373.0	– –	Peng et al. (2018)
Chitosan/waste coffee-grounds	Metamizol Acetylsalicylic acid Acetaminophen Caffeine	0.25–2.0 0.25–2.0 0.25–2.0 0.25–2.0	1.0–7.6 1.3–9.9 1.0–6.3 1.2–8.2	– – – –	Lessa et al. (2018))

8.6.2 Effect of Initial pH

pH is an imperative parameter in adsorption processes since it immensely affects the surface charge/binding sites of the adsorbent and the ionization degree of the pharmaceutical compounds (Mirjavadi et al. 2019). For instance, it was proved that the removal of sulfamethoxazole, carbamazepine, and ketoprofen by a phosphorized carbonaceous adsorbent was greatly dependent on the operational pH value (Turk Sekulic et al. 2019a). Similarly, it was claimed that the efficiency of the chitosan-based magnetic composite particles in terms of diclofenac sodium and tetracycline hydrochloride elimination from water was relied on the pH of solution (Zhang et al. 2016b). It seems that pH controls the pharmaceuticals adsorption since it influences the dominant adsorbent/adsorbate interaction.

The point of zero charge (pH_{zpc}) has been extensively utilized in adsorption studies for the determination of the mechanism and behavior of the removal pathway. At pH equal to pH_{zpc}, the surface charge of the material is zero. However, at pHs greater and lower than pH_{zpc}, the synthesized adsorbent owns negative and positive charge, respectively (Beheshti et al. 2019). It is fair to suggest that pHs > pH_{zpc} is favorable for the adsorption of cationic pharmaceuticals. Contrary to this, anionic compounds are better removed at pHs < pH_{zpc}. Table 8.7 indicates the optimum pH value for the removal of pharmaceuticals with different adsorbents.

Turk Sekulic et al. (2019b) investigated the removal of ketoprofen by the phosphorous-doped microporous carbonous material at various pH values (Fig. 8.5). It was found that the synthesized carbonous adsorbent had the pH_{zpc} of 4.12. By increasing the pH value from 2 to 9, the adsorption capacity decreased. The reduction in adsorption capacity at high pH was attributed to the repulsion force between the negatively charged adsorbent and the negatively charged adsorbate. At pH < pH_{zpc}, the adsorption mechanism involved H-bonding and π-π interaction (Turk Sekulic et al. 2019b).

In another investigation, Reza et al. (2014) employed microwave-assisted activated bamboo waste as an adsorbent for ibuprofen elimination. It was claimed that the maximum ibuprofen removal was attained at pH 2-5 (removal efficiency 85.66–97.02%). Ibuprofen is an acidic drug and has the pK_a of 4.15, indicating that at pH higher than 4.15, it acquires a negative charge due to the dissociation of molecules of this compound. Hence, electrostatic repulsion between adsorbate anions and the negative surface of the adsorbent led to a reduction in adsorption efficiency at high pH value (Reza et al. 2014). Similar mechanism was reported by Zhang et al. (2017a) for adsorption of sulfadiazine by expanded graphite.

Gadipelly et al. (2018) examined the applicability of metal organic framework 5 (MOF5) for the effective adsorption of ciprofloxacin hydrochloride from aqueous solution over a pH ranged from 3 to 9. They observed that adsorption capacity increased with the increasing pH and attained a maximum of 87.7 mg/g at pH 5.5, thereby decreasing on further increment in the pH. Ciprofloxacin hydrochloride mainly exists as a cation below pH 5.5, an anion above pH 9, and is zwitterion between pH 5.5 and 9 due to its pK_a values of 5.9 and 8.9. MOF5 exhibited a nega-

Table 8.7 The optimum initial pH value for the maximum removal of pharmaceuticals with different adsorbents

Adsorbent	Pharmaceutical	Optimum pH	Sources
Modified mica and montmorillonite	Ibuprofen	4–9	Martín et al. (2019)
Polyamidoamine/silica Nanohybrid	Clofibric acid Diclofenac Ibuprofen Ketoprofen	9	Lotfi et al. (2019)
Multiwalled carbon nanotube composite	Potassium diclofenac	6	Pires et al. (2019)
Phosphorised carbonaceous	Sulfamethoxazole Carbamazepine	3–5 No effect	Turk Sekulic et al. (2019a)
Polypyrrole/Calotropis gigantea fiber	Enrofloxacin Ciprofloxacin Norfloxacin	6 8 8	Duan et al. (2019))
Powdered zeolites	Azithromycin Ofloxacin Sulfamethoxazole	8 6 2	de Sousa et al. (2018))
Magnetic graphene oxide	Tetracycline hydrochloride Oxytetracycline	3.3 3.3	Miao et al. (2019)
Montmorillonite	Propranolol	2–9	del Mar Orta et al. (2019)
Magnetite nanoparticles modified β-cyclodextrin polymer coupled with $KMnO_4$ oxidation	Acetaminophen	7	Shi et al. (2019)
Goethite	Diclofenac	5.23	Zhao et al. (2017)
Activated carbon from spent tea leaves	Acetaminophen	3–7	Wong et al. (2018)
Chitosan/waste coffee-grounds	Metamizol Acetylsalicylic acid Acetaminophen Caffeine	6	Lessa et al. (2018)
Polypyrrole functionalized Calotropis gigantea fiber	Enrofloxacin CiprofloxacinNorfloxacin	6	Duan et al. (2019)
Weathered microplastic polystyrene	Oxytetracycline	5	Zhang et al. (2018a)
Magnetic chitosan-g-poly(2-acrylamide-2-methylpropanesulfonic acid)	Tetracycline Chlorotetracycline	3.3	Lu et al. (2018)
Walnut shell-based activated carbon	Metronidazole Sulfamethoxazole	8 5.5	Teixeira et al. (2019))
Magnetic nanoparticles	Levofloxacin	6.5	Al-Jabari et al. (2019)
Chitosan/waste coffee-grounds	Metamizol Acetylsalicylic acid Acetaminophen Caffeine	6	Lessa et al. (2018)
Functional polyaniline/ multi-walled carbon nanotube	Meloxicam	2	Dutra et al. (2018)

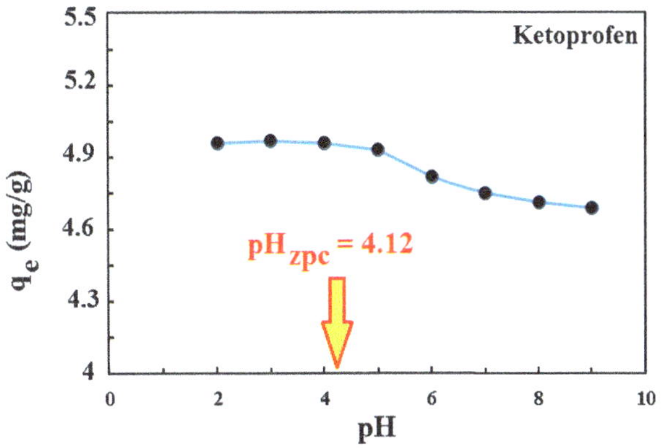

Fig. 8.5 Fluctuation of the adsorption capacity toward ketoprofen removal at different pH values

tive zeta potential over the studied pH values. Cationic species of the drug were dominantly removed by the used adsorbent. Electrostatic repulsion force was the main reason for the reduced adsorption capacity (Gadipelly et al. 2018). However, in another investigation regarding ciprofloxacin removal by modified waste grapefruit peel, it was expressed that ion exchange and π-π interaction were involved in the adsorption process as well as electrostatic attraction (Fu et al. 2019).

8.6.3 Effect of Temperature

Temperature is another significant factor which mainly influences the nature of the adsorption process, adsorbate/adsorbent interaction, pharmaceuticals movement, adsorption rate, and adsorption energy (Li et al. 2019b). Generally, adsorption processes are categorized into two different groups: (1) endothermic and (2) exothermic. If the adsorption capacity of the synthesized material increases with increasing temperature then the adsorption is an endothermic. On the other hand, if enhancement of adsorption capacity was observed by lowering the temperature, the process is exothermic. Table 8.8 shows the results of various studies on the effect of temperature on the pharmaceuticals adsorption by various adsorbents.

Al-Khateeb et al. (2014) studied the removal of the aspirin, acetaminophen, and caffeine onto graphene nanoplatelets. They inferred that higher adsorption capacity at lowered temperature may indicate the exothermic nature of the adsorption process. The percentage of adsorption of aspirin was decreased from 64.8 to 63.6 and 60.7%, when the solution temperature was raised from 296 to 308, and 323 K, respectively. For caffeine, the percentage of adsorption decreased from 98.5 to 96.0 and 93.2%. Adsorption of acetaminophen decreased from 90.3 to 84.9 and 79.7% (Al-Khateeb et al. 2014). Such behavior may be because of the adsorptive forces reduction between the adsorbent active sites and the pharmaceuticals species.

Table 8.8 The nature of the adsorption process in terms of temperature as well as thermodynamic parameters in each study

Adsorbent	Pharmaceutical	Temperature	Process	$\Delta H°$	$\Delta S°$	$\Delta G°$	Source
Multi-walled carbon nanotube	Potassium diclofenac	25–45	Endothermic	24.71	148.42	−19.52 to −22.47	Pires et al. (2019)
Polypyrrole/multi-walled carbon nanotube	Potassium diclofenac	25–45	Endothermic	52.67	246.0	−20.64 to −25.57	
Magnetic graphene oxide	Chlortetracycline	283–313	Endothermic	12.3666	46.4173	−0.7406 to −2.1331	Miao et al. (2019)
	Tetracycline hydrochloride	283–313	Endothermic	18.9181	67.5944	−0.2735 to −2.3013	
	Oxytetracycline	283–313	Endothermic	7.5256	30.0922	−0.9523 to −1.8551	
Commercial organoclay	Diclofenac sodium	15–50	Endothermic	112.3	452.6	−16.67 to −32.5	Maia et al. (2019)
Activated carbon	Chlorpheniramine	298–308	Endothermic	15.11	75.4	−7.35 to −7.85	(Ali et al. (2019)
Activated carbon	Ibuprofen	298–308	Endothermic	18.18	77.16	−4.81 to −5.40	
Oxidized activated carbon	Chlorpheniramine	298–308	Endothermic	23.52	105.2	−7.82 to −8.58	
Oxidized activated carbon	Ibuprofen	298–308	Endothermic	22.19	94.92	−6.10 to −6.82	
Ethylene diamine to produce basic surface	Chlorpheniramine	298–308	Endothermic	18.52	86.1	−7.14 to −7.74	
Ethylene diamine to produce basic surface	Ibuprofen	298–308	Endothermic	21.20	73.37	−0.66 to −1.35	
Ethylamine to produce hydrophobic carbonaceous surface	Chlorpheniramine	298–308	Endothermic	21.41	99.34	−8.19 to −8.88	
Ethylamine to produce hydrophobic carbonaceous surface	Ibuprofen	298–308	Endothermic	15.58	61.31	−2.69 to −3.20	
Functionalized bean husks	Ibuprofen	303–323		96.234	109.111	−26.678 to −16.287	Bello et al. (2019)
Pretreated oat hulls	Ciprofloxacin	288–318	Endothermic	+17	+163	−29 to −34	Movasaghi et al. (2019)

Adsorbent	Pharmaceutical	Temperature	Process	ΔH°	ΔS°	ΔG°	Source
Double-oxidized graphene oxide	Acetaminophen	283–313	Exothermic	−10.84	8.81	−13.36 to −13.64	Moussavi et al. (2016)
Graphene nanoplatelets	Aspirin Acetaminophen Caffeine	296-323	Exothermic	−5.42 −25.4 −46.6	−13.1 −67.3 −123.0	−1.54 −5.42 −10.2	Al-Khateeb et al. (2014)
Amorphous nano-carbon	Tetracycline Sulfadiazine	298–313	Endothermic	52.38 34.94	223.39 147.70	−14.19 to −17.54 −9.07 to −11.28	Wu et al. (2016)
	Sulfamethoxazole			33.43	144.10	−9.51 to −11.67	

Wu et al. (2016) studied the removal of tetracycline and sulfamethoxazole antibiotics on amorphous nano-carbon at different temperatures. They noticed that by increasing the temperature from 298 to 313 K, the adsorption capacity toward tetracycline and sulfamethoxazole increased from 79.85 and 96.67 mg/g to 129.54 and 113.56 mg/g, respectively. It can be deduced that the adsorption capacity was higher at greater temperature (Wu et al. 2016). This phenomenon may be attributed to the mobility of the pharmaceutical molecules and an increase in the number of available active sites of the adsorbent. Furthermore, the increased temperature reduces the thickness and mass transfer resistance of the boundary layer surrounding the nanoparticles (Xu et al. 2018).

Interestingly, some researches have expressed that the temperature variations during the adsorption process had no significant effect on the removal efficiency. Baccar et al. (2012) studied the adsorption of ibuprofen, ketoprofen, naproxen, and diclofenac onto a low-cost activated carbon, prepared at the laboratory scale from olive-waste cake. Identical results were also reported regarding ibuprofen adsorption by powered activated carbon (Mestre et al. 2007) and paracetamol adsorption by grape stalk (Villaescusa et al. 2011). It is believed that the process was athermic. One possible explanation of this behavior is that the molecule drugs are well solvated in their aqueous solution.

Aside from the effect of the temperature on the removal efficiency or the adsorption capacity, temperature results provide more information regarding thermodynamic parameters such as Gibbs free energy (ΔG°, kJ/mol), enthalpy (ΔH°, kJ/mol), and entropy (ΔS°, kJ/mol K) and can be determined by the following equations:

$$\Delta G^\circ = -RT \ln K_0$$

$$\Delta G^\circ = \Delta H^\circ - T\Delta S^\circ$$

$$\ln K_0 = \frac{\Delta S^\circ}{R} - \frac{\Delta H^\circ}{RT}$$

T—the system temperature (K)
R—the universal gas constant (8.314 J/mol K)
K_0—is the equilibrium constant

ΔG° is related to spontaneity of the adsorption process. ΔH° can possess two different values that express the adsorption process as an endothermic or exothermic process if its value is positive or negative, respectively. Also, the type of the adsorption, physisorption, or chemisorption might be determined by ΔH°. In fact, physical adsorption has enthalpy lower than 20 kJ/mol; however, chemical adsorption has the enthalpy ranging from 80 to 400 kJ/mol. Of course, removal processes having ΔH° values between 20 and 80 kJ/mol are combination of physical/chemical adsorption.

Miao et al. (2019) studied the adsorption of tetracycline using magnetic graphene oxide at 283, 298, and 313 K and reported that the adsorption capacity was highest at 313 K and lowest at 283 K. The results exhibited that the values of $\Delta G°$ were negative (spontaneous process) and keep decreasing with increasing temperature. The positive values of $\Delta H°$ implied that the treatment process was endothermic, favoring at higher temperatures. Moreover, the positive value of $\Delta S°$ indicated that the randomness at the adsorption solid–liquid interface increased during the adsorption process.

Nasseh et al. (2019) evaluated adsorption thermodynamics of metronidazole on a new $FeNi_3/SiO_2/CuS$ magnetic nanocomposite. The corresponding $\Delta G°$ values were from −19.88 to −26.20 kJ/mol for temperatures from 278 to 323 K, respectively. The values of $\Delta H°$ and $\Delta S°$ were 0.05 kJ/mol and 82.11 J/mol K, respectively. Endothermicity, spontaneity, and feasibility of the adsorption process were verified by positive enthalpy and negative Gibbs free energy. The positive value of entropy variations ($\Delta S°$) revealed that the degree of freedom is magnified in the solid-liquid interface during the adsorption (Nasseh et al. 2019).

Jodeh et al. (2016) studied the thermodynamic parameters of diclofenac adsorption from aqueous solution using *Cyclamen persicum* tubers-based activated carbon. The values of $\Delta G°$ (−3.04 to −1.15 kJ/mol for 288–318 K), $\Delta H°$ (−17.45 kJ/mol), and $\Delta S°$ (−50.08 J/mol K) exhibited that the process was exothermic and spontaneous. From the value of $\Delta H°$ which was less than 40 kJ/mol, it is fair to suggest that the adsorption was physical. In another point of view revealing the same outcome, the absolute magnitude of $\Delta G°$ was between 0 and 20 kJ/mol and the adsorption could be classified as physisorption type (Gereli et al. 2006; Ahmed 2017a). Negative value of $\Delta S°$ suggested that diclofenac removal was enthalpy driven.

Sun et al. (2019) examined the adsorption of nitroimidazole antibiotics from aqueous solutions on self-shaping porous biomass carbon foam pellets derived from *Vallisneria natans* waste. Thermodynamic parameters implied an endothermic adsorption process and indicated that the randomness of adsorbate at the solid-liquid interface was increased. Sun et al. (2019) calculated the activation energy of adsorption process utilizing the formula below proposed by Arrhenius:

$$\ln K = \ln K_0 - \frac{E_a}{RT}$$

K—pseudo-second-order rate constant (g/mg min)
E_a—the activation energy (kJ/mol)

The activating energy calculated from the experimental data done by Sun et al. (2019) was 15.33 kJ/mol, indicating that chemical adsorption was involved in the adsorption processes (Sun et al. 2019). Normally, the activating energy of physical adsorption is not more than 4.184 kJ/mol due to the weak van der Waals forces (Duranoğlu et al. 2012).

8.6.4 Effect of Adsorbent Dosage

Adsorbent dosage is a substantial parameter affecting the capacity of an adsorbent for a given amount of the adsorbent used during pharmaceuticals removal. In general, the removal efficiency of pharmaceutical drugs increase with increasing amount of adsorbent. This positive fact could be attributed to the number and availability of sorption sites enhancement by increasing adsorbent dosage (Wong et al. 2018).

Quercus Brantii(Oak) acorn was utilized as an inexpensive material for omission of two important drugs, acetaminophen and ibuprofen, from water ways (Nourmoradi et al. 2018). They noticed that the removal efficiency of acetaminophen increased from 10.67 to 89.55% by raising adsorbent dosage from 0.5 to 10 g/L. Similar behavior was observed for removal pathway of ibuprofen. The availability of higher surface area and empty cavities for drugs at higher adsorbent dosage were introduced as the main reason for such behavior (Nourmoradi et al. 2018). It was seen that increasing the adsorbent dosage from a certain value of more than 5.0 g/L, the removal percentage was not changed significantly (Nourmoradi et al. 2018). Ghemit et al. (2019a) tested the elimination of diclofenac and ibuprofen from aqueous media by organobentonites. They found that with the enhancement of the adsorbent mass from 10 to 100 mg the adsorbed amount improved with an increase in adsorbent dose. Similar observations were reported regarding ciprofloxacin removal by magnetic Fe_3O_4/red mud-nanoparticles (Fig. 8.6) (Aydin et al. 2019a).

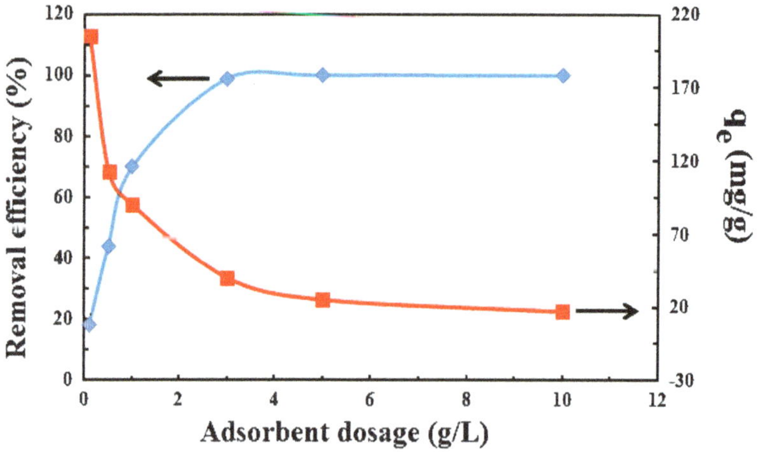

Fig. 8.6 The figure indicates that by increasing the adsorbent dosage, the removal efficiency continuously increased. However, the higher adsorption capacity was attained at the lower adsorbent dosage

Wang et al. (2020) attempted to remove low-concentration antibiotics from aqueous solution by regenerable $Mg(OH)_2$. They studied the effect of adsorbent dosage ranging from 0.1 to 1 g/L. The results indicated that the removal rate of ciprofloxacin decreased from 91.1 to 86%, and the adsorption capacity for ciprofloxacin decreased from 182.1 to 17.19 mg/g by increasing the adsorbent dosage, respectively. Excess usage of adsorbent may increase the overall cost of the operation along with lowering adsorption capacity and removal efficiency.

8.7 Adsorption Isotherms

Adsorption isotherm models are essential tools for the evaluation of adsorption behavior via adsorption. In another words, these models are employed to assess the surface properties of the adsorbents as well as the interaction between adsorbent and adsorbate. Adsorption isotherms are based on a relationship between the amount of pharmaceutical adsorbed per unit weight of the adsorbent and the pharmaceutical concentration at a constant temperature, under equilibrium state.

Accordingly, various models have been used in extensive studies (Syafiuddin et al. 2018; Neris et al. 2019). Among them, Langmuir, Freundlich, and Temkin models are the most common models of adsorption in so many studies.

Langmuir model, a theoretical model originally developed to the adsorption of gases on a solid material, is valid for monolayer adsorption onto a surface containing a finite number of identical sites. The model is based on the assumptions that maximum adsorption corresponds to a saturated monolayer of pharmaceuticals molecules on the adsorbent surface, that the energy of adsorption is constant. Also, there are no any lateral forces between the adsorbed molecules. The nonlinear Langmuir isotherm is given as:

$$q_e = \frac{q_m K_L C_e}{1 + K_L C_e}$$

q_e—the amount of adsorbate adsorbed at equilibrium (mg/g)
q_m—the maximum saturated monolayer adsorption capacity of an adsorbent (mg/g)
K_L—the constant related to the affinity between adsorbent/adsorbate (L/mg)
C_e—the adsorbate concentration at equilibrium (mg/L)

Since now, four linearized forms of Langmuir equation have been proposed. They are shown in:

$$\text{Langmuir} - 1\left(\text{Hanes} - \text{Woolf linearization}\right)$$

$$\frac{C_e}{q_e} = \frac{C_e}{q_m} + \frac{1}{q_m K_L} \quad \left(\frac{C_e}{q_e}\ \text{vs.}\ C_e\right)$$

$$\text{Langmuir} - 2(\text{Lineweaver} - \text{Burk linearization})$$
$$\frac{1}{q_e} = \frac{1}{q_m} + \frac{1}{q_m K_L} \frac{1}{C_e} \quad \left(\frac{1}{q_e} \text{vs.} \frac{1}{C_e} \right)$$

$$\text{Langmuir} - 3(\text{Eadie} - \text{Hoffsiee linearization})$$
$$q_e = q_m - \frac{1}{K_L} \frac{q_e}{C_e} \quad \left(q_e \text{vs.} \frac{q_e}{C_e} \right)$$

$$\text{Langmuir} - 4(\text{Scatchard linearization})$$
$$\frac{q_e}{C_e} = q_m K_L - K_L q_e \quad \left(\frac{q_e}{C_e} \text{vs.} q_e \right)$$

A mathematical expression developed by Webber and Chakkravorti has been used to test the favorability of the adsorption process:

$$R_L = \frac{1}{1 + K_L C_0}$$

R_L—separation factor and dimensionless
C_0—initial concentration of pharmaceutical compounds

The Freundlich isotherm, one of the earliest empirical equations, assumes that the adsorption occurs on a heterogeneous surface, and the amount that is adsorbed increases infinitely with an increase in concentration. The model cannot describe the behavior of the adsorption at very low or very high (saturated state) concentrate of adsorbate. The nonlinear and linear forms of the Freundlich model can be written as (Piri et al. 2019):

$$q_e = K_F C_e^{1/n}$$

$$\ln q_e = \ln K_F + \frac{1}{n} \ln C_e$$

K_F—the Freundlich constant (mg/g)/(mg/L)^n
n—the Freundlich intensity dimensionless parameter

For the various studies on the application of various adsorbents for pharmaceutical removal, it was observed that Langmuir model approximately exhibited better results than Freundlich models (Table 8.9). This means that drugs elimination via adsorption generally obey Langmuir assumption.

Wu et al. (2019a) studied the competitive adsorption of antibiotic tetracycline and ciprofloxacin on montmorillonite. They examined Freundlich, Langmuir, Temkin, and Dubinin-Radushkevich isotherm models. It was found that Langmuir isotherm model better described the experimental data in single and mixed system. The R^2

Table 8.9 The best-fitted isotherm and kinetic model for various studies using different adsorbents for the removal of distinctive pharmaceuticals

Adsorbent	Pharmaceuticals	Isotherm	Kinetic	Sources
Modified mica and montmorillonite	Ibuprofen	Langmuir/ Freundlich	Pseudo second order	Martín et al. (2019)
Polypyrrole/multiwalled carbon nanotube	Diclofenac	Langmuir–Freundlich	Pseudo second order	Martín et al. (2019)
Polypyrrole/Calotropis gigantea fiber	Ciprofloxacin Enrofloxacin	Redlich-Peterson	Pseudo second order	Duan et al. (2019)
Powdered zeolites	Azithromycin Ofloxacin Sulfamethoxazole	Freundlich	Pseudo second order	de Sousa et al. (2018)
Montmorillonite	Tetracycline Ciprofloxacin	Langmuir	Pseudo second order	Wu et al. (2019a)
Carbon aerogels	Terramycin	–	Pseudo second order	Tian et al. (2019)
Magnetic graphene oxide	Tetracycline	Freundlich	Pseudo second order	Miao et al. (2019))
Montmorillonite	Propranolol	Freundlich/ Dubinin-Radushkevich	Pseudo second order	del Mar Orta et al. (2019)
Magnetite nanoparticles modified β-cyclodextrin	Acetaminophen	Langmuir	Pseudo second order	Shi et al. (2019)
Commercial organoclay	Diclofenac sodium	Temkin/ Freundlich	Pseudo first order	Maia et al. (2019)
Goethite	Diclofenac	–	Pseudo second order	Zhao et al. (2017)
Activated carbon from spent tea leaves	Acetaminophen	Langmuir	Pseudo second order	Wong et al. (2018)
Polyamidoamine/silica Nanohybrid	Clofibric acid Diclofenac Ibuprofen Ketoprofen	Langmuir	Pseudo second order	(Lotfi et al. 2019)
Activated charcoal	Propafenone Diphenhydramine Propranolol Thioridazine	Langmuir	–	Zhao et al. (2018)

(continued)

Table 8.9 (continued)

Adsorbent	Pharmaceuticals	Isotherm	Kinetic	Sources
Functionalized carbonaceous adsorbent	Sulfamethoxazole Carbamazepine Ketoprofen	Langmuir	Pseudo second order	Turk Sekulic et al. (2019a)
Microwave functionalized biochar derived from lignocellulosic waste biomass	Naproxen	Langmuir	Pseudo second order	Paunovic et al. (2019)
Fe_3O_4@SiO_2/SiCRG	Metoprolol	Langmuir/Toth	–	Soares et al. (2016)
Activated carbon of babassu coconut mesocarp	Acetylsalicylic acid	Sips	Pseudo second order	Hoppen et al. (2019)
Iron oxide-mesoporous silica MCM-41	Acetylsalicylic acid	Langmuir	–	Teo et al. (2016)
Metal-organic frameworks	Acetic acid	Freundlich	Pseudo second order	Zhang et al. (2016a)
Activated carbons	Acetaminophen	Langmuir	–	Rey-Mafull et al. (2014)
Activated carbon prepared from rice husk	Paracetamol	Langmuir	Pseudo second order	George Nche et al. (2017)
Single walled carbon nanotubes	Amoxicillin	Langmuir	Pseudo second order	Balarak et al. (2016)
Multi-walled carbon nanotubes with different oxygen contents	Ciprofloxacin	Dubinin-Radushkevich	Pseudo second order	Yu et al. (2016)
Magnetic multi-walled carbon nanotube	Amoxicillin	Langmuir	–	Fazelirad et al. (2015)
Graphene oxide	Tetracycline	Langmuir	Pseudo second order	Gao et al. (2012)
Magnetic chitosan-g-poly(2-acrylamide-2-methylpropanesulfonic acid)	Tetracycline Chlorotetracycline	Langmuir	Pseudo second order	Lu et al. (2018)
Pretreated oat hulls	Ciprofloxacin	Freundlich	Pseudo second order	Movasaghi et al. (2019)
A novel core-shell-structure activated carbon	Sulfamethoxazole	Freundlich	Pseudo second order	Pamphile et al. (2019)
Auricularia-based magnetic porous carbon		Langmuir	Pseudo second order	Xie et al. (2019)
Nitrogen-functionalized carbon	Carbamazepine Clofibric acid Oxytetracycline	Langmuir Freundlich Freundlich	Pseudo second order	Prarat et al. (2019)

value, the maximum saturated monolayer adsorption capacity, and experimental maximum adsorption capacity for tetracycline were 0.99, 1.06, and 1.00 mmol/g, respectively. Freundlich isotherm was also reported in the literature. It was remarked that tetracycline removal by magnetic graphene oxide was well-described by the Freundlich model than Langmuir model (Miao et al. 2019). It seems that the type of adsorbent is important for the determination of the best isotherm model.

Moreover, some researches have expressed more than one isotherm as the best model describing the removal process. Martin et al. (2019), for instance, believed that omission of ibuprofen from aqueous media by montmorillonite is fitted well with both Freundlich and Langmuir equations since they revealed high R^2 values (greater than 0.999). Maia et al. (2019) showed that the removal pathway of diclofenac sodium by commercial organoclay was governed by Freundlich at 15 and 30 °C; however, at 50 °C Temkin was more practical in adsorption process description. Chemisorption was assumed as the main removal mechanism.

Dubinin-Radushkevich model was also employed in some studies. Dubinin-Radushkevich, a semiempirical temperature-dependent model, is based on the fact that adsorption mechanism is pore filling, following Gaussian energy distribution onto heterogeneous surfaces. Dubinin-Radushkevich isotherm is expressed as follows:

$$q_e = q_{DR} e^{-K_{DR}\varepsilon^2}$$

The linear form of the presented model is:

$$\ln q_e = -K_{DR}\varepsilon^2 + \ln q_{DR}$$

$$\varepsilon = RT \ln\left(1 + \frac{1}{C_e}\right)$$

q_{DR}—the adsorption capacity (mg/g)
K_{DR}—is a constant related to the sorption energy (mol^2/kJ^2)
ε—the Polanyi potential
R—the gas constant (8.314)

Yu et al. (2016) employed multi-walled carbon nanotubes with different oxygen contents for ciprofloxacin removal. They stated that Dubinin-Radushkevich and Langmuir model were revealed as the best model for describing the adsorption process (Yu et al. 2016). Also, they calculated the average free energy of adsorption by using the below formula.

$$E = \frac{1}{\sqrt{2K_{DR}}}$$

E—the average (mean) adsorption energy (kJ/mol)

Yu et al. (2016) believed that the removal process was chemisorption since the adsorption energy was in the range of 20–40 kJ/mol. Generally, physical adsorption exists in the adsorption process at the adsorption energy of lower than 8 kJ/mol, while chemical adsorption such as ion exchange owns the adsorption energy of 8–16 kJ/mol (Tahir and Rauf 2006). In a similar study concerning diclofenac sodium removal by commercial organoclay, the adsorption energy at 30 and 50 °C was found to be 8.53 and 10.96 kJ/mol, suggesting a chemical adsorption process (Maia et al. 2019).

8.8 Adsorption Kinetics

For the implementation of an industrial scale adsorption unit, it is necessary to reveal the controlling mechanism for the pollutant removal by means of kinetics models. The adsorption kinetics exhibit the rate of adsorbate attachment to the surface of the adsorbent. Many models have been suggested to find the best order of the reaction in order to design an optimum treatment unit. Accordingly, Lagergren pseudo-first-order model and Lagergren pseudo-second-order model are the two most studied models in countless studies.

The Lagergren pseudo-first-order kinetic rate equation is based on the premise that the adsorption rate is corresponding to the difference between the adsorbed amount and the equilibrium adsorption capacity and can be written as follows:

$$\frac{dq_t}{dt} = K_1\left(q_e - q_t\right)$$

K_1—the rate constant of the pseudo first-order model (min^{-1})

Integrating the equation for boundary conditions ($t = 0$, $q_t = 0$ and $t = \text{t}$, $q_e = q_t$) leads to the following linear equation:

$$\ln\left(q_e - q_t\right) = \ln q_e - K_1 t \quad \left(\text{Or } \log\left(q_e - q_t\right) = \log q_e - \frac{K_1 t}{2.303}\right)$$

which can be rearranged in a nonlinearized form:

$$q_t = q_e\left(1 - e^{K_1 t}\right)$$

The differential equation of Lagergren pseudo-second-order model is generally known and described as:

$$\frac{dq_t}{dt} = K_2\left(q_e - q_t\right)^2$$

K_2—the rate constant of the pseudo second-order model (g/mg min)

The integration of this equation for the boundary conditions ($t = 0$, $q_t = 0$ and $t = t$, $q_e = q_t$) gives the formula below:

$$q_t = \frac{q_e^2 K t}{1 + q_e t}$$

Equation below can be rearranged to obtain the linear form:

$$\frac{t}{q_t} = \frac{1}{K q_e^2} + \frac{t}{q_e}$$

Table 8.9 exhibits the best fitting model for the removal of pharmaceutical compounds by various types of adsorbents. As can be observed from Table 8.9, adsorption kinetics of almost all of the pharmaceuticals is nicely described by pseudo second order. For instance, Tian et al. (2019) used carbon aerogel to remove antibiotics (Tian et al. 2019). They expressed that the adsorption kinetics was satisfactory followed by pseudo second order and adsorption process was mainly based on the chemical adsorption with the assistance of physical adsorption. Lotfi et al. (2019) believed that pseudo second order was more suitable than pseudo first order and intraparticle diffusion to expound the kinetics of clofibric acid, diclofenac, ibuprofen, and ketoprofen by polyamidoamine/silica nanohybrid. Other studies complying pseudo second order are listed in Table 8.9. These evidences propose that the adsorption of pharmaceuticals is chiefly governed by chemisorption via exchanging or sharing electrons between adsorbate and adsorbent.

8.9 Adsorbents for Pharmaceuticals Removal

To date, so many adsorbents with various advantages and disadvantages have been proposed for removal of pollutants. In this review, the following materials are considered since they are approximately famous and extensively employed in countless studies.

8.9.1 Activated Carbon

Among many materials that have been suggested for adsorption processes, activated carbon owing to its large specific surface, its micropore structure, unique surface chemistry, tunable pore structure, good thermostability at high temperatures in inert or reduction atmospheres, low acid-base reactivity and high adsorption capacity offers a great potential for the removal of pharmaceutical compounds (Ao et al. 2018). Table 8.10 shows several studies used different activated carbon for pharmaceutical removal as well as the maximum adsorption capacity. Figure 8.7 shows a simple procedure for production of activated carbon.

Table 8.10 The utilization of different activated carbon for the removal of various pharmaceutical compounds

Adsorbents	Origin	Pharmaceuticals	Adsorption capacity (mg/g)	Sources
Commercial activated carbon	–	Diclofenac Naproxen	50 80	Varga et al. (2019)
Pulverized activated carbon	–	Diclofenac Naproxen	320 280	Varga et al. (2019)
Powdered activated carbon	Vegetable	Carbamazepine Sildenafil citrate	242 341	Delgado et al. (2019)
Activated carbon	–	Ibuprofen Naproxen Diclofenac	160 241 188	Bahamon and Vega (2017)
NiFe2O4/activated carbon	Vegetable	Ibuprofen Ketoprofen	261.35 97.75	Fröhlich et al. (2019)
Activated carbon with steam preparation	Shell of *Aegle Marmelos*	Ibuprofen	12.7	Chakraborty et al. (2018)
Thermally activated carbon preparation	Bovine bone	Ibuprofen	56.8	Cazetta et al. (2016)
Chemically activated high grade nanoporous carbons	Sisal (Agave sisalana) residues biomass	Ibuprofen Iopamidol	325 1097	Mestre et al. (2019)
Surface functionalized activated carbons	Date palm leaflets	Chlorpheniramine	455	Ali et al. (2019)
A novel core-shell-structure activated carbon	A coal power plant	Sulfamethoxazole	95.5	Pamphile et al. (2019)
Activated carbon-decorated polyacrylonitrile nanofibers	Waste tires	Ciprofloxacin Danofloxacin Enrofloxacin	93 99 112	Mogolodi Dimpe and Nomngongo (2019)
Auricularia-based magnetic porous carbon	Auricularia	Tetracycline	397.25	Xie et al. (2019)
Ultrafine magnetic biochar	–	Carbamazepine tetracycline	62.7 94.2	Shan et al. (2016)
Ultrafine magnetic Activated carbon	–	Carbamazepine tetracycline	135.1 45.3	Shan et al. (2016)
Nitrogen-functionalized carbon	Pomelo peel waste	Carbamazepine Clofibric acid Oxytetracycline	216.2 19.4 64.9	Prarat et al. (2019)
Activated carbon	Bamboo	Ciprofloxacin	615	Wang et al. (2015)

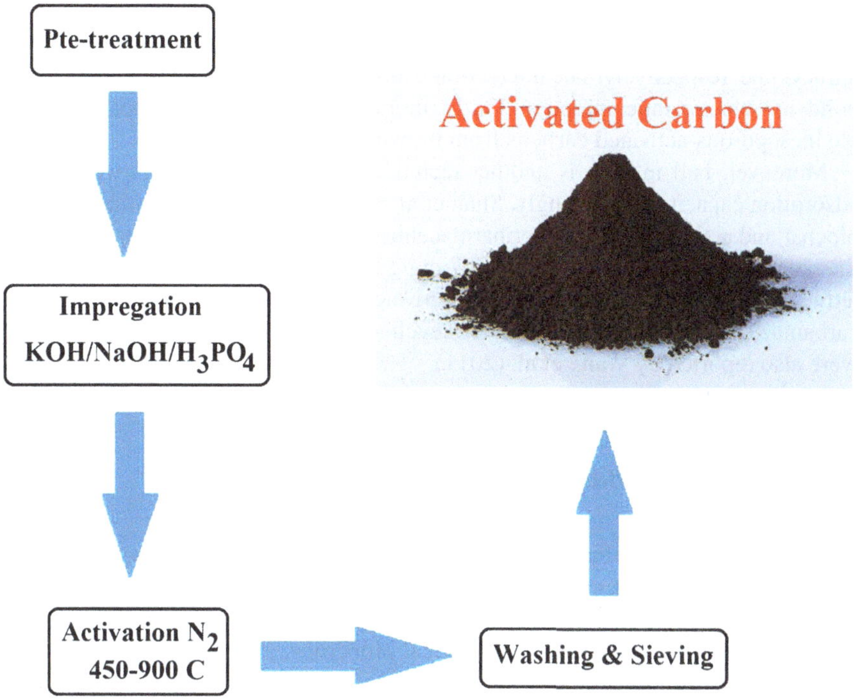

Fig. 8.7 It shows that the production of the activated carbon is constituted of four steps that the most important ones are impregnation and activation

Varga et al. (2019) employed commercial activated carbon for pharmaceutical adsorption. In their studies, the experimental maximal adsorption capacity of 50 and 80 mg/g was found for diclofenac and naproxen, respectively. Furthermore, they pulverized and sieved activated carbon through a 0.1 mm mesh size and reach the adsorption capacity of 320 and 280 mg/g for diclofenac and naproxen, respectively. Availability of the pores of granules was one of the main factors influencing adsorption (Varga et al. 2019). Delgado et al. (2019) reported that powdered active carbon is more effective that granular active carbon for pharmaceuticals removal since it has greater surface area with the value of 1328 m^2/g and 956 m^2/g, respectively. Also, micro-grain activated carbon, characterized by having a particle size of 200–600 μm (between PAC (<100 μm) and GAC (>800 μm)), was utilized for drug removal (Alves et al. 2018).

Magnetic adsorbents have recently been proposed in adsorption process due to their easy separation and better adsorption performance (Mollahosseini et al. 2019). Fröhlich et al. (2019) stated $NiFe_2O_4$/activated carbon composite as a promising magnetic adsorbent for elimination of ibuprofen and ketoprofen pharmaceuticals from water. The highest adsorption capacity according to Sips model was found to be 261.35 for ibuprofen and 97.75 for ketoprofen with initial concentration of 0–100 ppm. Also, auricularia-based magnetic porous carbon was developed by Xie et al. (2019) via facile simultaneous activation and magnetization strategy. They attained the surface area and adsorption capacity of 823.2 m^2/g and 397.25 mg/g for tetracycline, respectively. The interaction mainly involved π-π interaction, hydrogen bond, and electrostatic attraction. Furthermore, Jiang et al. (2019) prepared separable mesoporous-activated carbons from brown coal with Fe_3O_4.

Moreover, ball milling is another technique to prepare materials with greater adsorption capacity. Accordingly, Shan et al. (2016) reported an ultrafine magnetic biochar and activated carbon for pharmaceutical adsorption. The biochar/Fe_3O_4 had the greatest adsorption capacity of 62.7 mg/g for carbamazepine and 94.2 mg/g for tetracycline, while values obtained for activated carbon/Fe_3O_4 were 135.1 mg/g for carbamazepine and 45.3 mg/g for tetracycline, respectively. Similar observations were also reported by Wang et al. (2011).

8.9.2 Graphene-Based materials

Graphene, a one-atom-thick and two-dimensional (2D) layer of sp^2-bonded carbon with a molecule bond length of 0.142 nm and an interplanar spacing of 0.123 nm, has gained many researchers attention due to its significant properties such as excellent mechanical, electrical, thermal, optical properties, and very high specific surface area (Al-Khateeb et al. 2014; Iannazzo et al. 2018).

Graphene atoms are able to contact with pollutant molecules from both sides that is favorable for adsorption processes. However, surface hydrophobicity and easy agglomeration in aqueous solution are the main limitations of graphene that necessitate the modification of this material (Li et al. 2019b). Zhu et al. (2015) studied the removal of ciprofloxacin by graphene and found that the maximum adsorption capacity of 322.6 mg/g was achieved at pH 4 and equilibrium time of 0.05 h. The removal process followed Langmuir and pseudo second order assumptions.

Graphene oxide (GO) and reduced graphene oxide (rGO) are of two main derivations of graphene. rGO is obtained from GO reduction. GO was employed for removal of sulfamethoxazole, ciprofloxacin, levofloxacin, and tetracycline that revealed the maximum adsorption capacity of 240, 379, 256.6, and 313 mg/g, respectively (Çalışkan and Göktürk 2010; Gao et al. 2012; Dong et al. 2016).

The combination of magnetic nanoparticles and graphene has recently attracted great interest. It has been proven that the introduction of Fe_3O_4 nanoparticles in graphene structure results in better performance. It was observed that the adsorption capacity of graphene/Fe_3O_4 (423 mg/g) was 1.35 times higher than graphene (313 mg/g) for tetracycline removal (Zhang et al. 2017b). In another study, magnetically separable Fe/Cu bimetallic nanoparticles were supported by GO for removal of pharmaceuticals compound (Tabrizian et al. 2019). Figure 8.8 shows the adsorption procedure and SEM images of the prepared materials. Under the optimum conditions (pH 6.5, contact time 0.25 h, adsorbent dosage 0.25 g/L), the maximum adsorption capacity of 201.9 mg/g was achieved, following Freundlich isotherm equation.

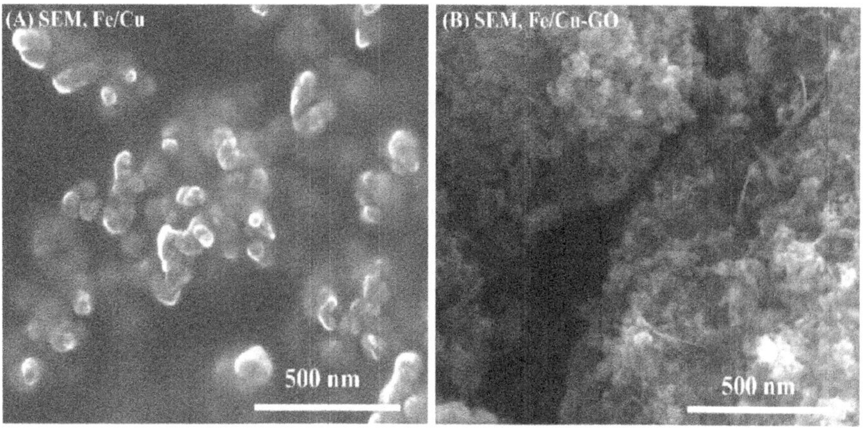

Fig. 8.8 The SEM images of F/Cu and Fe/Cu-graphene oxide composite. Reprinted with permission of Elsevier from (Tabrizian et al. 2019)

The utilization of the magnetic graphene-based adsorbents was further studied by the researchers. In paper manufacturing, textile printing, photographic, and leather processing industries, thiourea dioxide has been widely employed as a strong reductant. Yang et al. (2017) used thiourea dioxide/reduced magnetic graphene oxide for the decontamination of tetracycline. They observed the maximum adsorption capacity of 1233 mg/g under pH 4, and adsorbent dosage 0.07 g/L. The removal process was governed by pseudo second-order and Langmuir models. Results indicated that higher temperature favored pharmaceutical adsorption, supporting that the adsorption was spontaneous and endothermic. Also, the presence of NaCl as the ionic strength in the solution facilitated the adsorption process, and the optimum adsorption capacity was obtained when the NaCl concentration was higher than 0.001 M.

8.9.3 Chitosan and Its Composites

Chitosan, composed of randomly distributed (1–4)-linked 2-amino-2-deoxy-β-D-glucopyranose units, is the second most abundant naturally available biopolymer, and has high affinity to adsorb different pollutants from aqueous solution. Properties such as low price and abundance, antibacterial property, non-toxicity, biocompatibility, biodegradability, macromolecular structure, hydrophilicity, cationicity, active sites, and high adsorption capability have made it as a suitable alternative (Vakili et al. 2019). Chitosan can be obtained by simple deacetylation of chitin as shown in Fig. 8.9. Since chitosan has high contents of amino and hydroxy functional groups, electrostatic interaction, ion exchange, chelation, and ion pair formation might be introduced as the main adsorption mechanisms. Table 8.11 reveals

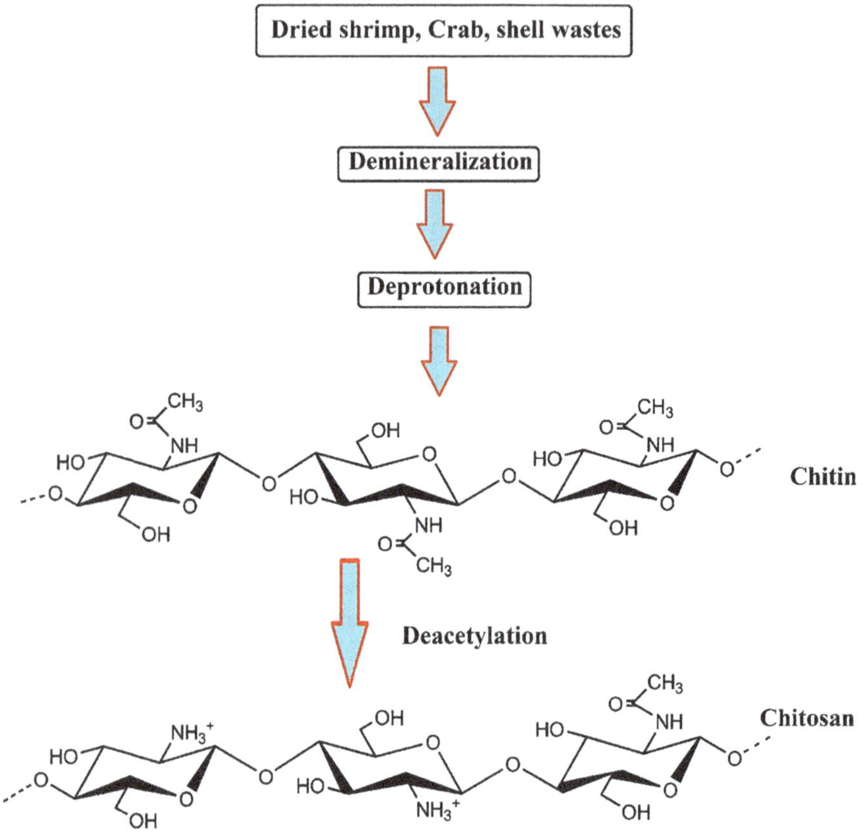

Fig. 8.9 It shows the procedure of chitosan from shrimps, crab, and shell wastes. Briefly, chitosan is obtained from the deacetylation of chitin

the utilization of chitosan and its composites for the removal of various pharmaceuticals.

However, separation of powdery chitosan from water requires high-speed centrifugation, which limits its large-scale applications. Liang et al. (2019) developed a magnetic composite based on amine-functionalized chitosan and magnetic nanoparticles to remove diclofenac sodium from water (Liang et al. 2019). TEM results revealed that the adsorbent had the mean diameter of 25 nm. The maximum adsorbent capacity of 469.48 mg/g was attained and the removal process was governed by Langmuir and pseudo second-order models. It was observed that hydrogen bonding was an important removal pathway.

Sometimes raw biopolymers including chitosan do not have an acceptable performance for particular pollutants unless they are modified. In this direction, chitosan/biochar hydrogel beads composite was employed for drug removal (Afzal et al. 2018). Figure 8.10 shows the SEM image of biochar before modification (a) and

Table 8.11 The utilization of chitosan and its composites for the removal of various pharmaceutical compounds

Adsorbents	Pharmaceuticals	Adsorption capacity (mg/g)	Sources
Magnetic amine-functionalized chitosan	Diclofenac sodium	469.48	Liang et al. (2019)
Magnetic activated carbon/chitosan	Ciprofloxacin Erythromycin Amoxicillin	90.10 178.57 526.31	Danalıoğlu et al. (2017)
Chitosan/biochar hydrogel beads	Ciprofloxacin	36.72	Afzal et al. (2018)
Chitosan derived granular hydrogel	Ciprofloxacin Enrofloxacin	267.7 387.7	Wang et al. (2019a)
Chitosan/waste coffee-grounds	Acetylsalicylic acid Acetaminophen Caffeine	9.92 7.52 8.21	Lessa et al. (2018)
Chitosan–ethylene glycol hydrogel	Perfluorooctanoic acid	1275.9	Long et al. (2019)
$ZnFe_2O_4$/chitosan	Diclofenac	188	dos Santos et al. (2019)
Bilayer amino-functionalized cellulose nanocrystals/chitosan	Diclofenac	444.44	Hu et al. (2019)
Magnetic genipin-cross-linked chitosan/graphene oxide-SO_3H	Tetracycline	473.28	Liu et al. (2019)

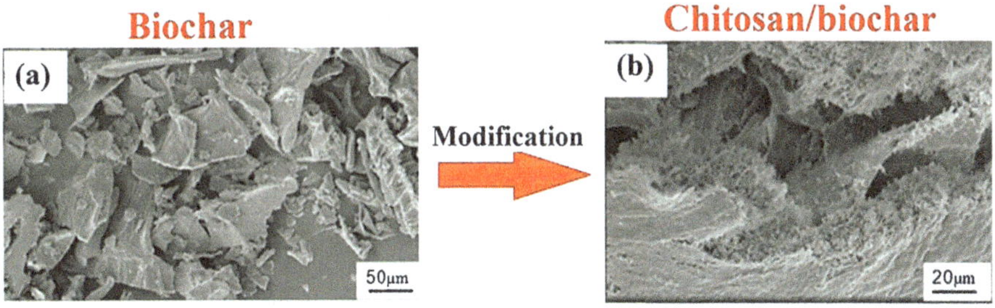

Fig. 8.10 The SEM images of biochar and chitosan/biochar composite. Reprinted with permission of Elsevier from (Afzal et al. 2018)

after modification with chitosan (b). It is obvious that various pores and sites are formed after modification which is beneficial in terms of pollutant removal. The maximum uptake capacity of 36.72 mg/g for ciprofloxacin was achieved thorough π-π electron donor-acceptor interaction, hydrogen bonding, and hydrophobic interaction at pH 3.

In view of environmental decontamination and protection, chitosan-based wastes have also been suggested. Waste coffee-grounds, for instance, remained after coffee dripping that need large quantity of oxygen for full decomposition. Lessa et al.

(2018), for the first time, utilized chitosan/waste coffee-grounds composite for pharmaceutical adsorption. Besides the fact that the synthesized adsorbent exhibited a suitable performance and it was cost-effective and eco-friendly, the mechanical properties of chitosan were enhanced after modification. Formerly, the application of chitosan was limited because of its low mechanical strength and deformation (Crini 2006).

Chitosan has low stability at strong pH changes that impede the applications of this material. To overcome this issue as well as mechanical problems, crosslinking, impregnation, and grafting techniques were suggested by the researchers. Glutaraldehyde, epichlorohydrin, ethylene glycol diglycidyl ether, and tripolyphosphate are the well-known cross-linking agents. dos Santos et al. (2019) used zinc ferrite as the core material and chitosan as the external layer by means of glutaraldehyde.

8.9.4 *Carbon Nanotube*

Carbon nanotubes (CNTs) have gained considerably high attention since its invention by Iijima in 1991. CNTs possess unique properties such as high surface area, porous structure, hydrophobic nature, tunable surface chemistry, thermal and chemical stability and mechanical properties make CNTs as a distinctive material for water purification (Anjum et al. 2019). They consist of one or more graphene sheets wrapped around themselves to form a cylindrical shape with a length of more than 20 μm and a radius of less than 100 nanometer (Fiyadh et al. 2019). Depending on the side walls configuration, CNTs have two types: single-walled carbon nanotubes (SWCNTs, contain one graphene sheet) and multi-walled carbon nanotubes (MWCNTs, contain more than one graphene sheet). Figure 8.11 illustrates the SWCNTs and MWCNTs.

Since now, various pharmaceuticals compounds have been removed by CNTs especially with MWCNTs (Table 8.12). Ncibi and Sillanpää (2017) compared the removal of carbamazepine and dorzolamide from aqueous solutions using mesoporous-activated carbons and MWCNTs. They found that for both drugs MWCNTs exhibited higher adsorption capacity of 224.6 and 78.8 mg/g, respectively. Also, MWCNTs had better performance than bentonite clay, zeolite, rice straw, Fe_3O_4 nanoparticles, sewage sludge/fish waste, and activated carbons from potato peels.

Similar to chitosan and activated carbon, CNTs were also integrated with magnetic nanoparticles. Bhatia et al. (2019) conducted a research to compare the applicability of MWCNTs and MWCNTs/Fe_3O_4 nanocomposite. They reported that the adsorption capacity of 671 and 1234 mg/g for MWCNT and MWCNT/Fe_3O_4 was found for isonicotinic acid removal, respectively. In fact, the authors proved that the presence of magnetic particles in the nanocomposite not only provides a better situation for separation of the spent adsorbent but also increases the adsorption capacity.

In recent years, the modification of CNTs with conducting polymers has been conducted in several investigations. Pires et al. (2019) formed composite named

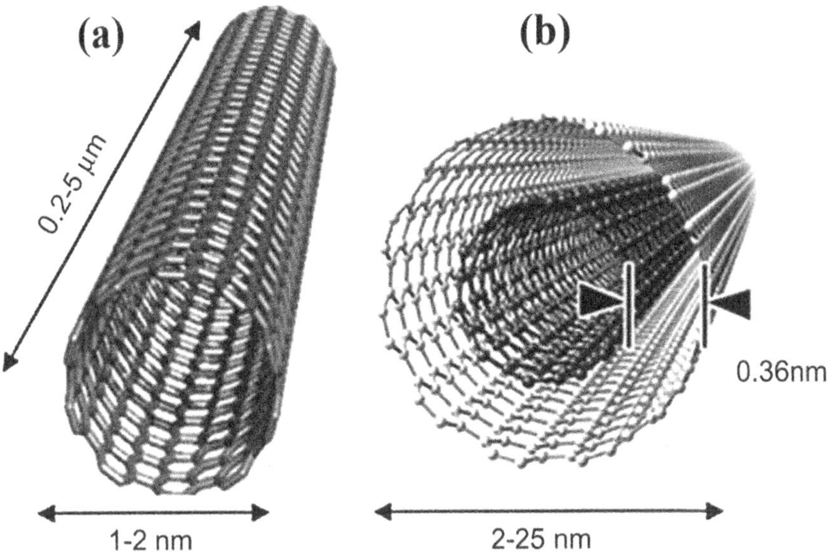

Fig. 8.11 The structure of single-walled carbon nanotubes and multi-walled carbon nanotubes

polypyrrole/MWCNTs for the removal of a non-steroidal anti-inflammatory drug. Some features of polypyrrole such as low cost, environmental stability, non-toxic nature, and facile synthesis via chemical and electrochemical methods make it stands out among the other conducting polymers. The BET analysis proved that the surface area of MWCNTs increased from 277.5 to 541.2 m^2/g after the addition of polypyrrole. In the case of adsorption capacity, the polypyrrole/MWCNTs (229.39 mg/g) revealed a higher value than pristine MWCNTs (59.67 mg/g).

In a similar study concerning CNTs modification with conducting polymers, Dutra et al. (2018) used functional polyaniline/multi-walled carbon nanotube composite for pharmaceutical elimination. Maximum meloxicam removal was achieved at pH 2 and contact time 6 min. Interestingly, Avelar Dutra et al. (2017) also did another research for removal of meloxicam by polyaniline-deposited cellulose fiber composite and reported that the highest adsorption capacity was arrived at 4 min of contact time. It is fair to suggest that polyaniline may lead to a rapid adsorption.

Metal-organic frameworks (MOFs), known as coordination polymers and a class of highly ordered porous crystalline hybrid material, were also used for CNTs modification with the aim of greater CNTs application. Xiong et al. (2018) proposed multi-walled carbon nanotube functionalized MIL-53(Fe) as a new adsorbent for tetracycline antibiotics removal. BET analysis showed that the surface area of pristine MIL-53 (Fe) and MWCNTs was 52.18 and 136.99 m^2/g, respectively; however, of MWCNTs/MIL-53 (Fe) it was 60.17 m^2/g. The results showed that adsorption capacity of MWCNTs/MIL-53 (Fe) for tetracycline antibiotics was at least 1.2 times higher than single MWCNTs. In addition, π–π interaction between adsorbent and adsorbate played an important role in the adsorption process.

Table 8.12 The utilization of carbon nanotubes and its composites for the removal of various pharmaceutical compounds

Adsorbents	Pharmaceuticals	Adsorption capacity (mg/g)	Sources
Multiwalled carbon nanotubes	Carbamazepine Dorzolamide	224.6 78.8	Ncibi and Sillanpää (2017)
Single walled carbon nanotube	Carbamazepine	130	Lerman et al. (2013)
Multiwalled carbon nanotubes (10–20 nm outer diameters)	Carbamazepine	441.4	Oleszczuk et al. (2009)
Multiwalled carbon nanotubes	Isonicotinic acid	671	Bhatia et al. (2019)
Multiwalled carbon nanotubes/Fe3O4	Isonicotinic acid	1234	
Multiwalled carbon nanotubes	Ciprofloxacin	200	Yu et al. (2016)
Carbon dot magnetic carbon nanotubes	Carbamazepine	65	Deng et al. (2019)
Functional polyaniline/ multiwalled carbon nanotube	Meloxicam	221.2	Dutra et al. (2018)
Multiwalled carbon nanotube	Potassium diclofenac	59.67	Pires et al. (2019)
Polypyrrole/multiwalled carbon nanotube	Potassium diclofenac	229.39	
Multi-walled carbon nanotube functionalized MIL-53(Fe)	Tetracycline Oxytetracycline chlortetracycline	364.37 325.59 180.68	Xiong et al. (2018)
Single-walled Carbon nanotube	Oxytetracycline Ciprofloxacin	554 721	Ncibi and Sillanpää (2015)
Double-walled carbon nanotube	Oxytetracycline Ciprofloxacin	507 605	
Multi-walled carbon nanotube	Oxytetracycline Ciprofloxacin	391 475	
Multiwall carbon nanotubes	Sulfamethoxazole	3.03	Wang et al. (2017)

8.9.5 *Clay and Zeolite Nanocomposites*

Clay, a small fine-grained natural particle on the surface of the earth (<2 μm), is mainly composed of silica, alumina, water, and weathered rock having different layered structures (Uddin 2017). Kaolinite, montmorillonite, and bentonite are the main types of clay minerals. The popularity of clay minerals is because of their inexpensiveness, accessibility, vast effective area, mechanical strength, high swelling capacity, layered structure, harmlessness, and ion exchange capability. Ions H^+, K^+, Na^+, Ca^{2+}, Mg^{2+}, NH^+_4, Cl^-, SO_4^{2-}, PO_4^{3-}, and NO_3^- are usually found on the surface of the clay (Ngulube et al. 2017). Commonly, pollutant elimination by clay-based adsorbents is eventuated by the contribution of the surface of the adsorbent, the edge of the plates and the available active sites situated between the layers (Chen et al. 2016).

Waste-based materials have gained much attention recently because they are readily available and reuse of them is profitable in terms of clean environment. Ashiq et al. (2019) tried to remove ciprofloxacin from water ways by developing a composite based on the municipal solid waste biochar/bentonite. They reported the maximum adsorption capacity of 190 mg/g, which was approximately 40% higher than raw biochar. They proposed the following equations for ciprofloxacin removal at acidic conditions:

$$H^+ + R \quad NH \rightarrow R \quad NH_2^+$$

$$R - NH_2^+ + M^- \rightarrow RNH_2M$$

R— the group represented of ciprofloxacin
M—bentonite group

In another study, norfloxacin removal was studied by attapulgite-biochar composites derived from potato stem and natural attapulgite (Li et al. 2017). Figure 8.12 shows the synthesis procedure and SEM images of the prepared material. Batch sorption experiment indicated that the maximum adsorption capacity of attapulgite-biochar composites (5.24 mg/g) was 1.68 times higher than pristine biochar, so modification process was successful. It is believed that the attapulgite particles on the biochar surface served as available vacant sorption sites and interacted with the drug through electrostatic interactions. In addition, norfloxacin was studied by montmorillonite-biochar and had the adsorption capacity of 25.53 mg/g (Zhang et al. 2018b). Higher capacity of montmorillonite-biochar than attapulgite-biochar may be attributed to N_2 adsorption-desorption analysis. Montmorillonite/biochar has the surface area and pore volume of 112.6 m^2/g and 0.604 cm^3/g; however of attapulgite-biochar they were 90.40 m^2/g and 0.1225 cm^3/g, respectively. Also, montmorillonite itself was employed for propranolol adsorption (del Mar et al. 2019).

Some researchers have focused on montmorillonitemodification to increase the applicability of these materials for a wider range of pollutant removal. Hydroxyapatite ($Ca_{10}(OH)_2(PO_4)_6$), similar to inorganic components of the bone matrix and the teeth, has been used for pollutants adsorption. Accordingly, hydroxyapatite/clay was synthesized with the aim of tetracycline removal (Ersan et al. 2015). Hydroxyapatite/clay showed the surface area of 43.07 m^2/g and total pore volume of 0.33 cm^3/g with the maximum adsorption capacity of 76.02 mg/g which was higher than similar studies with various adsorbents like montmorillonite (54 mg/g) and kaolinite (4.32 mg/g) (Figueroa et al. 2004; Li et al. 2010).

Zeolites are environmentally microporous compatible crystalline alumino-silicates in a tetrahedral structure, which are linked together by oxygen atoms. The pores of the molecular size are between 0.5 and 1.2 nm. Zeolites are widely used in water/wastewater treatment due to their properties such as low cost, availability, cation exchange capability, high degree of hydration/dehydration, low density, and good crystal stability (Wen et al. 2018). Concerning the ratio of the silica to aluminum ratio (Si/Al ratio), the low-silica and high-silica zeolites are prepared with Si/

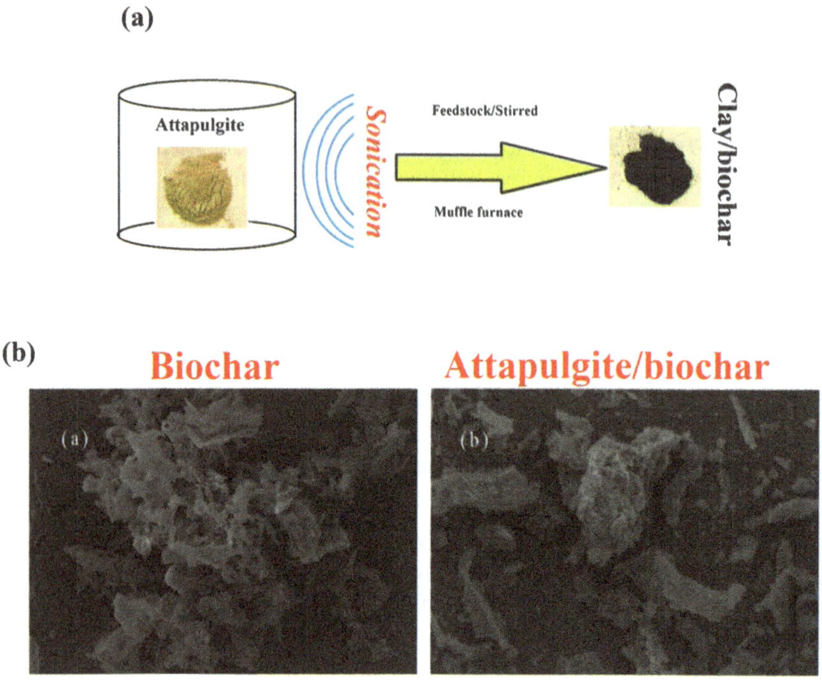

Fig. 8.12 (**a**) The synthesis procedure of clay/biochar composite production and (**b**) the SEM images of the prepared materials. Reprinted with permission of Elsevier from (Li et al. 2017)

Al lower than 2 and up to several thousands, respectively (Jiang et al. 2018). There are 194 unique zeolite frameworks and over 40 naturally occurring zeolites (Ozekmekci et al. 2015).

Researches on the synthesis of magnetic nanoparticles based on zeolite have been undertaken. Natural zeolite and natural zeolite coated with magnetic Fe_3O_4 nanoparticles were investigated in terms of cephalexin removal (Mohseni-Bandpi et al. 2016). Zeolite/Fe_3O_4 (24.9 mg/g) was found to have a higher adsorption capacity than pristine natural zeolite (14.1 mg/g). It was observed that the adsorption process was non-spontaneous and exothermic, following Langmuir and pseudo second-order models. In comparison with other adsorbents for cephalexin adsorption, zeolite/Fe_3O_4 revealed satisfactory results. For instance, activated carbon, bentonite, activated carbon nanoparticles, and MnO_2-coated zeolite showed the maximum adsorption capacity of 17.361 mg/g (Al-Khalisy et al. 2010), 10.384 mg/g (Al-Khalisy et al. 2010), 7.080 mg/g (Pouretedal and Sadegh 2014), and 24.5 mg/g (Samarghandi et al. 2015), respectively.

In another study, natural clinoptilolite zeolite was treated by mechanochemistry (Chen et al. 2019c). In fact, mechanochemistry is a facile free-solvent method to change the physical and chemical properties of the material without the need of special tools or reagents (Majano et al. 2014). The procedure was initiated by milling the humic acid and natural clinoptilolite zeolite in a planetary ball mill at room temperature. The maximum adsorption capacity of 47.68 mg/g was attained. Table 8.13 shows the usage of clay, zeolite, and their composites for the removal of various pharmaceutical compounds.

Table 8.13 The utilization of clay, zeolite, and their composites for removal of various pharmaceutical compounds

Adsorbents	Pharmaceuticals	Adsorption capacity (mg/g)	Sources
Magnetic polymer clay	Atenolol Ciprofloxacin Gemfibrozil	15.6 39.1 24.8	Arya and Philip (2016)
Montmorillonite	Tetracycline ciprofloxacin	1 mmol/g 1.01 mmol/g	Wu et al. (2019a)
Clay/biochar	Norfloxacin	5.24	Li et al. (2017)
Montmorillonite-biochar	Norfloxacin	25.53	Zhang et al. (2018b)
Municipal solid waste derived biochar/bentonite clay	Ciprofloxacin	190	Ashiq et al. (2019)
Fe-pillared montmorillonite	Levofloxacin	48.61	Liu et al. (2015)
Nanotube structured halloysite	Niprofloxacin	21.7	Duan et al. (2018)
Hydroxyapatite/clay	Tetracycline	76.02	Ersan et al. (2015)
Organobentonites	Diclofenac Ibuprofen	600.6 194.9	Ghemit et al. (2019b)
2:1 layered non-swelling clay mineral illite	Tetracycline	32	Chang et al. (2012)
Na-montmorillonite	Tetracycline	0.92 mmol/g	Wu et al. (2019b)
Commercial organoclay	Diclofenac sodium	0.133 Mol/g	Maia et al. (2019)
Montmorillonite	Ciprofloxacin	Up to 330	Roca Jalil et al. (2015)
Palygorskite/montmorillonite	Bisphenol A ciprofloxacin	77.3 mg/kg 107,000 mg/kg	Berhane et al. (2016))
Mechanochemistry treated zeolite	Levofloxacin	47.68	Chen et al. (2019c)
Zeolite-hydroxyapatite-activated oil palm ash	Tetracycline	244.63	Khanday and Hameed (2018)
NaY zeolite from wheat straws ash	Doxycycline	269.75	Ali and Ahmed (2017)
Struvite loaded zeolite	Tetracycline	197.53 mmol/kg	Wang et al. (2019c)
Natural zeolite Zeolite coated with manganese oxide nanoparticles	Cephalexin	16.1 24.5	Samarghandi et al. (2015)
Natural zeolite Natural zeolite/Fe_3O_4	Cephalexin	16.1 24.9	Mohseni-Bandpi et al. (2016)
Natural clinoptilolite Natural clinoptilolite/triton X-100 Natural clinoptilolite/tween 80	Apramycin	0.089 0.091 0.092	Saadi et al. (2019)

8.10 Conclusion

Pharmaceutical compounds are complex biological persistent substances that have been employed for the treatment of humans and animals. However, excessive consumption of drugs has consequently led to the presence and detection of these hazardous substances in water media including surface water, ground water, and potable water at low and moderate concentrations. Most of the conventional WWTPs that consist of preliminary, primary, and secondary treatments are not capable of complete removal of all pharmaceuticals. Similar observation was reported for WTPs. The fluctuation of the drug concentration in a year, temperature, pH, and type of the pharmaceutical were among the most parameters that affected the final efficiency. Many studies have revealed that conventional WWTPs are the main entry point of pharmaceutical compounds to the environment.

In this review, the authors have described the recent works on adsorption processes regarding pharmaceutical compound removal. In the first part, the influential parameters including contact time, initial pharmaceutical concentration, pH, adsorbent dosage, and temperature affecting the adsorbent/adsorbate interaction were discussed. It seems that pH was the most important factor since it affects the surface charge/binding sites of the adsorbent and the ionization degree of the pharmaceutical compounds. Thermodynamic parameters exhibited that the adsorption process was endothermic, suggesting that higher temperature is favored for pharmaceuticals removal. Furthermore, it was seen that the Langmuir model was better to describe the equilibrium data, which indicated homogenous monolayer coverage of drugs on the adsorbent surface. Also, the kinetics data were well described by pseudo-second-order model which confirmed the chemisorption nature of the adsorption process in many studies.

This review presented the most used adsorbents such as activated carbon, graphene-based adsorbents, chitosan, carbon nanotubes, clay, and zeolites in the removal of drugs from aqueous media. Intending the faults in the pristine materials, many studies focused on the modification of the raw material with the aim of higher applicability. Results proved that the modification processes were almost successful in terms of pharmaceutical removal with higher adsorption capacity. Magnetic nanoparticles were highly used in the prepared composites since its easy and facile separation after completion of the treatment process.

Still, the prepared composites own intrinsic disadvantages like low adsorption capacity, experimental-based usage, hard synthesis procedure with hazardous chemicals, and expensive that may hinder the utilization of the adsorbents. In fact, more work is necessary to resolve these issues and gaps in future studies.

References

Abazari R, Mahjoub AR, Shariati J (2019) Synthesis of a nanostructured pillar MOF with high adsorption capacity towards antibiotics pollutants from aqueous solution. J Hazard Mater 366:439–451. https://doi.org/10.1016/j.jhazmat.2018.12.030

Afzal MZ, Sun X-F, Liu J et al (2018) Enhancement of ciprofloxacin sorption on chitosan/biochar hydrogel beads. Sci Total Environ 639:560–569. https://doi.org/10.1016/j.scitotenv.2018.05.129

Ahmed MJ (2017a) Adsorption of non-steroidal anti-inflammatory drugs from aqueous solution using activated carbons: review. J Environ Manage 190:274–282. https://doi.org/10.1016/j.jenvman.2016.12.073

Ahmed MJ (2017b) Adsorption of quinolone, tetracycline, and penicillin antibiotics from aqueous solution using activated carbons: review. Environ Toxicol Pharmacol 50:1–10. https://doi.org/10.1016/j.etap.2017.01.004

Al-Faham Z, Habboub G, Takriti F (2011) The sale of antibiotics without prescription in pharmacies in Damascus, Syria. J Infect Dev Ctries 5. https://doi.org/10.3855/jidc.1248

Ali MMM, Ahmed MJ (2017) Adsorption behavior of doxycycline antibiotic on NaY zeolite from wheat (*Triticum aestivum*) straws ash. J Taiwan Inst Chem Eng 81:218–224. https://doi.org/10.1016/j.jtice.2017.10.026

Ali SNF, El-Shafey EI, Al-Busafi S, Al-Lawati HAJ (2019) Adsorption of chlorpheniramine and ibuprofen on surface functionalized activated carbons from deionized water and spiked hospital wastewater. J Environ Chem Eng 7:102860. https://doi.org/10.1016/j.jece.2018.102860

Al-Jabari MH, Sulaiman S, Ali S et al (2019) Adsorption study of levofloxacin on reusable magnetic nanoparticles: kinetics and antibacterial activity. J Mol Liq 291:111249. https://doi.org/10.1016/j.molliq.2019.111249

Al-Khalisy RS, Al-Haidary AMA, Al-Dujaili AH (2010) Aqueous phase adsorption of cephalexin onto bentonite and activated carbon. Sep Sci Technol 45:1286–1294. https://doi.org/10.1080/01496391003689017

Al-Khateeb LA, Almotiry S, Salam MA (2014) Adsorption of pharmaceutical pollutants onto graphene nanoplatelets. Chem Eng J 248:191–199. https://doi.org/10.1016/j.cej.2014.03.023

Al-Maadheed S, Goktepe I, Latiff ABA, Shomar B (2019) Antibiotics in hospital effluent and domestic wastewater treatment plants in Doha, Qatar. J Water Process Eng 28:60–68. https://doi.org/10.1016/j.jwpe.2019.01.005

Al-Riyami IM, Ahmed M, Al-Busaidi A, Choudri BS (2018) Antibiotics in wastewaters: a review with focus on Oman. Appl Water Sci 8:199. https://doi.org/10.1007/s13201-018-0846-z

Alves TC, Cabrera-Codony A, Barceló D et al (2018) Influencing factors on the removal of pharmaceuticals from water with micro-grain activated carbon. Water Res 144:402–412. https://doi.org/10.1016/j.watres.2018.07.037

Anjum H, Johari K, Gnanasundaram N et al (2019) A review on adsorptive removal of oil pollutants (BTEX) from wastewater using carbon nanotubes. J Mol Liq 277:1005–1025. https://doi.org/10.1016/j.molliq.2018.10.105

Ansari F, Erntell M, Goossens H et al (2009) The European surveillance of antimicrobial consumption (ESAC) point-prevalence survey of antibacterial use in 20 European hospitals in 2006. Clin Infect Dis 49:1496–1504. https://doi.org/10.1086/644617

Ao W, Fu J, Mao X et al (2018) Microwave assisted preparation of activated carbon from biomass: a review. Renew Sustain Energy Rev 92:958–979. https://doi.org/10.1016/j.rser.2018.04.051

Arya V, Philip L (2016) Adsorption of pharmaceuticals in water using Fe_3O_4 coated polymer clay composite. Micropor Mesopor Mater 232:273–280. https://doi.org/10.1016/j.micromeso.2016.06.033

Ashiq A, Adassooriya NM, Sarkar B et al (2019) Municipal solid waste biochar-bentonite composite for the removal of antibiotic ciprofloxacin from aqueous media. J Environ Manage 236:428–435. https://doi.org/10.1016/j.jenvman.2019.02.006

Avelar Dutra FV, Pires BC, Nascimento TA et al (2017) Polyaniline-deposited cellulose fiber composite prepared via in situ polymerization: enhancing adsorption properties for removal of meloxicam from aqueous media. RSC Adv 7:12639–12649. https://doi.org/10.1039/C6RA27019K

Aydin S, Aydin ME, Beduk F, Ulvi A (2019a) Removal of antibiotics from aqueous solution by using magnetic Fe_3O_4/red mud-nanoparticles. Sci Total Environ 670:539–546. https://doi.org/10.1016/j.scitotenv.2019.03.205

Aydin S, Aydin ME, Ulvi A, Kilic H (2019b) Antibiotics in hospital effluents: occurrence, contribution to urban wastewater, removal in a wastewater treatment plant, and environmental risk assessment. Environ Sci Pollut Res Int 26:544–558. https://doi.org/10.1007/s11356-018-3563-0

Azuma T, Ishiuchi H et al (2015) Occurrence and fate of selected anticancer, antimicrobial, and psychotropic pharmaceuticals in an urban river in a subcatchment of the Yodo River basin, Japan. Environ Sci Pollut Res Int 22(23): 18676–18686
Baccar R, Sarrà M, Bouzid J et al (2012) Removal of pharmaceutical compounds by activated carbon prepared from agricultural by-product. Chem Eng J 211–212:310–317. https://doi.org/10.1016/j.cej.2012.09.099
Bahamon D, Vega LF (2017) Pharmaceuticals removal from water effluents by adsorption in activated carbons using Monte Carlo simulations. In: Espuña A, Graells M, Puigjaner LBT-CACE (eds) 27 European symposium on computer aided process engineering. Elsevier, pp 2695–2700
Balarak D, Mahdavi Y, Maleki A et al (2016) Studies on the removal of amoxicillin by single walled carbon nanotubes, JPRI 10:1–9. https://doi.org/10.9734/BJPR/2016/24150
Bean TG, Rattner BA, Lazarus RS et al (2018) Pharmaceuticals in water, fish and osprey nestlings in Delaware River and Bay. Environ Pollut 232:533–545
Beheshti F, RMA T, Khadir A (2019) Sulfamethoxazole removal by photocatalytic degradation utilizing TiO_2 and WO_3 nanoparticles as catalysts: analysis of various operational parameters. Int J Environ Sci Technol 16:7897–7996. https://doi.org/10.1007/s13762-019-02212-x
Bello OS, Alao OC, Alagbada TC, Olatunde AM (2019) Biosorption of ibuprofen using functionalized bean husks. Sustain Chem Pharm 13:100151. https://doi.org/10.1016/j.scp.2019.100151
Berhane TM, Levy J, Krekeler MPS, Danielson ND (2016) Adsorption of bisphenol A and ciprofloxacin by palygorskite-montmorillonite: effect of granule size, solution chemistry and temperature. Appl Clay Sci 132–133:518–527. https://doi.org/10.1016/j.clay.2016.07.023
Bhatia D, Datta D, Joshi A et al (2019) Adsorption of isonicotinic acid from aqueous solution using multi-walled carbon nanotubes/Fe_3O_4. J Mol Liq 276:163–169. https://doi.org/10.1016/j.molliq.2018.11.127
Bin Abdulhak AA, Al Tannir MA, Almansor MA et al (2011) Non prescribed sale of antibiotics in Riyadh, Saudi Arabia: a cross sectional study. BMC Public Health 11:538. https://doi.org/10.1186/1471-2458-11-538
Bound JP, Nikolaos V (2005) Household disposal of pharmaceuticals as a pathway for aquatic contamination in the United Kingdom. Environ Health Perspect 113:1705–1711. https://doi.org/10.1289/ehp.8315
Brown KD, Kulis J, Thomson B et al (2006) Occurrence of antibiotics in hospital, residential, and dairy effluent, municipal wastewater, and the Rio Grande in New Mexico. Sci Total Environ 366:772–783. https://doi.org/10.1016/j.scitotenv.2005.10.007
Çalışkan E, Göktürk S (2010) Removal of sulfamethoxazole and ciprofloxacin from aqueous solutions. Sep Sci Technol 45:244–255. https://doi.org/10.1080/01496390903409419
Carolin CF, Kumar PS, Saravanan A et al (2017) Efficient techniques for the removal of toxic heavy metals from aquatic environment: a review. J Environ Chem Eng 5:2782–2799
Castiglioni S, Bagnati R, Fanelli R et al (2006) Removal of pharmaceuticals in sewage treatment plants in Italy. Environ Sci Technol 40:357–363
Causanilles A, Ruepert C, Ibáñez M et al (2017) Occurrence and fate of illicit drugs and pharmaceuticals in wastewater from two wastewater treatment plants in Costa Rica. Sci Total Environ 599–600:98–107. https://doi.org/10.1016/j.scitotenv.2017.04.202
Cazetta AL, Martins AC, Pezoti O et al (2016) Synthesis and application of N–S-doped mesoporous carbon obtained from nanocasting method using bone char as heteroatom precursor and template. Chem Eng J 300:54–63. https://doi.org/10.1016/j.cej.2016.04.124
Chakraborty P, Banerjee S, Kumar S et al (2018) Elucidation of ibuprofen uptake capability of raw and steam activated biochar of *Aegle marmelos* shell: isotherm, kinetics, thermodynamics and cost estimation. Process Saf Environ Prot 118:10–23. https://doi.org/10.1016/j.psep.2018.06.015
Chang P-H, Li Z, Jean J-S et al (2012) Adsorption of tetracycline on 2:1 layered non-swelling clay mineral illite. Appl Clay Sci 67–68:158–163. https://doi.org/10.1016/j.clay.2011.11.004
Chen L, Zhou CH, Fiore S et al (2016) Functional magnetic nanoparticle/clay mineral nanocomposites: preparation, magnetism and versatile applications. Appl Clay Sci 127–128:143–163. https://doi.org/10.1016/j.clay.2016.04.009

Chen C, Pankow CA, Oh M et al (2019a) Effect of antibiotic use and composting on antibiotic resistance gene abundance and resistome risks of soils receiving manure-derived amendments. Environ Int 128:233–243. https://doi.org/10.1016/j.envint.2019.04.043

Chen H, Bai X, Jing L et al (2019b) Characterization of antibiotic resistance genes in the sediments of an urban river revealed by comparative metagenomics analysis. Sci Total Environ 653:1513–1521. https://doi.org/10.1016/j.scitotenv.2018.11.052

Chen Z, Ma W, Lu G et al (2019c) Adsorption of levofloxacin onto mechanochemistry treated zeolite: modeling and site energy distribution analysis. Sep Purif Technol 222:30–34. https://doi.org/10.1016/j.seppur.2019.04.010

Crini G (2006) Non-conventional low-cost adsorbents for dye removal: a review. Bioresour Technol 97:1061–1085

Cosenza A, Maida CM, Piscionieri D, Fanara S, Di Gaudio F, Viviani G (2018) Occurrence of illicit drugs in two wastewater treatment plants in the South of Italy, Chemosphere. 198:377–385. https://doi.org/10.1016/j.chemosphere.2018.01.158

Daglioglu N, Guzel EY, Kilercioglu S (2019) Assessment of illicit drugs in wastewater and estimation of drugs of abuse in Adana Province, Turkey. Forensic Sci Int 294:132–139. https://doi.org/10.1016/j.forsciint.2018.11.012

Danalıoğlu ST, Bayazit ŞS, Kerkez Kuyumcu Ö, Salam MA (2017) Efficient removal of antibiotics by a novel magnetic adsorbent: magnetic activated carbon/chitosan (MACC) nanocomposite. J Mol Liq 240:589–596. https://doi.org/10.1016/j.molliq.2017.05.131

Danner M-C, Robertson A, Behrends V, Reiss J (2019) Antibiotic pollution in surface fresh waters: occurrence and effects. Sci Total Environ 664:793–804. https://doi.org/10.1016/j.scitotenv.2019.01.406

Dawood S, Sen T (2014) Review on dye removal from its aqueous solution into alternative cost effective and non-conventional adsorbents. J Chem Process Eng 1:1–11

Delgado N, Capparelli A, Navarro A, Marino D (2019) Pharmaceutical emerging pollutants removal from water using powdered activated carbon: study of kinetics and adsorption equilibrium. J Environ Manage 236:301–308. https://doi.org/10.1016/j.jenvman.2019.01.116

Deng Y, Ok YS, Mohan D et al (2019) Carbamazepine removal from water by carbon dot-modified magnetic carbon nanotubes. Environ Res 169:434–444. https://doi.org/10.1016/j.envres.2018.11.035

Deo RP (2014) Pharmaceuticals in the surface water of the USA: a review. Curr Environ Health Rep 1:113–122. https://doi.org/10.1007/s40572-014-0015-y

Dhiman N, Sharma N (2019) Removal of pharmaceutical drugs from binary mixtures by use of ZnO nanoparticles: (Competitive adsorption of drugs). Environ Technol Innov 15:100392. https://doi.org/10.1016/j.eti.2019.100392

Dong S, Sun Y, Wu J et al (2016) Graphene oxide as filter media to remove levofloxacin and lead from aqueous solution. Chemosphere 150:759–764. https://doi.org/10.1016/j.chemosphere.2015.11.075

dos Santos JMN, Pereira CR, Foletto EL, Dotto GL (2019) Alternative synthesis for $ZnFe_2O_4$/chitosan magnetic particles to remove diclofenac from water by adsorption. Int J Biol Macromol 131:301–308. https://doi.org/10.1016/j.ijbiomac.2019.03.079

Duan W, Wang N, Xiao W et al (2018) Ciprofloxacin adsorption onto different micro-structured tourmaline, halloysite and biotite. J Mol Liq 269:874–881. https://doi.org/10.1016/j.molliq.2018.08.051

Duan W, Li M, Xiao W et al (2019) Enhanced adsorption of three fluoroquinolone antibiotics using polypyrrole functionalized *Calotropis gigantea* fiber. Colloids Surf A Physicochem Eng Asp 574:178–187. https://doi.org/10.1016/j.colsurfa.2019.04.068

Duranoğlu D, Trochimczuk AW, Beker U (2012) Kinetics and thermodynamics of hexavalent chromium adsorption onto activated carbon derived from acrylonitrile-divinylbenzene copolymer. Chem Eng J 187:193–202

Dutra FVA, Pires BC, Nascimento TA, Borges KB (2018) Functional polyaniline/multiwalled carbon nanotube composite as an efficient adsorbent material for removing pharmaceuticals from aqueous media. J Environ Manage 221:28–37. https://doi.org/10.1016/j.jenvman.2018.05.051

Ersan M, Guler UA, Acıkel U, Sarioglu M (2015) Synthesis of hydroxyapatite/clay and hydroxy-apatite/pumice composites for tetracycline removal from aqueous solutions. Process Saf Environ Prot 96:22–32. https://doi.org/10.1016/j.psep.2015.04.001

Faleye AC, Adegoke AA, Ramluckan K et al (2019) Concentration and reduction of antibiotic residues in selected wastewater treatment plants and receiving waterbodies in Durban, South Africa. Sci Total Environ 678:10–20. https://doi.org/10.1016/j.scitotenv.2019.04.410

Fazelirad H, Ranjbar M, Taher MA, Sargazi G (2015) Preparation of magnetic multi-walled carbon nanotubes for an efficient adsorption and spectrophotometric determination of amoxicillin. J Ind Eng Chem 21:889–892. https://doi.org/10.1016/j.jiec.2014.04.028

Figueroa RA, Leonard A, MacKay AA (2004) Modeling tetracycline antibiotic sorption to clays. Environ Sci Technol 38:476–483. https://doi.org/10.1021/es0342087

Fiyadh SS, AlSaadi MA, Jaafar WZ et al (2019) Review on heavy metal adsorption processes by carbon nanotubes. J Clean Prod 230:783–793. https://doi.org/10.1016/j.jclepro.2019.05.154

Fröhlich AC, Foletto EL, Dotto GL (2019) Preparation and characterization of NiFe2O4/activated carbon composite as potential magnetic adsorbent for removal of ibuprofen and ketoprofen pharmaceuticals from aqueous solutions. J Clean Prod 229:828–837. https://doi.org/10.1016/j.jclepro.2019.05.037

Fu Y, Yang Z, Xia Y et al (2019) Adsorption of ciprofloxacin pollutants in aqueous solution using modified waste grapefruit peel. Energy Sources A Recover Util Environ Eff:1–10. https://doi.org/10.1080/15567036.2019.1624877

Gadipelly CR, Marathe KV, Rathod VK (2018) Effective adsorption of ciprofloxacin hydrochloride from aqueous solutions using metal-organic framework. Sep Sci Technol 53:2826–2832. https://doi.org/10.1080/01496395.2018.1474225

Gao Y, Li Y, Zhang L et al (2012) Adsorption and removal of tetracycline antibiotics from aqueous solution by graphene oxide. J Colloid Interface Sci 368:540–546. https://doi.org/10.1016/j.jcis.2011.11.015

Gelband H, Duse AG (2011) Executive summary. SAMJ South African Med J 101:551

George Nche N-A, Bopda A, Donald Raoul T, et al (2017) Removal of paracetamol from aqueous solution by adsorption onto activated carbon prepared from rice husk. J Chem Pharm Res 9(3):56–6

Gereli G, Seki Y, Kuşoğlu İM, Yurdakoç K (2006) Equilibrium and kinetics for the sorption of promethazine hydrochloride onto K10 montmorillonite. J Colloid Interface Sci 299:155–162

Ghemit R, Makhloufi A, Djebri N et al (2019a) Adsorptive removal of diclofenac and ibuprofen from aqueous solution by organobentonites. Groundw Sustain Dev 8:520–529. https://doi.org/10.1016/j.gsd.2019.02.004

Ghemit R, Makhloufi A, Djebri N et al (2019b) Adsorptive removal of diclofenac and ibuprofen from aqueous solution by organobentonites: study in single and binary systems. Groundw Sustain Dev 8:520–529. https://doi.org/10.1016/j.gsd.2019.02.004

Ghenaatgar A, Tehrani RMA, Khadir A (2019) Photocatalytic degradation and mineralization of dexamethasone using WO3 and ZrO2 nanoparticles: Optimization of operational parameters and kinetic studies. J Water Process Eng 32:100969. https://doi.org/10.1016/j.jwpe.2019.100969

Greenham RT, Miller KY, Tong A (2019) Removal efficiencies of top-used pharmaceuticals at sewage treatment plants with various technologies. J Environ Chem Eng:103294. https://doi.org/10.1016/j.jece.2019.103294

Gros M, Rodríguez-Mozaz S, Barceló D (2013) Rapid analysis of multiclass antibiotic residues and some of their metabolites in hospital, urban wastewater and river water by ultra-high-performance liquid chromatography coupled to quadrupole-linear ion trap tandem mass spectrometry. J Chromatogr a 1292:173–188. https://doi.org/10.1016/j.chroma.2012.12.072

Homem V, Santos L (2011) Degradation and removal methods of antibiotics from aqueous matrices–a review. J Environ Manag 92:2304–2347

Hoppen MI, Carvalho KQ, Ferreira RC et al (2019) Adsorption and desorption of acetylsalicylic acid onto activated carbon of babassu coconut mesocarp. J environ Chem Eng 7:102862. https://doi.org/10.1016/j.jece.2018.102862

Hu D, Huang H, Jiang R et al (2019) Adsorption of diclofenac sodium on bilayer amino-functionalized cellulose nanocrystals/chitosan composite. J Hazard Mater 369:483–493. https://doi.org/10.1016/j.jhazmat.2019.02.057

Huang F, Zou S, Deng D et al (2019a) Antibiotics in a typical karst river system in China: spatiotemporal variation and environmental risks. Sci Total environ 650:1348–1355. https://doi.org/10.1016/j.scitotenv.2018.09.131

Huang Y-H, Liu Y, Du P-P et al (2019b) Occurrence and distribution of antibiotics and antibiotic resistant genes in water and sediments of urban rivers with black-odor water in Guangzhou, South China. Sci Total environ 670:170–180. https://doi.org/10.1016/j.scitotenv.2019.03.168

Huo P, Zhou M, Tang Y et al (2016) Incorporation of N–ZnO/CdS/Graphene oxide composite photocatalyst for enhanced photocatalytic activity under visible light. J alloys Compd 670:198–209. https://doi.org/10.1016/j.jallcom.2016.01.247

Iannazzo D, Pistone A, Ziccarelli I, Galvagno S (2018) Graphene-based materials for application in pharmaceutical nanotechnology. In: AMBT-F (ed) Grumezescu graphenes and nanotubes. William Andrew Publishing, pp 297–329

Jiang N, Shang R, Heijman SGJ, Rietveld LC (2018) High-silica zeolites for adsorption of organic micro-pollutants in water treatment: a review. Water res 144:145–161. https://doi.org/10.1016/j.watres.2018.07.017

Jiang Y, Xie Q, Zhang Y et al (2019) Preparation of magnetically separable mesoporous activated carbons from brown coal with Fe3O4. Int J min Sci Technol 29:513–519. https://doi.org/10.1016/j.ijmst.2019.01.002

Jodeh S, Abdelwahab F, Jaradat N et al (2016) Adsorption of diclofenac from aqueous solution using Cyclamen persicum tubers based activated carbon (CTAC). J Assoc Arab Univ basic Appl Sci 20:32–38. https://doi.org/10.1016/j.jaubas.2014.11.002

Kanakaraju D, Glass BD, Oelgemöller M (2018) Advanced oxidation process-mediated removal of pharmaceuticals from water: a review. J Environ Manag 219:189–207

Kaya Y, Bacaksiz AM, Bayrak H et al (2019) Investigation of membrane fouling in an anaerobic membrane bioreactor (AnMBR) treating pharmaceutical wastewater. J Water Process Eng 31:100822. https://doi.org/10.1016/j.jwpe.2019.100822

Khanday WA, Hameed BH (2018) Zeolite-hydroxyapatite-activated oil palm ash composite for antibiotic tetracycline adsorption. Fuel 215:499–505. https://doi.org/10.1016/j.fuel.2017.11.068

Khanday WA, Ahmed MJ, Okoye PU et al (2019) Single-step pyrolysis of phosphoric acid-activated chitin for efficient adsorption of cephalexin antibiotic. Bioresour Technol 280:255–259. https://doi.org/10.1016/j.biortech.2019.02.003

Khadir A, Negarestani M, Mollahosseini A (2020) Sequestration of a non-steroidal anti-inflammatory drug from aquatic media by lignocellulosic material (Luffa cylindrica) reinforced with polypyrrole: Study of parameters, kinetics, and equilibrium. J Environ Chem Eng 103734. https://doi.org/10.1016/j.jece.2020.103734

Kim SH, Shon HK, Ngo HH (2010) Adsorption characteristics of antibiotics trimethoprim on powdered and granular activated carbon. J Ind Eng Chem 16:344–349. https://doi.org/10.1016/j.jiec.2009.09.061

Klein EY, Van Boeckel TP, Martinez EM et al (2018) Global increase and geographic convergence in antibiotic consumption between 2000 and 2015. Proc Natl Acad Sci 115:E3463–E3470

Kołecka K, Gajewska M, Stepnowski P, Caban M (2019) Spatial distribution of pharmaceuticals in conventional wastewater treatment plant with sludge treatment reed beds technology. Sci Total Environ 647:149–157. https://doi.org/10.1016/j.scitotenv.2018.07.439

Kumar R, Tscharke B, O'Brien J et al (2019) Assessment of drugs of abuse in a wastewater treatment plant with parallel secondary wastewater treatment train. Sci Total Environ 658:947–957. https://doi.org/10.1016/j.scitotenv.2018.12.167

Kümmerer K (2001) Drugs in the environment: emission of drugs, diagnostic aids and disinfectants into wastewater by hospitals in relation to other sources—a review. Chemosphere 45:957–969

Kurasam J, Sihag P, Mandal PK, Sarkar S (2018) Presence of fluoroquinolone resistance with persistent occurrence of gyrA gene mutations in a municipal wastewater treatment plant in India. Chemosphere 211:817–825. https://doi.org/10.1016/j.chemosphere.2018.08.011

La Lien T, Hoa QN, Chuc TN et al (2016) Antibiotics in wastewater of a rural and an Urban Hospital before and after wastewater treatment, and the relationship with antibiotic use—a one year study from Vietnam. Int J Environ Res Publ Health 13(6):588. https://doi.org/10.3390/ijerph13060588

Le T-H, Ng C, Chen H et al (2016) Occurrences and characterization of antibiotic-resistant bacteria and genetic determinants of hospital wastewater in a tropical country. Antimicrob Agents Chemother 60:7449–7456. https://doi.org/10.1128/AAC.01556-16

Lee S-H, Kim K-H, Lee M, Lee B-D (2019) Detection status and removal characteristics of pharmaceuticals in wastewater treatment effluent. J Water Process Eng 31:100828. https://doi.org/10.1016/j.jwpe.2019.100828

Lerman I, Chen Y, Xing B, Chefetz B (2013) Adsorption of carbamazepine by carbon nanotubes: effects of DOM introduction and competition with phenanthrene and bisphenol A. Environ Pollut 182:169–176. https://doi.org/10.1016/j.envpol.2013.07.010

Lessa EF, Nunes ML, Fajardo AR (2018) Chitosan/waste coffee-grounds composite: an efficient and eco-friendly adsorbent for removal of pharmaceutical contaminants from water. Carbohydr Polym 189:257–266. https://doi.org/10.1016/j.carbpol.2018.02.018

Leung HW, Minh TB, Murphy MB et al (2012) Distribution, fate and risk assessment of antibiotics in sewage treatment plants in Hong Kong, South China. Environ Int 42:1–9. https://doi.org/10.1016/j.envint.2011.03.004

Li WC (2014) Occurrence, sources, and fate of pharmaceuticals in aquatic environment and soil. Environ Pollut 187:193–201. https://doi.org/10.1016/j.envpol.2014.01.015

Li S-W, Lin AY-C (2015) Increased acute toxicity to fish caused by pharmaceuticals in hospital effluents in a pharmaceutical mixture and after solar irradiation. Chemosphere 139:190–196. https://doi.org/10.1016/j.chemosphere.2015.06.010

Li Z, Chang P-H, Jean J-S et al (2010) Interaction between tetracycline and smectite in aqueous solution. J Colloid Interface Sci 341:311–319. https://doi.org/10.1016/j.jcis.2009.09.054

Li Y, Wang Z, Xie X et al (2017) Removal of Norfloxacin from aqueous solution by clay-biochar composite prepared from potato stem and natural attapulgite. Colloids Surf A Physicochem Eng Asp 514:126–136. https://doi.org/10.1016/j.colsurfa.2016.11.064

Li S, Shi W, Li H et al (2018) Antibiotics in water and sediments of rivers and coastal area of Zhuhai City, Pearl River estuary, South China. Sci Total Environ 636:1009–1019. https://doi.org/10.1016/j.scitotenv.2018.04.358

Li L, Zou D, Xiao Z et al (2019a) Biochar as a sorbent for emerging contaminants enables improvements in waste management and sustainable resource use. J Clean Prod 210:1324–1342. https://doi.org/10.1016/j.jclepro.2018.11.087

Li M, Liu Y, Zeng G et al (2019b) Graphene and graphene-based nanocomposites used for antibiotics removal in water treatment: a review. Chemosphere 226:360–380. https://doi.org/10.1016/j.chemosphere.2019.03.117

Li S, Shi W, You M et al (2019c) Antibiotics in water and sediments of Danjiangkou reservoir, China: spatiotemporal distribution and indicator screening. Environ Pollut 246:435–442. https://doi.org/10.1016/j.envpol.2018.12.038

Liang P, Wu S, Zhang C et al (2018) The role of antibiotics in mercury methylation in marine sediments. J Hazard Mater 360:1–5. https://doi.org/10.1016/j.jhazmat.2018.07.096

Liang XX, Omer AM, Hu Z et al (2019) Efficient adsorption of diclofenac sodium from aqueous solutions using magnetic amine-functionalized chitosan. Chemosphere 217:270–278. https://doi.org/10.1016/j.chemosphere.2018.11.023

Lindberg R, Jarnheimer P-Å, Olsen B et al (2004) Determination of antibiotic substances in hospital sewage water using solid phase extraction and liquid chromatography/mass spectrometry and group analogue internal standards. Chemosphere 57:1479–1488. https://doi.org/10.1016/j.chemosphere.2004.09.015

Liu Y, Dong C, Wei H et al (2015) Adsorption of levofloxacin onto an iron-pillared montmorillonite (clay mineral): kinetics, equilibrium and mechanism. Appl Clay Sci 118:301–307. https://doi.org/10.1016/j.clay.2015.10.010

Liu Y, Liu R, Li M et al (2019) Removal of pharmaceuticals by novel magnetic genipin-crosslinked chitosan/graphene oxide-SO3H composite. Carbohydr Polym 220:141–148. https://doi.org/10.1016/j.carbpol.2019.05.060

Long L, Hu X, Yan J et al (2019) Novel chitosan–ethylene glycol hydrogel for the removal of aqueous perfluorooctanoic acid. J Environ Sci 84:21–28. https://doi.org/10.1016/j.jes.2019.04.007

Lotfi R, Hayati B, Rahimi S et al (2019) Synthesis and characterization of PAMAM/SiO2 nanohybrid as a new promising adsorbent for pharmaceuticals. Microchem J 146:1150–1159. https://doi.org/10.1016/j.microc.2019.02.048

Lu T, Zhu Y, Qi Y et al (2018) Magnetic chitosan–based adsorbent prepared via Pickering high internal phase emulsion for high-efficient removal of antibiotics. Int J Biol Macromol 106:870–877. https://doi.org/10.1016/j.ijbiomac.2017.08.092

Madikizela LM, Tavengwa NT, Chimuka L (2017) Status of pharmaceuticals in African water bodies: occurrence, removal and analytical methods. J Environ Manage 193:211–220. https://doi.org/10.1016/j.jenvman.2017.02.022

Maia GS, de Andrade JR, da Silva MGC, Vieira MGA (2019) Adsorption of diclofenac sodium onto commercial organoclay: kinetic, equilibrium and thermodynamic study. Powder Technol 345:140–150. https://doi.org/10.1016/j.powtec.2018.12.097

Majano G, Borchardt L, Mitchell S et al (2014) Rediscovering zeolite mechanochemistry – a pathway beyond current synthesis and modification boundaries. Micropor Mesopor Mater 194:106–114. https://doi.org/10.1016/j.micromeso.2014.04.006

del Mar Orta M, Martín J, Medina-Carrasco S et al (2019) Adsorption of propranolol onto montmorillonite: kinetic, isotherm and pH studies. Appl Clay Sci 173:107–114. https://doi.org/10.1016/j.clay.2019.03.015

Martín J, Orta M del M, Medina-Carrasco S et al (2019) Evaluation of a modified mica and montmorillonite for the adsorption of ibuprofen from aqueous media. Appl Clay Sci 171:29–37. https://doi.org/10.1016/j.clay.2019.02.002

Mestre AS, Pires J, Nogueira JMF, Carvalho AP (2007) Activated carbons for the adsorption of ibuprofen. Carbon N Y 45:1979–1988. https://doi.org/10.1016/j.carbon.2007.06.005

Mestre AS, Hesse F, Freire C et al (2019) Chemically activated high grade nanoporous carbons from low density renewable biomass (*Agave sisalana*) for the removal of pharmaceuticals. J Colloid Interface Sci 536:681–693. https://doi.org/10.1016/j.jcis.2018.10.081

Miao J, Wang F, Chen Y et al (2019) The adsorption performance of tetracyclines on magnetic graphene oxide: a novel antibiotics absorbent. Appl Surf Sci 475:549–558. https://doi.org/10.1016/j.apsusc.2019.01.036

Mirjavadi ES, Tehrani RMA, Khadir A (2019) Effective adsorption of zinc on magnetic nanocomposite of Fe_3O_4/zeolite/cellulose nanofibers: kinetic, equilibrium, and thermodynamic study, Environ Sci Pollut Res. https://doi.org/10.1007/s11356-019-06165-z

Mogolodi Dimpe K, Nomngongo PN (2019) Application of activated carbon-decorated polyacrylonitrile nanofibers as an adsorbent in dispersive solid-phase extraction of fluoroquinolones from wastewater. J Pharm Anal 9:117–126. https://doi.org/10.1016/j.jpha.2019.01.003

Mohseni-Bandpi A, Al-Musawi TJ, Ghahramani E et al (2016) Improvement of zeolite adsorption capacity for cephalexin by coating with magnetic Fe_3O_4 nanoparticles. J Mol Liq 218:615–624. https://doi.org/10.1016/j.molliq.2016.02.092

Mollahosseini A, Khadir A, Saeidian J (2019) Core–shell polypyrrole/Fe_3O_4 nanocomposite as sorbent for magnetic dispersive solid-phase extraction of Al^{3+} ions from solutions: investigation of the operational parameters. J Water Process Eng 29. https://doi.org/10.1016/j.jwpe.2019.100795

Moussavi G, Hossaini Z, Pourakbar M (2016) High-rate adsorption of acetaminophen from the contaminated water onto double-oxidized graphene oxide. Chem Eng J 287:665–673. https://doi.org/10.1016/j.cej.2015.11.025

Movasaghi Z, Yan B, Niu C (2019) Adsorption of ciprofloxacin from water by pretreated oat hulls: equilibrium, kinetic, and thermodynamic studies. Ind Crops Prod 127:237–250. https://doi.org/10.1016/j.indcrop.2018.10.051

Nasseh N, Barikbin B, Taghavi L, Nasseri MA (2019) Adsorption of metronidazole antibiotic using a new magnetic nanocomposite from simulated wastewater (isotherm, kinetic and thermodynamic studies). Compos B Eng 159:146–156. https://doi.org/10.1016/j.compositesb.2018.09.034

Ncibi MC, Sillanpää M (2015) Optimized removal of antibiotic drugs from aqueous solutions using single, double and multi-walled carbon nanotubes. J Hazard Mater 298:102–110. https://doi.org/10.1016/j.jhazmat.2015.05.025

Ncibi MC, Sillanpää M (2017) Optimizing the removal of pharmaceutical drugs carbamazepine and Dorzolamide from aqueous solutions using mesoporous activated carbons and multi-walled carbon nanotubes. J Mol Liq 238:379–388. https://doi.org/10.1016/j.molliq.2017.05.028

Neris JB, Luzardo FHM, da Silva EGP, Velasco FG (2019) Evaluation of adsorption processes of metal ions in multi-element aqueous systems by lignocellulosic adsorbents applying different isotherms: a critical review. Chem Eng J 357:404–420. https://doi.org/10.1016/j.cej.2018.09.125

Ngulube T, Gumbo JR, Masindi V, Maity A (2017) An update on synthetic dyes adsorption onto clay based minerals: a state-of-art review. J Environ Manage 191:35–57. https://doi.org/10.1016/j.jenvman.2016.12.031

Nourmoradi H, Moghadam KF, Jafari A, Kamarehie B (2018) Removal of acetaminophen and ibuprofen from aqueous solutions by activated carbon derived from *Quercus Brantii* (Oak) acorn as a low-cost biosorbent. J Environ Chem Eng 6:6807–6815. https://doi.org/10.1016/j.jece.2018.10.047

Ohlsen K, Ternes T, Werner G et al (2003) Impact of antibiotics on conjugational resistance gene transfer in *Staphylococcus aureus* in sewage. Environ Microbiol 5:711–716

Oleszczuk P, Pan B, Xing B (2009) Adsorption and desorption of oxytetracycline and carbamazepine by multiwalled carbon nanotubes. Environ Sci Technol 43:9167–9173. https://doi.org/10.1021/es901928q

Ozekmekci M, Salkic G, Fellah MF (2015) Use of zeolites for the removal of H_2S: a mini-review. Fuel Process Technol 139:49–60. https://doi.org/10.1016/j.fuproc.2015.08.015

Palli L, Spina F, Varese GC et al (2019) Occurrence of selected pharmaceuticals in wastewater treatment plants of Tuscany: an effect-based approach to evaluate the potential environmental impact. Int J Hyg Environ Health 222:717–725. https://doi.org/10.1016/j.ijheh.2019.05.006

Pamphile N, Xuejiao L, Guangwei Y, Yin W (2019) Synthesis of a novel core-shell-structure activated carbon material and its application in sulfamethoxazole adsorption. J Hazard Mater 368:602–612. https://doi.org/10.1016/j.jhazmat.2019.01.093

Patrolecco L, Rauseo J, Ademollo N et al (2018) Persistence of the antibiotic sulfamethoxazole in river water alone or in the co-presence of ciprofloxacin. Sci Total Environ 640–641:1438–1446. https://doi.org/10.1016/j.scitotenv.2018.06.025

Paunovic O, Pap S, Maletic S et al (2019) Ionisable emerging pharmaceutical adsorption onto microwave functionalised biochar derived from novel lignocellulosic waste biomass. J Colloid Interface Sci 547:350–360. https://doi.org/10.1016/j.jcis.2019.04.011

Pena A, Paulo M, Silva LJG et al (2010) Tetracycline antibiotics in hospital and municipal wastewaters: a pilot study in Portugal. Anal Bioanal Chem 396:2929–2936. https://doi.org/10.1007/s00216-010-3581-3

Peng G, Zhang M, Deng S et al (2018) Adsorption and catalytic oxidation of pharmaceuticals by nitrogen-doped reduced graphene oxide/Fe_3O_4 nanocomposite. Chem Eng J 341:361–370. https://doi.org/10.1016/j.cej.2018.02.064

Peng J, Wang X, Yin F, Xu G (2019) Characterizing the removal routes of seven pharmaceuticals in the activated sludge process. Sci Total Environ 650:2437–2445. https://doi.org/10.1016/j.scitotenv.2018.10.004

Pires BC, do Nascimento TA, FVA D, Borges KB (2019) Removal of a non-steroidal anti-inflammatory by adsorption on polypyrrole/multiwalled carbon nanotube composite—study of kinetics and equilibrium in aqueous medium. Colloids Surf A Physicochem Eng Asp 578:123583. https://doi.org/10.1016/j.colsurfa.2019.123583

Piri F, Mollahosseini A, Khadir A, Milani Hosseini M (2019) Enhanced adsorption of dyes on microwave-assisted synthesized magnetic zeolite-hydroxyapatite nanocomposite. J Environ Chem Eng 103:338. https://doi.org/10.1016/j.jece.2019.103338

Pouretedal HR, Sadegh N (2014) Effective removal of amoxicillin, cephalexin, tetracycline and penicillin G from aqueous solutions using activated carbon nanoparticles prepared from vine wood. J Water Process Eng 1:64–73. https://doi.org/10.1016/j.jwpe.2014.03.006

Prarat P, Hadsakunnee K, Padejepan L et al (2019) Pharmaceuticals and personal care products removal from aqueous solution by nitrogen-functionalized carbon adsorbent derived from pomelo peel waste. IOP Conf Ser Earth Environ Sci 257:12019. https://doi.org/10.1088/1755-1315/257/1/012019

Praveena SM, Shaifuddin SNM, Sukiman S et al (2018) Pharmaceuticals residues in selected tropical surface water bodies from Selangor (Malaysia): occurrence and potential risk assessments. Sci Total Environ 642:230–240

Qiu W, Sun J, Fang M et al (2019) Occurrence of antibiotics in the main rivers of Shenzhen, China: association with antibiotic resistance genes and microbial community. Sci Total Environ 653:334–341. https://doi.org/10.1016/j.scitotenv.2018.10.398

Quesada HB, Baptista ATA, Cusioli LF et al (2019) Surface water pollution by pharmaceuticals and an alternative of removal by low-cost adsorbents: a review. Chemosphere 222:766–780. https://doi.org/10.1016/j.chemosphere.2019.02.009

Reis EO, Foureaux AFS, Rodrigues JS et al (2019) Occurrence, removal and seasonal variation of pharmaceuticals in Brasilian drinking water treatment plants. Environ Pollut 250:773–781. https://doi.org/10.1016/j.envpol.2019.04.102

Rey-Mafull CA, Tacoronte JE, Garcia R et al (2014) Comparative study of the adsorption of acetaminophen on activated carbons in simulated gastric fluid. Springerplus 3:48. https://doi.org/10.1186/2193-1801-3-48

Reza RA, Ahmaruzzaman M, Sil AK, Gupta VK (2014) Comparative adsorption behavior of ibuprofen and clofibric acid onto microwave assisted activated bamboo waste. Ind Eng Chem Res 53:9331–9339. https://doi.org/10.1021/ie404162p

Rivera-Jaimes JA, Postigo C, Melgoza-Alemán RM et al (2018) Study of pharmaceuticals in surface and wastewater from Cuernavaca, Morelos, Mexico: occurrence and environmental risk assessment. Sci Total Environ 613:1263–1274

Roca Jalil ME, Baschini M, Sapag K (2015) Influence of pH and antibiotic solubility on the removal of ciprofloxacin from aqueous media using montmorillonite. Appl Clay Sci 114:69–76. https://doi.org/10.1016/j.clay.2015.05.010

Saadi Z, Fazaeli R, Vafajoo L, Naser I (2019) Adsorptive removal of apramycin antibiotic from aqueous solutions using Tween 80-and Triton X-100 modified clinoptilolite: experimental and fixed-bed modeling investigations. Int J Environ Health Res:1–26. https://doi.org/10.1080/09603123.2019.1612039

Sabri NA, Schmitt H, Van der Zaan B et al (2018) Prevalence of antibiotics and antibiotic resistance genes in a wastewater effluent-receiving river in the Netherlands. J Environ Chem Eng. https://doi.org/10.1016/j.jece.2018.03.004

Salleh MAM, Mahmoud DK, Karim WAWA, Idris A (2011) Cationic and anionic dye adsorption by agricultural solid wastes: a comprehensive review. Desalination 280:1–13

Samarghandi MR, Al-Musawi TJ, Mohseni-Bandpi A, Zarrabi M (2015) Adsorption of cephalexin from aqueous solution using natural zeolite and zeolite coated with manganese oxide nanoparticles. J Mol Liq 211:431–441. https://doi.org/10.1016/j.molliq.2015.06.067

Sanganyado E, Gwenzi W (2019) Antibiotic resistance in drinking water systems: occurrence, removal, and human health risks. Sci Total Environ 669:785–797. https://doi.org/10.1016/j.scitotenv.2019.03.162

Santoro DO, Cardoso AM, Coutinho FH et al (2015) Diversity and antibiotic resistance profiles of pseudomonads from a hospital wastewater treatment plant. J Appl Microbiol 119:1527–1540. https://doi.org/10.1111/jam.12936

Shan D, Deng S, Zhao T et al (2016) Preparation of ultrafine magnetic biochar and activated carbon for pharmaceutical adsorption and subsequent degradation by ball milling. J Hazard Mater 305:156–163. https://doi.org/10.1016/j.jhazmat.2015.11.047

Shi Y, Zhang Y, Cui Y et al (2019) Magnetite nanoparticles modified β-cyclodextrin polymer coupled with $KMnO_4$ oxidation for adsorption and degradation of acetaminophen. Carbohydr Polym 222:114972. https://doi.org/10.1016/j.carbpol.2019.114972

Shokoohi R, Leili M, Dargahi A et al (2017) Common antibiotics in wastewater of Sina and Besat hospitals, Hamadan, Iran. Arch Hyg Sci 6:152–159. https://doi.org/10.29252/ArchHygSci.6.2.152

Soares SF, Simões TR, António M et al (2016) Hybrid nanoadsorbents for the magnetically assisted removal of metoprolol from water. Chem Eng J 302:560–569. https://doi.org/10.1016/j.cej.2016.05.079

Song L, Li L, Yang S et al (2016) Sulfamethoxazole, tetracycline and oxytetracycline and related antibiotic resistance genes in a large-scale landfill, China. Sci Total Environ 551:9–15

de Sousa DNR, Insa S, Mozeto AA et al (2018) Equilibrium and kinetic studies of the adsorption of antibiotics from aqueous solutions onto powdered zeolites. Chemosphere 205:137–146

Spataro F, Ademollo N, Pescatore T et al (2019) Antibiotic residues and endocrine disrupting compounds in municipal wastewater treatment plants in Rome, Italy. Microchem J 148:634–642. https://doi.org/10.1016/j.microc.2019.05.053

Sun L, Wan S, Yuan D, Yu Z (2019) Adsorption of nitroimidazole antibiotics from aqueous solutions on self-shaping porous biomass carbon foam pellets derived from *Vallisneria natans* waste as a new adsorbent. Sci Total Environ 664:24–36. https://doi.org/10.1016/j.scitotenv.2019.01.412

Syafiuddin A, Salmiati S, Jonbi J, Fulazzaky MA (2018) Application of the kinetic and isotherm models for better understanding of the behaviors of silver nanoparticles adsorption onto different adsorbents. J Environ Manage 218:59–70. https://doi.org/10.1016/j.jenvman.2018.03.066

Szekeres E, Baricz A, Chiriac CM et al (2017) Abundance of antibiotics, antibiotic resistance genes and bacterial community composition in wastewater effluents from different Romanian hospitals. Environ Pollut 225:304–315. https://doi.org/10.1016/j.envpol.2017.01.054

Szymonik A, Lach J, Malińska K (2017) Fate and removal of pharmaceuticals and illegal drugs present in drinking water and wastewater. Ecol Chem Eng S 24:65–85. https://doi.org/10.1515/eces-2017-0006

Tabrizian P, Ma W, Bakr A, Rahaman MS (2019) pH-sensitive and magnetically separable Fe/cu bimetallic nanoparticles supported by graphene oxide (GO) for high-efficiency removal of tetracyclines. J Colloid Interface Sci 534:549–562. https://doi.org/10.1016/j.jcis.2018.09.034

Tahir SS, Rauf N (2006) Removal of a cationic dye from aqueous solutions by adsorption onto bentonite clay. Chemosphere 63:1842–1848. https://doi.org/10.1016/j.chemosphere.2005.10.033

Teixeira S, Delerue-Matos C, Santos L (2019) Application of experimental design methodology to optimize antibiotics removal by walnut shell based activated carbon. Sci Total Environ 646:168–176. https://doi.org/10.1016/j.scitotenv.2018.07.204

Teixeira-Lemos E, Teixeira-Lemos L, Oliveira J, Pais do Amaral J (2019) Pharmaceuticals in the environment. focus on drinking water, 325–335. https://doi.org/10.1016/B978-0-12-409547-2.13941-1

Teo HT, Siah WR, Yuliati L (2016) Enhanced adsorption of acetylsalicylic acid over hydrothermally synthesized iron oxide-mesoporous silica MCM-41 composites. J Taiwan Inst Chem Eng 65:591–598. https://doi.org/10.1016/j.jtice.2016.06.006

Ternes T, Joss A (2007) Human pharmaceuticals, hormones and fragrances. IWA Publishing, London

Thomas KV, Dye C, Schlabach M, Langford KH (2007) Source to sink tracking of selected human pharmaceuticals from two Oslo city hospitals and a wastewater treatment works. J Environ Monit 9:1410–1418. https://doi.org/10.1039/b709745j

Tian X, Liu J, Wang Y et al (2019) Adsorption of antibiotics from aqueous solution by different aerogels. J Non Cryst Solids 505:72–78. https://doi.org/10.1016/j.jnoncrysol.2018.10.033

Timraz K, Xiong Y, Al Qarni H, Hong P-Y (2017) Removal of bacterial cells{,} antibiotic resistance genes and integrase genes by on-site hospital wastewater treatment plants: surveillance

of treated hospital effluent quality. Environ Sci Water Res Technol 3:293–303. https://doi.org/10.1039/C6EW00322B
Tiwari B, Sellamuthu B, Ouarda Y et al (2017) Review on fate and mechanism of removal of pharmaceutical pollutants from wastewater using biological approach. Bioresour Technol 224:1–12. https://doi.org/10.1016/j.biortech.2016.11.042
Turk Sekulic M, Boskovic N, Milanovic M et al (2019a) An insight into the adsorption of three emerging pharmaceutical contaminants on multifunctional carbonous adsorbent: mechanisms, modelling and metal coadsorption. J Mol Liq 284:372–382. https://doi.org/10.1016/j.molliq.2019.04.020
Turk Sekulic M, Boskovic N, Slavkovic A et al (2019b) Surface functionalised adsorbent for emerging pharmaceutical removal: adsorption performance and mechanisms. Process Saf Environ Prot 125:50–63. https://doi.org/10.1016/j.psep.2019.03.007
Uddin MK (2017) A review on the adsorption of heavy metals by clay minerals, with special focus on the past decade. Chem Eng J 308:438–462. https://doi.org/10.1016/j.cej.2016.09.029
Vakili M, Deng S, Cagnetta G et al (2019) Regeneration of chitosan-based adsorbents used in heavy metal adsorption: a review. Sep Purif Technol 224:373–387. https://doi.org/10.1016/j.seppur.2019.05.040
Van Boeckel TP, Brower C, Gilbert M et al (2015) Global trends in antimicrobial use in food animals. Proc Natl Acad Sci 112:5649–5654
Varela AR, Nunes OC, Manaia CM (2016) Quinolone resistant Aeromonas spp. as carriers and potential tracers of acquired antibiotic resistance in hospital and municipal wastewater. Sci Total Environ 542:665–671. https://doi.org/10.1016/j.scitotenv.2015.10.124
Varga M, ELAbadsa M, Tatár E, Mihucz VG (2019) Removal of selected pharmaceuticals from aqueous matrices with activated carbon under batch conditions. Microchem J 148:661–672. https://doi.org/10.1016/j.microc.2019.05.038
Verlicchi P, Al Aukidy M, Galletti A et al (2012) Hospital effluent: investigation of the concentrations and distribution of pharmaceuticals and environmental risk assessment. Sci Total Environ 430:109–118
Verlicchi P, Zambello E, Al Aukidy M (2013) Removal of pharmaceuticals by conventional wastewater treatment plants. In: Petrovic M, Barcelo D, Pérez SBT-CAC (eds) Analysis, removal, effects and risk of pharmaceuticals in the water cycle. Elsevier, pp 231–286
Verlicchi P, Al Aukidy M, Zambello E (2015) What have we learned from worldwide experiences on the management and treatment of hospital effluent?—an overview and a discussion on perspectives. Sci Total Environ 514:467–491
Villaescusa I, Fiol N, Poch J et al (2011) Mechanism of paracetamol removal by vegetable wastes: the contribution of π–π interactions, hydrogen bonding and hydrophobic effect. Desalination 270:135–142. https://doi.org/10.1016/j.desal.2010.11.037
Wang J, Wang S (2016) Removal of pharmaceuticals and personal care products (PPCPs) from wastewater: a review. J Environ Manag 182:620–640
Wang Y, Rao GY, Hu JY (2011) Adsorption of EDCs/PPCPs from drinking water by submicron-sized powdered activated carbon. Water Sci Technol Water Supply 11:711–718
Wang YX, Ngo HH, Guo WS (2015) Preparation of a specific bamboo based activated carbon and its application for ciprofloxacin removal. Sci Total Environ 533:32–39. https://doi.org/10.1016/j.scitotenv.2015.06.087
Wang F, Ma S, Si Y et al (2017) Interaction mechanisms of antibiotic sulfamethoxazole with various graphene-based materials and multiwall carbon nanotubes and the effect of humic acid in water. Carbon N Y 114:671–678. https://doi.org/10.1016/j.carbon.2016.12.080
Wang Q, Wang P, Yang Q (2018) Occurrence and diversity of antibiotic resistance in untreated hospital wastewater, Sci Total Environ 621: 990–999. https://doi.org/10.1016/j.scitotenv.2017.10.128
Wang N, Xiao W, Niu B et al (2019a) Highly efficient adsorption of fluoroquinolone antibiotics using chitosan derived granular hydrogel with 3D structure. J Mol Liq 281:307–314. https://doi.org/10.1016/j.molliq.2019.02.061

Wang Y, Jiao W-B, Wang J-T et al (2019b) Amino-functionalized biomass-derived porous carbons with enhanced aqueous adsorption affinity and sensitivity of sulfonamide antibiotics. Bioresour Technol 277:128–135. https://doi.org/10.1016/j.biortech.2019.01.033

Wang Y, Wang X, Li J et al (2019c) Coadsorption of tetracycline and copper(II) onto struvite loaded zeolite—an environmentally friendly product recovered from swine biogas slurry. Chem Eng J 371:366–377. https://doi.org/10.1016/j.cej.2019.04.058

Wang Y, Lin J, Wang Y et al (2020) Highly efficient and selective removal of low-concentration antibiotics from aqueous solution by regenerable $Mg(OH)_2$. J Environ Sci 87:228–237. https://doi.org/10.1016/j.jes.2019.06.017

Wei R, Ge F, Huang S et al (2011) Occurrence of veterinary antibiotics in animal wastewater and surface water around farms in Jiangsu Province, China. Chemosphere 82:1408–1414. https://doi.org/10.1016/j.chemosphere.2010.11.067

Wen J, Dong H, Zeng G (2018) Application of zeolite in removing salinity/sodicity from wastewater: a review of mechanisms, challenges and opportunities. J Clean Prod 197:1435–1446. https://doi.org/10.1016/j.jclepro.2018.06.270

Wong S, Lim Y, Ngadi N et al (2018) Removal of acetaminophen by activated carbon synthesized from spent tea leaves: equilibrium, kinetics and thermodynamics studies. Powder Technol 338:878–886. https://doi.org/10.1016/j.powtec.2018.07.075

Wong JKH, Tan HK, Lau SY et al (2019) Potential and challenges of enzyme incorporated nanotechnology in dye wastewater treatment: a review. J Environ Chem Eng 7:103261. https://doi.org/10.1016/j.jece.2019.103261

Wu Y, Xi B, Hu G et al (2016) Adsorption of tetracycline and sulfonamide antibiotics on amorphous nano-carbon. Desalin Water Treat 57:22682–22694. https://doi.org/10.1080/19443994.2015.1135407

Wu M, Zhao S, Jing R et al (2019a) Competitive adsorption of antibiotic tetracycline and ciprofloxacin on montmorillonite. Appl Clay Sci 180:105175. https://doi.org/10.1016/j.clay.2019.105175

Wu M, Zhao S, Tang M et al (2019b) Adsorption of sulfamethoxazole and tetracycline on montmorillonite in single and binary systems. Colloids Surf Physicochem Eng Asp 575:264–270. https://doi.org/10.1016/j.colsurfa.2019.05.025

Xie A, Cui J, Chen Y et al (2019) Simultaneous activation and magnetization toward facile preparation of auricularia-based magnetic porous carbon for efficient removal of tetracycline. J Alloys Compd 784:76–87. https://doi.org/10.1016/j.jallcom.2018.12.375

Xiong W, Zeng G, Yang Z et al (2018) Adsorption of tetracycline antibiotics from aqueous solutions on nanocomposite multi-walled carbon nanotube functionalized MIL-53(Fe) as new adsorbent. Sci Total Environ 627:235–244. https://doi.org/10.1016/j.scitotenv.2018.01.249

Xu J, Cao Z, Zhang Y et al (2018) A review of functionalized carbon nanotubes and graphene for heavy metal adsorption from water: preparation, application, and mechanism. Chemosphere 195:351–364. https://doi.org/10.1016/j.chemosphere.2017.12.061

Yang Y, Hu X, Zhao Y et al (2017) Decontamination of tetracycline by thiourea-dioxide–reduced magnetic graphene oxide: effects of pH, ionic strength, and humic acid concentration. J Colloid Interface Sci 495:68–77. https://doi.org/10.1016/j.jcis.2017.01.075

Yang Y-Y, Zhao J-L, Liu Y-S et al (2018a) Pharmaceuticals and personal care products (PPCPs) and artificial sweeteners (ASs) in surface and ground waters and their application as indication of wastewater contamination. Sci Total Environ 616:816–823

Yang Y, Song W, Lin H et al (2018b) Antibiotics and antibiotic resistance genes in global lakes: a review and meta-analysis. Environ Int 116:60–73. https://doi.org/10.1016/j.envint.2018.04.011

Yu F, Sun S, Han S et al (2016) Adsorption removal of ciprofloxacin by multi-walled carbon nanotubes with different oxygen contents from aqueous solutions. Chem Eng J 285:588–595. https://doi.org/10.1016/j.cej.2015.10.039

Yu Q, Geng J, Ren H (2019) Occurrence and fate of androgens in municipal wastewater treatment plants in China. Chemosphere 237:124371. https://doi.org/10.1016/j.chemosphere.2019.124371

Zeng Z, Tan X, Liu Y et al (2018) Comprehensive adsorption studies of doxycycline and ciprofloxacin antibiotics by biochars prepared at different temperatures. Front Chem 6:80. https://doi.org/10.3389/fchem.2018.00080

Zeng Q, Sun J, Zhu L (2019) Occurrence and distribution of antibiotics and resistance genes in greenhouse and open-field agricultural soils in China. Chemosphere 224:900–909. https://doi.org/10.1016/j.chemosphere.2019.02.167

Zhang Y, Geißen S-U, Gal C (2008) Carbamazepine and diclofenac: removal in wastewater treatment plants and occurrence in water bodies. Chemosphere 73:1151–1161. https://doi.org/10.1016/j.chemosphere.2008.07.086

Zhang Q-Q, Ying G-G, Pan C-G et al (2015) Comprehensive evaluation of antibiotics emission and fate in the river basins of China: source analysis, multimedia modeling, and linkage to bacterial resistance. Environ Sci Technol 49:6772–6782

Zhang H, Lan X, Bai P, Guo X (2016a) Adsorptive removal of acetic acid from water with metal-organic frameworks. Chem Eng Res Des 111:127–137. https://doi.org/10.1016/j.cherd.2016.04.020

Zhang S, Dong Y, Yang Z et al (2016b) Adsorption of pharmaceuticals on chitosan-based magnetic composite particles with core-brush topology. Chem Eng J 304:325–334. https://doi.org/10.1016/j.cej.2016.06.087

Zhang L, Wang Y, Jin S et al (2017a) Adsorption isotherm, kinetic and mechanism of expanded graphite for sulfadiazine antibiotics removal from aqueous solutions. Environ Technol 38:2629–2638. https://doi.org/10.1080/09593330.2016.1272637

Zhang Y, Jiao Z, Hu Y et al (2017b) Removal of tetracycline and oxytetracycline from water by magnetic Fe_3O_4@graphene. Environ Sci Pollut Res 24:2987–2995. https://doi.org/10.1007/s11356-016-7964-7

Zhang H, Wang J, Zhou B et al (2018a) Enhanced adsorption of oxytetracycline to weathered microplastic polystyrene: kinetics, isotherms and influencing factors. Environ Pollut 243:1550–1557. https://doi.org/10.1016/j.envpol.2018.09.122

Zhang J, Lu M, Wan J et al (2018b) Effects of pH, dissolved humic acid and Cu^{2+} on the adsorption of norfloxacin on montmorillonite-biochar composite derived from wheat straw. Biochem Eng J 130:104–112. https://doi.org/10.1016/j.bej.2017.11.018

Zhao Y, Liu F, Qin X (2017) Adsorption of diclofenac onto goethite: adsorption kinetics and effects of pH. Chemosphere 180:373–378. https://doi.org/10.1016/j.chemosphere.2017.04.007

Zhao Y, Choi J-W, Bediako JK et al (2018) Adsorptive interaction of cationic pharmaceuticals on activated charcoal: experimental determination and QSAR modelling. J Hazard Mater 360:529–535. https://doi.org/10.1016/j.jhazmat.2018.08.039

Zhao F, Yang L, Chen L et al (2019) Soil contamination with antibiotics in a typical peri-urban area in eastern China: seasonal variation, risk assessment, and microbial responses. J Environ Sci 79:200–212. https://doi.org/10.1016/j.jes.2018.11.024

Zhou Y, Niu L, Zhu S et al (2017) Occurrence, abundance, and distribution of sulfonamide and tetracycline resistance genes in agricultural soils across China. Sci Total Environ 599:1977–1983

Zhu X, Tsang DCW, Chen F et al (2015) Ciprofloxacin adsorption on graphene and granular activated carbon: kinetics, isotherms, and effects of solution chemistry. Environ Technol 36:3094–3102. https://doi.org/10.1080/09593330.2015.1054316

Chapter 9
Treatment of Dairy Byproducts with the Conversion of Useful Bio-Products

Rajesh K. Srivastava

Contents

Abbreviations

ASBR Anaerobic mode sequencing batch reactor
BOD Biological oxygen demand
CSIR Continuous flow stirred tank reactor
CWP Cheese whey powder
DO Dissolve oxygen
ETP Effluent treatment plant
GC Gas chromatography
HAD Hybrid anaerobic digesters
HAS Human serum albumin
kHz Kilo hertz
LC Liquid chromatography

R. K. Srivastava (✉)
Department of Biotechnology, GIT, Gandhi Institute of Technology and Management (GITAM) Deemed to be University, Visakhapatnam, AP, India

Inamuddin, A. M. Asiri (eds.), *Sustainable Green Chemical Processes and their Allied Applications*, Nanotechnology in the Life Sciences,
https://doi.org/10.1007/978-3-030-42284-4_9

LOD	Limit of detection
LOQ	Limit of quantification
$m\ s^{-1}$	Meter per second
m^{-1}	Per meter
MS	Mass spectroscopy
NPN	Non-protein nitrogen
Rf	Retardation factor
RI	Resistance index
SCM	Sweetened condensed milks
TFA	Trifluoroacetic acid
TSS	Total soluble solids
UFSB	Up-flow sludge blanket
UF	Ultrafiltration
ZOSS	Zirconium oxide stainless steel

9.1 Introduction: Basic Concept of Milk Food

Milk is found as white liquid and it is produced by the mammary glands of cow, buffalos, goat, camel, human being, and other mammal species. And it has helped for feeding to their offspring till they can easily consume solid food. Milk contains valuable nutrients via offering a range of health benefit components (such as calcium can help in the prevention of osteoporosis disease). It has found that some people are unable to digest for lactose due to not enough of an enzyme production (lactase) and it is milk sugar and it is necessary for an infant or other young mammal as food. Lactose intolerance and milk allergies people can take the range of substitute milks (such as almond and soy milk) (Thorning et al. 2016).

A food can be synthesized from milk or its constituents and it is very important, necessary and preferred food which is taken by all aged small children as well as elder and older. Various types of milk products can be prepared by addition of byproducts (milk processing) or non-dairy components. Fats of non-dairy origin are used in the production process as individual ingredients. Dairy products are not applied for nutrition of infants and babies but this can be added for purposes other than substitution of milk constituents. Final milk products are found to contain milk constituents (above 50%), ice cream and sweet milk processing products (above 40%). There are various dairy products which needs different milk processing. Some dairy products are cultured product, cream product, cream ice, cream cheese butter, cream-vegetable rendered mixture, curd and curd mass bar etc. (Allwyn Sundarraj et al. 2018). Dairy-based products are probiotic food products and it is reported as rapidly growing area of functional food and it is synthesized by microbial inoculants via enhancing properties (i.e. tastes, aromas or flavors, consistency, shelflife or duration of products quality stability, and nutritional values of food or dairy or milk products. It has found the big challenges or issues for engineers, microbiological scientists or technologist for disposal of dairy foods or its byproducts or wastes. Food industries can develop the techniques or procedure for milk waste products utilization for new dairy products or other products processing tasks. They need to find the best or sustainable ways for these products destruction or degradation via application of suitable enzymes or microbial system activity without escaping proper treatment processes by various dairy industries (De Cortes-Sanchez et al. 2015).

Discarding of the wastes milk is reported as important problem for environment via causing soil or water pollution. Nutriment materials in the wasted milk from milk factory are found for increasing and developing microorganisms which could be good nutriment source for flies, insects and gnawings. Toxic components are also synthesized with pathogen organisms in them. Waste of milk factory is reported for potential contamination source for people and other alive (Tawfika et al. 2008). Milk can be produced, transported, stored and sold as solid, fluid and gas. Milk wastes are discarded without any process, causing damage to all alive. Milk wastes contain animal protein, oil, sugar and mineral components and are evaluated in different ways by a lot of country in world. Whey in cheese and casein technology is important side products and can cause important problem via spoil easily in a few modern factory in our country. It can be given to animal or poured to fields or discarded to canals and rivers. Many countries do not go for squander milk, milk products and their wastes (Raghunath et al. 2016).

Various modes of aerobic technologies are utilized as treatment methods and some treatment methods of aerobic modes are activated sludge mode reactor, sequencing mode of batch reactor and rotating mode of biological contractor reactor with trickling filter as good examples and utilized for dairy waste treatment processes. Upflow anaerobic sludge blanket mode reactor (UASB), anaerobic mode sequencing batch reactors (ASBR), continuous flow stirred tank reactor (CSTR), hybrid mode of anaerobic digestor (HAD) contact reactor flow anaerobic mode with filter with two phases or stages processing systems are employed for waste treatment purposes and most of these treatments are completed in anaerobic conditions under influences and separation of acid forming and methane biofuel forming microbial system (Kushwaha et al. 2011). It has reported for combined of aerobic and anaerobic processes as single unit system for waste treatment with application of membrane technology.

These arrangements can provide an alternative option for bio-mode of waste treatment via helping in reduction of biological oxygen demand (BOD) in dairy wastewater with replacement of secondary clarifier system in the waste treatment with membrane filter. Wastewater treatment in anaerobic mode or conditions is found to more viable and economical feasible with alternatives option for reducing high BOD values via comparing to conventional aerobic mode of waste treatment (Kushwaha et al. 2011; Birwal et al. 2017). The developed model has helped in identification of the optimal concentration of products in portfolio composition of milk via the optimal products formation with portfolio compositions via allocating the raw milk materials to most profitable dairy products. This model can help in accounting for critical and important constraints (such as recipes preparation of milk products, composition variation, interdependency modes of dairy products formation, seasonal variation, demands, supplies, capacities and transport flow of milk or its products) (Mula et al. 2010).

All the relevant constraints in waste products treatment can provide the easy ways of understanding the dynamics mode of dairy products synthesis in comprehensive models. Structure and components of the model and its outputs has been studied and analyzed in multiples sessions with its proper approval for implementation (relevant to employee of Friesland, Campina). The elements or components in this model are reported with its necessarily for optimal valorization of raw milk. Comprehensiveness and functionality of this model has been illustrated for analysis of the effect of seasonality on milk valorization. This model has helped in understanding of large difference in profits with shifting in allocation of milk in different seasons with consideration of impact on volarization of raw milk (Banaszewska et al. 2013).

Along with many food industries, dairy industry is found one of the most polluting source and dairy product components is responsible for water pollution. Due to increased or enhanced demand of milk or its products, the dairy industries are found to grow rapidly in the world and most quantity of waste generation is responsible for environmental problems. Due to poor or minimum efforts, untreated wastewaters are found to pollutant concentration in high levels. Treatment system with poor design and operation conditions in dairy industries are reported to create major environmental problems due to their effluent discharge to the land surfaces or water body resources. Various operations like pasteurization for cream, cheese or milk powder development from milk is reported in various dairy industries. Effluent Treatment Plant (ETP) is applied evaluation of pH, value of chemical oxygen demand (COD), BOD and total soluble solids (TSS) content or compounds from each sample taken (from inlet chamber, aeration tank, equalization tank, and clarifier outlet) during the dairy components containing wastewater treatment or its waste compounds or byproducts. The COD (94%), BOD (95%) at 27 °C and TSS (93%) are reported in removal efficiency of ETP and shown in Fig. 9.1. Author will discuss the different type of milk products, their processing and respective wastes generation as well as effective treatment methods.

9.2 Milk and Its Production

Milk production from cow cattle is reported as important or crucial sources of protein, needed in human nutrition and it is reported to consist of protein (3.5%) especially abundance in casein or various varieties of whey proteins of whey proteins (i.e. β-lactoglobulin, α-lactalbumin or serum proteins). These proteins can be identified by various types of allelic variants or posts translated processed modified proteins or its products and are reported in variation in breed, individual species, and stages of lactation with healthy or nutritional status of the milk producing animals. Reliable analysis methods (i.e. liquid chromatography~LC or ultra-violet detection methods) are applied for its detection and quantification of these proteins. Several

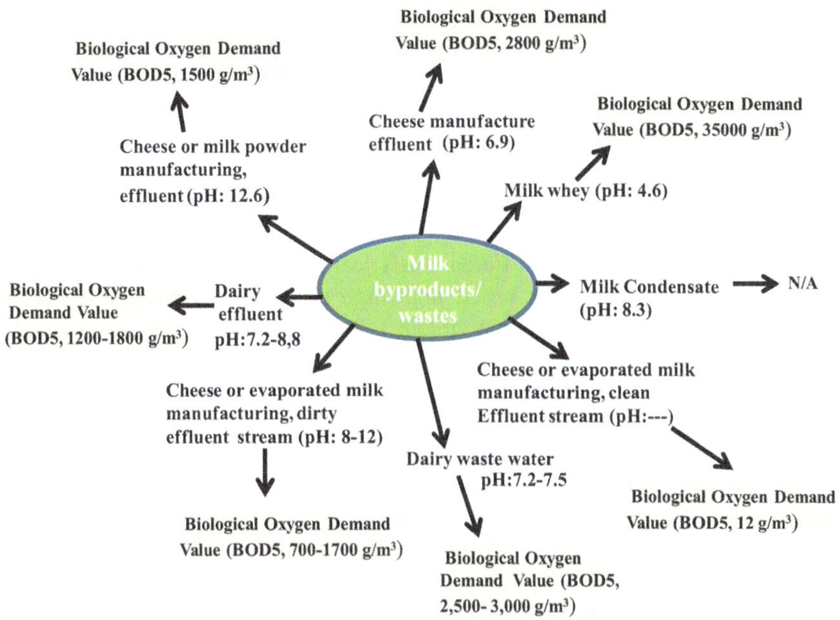

Fig. 9.1 Water pollution (in term of BOD values) reported due to disposal of untreated milk byproducts from dairy industries (Watkins and Nash 2010)

other analysis i.e. reverse phase- high performance liquid chromatography (RP-HPLC) and mass spectroscopy (MS) with its detection parameters are used to optimize the analysis or detection of intact bovine proteins quantity in raw milk.

An optimal method of detection via RP-HPLC technique is reported to use the 20, 28 or 40% phase gradient B component (in different quantity in %) with trifluoroacetic acids (0.02%) as combination of two or both mobile phases. In this technique, optimal flow rate (0.2 mL/min) and temperature (75 °C) is maintained on C8 column with scanning facility of every 3 s at 600–3000 *m/z* window screen. This technique has good ionization efficiency, linearity of calibration (LOC), limit of detection (LOD) or limit of quantification (LOQ). Further this technique has better sensitivity, precision, reproducibility and mass accuracy as better property. This technique can be easily proceded with optimization of process detection with external standards, purchased commercially (Vincent et al. 2016).

Fractionation of the mixture of proteins is reported to getting individual dairy protein fractions from milk and whey in form pure. These proteins can provide the special nutritional needs to our diet via improving healthy conditions. The processes of ion exchange chromatography are applied for obtaining the respective component fractions in mixture of protein solution via using selection elution modes. It can help in releasing these proteins in separate fractions for several proteins which is bound from whey. Bound proteins can be released all after isolation of whey protein. These proteins are reported in prototype beverages and are gone for examination of clarity before and after thermal processing steps. Whey proteins in beverages food are isolated at pH range (2–7) before heating and after thermal processing temperature (88 °C for 120 s) with separation of beverage (at pH $\leq$ 3.0) (Etzel 2004).

Buttermilk from natural fermentation processes is prepared from souring of cream or liquid milk and is collected during two seasons in northern Ethiopia country. Buttermilk has been studied for variation in chemical composition and favoring compounds.. From this milk products, protein, fats, organic acids, or volatile compounds has been quantified by different analytical techniques i.e. Kjeldahl method, techniques such as Kjeldahl, Gerber, high high pressure liquid chromatography (HPLC) and head space gas chromatography (GC) methods. It has reported that difference in concentration of organic acids or volatile compounds are reported the different values among samples analysis. Introduction or application of the standardization processes is used to supply of products of homogenous buttermilk products in markets (Gebreselassie et al. 2016).

Casein is found only in the milk on nature. Casein protein is reported around 80% of milk proteins. And casein and caseinates are two products, reported to wide use in many areas. Production of high quality casein can be achieved by using skimmed milk as raw material with good quality. And lactose conversion by lactic acid can affect the quality of casein via increasing of acidity due to more lactic acid formation. Excessive heating of milk before precipitation or manufacturing process can produce undesirable color with of reduction of casein yield (Badem and Uçar 2017). More example of milk products are shown in Table 9.1 and also in Fig. 9.2.

9.3 Milk Processing Steps for Milk Products

Vacreation is short-time pasteurization via continuous flow of high-temperature to milk or milk products. This process is completed in vacreator equipment which includes three successive stages, each of which operates under a pressure lower than that of the atmosphere. Application of various modification processes are found and some of these processes are hydrogenation, chemical interesterication (with sodium methanolate or sodium methoxide at low temperature of 100 °C) and fractionation processes and responsible for alteration of chemical compositions of milk fats) Modification of physical or chemical properties or characters of milk fat is found more suitable for market purposes with wider applications or for tailored specific tasks or end uses (Sichien et al. 2009). There are some common milk processing steps which are shown below.

- Cleaning of milk equipments
- Enzymes addition or utilization in Dairy
- Heat Treatment process of Milk (pasteurization, sterilization or ultra heat treatment)

Table 9.1 Various milk products reported from dairy industries

Milk products	Source	Processes	References
1. Albumin (concentrate of milk whey proteins)	Milk whey transgenic cattle	Fractionation of the mixture of proteins is done via thermal processing and high levels of recombinant human serum albumin (HSA) in transgenic cattle milk due to phiC31 integrase-mediated gene delivery	Luo et al. (2015), Vincent et al. (2016), Etzel (2004)
2. Butter (dairy composite product)	Cow milk	Churning fresh or fermented cream or milk as emulsion-and-fat basis	Deosarkar et al. (2016)
3. Buttermilk (byproduct)	Cow milk	Variable fermentation process and standardization process for the fermented milk	Gebreselassie et al. (2016)
4. Butter paste (dairy composite product)	Cow milk cream	Emulsion-and-fat basis with specific fractions to improve butter spreadability	Henning et al. (2006)
5. Canned milk (packaged and concentrated or condensed)	Cow or buffalos milk	Removal or evaporation of water from milk and it contains crystal lactose, color and flavor and age gelation	Nieuwenhuijse (2011)
6. Canned composite milk (packaged and composite milk)	Cow or buffalos milk	Commercial sterilizing condition (109.3–121.1 °C under pressure of 5–15 psi)	Chiewchan et al. (2006)
7. Casein (major fraction of milk protein)	Skimmed milk	Clotting agents (are called rennet casein and acid caseins) are utilized.	Badem and Uçar (2017)
8. Caseinate (sodium, potassium or calcium, etc.)	Skimmed milk	Casein by-process with the acid casein coagulant or solutions of alkaline metal hydroxides. It is formed with casein fractions to appropriate salt form at pH 6–7.	Jacob et al. (2010)
9. Cheese	Milk or milk products	Milk processing byproducts. Clarification, separation and standardization	Bergamaschi and Bittante (2018)
10. Cheese product (Mozzarella cheese)	Buffalo milk	Cheese making technology has helped via action of lactic acid on dicalcium-para-caseinate	Jana and Mandal (2011)
11. Concentrate of whey proteins (protein contents in the range 35–65%)	Milk whey	Prepared by concentration or ultra-filtration by removal of lactose, minerals and non-protein nitrogen (NPN), leaving the whey proteins to be spray-dried	Early (2012)

(continued)

Table 9.1 (continued)

Milk products	Source	Processes	References
12. Condensed, or evaporated milk	Fresh milk	Prepared by partial water removal and contains solids not below 20%. Fresh milk is clarified and standardized to a suitable level of fat, and it is then heat treated at 85–90 °C for several seconds.	Kalyankar et al. (2016)
13. Cream (emulsion of fat and milk plasma)	Milk and or milk products	Revalorization of milk fat is done and at least 10% fat mass fraction with poor spreadability at refrigeration temperature and its high content in saturated fatty acids	Lubary et al. (2011)
14. Condensed whole milk with sugar	Milk of crossbred cow's	28.5% milk solids, 8.5% milk fat 34% fat-free milk solids. Cooking temperature of the caramel mix and processing conditions	Dey and De (2014)
15. Condensed cream with sugar	Butter rich milk	37% Milk solids milk protein, 34% fat-free milk solids, 19% milk fat by evaporation processes. Homogenization, stabilization, and heat sterilization for evaporated milk and sugar addition, lactose crystallization for sweetened condensed milks (SCMs)	Sharma et al. (2015)

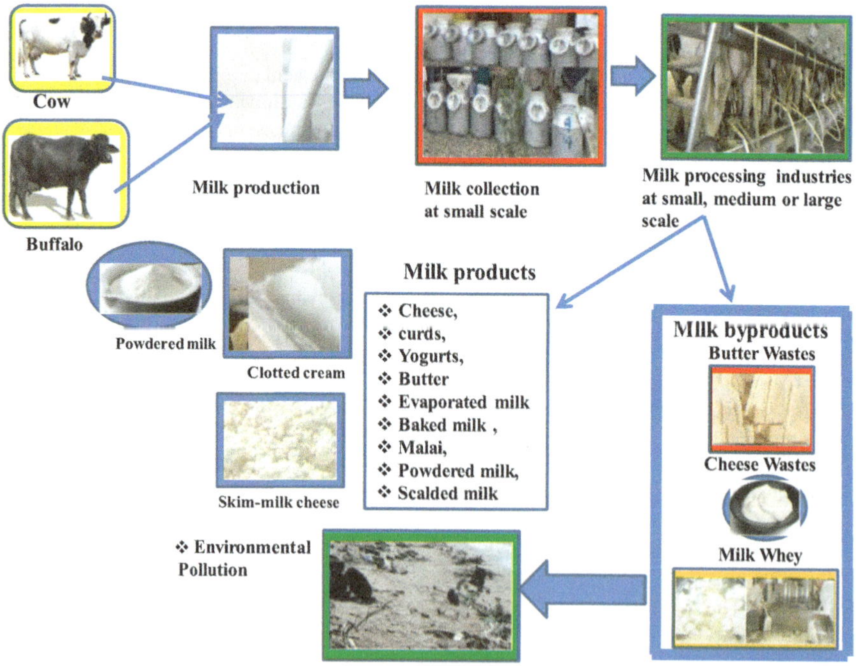

Fig. 9.2 Schematic diagram for milk production and milk processing on environment

- Processing of Whey (Fines removal, Whey cream removal, Pasteurization, Bleaching (optional) and storage
- Fermented/sour milk (Itittuu) processes
- Cottage cheese (Ayib) making and utilization
- Butter making and preservation
- Extraction of Casein from Milk

Advances in development processing for cream separation tasks Separators for improved separation development Heat treatment under vacuum condition in stainless steel chamber is proceeded under vacreator process. Sprays drying process applications is reported in cheese preparation industry via processing of milk whey component and are reported for different characters or properties in different type of whey component, produced by spray drying processes. In this milk processing, whey property is studied with analytical effects of spray –dryer operation parameters (such as flow rate of feed, type of atomizer and temperature of inlet or outlet air).

Physical properties of milk powder is studied and analyzed and some of these are bulk liquid density, sizes of particle, moisture content or quantity, insoluble solids amounts, wetting ability or morphology or physical appearance of powder particles. Spray dried whey product are reported for easy storage, handling or transport tasks. And this technique produced whey quality is quiet dependent on the parameters of operation of spray dryer equipment or vessel (i.e. conditions of spray drying) (Chegini and Taheri 2013).

Heat treatment is milk processing technology in the dairy industry via destroying several microbial agents such pathogenic bacteria or yeast cell. These processes can reduce the spoilage processing via ensuring the milk safety and sufficient periods of shelflife. Alternative milk processing development are reported and high pressure processing, or pulsed field processing technology are used for destroying the microbial cells contamination as non-thermal processing, employed mostly as bactericidal treatment in many dairy industries.. This choice can be attributed to the tremendous success processes for pasteurization of milk via avoiding heat treatment for wine and beer via preventing spoilage (Takeda et al. 2015).

Fermentative properties of *Streptococcus thermophilus* can be reported in dairy product formation as alone or single microbial source or combination of other microbial strain (with *Lactobacillus belbrueckii* subsp). *Lactobacillus belbrueckii* has been evaluated in skim milk products via subjection of distinct sterilization conditions. Microbial growth with organic acids and extracellular polysaccharides (EPS) production are studied with determination of its quantity in pasteurized or sterilized milk via autoclave processes. There are reports that *S. thermophilus* microbial strain can be cultivated or grown in pasteurized skim milk with its good growth rate. But *Sreptococcus thermophilus* growth can be accelerated or enhanced in autoclaved milk liquid. Accelerated in autoclaved milk. EPS product formation is reported by *Lactobacillus belgaricus* without affecting its production with combination of *S. thermophilus* strain. Milk fermented in yogurt form is reported by *Lactobacillus belgaricus* and it is minimally affected with combination of *S. thermophilus* strain under a constant environment via facilating the fermentation processes for pasteurized raw milk in dairy industries (Nishimura et al. 2015).

High-amplitude ultrasound-promoted pasteurization can be worked by developing of intense acoustic cavitations and it generates violently imploding vacuum bubbles via causing micro-jets for lethal to microorganisms. Freshly prepared water and milk-based fruit smoothies has been processed by this approach and it can be applied for foods and beverages an emerging technique. It can achieve the required levels of microbial inactivation at much lower temperatures (about 40 °C), without deterioration of the quality of the products. Ultrasound waves can produce longitudinal sound wave^{s} (frequency of 20 kHz or more). These waves do not significantly modify the material, confirmed via analysis of the material property. High amplitude of ultrasonic waves is reported for food processing task with the capacity of altering the food material structure by cavitations and increasing amounts being used in the dairy industry (Ghosh 2017).

Numbers of viable cells for *Escherichia coli* are are reported to reduce by 100% after 10.0 min of ultrasonication treatment. Viable counts of *Pseudomonas fluorescens* also reported for its reduction up to 100% after treatment for 6.0 min and 99% reduction is reported for *Listeria monocytogenes* with treatment for 10 min. The equipment of infra-red (IR) has been applied for analysis of raw and pasteurized milk after treatment via ultrasonic waves (Cameron et al. 2009). Here we will discuss the different type of milk wastes or byproducts in next section and shown in Fig. 9.1.

9.4 Dairy Wastes and Byproducts

Dairy factory has generated amount of wastewaters, containing milk or other products or byproducts during processing operations cleaning products and various additives. Bovine milk is reported to contain the different components quantity such as water (87%), fat component (4%), protein (3.5), lactose sugar (4.7%) and ashes compounds (0.8%). Triglycerides (98%) is reported as major fraction in milk fat content where as other fats components are phospholipids quantity (0.5–1.0) or sterol component (0.2–0.5%) for total fat quantity. Terpene as organic compounds is reported on the diet of the cows with plants origin (De Noni and Battelli 2008). Untreated dairy wastewater is reported for good nutrient-rich source via providing beneficial nitrogen and phosphorus for grass covered land via irrigation. This wastewater is responsible for adverse or undesirable changes to property of soil and its structure (porosity nature) and infiltration rate of soil. Further downstream processes of milk products contamination for streams or ground water resource can be depressed the plant growth for using wastewater for irrigation task (Halliwell et al. 2001).

Irrigation from this wastewater can be achieved via completion of proper treatment which can contain lesser quantity of some elements such as sodium (Na), calcium (Ca), magnesium (Mg), phosphorus (P), nitrogen (N) or some carbon waste in land area irrigated with wastewater uses and these lands can make worthy via using bioremidation processes that now under consideration. System management with poor treatment facility or technique can not able to remove excess quantity of waste nutrient (i.e. nitrates) and it can result for leaching through soil profile via making the contamination of underground water resources (Chaiudhari and Dhoble 2010). Acceleration of eutrophication processes in streams or lakes are found due to more quantity of nitrogen and phosphorus via exporting off site in surface runoff. Treated of milk wastewater can contain fewer quantity of nutrients and can be resulted with lower risks of contamination of underground or surface water resources (Watkins and Nash 2010).

Cheese or milk powder, cheese or evaporated milk cheese, fat content products and other milk fraction can be found in milk wastewater effluent and are responsible for increased value of BOD and COD in water bodies (river, pond or lake) which has contained disposal milk wastewater and BOD value is reported in Figs. 9.1 and 9.3. Dairy effluents generated from various dairy industries are reported to contain the various components and most common components are soluble organic compounds, suspended solids (inert or reactive nature) with trace elements and all these components are responsible for contributing the major environmental problems such as high or more values of biological oxygen demand (BOD) or chemical oxygen demand (COD) (Geilman et al. 1992). These dairy wastes are responsible for changing color (i.e. white) and also responsible for high pH value (slightly alkaline in nature) and later these wastes can responsible for lower pH value of water (acidic nature) at rapid rate due to formation of lactic acid from lactose sugar during fermentation processes in water bodies. More quantity of solid matters from milk waste from cheese industries is reported with fine curd form and these can responsible for pollution effect in water resource due to high values of BOD (Tawfik et al. 2008).

Further, decomposition of casein protein in milk wastes is responsible for heavy black sludge formation and also reported for strong or sharp butyric acod odors due to presence of milk waste in water pollution. In this polluted water bodies or resource has changed temperature, pH (6.5–8.0), color, dissolved oxygen (DO), with increased value of BOD and COD. These water bodies have more amount suspended solids, chlorides, sulphates, oil content or grease components (Quasim and Mane 2013; Tsachev 1982). These components quantity is dependent on quantity or types of processed milk products or its constituents and casein, inorganic salts, with detergent or sanizers compounds during washing task Sodium element from caustic soda uses for cleaning is reported in polluted water (Slavov 2017; Shete and Shinkar 2013). More examples of milk waste products are discussed in Table 9.2 and shown in Fig. 9.3.

9.5 Products from Milk Byproducts Utilization

Formations of important byproducts compounds from dairy industry are reported with containing of milk whey components ((i.e. lactose and proteins). The processes of co-utilization for lactose or protein components (in milk whey fraction or portion) is reported for biochemical products formation via utilization of efficient microbial systems or its bioprocesses and comprehensive utilization of cheese whey

Table 9.2 Milk waste products or byproducts reported from dairy industry (Shete and Shinkar 2013)

Milk waste type	BOD (mg/L) and COD (mg/L) values respectively	pH and TSS (mg/L) values	References
1. Milk and dairy products factory	1.0×10^3 and 4.8×10^3	8.34 and 5.8×10^3	Cristian (2010), Janczukowicz et al. (2008)
2. Dairy effluent	$1.9–2.7 \times 10^3$ and $1.2–1.8 \times 10^3$	7.2–8.8 and 5.00–7.4×10^2	Deshannavar et al. (2012), Kocadagistan et al. (2005)
3. Arab dairy factory	3.38×10^3 and 1.9×10^3	7.9 and 8.31×10^2	Tawfik et al. (2008), Kisielewska et al. (2018)
4. Dairy waste water	$2.5–3.00 \times 10^3$ and $1.30–1.60 \times 10^3$	7.2–7.5 and $7.2–8.0 \times 10^4$	Qazi et al. (2011), Shi et al. (2011)
5. Dairy effluent	$1.12–3.36 \times 10^3$ and $3.20–17.5 \times 10^2$	5.6–8 and 0.28–1.9×10^2	Slavov (2017), Karadag et al. (2015)
6. Whey (Total solids: 56782 mg/liter and suspended solids: 22050 mg/liter)	7.15×10^4 and 2.0×10^4	4.1 and 2.2×10^4	Deshpande et al. (2012), Sankar Ganesh and Ramasamy (2007)
7. Bhandara co-operative dairy industry wastewater	$1.4–2.5 \times 10^3$ and $0.800–1.0 \times 10^3$	7.1–8.2 and 1.04–1.8×10^3	Powar et al. (2013), Hossein and Jalal (2011)
8. Cheese whey pressed (organic and saline effluent)	$8.0–9.0 \times 10^4$ and $1.20–1.35 \times 10^5$	6 and 0.800–1.1×10^4	Kabbout et al. (2011), Carvalho et al. (2013)
9. Aavin dairy industry washwater	– and $2.5–3.30 \times 10^3$	6.4–7.1and $6.3–7.3 \times 10^2$	Sathyamoorthy and Saseetharan (2012), Shete and Shinkar (2013)
10. Dairy industry wastewater	2.1×10^3 and 1.04×10^3	7–8 and $1.2–2.5 \times 10^3$	Arumugam and Sabarethinam (2008), Keffala et al. (2017)
11. Mixed dairy	0.24–5.9 and 0.5–10.4	4–11 and 0.06–5.80	Tawfik et al. (2008), Sarkar et al. (2006)
12. Milk reception	0.8 and 2.54	7.18 and 0.65	Janczukowicz et al. (2008), Brião and Granhen Tavares (2007)
13. Whey processing effluent	0.59–1.21 and 1.07–2.18	10.37 and 3.82	Rivas et al. (2010), Chatzipaschali and Stamatis (2012)
14. Dairy and sewage (7:3)	1.08–2.81 and 2.04–4.73	9.1 ± 6.7 and 0.53–1.13	Tawfik et al. (2008)
15. Fluid milk	0.5–1.3 and 0.95–2.4	5–9.5 and 0.09–0.45	Demirel et al. (2005)
16. Yoghurt	– and 6.5	4.53 and –	Un and Ozel (2013)
17. Butter	0.22–2.65 and 8.93	12.08 and 0.7–5.07	Janczukowicz et al. (2008), Saddoud et al. (2007)

(continued)

Table 9.2 (continued)

Milk waste type	BOD (mg/L) and COD (mg/L) values respectively	pH and TSS (mg/L) values	References
18. Washing wastewater	3.47 and 14.64	10.37 and 3.82	Janczukowicz et al. (2008), Guimarães et al. (2010)
19. Cottage cheese whey	– and 79	4.5 and –	Carvalho et al. (2013), Najafpour et al. (2008)
20. Hard cheese whey	9.48 73.45	5.8 and 7.15	Janczukowicz et al. (2008), Beaulieu et al. (2006)

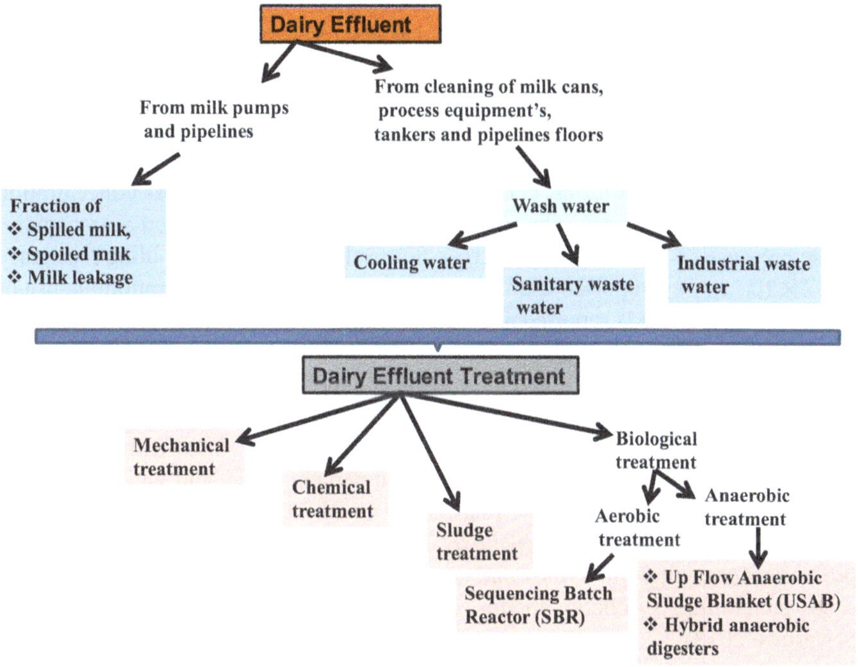

Fig. 9.3 Dairy wastes sources and its treatment methods (Birwal et al. 2017)

protein (CWP) is found for lactic acid production as low cost and labor saving methods. In lactic acid production has reported by application of *Lactobacillus bulgaricus* under optimized process condition. *Lactobacillus bulgaricus* CGMCC 1.6970 strain is reported for consumption of CWP substrate for lactic acid production without addition of any extra nutrients solution or substrates.

This microbial strain has ability for CWP substrate with addition of different type of proteases enzymes for CWP hydrolysis and it has shown highest lactose yield from neutral nature of proteases enzymes with pretreated CPW at an optimal condition. From batch fermentation processes, high lactose concentration

(71.0 g L^{-1}) with average productivity (1.5 g L^{-1} h^{-1}) is reported from CWP substrate and fed-batch fermentation is reported for better lactose concentration (113 g L^{-1}) with average productivity (2.4 g L^{-1} h^{-1}) at open condition without sterilization processes compared to batch data. This type of biological approaches for biosynthesis of lactic acid is shown as simple or more economic strategy or plan for CWP consumption or processes of valorization for lactic acid biosynthesis (Liu et al. 2018).

Membrane technology such as cross-flow and pressure mediated ultrafiltration (UF) is applied for whey fermentation using a bioreactor fermentation to synthesize lactic acid and it has been alignment with equipment of fitting of membrane filter of zirconium oxide stainless steel (ZOSS) for waste product removal tasks. Various types of filtration resistance factor (i.e. retardation factor ~ Rf and resistance index ~ RI values) for waste components has been studied for better understanding the interaction between membrane filter and fermentation feed solution.

Various waste nature raw materials (i.e. deproteinated form of synthetic whey (S1) deproteinated synthetic whey (S1), synthetic whey protein (S2) or dairy whey permeate (S3) components are utilized as feed solution for UF fractions tests under constant operating conditions. Permeate solution with highest lactose concentration (0.7%) with good permeate flux (5.5×10^{-6} m^3 m^{-2} s^{-1}) is reported at transmembrane pressure (2.1×10^3 Pa) with cross-flow velocity (1 m s^{-1}. Fouling resistance (6.87×10^{11} m^{-1}) is reported as minimal value with bioreactor operation via use of ZOSS membranes. It is reported as promise concept for production lactic acid with its filtration facility in the dairy industry (Wojtyniak et al. 2016).

Whey as the substrate is utilized for lactic acid production using batch fermentation process and *Lactobacillus delbruekii*-911 strain at optimum conditions (i.e. agitation rate (400 rpm), temperature (47 °C) or pH (5.5–6.0)) with addition of yeast extract as nitrogen source. Ninety-seven percent lactic acid yield is reported as maximum value with *Lactobacillus delbruekii*-911 as more advantageous microbial strain than *Lactobacillus helveticus*. Inoculum was propagated in fermentation processes containing milk whey and compared with propagation of culture in glucose containing synthetic media (Satyanarayana and Venkateshwar 2004; Srivastava 2018).

Assessment processes for microbial efficiency has done for the potential *Lactobacillus delbrueckii* spp. *bulgaricus* strain for dairy wastewater utilization and biomethane production via improved characteristics identification via nuclear magnetic resonance utility. Assessment study is done for waste metabolite presence in the unprocessed wastewater system and it is good sources of value added products (bioenergy) in bio refinery or other relevant compounds synthesis. Dairy wastes have more a potential for valuable nutrient source for utilization in fermentation processes *f*. The chemical oxygen demand (COD ~ 97%) or biological oxygen demand (BOD ~ 98%) are reported for decreased values and tested for one or two stages fermentation processes. Lactic acid bacteria are not added in one stage fermentation of wastewater containing system and reported methane production (up to 38% yield). In two stages fermentation of wastewater system, acetogenesis and methanogenesis processes is reported with *L. delbduekii* strain and 76% (W/Ws) of methane yield is reported during acetogenesis stage. *L. delbrueckii* is shown its potential for fermentation of dairy WWs and methane production. It has decreased the polluting effect from waste streams with reduction value in the chemical oxygen demand (0.20 g L^{-1}) and biological oxygen demand (0.03 g L^{-1}) (Juodeikiene et al. 2017).

Hydrogen and methane production is reported from diversity of microbial communities with intermediate valuable products synthesis. Culture independent techniques or methods are applied as tool for exploration of engineered microbial communities system for exhibition of new insights or information's from their structural and functional tasks study. It showed potential applications in various industries. Macrobiotic with anaerobic process is applied for specific waste and byproducts treatment (Yang et al. 2007).

Populations of microbial strains system with their biochemical reactions intermediate compounds and its final products are reported at large scale anaerobic digestion or degradation of mixture of mozzarella cheese, processed whey or buttermilk components. These are needed for amendment with industrial useful animal manure pellets (5%w/v). Culture independent techniques are employed for high-throughput sequencing for microbial enumerations for lactic acid bacteria family (*Lactobacillaceae* or *Streptococcaceae*) (Yang et al. 2007). Enhanced quantity of hydrogen biofuel production (10 mL H_2 g.VS^{-1} from 13, 50, 14 days periods) is reported and during incubation periods, a gradual decreasing number of lactic acid bacteria is detected with simultaneous increasing number of clostridia bacterial group (Clostridiacease and Tissierellaceae family). Archaeal bacterial populations (0.1%) in the biosystem are reported as inoculums with non-inoculated samples from dairy waste mixture. Methanogenic archaea (*Methanoculleus*) is reported for methane gas production via utilization of hydrogentrophic pathway with dairy waste compounds in the bioreactor (Yang et al. 2007; Yang et al. 2007; Pagliano et al. 2019) and few more examples of milk waste compound utilization is shown in Table 9.3.

9.6 Conclusions

Milk as nutritious food is reported as complete sets of nutrients for maintaining good health to young children, elders and old people and it is biosynthesized mainly from cow and buffalos other mammals. For making of different products of milk, milk needs proper collection, handling and processing processes. During milk processing steps, different type of milk byproducts or wastes are generated. Most of dairy industries, they dispose it to soil or water bodies sources which cause the different types of pollution problem (such as increased value of COD and BOD) which is main reason for threatening to water habitat living organism. Now due to more awareness of pollutions problem, dairy industries have started to proper processing of wastes compounds or byproducts and this steps has helped in reduction of environmental problem. Most of research has been discussed to utilize the waste

Table 9.3 Valuable products for milk waste compounds or byproducts utilization

Milk wastes	Valuable milk products	References
1. Milk whey (25%)	Papaya based beverage with whey incorporation with oBrix TSS (15.1%), acidity (0.30%), reducing sugar (5.37 g/100 g), total sugar (14.1 g/100 g) and ascorbic acid (5.60 mg/100 g) for ready-to-serve (RTS) food	Panghal et al. (2017)
2. Milk whey	Production of liquid whey protein concentrates by ultrafiltration followed by thermal denaturation and homogenization of the ultrafiltrated concentrate as well as on the production of ultrafiltrated permeates concentrated by reverse osmosis. It has been inoculated with kefir grains (fresh and thawed) and/or commercial probiotic bacteria to get fermentative products with lactococci and lactobacilli (7 log CFU/mL) and yeasts (6 log CFU/mL) after 14 days of refrigerated storage	Pereira et al. (2015)
3. Cheese whey wastewater	**Hydrogen** production is reported under thermophilic microbial communities anaerobic conditions	Azbar et al. (2009)
4. Dairy wastes	Methane production in the reactors	Pagliano et al. (2019)
5. Dairy wastewater (WWs)	Methane yield (76%) from *L. delbrueckii* with reduction in reduction value of chemical oxygen demand and biological oxygen demand in the two-stage fermentation	Juodeikiene et al. (2017)
6. Cheese whey powder	D-lactic acid production (113 g L^{-1}) is reported under optimized *Lactobacillus bulgaricus* CGMCC 1.6970 is utilized for CWP consumption via fed-batch fermentation	Liu et al. (2018)
7. Specific milk waste and byproducts treatment	Hydrogen and methane production from diversity of microbial communities	Yang et al. (2007)
8. Acid whey	n-Caproate production (134.3 mmol $CL^{-1}day^{-1}$ after 37 days) from Coriobacteriaceae, Ruminococcaceae and Prevotellaceae in in a UASB bioreactor	Duber et al. (2018)
9. Dairy industry anaerobic sludge	*Acinetobacter junii* AH4-A strain is used for bio-hydrogen production (concentration ~945.7 mL/L and H_2 yield ~1.35 mol H_2/mol glucose from dairy industry anaerobic sludge (10–100%)	Murugan et al. (2018)
10. Dairy industrial coproducts (cheese whey and whey permeate)	Yeast *Kluyveromyces marxianus* produced ethanol yield (η ~ 76%), with volumetric productivity (QP) (0.81 g L^{-1} h^{-1}) in 10 h in under fermentation anaerobic conditions and biomass production (11.48 g L^{-1}) and productivity (QX 0.75 g L^{-1} h^{-1}) in, in 14 h, with a specific growth rate (0.253 h^{-1}) with COD (86%) reduction at the end of the aerobic process	Murari et al. (2019)

compounds or byproducts for conversion into valuable bio-products such as ethanol or other biofuels, lactic acid and new products development. This information has been shown in tables. Number of processing steps has been applied for reduction of dairy wastes compounds concentration from wastewater source or soil. This chapter has focused all new sight for dairy wastes conversion into new and useful products via minimizing the pollution issues in our environment.

References

Allwyn Sundarraj A, Angeline Rajathi A, Vishaal SC, Rohit D, Saran Prakash M, Alexander Sam A, Seihenbalg SS (2018) Food biotechnology applications in dairy and dairy products. J Pharm Res 12(4):520–525. http://jprsolutions.info/files/final-file-5ac459ba01e 826.42478943.pdf

Arumugam A, Sabarethinam PL (2008) Performance of a three phase fluidized bed reactor with different support particles in treatment of dairy wastewater. ARPN J Eng Appl Sci 3(5):42–44. ISSN 1819-6608. doi:10.1.1.567.1560

Azbar N, Dokgöz FTÇ, Gundogdu TK, Korkmaz KS (2009) Continuous fermentative hydrogen production from cheese whey wastewater under thermophilic anaerobic conditions. Int J Hydrogen Energy 34(17):7441–7447. https://doi.org/10.1016/j.ijhydene.2009.04.032

Badem A, Uçar G (2017) Production of caseins and their usages. Int J Food Sci Nutr 2(1):4–9. ISSN: 2455-4898. www.academia.edu/30849441/Production_of_caseins_and_their_usages

Banaszewska A, Cruijssen F, van der Vorst JGAJ, Claassen GDH, Kampman JL (2013) A comprehensive dairy valorization modelJ. J Dairy Sci 96:761–779. https://doi.org/10.3168/jds.2012-5641. Epub 2012 Nov 29

Beaulieu J, Dupont C, Lemieux P (2006) Whey proteins and peptides: beneficial effects on immune health. Therapy 3:69–78. https://doi.org/10.1586/14750708.3.1.69

Bergamaschi M, Bittante G (2018) From milk to cheese: evolution of flavor fingerprint of milk, cream, curd, whey, ricotta, scotta, and ripened cheese obtained during summer Alpine pasture. J Dairy Sci 101(5):3918–3934. https://doi.org/10.3168/jds.2017-13573. Epub 2018 Feb 14

Birwal P, Deshmukh G, Priyanka SSP (2017) Advanced technologies for dairy effluent treatment. J Food Nutr Popul Health 1(1):7. http://www.imedpub.com/food-nutrition-and-population-health/

Brião VB, Granhen Tavares CR (2007) Effluent generation by the dairy industry: preventive attitudes and opportunities. Braz J Chem Eng 24(4):487–487. http://www.scielo.br/pdf/bjce/v24n4/a03v24n4.pdf

Cameron M, McMaster LD, Britz TJ (2009) Impact of ultrasound on dairy spoilage microbes and milk components. Dairy Sci Technol 89(1):83–98. https://doi.org/10.1051/dst/2008037

Carvalho F, Prazeres AR, Rivas J (2013) Cheese whey wastewater: characterization and treatment. Sci Total Environ 445–446:385–396. https://doi.org/10.1016/j.scitotenv.2012.12.038. Epub 2013 Jan 29

Chaiudhari DH, Dhoble RM (2010) Performance evaluation of effluent treatment plant of dairy industry. Curr World Environ 5(2):373–378. https://doi.org/10.12944/CWE.5.2.26

Chatzipaschali AA, Stamatis AG (2012) Biotechnological utilization with a focus on anaerobic treatment of cheese whey: current status and prospects. Energies 5(9):3492–3525. https://doi.org/10.3390/en5093492

Chegini GR, Taheri M (2013) Whey powder: process technology and physical properties: a review. Middle East J Sci Res 13(10):1377–1387. https://doi.org/10.5829/idosi.mejsr.2013.13.10.1239

Chiewchan N, Phungamngoen C, Siriwattanayothin S (2006) Effect of homogenizing pressure and sterilizing condition on quality of canned high fat coconut milk. J Food Eng 73(1):38–44. https://doi.org/10.1016/j.jfoodeng.2005.01.003

Cristian O (2010) Characteristics of the untreated wastewater produced by food industry Analele Universităţii din Oradea. Fascicula Protecţia Mediului 15:709–714. ISSN 1224-6255. https://www.cabdirect.org/cabdirect/abstract/20113176344

De Cortes-Sanchez AJ, Valle-Gonzalez ER, Salazar-Flores RD, Ashutosh S (2015) Biotechnological alternatives for the utilization of dairy industry waste products. Adv Biosci Biotechnol 6:223–235. https://doi.org/10.4236/abb.2015.63022

De Noni I, Battelli G (2008) Terpenes and fatty acid profiles of milk fat and "Bitto" cheese as affected by transhumance of cows on different mountain pastures. Food Chem 109:299–309. https://doi.org/10.1016/j.foodchem.2007.12.033. Epub 2008 Feb 1

Demirel B, Yenigun O, Onay TT (2005) Anaerobic treatment of dairy wastewaters: a review. Process Biochem 40:2583–2595. https://doi.org/10.1016/j.procbio.2004.12.015

Deosarkar SS, Khedkar CD, Kalyankar SD (2016) Butter: manufacture. In: Caballero B, Finglas P, Toldrá F (eds) The encyclopedia of food and health, vol 1. Academic Press, Oxford, pp 529–534
Deshannavar UB, Basavaraj RK, Naik NM (2012) High rate digestion of dairy industry effluent by upflow anaerobic fixed-bed reactor. J Chem Pharm Res 4(6):2895–2899. http://www.jocpr.com/articles/high-rate-digestion-of-dairy-industryeffluent-by-upflow-anaerobic-fixedbed-reactor.pdf
Deshpande DP, Patil PJ, Anekar SV (2012) Biomethanation of dairy waste. Res J Chem Sci 2(4):35–39. ISSN 2231-606X. https://www.semanticscholar.org/paper/Biomethanation-of-dairy-waste.-Deshpande-Patil/2773b9db0089bacda4df02759146567266a42cf
Dey A, De PS (2014) Influence of condensed tannins from *Ficus bengalensis* leaves on feed utilization, milk production and antioxidant status of crossbred cows. Asian-Australas J Anim Sci 27(3):342–348. https://doi.org/10.5713/ajas.2013.13295
Duber A, Jaroszynski L, Zagrodnik R, Chwialkowska J, Juzwa W, Ciesielskid S, Oleskowicz-Popiel P (2018) Exploiting the real wastewater potential for resource recovery – n-caproate production from acid whey. Green Chem 20:3790–3803. https://doi.org/10.1039/C8GC01759J
Early R (2012) Dairy products and milk-based food ingredients. In Baines D, Seal R (eds), Natural food additives, ingredients and flavourings, a volume in Woodhead Publishing series in food science, technology and nutrition, pp 417–445. https://doi.org/10.1533/9780857095725.2.417
Etzel MR (2004) Manufacture and use of dairy protein fractions. J Nutr 134(4):996S–1002S. https://doi.org/10.1093/jn/134.4.996S
Gebreselassie N, Abrahamsen RK, Beyene F, Abay F, Narvhus JA (2016) Chemical composition of naturally fermented buttermilk. Int J Dairy Technol 69(2):200–208. https://doi.org/10.1111/1471-0307.12236
Geilman WG, Schmitd D, Herfurth-Kennedy C, Path J, Cullor J (1992) Production of an electrolyte beverage from milk permeate. J Dairy Sci 75:2364–2369. https://doi.org/10.3168/jds.S0022-0302(92)77996-X
Ghosh S (2017) Thermosonication as an upcoming technology in the dairy industry: an overview. J Adv Dairy Res 5:189. https://doi.org/10.4172/2329-888X.1000189
Guimarães PMR, Teixeira JA, Domingues L (2010) Fermentation of lactose to bio-ethanol by yeasts as part of integrated solutions for the valorisation of cheese whey. Biotechnol Adv 28:375–384. https://doi.org/10.1016/j.biotechadv.2010.02.002. Epub 2010 Feb 11
Halliwell DJ, Barlow KM, Nash DM (2001) A review of the effects of wastewater sodium on soil physical properties and their implications for irrigation systems. Aust J Soil Res 39:1259–1267. https://doi.org/10.1071/SR00047
Henning DR, Baer RJ, Hassan AN, Dave R (2006) Major advances in concentrated and dry milk products, cheese, and milk fat-based spreads. J Dairy Sci 89(4):1179–1188. https://doi.org/10.3168/jds.S0022-0302(06)72187-7
Hossein H, Jalal S (2011) Optimizing OLR and HRT in a UASB reactor for pretreating high-strength municipal wastewater. Chem Eng Transact 24:1285–1290. https://doi.org/10.3303/CET1124215
Jacob M, Jaros D, Rohm H (2010) Recent advances in milk clotting enzymes. Int J Dairy Technol 63:1–18. https://doi.org/10.1111/j.1471-0307.2010.00633.x
Jana AH, Mandal PK (2011) Manufacturing and quality of mozzarella cheese: a review. Int J Dairy Sci 6(4):199–226. https://doi.org/10.3923/ijds.2011.199.226
Janczukowicz W, Zieliński M, Debowski M (2008) Biodegradability evaluation of dairy effluents originated in selected sections of dairy production. Bioresour Technol 99(10):4199–4205. https://doi.org/10.1016/j.biortech.2007.08.077
Juodeikiene G, Cizeikiene D, Glasner C, Bartkiene E, Dikiy A, Shumilina E, Ilić N, Berardino SD, Foncesa C (2017) Evaluation of the potential of utilizing lactic acid bacteria and dairy wastewaters for methane production. Energ Explor Exploit 35(3):388–402. https://doi.org/10.1177/0144598717698081

Kabbout R, Baroudi M, Dabboussi F, Halwani J, Taha S (2011) Characterization, physicochemical and biological treatment of sweet whey (major pollutant in dairy effluent). In: 2011 international conference on biology, environment and chemistry IPCBEE, vol 2. IACSIT Press, Singapore, pp 123–127

Kalyankar SD, Deshmukh MA, Khedkar CD, Deosarkar SS, Sarode AR (2016) Condensed milk. In: Caballero B, Finglas P, Toldrá F (eds) The encyclopedia of food and health, vol vol. 2. Academic Press, Oxford, pp 291–295

Karadag D, Köroğlu OE, Ozkaya B, Cakmakci M (2015) A review on anaerobic biofilm reactors for the treatment of dairy industry wastewater. Process Biochem 50:262–271. https://doi.org/10.1016/j.procbio.2014.11.005

Keffala C, Zouhir F, Abdallah KBH, Kammoun S (2017) Use of bacteria and yeast strains for dairy wastewater treatment. Int J Res Eng Technol 6(4):108–113. https://orbi.uliege.be/bitstream/2268/226062/1/IJRET20170603019.pdf

Kisielewska M, Dębowski M, Zieliński M, Krzemieniewski M (2018) Enhancement of dairy wastewater treatment in a combined anaerobic baffled and biofilm reactor with magneto-active packing media. J Ecol Eng 19(5):165–171. https://doi.org/10.12911/22998993/89816

Kocadagistan B, Kocadagistan E, Topcu N, Demircioglu N (2005) Wastewater treatment with combined upflow anaerobic fixed-bed and suspended aerobic reactor equipped with a membrane unit. Process Biochem 40:177–182. https://doi.org/10.1016/j.procbio.2003.11.055

Kushwaha JP, Srivastava VC, Mall ID (2011) An overview of various technologies for the treatment of dairy wastewaters. Food Sci Nutr 51:442–452. https://doi.org/10.1080/10408391003663879

Liu P, Zheng Z, Xu Q, Qian Z, Liu J, Ouyan J (2018) Valorization of dairy waste for enhanced D-lactic acid production at low cost. Process Biochem 71:18–22. https://doi.org/10.1016/j.procbio.2018.05.014

Lubary M, Hofland GW, ter Horst JH (2011) The potential of milk fat for the synthesis of valuable derivatives. Eur Food Res Technol 232(1):1–8. https://link.springer.com/article/10.1007/s00217-010-1387-3

Luo Y, Wang Y, Liu J, Lan H, Shao M, Yu Y, Quan F, Zhang Y (2015) Production of transgenic cattle highly expressing human serum albumin in milk by phiC31 integrase-mediated gene delivery. Transgenic Res 24(5):875–883. https://doi.org/10.1007/s11248-015-9898-0. Epub 2015 July 22

Mula J, Peidro D, Diaz-Madronero M, Vicens E (2010) Mathematical programming models for supply chain production and transport planning. Eur J Oper Res 204:377–390. https://doi.org/10.1016/j.ejor.2009.09.008

Murari CS, Niz da Silva DCM, Schuina GL, Mosinahti EF, Bianchi VLD (2019) Bioethanol production from dairy industrial coproducts. Bioenergy Res 12(1):112–122. https://doi.org/10.1007/s12155-018-9949-5

Murugan RS, Dinesh GH, TRA S, Boobalan T, Jothibasu M, Manimaran PS, Selvakumar G, Arun A (2018) *Acinetobacter junii* AH4-a potential strain for bio-hydrogen production from dairy industry anaerobic sludge. J Pure Appl Microbiol 12(4):1761–1769. https://doi.org/10.22207/JPAM.12.4.09

Najafpour GD, Hashemiyeh BA, Asadi M, Ghasemi MB (2008) Biological treatment of dairy wastewater in an upflow anaerobic sludge-fixed film bioreactor. Am Eurasian J Agric Environ Sci 4:251–257. ISSN 1818-6769. https://pdfs.semanticscholar.org/3851/86c3fd1eae864b2338b45a582dfca488792a.pdf

Nieuwenhuijse H (2011) Concentrated dairy products: sweetened condensed milk. In Fuquay JW, McSweeney PLH, Fox PF (eds), Encyclopedia of dairy sciences, 2nd edn, kindle edn. Academic Press, pp 869–873. https://doi.org/10.1016/B978-0-08-100596-5.00697-1

Nishimura J, Makino S, Kimura K, Isogai E, Saito T (2015) Influence of different sterilization conditions on the growth and exopolysaccharide of *Streptococcus thermophilus* and co-cultivation with *Lactobacillus delbrueckii* subsp. *bulgaricus* OLL1073R-1. Adv Microbiol 5:760–767. https://doi.org/10.4236/aim.2015.511080

Pagliano G, Ventorino V, Panico A, Romano I, Pirozzi F, Pepe O (2019) Anaerobic process for bioenergy recovery from dairy waste: meta-analysis and enumeration of microbial community related to intermediates production. Front Microbiol 9:3229. https://doi.org/10.3389/fmicb.2018.03229

Panghal A, Kumar V, Dhull SB, Gat Y, Chhikara N (2017) Utilization of dairy industry waste-whey in formulation of papaya RTS beverage. Curr Res Nutr Food Sci 5(2):168–174. https://doi.org/10.12944/CRNFSJ.5.2.14

Pereira C, Henriques M, Gomes D, Gomez-Zavaglia A, de Antoni G (2015) Novel functional whey-based drinks with great potential in the dairy industry. Food Technol Biotechnol 53(3):307–314. https://doi.org/10.17113/ftb.53.03.15.4043

Powar MM, Kore VS, Kore S, Kulkarni GS (2013) Review on applications of UASB technology for wastewater treatment. Int J Adv Sci Eng Technol 2(2):125–133. ISSN 2319-5924. https://pdfs.semanticscholar.org/65ac/3f1bf74400b0836cbb7c5f37392a5ca3ca4f.pdf

Qazi JI, Nadeem M, Baig SS, Baig S, Syed Q (2011) Anaerobic fixed film biotreatment of dairy wastewater. Middle-East J Sci Res 8(3):590–593. https://www.researchgate.net/publication/267688317_Anaerobic_Fixed_Film_Biotreatment_of_Dairy_Wastewater

Quasim W, Mane AV (2013) Characterization and treatment of selected food industrial effluents by coagulation and adsorption techniques. Water Res Ind 4:1–12. https://doi.org/10.1016/j.wri.2013.09.005

Raghunath BV, Punnagaiarasi A, Rajarajan G, Irshad A, Elango A, Mahesh Kumar G (2016) Impact of dairy effluent on environment—a review. In: Prashanthi M, Sundaram R (eds) Integrated waste management in India: status and future prospects for environmental sustainability. Springer International Publishing, Cham, pp 229–249. https://doi.org/10.1007/978-3-319-27228-3_22

Rivas J, Prazeres AR, Carvalho F, Beltrán F (2010) Treatment of cheese whey wastewater: combined coagulation-flocculation and aerobic biodegradation. Agric Food Chem 58(13):7871–7877. https://doi.org/10.1021/jf100602j

Saddoud A, Hassaïri I, Sayadi S (2007) Anaerobic membrane reactor with phase separation for the treatment of cheese whey. Bioresour Technol 98:2102–2108. https://doi.org/10.1016/j.biortech.2006.08.013

Sankar Ganesh P, Ramasamy EV (2007) Studies on treatment of low-strength effluents by UASB reactor and its application to dairy industry wash waters. Ind J Biotechnol 6:234–238. ISSN 0972-5849. http://hdl.handle.net/123456789/3012

Sarkar B, Chakrabarti PP, Vijaykumar A, Kale V (2006) Wastewater treatment in dairy industries—possibility of reuse. Desalination 195:141–152. https://doi.org/10.1016/j.desal.2005.11.015

Sathyamoorthy GL, Saseetharan MK (2012) Dairy wastewater treatment by anaerobic hybrid reactor—a study on the reactor performance and optimum percentage of inert media fill inside reactor. Res J Chem Environ 16(1):51–56. ISSN 0972-0626. https://www.tib.eu/en/search/id/BLSE%3ARN312609949/Dairy-Wastewater-Treatment-by-Anaerobic-Hybrid/

Satyanarayana D, Venkateshwar S (2004) Lactic acid production from dairy waste by fermentation using *Lactobacillus delbruekii*. Asian J Microbiol Biotechnol Environ Sci 6(1):139–140. http://www.envirobiotechjournals.com/article_abstract.php?aid=3924 &iid=142& jid=1

Sharma P, Patel H, Patel A (2015) Evaporated and sweetened condensed milks. In: Chandan RC, Kilara A, Shah NP (eds) Dairy processing and quality assurance. Wiley, pp 310–332. https://doi.org/10.1002/9781118810279.ch13

Shete MBS, Shinkar NP (2013) Comparative study of various treatments for dairy industry wastewater. IOSR J Eng (IOSRJEN) 3(8):42–47. ISSN 2250-3021. https://doi.org/10.9790/3021-03844247

Shi R, Xu H, Zhang Y (2011) Enhanced treatment of wastewater from the vitamin C biosynthesis industry using a UASB reactor supplemented with zero-valent iron. Environ Technol 32:1859–1865. https://doi.org/10.1080/09593330.2011.566583

Sichien M, Thienpont N,. Fredrick E, Trung Le T, Van Camp J, Dewettinck K (2009) Processing means for milk fat fractionation and production of functional compounds. In Corredig M (eds),

Dairy-derived ingredients food and nutraceutical uses. Woodhead Publishing Series in Food Science, Technology and Nutrition, pp 68–102. https://doi.org/10.1533/9781845697198.1.68
Slavov AK (2017) General characteristics and treatment possibilities of dairy wastewater—a review. Food Technol Biotechnol 55(1):14–28. https://doi.org/10.17113/ftb.55.01.17.4520
Srivastava RK (2018) Utilization of milk whey (sustainable substrate) in microbial biosynthesis of lactic acid. SFJ Ferm Micro Technol 1:2. http://scifedpublishers.com/fulltext/utilization-of-milk-whey-sustainable-substrate-in-microbial-biosynthesisof-lactic-acid/22434
Takeda Y, Shimada M, Ushida Y, Saito H, Iwamoto H, Okawa T (2015) Effects of sterilization process on the physicochemical and nutritional properties of liquid enteral formula. Food Sci Technol Res 21(4):573–581. https://doi.org/10.3136/fstr.21.573
Tawfik A, Sobhey M, Badawy M (2008) Treatment of a combined dairy and domestic wastewater in an up-flow anaerobic sludge blanket (UASB) reactor followed by activated sludge (AS system). Desalination 227:167–177. https://doi.org/10.1016/j.desal.2007.06.023
Tawfika A, Sobheyb M, Badawya M (2008) Treatment of a combined dairy and domestic wastewater in an up-flow anaerobic sludge blanket (UASB) reactor followed by activated sludge (AS system). Desalination 227:167–177. https://doi.org/10.1016/j.desal.2007.06.023
Thorning TK, Raben A, Tholstrup T, Soedamah-Muthu SS, Givens I, Astrup A (2016) Milk and dairy products: good or bad for human health? An assessment of the totality of scientific evidence. Food Nutr Res 60:32527. https://doi.org/10.3402/fnr.v60.32527
Tsachev T (1982) Dairy industry wastewater treatment. In: Industrial wastewater treatment. Sofia: State Publishing House Technique (in Bulgarian). pp 239–241
Un UT, Ozel E (2013) Electrocoagulation of yoghurt industry wastewater and the production of ceramic pigments from the sludge. Sep Purif Technol 120:386–391. https://doi.org/10.1016/j.seppur.2013.09.031
Vincent D, Elkins A, Condina MR, Ezernieks V, Rochfort S (2016) Quantitation and identification of intact major milk proteins for high-throughput LC-ESI-Q-TOF MS analyses. PLoS One 11(10):e0163471
Watkins M, Nash D (2010) Dairy factory wastewaters, their use on land and possible environmental impacts—a mini review. Open Agric J 4:1–9. https://doi.org/10.2174/1874331501004010001
Wojtyniak B, Kołodziejczyk J, Szaniawsk D (2016) Production of lactic acid by ultrafiltration of fermented whey obtained in bioreactor equipped with ZOSS membrane. Chem Eng J 305:28–36. https://doi.org/10.1016/j.cej.2016.01.048
Yang P, Zhang R, McGarvey JA, Benemann JR (2007) Biohydrogen production from cheese processing wastewater by anaerobic fermentation using mixed microbial communities. Int J Hydrogen Energy 32:4761–4771. https://doi.org/10.1016/j.ijhydene.2007.07.038

Chapter 10
Role of Polyelectrolytes in the Treatment of Water and Wastewater

Shagufta Jabin and J. K. Kapoor

Contents

S. Jabin (✉)
Faculty of Engineering and Technology, Department of Chemistry, Manav Rachna International Institute of Research and Studies, Faridabad, Haryana, India

J. K. Kapoor (✉)
Department of Chemistry, National Institute of Technology, Kurukshetra, Haryana, India

Inamuddin, A. M. Asiri (eds.), *Sustainable Green Chemical Processes and their Allied Applications*, Nanotechnology in the Life Sciences,
https://doi.org/10.1007/978-3-030-42284-4_10

10.1 Introduction

There are problems which many countries are facing irrespective of their economic system and level of development such as environmental development, rapid deforestation, land degradation, different types of pollution, water shortage and contamination of water are common problems to Asia, Europe and America. The water related problems are caused by growing world population which consumes large amount of water and produces large amount of pollutants. In order to provide clean water, it has to be purifies before getting released to the environment (Schmidt and Morche 2006).

As per literature, "21 % of the people do not have access to safe water" (Cairncross and Petach 2013). Due to over population, availability of quality water will decrease in the coming century. UN Report indicates that "2.6 billion people are still suffering from poor quality water around the world" (United Nation Report 2010). According to literature, the main sources of water are rain water, rivers, ponds and unprotected and uncovered wells. It was further noted that problem is similar for both rural and urban; since rural people have to deal with bad quality of water from different sources mentioned above, whereas urban population has to deal with industrial wastewater pollution. Water related health problems are growing and according to WHO report in 2004, "it kills more than 5 million people in a year with infants mostly affected" (WHO/UNICEF Joint Monitoring Programme for Water Supply and Sanitation 2004). This figure is more as compared to those killed in wars and other natural disasters. The problem is of great significance in developing countries like India where polluted water, water shortage and unsanitary conditions prevail. WHO/UNICEF said in 2004 that "Access to water has improved greatly but access to safe water is still a major issue (WHO/UNICEF Joint Monitoring Programme for Water Supply and Sanitation 2004). To meet the 2015 target of united nation millennium development goals (MDGS) on access to safe drinking water, countries need awareness and well researched plan to manage water quality in rivers, ponds, canals, ground water and industrial wastewater. Most of contaminants in waste water can be related to particles since it is absorbed or incorporated in particulate matter (Akpor 2011)

10.2 Characteristics of Suspension

The character of water and wastewater impurities play important role in deciding the selection process (Moussas and Zouboulis 2009). The literature data reveals that most of the impurities are associated with suspension (Franceschi et al. 2002). So it is very necessary to study the details of suspension in water and waste water. As we all know that water is containing different types of impurities. Water impurities range from suspended matter to dissolved matter.

Suspended matter can be both organic and inorganic in nature. The fine fractions of suspended solids carry significantly high pollutant loads as compared to other water pollutants (James et al. 2003). Thus, high concentration of suspended solids can contribute to various adverse effects such as high turbidity and high total suspended solids (James and Ebeling 2004). Turbidity is one of the most unaesthetic parts of any type of water; whether it is an industrial water, ground water or surface water. The variation in the turbidity of water at different levels makes it difficult to handle. Many other pollutants have strong affinity to TSS and therefore such other pollutants present in water are removed with TSS (Rout et al. 1999). The literature data reveals that most of the impurities are associated with suspension (Beckett and Le 1990). So it is very necessary to study the details of suspension in different kind of water. The particulate matters have been categorized and the whole range of impurities was divided in three different categories:

- Settleable suspended matter
- Non settleable suspended matter
- Dissolved matter

The size of these matters is as follows:

- Soluble matter: 0–0.08 μm
- Non settleable suspended matter: 0.08–100 μm
- Settleable suspended matter >100 μm

Later on non settleable matters have been further classified by researchers into two categories.

- Supra-colloidal
- Colloidal

Odegard demonstrated that chemical oxygen demand (COD) is having only 25% dissolved matter where as others are included in suspension (Odegard 1987). Munch et al. in 1980 discussed and summarized the findings of different scientists and classified the contaminants according to their size and the same has been tabulated in Table 10.1 (Munch et al. 1980).

Table 10.1 Classification of contaminants in different water

	Classification			
Parameters	Dissolved (<0.025 μm)	Colloidal (0.025–3 μm)	Supra-colloidal (3–106 μm)	Settleable > 106 μm
BOD (%)	17	16	46	21
COD (%)	12	15	30	43
Total phosphorus (%)	63	3	12	22
Turbidity (%)	7	62	10	21
TSS (%)	–	–	52	48

Table 10.2 Average water composition given by scientist "Nieuwenhuijzen" in 2004

	Fractions					
Parameters	Dissolved	Supra-dissolved	Colloidal	Supra-colloidal	Suspended	Settleable
TSS (%)	–	–	–	–	52(±18)	48(±18)
Turbidity (%)	–	–	7(±4)	10(±7)	62(±17)	21(±12)
BOD (%)	48(±12)	–	–	14(±6)	30(±8)	8(±4)
COD (%)	36(±10)	3(±4)	2(±2)	11(±6)	27(±11)	21(±9)
N_{total} (%)	83(±25)	1(±1)	3(±3)	4(±2)	5(±3)	4(±4)
P_{total} (%)	53(±18)	3(±3)	3(±1)	5(±2)	30(±12)	6(±3)

The results demonstrate that phosphorus is mostly present in water in soluble form, whereas all other impurities such as chemical oxygen demand (COD), biological oxygen demand (BOD), turbidity and total suspended solids (TSS) are more significantly present in suspension form. Later on, Nieuwenhuijzen in 2004 categorized the different impurities according to their size (Nieuwenhuijzen et al. 2004).

It can be seen from the Table 10.2 that total phosphorus are mostly present in dissolved form; whereas TSS, turbidity, COD and BOD are mostly present in settleable and colloidal suspension form.

So it is very necessary to remove the suspension from different kinds of water, whether it is settleable suspended matter or non settleable suspended matter. Coagulation-flocculation is an important physicochemical operation for removal of suspension (Abu Hassan et al. 2009; Zainol et al. 2011) The mechanism of coagulation flocculation is the subject of continual review. The use of inorganic coagulants has its own limitations. A large amount of coagulants is required and it generates a large volume of sludge. In contrast a proper use of polyelectrolytes can remove pollutants from different kind of wastewater successfully.

10.3 Chemistry of Polyelectrolytes

Polyelectrolytes are those polymers which are water soluble and carrying ionic charge. Organic polyelectrolytes were introduced into environment engineering because of its potentials. Polyelectrolytes are used successfully in water and wastewater treatment (Lee et al. 1998).

10.3.1 Properties of Polyelectrolytes

Polyelectrolytes show different kind of physical, chemical and mechanical characteristics due to chemical structure, geometrical structure, molecular weight and charge density (Vajihinejad et al. 2019)

10.3.1.1 Chemical Structure

Chemical structure of polyelectrolytes depends upon number of monomer units, homochain and heterochain polymers. This parameter relates to three dimensional aggregated structure and influences characters of polyelectrolytes.

10.3.1.2 Geometrical Structure

Geometrical structure depends upon spatial arrangement of monomeric units with respect to each other. Several configurations are possible in polymer structure e.g. linear, branched and crossed linked structure.

Functional group is an important part of the structure of polyelectrolytes. Functional group of polyelectrolytes along with its charge has been shown in Fig. 10.1. They have charged sites with mobile counter ions. Anionic groups include carboxylate and sulphonate, while cationic group is quaternary amine. The structure of amine is similar to NH_3 but is protonated to give NH_4^+ but with R group instead of H attached to nitrogen. R is any hydrocarbon.

10.3.1.3 Molecular Weight

Mechanical properties of polyelectrolytes increase with increase in molecular weight in non linear fashion. High molecular weight gives increased mechanical strength to polyelectrolytes. Molecular weight of polyelectrolytes is in the order of a few thousand to tens of millions. Polyelectrolytes are considered as low, medium and high molecular weight if their molecular weight happens to be $<10^5$, 10^5–10^6 and $>10^6$ Dalton respectively. According to Gregory and Bolto, polymers generally

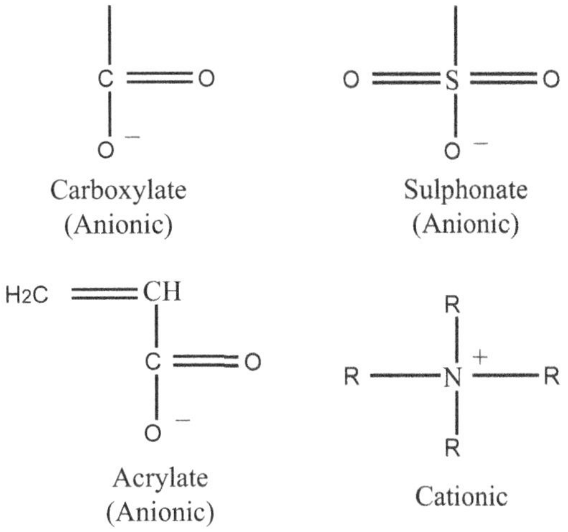

Fig. 10.1 Functional group of polyelectrolytes with anionic and cationic chargc

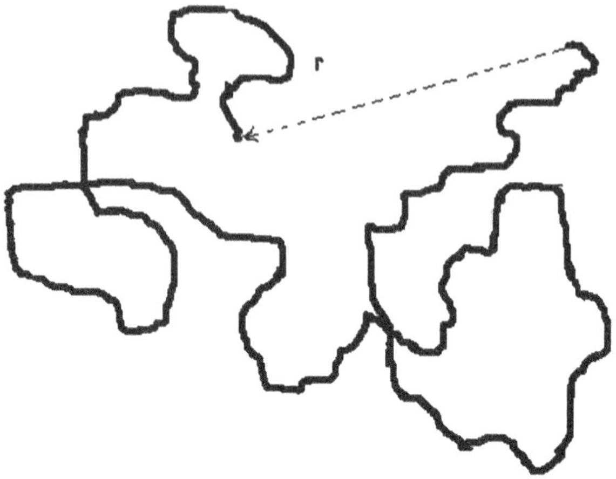

Fig. 10.2 Random coil configuration of a polymer chain with distance '*r*'. Modified after Bolto and Gregory (2007)

have a random coil configuration in aqueous solution as shown in Fig. 10.2 (Bolto and Gregory 2007). The size of coil is found to be in the range of a few hundred nm in case of very high molecular weight polyelectrolytes. The size is proportional to square root of molecular weight. The diameter of the coil of a polyelectrolyte is taken as root mean square value of distance r measured end to end as shown in Fig. 10.2. The end to end distance "*r*" can be in the range of 10 μm or more if polymer chain is fully stretched. But fully stretched polymer chain configuration is seldom found, as the most common configuration is the random coil configuration.

The size of random coil in a polyelectrolyte is dependent upon the kind of interaction between polymer segments. Sufficient repulsion takes place between different segments then the coil expands. This is particularly appreciable where the segments of polyelectrolytes are charged. The polymer coils get expanded due to ionic strength effects in such cases. However, the repulsion between the charged segments gets reduced by ions in solution at high ionic strength, and the coil expansion also gets correspondingly reduced. Repulsion between coil segments becomes more pronounced as the salt concentration is reduced, and the polymer coil acquires expanded configuration as shown in Fig. 10.3.

As per Bolto, the volume of a polyelectrolyte chain in random coil configuration is much larger than the same number of isolated monomer units taken together (Bolto and Gregory 2007). It is due to space occupied by aqueous solution in the coil. That's why polyelectrolyte solution shows higher viscosity than water. This effect can be used to estimate molecular weight of polyelectrolytes. Effectiveness of flocculants depends upon their molecular weight to a large extent (Owen et al. 2008).

When polyelectrolyte solution encounters high shear stain, the polymer chain gets significantly extended. In the process of extension it can get ruptured at the center of the chain where strain happens to be the maximum, causing reduction in molecular weight. This causes considerable low value of viscosity, which can be used for flocculation under condition of turbulent (Tripathy and Rajan De 2006).

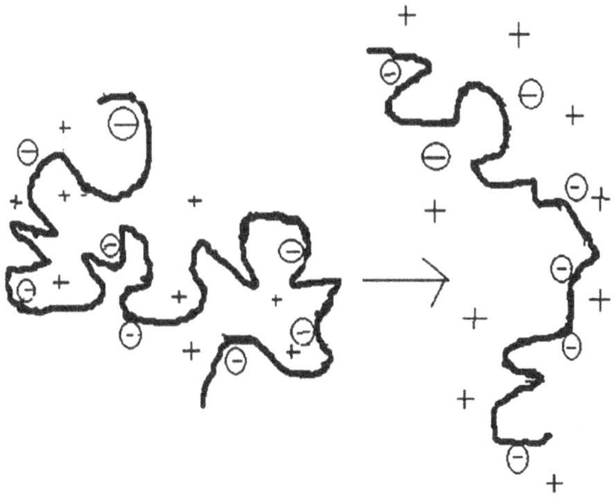

Fig. 10.3 Chain of Polyelectrolyte expansion with decrease in ionic strength. Modified after Bolto and Gregory (2007)

Low molecular weight polyelectrolytes generally react with soluble organics matter, whereas high molecular weight polyelectrolytes at low dosages preferentially react with suspended matter.

10.3.1.4 Crystallinity

High crystallinity is generally associated with those polyelectrolytes which have simple unit structure and exhibit a high degree of long range molecular weight. Further heat, UV and other radiations may affect the polymer properties by changing physical and chemical nature of polyelectrolytes. If effect is undesirable, then degradation is commonly used which may cause additional cross linking substitution, hydrolysis leading to different types of polyelectrolytes. Polyelectrolytes having aromatic groups are more resistant to chemical attack. Thermal as well as light energy alone or in the presence of oxygen or moisture can cause bond scission. Polyelectrolyte containing –CN, –CO, –C=C–, –OH and –C–Cl may easily be degraded by light or heat, whereas polyelectrolytes with aromatic backbone show excellent thermal stability owing to inertness (Buchhammer et al. 2000)

10.3.1.5 Charge Density

Charge density is measured experimentally by colloid titration method (Yu and Somasundaran 1996). Odegard demonstrated that "charge density is more important factor than molecular weight in selecting the polyelectrolyte" (Odegaard 1987). Charge density is measured as mole percent of charged groups i.e. milli equivalent per gram (meq/g). According to Gregory AND Bolto, charge density is classified in the following ways as shown in Table 10.3.

Table 10.3 Range of charge density of polyelectrolytes (Bolto and Gregory 2007)

Mol% of ionic group	Range of charge density
<25%	Low
25–50%	Medium
50–100%	High

10.3.2 Classification

According to their ionic charge, polyelectrolytes have been classified into four different categories. Cationic and anionic polyelectrolytes are more commonly used in water treatment. Polymers with no charged group are known as non-ionic polymers. Literature review reveals that polyampholytes have very little utility for water treatment (Bratby 2006). According to literature, cationic polyelectrolytes have been most widely used as primary coagulants, however non-ionic and anionic polyelectrolytes are mostly used as flocculants (Bratby 2006).

10.3.2.1 Synthetic Cationic Polyelectrolytes

Different types of cationic polyelectrolytes are available commercially. Most of them have been reviewed by Bolto and Gregory (Bolto and Gregory 2007). Strong polyelectrolytes have positive charge irrespective of pH (Bing-Hui et al. 2005). Polyelectrolytes are available in synthetic and natural forms.

Polydiallyl Dimethyl Ammonium Chloride

It is commonly referred as poly DADMAC. It is obtained from polymerization of diallyldimethyl ammonium chloride. Its molecular weight is low to medium and consists of five membered pyrrolidinium moiety (Choi et al. 2001). It is water soluble polyelectrolyte and has high charge density. The structure of PDADMAC is shown in Fig. 10.4.

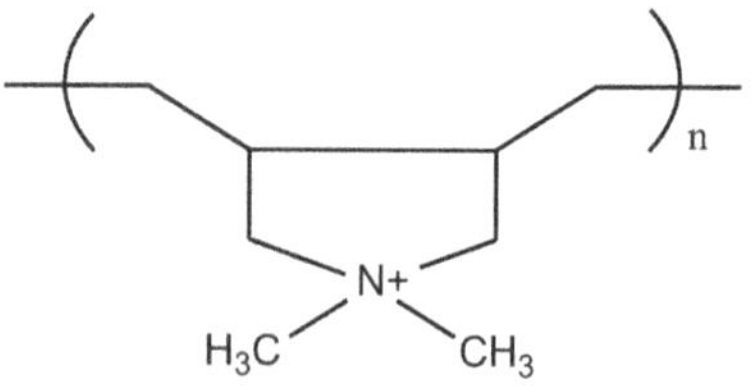

Fig. 10.4 Repeating unit of PDADMAC

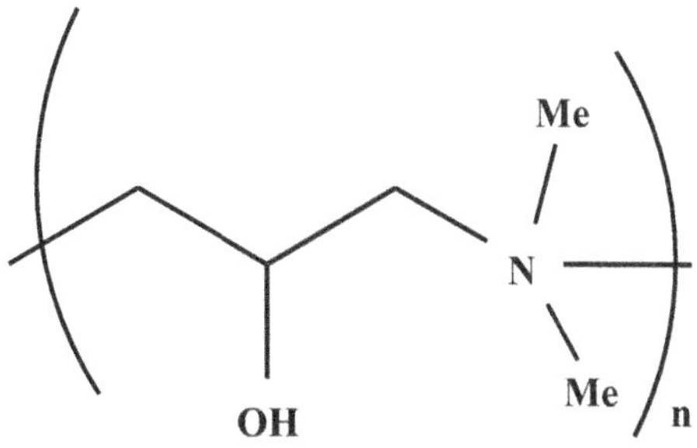

Fig. 10.5 Repeating unit of polyamines

Polyamines

Polyamines are those polyelectrolytes which are prepared by polymerization of epichlorohydrin with primary or secondary amine. Polyelectrolytes generally have low molecular weight and high charge density where all the nitrogen are in the quaternary ammonium form and located within the main chain as shown in Fig. 10.5 (Duk Jong et al. 2003).

Cationic Polyacrylamide

Cationic polyacrylamide is heterogenic group of polymer obtained from modification of polyacrylamide or co-polymerization of dimethylaminoethyl acrylate or methacrylate as shown in Fig. 10.6. The cationic part in cationic PAMS is roughly in the range of 10–80% (Rinaudo 2006).

It should be noted that copolymers of acrylamide may lose their cationic group in a hydrolysis of ester group taking place under alkaline condition roughly above pH 6.

$$\mathrm{Pol-COOCH_2CH_2N\,Me_3^{+} + OH^{-} \rightarrow Pol-COO^{-} + HO-CH_2CH_2NMe_3^{+}}$$

The hydrolysis has been found to be dependent on pH and charge density.

10.3.2.2 Natural Cationic Polyelectrolytes

Many polyelectrolytes are found in nature. Some of them inherit cationic properties while others are modified to produce cationic polyelectrolyte. Chitosan is the most popular amongst them. Chitosan is biopolymer produced from chitin, the water insoluble structure polymer. It is a low to medium molecular weight hetero polymer consisting of *N*-acetyl glucose amine or weakly charged glucose amine obtained from full or partial deacetylation of *N*-acetyl glucose amine groups of chitin (Guibal et al. 2006). The commercial product is having medium molecular weight and

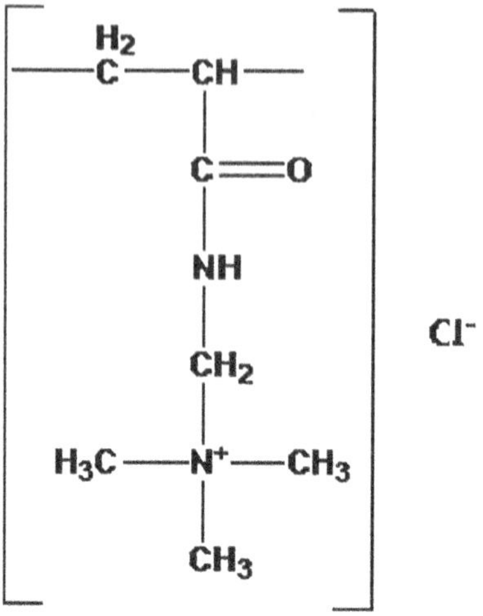

Fig. 10.6 Structure of cationic polyacrylamide (CPAM)

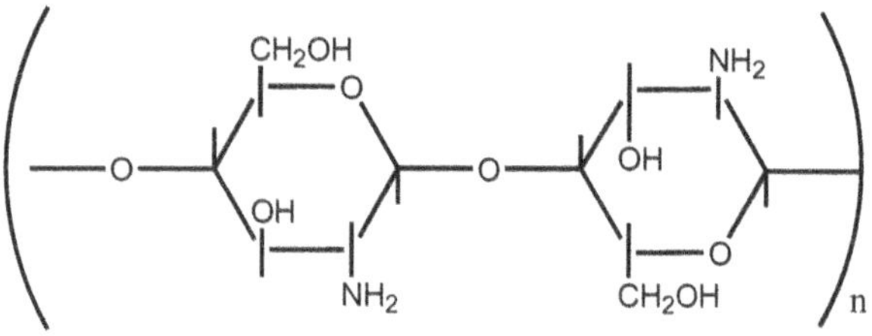

Fig. 10.7 Repeating unit of Chitosan

charge density that is pH dependent. Chitosan is having significance role in water purification and reviewed on a large scale (Abd-Elhakeem and Alkhulaqi 2014; Ting et al. 2019). At pH 7 the amino groups are un-protonated rendering the molecule insoluble (the pK_a value of amino group in chitosan is 6.6). Being non toxic in nature, chitosan can also be safely used as a primary coagulant in water treatment (Caulfield et al. 2003). The structure of chitosan is given in Fig. 10.7. The charge densities of cationic polyelectrolytes have been tabulated in Table 10.4.

10.3.2.3 Synthetic Anionic Polyelectrolytes

The most popular anionic polyelectrolyte contains weak carboxylic acid group, so its charge density depends upon pH (Lourenço et al. 2018).

Anionic Polyacrylamide (PAM)

High molecular weight carboxylic acid polymers derived from polyacrylamide are used as flocculant in water industries. Copolymers with anionic structure are prepared by copolymerization of acrylic acid and acrylamide or its salts, or by polym-

Table 10.4 Formulation and charge density of some cationic polyelectrolytes

Mol. formula	Charge density (mol%)	Charge density (meq/g)
$C_8H_{16}NCl$	100	6.2
$C_5H_{12}ONCl$	100	7.3
$C_8H_{16}O_2NCl$	100	5.2
$(C_8H_{16}O_2NCl)_{0.5}(C_3H_5ON)_{0.5}$	50	3.8
$(C_8H_{16}O_2NCl)_{0.5}(C_3H_5ON)_{0.5}$	25	2.5
Chitosan	100	5.2

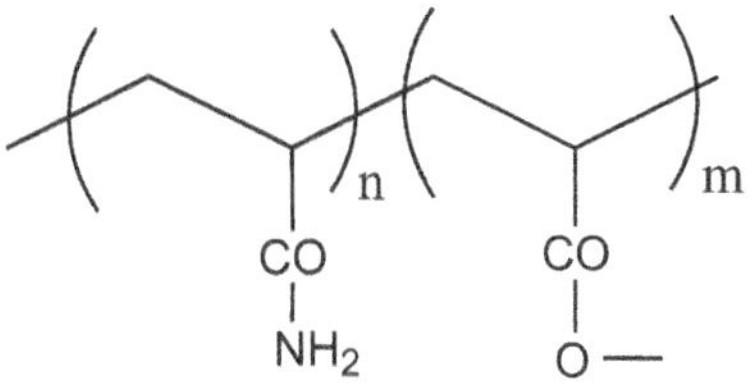

Fig. 10.8 Repeating unit of anionic PAM

erization of acryl amide followed by partial hydrolysis as shown in Fig. 10.8 (Haydar and Aziz 2009). The former method provides random copolymers, while the latter method gives some clusters of anionic groups.

10.3.2.4 Natural Anionic Polyelectrolytes

Many sulphated polysaccharides are available as their derivatives. But their applications are mainly restricted to medical purpose only. Tannin has received attention but it is also pH dependent (Bache and Gregory 2007). It works at pH more than 12 only.

Charge density of anionic polyelectrolytes has been tabulated in Table 10.5.

10.3.2.5 Non-ionic Polyelectrolytes

Natural non-ionic polyelectrolytes are starch, cellulose derivatives, gelatins and glues. Their structure, molecular weight, biodegradability and ease of dissolution are found to be different. Synthetic polyelectrolyte namely non-ionic PAM has also been used in water treatment (Selvapathy and Reddy 1992). It consists of 1–3% anionic group as in case of PAM, where they are prepared by hydrolysis of amide groups under controlled conditions. As per literature survey, the most reliable method of finding the degree of hydrolysis of concentrated PAM sample is ^{13}C NMR spectroscopy (Teh et al. 2016).

Table 10.5 Formulation and charge density of some anionic polyelectrolytes (Bolto and Gregory 2007)

Mol. formula	Charge density (mol%)	Charge density (meq/g)
$C_3H_3O_2Na$	100	10.6
$(C_3H_3O_2Na)_{0.75}(C_3H_5ON)_{0.25}$	75	8.5
$(C_3H_3O_2Na)_{0.50}(C_3H_5ON)_{0.50}$	50	6.1
$(C_3H_3O_2Na)_{0.25}(C_3H_5ON)_{0.75}$	25	3.3
$(C_3H_3O_2Na)_{0.1}(C_3H_5ON)_{0.9}$	10	1.4

10.3.3 Polyelectrolyte Selection

The selection of type of polyelectrolyte for use as coagulant and flocculant aid is generally an important task. Due to the great variation different raw water types and difference in experimental protocols, it is very challenging to predict the type of polyelectrolyte to be used in a particular kind of water. Confirmation of type of polyelectrolytes and its dosages can only be determined by jar tests (Song et al. 2004). There are complex inter-relations involving polyelectrolyte structure, charge density, dosage, mixing conditions, quantum and category of impurities and organic matter. The charge density and molecular weight are two important factors which affect the rate of adsorption (Mende et al. 2007). Of these two, charge density is crucial in obtaining the optimum dosage while charge neutralization mechanism prevails (Lee et al. 2014). High molecular weight is essential factor when reaction occurs by polymer bridging. Performance optimization using coagulation-flocculation method for the treatment of industrial wastewater has been reported in several studies, which inter-alia includes finding the most suitable polyelectrolyte, appropriate conditions for experiment, suitable range of pH and adsorption of proper flocculation mechanism (Cho et al. 2006). Water can be treated effectively by utilizing natural coagulants with low stress on environment.

10.4 Flocculation Mechanism of Polyelectrolytes

The flocculation mechanism of particles can be broadly divided into two categories: polymer bridging and charge neutralization.

10.4.1 Bridging Mechanism

A small dosage of long chain polymers added to a suspension of colloidal particles adsorb on to them in such a way that each chain gets attached to two or more particles forming a bridge as shown in Fig. 10.9a. This mechanism is called polymer

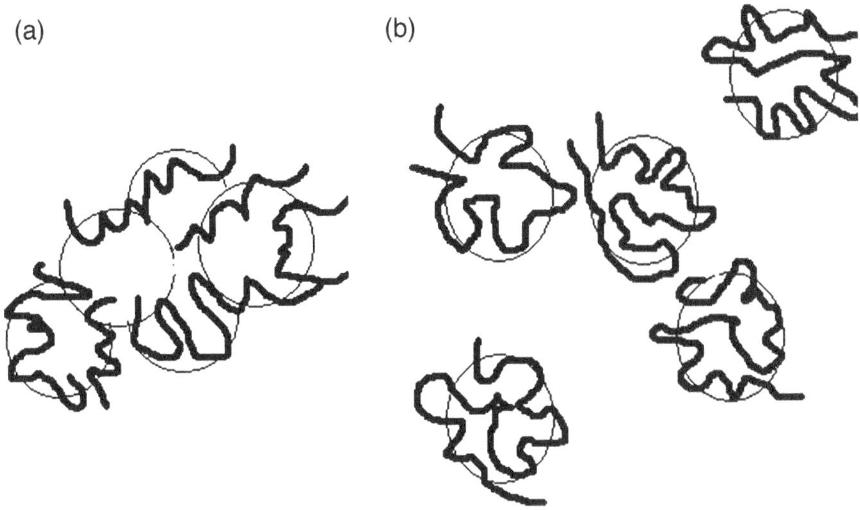

Fig. 10.9 Schematic diagram of (**a**) polymer bridging (**b**) restabilzation by adsorbed polymer chain. Modified after Bolto and Gregory (2007)

bridging. Interestingly, flocculation by bridging takes place up to a certain dosage of polymer only, which starts diminishing thereafter. This phenomenon is called steric stabilization. Fundamental requirement for polymer bridging to start is that there has to be enough vacant space on particles surface so that chains of polymer segments get attached to the particles and that the span of each pair of particles should be such as to promote inter particle repulsion (Bolto and Gregory 2007). On the other hand adsorbed amount should be low, otherwise surface of particle will be completely covered and insufficient adsorption site will be available. In this case, particles get restablize (Fig. 10.9b). The aggregates (flocs) formed by bridge flocculation is much stronger as compared to the ones formed by the addition of salts. Whereas, stronger flocs are produced by the bridging mechanism but if it is broken, it may not re-form again (Hjorth and Jorgensen 2012).

10.4.2 Electrostatic Patch Mechanism

It has been observed that in case of flocculants with high cationic charges in anionic colloidal suspensions, the involvement of high energy of attraction supports a flattened adsorbed surface which considerably diminishes the chance of formation of trains and loops to bridge the suspended particles (Wang et al. 2015). In this circumstance, each charged location on the particles surface will not be neutralized by oppositely charged polymer segment due to geometric restriction, which results in polymer chains to adsorb to produce "patches" of charge surrounded by areas of opposite charge as shown in Fig. 10.10. This gives rise to strong attraction between oppositely charged adsorbed patches and surrounded areas (Narkis et al. 1991).

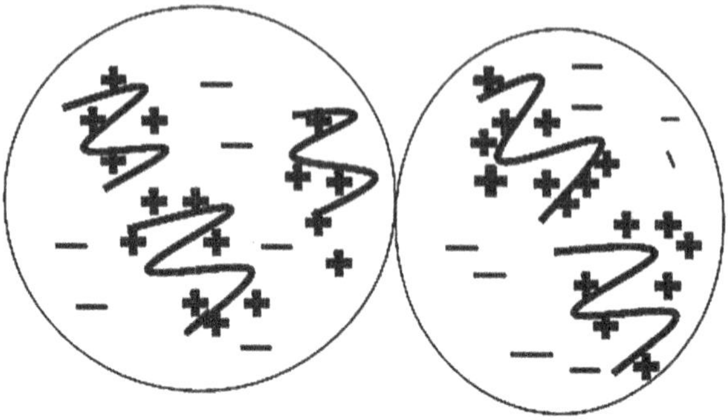

Fig. 10.10 Electrostatic patch model for flocculation. Modified after Bolto and Gregory (2007)

It is evident from above that at low particle concentration with high cationic charge on the flocculants and anionic colloidal solids, electrostatic patch mechanism would prevail. However, at high particle concentration for collisions to take place on a time scale similar to the one required for the polymer to attain a flattened configuration, bridging mechanism would be predominant (Kozlova and Santore 2006). Therefore, controlling mechanism of flocculation is determined by flocculants concentration.

Dosing and mixing prominently affects the level of flocculation. It has been found that flocculation takes place rapidly at high solid concentration and low polymer dosage, but the flocs formed by this way are not stable and get broken even at moderate stirring. So the floc size can be maintained by decreasing the rate of stirring shortly after polymer dosing. This is however, extremely difficult in practice due to precise control required. It has been observed that optimum flocculation takes place when half the area is covered with polyelectrolytes (Petzold et al. 2003). As the concentration increases from this level, the degree of flocculation decreases as the particles may be completely covered by the adsorbed polymer layer. Therefore overdosing may result in creation of a well-established suspension, which is not desirable. However in principle a sufficient level of flocculation can be achieved with much lower polymer dose.

10.5 Kinetic Aspects of Flocculation

Addition of polyelectrolyte in water containing suspended particles gives rise to a series of processes. These processes are discussed hereinafter:

10.5.1 Mixing

This is the first process wherein polymer is distributed evenly throughout the suspension (Gregory and Guibai 1991). The process of mixing is generally carried out swiftly else extra concentration may cause uneven adsorption and some particles

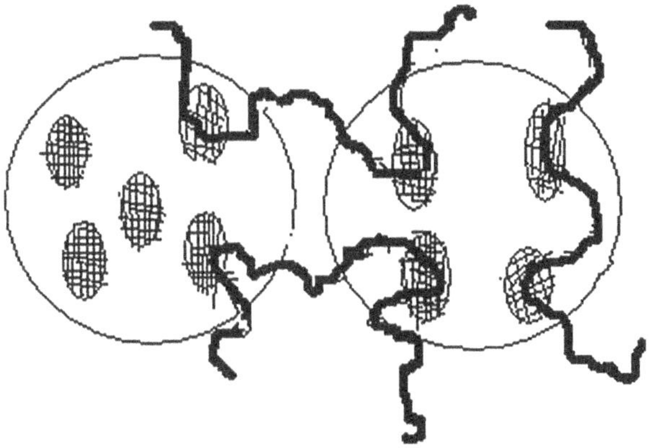

Fig. 10.11 Mode of flocculation and adsorption by anionic polymer on particle with cationic patches (dual polymer system). Modified after Bolto and Gregory (2007)

may re-establish as a result of adsorbing more quantity of polyelectrolyte. Improper mixing generally results in cloudy appearance in water after polymer flocculation (Oruc and Sabah 2006). It is desirable to have considerable amount of turbulence for proper mixing, although polymer chains may encounter some scission in this condition. Rapid mixing generally has direct influence on the flocculation process. Mixing effect is particularly more pronounced for concentrated suspension (Mende et al. 2007). Figure 10.11 shows mode of flocculation and adsorption by anionic polymer on particle with cationic patches.

10.5.2 Adsorption

Concentration is the determining factor for the rate of attachment of polymer chains to particles. As the particle concentration increases, so the optimum polymer dosage will also increase. Diffusion is the determining parameter for adsorption in case of low molecular weight polymers, however collisions between polymer molecules and particles play a major role in case of high molecular weight polymers. Mixing conditions determine the rate of polymer adsorption. The time required for adsorption of polymers can be several minutes for low turbid water, while it will be less than 1 s for high turbid water.

10.5.3 Rearrangement of Adsorbed Chain

A polymer chain adsorbed on a particle is required to be rearranged. A polymer chain of adsorbed configuration attains equilibrium with distinguished loops, trains and tails distribution as shown in Fig. 10.12. Rearrangement time of adsorbed

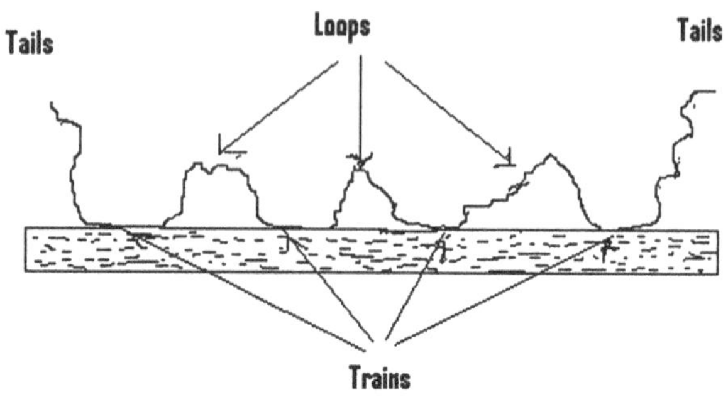

Fig. 10.12 A model of adsorbed polymer chain. Modified after Bolto and Gregory (2007)

chains depend upon several factors, although little information is available in literature about rearrangement of adsorbed chain (Pelssers et al. 1990). It takes several seconds for rearrangement in case of high molecular weight polymers. During the period of rearrangement adsorbed polymers have more extended configuration than the one in final equilibrium state, thus easily forming bridge contacts. A kinetic model with active and inactive adsorbed polymer chains has been developed by Pelssers et al. (1990). In this model active polymer chain has extended configuration, while inactive one is flattened on the surface. Flocculation by a high molecular weight polymer gets considerably enhanced if particles pre-adsorb low molecular weight polymer to cover around 50% of the site. This may be possible because pre-adsorbed polymer captures several surface sites making them unavailable to the second polymer, which in turn facilitates the second polymer to acquire a more enlarged adsorbed configuration. A model of polymer chain with adsorption is shown in Fig. 10.12.

10.5.4 Flocculation

When particles have acquired enough adsorption of polymer to become destabilized then collision results in attachment, either by bridging or by electrostatic mechanism. For high solid concentration, flocculation rates may become very high. All flocculation processes are done under applied shear in the tank (Crittenden et al. 2005). In case of high molecular weight polymers, particle collision time is shorter than adsorption time. It generally occurs in more concentrated suspension. Bridging interaction most likely gives the possibility of non equilibrium flocculation. In case of cationic polyelectrolytes and negative charged particles, bridging may occur however possibility of electrostatic patch flocculation would be more if the adsorbed chains are in a flatter configuration

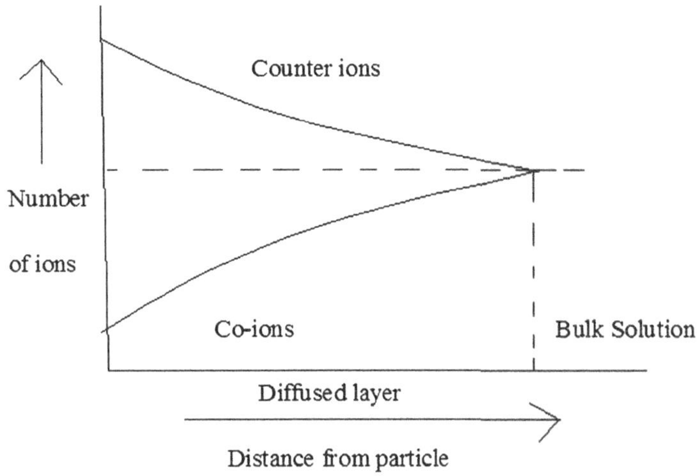

Fig. 10.13 Charge system in a colloidal suspension

10.6 Colloidal Stability of Particles of Water

Colloidal particles are generally stable if their concentration does not change with time. One of the most important factors contributing to stability of colloidal suspension is excessively large surface to volume ratio due to very small size (Bache and Gregory 2007). Surface phenomenon predominates over mass phenomenon. Ions present in water near the colloid will be affected by the charged surface. The charge system in a colloidal suspension has been shown in Fig. 10.13. When two colloids come in close proximity then two forces act on them. One is electrostatic force and the other one is van der waal force. Brownian movement may produce enough momentum for particles to collide with each other (Soos et al. 2008).

The stability of hydrophilic colloids is much greater than hydrophobic ones (Snoeyink and Jenkins 1980). The charge is reflected by electrophoretic movement of colloids under influence of electric field. Addition of cationic polyelectrolytes to hydrophilic colloids results in fixation of cation on negatively charged groups of colloids. The fixation may be electrostatic or chemical interaction. The hydrophilic colloids have strong tendency to bind or absorb water (Snoeyink and Jenkins 1980).

Hydrophilic colloids are stabilized by hydration of dispersed particles and electrical repulsion between particles. That's why their free energy decreases and entropy increases, which results in thermodynamic stability. Therefore, Gibbs free energy of system is negative.

$$\Delta G = \Delta H - T\Delta S < 0$$

The stability of hydrophilic colloids can be reduced by lowering solubility of colloids in water and change in pH value. Solution of hydrophilic colloids even at moderate concentration can have different properties from water because of large size of molecules. Hydrophilic colloids are true solution. They have some characteristics of particles because of large molecular size. They can scatter light and are unable to pass through dialysis membrane unlike smaller particles.

On the other hand hydrophobic colloids are thermodynamically unstable, since the energy input to the dispersed hydrophobic colloids is very high. Therefore, Gibbs free energy of system is positive.

$$\Delta G = \Delta H - T\Delta S > 0$$

Despite unstable nature, hydrophobic colloids can remain in finely dispersed phase for a very long time. It may be because of kinetic stability. The unstable hydrophobic colloidal particles maintain their stability through interactions/forces at their surface. These forces are called colloid forces or colloid interaction. Particles in water collide with each other and have ample opportunity to form aggregates.

10.7 Conclusion

The technology for suspension removal has great degree of variation so far in terms of removal efficiencies of water and industrial wastewater. The suspended solids are having high level of pollutants as compared to soluble water pollutants. So there is a requirement for green technology for its removal. Polyelectrolytes can be used for treating different kind of industrial water with great success. Sludge quantity is much reduced when polyelectrolytes are used in removal of suspension. It is concluded that low dosages of polyelectrolytes are required for coagulation- flocculation process so the question of toxicity in treated water is ruled out. That's why it should be understood that it is eco friendly and green technology for removal of suspension from water and industrial wastewater.

It has also been concluded that molecular weight of polyelectrolytes is more important in case of polymer bridging however polyelectrolytes having high charge density is more effective in case of charge neutralization. A combination of polyelectrolytes can also be used for complex industrial wastewater. Polyelectrolytes are having many advantages in water chemistry. It is quite reasonable to believe that the polyelectrolyte with maximum charge density is used more efficiently in industrial wastewater. However more quantitative study is required on this aspect to explore its further application in industrial wastewater.

References

Abd-Elhakeem MA, Alkhulaqi TA (2014) Simple, rapid and efficient water purification by chitosan coated magnetite nanoparticles. J Environ Nanotechnol 3(4):17–20

Abu Hassan MA, Tan PL, Zainon Noor Z (2009) Coagulation and flocculation treatment of wastewater in textile industry using chitosan. J Chem Nat Resour Eng 4(1):43–53. https://doi.org/10.1016/j.sbspro.2015.06.234

Akpor OB (2011) Wastewater effluent discharge: effects and treatment processes. In: Third international conference on chemical, biological and environmental engineering, Thailand, vol. 20, pp 85–90

Bache DB, Gregory R (2007) Flocs in water treatment. IWA Publishing, London

Beckett R, Le NP (1990) Electrokinetics of natural and synthetic calcite suspensions. Colloid Surf A Physicochem Eng Aspects 44:35–49. https://doi.org/10.1016/S0927-7757(97)00179-90

Bing-Hui T, Bin F, Xian-Jia P, Zhao-Kun L (2005) A cleaner two-step synthesis of high purity diallyldimethylammonium chloride monomers for flocculant preparation. J Environ Sci 17:798–801

Bolto B, Gregory J (2007) Organic polyelectrolytes in water treatment. Water Res 41(11):2301–2324. https://doi.org/10.1016/j.watres.2007.03.012

Bratby J (2006) In: Introduction in coagulation and flocculation in water and wastewater treatment, Chapter 1. IWA, London, pp 1–8

Buchhammer HM, Petzold G, Lunkwitz K (2000) Nanoparticles based on polyelectrolyte complexes: effect of structural and net charge on the sorption capability for solved organic molecules. Colloid Polym Sci 278(9):841–847. https://doi.org/10.1007/S003960000330

Cairncross S, Petach H (2013) J Water Sanit Hyg Dev 3(4):479–480. https://doi.org/10.1016/j.colsurfa.2006.03.029

Caulfield MJ, Hao X, Qiao GG, Solomon DH (2003) Degradation on polyacrylamides. Polymer 44(14):3817–3826

Cho BU, Garnier G, van de Ven TGM, Perrier M (2006) A bridging model for the effects of a dual component flocculation system on the strength of fiber contacts in flocs of pulp fibers: implications for control of paper uniformity. Colloids Surf A Physicochem Eng Asp 287(1–3):117–125

Choi JH, Shin WS, Lee SH (2001) Application of synthetic polyamine flocculants for dye wastewater treatment. Sep Sci Technol 36:2945–2968. https://doi.org/10.1081/SS-100107638

Crittenden JC, Trussel RR, Howe KJ, Tchobanoglous G (2005) Coagulation, mixing and flocculation. In: Water treatment: principles and design, 2nd edn. Wiley, Hoboken, pp 643–779

Duk Jong J, Won Sik S, Young-Hun K, Ji Hoon K, Jeong Hak C, Sang Jun C (2003) Effect of polyamine flocculant types on dye wastewater treatment. Separ Sci Technol 38(3):661–678. https://doi.org/10.1081/SS-120016657

Franceschi MA, Girou AM, Carro-Diaz MT, Maurette E, Puech-Costesb E (2002) Optimisation of the coagulation-flocculation process of raw water by optimal design method. Water Res 36(14):3561–3567

Gregory J, Guibai L (1991) Effects of dosing and mixing conditions on polymer flocculation of concentrated suspensions. Chem Eng Commun 108(1):3–21. https://doi.org/10.1080/00986449108910948

Guibal E, Van Vooren M, Dempsey BA (2006) A review of the use of chitosan for the removal of particulate and dissolved contaminants. Sep Sci Technol 41:2487–2514. https://doi.org/10.1080/01496390600742807

Haydar S, Aziz JA (2009) Coagulation–flocculation studies of tannery wastewater using combination of alum with cationic and anionic polymers. J Hazard Mater 168:1035–1040. https://doi.org/10.1016/j.jhazmat.2009.02.140

Hjorth M, Jørgensen BU (2012) Polymer flocculation mechanism in animal slurry established by charge neutralization. Water Res 46:1045–1051. https://doi.org/10.1016/j.watres.2011.11.078

James M, Ebeling S (2004) Application of chemical coagulation aids for the removal of suspended solids (TSS) and phosphorus from the microscreen effluent discharge of an intensive recirculating aquaculture system. N Am J Aquacult 66:198–207. https://doi.org/10.1577/A03-056.1

James ME, Sibrell PL, Ogden SR, Summerfelt ST (2003) Evaluation of chemical coagulation-flocculation aids for the removal of suspended solids and phosphorus from intensive recirculating aquaculture effluent discharge. Aquacult Eng 29:23–42. https://doi.org/10.1016/S0144-8609(03)00029-3

Kozlova N, Santore MM (2006) Surface pattern recognition by colloids particles. Langmuir 22(3):1135–1142. https://doi.org/10.1021/la0006059

Lee JF, Liao PM, Tsen DH, Wen PT (1998) Behavior of organic polymers in drinking water purification. Chemosphere 37(6):1045–1061

Lee CS, Robinson J, Chong MF (2014) A review on application of flocculants in wastewater treatment. Process Saf Environ Prot 92:489–508. https://doi.org/10.1016/j.psep.2014.04.010

Lourenço A, Arnold J, Gamelas JAF, Cayre OJ, Rasteiro MG (2018) Anionic polyelectrolytes synthesized in an aromatic-free-oils process for application as flocculants in dairy-industry-effluent treatment. Ind Eng Chem Res 57(49):16884–16896. https://doi.org/10.1021/acs.iecr.8b03546

Mende M, Schwarz S, Petzol G, Jaeger W (2007) Destabilization of model silica dispersions by polyelectrolyte complex particles with different charge excess, hydrophobicity, and particle size. J Appl Polym Sci 103(6):3776–3784. https://doi.org/10.1002/app.25573

Moussas PA, Zouboulis AI (2009) Enhanced coagulation-flocculation performance of iron-based coagulants: Effects of PO_4^- and SiO_3^{2-} modifiers. Water Res 43(14):3511–3524

Munch R, Hwang CP, Lackie TH (1980) Wastewater fractions add to total treatment picture. Water Sewage Works 127(12):49–54

Narkis N, Ghattas B, Rebhun M, Rubin AJ (1991) The mechanism of flocculation with aluminium salts in combination with polymeric flocculants as flocculant aids. Water Supply 9(1):37–44

Nieuwenhuijzen AF, Van der Graaf J, Kampschreur M, Mela A (2004) Particle related fractionation and characterization of municipal wastewater. Water Sci Technol 50(12):125–132

Odegaard H (1987) Optimised particle separation in the primary step of wastewater treatment. Water Sci Technol 37(10):43–53

Oruç F, Sabah E (2006) Effect of mixing conditions on flocculation performance of fine coal tailings. In: XXIII international mineral processing congress, 3–8 September, Istanbul-Turkey

Owen AT, Fawell PD, Swift JD, Labbett DM, Benn FA, Farrow JB (2008) The preparation and ageing of acrylamide/acrylate copolymer flocculant solutions. Int J Miner Process 87:90–99. https://doi.org/10.1016/j.minpro.2007.05.003

Pelssers EGM, MAC S, Fleer GJ (1990) J Chem Soc Faraday Trans 86(9):1355–1361. https://doi.org/10.1039/ft9908601355

Petzold G, Schwarz S, Lunkwitz K (2003) Higher efficiency in particle flocculation by using combinations of oppositely charged polyelectrolytes. Chem Eng Technol 26(1):48–53. https://doi.org/10.1002/ceat.200390006

Rinaudo M (2006) Chitin and chitosan: Properties and applications. Prog Polym Sci 31(7):603–632. https://doi.org/10.1016/j.progpolymsci.2006.06.001

Rout D, Verma R, Agarwal SK (1999) Conditioning process and characterization of fresh activated sludge. Water Sci Technol 40(2):137–141. https://doi.org/10.2166/wst.1999.0106

Schmidt KH, Morche (2006) Sediment output and effective discharge in two small high mountain catchments in the Bavarian Alps, Germany. Geomorphology 80:131–145. https://doi.org/10.1016/j.geomorph.2005.09.013

Selvapathy P, Reddy MJ (1992) Effect of polyelectrolytes on turbidity removal. Water Supply 10(4):175–178

Snoeyink VL, Jenkins D (1980) Water chemistry. Wiley, New York

Song Z, William CJ, Edyvean RGJ (2004) Treatment of tannery wastewater by chemical coagulation. Deaslination 164:249–259. https://doi.org/10.1016/S0011-9164(04)00193-6

Soos M, Moussa AS, Ehrl L (2008) Effect of shear rate on aggregate size and morphology investigated under turbulent conditions in stirred tank. J Colloid Interface Sci 319:577–589. https://doi.org/10.1016/j.jcis.2007.12.005

Teh CY, Budiman PM, Shak KPY, Wu TY (2016) Recent advancement of coagulation-flocculation and its application in wastewater treatment. Ind Eng Chem Res 55:4363–4389. https://doi.org/10.1021/acs.iecr.5b04703

Ting L, Luo C, Qi D, Zhang D, Zhao H (2019) Efficient treatment of emulsified oily wastewater by using amphipathic chitosan-based flocculant. React Funct Polym 139:133–141. https://doi.org/10.1016/j.reactfunctpolym.2019.03.019

Tripathy T, Rajan De B (2006) Flocculation: a new way to treat waste water. J Phys Sci 10:93–127

United Nation Report (2010) The human right to water and sanitation, Fact sheet No-35

Vajihinejad V, Gumfekar SG, Bazoubandi B, Najafabadi ZR, João BP (2019) Water soluble polymer flocculants: synthesis, characterization, and performance assessment. Macromol Mater Eng 304(2):1800526. https://doi.org/10.1002/mame.201800526

Wang C, Alpatova A, McPhedran KN, El-Din MG (2015) Coagulation/flocculation process with polyaluminum chloride for the remediation of oil sands process-affected water: performance and mechanism study. J Environ Manage 160:254–262. https://doi.org/10.1016/j.jenvman.2015.06.025

WHO/UNICEF Joint Monitoring Programme for Water Supply and Sanitation (2004) Meeting the management drinking water and sanitation target—a midterm assessment of progress. ISBN-9241562781

Yu X, Somasundaran P (1996) Role of polymer conformation in interparticle-bridging dominated flocculation. J Colloid Interface Sci 177(2):283–287. https://doi.org/10.1006/jcis.1996.0033

Zainol NA, Aziz HA, Yusoff MS, Umar M (2011) The use of polyaluminum chloride for the treatment of landfill leachate via coagulation and flocculation processes. Res J Chem Sci 1(3):52–57

Chapter 11
Green Technologies for the Treatment and Utilisation of Dairy Product Wastes

Shivani Garg, Nelson Pynadathu Rumjit, Paul Thomas, Sikander, Chin Wei Lai, and P. J. George

Contents

S. Garg
Institute of Environmental Studies, Kurukshetra University, Kurukshetra, Haryana, India

N. P. Rumjit · P. Thomas · C. W. Lai (✉)
Nanotechnology and Catalysis Research Centre (NANOCAT), Institute for Advanced Studies (IAS), University of Malaya (UM), Kuala Lumpur, Malaysia
e-mail: cwlai@um.edu.my

Sikander
Department of Geology, Kurukshetra University, Kurukshetra, Haryana, India

P. J. George
Sam Higginbottom University of Agriculture, Technology and Sciences, Allahabad, Uttar Pradesh, India

Inamuddin, A. M. Asiri (eds.), *Sustainable Green Chemical Processes and their Allied Applications*, Nanotechnology in the Life Sciences,
https://doi.org/10.1007/978-3-030-42284-4_11

Abbreviations

BOD	Biochemical oxygen demand
CFC	Chloride fluoride carbon
CMC	Carboxymethylcellulose
COD	Chemical oxygen demand
FAO	Food and agriculture organization
SBR	Sequencing batch reactors
TDS	Total dissolved solids

11.1 Introduction

Industrialisation plays a keen role in economic development in any nation. Pollution caused by industries is one of the primary concerns in the whole world, and of all industrial sectors, the food industry is one of the leading industrial sectors requiring excessive usage of water. The dairy industry is a type of food industry and produces many products such as milk and its by-products include butter, cheese, cream and yoghurt along with produce wastes (solid and liquid) (Jaganmai and Jinka 2017). During the last few decades, increased demand for milk and dairy products in different parts of the world has directed to the immense growth of the dairy sector (Chokshi et al. 2016). The demand of dairy products surges dairy productivity; however, it also increases the release of toxic effluents to the environment. This toxic release of effluents causes a severe impact on the ecosystem and organisms. The presence of high organic content in dairy effluents is a real threat to the ecosystem. It is estimated that per year about 4–11 million tonnes of dairy effluents are released directly and indirectly to the environment.

The dairy industry contributes 15% of the global share with different types of milk form (pasteurized, powdered milk, ghee, ice cream, butter, sweetened milk and cheese) (Chaudhry and Garg 2018). All the dairy products produce some valuable by-products during preparation, viz. whey and buttermilk that are generated in huge amount (Gopinathan and Thirumurthy 2012). Approximately 29 million tonnes of dairy products are wasted every year in Europe alone, and from Indian dairy industries, 110 million tonnes and 275 million tonnes of milk and contaminated water are discharged every year, respectively (Kushwaha et al. 2011).

For processing 1 L of milk, 1–3 L of clean water with the sanitizing agent is used, which produces approximately 2.0–2.5 L of wastewater (Kavitha et al. 2013). Raw wastewater discharge from the dairy industry causes reduction of dissolved oxygen in the environment. The presence of fat in dairy industry effluents forms a film layer on water surface that acts as a barrier preventing oxygen transfer and therefore causing a severe threat to the survival of aquatic living organisms and ecosystem. As the water supply is the main component for the operation of the dairy industry, the management of the wastewater is of significant concern.

The standard techniques or conventional methods are common in the management of dairy industry effluent. However, the costs and efficacy in the removal of COD associated with physiochemical techniques forced to look for other alternative sustainable solutions. Biotechnological management methods utilised for the management of effluents from dairy industries had been considered as good options (Faria et al. 2017). The biological treatment option could be able to reduce the discharge of organic load from the dairy industry. Many studies have been done for the treatment and utilisation of dairy wastes (Ganju and Gogate 2017; Chandra et al. 2018). Various dairy product wastes can be recovered and used as the primary material (raw) for the manufacture of other products or as fuel for energy production (Chandra et al. 2018; Wong et al. 2019). Previous studies pointed out that discharging wastewater with a high concentration of nutrients may lead to eutrophication in the water bodies (Abdel-Raouf et al. 2012).

The dairy industry effluent contains a high concentration of nutrient (Singh et al. 2014). Many regulatory governmental and non-governmental organizations in several countries like the United States have enclosed regulations and guidelines for the effluent discharge and marked with lower and upper limits of particular nutrients in the wastewaters.

Therefore it becomes crucial for these industries to develop more efficient and economically profitable ways of dairy waste management. This chapter highlights the characteristics of dairy product wastes and their management methods, which are more environment friendly and help in the reuse of dairy product wastes.

11.1.1 Outline of Dairy Products

Milk is an essential material for trades and required in daily life. So milk is processed in the dairy industry in many types of products (Burke et al. 2018). For the basic public needs and health consideration, the products should be of good quality, pure and hygienic. Quality control measures have to be carried out at all stages of production to maintain product quality. The dairy industries are rapidly rising all over the world. To understand the problems of wastes produced from this industry, it is required to study the different processes used for production in the industries. Many dairy industries produce pasteurized milk, making ghee and butter. In other industries, the supply of milk is more abundant; powdered milk, ice cream, cheese, yoghurt and condensed milk are also produced. Major dairy products produced in dairy industry are shown in Fig. 11.1.

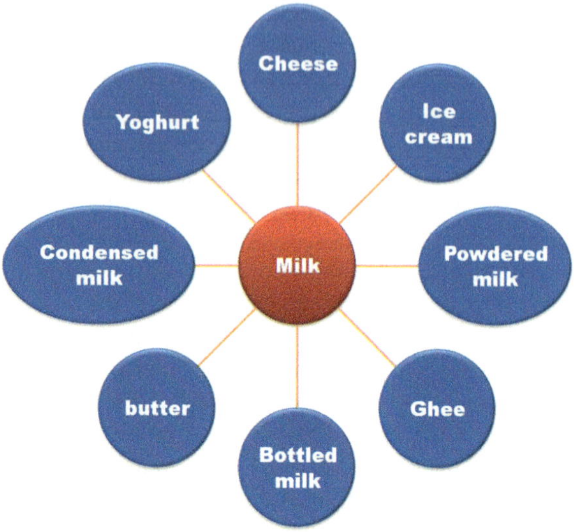

Fig. 11.1 Milk and different products processed from the raw milk

The chain of operations involved in dairy industries are homogenization, standardization/clarification (removal of solids), separation (cream removal for standardization of milk to maintain needed fats in the products), and pasteurization (heating at 62–71 °C for 15 s to 30 min followed by chilling to 4 °C) (Sfakianakis and Tzia 2014).

Cheese is made from milk in which protein and fat ratio is adjusted in the cheese production. In ice cream production, mixing of milk, additive, thickeners and sugar is done. After pasteurisation and cooling, different flavours and aromas are added, and then packaging is done. In the production of condensed milk, hot milk is homogenised and evaporated to form sugar-free and sweet condensed milk. Milk powder is prepared by vacuum evaporation and steam drying.

11.2 Sources and Composition of Wastes

In the dairy industry, operating and production methods, type of product being processed and produced, waste management and processing plant design affect the composition of dairy wastes (Shete and Shinkar 2013).

Waste generated from dairy industries is divided into three categories, such as solid (sludge), liquid (effluent), and gases (greenhouse gases) (Slavov 2017). Solid waste is also produced in the form of by-products(whey) from cheese whey. These wastes were utilised as the primary materials for bioenergy production. Dairy wastes are rich in SS (suspended solids) and organic matters, along with the presence of

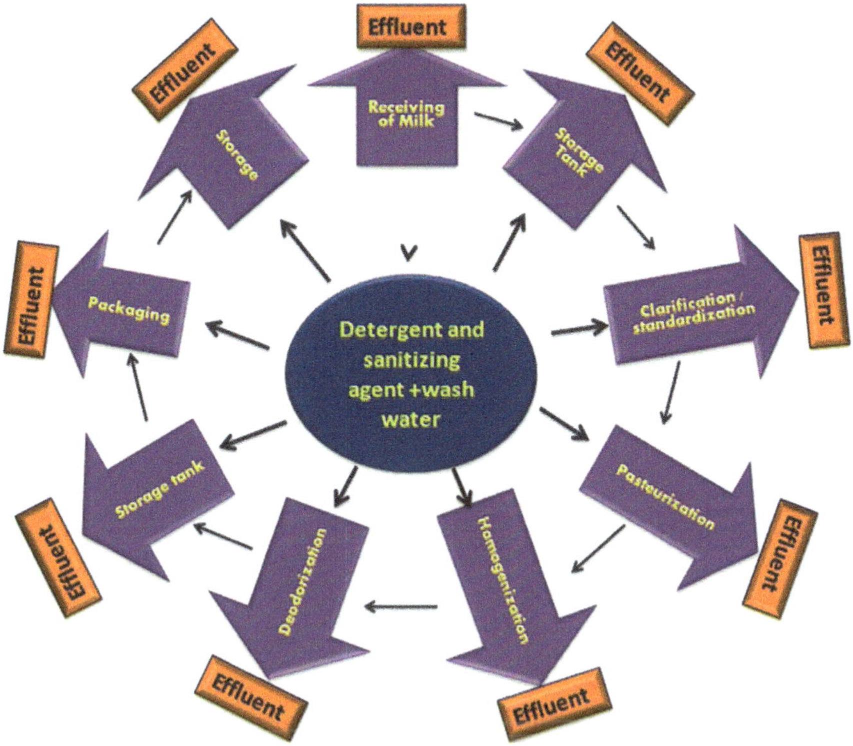

Fig. 11.2 Types of effluents produced during the processing of the dairy industry

grease and oil and high concentration of nitrogen and phosphorous. High quantity of water is needed for different purposes in the processing of dairy products. The three types of categories of wastewater are sanitary wastewater, cleansing wastewater and processing wastewater generated depending upon their origin and composition from the dairy industry (Britz et al. 2006). Dairy effluent comprises nutrients, fats, chlorides, sulphates, total solid organic components and lactose, which are typically characterised by high BOD and COD (Yonar et al. 2018). The different types of effluents produced during the processing of the dairy industry are shown in Fig. 11.2.

11.2.1 Solid Waste from Dairy Industries and Their Utilisation

Different solid wastes are produced from the dairy industries having many types of components. Solid wastes generally comprise damaged products and out-of-date products. These are described as follows:

11.2.1.1 Butter Wastes

Butter wastes produced during butter production have four groups:

- Milk without fat when butter directly processed from milk
- The residue from butter prepared from yoghurt
- Milk without fat when butter is formed by separating milk
- Cream as a residue by churning

Milk without fat can be used as a side product in milk powder factories and as a nutritious supplement. It is also used in yoghurt and kefir production. The waste produced during the churning of butter plays a beneficial role in preventing bacterium growth that causes a foul smell.

11.2.1.2 Whey

Whey is generated as a by-product during cheese and casein production. Whey waste contains carbohydrates, minerals, proteins as well as lipids (De Jesus et al. 2015). After the process of sterilisation, pasteurisation or spray drying, the natural proteins and vitamins of milk are denatured. So, to obtain the concentrates of lactose, proteins, vitamins and minerals from whey, low degree of microbial purity of whey is performed.

Lactose is used for the preparation of media for bacterial growth and in infant milk (Linares et al. 2017). The denatured whey proteins stimulate immune-regulatory properties for cancer treatment and antioxidants (Falkowski et al. 2018). These whey proteins also have beneficial effects on cancer prevention (Kassem 2015).

11.2.1.3 Sludge

Sludge is produced during the treatment of wastewater from the production unit containing a high amount of organic compounds, with a considerable concentration of nitrogen compounds, phosphorus, potassium, lactose, casein, oil and grease and fatty acids (Singh et al. 2013). Sludge is treated and utilised for different purposes that will be discussed in the further section ahead.

11.2.2 Liquid Wastes from Dairy Industries (Effluent)

Liquid waste is generated in the form of effluent from washing (milk transport vehicles before entering into the production area), cleaning, milk spills, cheese whey and other processing units. The effluent characteristics are based on the nature of product processed from milk in the dairy industry (Carvalho et al. 2013). Dairy

Table 11.1 Physicochemical parameters and quality standards of dairy industry wastewater discharge into the environment

Parameters	Dairy industry influent (Al-Wasify et al. 2018)	World Bank report	Indian effluent standards (CPCB 2006)
pH	9.8	6–9	5.5–9
BOD (mg/L)	650	50	100
COD (mg/L)	1448	250	250
TDS (mg/L)	1222	–	2100
Temperature (°C)	34	–	40
Appearance	White to grey	Normal	Normal
Odour	Unpleasant	Odourless	Odourless

effluent has high quantities of milk constituents, lactose, inorganic salts, fat, minerals (N, P, K), organic matter, detergent and sanitisers (Singh et al. 2013). Dairy wastewater is differentiated by BOD, COD and pH value variations. The pH variation is due to the presence of sanitising agents and detergents used in cleaning along with a wide range of acids (Britz et al. 2006). Table 11.1 lists the physicochemical parameters and quality standards of dairy industry wastewater discharge into the environment. Dairy wastewater has an unpleasant smell, turbid character and white colour (Kolev Slavov 2017).

The effluent temperature of dairy wastewater is higher than in other industries and sewage wastewater due to the faster biodegradation. There is nearly 5–6 °C temperature difference in the dairy effluent during summer and winter. The total suspended solids (TSS) are also essential constituents of dairy effluent (Porwal et al. 2015).

11.2.3 Gaseous Waste

Dairy industries produce air emissions having combustion gases from different generators, milk powder dust and refrigerant gases (Milani et al. 2011). Dairy effluent consists of organic matters that in anaerobic conditions produce methane, carbon dioxide, oxides of sulphur and nitrogen gases.

11.3 Environmental Impacts of Dairy Product Wastes

Dairy industry poses a pivotal role in polluting surface water, soil and biodiversity (Deshpande et al. 2012; Chen et al. 2018). The main dairy waste is in the form of wastewater that comprises many organic compounds, oil and grease, minerals and dissolved solids (Raghunath et al. 2016). When effluent with high organic load is discharged, the degradation of dissolved oxygen level takes place and leads to the growth of mosquitoes and flies in the water body (Al-Wasify et al. 2018).

Table 11.2 Physicochemical characteristics associated with dairy effluents

Waste	pH	BOD (mg/L)	COD (mg/L)	TDS (mg/L)	References
Dairy effluent	7.2–8.8	1200–1800	1900–2700	500–740	Sukhadev Shivsharan et al. (2013)
Whey wastewater	4.1	20,000	71,526	22,050	Deshpande et al. 2012
Cheese whey factory	6	>1,20,000	>80,000	>8000	Tikariha and Sahu (2014)
Milk-processing plants	8.34	4840	10,251	5802	Cristian (2010)

Proteins, lactose and fats are the main organic compounds found in dairy wastewater. These organic components increase BOD in water bodies (Raghunath et al. 2016). Organic wastes in dairy wastewater are biodegradable. In aerobic conditions, the organic matter is biodegraded by aerobic bacteria as

$$\text{Organic compounds}(\text{carbon compounds}) + O_2 \rightarrow CO_2 + H_2O + \text{bacteria}$$

The dairy wastewater promotes the growth of bacteria and algae in water bodies after discharge and eutrophication in water bodies, decreases oxygen availability and results in anaerobic conditions. The bacteria convert the nitrogen compounds into ammonium, nitrite and nitrate ions that cause gradual death of fishes (Shete and Shinkar 2013).

Whey is considered as the most polluting constituent of dairy wastewater that alters the physical and chemical composition of water (De Jesus et al. 2015). Table 11.2 lists various physicochemical characteristics associated with dairy effluents. The lactose in the effluent from dairy industry causes the growth of fungus as it acts as a source of sugar (Raghunath et al. 2016). The effluent is generally of white to grey colour as it contains organic compounds. Dairy wastewater has milk waste causing the turbidity in discharged water. The industrial wastewater causes degradation of the environment, which adversely affects living organisms as well as agriculture (Verma and Singh 2017).

The dairy industry effluent has dissolved salts that can unfavourably affect soil structure if wastewater is used for irrigation. High salt concentration in soil affects the growth of vegetation. Wastewater can leach into groundwater bodies; the leachate could contaminate the groundwater sources (Khalid et al. 2018).

11.4 Treatment Techniques and Utilisation of Dairy Waste

Dairy product processing affects the environment by producing wastes with pH variation and high organic and nutrient concentrations. The hight temperature of effluent causes the death of biota of water bodies (Porwal et al. 2015). The primary

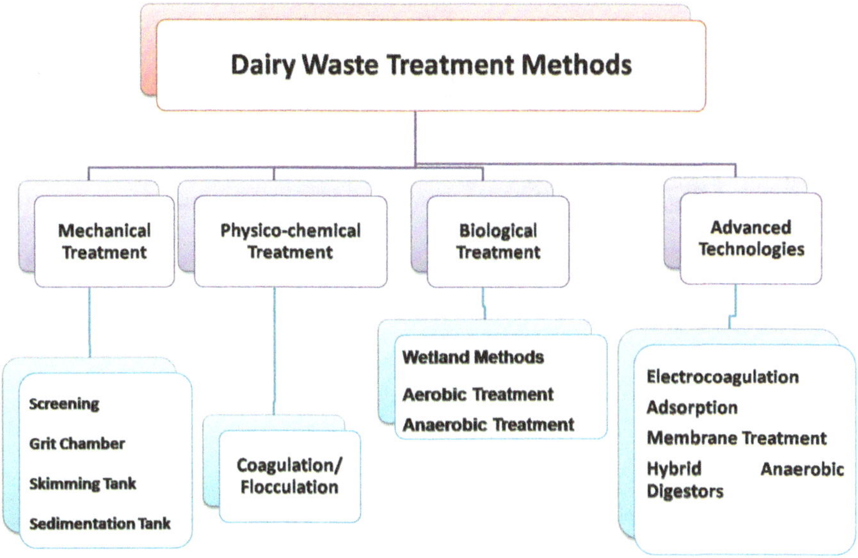

Fig. 11.3 Different types of management methods for wastewater treatment

treatment process for effluent from dairy industries is typical at the beginning of treatment. The different types of treatment methods for wastewater management are shown in Fig. 11.3.

The dairy industry wastes constitute different types of components, so methods of treatment are listed below:

11.4.1 Mechanical Methods

The mechanical techniques include screening, grit chambers, skimming tank or sedimentation tank process. This is also called pretreatment of wastewater. Wastewater contains floatable grit, oil and grease that causes a problem during treatment processes so these should be removed during the initial stage of operation.

Figures 11.4 and 11.5 illustrate the mechanical treatment route. These methods reduce the organic components in wastewater (Zheng et al. 2013). It constitutes the following steps:

- *Screening*: Screens are designed to remove the extensive-size floating material from wastewater during treatment. Screens have uniform-size pores that recollect the solids of more massive size than the pore size of the screen (Bhargava 2016). Generally, two types of screens are used which have an opening of one-fourth inch and less than one-fourth inch. The automatically operated rakes are used to clean screens.

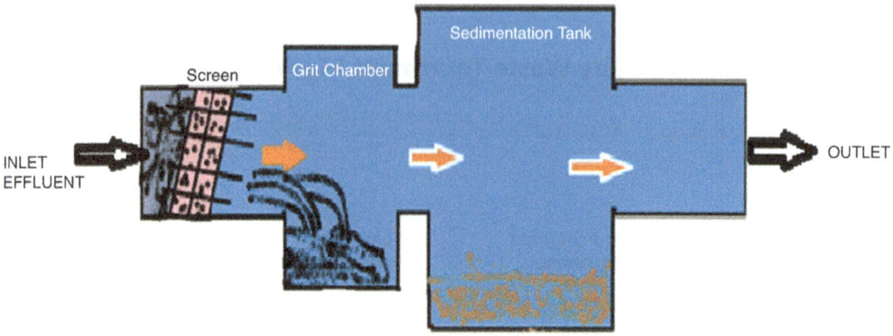

Fig. 11.4 The general process of mechanical treatment of dairy wastewater

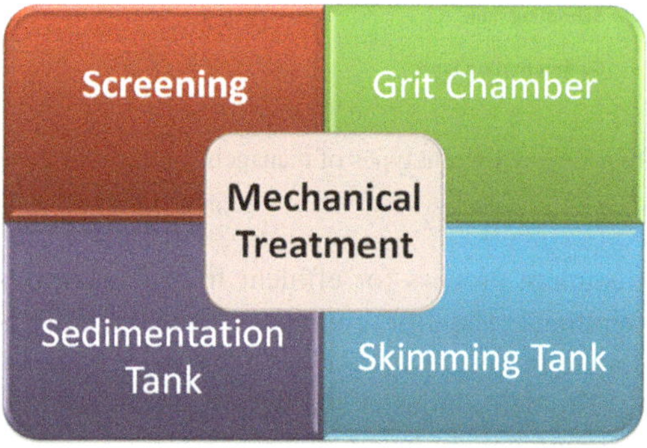

Fig. 11.5 Flow chart of the process of pretreatment or mechanical treatment of dairy wastewater

- *Grit Chamber*: The second unit of the treatment process is the grit chamber that removes suspended heavier inorganic components (sand and grit)and passes lighter organic material. This chamber removes suspended inorganic solids, and suspended solids are removed in primary sedimentation tanks. The velocity of flow of waste water in the chamber called differential sedimentation and differential scouring velocity as the flow of wastewater through the chambers neither too low nor should be high so that it doesn't allow the settling of lighter material on the floor. Grit waste can be discharged after wash to remove large-size organic matter in grit. Solids collected from the sedimentation tank can be disposed of after the treatment. Grit chamber is classified as mechanically and manually cleaned.
- Mechanically cleaned grit chamber has scraper blades that collect grit on the floor and are elevated to the ground by different mechanisms. Manually cleaned grit chambers are cleaned once a week by different methods, and the simplest method is by shovel.
- *Skimming Tank*: Skimming tank helps in the removal of floating matter (oil, fat and grease) and the rest of the liquid flow for further treatments. The artificial

aeration is provided at the bottom of the chamber so that the floating material rises and remains on the surface for easy removal mechanically or by hands.

- *Sedimentation Tank*: This process is used for SS removal from effluent by the principle of gravity (Cheremisinoff 2002). This process is used before the purification and disinfectant methods. It helps in the removal of small particulate matter and some microorganisms by gravity that settle down on the bottom. The success of the sedimentation process is based on the longevity of the undisturbed effluent in the tank; the longer the undisturbed water, the maximum the settling of the sediments at the bottom. Sedimentation process can be accelerated with the help of natural coagulants or some particular chemicals such as aluminium sulphate, ferric sulphate and poly-aluminium chloride (liquid alum).

11.4.2 Physicochemical Treatment

Dairy wastewater contains milk fat and protein colloids that can be treated by physicochemical methods. The most common physicochemical methods are coagulation and flocculation (Carvalho et al. 2013). The methods help in the reduction of colloidal particles from water that is responsible for the turbidity of the water. These processes reduce the organic compounds responsible for increasing the COD and BOD contents. The processes are explained as follows:

- *Coagulation or flocculation*: The solids that float on the surface of effluent are settled by specific adding material called coagulants, and the process is called coagulation (Kweinor Tetteh and Rathilal 2019). Figure 11.6 shows the pictorial representation of the coagulation/flocculation process. Coagulants destabilise the particulate matter with the collision of particles and floc formation results in sedimentation or floatation (Sahu and Chaudhari 2013). The coagulants neutralise the charged non-settable particles. After the collision, these small-size particles combine and stick together to form larger particles called macro flocs. As the

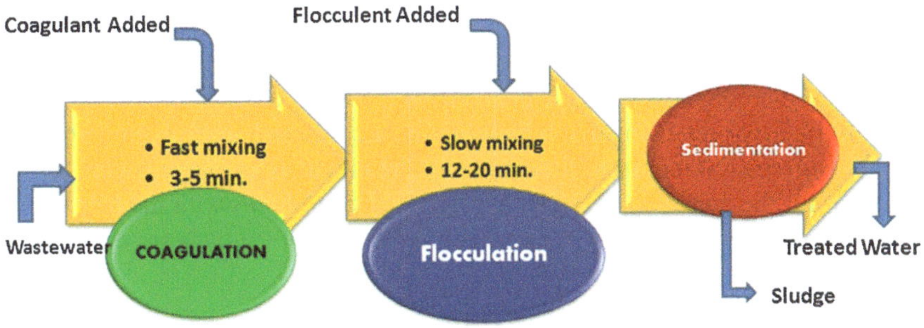

Fig. 11.6 Pictorial representation of coagulation/flocculation process

floc is formed with its optimum size and strength, the sedimentation or filtration process is initiated. The activated charcoal (powdered) after chitosan treatment for wastewater treatment is useful in the reduction of colour and odour (Ahmad et al. 2019). The reduction of COD occurs by the usage of lactic acid bacteria with chitosan by 49–82%, while carboxymethylcellulose usage results in a reduction of COD by 65–78% (Slavov 2017).

- Use of inorganic coagulant, for example, alum and ferric sulphate:

$$Al_2(SO_4)_3 + 3Ca(HCO_3)_2 \rightarrow 2Al(OH)_3 + 3CaSO_4 + 6CO_2$$

$$Fe_2(SO4)_3 + 3Ca(HCO_3)_2 \rightarrow 2Fe(OH)_3 + 3CaSO_4 + 6CO_2$$

- The common coagulant reagents are talcum, activated silica and carbon, mineral and organic coagulants (iron and aluminium salt), anionic or cationic flocculants and pH control reagents (acids and bases). The coagulant tannin is very significant for dairy industry wastewater treatment (Dela Justina et al. 2018). Recent studies showed the flocs produced using coagulant tannin observed to be have better stability even at high-speed gradients and tannin observed to be a promising coagulant in the treatment of dairy wastewater effluents.
- *Oxidation treatment*: The oxidation pretreatment techniques with hydrogen peroxide and ferrous sulphate result in the removal of fat present in dairy effluents. The waste produced from the oxidation treatment techniques is biodegradable as compared to the inorganic coagulants.

11.4.3 Green Treatment Technologies

Biological treatment methods are considered the most promising green techniques for the management of dairy effluent that includes many processes that are shown in Fig. 11.7. It is the most suitable methods for the treatment of organic matter from dairy waste (Carvalho et al. 2013).

During aerobic biodegradation, the formation of sludge may cause disposal problems which are expensive due to the sludge treatment (Dąbrowski et al. 2017). Biological treatment can transform the organic matter and absorb heavy metals with the help of microorganisms (Mosa et al. 2016).

Table 11.3 lists the comparison between the aerobic and anaerobic treatment of dairy wastewater; sometimes, the anaerobic treatment is employed with aerobic treatment for the reduction of BOD, nitrogen and phosphorus with the help of biological nutrient removal (BNR). The biological treatment methods are discussed as follows:

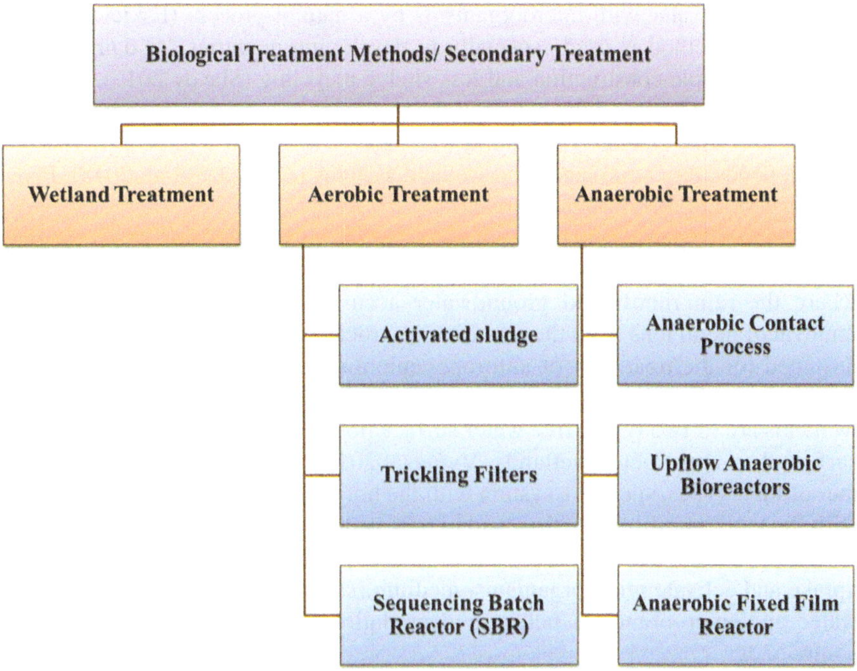

Fig. 11.7 Different types of biological/secondary treatment methods for dairy waste treatment

Table 11.3 Comparison between the aerobic and anaerobic treatment of dairy wastewater

Factors	Aerobic treatment	Anaerobic treatment
Process	Activated sludge, trickling filters, SBR	UASB, anaerobic contact process, fixed-film reactors
Size	Large-size reactors	Small-size reactors
Biomass yield	The large volume of biomass produced	Lower biomass produced
Energy required	High	Low
Sludge formation	High	Low
Aeration	Oxygen required for activity	No aeration required

11.4.3.1 Wetland Treatment Method

Wetlands known as sustainable wastewater treatment methods are economical, energy efficient and environmentally friendly as compared to other conventional treatment processes (Stefanakis et al. 2014). This method is used all over the world for many years (Hunt and Poach 2001). These treatment methods include the treatment of wastewater with microorganisms by a natural process (Eskicioglu et al. 2018). This method is used especially in developing countries for dairy effluent because of simple construction and less sludge recycling (Slavov 2017). Wetlands are of two types based on the nature of the treatment of wastewater, namely natural and constructed wetlands. The natural wetlands are available in three forms such as aquatic ponds, terrestrial lands and wetland systems (Almuktar et al. 2018). Plants play an essential role during treatment.

They absorb the nutrients and minerals in the wastewater and use it for their physiological processes. Natural wetlands are generally found in low-lying areas where the rain runoff and groundwater accumulate, and these are good for removal of metal ions from the wastes. Constructed wetlands (CWs) are specially designed for the treatment of anthropogenic discharge. These are semi-natural, economical and designed to treat wastewater by biological processes. Constructed wetlands are of two types: free water surface flow-constructed wetlands and subsurface flow-constructed wetlands (Vymazal 2010). The subsurface CWs are further of three types, such as a system with the horizontal subsurface flow, a system with the vertical subsurface flow and hybrid systems. Constructed wetlands work with three natural processes as follows: plants purify the wastewater by nutrient uptake and activate microorganisms, medium (sand and gravel) filters the wastewater and microorganism helps in degradation of organic matter and other compounds.

Free Water Surface Wetlands

Wetlands are designed with open water areas with floating vegetation and emergent plants. Topographical and environmental factors also affect the construction and design of the wetlands. These are the same as natural wetlands in designs. These are used to treat wastewater from lagoons and trickling filters (Andersson et al. 2005).

Horizontal Flow-Constructed Wetlands

Wetlands are designed with gravel and soil beds planted with vegetation. The wastewater flows below the surface of media so that exposure of wastewater to mosquito, pathogens and other organisms is minimised. These are designed for primary effluent treatment (Zidan et al. 2015). It contains influent piping, coarse media, a synthetic liner to prevent leaching and outlet piping with water level control.

Figure 11.8 illustrates the diagrammatical representation of horizontal flow-constructed wetlands. These generally have low-cost maintenance and are used for small-scale industries. The flow of water occurs horizontally in these wetlands. The filter material filters out the particles and microorganisms degrade the organic matter that is utilised by the vegetation.

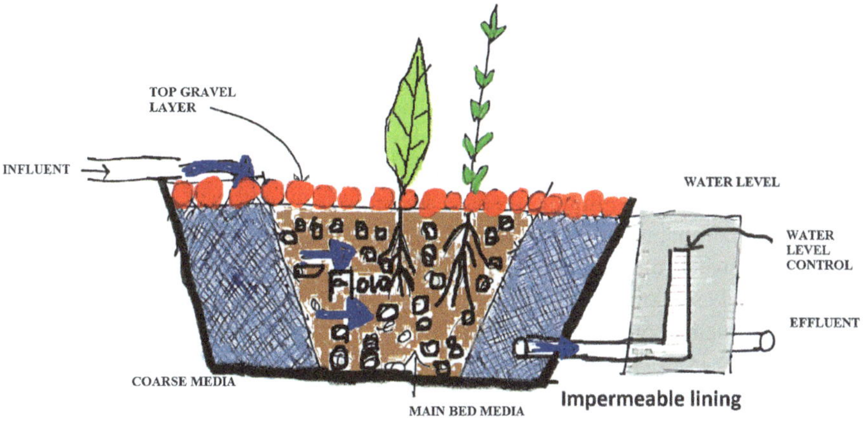

Fig. 11.8 A diagrammatical representation of horizontal flow-constructed wetlands

Vertical Flow-Constructed Wetlands

In this type of wetlands, oxygen supply is provided, and the flow of water is vertical. They can reduce the BOD of the wastewater. These consist of gravel and sand filters with aquatic vegetation. The intermittent pumps of wastewater are lined over the filter surface with a high injection of oxygen. The wastewater drains vertically through the filters (sand and gravel) towards drainage or effluent systems (Tsihrintzis 2017).

Plants used in constructed wetlands are sedges, water hyacinth, common cattail, spatterdock and waterweed. Dissolved metals are removed by the macrophytes, which cause phytotoxic symptoms.

Many dissolved metals like zinc, copper and lead react with sulphides; retain the sediment in wetlands; and cannot be used for agriculture purposes. The main disadvantages of the treatment by wetlands are a requirement of a large area for construction, the threat of leaching at the surface and in groundwater, growth of insects and presence of volatile material (Slavov 2017).

11.4.3.2 Aerobic Treatment Methods

Aerobic treatment processes work on the availability of oxygen and growth of microorganisms, which oxidises the organic material into water, cellular material and carbon dioxide (Britz et al. 2006). The dairy effluent treatment with aerobic treatment processes includes trickling filters, activated sludge and aerated lagoons (Samer 2015). The advanced aerobic treatment methods include sequential batch reactors (SBR). Aerobic processes are generally preceded at low DO (dissolved oxygen) concentration as the microorganism activity depends upon the environmental conditions of the area. The main processes used in the dairy wastewater treatment aerobically are explained as below:

Activated Sludge Treatment

Aerobic activated sludge treatment has not been used commonly as it produces stinking air (Kiesewetter et al. 2012). The activated sludge treatment systems include a tank with aeration with solid settler and solid removal line. The wastewater flows under constant aeration in the presence of activated sludge. The chemicals ferric chloride and aluminium sulphate are utilised for the sludge treatment. The most common chemical used is $Al_2(SO_4)\cdot nH_2O$ (alum) and it acts as a coagulant during wastewater management (Sahu and Chaudhari 2013).The composed sludge can be reused as a soil conditioner as carbon, nitrogen and phosphorous with other essential elements present in the sludge.

The sludge in dried form is used for landfill. The flow diagram of the working procedure of a conventional activated sludge treatment system is illustrated in Fig. 11.9. The extended aeration system is used to minimise the sludge formation as oxidation of organic material occurs. The retention time also increases for the consumption of sludge through endogenous respiration by 1–2 days. The sludge produced in extended aeration-activated sludge systems is non-biodegradable, very light in weight and settles with difficulty, so the retention time in settling tank increases.

Trickling Filters

The trickling filters are also called biological air filters. These are designed as a bed of permeable material with microorganisms and through which effluent flows (Zylka et al. 2018). The filter material consists of rocks or a range of plastic material.

The wastewater is spread over the filter bed with the help of rotary distributor and wastewater is retained there for some time so as it comes in contact with filter material. Figure 11.10 shows the pictorial representation of the operation of trickling filters.

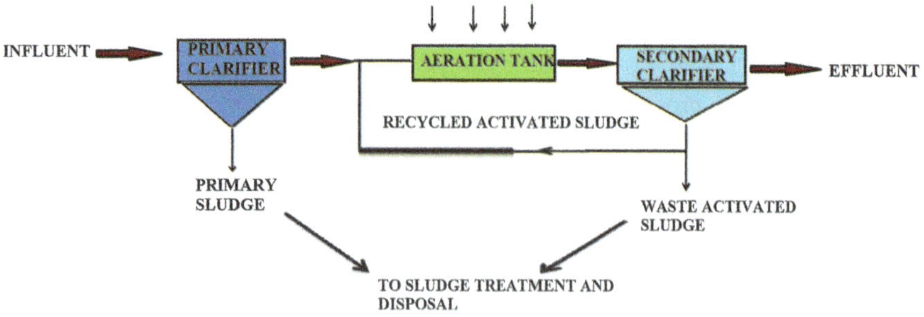

Fig. 11.9 The flow diagram of the working procedure of a conventional activated sludge treatment system

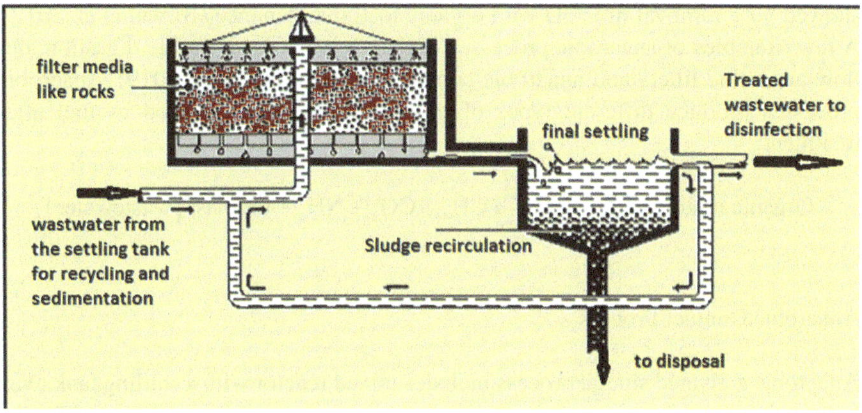

Fig. 11.10 Pictorial representation of the operation of trickling filters

They collect water that flows to the settling tank in which the solids settle down and are further treated. Usually, the collected wastewater from the settling tank is recycled in trickling tank to dilute the incoming wastewater from the industry. The disadvantages of trickling filters are long operation period and high incidence of clogging.

Sequence Batch Bioreactors

SBR is more preferred for dairy effluent treatment due to its effluent flexibility and different loading capability (Slavov 2017). The operation procedure of SBR and activated sludge treatment system is the same. The process aeration and sedimentation are common in both the systems.

The main difference between both the systems is that the processes equalisation, aeration and sedimentation are carried out in SBR (Dutta and Sarkar 2015). It is observed that in SBR after treatment, the reduction of COD, total solids, volatile solids and total nitrogen occurs (Britz et al. 2006).

The performance of intermittent aeration SBR improved because of the remediation limitation of nutrients by aerobic technologies as high organic material present in wastewater (Gil-Pulido et al. 2018).

11.4.3.3 Anaerobic Treatment

Anaerobic processes are more suitable processes for dairy wastewater treatment than aerobic process because these are more suitable for the treatment of high organic content in wastewater (Bharati et al. 2013). Anaerobic processes do not require aeration and produce low sludge and low area demand. According to a report, the discharge after treated raw milk by industrial scale anaerobic filters showed 90% removal of COD with organic load maintenance (Birwal et al. 2017). A few examples of anaerobic processes of dairy waste are anaerobic digestion, up-flow anaerobic filters and anaerobic contact process (Goli et al. 2019). Anaerobic processes produce different types of gases which can be utilised as fuel after treatment:

$$\text{Organic material} + \text{nutrients} \rightarrow CH_4 + CO_2 + NH_4 + \text{biomass}(\text{wastewater})$$

Anaerobic Contact Process

Anaerobic activated sludge process includes mixed reactor with a settling tank (Van Lier et al. 2008). Aeration is not required in this process. ACP was designed for the dilute wastewater from meatpacking plants which remove the flocs.

However, these are also suitable for the wastewater that contains suspended solids with microorganisms forming settleable flocs. Degasifier is used to remove the gas bubbles formed during operation to prevent floating of sludge on the surface. The HRT of this reactor is 0.5–5 days and the organic loading rate is 1.0–8 kg COD/m^3/day. Figure 11.11 illustrates the procedure of the anaerobic contact process.

Up-Flow Anaerobic Sludge Bioreactor

For the treatment of heavy load of effluent, having COD value higher than 42 g/L, UASB treatment is considered as suitable. UASB is designed as a biological reaction zone with a sedimentation zone. The organic materials of effluent are converted into CH_4 (methane) and CO_2 (carbon dioxide); in the biological reaction zones the wastewater flows upward through the activated sludge bed (Birwal 2017). The gas and sludge are separated in different separator zones. The substrate degradation occurs in the lower part because of activated sludge having anaerobic bacteria. For dairy wastewater treatment, UASB reactor is used (Kavitha et al. 2013).

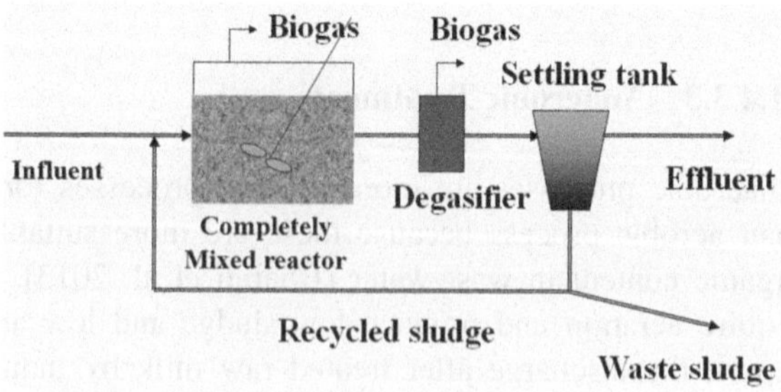

Fig. 11.11 The pictorial diagram of the procedure of the anaerobic contact process

Anaerobic Fixed-Film Reactors

Anaerobic fixed-film systems are a porous medium that acts as a bed to support the biomass which digests the wastes in the wastewater (Yousefzadeh et al. 2017). The medium may be rocks, plastic, wood or other natural or synthetic material. Generally, two types of fixed-film reactors are employed during treatment: one is fixed medium relative to fluid flow, such as trickling filters, and the other medium is in motion, such as rotating biological reactor. Trickling filter systems are designed as a media for beds across which wastewater flows. The wastewater flows through the media by gravity. The rotating disc helps in the movement of solids in the wastewater from which biomass is sloughed. Fixed-film reactors are used to treat low-strength waste at a fixed temperature. Fixed-film reactors are not suitable for wastewater to have a large amount of particulate matter as it creates clogging in the reactors. The larger the surface area of the reactor for the growth of bacteria on the media, the faster the digestion of the material.

11.4.4 Advanced Treatment Methods

Advanced techniques are used to purify and disinfect the treated wastewater for further utilisation and recycling. These are categorised under tertiary treatment in wastewater treatment operations. These treatments help in the removal of pathogens from the secondary treated effluents. The following techniques are classified under advanced treatment techniques:

11.4.4.1 Electrocoagulation

Electrocoagulation process is a technique of electrolysis that removes colouring matter, dissolved organic waste and turbidity by flowing an electrical current through wastewater with the help of electrodes. This process is primarily used to remove suspended colloidal particles (Reilly et al. 2019).

Metals, inorganic and organic wastes and colloidal particles possess charge in water and when the electrical charge is applied to contaminant water, the charges destabilise and separate from the wastewater. Oil and COD are removed from dairy effluent by using electrocoagulation (Lumina and Pavithra 2018).

Figure 11.12 shows the diagrammatical representation of electrocoagulation reactor. It is observed that approximately 98% COD and 97% turbidity are removed from dairy wastewater by applying electrocoagulation technique (Shivayogimath and Naik 2014).

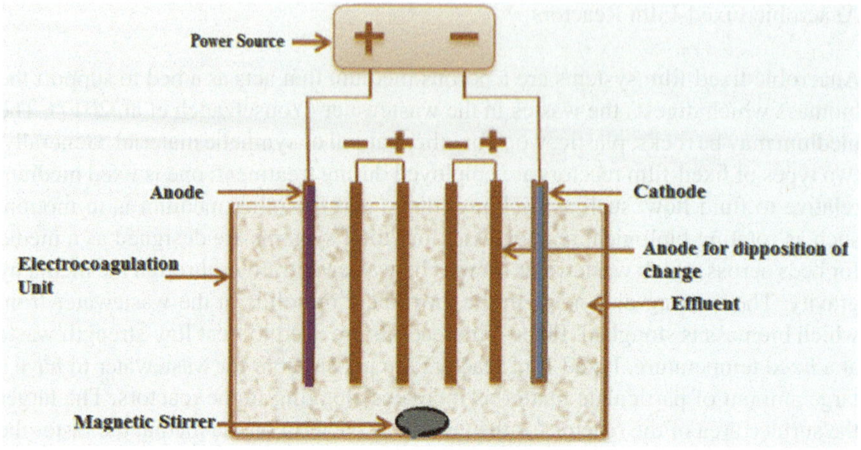

Fig. 11.12 The diagrammatical representation of electrocoagulation reactor

11.4.4.2 Adsorption

Adsorption is a process in which a solid material called adsorbent, e.g., activated carbon, helps in eliminating a massive amount of non-biodegradable organic matter from wastewater. Other examples of adsorbent are fly ash, rice husk and straw dust. Rice husk adsorbent is used for the removal of organic pollutants from dairy wastewater (Pathak et al. 2016). Activated charcoal is also used for the treatment of dairy wastewater, which removes approximately 65% of COD and 67% of BOD (Madhavrao Kanawade and Bhusal 2015).

11.4.4.3 Membrane Processes

Membrane processes use the membrane for the filtration of wastewater, so it is able to be utilised for drinking and other purposes (Nqombolo et al. 2018). Many types of membrane filtration process available nowadays use different types and sizes of pores in the membrane. Membrane processes are classified into ultrafiltration, nanofiltration and reverse osmosis, which use pressure for transportation of treated wastewater through the membrane (Frenkel 2008). Reverse osmosis membrane filtration is an extensively used process that removes ions and large particles from effluent. The demineralisation of water can also be done with the help of this process.

The material of the membrane used in RO systems is cellulose acetate, and the thickness of the membrane is 100 μm. In Andhra Pradesh, Gujrat, Tamil Nadu and Rajasthan, many membrane-based brackish water desalination plants with membrane filters have been set up for supply of safe and clean water for drinking to the villagers of states since the early 1990s (AGM 2010). In nanofiltration, the size of the pore of the membrane varies in between 0.5 and 1.5 nm. Nanofiltration allows infiltration of salts; reduces TDS, hardness, odour and colour; and removes heavy metals from the water (Zolfaghari and Kargar 2019).

The ultrafiltration process helps in the removal of microorganisms, such as bacteria and viruses. The ultrafiltration acts as a barrier for the complete exclusion of *E. coli* and total coliforms in most of the cases (Falsanisi et al. 2010).

11.4.4.4 Hybrid Anaerobic Digesters

The combination of UASB and an anaerobic filter is utilised for the treatment of dairy wastewater on a lab scale. Anaerobic digesters produce methane and remove COD from the wastewater (Musa et al. 2018). Anaerobic digesters digest the organic sludge by microbes in anaerobic conditions. The microbes used in the process are facultative anaerobes; for example, *Bacteroides, Clostridium, Butyrivibrio, Eubacterium, Bifidobacterium and Lactobacillus* are the most common for fermentation of wastewater.

11.5 Value-Added Products from Dairy Wastes

The concept is a new green management technology in the conversion of waste to valuable by-products. Biotechnological processes, for example, aerobic and anaerobic treatment processes, are used for the treatment of dairy wastes. The combination of biotechnological and physicochemical processes is utilised to produce valuable goods (Chen et al. 2018). Some of the examples of products produced from dairy wastes are given below:

11.5.1 Biofertiliser

The sludge produced during the treatment of wastewater in different stages is found to be useful for fast-growing rhizobia. Production of biofertilisers will help in the reduction of sludge from the dairy industry (Singh et al. 2013). Anaerobic digestion of cattle and dairy waste is already an established technology in the production of biocomposite.

The biocomposite fertiliser could act as a soil conditioner to improve the nutrient characteristics of the soil.

11.5.2 Biofuels

The dairy waste is used for the production of ethanol by using yeast, for example, *Kluyveromyces fragilis* as this yeast hydrolyses the lactose of whey (Pasotti et al. 2017) and *Candida* inconspicuous W16 immobilised in alginate calcium gel that produces a high amount of ethanol from the whey as substrate (Dahiya and Vij 2012). For the production of biodiesel production and nutrient removal, *Chlorella sp.* culture on dairy wastewater has been reported to have an enormous potential of the vast amount of biodiesel (Choi and Choi 2016).

11.5.3 Single-Cell Protein

Single-cell proteins are the products that have high protein content. The different types of microorganisms called generally recognised as safe are utilised for the manufacturing of SCP (Spalvins et al. 2018). Generally, yeast such as *Candida* (*C. pseudotropicalis*), *Kluyveromyces* (*K. lactis and K. marxianus*) and *Trichosporon* are preferred for the production of SCP as these are cheap, easy to grow and larger (Kasmi 2018).

11.5.4 Organic Acid Production

Milk whey can be used to produce different organic acids by using microorganisms. Succinic acid can be manufactured by fermenting milk whey in the presence of *Actinobacillus succinogenes* 130Z (Wan et al. 2008). Lactic acid can be produced with fermenting yoghurt whey having glucose and lactose by *Lactobacillus* (*L. casei* ATCC393) (Dimitrellou et al. 2014).

11.6 Biorefinery: New Green Management Route

The dairy biorefinery concept evolves as a new green management technique for change in precipitation from waste management to the production of valuable by-products (Dugmore et al. 2017). The waste to economic opportunity management integrates the process that strengthens biofuels and other valuable products recovery by production in cascading route, so these by-products in turn used as inputs for new value-added products. Figure 11.13 shows the schematic diagram of a biorefinery based on dairy waste effluents (Ahmad et al. 2019). The anaerobic digester is the primary technique employed in biorefineries; advanced alternative techniques are added downstream to yield other valuable products which could reimburse the operational and capital cost with profit.

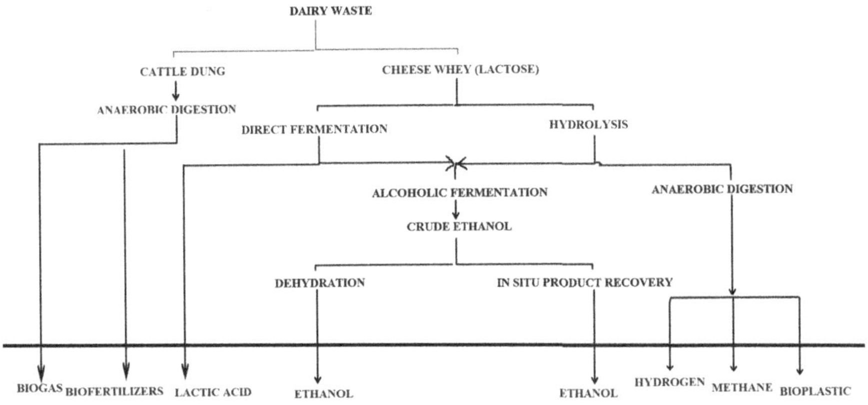

Fig. 11.13 The schematic diagram of a biorefinery based on dairy waste effluents

New biorefinery design utilises dairy waste for the production of other valuable by-products such as hydrogen, ethanol, fertilisers, lactic acid and volatile fatty acids (Morales-Polo et al. 2018). Whey and cattle dung are the significant waste produced from the dairy industry. Whey wastewater is the by-product produced during cheese processing, which is rich in lipids, proteins, mineral salts and lactose (Slavov 2017).

The oligosaccharides such as glucose as a by-product are derived from lactose that can be fermented for ethanol production. Similarly, the cheese whey waste can be fermented to ethanol for the production of biofuel (Shen et al. 2019). The ethanol that is produced as secondary product is further processed using purification route through pervaporation, membrane separation, distillation and precipitation. The carbohydrate residues derived through these processes can be processed for hydrogen production, methane and volatile fatty acids as the attractive environmental option.

The volatile fatty acids can be used as raw chemicals for the manufacturing of bioplastic with the aerobic process. The biorefinery utilises anaerobic digestion for fertiliser production and methane in the biorefinery route has an additional advantage that no greenhouse gas emission is associated with biorefinery processing route.

11.7 Conclusion and Future Prospects

The dairy industry plays a significant role in polluting air, soil and water environments as these produce a vast amount of waste with high COD, BOD and organic material in wastewater, solid and liquid by-product waste form. These wastes can cause an environmental problem and health hazards to human, flora and fauna of water resources.

The techniques utilised for the management of this type of waste are of a wide range and must be inexpensive. Biological treatment methods for dairy wastes are considered to be most effective because of the high efficiency of reduction of organic matter, COD and BOD in a combination of other advanced technologies. The aerobic and anaerobic processes are used to treat wastewater through different methods. However, both the processes are used together in order that the wastewater reaches the effluent discharge limits. Aerobic processes remove contaminants from wastewater, and anaerobic processes help in the disposal and further utilisation of treated waste products. Biotechnological approaches help in the utilisation of waste for useful purposes.

These approaches reduce the cost of treatment and load of pollution on the environment and strengthen the economy of the industry. All the methods of treatment of dairy waste are having some disadvantages when used individually, but in the combination process, the efficiency is increased. More studies are required to customise and optimise the processes and reduce the cost of treatment in the industry as individually, and these processes are time consuming and costly. Dairy wastewaters during anaerobic treatment produce methane, carbon dioxide and ammonia which can be collected and used as fuel. More studies are also required to combine the treatment of wastewater and the utilisation of the different generated waste products for further use. Bioremediation technology is also an alternative for the treatment processes that include chemicals or are time consuming.

References

Abdel-Raouf N, Al-Homaidan AA, Ibraheem IBM (2012) Microalgae and wastewater treatment. Saudi J Biol Sci 19:257–275. https://doi.org/10.1016/j.sjbs.2012.04.005

Ahmad T, Aadil RM, Ahmed H, Rahman U, Soares BCV, Souza SLQ, Pimentel TC, Scudino H, Guimarães JT, Esmerino EA, Freitas MQ, Almada RB, Vendramel SMR, Silva MC, Cruz AG (2019) Trends in food science & technology treatment and utilization of dairy industrial waste: a review. Trends Food Sci Technol 88:361–372. https://doi.org/10.1016/j.tifs.2019.04.003

Almuktar SAAAN, Abed SN, Scholz M (2018) Wetlands for wastewater treatment and subsequent recycling of treated effluent: a review. Environ Sci Pollut Res 25:23595–23623. https://doi.org/10.1007/s11356-018-2629-3

Al-Wasify RS, Ali MN, Hamed SR (2018) Application of different magnetic intensities for the treatment of landfill leachate in Egypt. Cogent Eng 5. https://doi.org/10.1080/23311916.2018.1436114

Andersson JL, Bastviken SK, Tonderski KS (2005) Free water surface wetlands for wastewater treatment in Sweden: nitrogen and phosphorus removal. Water Sci Technol 51:39–46. https://doi.org/10.2166/wst.2005.0283

Bharati M, Shete S, Shinkar NP, Kamaltai Gawai S (2013) Comparative study of various treatments for dairy industry wastewater. IOSR J Eng, 3, 42-47. https://doi.org/10.9790/3021-03844247

Bhargava A (2016) Physico-chemical waste water treatment technologies. An overview. Int J Sci Res Educ 4:5308–5319. https://doi.org/10.18535/ijsre/v4i05.05

Birwal P (2017) Advanced technologies for dairy effluent treatment. J Food Nutr Popul Health 1:1

Birwal P, Deshmukh G, Priyanka S, Saurabh SP (2017) Advanced technologies for dairy effluent treatment. J Food Nutr Popul Health 1:1

Britz TJ, Van Schalkwyk C, Hung YT (2006) Treatment of dairy processing wastewaters. Waste treatment in the food processing industry, 1–28. https://doi.org/10.12691/jaem-2-1-4

Burke N, Zacharski KA, Southern M, Hogan P, Ryan MP, Adley CC (2018) The Dairy Industry: Process, Monitoring, Standards, and Quality. In Descriptive Food Science. IntechOpen. https://doi.org/10.5772/intechopen.80398
Carvalho F, Prazeres AR, Rivas J (2013) Cheese whey wastewater: characterization and treatment. Sci Total Environ 445–446:385–396. https://doi.org/10.1016/J.SCITOTENV.2012.12.038
Chandra R, Castillo-Zacarias C, Delgado P, Parra-Saldívar R (2018) A biorefinery approach for dairy wastewater treatment and product recovery towards establishing a biorefinery complexity index. J Clean Prod 183:1184–1196. https://doi.org/10.1016/J.JCLEPRO.2018.02.124
Chaudhry S, Garg S (2018) Industrial wastewater pollution and advanced treatment techniques. In: Advanced treatment techniques for industrial wastewater. Hershey, IGI Global, pp 74–97. https://doi.org/10.4018/978-1-5225-5754-8.ch006
Chen GQ, Talebi S, Gras SL, Weeks M, Kentish SE (2018) A review of salty waste stream management in the Australian dairy industry. J Environ Manag 224:406–413. https://doi.org/10.1016/J.JENVMAN.2018.07.056
Cheremisinoff NP (2002) Handbook of water and wastewater treatment technologies. Butterworth-Heinemann, Oxford
Choi H-J, Choi H-J (2016) Dairy wastewater treatment using microalgae for potential biodiesel application. Environ Eng Res 21:393–400. https://doi.org/10.4491/eer.2015.151
Chokshi K, Pancha I, Ghosh A, Mishra S (2016) Microalgal biomass generation by phycoremediation of dairy industry wastewater: an integrated approach towards sustainable biofuel production. Bioresour Technol 221:455–460. https://doi.org/10.1016/J.BIORTECH.2016.09.070
Cristian O (2010) Characteristics of the untreated wastewater produced by food industry. An Univ Oradea Fasc Prot Med 15:709–714
Central Pollution Control Board (CPCB) (2006) Performance Status of Common Effluent Treatment Plants in India, Control of Urban Pollution Series, CUPS 60/2005-06.
Dąbrowski W, Żyłka R, Malinowski P (2017) Evaluation of energy consumption during aerobic sewage sludge treatment in dairy wastewater treatment plant. Environ Res 153:135–139. https://doi.org/10.1016/j.envres.2016.12.001
Dahiya M, Vij S (2012) Comparative analysis of bioethanol production from whey by different strains of immobilized thermotolerant yeast. Int J Sci Res Publ 2:1–5
De Jesus C-SA, Elba Ruth V-G, Daniel S-FR, Sharma A, De Jesus C-SA, Ruth V-GE, Daniel S-FR, Sharma A (2015) Biotechnological alternatives for the utilization of dairy industry waste products. Adv Biosci Biotechnol 06:223–235. https://doi.org/10.4236/abb.2015.63022
Dela Justina M, Rodrigues Bagnolin Muniz B, Mattge Bröring M, Costa VJ, Skoronski E (2018) Using vegetable tannin and polyaluminium chloride as coagulants for dairy wastewater treatment: a comparative study. J Water Process Eng 25:173–181. https://doi.org/10.1016/J.JWPE.2018.08.001
Deshpande DP, Patil PJ, Anekar SV (2012) Biomethanation of dairy waste. Res J Chem Sci 2:35–39
Dimitrellou D, Kandylis P, Sidira M, Koutinas AA, Kourkoutas Y (2014) Free and immobilized Lactobacillus casei ATCC 393 on whey protein as starter cultures for probiotic Feta-type cheese production. J Dairy Sci 97:4675–4685. https://doi.org/10.3168/jds.2013-7597
Dugmore TIJ, Clark JH, Bustamante J, Houghton JA, Matharu AS (2017) Valorisation of biowastes for the production of green materials using chemical methods. Top Curr Chem 375(2):46
Dutta A, Sarkar S (2015) Sequencing batch reactor for wastewater treatment: recent advances. Curr Pollut Rep 1:177–190. https://doi.org/10.1007/s40726-015-0016-y
Eskicioglu C, Galvagno G, Cimon C (2018) Approaches and processes for ammonia removal from side-streams of municipal effluent treatment plants. Bioresour Technol 268:797–810. https://doi.org/10.1016/J.BIORTECH.2018.07.020
Falkowski M, Maciejczyk M, Koprowicz T, Mikołuć B, Milewska A, Zalewska A, Car H (2018) Whey protein concentrate WPC-80 improves antioxidant defense systems in the salivary glands of 14-month Wistar rats. Nutrients 10. https://doi.org/10.3390/nu10060782

Falsanisi D, Liberti L, Notarnicola M, Falsanisi D, Liberti L, Notarnicola M (2010) Ultrafiltration (UF) pilot plant for municipal wastewater reuse in agriculture: impact of the operation mode on process performance. Water 2:872–885. https://doi.org/10.3390/w2040872

Faria A, Gonçalves L, Peixoto JM, Peixoto L, Brito AG, Martins G (2017) Resources recovery in the dairy industry: bioelectricity production using a continuous microbial fuel cell. J Clean Prod 140:971–976. https://doi.org/10.1016/j.jclepro.2016.04.027

Frenkel VS (2008) Membranes in water and wastewater treatment. In: World environmental and water resources congress 2008. American Society of Civil Engineers, Reston, VA, pp 1–10

Ganju S, Gogate PR (2017) A review on approaches for efficient recovery of whey proteins from dairy industry effluents. J Food Eng 215:84–96. https://doi.org/10.1016/J.JFOODENG.2017.07.021

Gil-Pulido B, Tarpey E, Almeida EL, Finnegan W, Zhan X, Dobson ADW, O'Leary N (2018) Evaluation of dairy processing wastewater biotreatment in an IASBR system: aeration rate impacts on performance and microbial ecology. Biotechnol Rep 19:e00263. https://doi.org/10.1016/J.BTRE.2018.E00263

Goli A, Shamiri A, Khosroyar S, Talaiekhozani A, Sanaye R, Azizi K (2019) A review on different aerobic and anaerobic treatment methods in dairy industry wastewater. J Environ Treat Tech 6(1):113–141

Gopinathan M, Thirumurthy M (2012) Feasibility studies on static pile co-composting of organic fraction of municipal solid waste with dairy waste water. Environ Res Eng Manag 60:34–39. https://doi.org/10.5755/j01.erem.60.2.963

Hunt PG, Poach ME (2001) State of the art for animal wastewater treatment in constructed wetlands. Water Science and Technology, 44(11–12), 19–25. https://doi.org/10.2166/wst.2001.0805

Jaganmai G, Jinka R (2017) Production of lipases from dairy industry wastes and its applications. Int J Curr Microbiol Appl Sci 5:67–73

Kasmi M (2018) Biological processes as promoting way for both treatment and valorization of dairy industry effluents. Waste Biomass Valorization 9:195–209. https://doi.org/10.1007/s12649-016-9795-7

Kassem JM (2015) Future challenges of whey proteins. Int J Dairy Sci 10:139–159. https://doi.org/10.3923/ijds.2015.139.159

Kavitha RV, Kumar S, Krishnamurthy V (2013) Performance evaluation and biological treatment of dairy waste water treatment plant by upflow anaerobic sludge blanket reactor. Int J Chem Petrochem Technol 3:9–20

Khalid S, Shahid M, Natasha BI, Sarwar T, Shah AH, Niazi NK (2018) A review of environmental contamination and health risk assessment of wastewater use for crop irrigation with a focus on low and high-income countries. Int J Environ Res Public Health 15. https://doi.org/10.3390/ijerph15050895

Kiesewetter J, Kraakman B, Cesca J, Trainor S, Witherspoon J (2012) Expanding the use of activated sludge at biological waste water treatment plants for odor control. Proc Water Environ Fed 2012:600–614. https://doi.org/10.2175/193864712811700363

Kolev Slavov A (2017) Dairy wastewaters – general characteristics and treatment possibilities—a review. Food Technol Biotechnol 55:14–28. https://doi.org/10.17113/ftb.55.01.17.4520

Kushwaha JP, Srivastava VC, Mall ID (2011) An overview of various technologies for the treatment of dairy wastewaters. Crit Rev Food Sci Nutr 51:442–452. https://doi.org/10.1080/10408391003663879

Kweinor Tetteh E, Rathilal S (2019) Application of organic coagulants in water and wastewater treatment. In: Organic polymers [Working title]. IntechOpen

Linares DM, Gómez C, Renes E, Fresno JM, Tornadijo ME, Ross RP, Stanton C (2017) Lactic acid bacteria and Bifidobacteria with potential to design natural biofunctional health-promoting dairy foods. Front Microbiol 8:846. https://doi.org/10.3389/fmicb.2017.00846

Lumina P, Pavithra MP (2018) Treatability studies of dairy wastewater by electrocoagulation process. Int J Appl Eng Res 13:249–252

Madhavrao Kanawade S, Bhusal VC (2015) Adsorption on dairy industrial wastewater by using activated charcoal as adsorbent. Int J Chem Mater Sci 3:25–032

Milani FX, Nutter D, Thoma G (2011) Invited review: environmental impacts of dairy processing and products: a review. J Dairy Sci 94:4243–4254. https://doi.org/10.3168/JDS.2010-3955

Morales-Polo C, del Mar Cledera-Castro M, Moratilla Soria BY (2018) Reviewing the anaerobic digestion of food waste: from waste generation and anaerobic process to its perspectives. Appl Sci 8:35. https://doi.org/10.3390/app8101804

Mosa KA, Saadoun I, Kumar K, Helmy M, Dhankher OP (2016) Potential biotechnological strategies for the cleanup of heavy metals and metalloids. Front Plant Sci 7:303. https://doi.org/10.3389/fpls.2016.00303

Musa MA, Idrus S, Hasfalina CM, Daud NNN (2018) Effect of organic loading rate on anaerobic digestion performance of mesophilic (UASB) reactor using cattle slaughterhouse wastewater as substrate. Int J Environ Res Public Health 15. https://doi.org/10.3390/ijerph15102220

Nqombolo A, Mpupa A, Moutloali RM, Nomngongo PN (2018) Wastewater treatment using membrane technology. In: Wastewater and water quality. InTech

Pasotti L, Zucca S, Casanova M, Micoli G, Cusella De Angelis MG, Magni P (2017) Fermentation of lactose to ethanol in cheese whey permeate and concentrated permeate by engineered *Escherichia coli*. BMC Biotechnol 17:48. https://doi.org/10.1186/s12896-017-0369-y

Pathak U, Das P, Banerjee P, Datta S (2016) Treatment of wastewater from a dairy industry using rice husk as adsorbent: treatment efficiency, isotherm, thermodynamics, and kinetics modelling. J Thermodyn 2016:1–7. https://doi.org/10.1155/2016/3746316

Porwal HJ, Mane AV, Velhal SG (2015) Biodegradation of dairy effluent by using microbial isolates obtained from activated sludge. Water Resour Ind 9:1–15. https://doi.org/10.1016/J.WRI.2014.11.002

Raghunath B V., Punnagaiarasi A, Rajarajan G, Irshad A, Elango A, Mahesh Kumar G (2016) Impact of dairy effluent on environment—a review. In: Integrated waste management in India, environmental science and engineering. Springer, pp 239–249

Reilly M, Cooley AP, Tito D, Tassou SA, Theodorou MK (2019) Electrocoagulation treatment of dairy processing and slaughterhouse wastewaters. Energy Procedia 161:343–351. https://doi.org/10.1016/J.EGYPRO.2019.02.106

Sahu OP, Chaudhari PK (2013) Review on chemical treatment of industrial waste water. J Appl Sci Environ Manag 17:241–257

Samer M (2015) Biological and chemical wastewater treatment processes. In: Wastewater treatment engineering. InTech

Sfakianakis P, Tzia C (2014) Conventional and innovative processing of milk for yogurt manufacture; development of texture and flavor: a review. Foods (Basel, Switzerland) 3:176–193. https://doi.org/10.3390/foods3010176

Shen J, Chen J, Jensen PR, Solem C (2019) Development of a novel, robust and cost-efficient process for valorizing dairy waste exemplified by ethanol production. Microb Cell Factories. https://doi.org/10.1186/s12934-019-1091-3

Shete BS, Shinkar NP (2013) Comparative study of various treatments for dairy industry wastewater. IOSR J Eng 3:42–47. https://doi.org/10.9790/3021-03844247

Shivayogimath CB, Naik VR (2014) Treatment of dairy industry wastewater using electrocoagulation technique. Int J Eng Res Technol 3:4

Singh AK, Singh G, Gautam D, Bedi MK (2013) Optimization of dairy sludge for growth of rhizobium cells. Biomed Res Int 2013:845264. https://doi.org/10.1155/2013/845264

Singh N, Singh R, Manzer M (2014) Waste water management in dairy industry: pollution abatement and preventive attitudes. Int J Sci Environ Technol 3:672–683

Slavov AK (2017) General characteristics and treatment possibilities of dairy wastewater—a review. Food Technol Biotechnol 55:14–28. https://doi.org/10.17113/ftb.55.01.17.4520

Spalvins K, Zihare L, Blumberga D (2018) Single cell protein production from waste biomass: comparison of various industrial by-products. Energy Procedia 147:409–418. https://doi.org/10.1016/J.EGYPRO.2018.07.111

Stefanakis A, Akratos CS, Tsihrintzis VA (2014) Vertical flow constructed wetlands—eco-engineering systems for wastewater and sludge treatment. Ecol Eng:392. https://doi.org/10.1016/j.ecoleng.2015.06.039

Sukhadev Shivsharan V, Kulkarni SW, Wani M (2013) Physicochemical characterization of dairy effluents. Int J Life Sci Biotechnol Pharma Res 2:1–12

Tikariha A, Sahu O (2014) Study of characteristics and treatments of dairy industry waste water. J Appl Environ Microbiol 2:16–22. https://doi.org/10.12691/jaem-2-1-4

Tsihrintzis VA (2017) The use of vertical flow constructed wetlands in wastewater treatment. Water Resour Manag 31:3245–3270. https://doi.org/10.1007/s11269-017-1710-x

Van Lier JB, Mahmoud N, Zeeman G (2008) Anaerobic wastewater treatment. IWA Publishing, London

Verma A, Singh A (2017) Physico-chemical analysis of dairy industrial effluent. Int J Curr Microbiol Appl Sci 6:1769–1775. https://doi.org/10.20546/ijcmas.2017.607.213

Vymazal J (2010) Constructed wetlands for wastewater treatment. Water 2:530–549. https://doi.org/10.3390/w2030530

Wan C, Li Y, Shahbazi A, Xiu S (2008) Succinic acid production from cheese whey using Actinobacillus succinogenes 130 Z. Appl Biochem Biotechnol 145:111–119. https://doi.org/10.1007/s12010-007-8031-0

Wong YM, Show PL, Wu TY, Leong HY, Ibrahim S, Juan JC (2019) Production of bio-hydrogen from dairy wastewater using pretreated landfill leachate sludge as an inoculum. J Biosci Bioeng 127:150–159. https://doi.org/10.1016/J.JBIOSC.2018.07.012

Yonar T, Sivrioglu O, Ozengin N (2018) Physico-chemical treatment of dairy industry wastewaters: a review. In: Technological approaches for novel applications in dairy processing. InTechOpen

Yousefzadeh S, Ahmadi E, Gholami M, Ghaffari HR, Azari A, Ansari M, Miri M, Sharafi K, Rezaei S (2017) A comparative study of anaerobic fixed film baffled reactor and up-flow anaerobic fixed film fixed bed reactor for biological removal of diethyl phthalate from wastewater: a performance, kinetic, biogas, and metabolic pathway study. Biotechnol Biofuels 10:139. https://doi.org/10.1186/s13068-017-0826-9

Zheng C, Zhao L, Zhou X, Fu Z, Li A (2013) Treatment technologies for organic wastewater. In: Water treatment. InTech

Zidan ARA, El-Gamal MM, Rashed AA, El-Hady Eid MAA (2015) Wastewater treatment in horizontal subsurface flow constructed wetlands using different media (setup stage). Water Sci 29:26–35. https://doi.org/10.1016/J.WSJ.2015.02.003

Zolfaghari G, Kargar M (2019) Nanofiltration and microfiltration for the removal of chromium, total dissolved solids, and sulfate from water. MethodsX 6:549–557. https://doi.org/10.1016/j.mex.2019.03.012

Zylka R, Dabrowski W, Gogina E, Yancen O (2018) Trickling filter for high efficiency treatment of dairy sewage. J Ecol Eng 19:269–275. https://doi.org/10.12911/22998993/89657

Chapter 12
Treatment of Textile Wastewater by Dual Coagulant from Fe(III) and Purple Okra (*Abelmoschus esculentus*) Waste

Thabata Karoliny Formicoli Souza Freitas, Elizangela Ambrosio, Fernando Santos Domingues, Henrique Cesar Lopes Geraldino, Maísa Tatiane Ferreira de Souza, Renata Padilha de Souza, and Juliana Carla Garcia

Contents

T. K. F. S. Freitas (✉) · E. Ambrosio · F. S. Domingues · H. C. L. Geraldino · M. T. F. de Souza · J. C. Garcia
GPDMA, Department of Chemistry, State University of Maringá, Maringá, Paraná, Brazil
e-mail: jucgarcia@ibest.com.br

R. P. de Souza
Department of Engineering and Biotechnology, Federal University of Technology-Paraná, Dois Vizinhos, Paraná, Brazil

Inamuddin, A. M. Asiri (eds.), *Sustainable Green Chemical Processes and their Allied Applications*, Nanotechnology in the Life Sciences,
https://doi.org/10.1007/978-3-030-42284-4_12

12.1 Introduction

The rapid progress in industrialization and urbanization contributes significantly to the release of organic and inorganic pollutants in soil, water, and air (Ren et al. 2017). Among the various industries, the textile segment is considered as one of the major contributors of water pollution since it consumes a high quantity of water and chemical, which contains various toxic agents, generating wastewater with an extremely complex matrix and that is heterogeneous (Souza et al. 2016; Bilińska et al. 2019; Khan et al. 2019).

The negative impacts due to the irregular release of textile wastewaters into waterbodies represent a major environmental concern. The highly colored wastewater not only causes aesthetic problems, as they changed the visual aspect of the aquatic bodies, but the presence of color inhibits the penetration of light, reducing the photosynthetic activity of aquatic plants and, consequently, the amount of dissolved oxygen available in the waterbodies (Verma et al. 2012; Chhabra et al. 2015; Lafi et al. 2018). Moreover, textile dyes or their decomposition derivatives have carcinogenic and/or mutagenic effect, being potentially toxic to aquatic and human life (Garcia et al. 2009; Ito et al. 2018; Sen et al. 2019). Therefore, textile wastewater, if discharged without appropriate prior treatment in natural waterbodies, poses a serious environmental hazard (Lafi et al. 2018; Khan et al. 2019).

Coagulation/flocculation technology is a process widely used for wastewater treatment in many industry fields (Choumane et al. 2017; Wei et al. 2018; Kim et al. 2019). This technique remains an advantageous process due to it is economical, simple, robust, efficient, easily operated, and consumes less energy compared to other wastewater treatment methods (Braga et al. 2018; Boulaadjoul et al. 2018; Mirbahoush et al. 2019). It not only removes suspended solid that causes turbidity in water, but also natural organic matter is eliminated (Huang et al. 2014).

To destabilize and aggregate the colloids and other substances dispersed, coagulant agents are added to wastewater. The opposite charges are neutralized, and, consequently, the repulsive potential of colloids is reduced, and microparticles are produced. From the collision of these microparticles with each other, larger particles are formed, called flocs, which are more easily removed. Coagulant aids are sometimes added after coagulation to accelerate particle aggregation, strengthen the flocs, and increase settling efficiency. These coagulant aids, commonly plant-based coagulants, act as bridges that adsorb and bind various colloids. Thereby, the size of floc increases and can be effectively removed by sedimentation. The latter process is called flocculation (de Souza et al. 2014; Wei et al. 2018; Lopes et al. 2019).

Although iron and aluminum inorganic coagulants are the most widely used reagents in the world (Sillanpää et al. 2018), the use of Al-based coagulants is questionable due to their toxicological potential. They can generate large volumes of sludge and a high level of residual aluminum in both treated water and sludge (Camacho et al. 2017; Zhu et al. 2011). Their accumulation in the brain has been associated with several neuropathological conditions such as Alzheimer's disease

(Ravi et al. 2018; Mirza et al. 2017), autism spectrum disorder (Mold et al. 2018b; Gamakaranage 2016), epilepsy (Mold et al. 2019), and multiple sclerosis (Mold et al. 2018a).

Another class of coagulants that is also available in the national and international market is the polymeric coagulants (synthetic origin). However, they also have some disadvantages compared to those of natural origin: (1) have low biodegradability sludge generation, (2) cannot be recycled as easily as those of natural origin (Hussain et al. 2019), and (3) may bring serious health hazards as they have carcinogenic, mutagenic, and neurotoxic effects (Antov et al. 2018; Mallevialle et al. 1984). One example is acrylamide (monomer of polyacrylamide used globally in water treatment) classified as neurotoxic in animals and humans, carcinogen and mutagen agent and reprotoxic (Shipp et al. 2006). It may highly dangerous and lethal at low concentrations (Xiong et al. 2018).

This way, new or improved alternatives are essential to reduce the negative impact on the environment and human health caused by using inorganic and synthetic coagulants. The use of natural coagulants has potential to be a possible solution (Antov et al. 2018; Megersa et al. 2019). Natural coagulants are derived from plants, microorganisms, and animals. Among them, plant-based source ones are the resources most available worldwide, becoming a viable alternative to inorganic and synthetic coagulants. For such reason, they receive an increasing importance over recent years (Choy et al. 2013).

Natural coagulants are cheap and economically feasible and have an environmentally friendly behavior, low toxicological risk (or inexistent), high biodegradability, and little environment impact, and they are safe (de Souza et al. 2016; Choy et al. 2016). They are also known for generating less sludge, reducing the quantity of residuals heavy metals, the amount of chemical coagulants, and the corrosive impact on the equipment caused by metals. Depending on their extraction source, may be harvested and processed locally (Bolto and Gregory 2007; Yin 2010; Antov et al. 2018).

The scientific community has shown great interest in studying the application of plant-based coagulants in wastewater treatment (Megersa et al. 2019; Tawakkoly et al. 2019; Antov et al. 2018), i.e., cassava (dos Santos et al. 2018), *Cereus peruvianus* cactus (de Souza et al. 2016), chayote waste (Almeida et al. 2017), chickpea (Lek et al. 2018), common oak (Antov et al. 2018), *Maerua subcordata* (Megersa et al. 2019), *Moringa oleifera* (Dotto et al. 2019), *Opuntia ficus-indica* cactus (Choudhary et al. 2019), orange industry residues (Kebaili et al. 2018), pine cone (Hussain et al. 2019), *Salvia hispanica* (Tawakkoly et al. 2019), starches (Choy et al. 2016), and tannin (Lopes et al. 2019). But there is no scientific research about applying purple okra residues on effluent treatment. This motivated us to study the use of purple okra waste as natural coagulant to intensify coagulation/flocculation treatment of textile wastewater, containing Fe(III) as primary coagulant, and to monitor acute toxicity to treated wastewater.

12.2 Materials and Methods

Textile wastewater was collected from stabilization pond in industrial laundry located in Brazil. Before proceeding to the stabilization pond, a pretreatment (mechanical sieving) was carried out to remove the coarse solids and floating bodies at wastewater treatment plant.

Purple okra pods were collected in West Parana State, Brazil. Darkened and wilted pods were chosen because their unappealing appearance and deprecated economic value disqualify them as fresh fruit for consumption. NaCl, $FeCl_3.6H_2O$ (FeIII) and $K_2Cr_2O_7$ were obtained from Synth (Brazil). Commercial polymers were donated by Faxon Química (Brazil), and they are available in the national market. Synthetic polyelectrolytes are high molecular weight polyacrylamide-based polymers. FXCS8®, FXCS4®, and FXCE2® are cationic, and FXNS1® and FXNS2® are non-ionic.

12.2.1 *Preparation of Gum and Mucilage from Purple Okra*

The pods were cleaned and washed with tap water up to room temperature (25 °C) to remove any dirt. The purple okra gum is deposited between exocarp (peel) and endocarp (Fig. 12.1). The pods were cut in half, seeds; septum and excess white fibers from endocarp were removed. Then, okra gum is easily removed from mesocarp by scraping with a spatula. Mucilage was prepared according to the methodology proposed by Okuda et al. (1999) and Freitas et al. (2015). Purple okra mucilage was denoted POM.

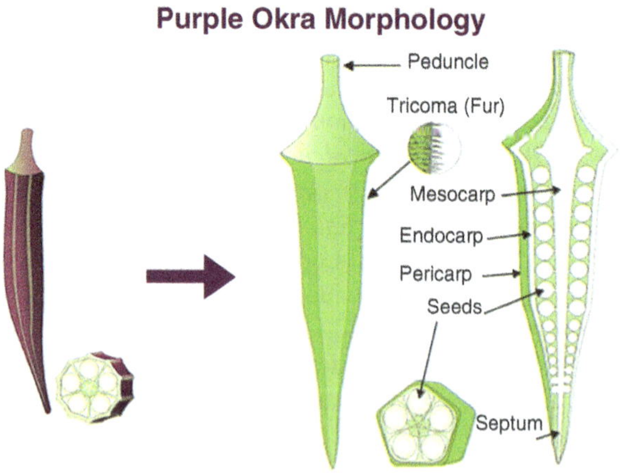

Fig. 12.1 Vegetal morphology of okra

12.2.2 *Coagulation/Flocculation Process*

Assays were performed with $FeCl_3.6H_2O$ as primary coagulant and POM as coagulant aids. All tests were carried out using 250 mL of textile wastewater in a jar tester apparatus (Millan—JT 203/6, Brazil) in triplicate. Experimental conditions for rapid mixing speed, slow mixing speed, and mixing time were described by de Souza et al. (2014). For analytical measurements, 30 mL of supernatant solution was collected after 1 h of sedimentation.

12.2.2.1 Process Using Inorganic Coagulant

CF process using only inorganic coagulant was used as control for comparison with the treatment using dual coagulant. The $FeCl_3.6H_2O$ at 320 mg L^{-1} was used and added to flasks before rapid mixing. Coagulation pH was adjusted to 6.0 (Freitas et al. 2015).

12.2.2.2 Process Using Dual Coagulant

When a natural polymer is used as coagulant aid and added after primary coagulant, it is called a dual coagulant (Wang et al. 2005). In this work, dual coagulant of mixed $FeCl_3.6H_2O$ and POM was denoted as Fe-POM.

Firstly, to evaluate the effect of pH (at temperature of 25 °C), medium pH was adjusted in the range of 3.0–8.0. 120 mg L^{-1} of $FeCl_3.6H_2O$ and 24.0 mg L^{-1} of POM were added to wastewater, respectively. The effect of inorganic coagulant was studied by varying the concentration of $FeCl_3.6H_2O$ (40.0–160.0 mg L^{-1}). Finally, the POM concentration was varied between 4.0 and 28.0 mg L^{-1} to assess the effect of natural coagulant.

12.2.2.3 Comparison Between Biopolymer from Purple Okra Waste and Synthetic Commercial Polymer

The comparison assays were carried out using 80.0 mg L^{-1} of $FeCl_3.6H_2O$ and 20.0 mg L^{-1} of POM or synthetic polymer and pH 6.0.

12.2.3 *Analytical Methods*

Turbidity (2130 B method), color (2120 B method), and chemical oxygen demand (COD, 5220 D method) were carried out according to methodologies described by APHA (1998). Absorbance at wavelength of 254 nm (ABS_{254nm}) was measured in a

UV-Vis spectrophotometer (Lambda 25, Perkin Elmer). Removal efficiency of coagulation/flocculation was calculated from Eq. 12.1. The graphics were plotted using OriginPro® 9.0.

$$\text{Removal}(\%) = \left(\frac{C_o - C_f}{C_o} \right) \times 100 \tag{12.1}$$

where C_o and C_f are the initial and final concentration, respectively.

12.2.4 Characterization of Purple Okra Mucilage

For the FTIR and ^{13}C CP/MAS measurements, the POM was dried by lyophilization and macerated. Two milligrams de POM was dispersed in 198 mg of potassium bromide powder (KBr, spectroscopic grade, Merck) and pressed to form KBr pellets at 3 ton. FTIR spectra were achieved at 20 °C under vacuum in range 400–4000 cm^{-1} using an FTIR spectrometer (Bruker Vertex, −70 V, Germany) with 64 scans at resolution of 4 cm^{-1}. Nuclear magnetic resonance of ^{13}C (NMR) spectrum was obtained using an NMR spectrometer (Varian Mercury Plus 300 model) coupled with solid probe CP/MAS.

12.2.5 Acute Toxicity Assays with A. salina

The nutritive solution was prepared using 23.0 g of NaCl in 1000 mL of mineral water, pH of this solution was adjusted between 8.0 and 9.0 with Na_2CO_3 (0.1 mol L^{-1}). Next, the cyst-like eggs were placed in this solution and hatched within 48 h under constant aeration at 25 °C. The most resistant nauplii were used as test organism. For this, artificial illumination (20 W) was placed on one of the sides of recipient to attract them.

Five milliliters of samples with the following concentration was added in a glass tube: 200, 400, 600, 800, and 1000 mL L^{-1}. Ten larvae were placed in each tube. Then, they were incubated for 24 h at 25 °C in the presence of artificial illumination. As a negative control, pure nutritive solution was used and, as a positive control, dichromate solution at 2.6 g L^{-1}. Mortality rate was calculated by Eq. 2 and LD_{50} using statistical software PriProbit with a confidence interval of 95%.

$$\text{Mortality rate}(\%) = \left(\frac{\text{dead nauplii}}{\text{total nauplii}} \right) \times 100. \tag{12.2}$$

12.3 Results and Discussion

12.3.1 Purple Okra Mucilage Characterization

The polymeric structures and chemical characteristics of natural coagulants, such as ionization and copolymerization degrees, as well as the number of substituted groups in the structure, are directly related to the degree of pollutant removal (Kakoi et al. 2016). FTIR (Fig. 12.2) and CP/MAS ^{13}C solid-state NMR (Fig. 12.3) spectra were accomplished to identify the nature of functional groups present in the POM.

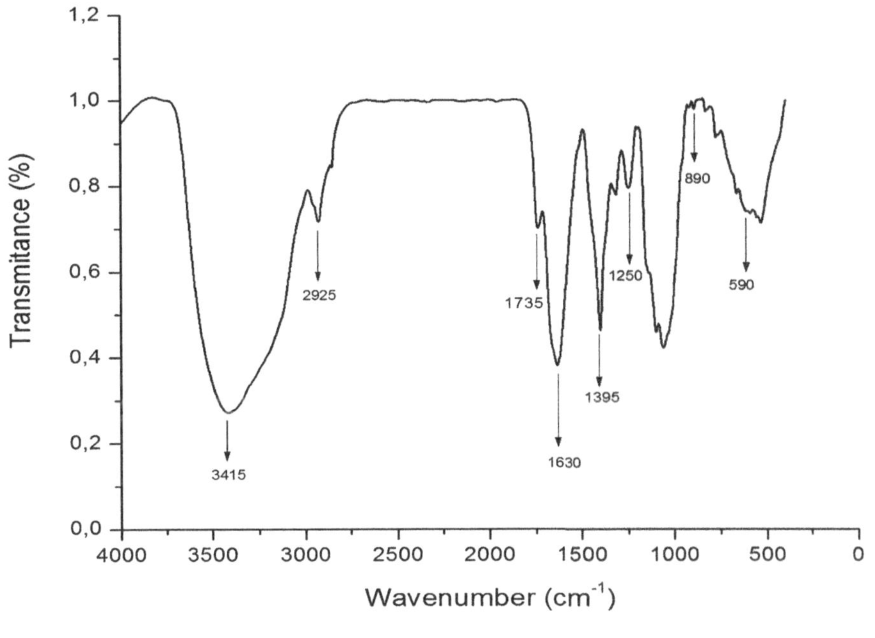

Fig. 12.2 FTIR spectrum of POM at range 400–4000 cm^{-1}

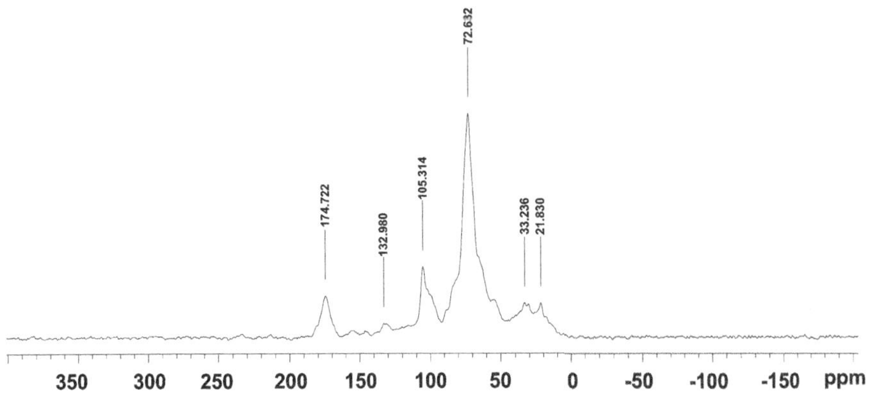

Fig. 12.3 ^{13}C CP/MAS NMR spectrum of purple okra mucilage

FTIR spectra showed intense and broad peak around 3415 cm^{-1} which corresponds to the hydroxyl groups (–OH). This group is typical in macromolecules such as carbohydrates due to the vibrational stretches associated with intermolecular and intramolecular bound free hydroxyl groups (Emeje et al. 2011). The asymmetric C–H characteristic peak, including CH, CH_2, and CH_3 stretching vibrations, was observed at 2925 cm^{-1} (Wu et al. 2012; Alba et al. 2015).

The absorption peaks at 1735 cm^{-1} and 1395 cm^{-1} indicate the presence of carboxylic groups (–COOH) in complex polysaccharides, which are characteristic of C–O–C and C–O stretching vibration (de Souza et al. 2016). A peak at 1250 cm^{-1} confirms the presence of ketones, aldehydes, and lactones or carboxyl groups (Kakoi et al. 2016).

C=O asymmetric stretching of double bond of deprotonated carboxylate functional groups (–COO) present on polysaccharide is characterized by intense peak around 1630 cm^{-1}. Polysaccharides containing pyrene monomers in their structures have specific bands at 1000–1200 cm^{-1} region. This region is superimposed by the C–O–C glycosidic band and stretching vibrations of C–OH side groups and may indicate certain similarities in the monosaccharide composition (Zhao et al. 2010; Zheng et al. 2014). The typical peak at 895 cm^{-1} and peak at 590 cm^{-1} confirm the β-glycosidic linkages between the monosaccharides (de Rosa et al. 2010).

^{13}C solid-state NMR spectrum detected signals typical of mono-, di-, and polysaccharides. The galacturonic acid structure, in the form of carboxylate anion (COO^-), is related by presence of carbon of $\underline{C}OOH$ group at 174.7 ppm. The carbons of acetyl groups ($O\underline{C}OCH_3$) are located in this same region. The signal at 105.3 and 72.6 ppm are attributed to the carbon of O–$\underline{C}$–O and $\underline{C}H$–O, relatively. The peaks of carbons from pyranoid rings are between 60 and 90 ppm. Spectra have signals at 21.8 and 33.2 ppm, which correspond to -CH_2 from $OCO\underline{C}H_2$ and –CH_3 from $OCO\underline{C}H_3$ (Nurdjanah et al. 2013; Zhu et al. 2014; Sinitsya et al. 1998).

The investigation of the main compounds (galacturonic acid) present in purple okra (*Abelmoschus esculentus*) mucilage is very important for a better understanding of its application in wastewater treatment as natural coagulant (Bathista et al. 2004). This typical polysaccharide (galacturonic acid) due to chemical characteristics promotes a better coagulation of colloidal particles of wastewater. The availability of different functional groups in POM indicates the presence of active sites for the retention of colloids onto natural coagulants through interparticle bridging and/or charge neutralization mechanism. Consequently, it promotes the destabilization of colloids and facilitates the removal of suspended and dissolved solids (Freitas et al. 2015; Kakoi et al. 2016).

The groups containing hydroxyl (–OH), mainly, phenols, carboxylic acids, and alcohols, exhibit a wide absorption range, leading to an improvement in coagulation capacity by charge neutralization and/or bridging. POM can be considered as a long-chain polymer because it has double bond hydrocarbons such as alkenes and alkanes. The interparticle bridging is also carried out by presence of a long-chain polymer (Dhivya et al. 2017).

Table 12.1 Physical-chemical characteristics of textile wastewater before and after coagulation/flocculation using Fe-POM (mean ± standard deviation)

Parameters	Unity	Raw TW	Treated water	Standards of release[a]
pH	–	7.2 ± 0.2	6.2 ± 0.2	5.0–9.0
Turbidity	NTU	145 ± 3	2.6 ± 1.5	100
COD	mg L^{-1}	884 ± 2	184 ± 4	200
$BOD_{5,20}$	mg L^{-1}	379 ± 6	36 ± 2	50
Temperature	°C	35 ± 2	24 ± 1	40
Color	mgPtCo L^{-1}	384 ± 1	12 ± 1	75
DO	mg L^{-1}	0.0	5.8	> 5.0
NOM	ABS	1.08	0.18	–

Determination in triplicate

TW textile wastewater, *COD* chemical oxygen demand, $BOD_{5,20}$ biochemical oxygen demand incubated for 5 days at 20 °C, *DO* dissolved oxygen, *NTU* nephelometric turbidity unit, *°C* Celsius degree, *mgPtCo* L^{-1} milligrams platinum-cobalt per liter, *NOM* natural organic matter represented by ABS_{254nm}, ABS_{254nm} absorbance at 254 nm

[a]Standards of release of textile wastewater established by environmental Brazilian legislation, Conama Resolution no. 430/2011 and CEMA Resolution no. 70/2009

12.3.2 Coagulation/Flocculation Treatment

Textile wastewater showed neutral pH, high color, moderate turbidity, low dissolved oxygen, and high organic matter expressed by high biochemical and chemical oxygen demand ($BOD_{5,20}$; COD) and high concentration of humic acid expressed by ABS_{254nm}, as shown in Table 12.1. According to Brazilian legislation in force, the observed data indicated that this textile wastewater must not be disposed into waterbody without proper treatment (Table 12.1). Unfortunately, with the exception of pH and temperature, all parameters fall outside Brazilian release standard. Coagulation/flocculation process using the novel dual coagulant (Fe-POM) removed 98.2%, 93.8%, 79.2%, and 83.2% of turbidity, color, COD, and ABS_{254nm}, respectively. Treatment with natural coagulant achieved textile wastewater release standards in all parameters.

12.3.3 Effect of pH

Coagulation/flocculation is a highly pH-dependent phenomenon (Stephenson and Duff 1996). It can change the charge of both coagulant and ionic polymers used in the process and, consequently, enhance or reduce the coagulation process efficiency (Somasundaran et al. 2005; Antov et al. 2018).The effect of pH (3.0–8.0) of textile wastewater on the coagulation/flocculation process efficiency is shown in Fig. 12.4.

Maximum coagulation is attained at both pH 5.0 and 6.0 with the removal of turbidity (98.2 and 95.3%), color (87.0 and 90.6%), COD (66.2 and 66.7%), and ABS_{254nm} (56.1 and 56.7%). It can be observed that all parameters, in pH 5.0 and

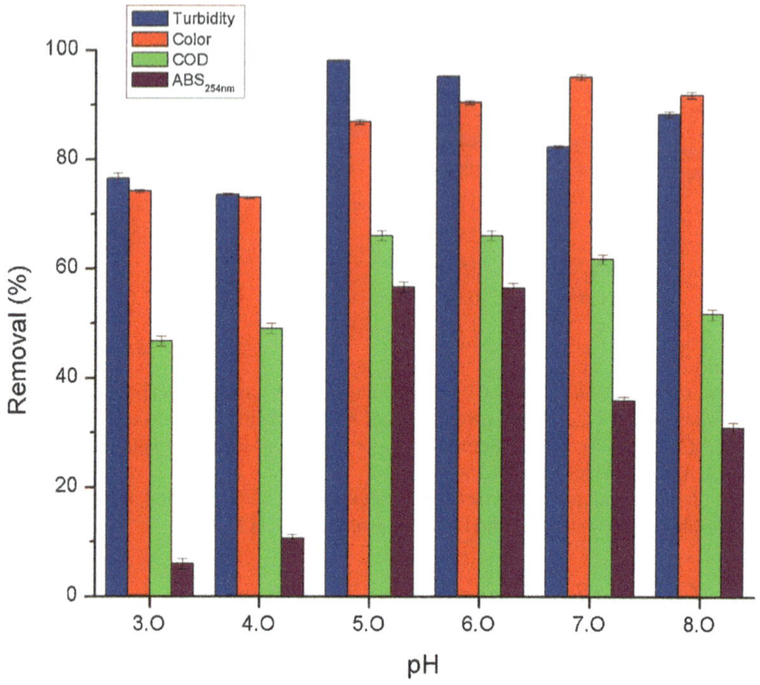

Fig. 12.4 The effect of pH on turbidity, COD, color, and ABS_{254nm} removal after addition of 120.0 mg L^{-1} Fe^{3+} and 24.0 mg L^{-1} POM (mean ± standard deviation)

6.0, exhibit similar behavior. Thus, pH 6.0 was considered optimum because it has maximum removal of organic matter (expressed both ABS_{254nm} and COD) and suspends solids (expressed both turbidity and color); it is mildly acidic and near neutrality. pH values far from neutrality have impacts on natural process of aquatic biota, thereby contributing to the growth of microorganisms (von Sperling 2014).

Agarwal et al. (2001, 2003) found the pH 7.0 as optimum using okra gum for sewage and tannery wastewater treatment. Freitas et al. (2015) used green okra gum for textile wastewater, and optimum pH was 6.0. Zhang et al. (2006) and de Souza et al. (2014) used cactus *Opuntia ficus* for treatment of water and textile wastewater, and, for both, optimum pH was 6.0.

When metallic coagulants are dissolved in water, two types of chemical equilibria, based on pH variation, may compete in the polymeric metal species formed: (1) if the pH is too low, some organic acids may not precipitate, resulting in poor pollutant removal, because there is a competition between the metallic ions and the excess protons for the organic ligands of the pollutants, and (2), at a higher pH, metallic ions may precipitate as metal hydroxides, and OH ion excess may compete with organic compounds for metallic adsorption sites (Stephenson and Duff 1996).

Highlighting the pH effect on the efficiency of Fe(III)-based coagulants, at pH 6.0, $Fe(OH)_2^+$ and $Fe(OH)_3$ are formed. In the same pH value the majority of colloids, including the dyes, are negatively charged and can be stable. Hence, Fe(III)-based coagulants are able to neutralize both the negatively charged dyes and organic matter without charge from textile wastewater in pH 6.0. These dyes and other polar molecules, with their oxygen-containing functional groups such as hydroxyl or carboxyl, react with the iron cations leading to colloid destabilization and consequent precipitation of flocs. While the dissolved organic compounds without charges are removed mainly by sorption onto the metal hydroxide surface (Duan and Gregory 2003). Thus, Fe(III)-based coagulants are able to neutralize the negatively charged dyes of textile wastewater in pH 6.0.

The possible active coagulant agent of POM is galacturonic acid (El-Mahdy and El-Sebaiy 1984; Miller et al. 2008; Alba et al. 2015). The galacturonic acid containing a carboxylic group on one of the polymeric chains is pH sensitive and can possess different conformations, molecular or ionized, with pH variation due to ionization of –COOH groups (Yu and Somasundaran 1996; Bolto and Gregory 2007). The greater efficacy of POM in the coagulation process occurred at pH 6.0, which can be attributed to the addition of protons (H^+) to some of the functional groups such as amino and carboxyl. As a result, a high positive charge density is formed on the POM that can neutralize or reduce the negative charges of colloids (Kakoi et al. 2016); as a consequence, there is a reduction in the repulsive force and an increase in the force of attraction (van der Waals forces) between POM and colloids to form settleable flocs (Ebeling et al. 2006).

12.3.4 Effect of Inorganic Coagulant Dosage

Coagulant dosage is essential for the process to successfully achieve the quality standards of treated wastewater. Often, a wrong coagulant dose leads to unsatisfactory results mainly, because of the lack of technique ajusted or poor optimization project (Choy et al. 2013). The effect of Fe(III)-based coagulant dosage on the coagulation/flocculation process is shown in Fig. 12.5.

The greatest removal of 98.0% in turbidity, 92.9% in color, 73.8% in COD, and 76.7% in ABS_{254nm} was found at 80 mg L^{-1} of Fe^{3+}. But, for removing only suspended and dissolved solids and natural organic matter (expressed by ABS_{254nm}) could be used the dosages of 40, 60, and 80 mg L^{-1} because they do not present significant differences ($P > 0.05$).

For an adequate organic matter removal, it is necessary to use the correct coagulant dosage, so coagulation/flocculation and sedimentation of flocs can occur with maximum effectiveness and minimum time. For ideal condition of coagulation must gradually increase the inorganic coagulant dosage until the optimum range has positive impact on increasing organic matter removal. At low dosage, there are not enough hydrolyzed species that are positively charged to form flocs, and coagulation/flocculation is incomplete. Alternately, when there is an excess of metallic coagulant, in the case of Fe^{3+}, there is an increase in the net positive charge, creating repulsive forces between colloids and ferric coagulant (Ebeling et al. 2006; Lenzi et al. 2012).

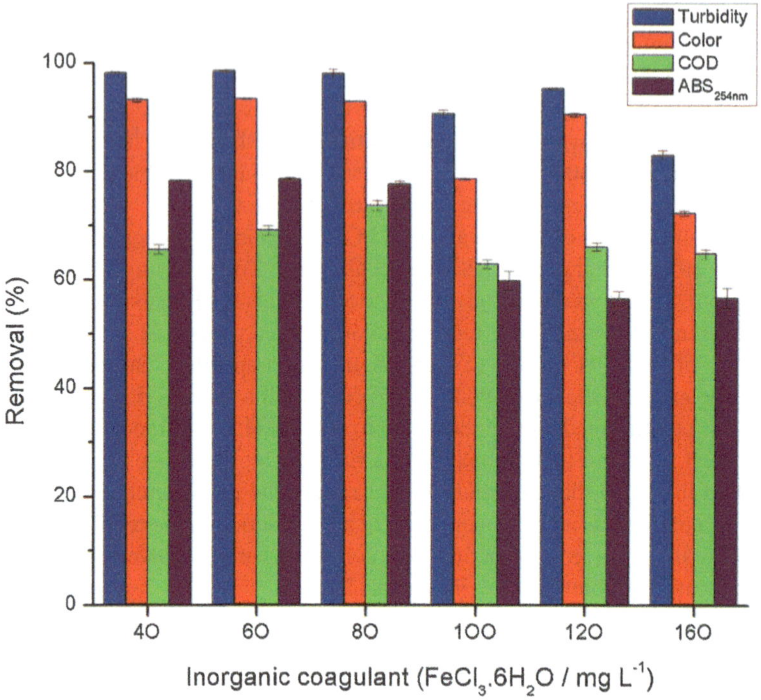

Fig. 12.5 Effect of inorganic coagulant dosage, $FeCl_3.6H_2O$, on the percentage removal of turbidity, color, COD, and ABS_{254nm} after addition of 24.0 mg L^{-1} POM at pH 6.0 (mean ± standard deviation)

12.3.5 Effect of Natural Coagulant Dosage

Even when natural coagulants were used in association with inorganic agents, POM dosage optimization is very important as well to increase the wastewater treatment efficiency (Bhole 1995). This effect is shown in Fig. 12.6.

Overall suspended and dissolved removal was high. Except for 4.0 and 8.0 mg L^{-1} of POM, there is not significant difference ($P > 0.05$) between the other dosages, meaning that POM dosages do not affect turbidity and color removal. For removal of ABS_{254nm} and COD, the efficiency varied significantly with different coagulant dosages. There is a simultaneous increase between dosage levels and COD removal. The highest efficacy was at 20 mg L^{-1} POM with 98.2% in turbidity, 93.8% in color, 79.2% in COD, and 83.2% in ABS_{254nm} removal.

When the coagulant dosage is optimum, maximum removal occurs. There is an effective linkage of multiple colloidal particles due to their destabilization followed by adsorption onto the coagulant, whose loops and tails extend throughout the solution by bridging with other particles. This requires sufficient unoccupied surface sites on a coagulant particle to attach segments of polymer chains adsorbed on other particles. If an excessive quantity of polymer is added, the coagulant particle entire surface becomes coated blocking sites that were available to "bridge" with other

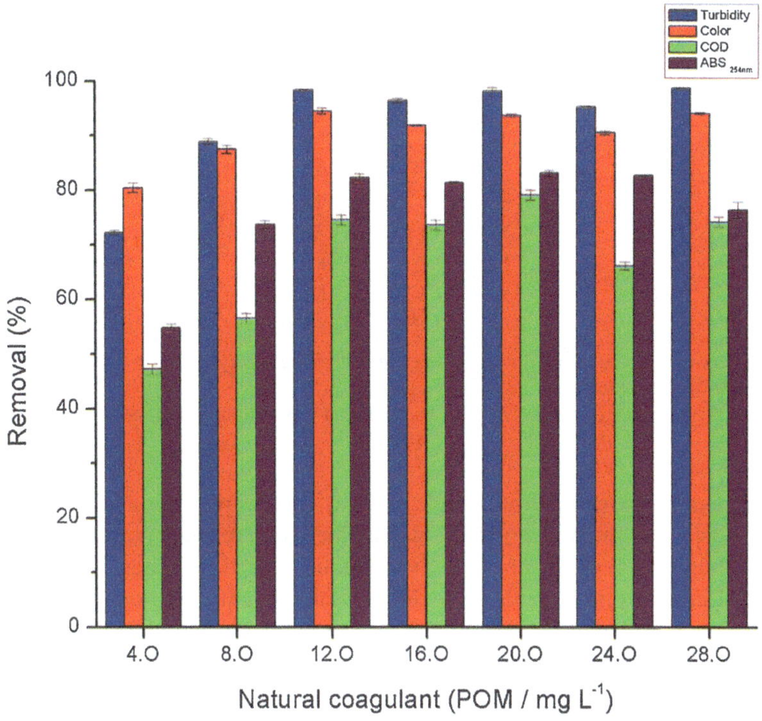

Fig. 12.6 Effect of natural coagulant dosage, POM, on percentage removal of turbidity, color, COD, and ABS_{254nm} of textile wastewater after addition of 80.0 mg L^{-1} Fe^{3+} at pH 6.0 (mean ± standard deviation)

particles. These excess levels can reestablish colloidal suspension and even break up the flocs. In contrast, when the coagulant dosage is too low, there is an incomplete and inefficient coagulation, and most of the colloids remain in suspension because bridging interactions are not enough to form the flocs (Sher et al. 2013; Lee et al. 2014).

12.3.6 Comparison Between Biopolymer from Purple Okra and Synthetic Commercial Polymer

The efficiency of the natural coagulant from purple okra and synthetic commercial, high molecular weight polyacrylamide-based polymers were evaluated on removal of ABS_{254nm}, color, COD, and turbidity which are shown in Table 12.2.

All polymers were efficient to remove suspended solids, except FXNI2®, because its turbidity removal was lower than natural and other commercial polymers. Only FXCE2® polymer was as efficient as natural polymer from purple okra on dissolved solid removal. Biopolymer presented higher removal values than synthetic commercial polymers for ABS_{254nm} and organic matter (COD).

Table 12.2 Efficiency of synthetic commercial polymer and biopolymer from purple okra on the percent removal of turbidity, color, COD, and ABS_{254nm} of textile wastewater (mean ± standard deviation)

		Removal (%)			
Polymer	Dosage/mg L^{-1}	Turbidity	Color	COD	ABS_{254nm}
FXCS8®	20.0	93.6 ± 0.5^{a}	87.6 ± 0.6^{a}	59.4 ± 1.0^{a}	69.2 ± 0.6^{a}
FXCS4®	20.0	95.8 ± 0.6^{b}	88.0 ± 0.4^{b}	68.6 ± 1.0^{b}	72.0 ± 0.1^{b}
FXCE2®	20.0	98.7 ± 0.4^{c}	94.7 ± 1.0^{c}	59.4 ± 1.0^{a}	$70.6 \pm 0.1^{a,b}$
FXNS1®	20.0	85.4 ± 0.7^{d}	81.8 ± 0.7^{d}	65.9 ± 1.0^{c}	75.9 ± 0.6^{c}
FXNS2®	20.0	91.3 ± 0.3^{e}	$89.6 \pm 0.4^{a,b}$	47.7 ± 1.0^{d}	78.3 ± 0.5^{d}
Fe-POM	20.0	98.2 ± 0.7^{b}	93.8 ± 0.3^{e}	79.2 ± 0.9^{e}	83.2 ± 0.6^{e}

Equal letters show there are no differences between them by Tukey Test ($P > 0.05$)
Determination in triplicate
COD chemical oxygen demand, ABS_{254nm} absorbance at wavelength at 254 nm

Biopolymers extracted from purple okra residues were better and more efficient than synthetic commercial polymer, placing them in a relevant position on the sustainable market, since synthetic polyacrylamide-based polymers are not easily degraded and they may generate toxic and carcinogenic by-products, e.g. acrylamide monomer, being harmful to animal and human health and the environment (Caulfield et al. 2002). Thus, the use of natural polymers from purple okra residues represents a green technique to wastewater treatment.

12.3.7 Toxicity Assays with Artemia salina*: Ecotoxicity*

The ecotoxiticy assay carried out before and after treatment of textile wastewater is shown in Fig. 12.7. To all the tests, the mortality rate was 0% for negative control (nutritive solution) and 100% for positive control (dichromate solution). The mortality rate of *A. salina* varied with an increase in wastewater concentration. Regardless of the wastewater concentration, mortality rate declined after the coagulation/flocculation. The maximum mortality rate occurred for untreated wastewater. Adding Fe-POM led to a performance improvement in relation to the treatment using only inorganic coagulant. Treatment using POM resulted in lower *A. salina* mortality rate for all wastewater concentration.

The highest LD_{50} value was found for untreated wastewater assays, noting that only 120 mL L^{-1} of wastewater was needed to kill half the population of *A. salina*. The coagulation/flocculation process leads to an increase in the LD_{50}. The LD_{50} obtained using only inorganic coagulant was of 800 mL L^{-1} of textile wastewater. The assays with Fe-POM showed that the mortality rate decreased significantly, and the LD_{50} achieved was 1500 mL L^{-1}, i.e., over than the undiluted treated wastewater (1000 mL L^{-1}), meaning it can be safely discharged without contamination risk.

The wastewater toxicity was decreased by coagulation/flocculation before its disposal in the aquatic environment. The coagulation/flocculation treatment using

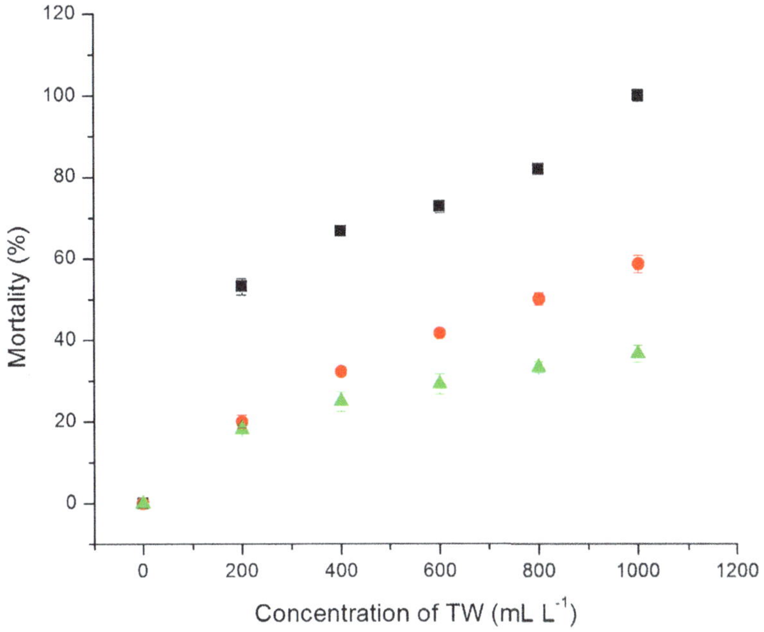

Fig. 12.7 Mortality rate of *A. salina* as a function of textile wastewater concentration: (■) in natura, (●) treated only with inorganic coagulant (Fe), and (▲) treated with dual coagulant (Fe-POM)

only inorganic coagulant is not suited to be used as single wastewater treatment system because their level of toxicity is moderate. However, an excellent and effective removal of toxicological contaminants and low ecotoxicity were carried out when the dual coagulant (Fe-POM) was used. This result suggests that their toxicity potential is very low, and this process can be used as wastewater treatment system.

Bandala et al. (2013) reported a toxicity reduction of petrochemical wastewater by using guar gum, locust bean gum, and *Opuntia* spp. as natural coagulant followed by aerobic biofilter. Toxicity tests were performed on *Lactuca sativa* L., and after treatment, the germinated seed index climbed to very toxic (−0.89 cm) from low toxic (−0.07 cm). Al-Matairi (2006) examined two coagulants, commercial polymer and aluminum sulfate in a slaughterhouse wastewater treatment. Toxicity assay was based on the Microtox. Polymer-treated wastewaters were more toxic than alum-treated at extremely low concentration.

12.4 Conclusion

POM is an efficient natural coagulant for the removal of both organic matter and suspended and dissolved solids from textile wastewater. The higher mortality rate of *A. salina* occurs in the untreated wastewater. The textile wastewater treatment by the coagulation/flocculation process leads to a decrease of toxicity in the aquatic environment. Using purple okra wastes as natural coagulant does not present risks to animal and human health, as they have an excellent safety profile (non toxics, non carcinogenic, and non irritantes) and are highly biodegradable and environmentally friendly. They are also completely available for an ecological and efficient treatment of textile wastewater. In addition, this paper strengthen the reutilization of food waste since purple okra pods were darkened and wilted and they had not actually any commercial value.

Acknowledgments The authors thank Dynamic's Industrial Laundry for providing textile wastewater and the State University of Maringá (Paraná, Brazil) and Chemistry Department for laboratorial support. We also thank FINEP, CNPq, COMCAP-UEM, CAPES, and Araucaria Foundation for funding.

References

Agarwal M, Srinivasan R, Mishra A (2001) Study on flocculation efficiency of okra gum in sewage waste water. Macromol Mater Eng 286:560–563. https://doi.org/10.1002/1439-2054(20010901)286:9<560::AID-MAME560>3.0.CO;2-B

Agarwal M, Rajani S, Mishra A, Rai JSP (2003) Utilization of okra gum for treatment of tannery effluent. Int J Polym Mater 52:1049–1057. https://doi.org/10.1080/714975900

Alba K, Laws AP, Kontogiorgios V (2015) Isolation and characterization of acetylated LM-pectins extracted from okra pods. Food Hydrocoll 43:726–735. https://doi.org/10.1016/j.foodhyd.2014.08.003

Al-Matairi NZ (2006) Coagulant toxicity and effectiveness in a slaughterhouse wastewater treatment plant. Ecotoxicol Environ Saf 65:74–83. https://doi.org/10.1016/j.ecoenv.2005.05.013

Almeida CA, de Souza MTF, Freitas TKFS, Ambrosio E, Geraldino HCL, Garcia JC (2017) Vegetable residue of chayote (Sechium edule SW.) as a natural coagulant for treatment of textile wastewater. Int J Energy Water Resour 1:37–46

Antov MG, Šćiba MB, Prodanović JM, Kukić DV, Vasić VM, Đorđević TR, Milošević MM (2018) Common oak (Quercus robur) acorn as a source of natural coagulants for water turbidity removal. Ind Crop Prod 117:340–346. https://doi.org/10.1016/j.indcrop.2018.03.022

APHA—American Public Health Association Standard (1998) Methods for the examination of water and wastewater, 20th edn. American Water Works Association, Water Environment Federation, Washington

Bandala ER, Tiro JB, Luján M, Camargo FJ, Sánchez-Salas JL, Reyna S, Moeller G, Torres LG (2013) Petrochemical effluent treatment using natural coagulants and an aerobic biofilter. Adv Environ Res 2(3):229–243. https://doi.org/10.12989/aer.2013.2.3.229

Bathista ALBS, Tavares MIB, Silva EO (2004) ^{13}C NMR study of *Albemoschus Esculentus* compounds. Ann Magn Reson 3:6–8

Bhole AG (1995) Relative evaluation of a few natural coagulants. J Water Supply Res Technol 44(6):284–290

Bilińska L, Blus K, Gmurek M, Ledakowicz S (2019) Coupling of electrocoagulation and ozone treatment for textile wastewater reuse. Chem Eng J 358:992–1001. https://doi.org/10.1016/j.cej.2018.10.093

Bolto B, Gregory J (2007) Organic polyelectrolytes in water treatment. Water Res 41:2301–2324. https://doi.org/10.1016/j.watres.2007.03.012

Boulaadjoul S, Zemmouri H, Bendjama Z, Drouiche N (2018) A novel use of Moringa oleifera seed powder in enhancing the primary treatment of paper mill effluent. Chemosphere 206:142–149. https://doi.org/10.1016/j.chemosphere.2018.04.123

Braga WLM, Roberto JA, Vaz C, Samanamud GRL, Loures CCA, França AB, Lofrano RCZ, Naves LLR, Gomes JHF, Naves FL (2018) Extraction and optimization of tannin from the flower of *Musa* sp. applied to the treatment of iron ore dump. J Environ Chem Eng 6:4310–4317. https://doi.org/10.1016/j.jece.2018.05.058

Camacho FP, Sousa VS, Bergamasco R, Teixeira MR (2017) The use of Moringa oleifera as a natural coagulant in surface water treatment. Chem Eng J 313:226–237. https://doi.org/10.1016/j.cej.2016.12.031

Caulfield JM, Qiao GG, Solomon DH (2002) Some aspects of the properties and degradation of polyacrylamides. Chem Rev 102:3067–3083. https://doi.org/10.1021/cr010439p

CEMA, Resolução n. 70 de 2009, Anexo 7 (2009) Condições e Padrões de Lançamento de Efluentes Líquidos Industriais. Instituto Ambiental do Paraná, Curitiba

Chhabra M, Mishra S, Sreekrishnan TR (2015) Combination of chemical and enzymatic treatment for efficient decolorization/degradation of textile effluent: high operational stability of the continuous process. Biochem Eng J 93:17–24. https://doi.org/10.1016/j.bej.2014.09.007

Choudhary M, Ray MB, Neogi S (2019) Evaluation of the potential application of cactus (Opuntia ficus-indica) as a bio-coagulant for pre-treatment of oil sands process-affected water. Sep Purif Technol 209:714–724. https://doi.org/10.1016/j.seppur.2018.09.033

Choumane FM, Benguella B, Maachou B, Saadi N (2017) Valorisation of a bioflocculant and hydroxyapatites as coagulation-flocculation adjuvants in wastewater treatment of the steppe in the wilaya of Saida (Algeria). Ecol Eng 107:152–159. https://doi.org/10.1016/j.ecoleng.2017.07.013

Choy SY, Prasad KMN, Wu TY (2013) A review on common vegetables and legumes as promising plant-based natural coagulants in water clarification. Int J Environ Sci Technol 12:367–390. https://doi.org/10.1007/s13762-013-0446-2

Choy SY, Prasad KMN, Wu TY, Raghunandan ME, Ramanan RN (2016) Performance of conventional starches as natural coagulants for turbidity removal. Ecol Eng 94:352–364. https://doi.org/10.1016/j.ecoleng.2016.05.082

CONAMA, Resolução n. 430 de 2011 (2011) Dispõe sobre as condições e padrões de lançamento de efluentes, complementa e altera a Resolução n. 357, de 17 de março, do CONAMA. Conselho Nacional de Meio Ambiente, Brasília

de Rosa IM, Kenny JM, Puglia D, Santulli C, Sarasini F (2010) Morphological, thermal and mechanical characterization of okra (*Abelmoschus esculentus*) fibres as potential reinforcement in polymer composites. Compos Sci Technol 70:116–122. https://doi.org/10.1016/j.compscitech.2009.09.013

de Souza MTF, Ambrosio E, de Almeida CA, Freitas TKFS, Santos LB, Almeida VC, Garcia JC (2014) The use of a natural coagulant (*Opuntia ficus-indica*) in the removal for organic materials of textile effluents. Environ Monit Assess 186:5261–5271. https://doi.org/10.1007/s10661-014-3775-9

de Souza MTF, de Almeida CA, Ambrosio E, Santos LB, Freitas TKFS, Manholer DD, de Carvalho GM, Garcia JC (2016) Extraction and use of *Cereus peruvianus* cactus mucilage in the treatment of textile effluents. J Taiwan Inst Chem E 67:174–183. https://doi.org/10.1016/j.jtice.2016.07.009

Dhivya S, Ramesh ST, Gandhimathi R, Nidheesh PV (2017) Performance of natural coagulant extracted from *Plantago ovata* seed for the treatment of turbid water. Water Air Soil Pollut 228:423. https://doi.org/10.1007/s11270-017-3592-1

dos Santos JD, Veit MT, Juchen PT, da Cunha Gonçalves G, Palácio SM, Fagundes-Klen M (2018) Use of different coagulants for cassava processing wastewater treatment. J Environ Chem Eng 6:1821–1827. https://doi.org/10.1016/j.jece.2018.02.039

Dotto J, Fagundes-Klen MR, Veit MT, Palácio SM, Bergamasco R (2019) Performance of different coagulants in the coagulation/flocculation process of textile wastewater. J Clean Prod 208:656–665. https://doi.org/10.1016/j.jclepro.2018.10.112

Duan J, Gregory J (2003) Coagulation by hydrolysing metal salts. Adv Colloid Interf Sci 100–102:475–502. https://doi.org/10.1016/S0001-8686(02)00067-2

Ebeling JM, Welsh CF, Rishel KL (2006) Performance evaluation of an inclined belt filter using coagulation/flocculation aids for the removal of suspended solids and phosphorus from microscreen backwash effluent. Aquacult Eng 35:61–77. https://doi.org/10.1016/j.aquaeng.2005.08.006

El-Mahdy AR, El-Sebaiy LA (1984) Preliminary studies on the mucilages extracted from okra fruits, taro tubers, Jew's mellow leaves and fenugreek seeds. Food Chem 14:237–249. https://doi.org/10.1016/0308-8146(84)90079-7

Emeje M, Isimi C, Byrn S, Fortunak J, Kunle O, Ofoefule S (2011) Extraction and physicochemical characterization of a new polysaccharide obtained from the fresh fruits of *Abelmoschus esculentus*. Iran J Pharm Res 10:237–246. PMID: 24250349

Freitas TKFS, Oliveira VM, de Souza MTF, Geraldino HCL, Almeida VC, Fávaro SL, Garcia JC (2015) Optimization of coagulation-flocculation process for treatment of industrial textile wastewater using okra (*A. esculentus*) mucilage as natural coagulant. Ind Crop Prod 76:538–544. https://doi.org/10.1016/j.indcrop.2015.06.027

Gamakaranage C (2016) Heavy metals and autism. J Heavy Met Toxicity Dis 1(3):12. https://doi.org/10.21767/2473-6457.100012

Garcia JC, Simionato JI, Almeida VC, Palacio SM, Rossi FL, Schneider MV, de Souza NE (2009) Evolutive follow-up of the photocatalytic degradation of real textile effluents in TiO_2 and TiO_2/H_2O_2 systems and their toxic effects on Lactuca sativa seedlings. J Braz Chem Soc 20:1589–1597. https://doi.org/10.1590/S0103-50532009000900005

Huang X, Gao B, Yue Q, Wang Y, Li Q (2014) Effect of Si/Ti molar ratio on enhanced coagulation performance, floc properties and sludge reuse of a novel hybrid coagulant:polysilicate titanium sulfate. Desalination 352:150–157. https://doi.org/10.1016/j.desal.2014.08.021

Hussain S, Ghouri AS, Ahmad A (2019) Pine cone extract as natural coagulant for purification of turbid water. Heliyon 5:e01420. https://doi.org/10.1016/j.heliyon.2019.e01420

Ito T, Shimada Y, Suto T (2018) Potential use of bacteria collected from human hands for textile dyes decolorization. Water Resour Ind 20:46–53. https://doi.org/10.1016/j.wri.2018.09.001

Kakoi B, Kaluli JW, Ndiba P, Thiong'o G (2016) Banana pith as a natural coagulant for polluted river water. Ecol Eng 95:699–705. https://doi.org/10.1016/j.ecoleng.2016.07.001

Kebaili M, Djellali S, Radjai M, Drouiche N, Lounici H (2018) Valorization of orange industry residues to form a natural coagulant and adsorbent. J Ind Eng Chem 64:292–299. https://doi.org/10.1016/j.jiec.2018.03.027

Khan S, Anas M, Malik A (2019) Mutagenicity and genotoxicity evaluation of textile industry wastewater using bacterial and plant bioassays. Toxicol Rep 6:192–201. https://doi.org/10.1016/j.toxrep.2019.02.002

Kim KW, Shon WJ, Oh MK, Yang D, Foster RI, Lee KY (2019) Evaluation of dynamic behavior of coagulation-flocculation using hydrous ferric oxide for removal of radioactive nuclides in wastewater. Nucl Eng Technol 51:738–745. https://doi.org/10.1016/j.net.2018.11.016

Lafi R, Gzara L, Lajimi RH, Hafiane A (2018) Treatment of textile wastewater by a hybrid ultrafiltration/electrodialysis process. Chem Eng Process 132:105–113. https://doi.org/10.1016/j.cep.2018.08.010

Lee S, Robinson J, Chong MF (2014) A review on application of flocculants in wastewater treatment. Process Saf Environ Prot 92:489–508. https://doi.org/10.1016/j.psep.2014.04.010

Lek BLC, Peter AP, Chong KHQ, Ragu P, Sethu V, Selvarajoo A, Arumugasamy SK (2018) Treatment of palm oil mill effluent (POME) using chickpea (Cicer arietinum) as a natural coagulant and flocculant: evaluation, process optimization and characterization of chickpea powder. Chem Eng 6:6243–6255. https://doi.org/10.1016/j.jece.2018.09.038

Lenzi E, Favero LOB, Luchese EB (2012) Introdução à Química da Água: Ciência, vida e sobrevivência. LTC, Rio de Janeiro

Lopes EC, Santos SCR, Pintor AMA, Boaventura RAR, Botelho CMS (2019) Evaluation of a tannin-based coagulant on the decolorization of synthetic effluents. J Environ Chem Eng 7:103125. https://doi.org/10.1016/j.jece.2019.103125

Mallevialle J, Brichet A, Fiessinger F (1984) How safe are organic polymers in water treatment. J Am Water Works Assoc 76:87–93
Megersa M, Gach W, Beyene A, Ambelu A, Triest L (2019) Effect of salt solutions on coagulation performance of Moringa stenopetala and Maerua subcordata for turbid water treatment. Sep Purif Technol 221:319–324. https://doi.org/10.1016/j.seppur.2019.04.013
Miller SM, Fugate EJ, Craver VO, Smith JA, Zimmerman JB (2008) Toward understanding the efficacy and mechanism of Opuntia ssp. as a natural coagulant for potential application in water treatment. Environ Sci Technol 42:4274–4279. https://doi.org/10.1021/es7025054
Mirbahoush SM, Chaibakhsh N, Moradi-Shoeili Z (2019) Highly efficient removal of surfactant from industrial effluents using flaxseed mucilage in coagulation/photo-Fenton oxidation process. Chemosphere 231:51. https://doi.org/10.1016/j.chemosphere.2019.05.118
Mirza A, King A, Troakes C, Exley C (2017) Aluminium in brain tissue in familial Alzheimer's disease. J Trace Elem Med Biol 40:30–36. https://doi.org/10.1016/j.jtemb.2016.12.001
Mold M, Chmielecka A, Rodriguez MRR, Thom F, Linhart C, King A, Exley C (2018a) Aluminium in brain tissue in multiple sclerosis. Int J Environ Res Public Health 15(8):1777. https://doi.org/10.3390/ijerph15081777
Mold M, Umar D, King A, Exley C (2018b) Aluminium in brain tissue in autism. J Trace Elem Med Biol 46:76–82. https://doi.org/10.1016/j.jtemb.2017.11.012
Mold M, Cottle J, Exley C (2019) Aluminium in brain tissue in epilepsy: a case report from Camelford. Int J Environ Res Public Health 16(12):2129. https://doi.org/10.3390/ijerph16122129
Nurdjanah S, Hook J, Paton J, Paterson J (2013) Galacturonic acid content and degree of esterification of pectin from sweet potato starch residue detected using 13C CP/MAS solid state NMR. Eur J Nutr Food Saf 3(1):16–37. https://doi.org/10.9734/ejnfs/2013/v3i127018
Okuda T, Baes AU, Nishijima W, Okada M (1999) Improvement of extraction method of coagulation active components from Moringa oleifera seed. Water Res 33:3373–3378. https://doi.org/10.1016/S0043-1354(99)00046-9
Ravi SK, Ramesh BN, Mundugaru R, Vincent B (2018) Multiple pharmacological activities of *Caesalpinia crista* against aluminium-induced neurodegeneration in rats: relevance for Alzheimer's disease. Environ Toxicol Pharmacol 58:202–211. https://doi.org/10.1016/j.etap.2018.01.008
Ren Z, Chon TS, Xia C, Li F (2017) The monitoring and assessment of aquatic toxicology. Biomed Res Int 1-2:9179728. https://doi.org/10.1155/2017/9179728
Sen SK, Patra P, Das CR, Raut S (2019) Pilot-scale evaluation of bio-decolorization and biodegradation of reactive textile wastewater: an impact on its use in irrigation of wheat crop. Water Resour Ind 21:100106. https://doi.org/10.1016/j.wri.2019.100106
Sher F, Malik A, Liu H (2013) Industrial polymer effluent treatment by chemical coagulation and flocculation. J Environ Chem Eng 1:684–689. https://doi.org/10.1016/j.jece.2013.07.003
Shipp A, Lawrence G, Gentry R, McDonald T, Bartow H, Bounds J, Macdonald N, Clewell H, Allen B, Van Landingham C (2006) Crit Rev Toxicol 36:481–608. https://doi.org/10.1080/10408440600851377
Sillanpää M, Ncibi MC, Matilainen A, Vepsäläinen M (2018) Removal of natural organic matter in drinking water treatment by coagulation: a comprehensive review. Chemosphere 190:54–71. https://doi.org/10.1016/j.chemosphere.2017.09.113
Sinitsya A, Čopíková J, Pavlíková H (1998) ^{13}C CP/MAS NMR spectroscopy in the analysis of pectins. J Carbohydr Chem 17:279–292. https://doi.org/10.1080/07328309808002328
Somasundaran P, Runkana V, Kapur PC (2005) Flocculation and dispersion of colloidal suspensions by polymers and surfactants: experimental and modeling studies. In: Stechemesser H, Dobia's B (eds) Coagulation and flocculation, 2nd edn. CRC Press, Boca Raton, pp 767–803. https://doi.org/10.1201/9781420027686.ch11
Souza RP, Freitas TKFS, Domingues FS, Pezoti O, Ambrosio E, Ferrari-Lima AM, Garcia JC (2016) Photocatalytic activity of TiO_2, ZnO and Nb_2O_5 applied to degradation of textile wastewater. J Photochem Photobiol A Chem 329:9–17. https://doi.org/10.1016/j.jphotochem.2016.06.013

Stephenson RJ, Duff SJB (1996) Coagulation and precipitation of a mechanical pulping effluent—I. Removal of carbon, colour and turbidity. Water Res 30(4):781–792. https://doi.org/10.1016/0043-1354(95)00213-8

Tawakkoly B, Alizadehdakhel A, Dorosti F (2019) Evaluation of COD and turbidity removal from compost leachate wastewater using Salvia hispanica as a natural coagulant. Ind Crop Prod 137:323–331. https://doi.org/10.1016/j.indcrop.2019.05.038

Verma AK, Dash RR, Bhunia P (2012) A review on chemical coagulation/flocculation technologies for removal of colour from textile wastewaters. J Environ Manag 93:154–168. https://doi.org/10.1016/j.jenvman.2011.09.012

von Sperling M (2014) Introdução à qualidade das águas e ao tratamento de esgotos. Editora UFMG, Belo Horizonte

Wang T, Gao BY, Yue QY, Zhou WZ, Chu YB (2005) On-line optical determination of floc size of Fe(III) coagulants. J Environ Sci 17:921–925

Wei H, Gao B, Ren J, Li A, Yang H (2018) Coagulation/flocculation in dewatering of sludge: a review. Water Res 143:608–631. https://doi.org/10.1016/j.watres.2018.07.029

Wu W, Zhu Y, Zhang L, Yang R, Zhou Y (2012) Extraction, preliminary structural characterization, and antioxidant activities of polysaccharides from Salvia miltiorrhiza Bunge. Carbohydr Polym 87:1348–1353. https://doi.org/10.1016/j.carbpol.2011.09.024

Xiong B, Loss RD, Shields D, Pawlik T, Hochreiter R, Zydney AL, Kumar M (2018) Polyacrylamide degradation and its implications in environmental systems. Clean Water 1–17

Yin CY (2010) Emerging usage of plant-based coagulants for water and wastewater treatment. Process Biochem 45:1437–1444. https://doi.org/10.1016/j.procbio.2010.05.030

Yu X, Somasundaran P (1996) Role of polymer conformation in interparticle-bridging dominated flocculation. J Colloid Interface Sci 177:283–287. https://doi.org/10.1006/jcis.1996.0033

Zhang J, Zhang F, Luo Y, Yang H (2006) A preliminary study on cactus coagulant in water treatment. Process Biochem 41:730–733. https://doi.org/10.1016/j.procbio.2005.08.016

Zhao L, Dong Y, Chen G, Hu Q (2010) Extraction, purification, characterization and antitumor activity of polysaccharides from Ganoderma lucidum. Carbohydr Polym 80:783–789. https://doi.org/10.1016/j.carbpol.2009.12.029

Zheng W, Zhao T, Feng W, Wang W, Zou Y, Zheng D, Takases M, Li Q, Wu H, Yang L, Wu X (2014) Purification, characterization and immunomodulating activity of a polysaccharide from flowers of Abelmoschus esculentus. Carbohydr Polym 106:335–342. https://doi.org/10.1016/j.carbpol.2014.02.079

Zhu G, Zheng H, Zhang Z, Tshukudu T, Zhang P, Xiang X (2011) Characterization and coagulation–flocculation behavior of polymeric aluminum ferric sulfate (PAFS). Chem Eng J 178:50–59. https://doi.org/10.1016/j.cej.2011.10.008

Zhu X, Liu B, Zheng S, Gao Y (2014) Quantitative and structure analysis of pectin in tobacco by ^{13}C CP/MAS NMR spectroscopy. Anal Methods 6:6407–6413. https://doi.org/10.1039/C4AY01156B

Chapter 13
Implementation Guidelines for Modelling Gasification Processes in Computational Fluid Dynamics: A Tutorial Overview Approach

João Cardoso, Valter Bruno Silva, and Daniela Eusébio

Contents

Abbreviations

ANN	Artificial neural network
CAD	Computer-aided design

J. Cardoso
Polytechnic Institute of Portalegre, Portalegre, Portugal

Instituto Superior Técnico, Universidade de Lisboa, Lisboa, Portugal
e-mail: jps.cardoso@ipportalegre.pt; joaoscardoso@tecnico.ulisboa.pt

V. B. Silva (✉) · D. Eusébio
Polytechnic Institute of Portalegre, Portalegre, Portugal
e-mail: valter.silva@ipportalegre.pt; danielafle@ipportalegre.pt

Inamuddin, A. M. Asiri (eds.), *Sustainable Green Chemical Processes and their Allied Applications*, Nanotechnology in the Life Sciences,
https://doi.org/10.1007/978-3-030-42284-4_13

CFD	Computational fluid dynamics
DDPM	Dense discrete phase model
DDPM-DEM	Dense discrete phase model-discrete element model
DDPM-KTGF	Dense discrete phase model-kinetic theory of granular flow
DNS	Direct numerical simulation
DO	Discrete ordinates
DPM	Discrete phase model
DTRM	Discrete transfer radiation model
HTML	Hypertext markup language
RANS	Reynolds-averaged Navier-Stokes simulations
RNG	Renormalization group
S2S	Surface-to-surface
SRS	Scale-resolving simulations
SST	Shear-stress transport
UDF	User-defined function

13.1 Introduction

Global warming and climate change stand as some of the greatest environmental, social and economic threats of our time. One solution to overcome this issue is by gradually employing renewable energy sources to replace fossil fuels envisioning an alternative to move towards sustainable development while mitigating environmental problems (Cai et al. 2011; Cardoso et al. 2019b; Couto et al. 2016a; Tarelho et al. 2011; Vicente et al. 2016).

Biomass and waste products carry great development potential since they can be easily stored and transported, and unlike other renewable energy sources, they can also be converted into biofuels thus increasing their applicability, contributing significantly to the energy independence of the region along with associated economic and environmental benefits (Ahmad et al. 2016; Cardoso et al. 2018b; Galadima and Muraza 2015; Pinto et al. 2014; Pio et al. 2017).

The economic and energetic performance of a solution turning biomass and wastes into energy depends on many variables, having the feedstock supply sustainability, the robust technical performance and the ability to predict the final products without carrying out expensive experimental tests, precedence in these lists (Cardoso et al. 2019a; Lamers et al. 2015; Leme et al. 2014; Lourinho and Brito 2015; Pereira et al. 2016). The correct assessment and approach to this and other concerns will determine the type of technology proper for each specific project. Selection of a preferred technology is complex and requires careful consideration of the fuel flexibility, type of biomass, global efficiency and performance, to mention a few.

It is then necessary to perform experimental and numerical work in different power scales comparing the most suitable technologies, typically combustion and gasification (Couto et al. 2017a; Maurer et al. 2014; Molino et al. 2016; Neves et al. 2011; Sanderson and Rhodes 2005). Both technologies are complex systems

Table 13.1 Brief pros and cons analysis of the available methodologies for thermochemical processes (Baruah and Baruah 2014; Fogler and Gurmen 2002; Sehrawat et al. 2015; Silva and Rouboa 2013)

	Pros	Cons
Equilibrium model	Simple, easy to implement and with quick convergence. Allows a practical description of the gasification processes with good approximation.	Poor process representation for lower operating temperatures. Inability to predict kinetics and hydrodynamic phenomena. Unfit for complex reactor designs.
Kinetic model	More accurate and detailed than equilibrium models. Capability to predict gas and temperature profiles inside the gasifier once it incorporates both reaction kinetics and reactor hydrodynamics.	Complex formulation and computationally expensive. Depends on reaction kinetics and type of gasifier.
Artificial neural network (ANN)	Ability to be self-thought from sampled experimental data (machine learning). Able to represent complex non-linear behaviours.	Slow convergence speed, less generalizing performance, arriving at a local minimum and over-fitting problems. Requires diligent work in implementing. Failure in case of limited data.
Flowsheet simulators	Helps to cut down on laboratory experiments and pilot plant runs. Can be used for risk-free analysis of various what-if scenarios.	Forces the user to think deeper about the problem at hands, in finding novel approaches to solve it, and to evaluate the assumptions closer. Process plants rarely operate entirely under steady-state conditions.
Computational fluid dynamics	Highly precise. Very often offers faster resolutions than physical modelling. Computational fluid dynamics (CFD) model studies are generally 20–40% less expensive than a comparable physical model effort.	It is generally run as a steady-state simulation. It carries a high degree of complexity; therefore, it requires incorporation of many engineering approximations, modelling shortcuts and real-world variabilities.

depending nonlinearly on a large number of parameters making the experimental work hard. Together with experimental characterization, the development of high-fidelity models is crucial for these technologies' evaluation (Cardoso et al. 2019c; Ramos et al. 2018; Sharma et al. 2015; Silva and Rouboa 2015; Vepsäläinen et al. 2013). Table 13.1 lists a set of different numerical approaches to handle thermochemical conversion processes.

A set of governing equations are applied behind computational fluid dynamics (CFD) models. These are built by resorting on conservation of mass equations, momentum, energy and species over a designated region of the reactor, capable of evaluating temperature and concentration, among numerous others parameters with a considerable precision rate (Couto et al. 2016b, 2017b; Olaofe et al. 2014; Silva and Rouboa 2015; Xue et al. 2012). Therefore, CFD is broadly recognized as an appropriate and useful tool to deal with thermochemical conversion processes, and for that reason, it will be the bottom line of this tutorial guide.

Since the late 1990s, we have witnessed the development of CFD studies applied to a large broad of chemical and physical problems. The trend towards the use of CFD solutions applied to thermochemical conversion processes was inevitable, often involving the prediction of products and frequently discussing the design of different configurations, becoming a major tool for energy-related researchers and scientists.

A multitude of options take place throughout the process solving and, in concert, lead to the overall necessity to be aware of the many pieces comprising the CFD simulation. It is not surprising that the full understanding of the mechanisms behind the simulation setup requires in-depth knowledge of the software fundamental steps and guidelines. The real value of this knowledge is many times jeopardized, and the researchers often focus their efforts only in the physical problem neglecting the best way to implement it in a computational environment.

In this chapter, one aims to provide a working background for the practical scientists, researchers and engineers who wish to apply a CFD simulation to thermochemical conversion procedures without needing to become a computational master. Throughout the chapter, we will focus on practical interpretations of common problems, based on somewhat simplified but effective approaches, introducing some of the necessary theory to understand the step-by-step process to assemble a full resolution. Although one shall focus on thermochemical conversion processes, most of the discussed methods herein apply to a large range of other physical and chemical problems.

13.2 Problem and Domain Identification

CFD is a powerful tool, but its application should be restricted to cases where its implementation is justifiable since it can easily become much more computationally expensive and time-consuming than a simpler approach capable of reaching similar results.

In some cases, the decision of using analytical methods or 1-D models is straightforward and provides the required insights for less accurate needs. Typical examples are the problems with an analytical solution, the ones in which the engineer is only required to acknowledge the trends concerning the problem in hands and also some margin of error is allowed. Furthermore, the analytical methods or simpler models always provide a good first shot and an effective way of comparison with more sophisticated approaches as CFD.

After carefully deciding on what type of results one wants to explore, and which is the most appropriate tool available, the user must focus on identifying the domain and isolate the intended section from the complete physical system. In the case of a gasification plant, the user can model a singular or a set of components such as the gasifier, the cleaning and feeding systems or even the heat exchangers. For the sake of simplicity, our attention here is devoted to the most important component in such setup, the gasifier. Figure 13.1 depicts a 75 kW_{th} pilot-scale bubbling fluidized bed gasifier.

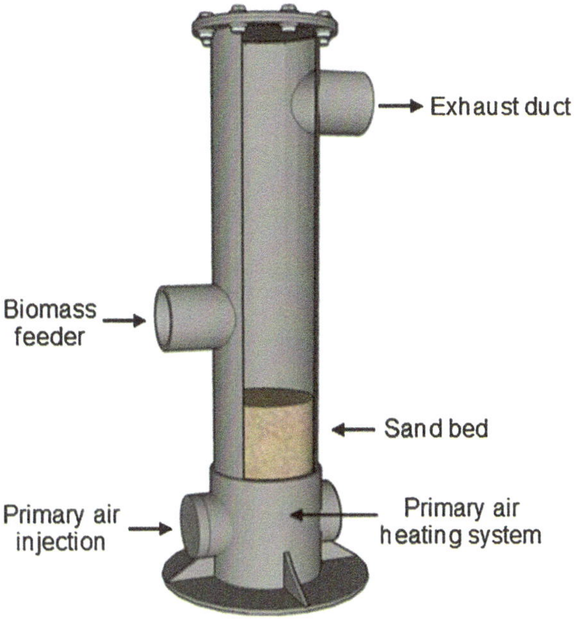

Fig. 13.1 Schematics of the 75 kW_{th} pilot-scale bubbling fluidized bed gasifier

Most of the studies found in the literature emphasize the simulation effort in the gasifier (Couto et al. 2015a, b; Dinh et al. 2017; Sharma et al. 2014; Xue et al. 2011). This is easily understandable because relevant outputs in a gasification process such as the syngas composition, the particle behaviour, the bed hydrodynamics, the flow patterns and the temperature distribution are uncovered in these simulations.

The simulation of an entire gasification facility is mostly applied to flowsheet simulators like ASPEN Plus, although this solution lacks to provide detailed results on particle behaviour or aiding to understand the complex hydrodynamics within.

13.3 Pre-processing

13.3.1 Geometry

The next step requires the geometry definition of the selected gasifier and the division of the geometry into small elements, the so-called mesh generation. The geometry can be imported from design software such as SolidWorks or generated under the ANSYS Fluent framework. Figure 13.2 shows the schematics of the 75 kW_{th} pilot-scale gasifier with all the respective inlets and outlet. Those are especially relevant at this stage because they allow defining the boundary conditions of such domain.

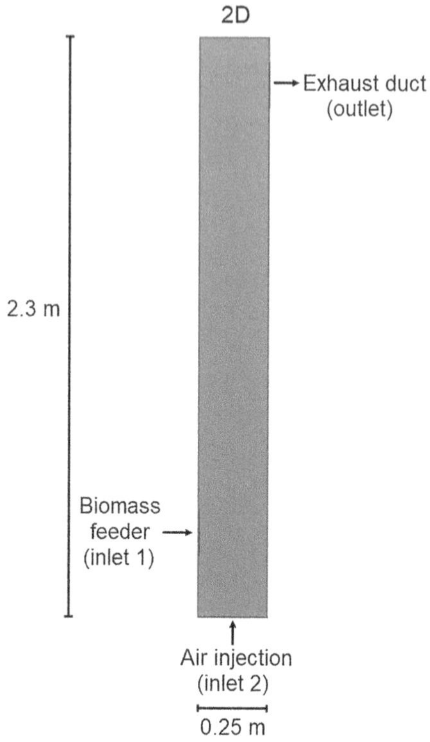

Fig. 13.2 Simplified geometry for the 75 kW_{th} pilot-scale fluidized bed reactor

The users can create a geometry from scratch or a pre-existing one from a computer-aided design (CAD) model. Starting with a pre-existing geometry can save some time and effort; however other challenges may arise such as how to extract the fluid region from a solid part. Also, trying to simplify or remove unnecessary features from an already defined solid can be tricky or time-consuming. When creating a geometry from scratch, users are able to analyse the problem beforehand, allowing them to remove features they deem unnecessary that would complicate meshing such as fillets or bolt heads, which frequently add no crucial information and can be assumed to have little to no impact on the final solution.

Some additional questions can figure out challenging issues. Should the user select a 2-D or a 3-D geometry? Can the user take advantage of the design symmetry? In fact, researchers usually tend to choose a 2-D geometry for easiness and moderate numerical effort. A more sophisticated approach of still using 2-D geometries takes advantage of design symmetry and uses 2-D axisymmetric problem setups (Xie et al. 2008). When the accuracy is of primary importance, the 3-D domains are an imposition, but some authors still use only a small and representative part of such domain to meet an equilibrium between accuracy and computational time. When describing fluidized bed systems, and in order to obtain fully developed turbulent velocity profiles at the air and/or biomass inlet, some authors include pre-inlet pipes for the oxidizer and/or substrate inlet. For simplification purposes, only the 2-D geometry scheme is addressed.

13.3.2 Mesh

As mentioned before, the geometry is split into small cells for numerical purposes. Here stands a major concern—the mesh must hold a good quality in several aspects, element distribution, element quantity and shape, and element smooth transition, to ensure that the results follow the pre-defined criteria of convergence. The process of developing a good mesh could be quite demanding, and a thumb rule is to generate a primary coarse mesh and then proceed by duplicating the number of elements in the following meshes (Cardoso et al. 2018a). When dealing with solid combustion or gasification, the user must guarantee a ratio within 5–12 between the maximum grid size and the particle diameter (Cardoso et al. 2018a). This rule may guide the user towards when to stop the mesh refinement. Lastly, the user must balance between the accuracy and the time needed for the simulation.

There is an assortment of parameters such as element quality, aspect ratio, skewness and orthogonal quality that are important indicators of the mesh quality. Element quality relates the ratio of the volume to the sum of the square of the edge lengths for 2-D elements, ranging from 0 to 1, in which higher values indicate higher element quality, 0 standing for null or negative volume element and 1 for perfect cube or square. Aspect ratio measures the stretching of a cell, and its acceptable range must be lower than 100. Skewness provides the level of distortion of the existing elements from standard or normalized elements; hence skewness metrics must be kept as low as possible, 0 for excellent and 1 for unacceptable. Orthogonal quality is determined by vectoring from the centre of an element to each of the adjacent elements, ranging from 0 to 1, where 0 claims the worst elements and 1 indicates high quality. When the user proceeds with a 2-D simulation, triangular- and quadrilateral-shaped elements are usually selected.

The user is advised to test with at least three or four meshes prior to the final decision. The chosen mesh must succeed at these quality parameters test, follow the implemented convergence criteria and reach the final solution in a reasonable amount of time.

An efficient way to refine a particular mesh is to identify and locate high gradients. This can be done by employing two possible routes: one is by manually setting a particular region where high gradients are expected, e.g. near walls, inlets/outlets, wall boundaries, smaller features and curved regions, and the other is to apply a mesh adaption feature, allowing the automatic mesh refinement in regions where the software sees fit without user interaction. However, in order to locally refine the mesh, the user must be able to recognize the areas where higher gradients are more likely to occur. ANSYS Meshing allows to speed up this process by implementing "Size Functions"; these controls automatically refine the mesh in the areas that will typically have higher gradients.

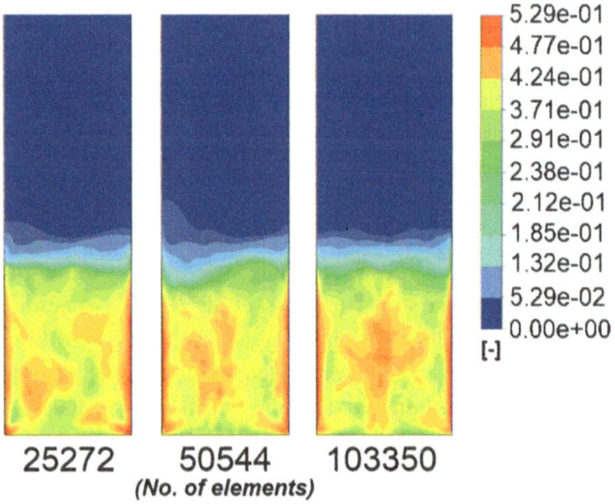

Fig. 13.3 2-D time-averaged solid volume fraction contours for each studied mesh resolution

Table 13.2 Mesh quality parameters

	Quality parameters	Coarse	Medium	Fine
2-D	Element size (cm)	0.9	0.6	0.4
	Number of elements	25,272	50,544	103,350
	Element quality	0.9281	0.9517	0.9981
	Aspect ratio	1.0686	1.0567	1.0010
	Skewness	$8.2534e^{-2}$	$5.6272e^{-2}$	$1.7071e^{-3}$
	Orthogonal quality	0.9895	0.9950	0.9999

13.3.3 Mesh Analysis

In order to allow readers acquiring a deeper understanding of the relationship between mesh density, results and computational cost, a mesh sensitivity analysis was conducted based on previous works developed by the authors (Cardoso et al. 2018a).

The simulations were carried out in a 2-D geometry and are shown in Fig. 13.3, holding the three mesh resolutions previously presented in Table 13.2. A mesh refinement ratio of 2 and a maximum grid spacing rule of 10–12 times the particle diameter were used to build all meshes in this study. All numerical simulations concern gasification experiments making use of eucalyptus wood as biomass, mean diameter of 5 mm, and quartz sand as bed material, mean diameter of 0.5 mm. Operating conditions were set at a superficial gas velocity of 0.25 m/s, time-averaged over a total of 3 s simulation time and operating temperature of 873 K. The geometry domain was created according to the real dimensions of the 75 kW_{th} fluidized bed reactor, 0.25 m width, 2.3 m height and static bed height of 0.23 m, so as to reproduce the established experimental operating conditions closer to a real scenario. At the bottom, atmospheric air is injected into the reactor, inlet, while at the top of the geometry, an opening is set to withdraw the produced syngas, outlet.

Additional information regarding the reactor configurations and simulation parameters can be found in Cardoso et al. (2017, 2018a).

Clearer solids and void distribution areas are provided by the finer meshes, 50,544 and 103,350 elements, while a rougher representation is granted by the coarser mesh, 25,272 elements, as it misinterprets solid presence throughout various areas along the bed. Indeed, different mesh densities result in representation dissimilarities. Convergence-wise, coarser meshes tend to fail to provide a realistic interpretation of the bed behaviour, delivering fallacious assumptions and results accentuating the need to perform a mesh sensitivity study. Finer meshes are capable of providing a more accurate solution; nevertheless, computational time increases as a mesh is made finer; therefore one must reasonably balance between accuracy and computational resource availability.

13.4 Settings

When the user initiates the ANSYS Fluent, several decisions must be made. The user should define the type of geometry, 2-D or 3-D; the level of accuracy, single or double precision; and the processing options, serial or parallel. If the user disposes of a multicore machine, parallel processing can actively speed up the simulation. Along this section, the most important setting steps for performing a simulation will be briefly discussed. For the sake of simplicity, the steps regarding the chemical model implementation will not be covered in this chapter, and its proper implementation will be later defined in the upcoming chapters due to their importance. Therefore, all attentions are devoted to the remaining settings.

In the setup general options, two types of solvers are available, "pressure-based", default option, and "density-based". The pressure-based solver is used for most cases, gasification processes included, such as handling problems with low Mach number, once being more accurate for incompressible subsonic flows. On the other hand, the density-based solver is more accurate for supersonic flow applications with higher Mach number, such as to capture interacting shock waves. Regarding the time dependence, the flow characteristics can be specified as either steady-state or transient.

An important step in the setup of the model is defining the materials and their physical properties. Within the materials section, the user can edit, or create, the properties of any material from the ANSYS database. Here, the user can input the relevant properties for the problem scope. ANSYS Fluent will automatically show the properties that need to be defined according to the models previously selected by the user. For solid materials in gasification processes, density, specific heat, thermal conductivity and viscosity must be defined. For each property, one may specify them as a constant, a linear or polynomial function, define it by a kinetic theory or even employ a user-defined function (UDF). UDF inclusion can be advantageous, allowing the user to customize the setup bringing the solution closer to its particular needs. In fact, ANSYS Fluent allows the user to customize a lot of its standard features by UDF inclusion.

With the materials once defined, one must set the phases and the interactions between them in the "Phases" dialogue box placed in the toolbars. The options here contained will vary regarding the type of multiphase model the user employs. Then, one must also set the parameters related to the operating conditions in the model, such as operating pressure, atmospheric pressure as the default value, temperature, density and gravity. To finish the main actions necessary to undertake in the setup stage, the user must now assign the boundary conditions to each previously designated zone and perform the required inputs for each boundary. Boundary conditions are a required and very important component of the mathematical model. These specify the boundary locations in the geometry, allowing to direct the motion of flow to enter and exit the solution domain. ANSYS Fluent provides various types of boundary conditions concerning the type of solution at hands and the physical models considered.

13.5 Mathematical Model Formulation

Combustion and gasification of waste materials include more than one species phase making the process more complicated to handle. They are typical cases of multiphase flows. In general, the mathematical treatment given in a multiphase flow describes the gas phase as a continuum approach, being the solid phase the one who differs in the approach.

The Eulerian-Eulerian method treats the solid phase as a continuum, while the Eulerian-Lagrangian numerical approach tracks the solid particles individually. Within the Eulerian-Eulerian method, the most implemented option used to handle the thermochemical waste conversion to energy is the Eulerian granular model, while regarding the Eulerian-Lagrangian methods, the discrete phase model (DPM) is the one who stands out. Table 13.3 depicts the most relevant features of both methods.

Table 13.3 Main characteristics of Euler granular and discrete phase model (Fan et al. 2018; Garzó et al. 2007; Godlieb et al. 2007)

Euler granular model	Discrete phase model
Treats continuous fluid (primary phase) as well as dispersed solids (secondary phase) as interpenetrating continua.	The discrete phase is modelled by the Lagrangian model, while the continuous phase is modelled by the Eulerian model.
Effects of particle-particle interactions are accounted for based on the kinetic theory of granular flow (KTGF).	The discrete and the continuous phases are coupled through source terms in the governing equations.
Applicable from dilute to dense particulate flows. Particle size distribution can also be accounted by assigning a separate secondary phase for each particle diameter.	It is recommended to keep a volume fraction inferior to 10%. On the other hand, mass loading can be rather large, in excess of 100%.
Compatible with species transport and homogeneous and heterogeneous reactions.	The discrete phase model (DPM) accounts for the effect of turbulence on the particle trajectories.

Despite being excellent options, they still present a bunch of limitations and drawbacks. As mentioned above, the Eulerian granular models treat both phases as a continuum, meaning that the information given from particle trajectories and size distribution is scarce. The treatment of polydisperse particles is also ineffective with this method (Garzó et al. 2007). The DPM method addresses these drawbacks by tracking every particle and the particle collision (Fan et al. 2018), but such power demands a terrible computational effort limiting the number of particles to 2×10^5, which is not ideal for treating combustors or gasifiers.

To solve the inability to handle with dense particulate flows, in recent years, a new Eulerian-Lagrangian method known as dense discrete phase model (DDPM) was brought in. This one shows better grid independence and turns the mathematical treatment of particle size distribution easier. The DDPM method can be divided into two approaches: the DDPM-kinetic theory of granular flow (KTGF) and the DDPM-discrete element model (DEM). The DDPM-KTGF approach is more suited for diluted to moderately dense particulate flows. Its main advantage is to allow for faster computations while predicting particle-particle collisions without full DEM. On the other hand, the DDPM-DEM approach is most suited for dense to near-packing limit particulate flows.

13.5.1 Turbulent Flow

The large percentage of flows in practical cases of engineering is turbulent. This means the flows are three-dimensional, irregular, aperiodic and with a broad range of length and time scales. This also means the mathematical treatment becomes more demanding, sometimes even exceptionally demanding. When a flow is under a turbulent regime, there are fluctuating velocity fields, which in turn affect the main hydrodynamic features.

The turbulence transfer between phases plays a predominant role in the case of gasification and combustion being crucial to be aware of the leading modelling approaches.

There are three basic approaches that are normally employed to compute a turbulent flow:

- Direct numerical simulation (DNS).
- Scale-resolving simulations (SRS).
- Reynolds-averaged Navier-Stokes simulations (RANS).

The DNS approach does not require any modelling to solve the full unsteady Navier-Stokes equations. This solution is known as being computationally demanding, and its application comes more as a research tool without any direct practical use in real industrial cases. In the SRS approach, there is a balance between modelling and resolution, with the smaller eddies than the grid being modelled and the bigger ones being directly resolved in the calculation. The traditional option in the industry relies on the use of the RANS models. It distinguishes itself from the

Table 13.4 Two-equation turbulence models and their respective recommended usage (ANSYS 2013)

Model	Recommended usage
1. Standard k-ε model	Default k-ε model and the most widely used engineering turbulence model for industrial applications. Suitable for exploring basic flow pattern and parametric studies. Appropriate for converging initial case before switching to other models.
2. RNG (renormalization-group) k-ε model	Recommended for complex shear flows involving rapid strain, moderate swirl, vortices and locally transitional flows.
3. Realizable k-ε model	Improved prediction for spreading rate of jets, superior ability to capture the mean flow of complex structures and for flows involving rotation, boundary layers under strong adverse pressure gradients, separation and recirculation. Delivers improved accuracy and easier convergence than RNG.
4. Standard k-ω model	Most widely adopted in the aerospace and turbo-machinery communities. Not recommended from an industrial standpoint except if the user disposes of good boundary conditions for k and ω.
5. SST (shear-stress transport) k-ω model	Recommended for high-accuracy boundary layer simulations. Wall-bounded flow (e.g. blades, airfoils, compressors and turbines, among others).

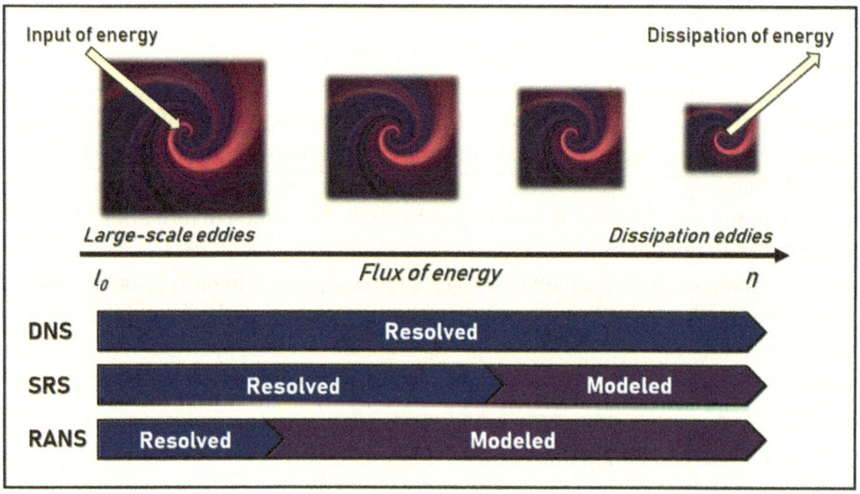

Fig. 13.4 Representation of the three main approaches used to compute a turbulent flow

previous methods by allowing a solution for steady-state simulations while modelling all the turbulence features. The RANS approach includes a large set of turbulence sub-models with specific features. Table 13.4 depicts the most relevant options within the RANS approach.

The right turbulence model to handle a real problem is still a very arguable question, and there is not a clear answer. Anyway, regarding combustion and

Table 13.5 Radiation models available in ANSYS fluent (ANSYS 2013)

Model	Applicability
Discrete ordinate (DO) model	The DO model is the most comprehensive radiation model but can easily become extremely computationally expensive.
Discrete transfer radiation model (DTRM)	The DTRM is a relatively simple model whose accuracy relies heavily on the number of rays; a large number of rays turn the problem-solving computationally intensive.
P-1 model	The radiative equations are solved with little computational cost but tend to overpredict radiative fluxes from localized heat sources.
Rosseland model	The Rosseland model can only be applied to optically thick media. It is faster than the P-1 model and requires less memory.
Surface-to-surface (S2S) model	The S2S model can be used in situations where there are no participating media (absorbs, emits or scatters a thermal ray as it travels through the medium); however, it cannot be used with periodic or symmetry boundary conditions.

gasification, the use of the standard k-ε model (ANSYS 2006; Zhang et al. 2015; Liu et al. 2013) still continues to be the primary option since it was first proposed by Launder and Spalding (1974). Figure 13.4 summarizes the three main approaches to compute a turbulent flow and corresponding features.

13.5.2 Radiation Model

The transfer of heat through electromagnetic energy is defined as radiation. Thermal radiation effects should be accounted whenever the heat radiation is at least equal or of greater magnitude than that of convective and conductive heat transfer rates, being of practical importance only at very high temperatures, above 800 K (Wong and Seville 2006). Radiation phenomena undergo complex interactions between the phases, so to accurately predict these interplays, computationally effective thermal radiation models are required to solve the radiative intensity transport equations. Table 13.5 describes the main radiation models available in ANSYS Fluent.

13.6 Solver Settings

Before proceeding with the solution calculation, the user must first set the solution methods. ANSYS Fluent provides multiple schemes to solve different types of solutions; however, only the available solution method settings for solving Eulerian multiphase flows will be addressed, and additional information on Lagrangian flows will be provided in the combustion tutorial chapter. For Eulerian multiphase flows, ANSYS Fluent solves the phase momentum equations, the shared pressure and the phasic volume fraction equations either by implementing a coupled or a segregated fashion. Figure 13.5 depicts an overview concerning the various features the user must engage to properly set the solver.

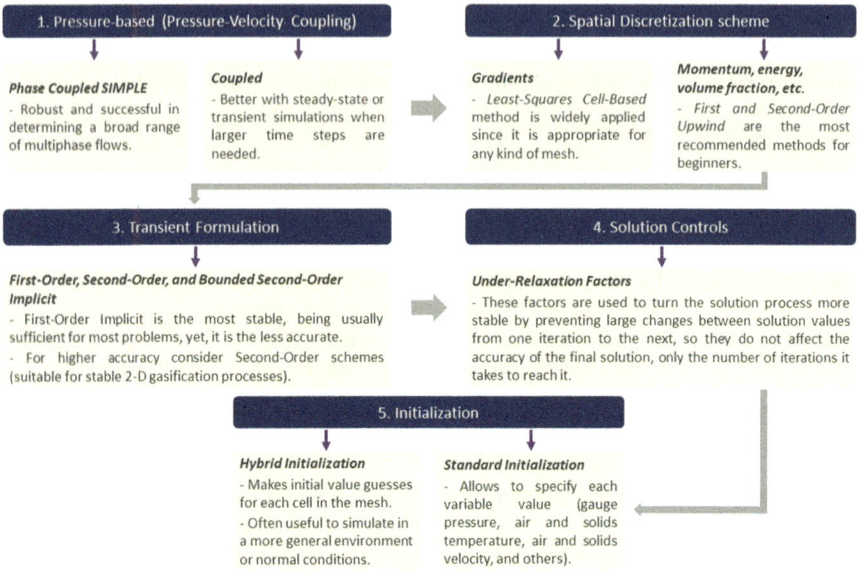

Fig. 13.5 Solver setting procedure overview

After the initialization, the "Patch" button becomes enabled. Setting the Patch values for individual variables in certain regions of the domain is an essential task while modelling multiphase flows and combustion problems. In order to do so, one must first create a domain region adaption, within the Adapt panel, which marks individual cells for refinement. In fluidized bed simulation, this marked cell region of interest is the area occupied by the static bed, delimited by the extremities of the reactor's domain and the bed height, and this selection is very important to set the different phase volume fractions in the region. Having defined the region for adaption, it is a good practice to display it so to visually verify if it encompasses the desired area. Following this procedure, the user may now return to the Patch panel and set the initial volume fraction of the solids, bed material and biomass in case of a binary mixture, in the marked bed region of the fluidized bed.

The "Run Calculation" allows to finally start the solver iterations. The available panel options in this task page vary concerning previous settings made. For a transient flow calculation, the user disposes of various options to determine the time step. Employing a "Fixed" time-stepping method allows the user to input the intended time step size, in seconds, and the number of time steps. If desired, the user may enable the "Extrapolate Variables" option, to estimate an initial guess for the next time step, allowing to speed up the transient solution by reducing required sub-iteration, and the "Data Sampling for Time Statistics", to compute the time average, mean and root mean square of the instantaneous values sampled during the calculation. All remaining options may be set as default.

Additionally, during the calculation process, the user may display contours, vectors, monitor plots or even mesh, for any desired quantity in the "Graphics and

Animations" dropdown list. Displaying the solid volume fraction contours in the reactor's domain is particularly useful for hydrodynamic analysis in fluidized bed gasification, allowing the user to follow the solid evolution being refreshed for every time step throughout the simulation time. Having finished all these solver settings, the user may now start the calculation.

- If the user desires to shorten the computational time for transient solutions, the phase-coupled SIMPLE scheme is more appropriate than the coupled scheme.
- In case the user chooses to enforce a couple scheme, a Courant number, 200 by default, and explicit relaxation factors for momentum and pressure, 0.75 by default, must be specified. The user may decrease the Courant number and the explicit relaxation factors if difficulties in reaching the convergence are encountered, whether being due to higher-order schemes or to the high complexity of the problem, such as in multiphase and combustion problems. These can later be heightened if the iteration process runs smoothly.
- Different numerical schemes may respond differently to the applied under-relaxation factors. For instance, setting lower under-relaxation factors for the volume fraction equation for the coupled scheme may lag the solution considerably, and values placed around 0.5 or above are considered acceptable. Contrarily, the phase-coupled SIMPLE generally requires a low under-relaxation for the volume fraction equation.
- When the solution solver requires higher-order numerical schemes or higher spatial discretization, it is recommended that the user initiates the solution by setting smaller time steps. These may be further increased after performing a few time steps so as to achieve a better approximation of the pressure field.
- For better convergence in gasification analysis, it is recommended to start the solution with a non-reacting flow and without radiation model. To do so, it is necessary to disable the chemical reactions, radiation equations and fluid-particle interactions. For instance, if the user intends to evaluate the solid particle behaviour within a fluidized bed, the solution must be initiated without the inclusion of the chemical reaction sub-model. This allows analysing, in a first stage, the hydrodynamics features by employing a simpler approach and determining if the results obtained are within tolerance and if proper behaviour is being achieved. After such validation, the chemical reactions model and the devolatilization sub-model can be securely added to the mathematical model.
- Lastly, the residuals stand as useful indicators of the iterative converge of the solution, quantifying the error in the solution of the equations system; thus it is important to monitor the residual behaviour during the calculation. Throughout this iterative process, the residuals are expected to progressively decay to smaller values, never reaching exact zero, up until they get levelled and substantial changes stop occurring. The lower the residual values are, the more numerically accurate the solution will be. If the residuals present an increasing behaviour within the first few iterations, one should consider lowering the under-relaxation factors and resume the calculation. Occasionally, the residuals may present a rather unstable behaviour showing huge fluctuations; on such occasions, the user

should proceed by reducing the under-relaxation factors. Yet, if the instabilities prevail, this might be a sign of previous misconfigurations; thereby, the user must re-check previous settings such as initial values, boundary conditions, mesh and fluid properties, in order to reach more stable residual curves.

13.7 Best Practices, Model Validation and Verification

After completing the whole simulation process and getting a solution, it is time to proceed with an in-depth analysis of the results. The overall accuracy of the numerical simulation depends on the magnitude of several factors:

- Round-off errors.
- Iteration errors.
- Discretization errors.
- Simplifications and assumptions.
- Differences between the numerical model and the real problem.

The first item emphasizes the errors that are coming up from the misrepresentation of real numbers due to incorrect truncation. Nowadays, with robust computational power, these errors are minimized, but a bunch of situations can lead to the source of the problem:

- Significant differences in length and time scales.
- Large range parameter variation.
- Major grid variations.

Some thumb rules are advised such as using a double-precision feature, defining target variables and comparing the double-precision results with single precision.

The iteration errors are intimately related with the user ability to reduce the presence of residuals in the numerical simulation. The best way to ensure a good convergence implies the right selection of residual criteria for the most critical parameters. It will be wise to plot and follow the residuals of these parameters through the convergence process. The user can impose tighter criteria in specific settings and must pay attention if the mass and energy balances are being respected.

As previously described, the use of the right mesh is crucial to get a good solution. There will always be differences between the solution found on the selected grid and an infinitely fine grid. The major goal here is to minimize such differences by simultaneously balancing the required time to run the numerical simulation. The reader should bear in mind that it is impossible to get an exact solution due to discretization errors but it is possible to keep them low by performing a mesh sensitivity analysis. Different discretization schemes in fine grids provide very similar results, while coarser meshes can lead to substantial dissimilarities. Anyway, the user should be aware of the most suitable discretization scheme for the problem in hands.

The last items seem obvious, but a significant contribution of the errors arises from inadequate simplifications, assumptions or incorrect use of mathematical models. The user can cut all the previous mistakes and still find substantial deviations between the experimental and numerical runs. Here, the user must be sure about the physics of the problem and test the impact of the riskiest assumptions. On the other hand, all the computational packages include a large number of models with their simplifications and assumptions, and the user must check the right adequacy for what it is intended.

In summary, the best practices to minimize such errors go through the following actions:

- Get a good grid quality by performing sensitivity analysis.
- Use the double-precision scheme and compare the results with single precision.
- Follow the residuals and check the mass and energy balances.
- Correct the defined residual criteria when needed.
- Use high-order discretization schemes.
- Check which one of the available models is suitable to reasonably depict the physics of the problem.

The listing of the previous errors and the ways to prevent it are a good source of inspiration for the next step. The user must verify and validate the results. It is essential to understand the meaning of verification and validation and the importance of performing both. By verification, one means the procedure to guarantee that the software package solves the mathematical problem with all the equations, boundary conditions and all the other computational settings. By validation, one means the procedure to ensure the employed model satisfies the experimental data in a broad range of conditions.

The validation procedure is sometimes hard because experimental data is not always available and when available rarely offers a large set of conditions for comparison. Anyway, the user must ensure that the model fits reasonably the experimental data at least in a set of conditions where the key parameters vary within an interesting range. If the model captures the key trends, there is a certain degree of confidence that the user can extrapolate conclusions in alternative geometries or under hazardous conditions that are prohibitive experimentally.

13.8 Post-processing

Having converged and validated the solution, it is now time to obtain the results first planned in the pre-analysis step. Researchers possess a wide range of options when it comes to post-processing such as contour plots, vectors, streamlines, iso-surfaces, video screenings, creating planes and lines to study particular solution regions, creating graphical representations and generate reports. Within the ANSYS framework, there are two possible routes to post-process the simulation results from Fluent, the Fluent post-processing integrated tools or the ANSYS CFD-Post application.

In the "Results" task page, one may find a set of the most common post-processing features, namely, contours, vectors, pathlines, particle tracks, animations, several types of plotting and reports. Supplementary post-processing tasks can be found within the "Postprocessing" banner in the toolbar tabs. Here, the user may access and create surface regions in the solution domain like points, lines and planes.

The other possible route for post-processing is to use the ANSYS CFD-Post. Both platforms are perfectly capable of addressing the most basic post-processing; however, contrary to the ANSYS Fluent post-processing built-in options, the CFD-Post provides far more powerful and sophisticated post-processing capabilities as 3D-viewer files, user variables, automatic HTML report generation and case comparison tools (Rüdisüli et al. 2012). ANSYS Fluent allows the user to send the case and data files to CFD-Post and to perform the various post-processing actions. For such, the user must select the quantities one desires to export by creating a "CFD-Post Compatible Automatic Export" within the "Calculation Activities" task page.

Regarding fluidized bed gasification, the use of CFD-Post is advantageous as it allows to produce visual data with higher quality, assisting to better visualize and understand the complex flow phenomena within the reactor. Indeed, the ANSYS CFD-Post options are immense and are best learned in a hands-on manner; the experience will lead researchers to take their result visualization and analysis to the next level by taking the maximum potential of such a broad application.

13.9 Conclusions

This chapter has summarized and discussed the main process steps one must endure to implement a model within the ANSYS Fluent framework. The implementation of chemistry sets will be later defined in the upcoming chapters due to their importance. At this stage, the user must have a clear idea on the theory behind how to manage the ANSYS Fluent software and how it interplays with gasification and combustion process implementation.

Acknowledgements The authors are deeply thankful to the Portuguese Foundation for Science and Technology (FCT) for the grant SFRH/BD/146155/2019 and for the projects IF/01772/2014 and CMU/TMP/0032/2017.

References

Ahmad A, Zawawi N, Kasim F, Inayat A, Khasri A (2016) Assessing the gasification performance of biomass: a review on biomass gasification process conditions, optimization and economic evaluation. Renew Sust Energ Rev 53:1333–1347. https://doi.org/10.1016/j.rser.2015.09.030

ANSYS (2006) Modeling turbulent flows

ANSYS (2013) ANSYS fluent theory guide

Baruah D, Baruah DC (2014) Modeling of biomass gasification: a review. Renew Sust Energ Rev 39:806–815. https://doi.org/10.1016/j.rser.2014.07.129

Cai Y, Huang G, Yeh S, Liu L, Li L (2011) A modeling approach for investigating climate change impacts on renewable energy utilization. Int J Energy Res 36:764–777. https://doi.org/10.1002/er.1831

Cardoso J, Silva V, Eusébio D, Brito P (2017) Hydrodynamic modelling of municipal solid waste residues in a pilot scale fluidized bed reactor. Energies 10:17773. https://doi.org/10.3390/en10111773

Cardoso J, Silva V, Eusébio D, Brito P, Tarelho L (2018a) Improved numerical approaches to predict hydrodynamics in a pilot-scale bubbling fluidized bed biomass reactor: a numerical study with experimental validation. Energy Convers Manag 156:53–67. https://doi.org/10.1016/j.renene.2014.11.089

Cardoso J, Silva V, Eusébio D, Brito P, Tarelho L (2018b) Comparative scaling analysis of two different sized pilot-scale fluidized bed reactors operating with biomass substrates. Energy 151:520–535. https://doi.org/10.1016/j.energy.2018.03.090

Cardoso J, Silva VB, Eusébio D (2019a) Techno-economic analysis of a biomass gasification power plant dealing with forestry residues blends for electricity production in Portugal. J Clean Prod 212:741–753. https://doi.org/10.1016/j.jclepro.2018.12.054

Cardoso J, Silva VB, Eusébio D (2019b) Process optimization and robustness analysis of municipal solid waste gasification using air-carbon dioxide mixtures as gasifying agent. Int J Energy Res 43:4715–4728. https://doi.org/10.1002/er.4611

Cardoso J, Silva VB, Eusébio D, Brito P, Boloy R, Tarelho SJ (2019c) Comparative 2D and 3D analysis on the hydrodynamics behaviour during biomass gasification in a pilot-scale fluidized bed reactor. Renew Energy 131:713–729. https://doi.org/10.1016/j.renene.2018.07.080

Couto N, Silva V, Monteiro E, Teixeira S, Chacartegui R, Bouziane K, Brito P, Rouboa A (2015a) Numerical and experimental analysis of municipal solid wastes gasification process. Appl Therm Eng 78:185–195. https://doi.org/10.1016/j.applthermaleng.2014.12.036

Couto N, Silva V, Monteiro E, Brito P, Rouboa A (2015b) Using an Eulerian-granular 2-D multiphase CFD model to simulate oxygen air enriched gasification of agroindustrial residues. Renew Energy 77:174–181. https://doi.org/10.1016/j.renene.2014.11.089

Couto N, Silva V, Monteiro E, Rouboa A, Brito P (2016a) Hydrogen-rich gas from gasification of Portuguese municipal solid wastes. Int J Hydrog Energy 41:10619–10630. https://doi.org/10.1016/j.ijhydene.2016.04.091

Couto N, Silva V, Bispo C, Rouboa A (2016b) From laboratorial to pilot fluidized bed reactors: analysis of the scale-up phenomenon. Energy Convers Manag 119:177–186. https://doi.org/10.1016/j.enconman.2016.03.085

Couto N, Silva V, Monteiro E, Rouboa A, Brito P (2017a) An experimental and numerical study on the miscanthus gasification by using a pilot scale gasifier. Renew Energy 109:248–261. https://doi.org/10.1016/j.renene.2017.03.028

Couto N, Silva V, Cardoso J, Rouboa A (2017b) 2nd law analysis of Portuguese municipal solid waste gasification using CO_2/air mixtures. J CO2 Util 20:347–356. https://doi.org/10.1016/j.jcou.2017.06.001

Dinh CB, Liao CC, Hsiau SS (2017) Numerical study of hydrodynamics with surface heat transfer in a bubbling fluidized-bed reactor applied to fast pyrolysis of rice husk. Adv Powder Technol 28:419–429. https://doi.org/10.1016/j.apt.2016.10.013

Fan H, Guo D, Dong J, Cui X, Zhang M, Zhang Z (2018) Discrete element method simulation of the mixing process of particles with and without cohesive interparticle forces in a fluidized bed. Powder Technol 327:223–231. https://doi.org/10.1016/j.powtec.2017.12.016

Fogler HS, Gurmen NM (2002) Aspen plus workshop for reaction engineering and design. The University of Michigan

Galadima A, Muraza O (2015) Waste to liquid fuels: potency, progress and challenges. Int J Energy Res 39:1451–1478. https://doi.org/10.1002/er.3360

Garzó V, Dufty JW, Hrenya CM (2007) Enskog theory for polydisperse granular mixtures. I. Navier-Stokes order transport. Phys Rev E Stat Nonlin Soft Matter Phys 76:031303. https://doi.org/10.1103/PhysRevE.76.031303

Godlieb W, Deen NG, Kuipers JAM (2007) A discrete particle simulation study of solids mixing in a pressurized fluidized bed. In: Paper presented at the 2007 ECI conference on the 12th international conference on fluidization—new horizons in fluidization engineering, Vancouver, Canada

Lamers P, Roni M, Tumuluru J, Jacobson J, Cafferty K, Hansen J, Kenney K, Teymouri F, Bals B (2015) Techno-economic analysis of decentralized biomass processing depots. Bioresour Technol 194:205–213. https://doi.org/10.1016/j.biortech.2015.07.009

Launder BE, Spalding DB (1974) The numerical computation of turbulent flows. Comput Methods Appl Mech Eng 3:269–289. https://doi.org/10.1016/0045-7825(74)90029-2

Leme M, Rocha M, Lora E, Venturini O, Lopes B, Ferreira C (2014) Techno-economic analysis and environmental impact assessment of energy recovery from municipal solid waste (MSW) in Brazil. Resour Conserv Recycl 87:8–20. https://doi.org/10.1016/j.resconrec.2014.03.003

Liu H, Elkamel A, Lohi A, Biglari M (2013) Computational fluid dynamics modeling of biomass gasification in circulating fluidized-bed reactor using the Eulerian–Eulerian approach. Ind Eng Chem Res 52:18162–18174. https://doi.org/10.1021/ie4024148

Lourinho G, Brito P (2015) Assessment of biomass energy potential in a region of Portugal (Alto Alentejo). Energy 81:189–201. https://doi.org/10.1016/j.energy.2014.12.021

Maurer S, Schildhauer TJ, Ruud van Ommen J, Biollaz SMA, Wokaun A (2014) Scale-up of fluidized beds with vertical internals: studying the sectoral approach by means of optical probes. Chem Eng J 252:131–140. https://doi.org/10.1016/j.cej.2014.04.083

Molino A, Chianese S, Musmarra D (2016) Biomass gasification technology: the state of the art overview. J Energy Chem 25:10–25. https://doi.org/10.1016/j.jechem.2015.11.005

Neves D, Thunman H, Matos A, Tarelho L, Barea AG (2011) Characterization and prediction of biomass pyrolysis products. Prog Energy Combust Sci 37:611–630. https://doi.org/10.1016/j.pecs.2011.01.001

Olaofe OO, Patil AV, Deen NG, van der Hoef MA, Kuipers JAM (2014) Simulation of particle mixing and segregation in bidisperse gas fluidized beds. Chem Eng Sci 108:258–269. https://doi.org/10.1016/j.ces.2014.01.009

Pereira S, Costa M, Carvalho M, Rodrigues A (2016) Potential of poplar short rotation coppice cultivation for bioenergy in southern Portugal. Energy Convers Manag 125:242–253. https://doi.org/10.1016/j.enconman.2016.03.068

Pinto F, André RN, Carolino C, Miranda M, Abelha P, Direito D, Perdikaris N, Boukis I (2014) Gasification improvement of a poor quality solid recovered fuel (SRF). Effect of using natural minerals and biomass wastes blends. Fuel 117:1034–1044. https://doi.org/10.1016/j.fuel.2013.10.015

Pio D, Tarelho L, Matos M (2017) Characteristics of the gas produced during biomass direct gasification in an autothermal pilot-scale bubbling fluidized bed reactor. Energy 120:915–928. https://doi.org/10.1016/j.energy.2016.11.145

Ramos A, Monteiro E, Silva V, Rouboa A (2018) Co-gasification and recent developments on waste-to-energy conversion: a review. Renew Sust Energ Rev 81:380–398. https://doi.org/10.1016/j.rser.2017.07.025

Rüdisüli M, Schildhauer TJ, Biollaz SMA, Ommen JR (2012) Scale-up of bubbling fluidized bed reactors—a review. Powder Technol 217:21–38. https://doi.org/10.1016/j.powtec.2011.10.004

Sanderson J, Rhodes M (2005) Bubbling fluidized bed scaling laws: evaluation at large scales. AIChE J 51:2686–2694. https://doi.org/10.1002/aic.10511

Sehrawat J, Patel M, Kumar B (2015) Gaussian process regression to predict incipient motion of Alluvial channel. In: Das K, Deep K, Pant M, Bansal JC, Nagar A (eds) Proceedings of fourth international conference on soft computing for problem solving, advances in intelligent systems and computing, vol 336. Springer, New Delhi. https://doi.org/10.1007/978-81-322-2220-0_35

Sharma A, Wang S, Pareek V, Yang H, Zhang D (2014) CFD modeling of mixing/segregation behavior of biomass and biochar particles in a bubbling fluidized bed. Chem Eng Sci 106:264–274. https://doi.org/10.1016/j.ces.2013.11.019

Sharma A, Wang S, Pareek V, Yang H, Zhang D (2015) Multi-fluid reactive modeling of fluidized bed pyrolysis process. Chem Eng Sci 123:311–321. https://doi.org/10.1016/j.ces.2014.11.019

Silva VB, Rouboa A (2013) Using a two-stage equilibrium model to simulate oxygen air enriched gasification of pine biomass residues. Fuel Process Technol 109:111–117. https://doi.org/10.1016/j.fuproc.2012.09.045

Silva VB, Rouboa A (2015) Combining a 2-D multiphase CFD model with a response surface methodology to optimize the gasification of Portuguese biomasses. Energy Convers Manag 99:28–40. https://doi.org/10.1016/j.enconman.2015.03.020

Tarelho L, Neves D, Matos M (2011) Forest biomass waste combustion in a pilot-scale bubbling fluidised bed combustor. Biomass Bioenergy 35:1511–1523. https://doi.org/10.1016/j.biombioe.2010.12.052

Vepsäläinen A, Shah S, Ritvanen J, Hyppänen T (2013) Bed Sherwood number in fluidised bed combustion by Eulerian CFD modelling. Chem Eng Sci 93:206–213. https://doi.org/10.1016/j.ces.2013.01.065

Vicente E, Tarelho L, Teixeira E, Duarte M, Nunes T, Colombi C, Gianelle V, Rocha G, Sanchez de la Campa A, Alves C (2016) Emissions from the combustion of eucalypt and pine chips in a fluidized bed reactor. J Environ Sci 42:246–258. https://doi.org/10.1016/j.jes.2015.07.012

Wong YS, Seville JPK (2006) Single-particle motion and heat transfer in fluidized beds. AIChE J 52:4099–4109. https://doi.org/10.1002/aic.11012

Xie N, Battaglia F, Pannala S (2008) Effects of using two- versus three-dimensional computational modeling of fluidized beds part I, hydrodynamics. Powder Technol 182:1–13. https://doi.org/10.1016/j.powtec.2007.07.005

Xue Q, Heindel TJ, Fox RO (2011) A CFD model for biomass fast pyrolysis in fluidized-bed reactors. Chem Eng Sci 66:2440–2452. https://doi.org/10.1016/j.ces.2011.03.010

Xue Q, Dalluge D, Heindel TJ, Fox RO, Brown RC (2012) Experimental validation and CFD modeling study of biomass fast pyrolysis in fluidized-bed reactors. Fuel 97:757–769. https://doi.org/10.1016/j.fuel.2012.02.065

Zhang Y, Lei F, Xiao Y (2015) Computational fluid dynamics simulation and parametric study of coal gasification in a circulating fluidized bed reactor. Asia Pac J Chem Eng 10:307–317. https://doi.org/10.1002/apj.1878

Chapter 14
Lignocellulosic Waste Materials for Industrial Water Purification

Fulga Tanasă, Carmen-Alice Teacă, and Marioara Nechifor

Contents

F. Tanasă (✉) · M. Nechifor
Department of Photochemistry and Polyaddition, "Petru Poni" Institute of Macromolecular Chemistry, Iasi, Romania
e-mail: ftanasa@icmpp.ro

C.-A. Teacă
Centre of Advanced Research on Bionanoconjugates and Biopolymers INTELCENTRU, "Petru Poni" Institute of Macromolecular Chemistry, Iasi, Romania

Inamuddin, A. M. Asiri (eds.), *Sustainable Green Chemical Processes and their Allied Applications*, Nanotechnology in the Life Sciences,
https://doi.org/10.1007/978-3-030-42284-4_14

14.1 Introduction

In the last decades, the industry has recorded a rapid development which entailed increased amounts of wastewater discharged in nature, and this has become a challenge not only for the scientists, but for the entire society as well, in order to limit the environmental pollution and human health hazards. There are many conventional methods known and in use, such as biological treatments (Fazal et al. 2015), flocculation (Jawad et al. 2015), membrane processes, chemical precipitation, ion exchange (Rosales et al. 2017), etc. Retention onto activated carbon is considered efficient, reliable, but not cost-effective, so that the research for another approach was encouraged despite the advantages of this method (high capacity of retention, as related to the high surface area and surface reactivity, as well as mechanical strength) (Mashhadi et al. 2016).

Basically, the mechanism of adsorption consists of a mass transfer with the accumulation at the interface of the two phases in direct contact. In the case of solid-liquid systems where adsorbed compounds are removed from their solution and linked onto the surface of the solid sorbent, both physisorption (weak and reversible) and chemisorption (strong binding through chemical bonds) occur and coexist rending stability to the process. Physisorption takes place at temperatures below or close to the critical temperature of the adsorbate and evolves with a decrease in free energy and entropy, and, thereby, the process is exothermic (Cooney 1999; De Gisi et al. 2016). On the other hand, the chemisorption occurs under the condition of a monolayer; hence the resulted bonds are hard to break. In time, a dynamic equilibrium is reached in relation to the adsorbate.

Adsorption onto lignocellulosic waste materials has been acknowledged as a viable alternative to the already existing technologies employed for the removal of various pollutants from the industrial wastewater. These biosorbents may come from agriculture and wood industry, and can originate in different parts of plants (bark, stem, leaves, roots, flowers, biomass, husk, hull, skin, shell, bran, kernel etc.), thus containing cellulose, hemicelluloses, and lignin in various ratios (Fig. 14.1) (Dai et al. 2018; Okoro and Okoro 2011).

Aside from the porosity of the materials, another characteristic of great relevance is the presence of reactive functional groups able to undergo reactions in solution and, therefore, bind pollutants such as heavy metal ions or organic small molecules by specific bonds (Nguyen et al. 2013; Chan et al. 2015; Heng et al. 2017; Zhang et al. 2016; Zhou et al. 2015). For example, most of the proton donor functional groups, namely, amine, amide or acetamide, carboxylic, phenolic, hydroxylic, etc., may interact with the metal ions in solution yielding in substitution compounds or even metallic complexes (Marin-Rangel et al. 2012; Saka et al. 2012). It has been shown that Cr(VI) was chemically bound to the adsorbent surface, and a certain amount of it was reduced to Cr(II) (Lin et al. 2018), as well as Cu, Ni, and Zn ions (Lee and Rowell 2004), while Pb(II) was retained by electrostatic attraction (Nguyen et al. 2013). As for the organic pollutants, adsorption was achieved through other types of bonds, such as hydrogen bonding (dyes), van der Waals bridging (pesticides), hydrophobic attraction (drugs), covalent bonds (oil), or even π-π stacking interactions (Zhou et al. 2015).

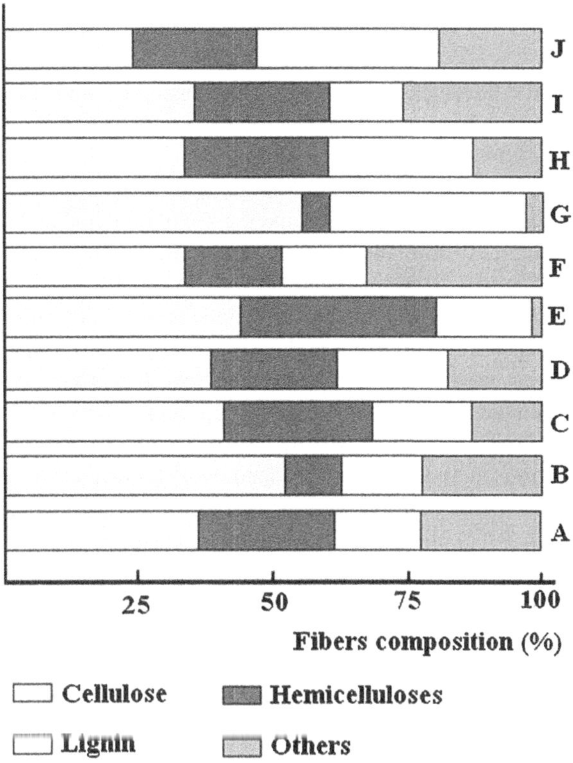

Fig. 14.1 Fiber composition (dry weight %) in correlation with their origin: A, rice straws; B, rice husk; C, wheat straws; D, corn stover; E, corncobs; F, rapeseed straws; G, sugarcane bagasse; H, sunflower stalks; I, sweet sorghum; J, palm kernel shell (redrawn after Dai et al. 2018)

There are factors influencing the purification process onto lignocellulosic materials that have to be considered especially when it comes to the retention effectiveness. Thus, the pH and temperature of the wastewater, the initial concentration of the pollutants and the pollutant-to-adsorbent ratio, and even the ionic strength may affect the process.

Another issue associated with the industrial water purification by means of lignocellulosic waste materials is the so-called secondary pollution caused by the environmental accumulation of waste adsorbents which can become, in their turn, source of further pollution. This problem must be dealt with, and there are some methods commonly applied: solvent extraction (it employs an appropriate solvent for the adsorbate in order to remove it from the sorbent material); calcination (the high temperature treatment yields in organic or inorganic volatiles that are released from the support); and biologic regeneration (specific microorganisms are used to release the adsorbent) (Dai et al. 2018).

In the following, some of the pollutants most commonly found in the wastewater, organic and inorganic, as well as the corresponding lignocellulosic sorbents employed for their efficient removal, will be reviewed highlighting the mechanism of retention onto the surface of the sorbent and factors that may influence the yield of the process.

14.2 Organic Pollutants

In the last decades, the highly increasing societal demand for food and goods has entailed large-scale environmental problems. The water, air, and soil pollution caused by the improper disposal of various by-products and wastes from agriculture and industry, including oil spills, associated with the lack of effective appropriate policies in some countries and the use of outdated technologies, among other causes, generated dramatic climate changes, wildlife extinction, desertification, etc. (O'Donnel 2018).

Wastewatermostcommonlycontainspesticides(suchas2,4-dichlorophenoxyacetic acid, metribuzin, and metalaxyl) and dyes (i.e., Congo red, rhodamine B, malachite green, methylene blue, crystal violet), but aromatic compounds (e.g., polycyclic aromatic hydrocarbons, phenol), oil, and drugs (such as tetracycline, caffeine, tylosin) were identified as well. Therefore, their removal was intensively studied, and various methods were developed in close correlation with specific pollutants: physical separation, sedimentation and filtration (Dai et al. 2018), flocculation (Guan and Tian 2017), membrane separation (Sarasidis et al. 2017), adsorption (Worch 2012), UV-vis treatment (Deng et al. 2017), and microwave-assisted purification (Gayathri et al. 2017); chemical, chemical, photochemical, and catalytic oxidation (Gayathri et al. 2017; Janssens et al. 2017); or even combinations of different methods (i.e., adsorption and photocatalysis on titanate nanotubes (Sandoval et al. 2017); membrane separation and oxidation, with simultaneous in situ regeneration of the sorbent) (Sarasidis et al. 2017), aiming at maximizing the yield of the process.

Unlike other methods, the adsorption onto lignocellulosic materials offers some specific advantages: simplicity, high removal rate, cost-effectiveness, an easy access to resources (agriculture and wood industry), and, of utmost importance, the presence of reactive functional groups able to bind specific organic pollutants – carboxyl (Bouguettoucha et al. 2016; Inam et al. 2015), carbonyl (Bouguettoucha et al. 2016; Zhu et al. 2016a, b, c, d), hydroxyl and amine (Inam et al. 2015; Shang et al. 2015), and amide (Sohrabi and Ameri 2015).

14.2.1 Pesticides

Pesticides are employed in agriculture and forestry in order to limit and prevent pests, fungi, and microorganism infestation, as well as weeds, and their extensive use affects the human health and environment.

Lignocellulosic waste sorbents proved to be highly effective for the retention of different pesticides (Fig. 14.2). Thus, the purification of wastewater containing **2,4-dichlorophenoxyacetic acid** (a well-known systemic pesticide, selective, and used as herbicide and plant growth regulator) attracted the interest mainly after being reported in 1987 by the International Agency for Cancer Research as having carcinogenic potential, as well as mutagenic (International Agency for Research on Cancer (IARC) 1987).

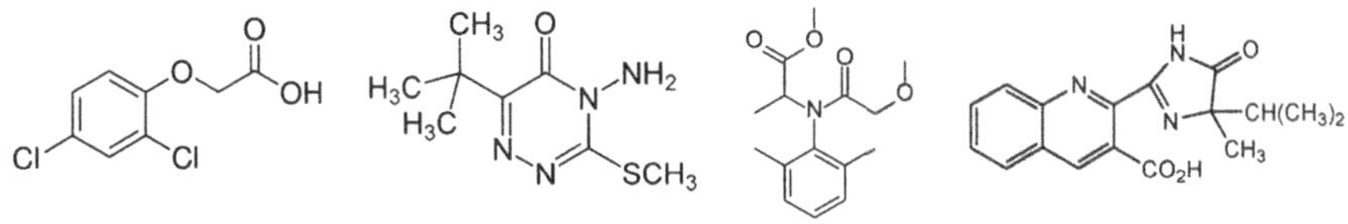

2,4-dichlorophenoxyacetic acid *metribuzin* *metalaxyl* *imazaquin*

Fig. 14.2 Some of most commonly used pesticides that can be retained onto lignocellulosic sorbents

Experiment conducted onto mustard plant ash (Trivedi et al. 2016) aimed at optimizing the purification process and factors affecting the adsorption, namely, the sorbent-to-sorbate ratio, initial pollutant concentration, contact time, and adsorption temperature, were investigated. Based on the experimental data, fitted to different kinetic and isothermal models, it was possible to indicate an optimum sorbent-to-pesticide ratio. The langsat empty fruits activated by phosphoric acid were also employed in the retention of 2,4-dichlorophenoxyacetic acid from wastewater (Njoku et al. 2015) with a reported efficiency of 261.1 mg/g, at 30 °C. The influence of the initial concentration of pollutant, shaking time, pH, and temperature was studied. It was also found the mechanism of sorption was the internal diffusion of particles. Reported data confirmed this material as an excellent sorbent. Another study employed fly ash originated from bagasse as a low-cost and effective sorbent (Deokar et al. 2016) using two different approaches: in-batch experiments and continuous packed-bed processes. The sorption process in batch depends on the initial amount of pesticide, contact time, pH, and temperature and the particle size of the sorbent, while the factors affecting the continuous packed-bed process are the concentration of the influent, flow rate, and bed height and model. It was thus possible to recommend optimal conditions for each study case. The sorbent particle size investigation showed the bigger the particles are, the higher is the amount of pollutant removed as a consequence of the greater BET surface.

Metribuzin is another pesticide of interest, as it is an herbicide used both pre- and post-emergence in crops (soy bean, potatoes, tomatoes, sugarcane, etc.), which acts by inhibiting photosynthesis of weeds. Since it was demonstrated that it contaminates the groundwater, its removal onto waste lignocellulosic materials was studied. A successful research reported on the reduction of water contamination by metribuzin using de-oiled two-phase olive mill waste (DW) as sorbent (Pena et al. 2016) in soil. The method consisted in preparing soils containing DW in various amounts and composted for different intervals. Thus, the study indicated that fresh DW retain lower amounts of metribuzin, as compared to aged DW. Therefore, the DW-composted soils may be a cost-effective solution for the sustainable decontamination of metribuzin-containing groundwater.

An interesting challenge was the removal of chiral pesticides such as **metalaxyl**, an acylalanine derivative (methyl N-(methoxyacetyl)-N-(2,6-xylyl)-DL-alaninate) fungicide with systemic function, and **imazaquin**, an imidazoline-based herbicide used to control a wide range of weed species. It was proven that adsorption onto lignocellulosic materials was applied only for metalaxyl (Gamiz et al. 2016) so far.

The study evidenced that soils amended with composted olive mill waste retain lower amount of pesticide than soils containing the corresponding biochar. At the same time, the enantiomer *R* was preferentially retained in soils containing composted olive mill waste, but biochar-amended soils adsorbed both enantiomers and blocked their leaching almost completely.

Despite the progress reported by researchers on the removal of pesticides, further studies are required in order to develop new sorbents based on agricultural waste able to purify wastewater more efficiently.

14.2.2 Dyes

Dyes represent a broad class of organic pollutants largely found in industrial wastewater. Due to their chemical structure (high content in aromatics, halogens, metal ions, etc.), they can severely damage the environment when improperly discharged in natural streams as they cause a decrease in water transparency, and alter the photosynthesis and water oxygenation, as well as self-cleaning processes (Tomczak and Tosik 2017; Dardouri and Sghaier 2017). Basically, dyes can be divided into groups: acid dyes (Congo red, eosin, etc.), basic (cationic) dyes (methylene blue, basic fuchsin, crystal violet, etc.), direct (substantive) dyes, mordants, and vat, reactive, and non-ionic (disperse) dyes.

Simple, low-cost, and easy-to-access sorbent materials obtained from lignocellulosic waste were successfully employed in industrial water decontamination when dyes were the pollutant. Thus, such materials removed cationic dyes (Amela et al. 2012; Zhu et al. 2016b), azoic dyes (Tomczak and Tosik 2017; Chebli et al. 2015), direct dyes (Karthick et al. 2017), and reactive dyes (Hong and Wang 2017; Tunç et al. 2009; Suteu et al. 2011; Suteu et al. 2015) from wastewater.

14.2.2.1 Anionic Dyes

Anionic dyes display a wide structural variety. According to some literature data (Zhou et al. 2015), acidic dyes, direct dyes, and reactive dyes can be included in the same group of anionic dyes.

Acid dyes have been typically applied to textiles under low pH conditions and are used to dye wool, silk, or nylon, but not cotton fabrics. They contain different chromophores (azo groups, anthraquinone, naphthol, etc.) in their structure and hence the broad palette of colors. Information on this specific subject is scarce. Still, it was reported that Naphthol Blue Black, Basic Blue 41, and Reactive Black 5 dyes were removed from water solution using the avocado peel residue as agriculture waste (Palma et al. 2016), and rice husk was employed to retain methyl orange (Hosseinzadeh and Mohammadi 2016), while modified barley straws were used to remove Acid Blue 40 (Oei et al. 2009) and the powder of the jackfruit leafs for Amido Black 10B (Ojha and Bulasara 2015). Other studies focused on the adsorption of Synolon Black

HWF-FS and Synolon Red 3HF onto mustard and linseed oil cakes (Safa 2016), when electrostatic interaction proved to be the sorption mechanism, and on the sorption of Acid Blue 113 onto the peel of overripe *Cucumis sativus* (Lee et al. 2015), which showed biosorption was endothermic, feasible, and spontaneous.

Direct dyes are water-soluble sodium salts of sulfonic acid bearing azoic groups as chromophores (some representative direct blue dyes are presented in Fig. 14.3). Despite the reports on the lignocellulosic materials originated from agricultural waste that can be used for the effective adsorption of direct dyes (Chebli et al. 2015), more research is still required.

Various lignocellulosic materials were tested for the sorption of direct dyes from industrial water. Thus, fibers of *Stipa tenacissima* were successfully used to remove the Congo red dye (Chebli et al. 2015) from aqueous solutions, mechanism of adsorption was identified (protonation, particle diffusion, electrostatic interactions), and an optimal set of parameters (pH = 4, temperature 25°) was recommended in order to reach the maximum of adsorption (7.93 mg/g). For the same dye, polyethyleneimine-modified wheat straws were employed (Shang et al. 2015), and it has been proven that the chemical modification of wheat straws enhanced their adsorption capacity. Another red dye, namely Direct Red 23, was effectively removed from wastewater using corn stover (Fathi et al. 2015).

Reactive dyes contain in their structure a chromophore bearing a substituent able to react with the substrate; thus, they bond it covalently during dyeing. The use of carbonaceous materials obtained from avocado skin for the removal of Reactive Black 5 was assessed (Palma et al. 2016) as related to a series of factors, such as carbonization parameters, and by comparison with other dyes. It has been shown that physical adsorption was the retention mechanism of Reactive Black 5 onto a new sorbent material based on banana pseudostem (Modenes et al. 2015). Modified barley straws turned out to be effective for the removal of reactive dyes (Oei et al. 2009) under certain conditions (sorbent-to-sorbate ratio, initial concentration of dye, solution pH, and temperature).

14.2.2.2 Cationic Dyes

Cationic dyes can dissociate and form positive ions in their corresponding aqueous solution and further interact with negatively charged groups on the fiber molecules. Representative dyes in this class are presented in Fig. 14.4.

Numerous and various recent reports have shown that agricultural waste can be turned into environmentally friendly adsorbent materials which can remove cationic dyes from industrial waters up to, or even beyond, satisfactory limits (Dai et al. 2018). Thus, *Cucumis sativus* was employed to remove such dyes, and the study on rhodamine B and crystal violet (Smitha et al. 2012) evidenced that the activation of the lignocellulosic waste using sulfuric acid enhanced the dye sorption, but the overall yield depended on the process parameters (dye concentration, sorbent-to-sorbate ratio, contact time, pH) as well.

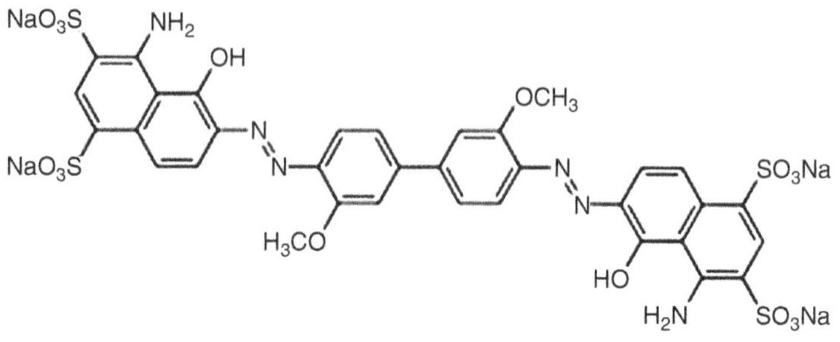

Direct Blue 1

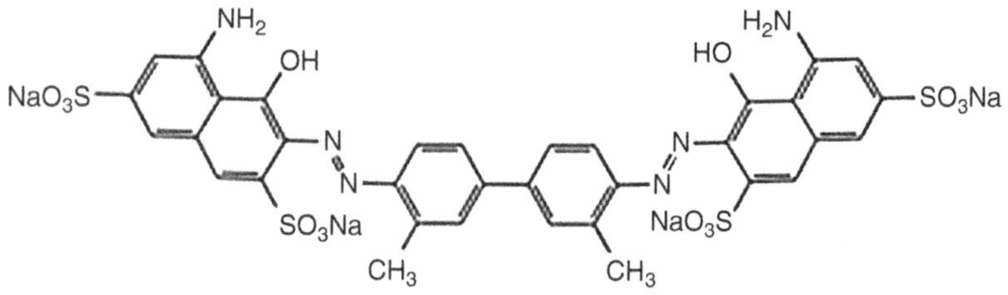

Direct Blue 14 (Trypan Blue)

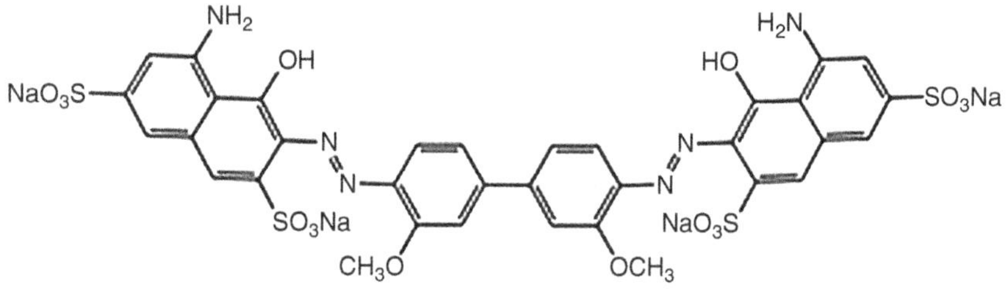

Direct Blue 15

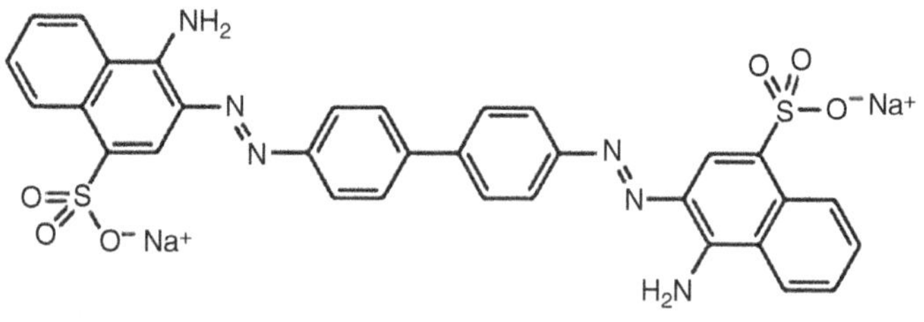

Congo Red

Fig. 14.3 Representative direct blue and red dyes

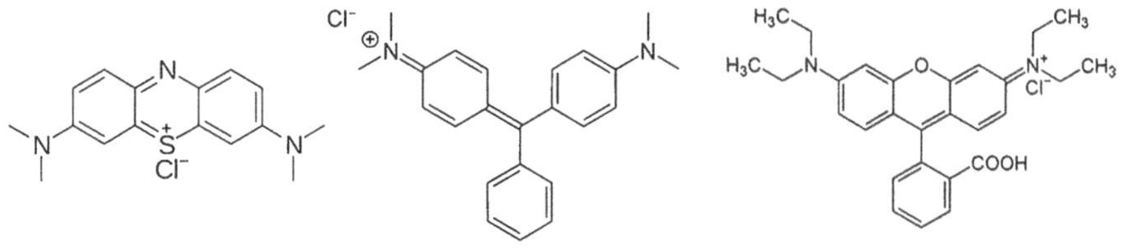

Methylene Blue Malachite Green Rhodamine B

Fig. 14.4 The most common cationic dyes

For the sorption of methylene blue, different lignocellulosic materials were tested as sorbents. Melon peel (Djelloul and Hamdaoui 2015) was set in fixed-bed columns for a dynamic retention of dye (where column parameters were estimated depending on the flow rate and initial concentration of dye), and the maximum of the adsorption process has been found to correspond to simultaneous high bed height, low flow rate, and high initial concentration of dye. Sumac leaves (Gülen et al. 2015) were also tested to remove methylene blue, as well as waste tea leaves chemically modified under mild conditions with citric acid (Zhou et al. 2016), material which can be regenerated after exhaustion up to 98.90% by cold atmospheric plasma treatment.

Malachite green was successfully retained on stems and leaves of *Solanum tuberosum* as powder through physisorption (Gupta et al. 2016), method which was also effective for methylene blue sorption, but to a greater extent.

14.2.2.3 Non-ionic Dyes

The term refers to disperse dyes, the only water-insoluble dyes that are fit for dyeing polyester and acetate fibers. As compared with other dyes, disperse dyes have the smallest molecules. Structurally, their molecule is based on an azobenzene or anthraquinone unit bearing nitro-, amino, hydroxyl, etc. groups.

The information on this subject is scarce, although the dye removal from wastewater using agriculture waste is a subject of high and constant interest (Dai et al. 2018; Zhou et al. 2015; Yagub et al. 2014; Worch 2012; Demirbas 2009). Still, there are some studies reporting on the retention of disperse dyes onto lignocellulosic waste materials. Disperse blue and red dyes commercially available were treated with oil palm ash made of oil palm waste from an oil manufacture, in batch and continuous flow experiments (Isa et al. 2007). It has been shown that the optimum adsorption was reached for pH = 2 and agitation time 60 min, for both dyes. Activated carbon made of bamboo stalk was employed for the removal of Disperse Red 167 (Zhou et al. 2015).

As for the adsorption mechanisms involved in the process of removal of dyes from industrial water using lignocellulosic waste, some synthetic data are presented in Table 14.1.

Table 14.1 The mechanism of adsorption of the most common dyes onto lignocellulosic waste sorbents

Dye	Lignocellulosic sorbent	Mechanism of adsorption	Reference
Methylene blue	*Camelina*	Intraparticle diffusion and surface adsorption	Sharma and Tiwari (2016)
	Sapindus seeds hulls	–//–	–//–
	Sumac (*Rhus coriaria* L.) leaves	Chemisorption, physisorption, electrostatic interactions	Gülen et al. (2015)
	Stem and leaves of *Solanum tuberosum*	Physisorption	Gupta et al. (2016)
	Water bamboo leaves	Intraparticle diffusion and electrostatic interactions	Zhu et al. (2016a, b, c, d)
	Walnut shells	Chemical reaction	Tang et al. (2017a)
	Rice straws (MW activated)	H-bonding electrostatic attraction	Mashhadi et al. (2016)
	Solanum tuberosum peel	Physisorption, electrostatic interactions	Guechi and Hamdaoui (2015a)
	Citric acid-modified peanut shells	Electrostatic interactions	Wang et al. (2015)
	Abelmoschus esculentus seed	Ion exchange, chemisorption	Nayak and Pal (2017)
	Chemically modified pine nut shells	Dipole-dipole interaction, π-π interaction	Naushad et al. (2015)
	Alfa grass	Ion exchange, chemical reaction, electrostatic interactions, external diffusion	Toumi et al. (2015)
	Coconut (*Cocos nucifera* L.) coir dust	Molecular adsorption by the van der Waals force, chemisorption	Etim et al. (2016)
Congo red	*Stipa tenassicima* fibers	Strong protonation, electrostatic interactions	Chebli et al. (2015)
	Polyethyleneimine-modified wheat straw	H-bonding, van der Waals, π-π conjugate, ion exchange	Shang et al. (2015)
Rhodamine B	Sulfuric acid-*Cucumis sativus*	Boundary layer diffusion, intraparticle diffusion	Smitha et al. (2012)
Chrystal violet	*Cucumis sativus*, sulfuric acid-*Cucumis sativus*	Boundary layer diffusion, intraparticle diffusion	Smitha et al. (2012)
Methyl orange	Aminated pumpkin seed	Electrostatic interactions	Subbaiah and Kim (2016)
Malachite green	Stem and leaves of *Solanum tuberosum*	Physisorption	Gupta et al. (2016)
Direct Red 23	Corn stalks	Electrostatic interactions	Fathi et al. (2015)
Acid Blue 113	*Cucumis sativus* peel	Chemisorption	Lee et al. (2015)
Reactive Blue 5G	Banana pseudostem	Ion exchange, physisorption	Modenes et al. (2015)
Reactive Red 141	Sesame waste	Electrostatic interactions	Sohrabi and Ameri (2015)
Synolon Black	Linseed oil cake	–//–	Safa (2016)
Synolon Red	Mustard oil cake	–//–	–//–

Removal of various classes of dyes from industrial water using lignocellulosic waste from agriculture remains an active field of research, as low-cost adsorbents are promising in terms of benefits in the future. Still, further investigation is needed in order to optimize the process (sorbent-to-sorbate ratio, size of dye molecules, particle size of the sorbent, functional groups present on the surface of the adsorbent, initial pH of the solution, batch or column system, processing temperature, etc.). But new approaches are also emerging since the nanotechnology opens alternative directions even in this domain.

14.2.3 Pharmaceuticals

Pharmaceuticals are considered as environmental pollutants as they have been often detected in city wastewater and even in reclaimed water due to the inefficient purification of industrial water in wastewater treatment plants.

Drugs, be they antibiotics or anti-inflammatory, painkillers or analgesics, psychotropics, antitumoral, or even disinfectants, as well as personal care products and steroids, may represent really hazardous agents when disposed of carelessly in the environment as they can degrade in water, air, and soil and, moreover, enter the metabolism of plants and animals.

Therefore, these organic compounds have to be removed from water in effective, harmless to the environment, ways. Adsorption onto lignocellulosic waste materials is one method that has been applied in this field in recent years with notable results (Blanch 2016; Portinho et al. 2017; Kyzas and Deliyanni 2015; de Andrade et al. 2018; Quesada et al. 2019).

Antibiotics are some of the most common organic pollutants, and their removal using agriculture waste was investigated. For the removal of ***tetracycline***, sorbent derived from rice straws (Wang et al. 2017) and animal waste (Wang et al. 2018) were studied. Experimental data have shown that sorption process was spontaneous and endothermic, but the retention was more effective on the rice straw-derived biochar. Sugarcane bagasse was used as precursor in order to obtain magnetic carbon composites, prepared via mild HTC and simple heat treatment, as sorbent for tetracycline (Rattanachueskul et al. 2016). The composite materials adsorbed rapidly the antibiotic due to their high adsorption *per* unit surface area. A hybrid material made of maize straws and goethite (an iron-bearing hydroxide mineral of the diaspore group) was successfully tested for the retention of ***tylosin*** (Yin et al. 2016). The results have indicated that a high temperature could favor the sorption of the antibiotic on both sorbents, and pH and ionic strength of the solution can influence the process. The sorption mechanism onto maize straw mainly involves electrostatic interactions and hydrophobic interactions, while the modified maize straws retain the antibiotic by electrostatic interactions, H-bonding, hydrophobic interactions, and surface complexation.

The removal of **anti-inflammatory drugs**, such as ***diclofenac***, ***ketoprofen***, ***naproxen***, ***nimesulide***, ***ibuprofen***, etc., was studied, and the corresponding retention mechanism was assessed (Zhou et al. 2015), as well as for other pharmaceuticals. For example, ***caffeine*** (a stimulant of the central nervous system, member of the methylxanthine class, considered the only psychoactive drug consumed worldwide) was retained from aqueous solution on sorbents based on grape stalks that were used in three different forms: raw, modified by phosphoric acid, and as activated carbon (Portinho et al. 2017). The best results have been obtained using the activated carbon material.

Pharmaceuticals, mainly antibiotics, and their use for the human health were a breakthrough in the history, but also entailed a series of risks which, without control, may lead to a global crisis. Therefore, their removal has to be of capital interest for academic and industrial R&D, and the recent experimental studies on various agricultural wastes as sorbents for drugs have indicated they are a viable option of great potential.

14.2.4 Oil and Other Organic Pollutants

Oil pollution came along with the industrial development, and it has been proven that oil spills are serious environmental concerns since they can easily spread onto water surface and form thin layers that hinder the oxygen transfer between air and water, thus severely affecting the marine biotope and the contiguous areas.

An increased attention is payed lately to specific sorbents made of lignocellulosic waste, such as barley straw (Ibrahim et al. 2010), coconut shells, garlic and onion peel (Zhou et al. 2015), and chemically modified sugarcane bagasse (Abdelwahab et al. 2017). In this latter case, the bagasse was modified by esterification with stearic acid and calcium oxide, and then coated with polyacrylonitrile, yielding in a hydrophobic adsorbent for the removal of diesel fuel from man-made seawater (mainly used in marine biology and in marine and reef aquaria). In the both stages of the chemical treatment, the resulted materials have acquired significantly increased sorption surface. The retention mechanism was chemisorption. It was also possible to assess the adsorption efficiency for different sorbates as follows: paraffin oil > vegetable oil > diesel oil > gasoline (Abdelwahab et al. 2017).

At the same time, the sorption of emulsified oil onto raw and modified bagasse and corn husk, by an MW-assisted procedure, has been also studied (Pachathu et al. 2016), when the maximum removal ratios were 98.07% (on modified bagasse) and 98.72% (on modified corn husk).

Other organic pollutants of high risk are the aromatic compounds, especially polycyclic aromatic ones, containing either separated or fused aromatic nuclei, as they have toxic, mutagenic, and carcinogenic potential. Despite the high interest of scientists on the removal of such pollutants from industrial water, mostly microbial, bacterial, and other biological approaches were investigated (Sherafatmand and Ng 2015; Kronenberg et al. 2017), while the information on sorption onto lignocellulosic materials is not so abundant (Zhou et al. 2015; De Gisi et al. 2016; Dai et al.

2018; Tran et al. 2015). Still, some research focused on obtaining activated carbon using waste banana peel (Gupta and Gupta 2015) when a sorption model has been established, and characteristic parameters of the adsorption isotherm were calculated.

The same situation is encountered for the retention of phenolic compounds using activated carbon from coconut shells (Karri et al. 2017), *Typha orientalis* (Feng et al. 2015), pine tree (Tonucci et al. 2015), and even zeolite-type materials made of bagasse fly ash (Shah et al. 2016). Nevertheless, these studies have indicated that the retention of phenolic compounds has been acquired by hydrogen bonding, hydrophobic and electrostatic interactions, π-π interactions, ion exchange, electronic interactions, diffusion at the interface, and internal diffusion of particles.

Further research is still needed in terms of expanding the range of lignocellulosic materials fit for the removal of organic pollutants from industrial water, increasing the volume of common agriculture waste used in such applications, and, last but not least, the possibility of using them in a less modified form.

14.3 Inorganic Pollutants

Most inorganic pollutants affecting the environment and, consequently, the wildlife and human health are heavy metals, and some phosphorus and nitrogen compounds, all resulted from industrial and agriculture activities. Industrial water containing such waste has to be specifically treated in order to remove these pollutants up to significant levels as they are toxic for the aquatic life and possess high mutagenic and carcinogenic character. Even more, once discharged in water, these pollutants enter the life cycle of plants and animals (fish included), which are subsequently used as food, thus generating a new cycle of pollution directly in the human body where they accumulate (entailing a toxic and inflammatory response) and cause genetic mutations.

Classic methods of removing heavy metals from wastewater are precipitation, flotation, ion exchange, and membrane processes, but all of them are energy intensive, use expensive equipment, require regeneration of the sorbents, and produce toxic sludge that is able to cause secondary pollution, and the level of retention is not as high as expected. Therefore, new solutions are required, and some of the most available are the lignocellulosic materials based on agriculture waste which have proved to be low-cost and non-aggressive toward the environment (Farhan et al. 2012; Du et al. 2016; Escudero-Oñate et al. 2017). The mechanisms involved in the retention of heavy metals from solution onto such adsorbent media include physisorption, chemisorption, ion exchange, membrane processes, particle diffusion, chelation, electrostatic interactions, surface complexation, ligand exchange, internal complexation, etc. (Nguyen et al. 2013; Agouborde and Navia 2009; Du et al. 2016). Some of them will be highlighted in the following. A schematic retention mechanism (ion exchange) is presented in Fig. 14.5.

There is a wide variety of reports on heavy metal retention on raw and modified agriculture waste, but only some of the most recent will be reviewed herein.

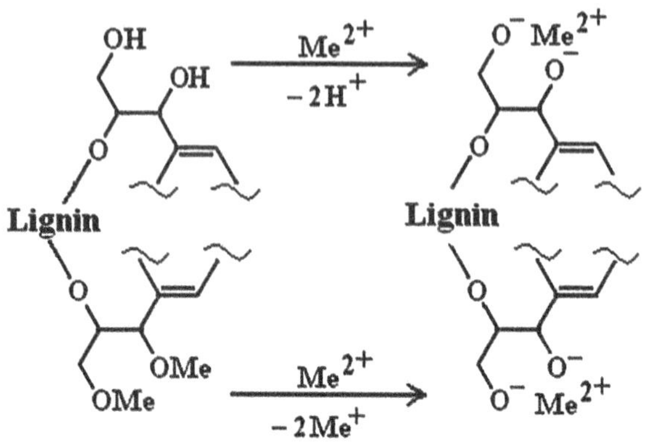

Fig. 14.5 Schematic representation of the ion exchange mechanism which can be applied for all heavy metal ions Me^{2+}

14.3.1 Lead

Lead (Pb) and its compounds (Pb(II) or Pb^{2+}) are toxic to human health, and its rapid accumulation into the body gravely affects the nervous, cardiovascular, digestive, and endocrine system, as well as the hematopoietic function of the body.

Rice husk has been identified as an effective sorbent for Pb(II) (Amer et al. 2017) as the maximum removal of 94% was reached under optimal conditions, namely: contact time = 30 min, pH = 5.5, particle size = 75–150 μm, and initial concentration = 4 g/L. A multiple mechanism was considered: complexation reactions of carboxylic and hydroxylic reactive groups, ion exchange with Ca^{2+} and Mg^{2+}, and physical attraction. It has been found that it is possible to regenerate the sorbent for further use. In a series of experiments on coconut shell, it was possible to prove that the raw sorbent retained a lower amount of Pb(II) than the corresponding modified coconut shell (activated with phosphoric acid), as the latter has a higher surface area and reactive functional groups (El-Deen and El-Deen 2016).

Other studies reported on the use of other lignocellulosic materials with different sorption capacity (mg/g): acid-modified rice straw, 18.98 (Guo et al. 2015); *Acacia nilotica* seed shell ash deposited onto magnetic nanoparticles, 37.6 (Omidvar-Hosseini and Moeinpour 2016); plum stone, 80.65 (Parlayıcı and Pehlivan 2017); *Citrus reticulata* waste biomass, 83.77 (Bhatti et al. 2010); and hazelnut husk-based activated carbon, 109.90 (Imamoglu et al. 2016).

14.3.2 Cadmium

Cadmium (Cd) is considered as extremely toxic (it is not biologic degradable) and can cause kidney failure upon accumulation into the human body. As pollutant, it may come from industry, and its accidental or careless disposal may entail the environmental accumulation and wildlife degradation due to its toxicity and mutagenic capacity.

The mechanism of removing Cd(II) by sorption onto raw and chemically modified walnut shells was the ion exchange, and it has been shown that the biosorption occurred spontaneously (Gondhalekar and Shukla 2015). By the chemical modification (alkaline treatment), the adsorption increased significantly for pH ranging from 2 to 6, and the desorption efficiency remained unmodified until the third cycle. The same mechanism was identified when succinyl-modified cellulosic biomass (*Abelmoschus esculentus*) was employed as sorbent for Cd(II) (Singha and Guleria 2014). The removal of Cd(II) from the aqueous solution (121.51 mg/g) depended on pH, contact time, temperature, and initial concentration of the metal ion. This sorbent was also effective in the case of other metal ions (adsorption capacity in mg/g): Cu^{2+}, 72.72; Zn^{2+}, 57.11; and Pb^{2+}, 273.97.

Another interesting study compared the retention capacity of sugarcane stalks over *Phragmites australis* (Farasati et al. 2015). The superior capacity of sorption of the sugarcane bagasse has been proven, and it was assessed that it is due to its extended contact surface. Activated carbons derived from waste of *Typha angustifolia* and *Salix matsudana* were prepared by phosphoric acid activation (Tang et al. 2017b) and then tested as sorbents for Cd(II) and Pb(II) in batch experiments. The adsorption process was spontaneous and endothermic and occurred through physical processes (intraparticle diffusion) and chemical reactions, both of them being limiting steps of the adsorption rate. At the same time, experimental data indicated that both sorbents can be satisfactorily regenerated and reused.

14.3.3 Copper

Copper and its alloys (brass and bronze) have been extensively used in a wide range of applications, ranging from common electric wiring to photovoltaic cells and phytotherapy. Cu(II) is a micronutrient, up to a certain concentration, necessary for the human and wildlife health. Its toxicity occurs upon accumulation in high amounts, and it affects mainly the liver and gall bladder. Retention of Cu(II) by adsorption is extensively used in wastewater treatment plants.

Considering the sorption of copper onto lignocellulosic materials, studies have shown that mechanisms involved were chemisorption, physisorption, electrostatic interactions, ion exchange, and particle diffusion (Dai et al. 2018). Thus, adsorption of Cu(II) onto potato peel was investigated in batch experiment and reached the maximum (84.74 mg/g) at 25 °C (Guechi and Hamdaoui 2015b), but depended on the initial concentration of ions.

Three different types of agriculture waste were studied as powders (banana stems, casuarina fruits, sorghum stems) (Mokkapati et al. 2016), and the corresponding adsorption capacity has been correlated with their microporous structure, given that all three sorbents have exhibited rough surface with some cavities. In the case of powders made of banana stems and casuarina fruits, the sorption mechanism was chemisorption, while for the sorghum stem powder, the mechanism occurred through physisorption and chemical attachment.

Raw pomegranate peel was employed for the adsorption of Cu(II) from its aqueous solutions (Ali et al. 2017). The highest sorption capacity was obtained at pH = 5.8, particle size = 630 μm, temperature = 40 °C, and contact time of 2 h, and by increasing the initial concentration of ion.

14.3.4 Nickel and Zinc

Nickel is essential for human life, but upon high concentrations (by absorption through the pores and sebaceous glands from the skin) it rouses an inflammatory response.

Different experimental studies have shown Ni(II) can be successfully retained onto sorbents made of lignocellulosic waste. For example, *Citrus limetta* peels (CLP) were employed for the sorption of Ni(II) ions from aqueous solution in batch system (Singh and Shukla 2017). The optimum values of process parameters were pH = 6, sorbent dose = 2 g/L, and contact time for equilibrium = 45 min. The maximum adsorption capacity was 27.78 mg/g. Experimental results have substantiated that the mechanism involved in sorption was the ion exchange, and by desorption, it was possible to reach almost complete recovery of Cu(II).

A complex study investigated the adsorbing capacity of different natural fibers (viz., spruce, coconut coir, sugarcane bagasse, kenaf bast, kenaf core, and cotton) as a function of their lignin content (Lee and Rowell 2004) in relation to copper, nickel, and zinc ions in aqueous solutions. The decreasing order of ion removal was as follows:

- Cu(II): kenaf bast > spruce > coconut coir > kenaf core > bagasse > cotton
- Zn(II): kenaf bast > kenaf core = spruce > coconut coir > bagasse > cotton
- Ni(II): kenaf bast > spruce > kenaf core > coconut coir > bagasse > cotton

All the fibers tested were more efficient in the adsorption of Cu(II) and Zn(II) than of Ni(II). Employing the extraction of the fibers with various reagents (diethyl ether, ethyl alcohol, hot water, and 1% sodium hydroxide), different extractives and cell wall components were removed, but ion sorption did not significantly increase. By removing the cell wall constituents, more lignin becomes available to contact with ion solution; however the study has indicated no consistent correlation between the increasing amount of available lignin and metal ion sorption.

Interesting results were also obtained with chemically modified kapok fibers (silky fibers that enclose the seeds of kapok trees *Ceiba pentandra*) (Zheng et al. 2015). By combined processes (chlorite-periodate oxidation), these fibers have demonstrated elevated capacity to adsorb heavy metal ions (mg/g): Pb, 93.55%; Cu, 91.83%; Cd, 89.75%; and Zn, 92.85% (Chung et al. 2008). The chemically oxidized kapok fibers can be also used as excellent sorbents for heavy metals, and their enhanced sorption may be due to the formation of carboxylic functional groups able to react chemically with the sorbate. Kapok fibers can be functionalized with diethylenetriaminepentaacetic acid (DTPA) in order to obtain sorbents with fast adsorption for heavy metals (Duan et al. 2013). Prior to this reaction, kapok fibers were washed with dichloromethane for the removal of the natural wax, and by further treatment with NaOH solution, fibers with a hollow structure and open ends were obtained, hence a larger specific surface area available for the reaction with DTPA. Maximum adsorption capacity for thusly modified kapok fibers (mg/g) was: Pb^{2+}, 310.6; Cd^{2+}, 163.7; and Cu^{2+}, 101.0. The sorption capacity of fibers remained high (over 90%) even after eight adsorption-desorption cycles.

Multi-ion sorbents were also investigated (Abdolali et al. 2014). A series of novel such multi-metal binding biosorbents (MMBB) was developed by combining in different formulations three lignocellulosic waste types (from agriculture and wood industry) for the effective removal of lead, cadmium, copper, and zinc from industrial water. The tested formulations ((1) tea waste, corncob, and sugarcane bagasse; (2) tea waste, corncob, and sawdust; (3) tea waste, corncob, and apple peel; (4) tea waste, corncob, and grape stalk) have proved to be more efficient for Cd(II), Cu(II), and Pb(II) than for Zn(II). The formulation 2 has maintained its excellent sorption capacity even after five sorption-desorption cycles.

14.3.5 Other Heavy Metals

Mercury (Hg(II)), chromium (Cr(VI)), and cobalt (Co(VI)) are highly toxic to human body; thus, their removal from industrial water is of utmost importance (Escudero-Oñate et al. 2017).

It has been found that rice husk and rice straw can be functionalized by the reaction with carbon disulfide (CS_2) under alkaline conditions, when S-containing xanthogenate-type reactive groups resulted ($(–O–CS_2]^-Na^+$), and these materials can adsorb Hg(II) by the ion exchange mechanism (Song et al. 2015). The sorption has been proved to be an endothermic and spontaneous process, and the sorbents have proved to be selective toward Hg(II), as the maximum capacity of adsorption was 119 mg/g for rice husk and 92 mg/g for rice straw.

Cr(VI) can be retained onto rice husk in relevant amounts (18.20 mg/g) at room temperature and pH = 6 (Ding et al. 2016), but after hydrothermal carbonization, the amount of adsorbed Cr(VI) was 31.1 mg/g. Experimental data have shown that the adsorption was prevented and the reduction increased when the pH of the solution was lower than 3.

Aminated rice straw-grafted poly(vinyl alcohol) (A-RS/PVA) was also employed to remove Cr(VI) from its aqueous solutions in batch experiments (Lin et al. 2018). This new sorbent sorption capacity was up to 140.39 mg/g, at initial pH = 2.0 and temperature = 60 °C, almost three times larger than that of the initial rice straw (34.90 mg/g). The process was spontaneous and endothermic, and the mechanism of sorption is based on the reduction reaction of Cr(VI) to Cr(III) which is less toxic.

Magnetic biochar prepared from *Melia azedarach* wood was studied as a possible sorbent for the removal of Cr(VI) (Zhang et al. 2018). It has proved a higher removal efficiency (99.8%) as compared with the biochar under the same conditions (concentration = 5 g/L, pH = 3.0, initial concentration of Cr(VI) 10 mg/L). The experimental data have indicated that the removal process consisted of a sequence of adsorption-reduction-adsorption steps: first, Cr(VI) was adsorbed onto the surface of the magnetic biochar and, subsequently, reduced to Cr(III) which was further adsorbed on the surface.

Other studies indicated more raw and modified materials as sorbents for Cr(VI) and Cr(III) (Miretzky and Cirelli 2010): bagasse, bark, bran, cake, coir, husk and hull, sawdust, biomass, etc.

Another relevant heavy metal, cobalt Co(II), was subjected to adsorption onto lignocellulosic sorbents, namely, lemon peel, in batch experiments, in order to be removed from its solutions (Bhatnagar et al. 2010). The maximum adsorption capacity of lemon peel was 22 mg/g at 25 °C. The adsorption process has been found to be exothermic. The cost efficiency study has indicated lemon peel as the most advantageous sorbent, given that its production price is ten times lower than that of activated carbons.

14.3.6 Nitrogen and Phosphorus

Nitrogen and phosphorus anions (NO_3^-; PO_4^{3-}; NH_4^+) become pollutants when various phytopharmaceuticals are excessively used. In principle, these anions are difficult to be retained onto the surface of common biochars due to their electroactive structure disparity.

Still, there are studies confirming that some lignocellulosic materials can retain certain anions (Dai et al. 2018). Thus, wood- and rice husk-based materials have proved to be able to adsorb ammonium ions in NH_4Cl solution and slurry solutions (Kizito et al. 2015). The mechanisms involved were chemisorption and physisorption, and the amount of sorbate was satisfactory. Other data have shown that the activated carbons based on avocado seed can bind ammonium ions (5.40 mg/g), at 298 K and pH = 5.0, by an ion exchange mechanism (Zhu et al. 2016d).

Phosphate anions (PO_4^{3-}) can be removed from polluted water using biochar made of lignocellulosic waste from different sources (oak, bamboo wood, maize, soybean, and peanut shell) (Jung et al. 2015).

It is generally accepted that there is, still, a great need for further research in this sensitive field, because heavy metal decontamination is a two-edge blade due to its high potential of secondary pollution, accumulation in living organisms, and cyclic contamination in an ascending spiral.

14.4 Concluding Remarks and Future Prospects

Purification of industrial water using lignocellulosic waste materials is a highly challenging theme of research.

On one hand, there is a significant amount of studies reporting on various materials and methods. On the other hand, most of the experimental data remain as literature because only very few of them are transferred to industry.

It is rather difficult to attempt to cover a wide range of inexpensive, locally available, and efficient materials as to be used instead of the commercially available activated carbons when it comes to removing various contaminants from industrial water. But little effort has been made to perform a cost comparison between these two classes of sorbents. This analysis is mainly needed in order to promote the large-scale use of lignocellulosic waste adsorbents so that their efficiency is maximized and the exploitation of local agriculture waste increases, as well as the technical feasibility, because, in the end, low-cost sorbents can provide promising benefits.

At this point, a few aspects of the ecological impact of the biobased materials should be mentioned. Concerns over global climate changes and the food insurance have been, at least partially, the driving force of the development of biobased materials. Still, this trend can be regarded as a challenge to academic and industry R&D scientists, and policy makers, as the manufacture of biobased materials requires land, sometimes the crops are used less for food, and it is associated with some adverse environmental effects. Recent data have indicated that one metric ton (t) of biobased materials may save, as compared to classic materials, 55 ± 34 GJ of primary energy and 3 ± 1 t carbon dioxide equivalents of greenhouse gases (Weiss et al. 2012). Still, the use of the biobased materials may also increase the eutrophication by 5 ± 7 kg phosphate equivalents/t and depletion of the ozone layer in the stratosphere by 1.9 ± 1.8 kg nitrous oxide equivalents/t. The additional land use may entail a partial loss of biodiversity, depletion of carbon in soil and soil erosion, possible deforestation, and greenhouse gas emission.

Nevertheless, future research must focus on:

- identifying solutions for local pollution problems, because, in spite of the global character of some ideas, their application is intrinsically linked to and limited by the local availability of materials;
- the design of new approaches by the use of combined methodologies (e.g., algae and microalgae, or combined bacterial and algal cultures applied directly in areas containing polluted water, followed by specific biologic decontamination and regeneration, if possible, in order to avoid secondary contamination);
- creating hybrid methods (such as the direct transfer of the removed dyes to assigned facilities for further decomposition up to non-toxic or low toxic compounds) and materials (i.e., lignocellulosic fibers can be incorporated into biopolymeric thermoplastic matrices, as chitosan, starch, or even cellulose and lignin, and, at the same time, various natural nanofillers with sorbent properties, such as clays and zeolites, can be added to the formulations);
- the research of new methods for the regeneration of sorbents without causing secondary pollution;
- the development of highly effective green modifiers so that raw materials would be harmlessly and cost-effectively transformed as to achieve an improved adsorption.

And the list may continue as only the human intelligence and creativity can limit the development of this field of research.

References

Abdelwahab NA, Shukry N, El-kalyoubi SF (2017) Preparation and characterization of polymer coated partially esterified sugarcane bagasse for separation of oil from seawater. Environ Technol 38:1905–1914. https://doi.org/10.1080/09593330.2016.1240243

Abdolali A, Ngo HH, Guo WS, Lee DJ, Tung KL, Wang XC (2014) Development and evaluation of a new multi-metal binding biosorbent. Bioresour Technol 160:98–106. https://doi.org/10.1016/j.biortech.2013.12.038

Agouborde L, Navia R (2009) Heavy metals retention capacity of a non-conventional sorbent developed from a mixture of industrial and agricultural wastes. J Hazard Mater 167:536–544. https://doi.org/10.1016/j.jhazmat.2009.01.027

Ali SB, Jaouali I, Souissi-Najar S, Ouederni A (2017) Characterization and adsorption capacity of raw pomegranate peel biosorbent for copper removal. J Clean Prod 142:3809–3821. https://doi.org/10.1016/j.jclepro.2016.10.081

Amela K, Hassen MA, Kerroum D (2012) Isotherm and kinetics study of biosorption of cationic dye onto banana peel. Energy Procedia 12:286–295. https://doi.org/10.1016/j.egypro.2012.05.208

Amer H, El-Gendy A, El-Haggar S (2017) Removal of lead (II) from aqueous solutions using rice straw. Water Sci Technol 76:1011–1021. https://doi.org/10.2166/wst.2017.249

Bhatnagar A, Minocha AK, Sillanpää M (2010) Adsorptive removal of cobalt from aqueous solution by utilizing lemon peel as biosorbent. Biochem Eng J 48:181–186. https://doi.org/10.1016/j.bej.2009.10.005

Bhatti HN, Bajwa II, Hanif MA, Bukhari IH (2010) Removal of lead and cobalt using lignocellulosic fiber derived from *Citrus reticulata* waste biomass. Korean J Chem Eng 27:218–227. https://doi.org/10.1007/s11814-009-0325-1

Blanch GL (2016) Removal of pharmaceuticals from WWTP streams by biological and physical processes. PhD Thesis, Universitat Autonome de Barcelona, Bellaterra, Spain. isbn: 9788449066733. https://www.tdx.cat/handle/10803/399517#page=1. Accessed 15 Aug 2019

Bouguettoucha A, Reffas A, Chebli D, Mekhalif T, Amrane A (2016) Novel activated carbon prepared from an agricultural waste, *Stipa tenacissima*, based on $ZnCl_2$ activation—characterization and application to the removal of methylene blue. Desalin Water Treat 57:24056–24069. https://doi.org/10.1080/19443994.2015.1137231

Chan YH, Yusup S, Quitain AT, Tan RR, Sasaki M, Lam HL, Uemura Y (2015) Effect of process parameters on hydrothermal liquefaction of oil palm biomass for bio-oil production and its life cycle assessment. Energy Convers Manag 104:180–188. https://doi.org/10.1016/j.enconman.2015.03.075

Chebli D, Bouguettoucha A, Mekhalef T, Nacef S, Amrane A (2015) Valorization of an agricultural waste, *Stipa tenassicima* fibers, by biosorption of an anionic azo dye, Congo red. Desalin Water Treat 54:245–254. https://doi.org/10.1080/19443994.2014.880154

Chung BY, Cho JY, Lee MH, Wi SG, Kim JH, Kim JS, Kang PH, Nho YC (2008) Adsorption of heavy metal ions onto chemically oxidized *Ceiba pentandra* (L.) Gaertn. (kapok) fibers. J Appl Biol Chem 51:28–35. https://doi.org/10.3839/jabc.2008.006

Cooney DO (1999) Adsorption design for wastewater treatment. Lewis Publishers, CRC Press LLC, Boca Raton, p 182

Dai Y, Sun Q, Wang W, Lu L, Liu M, Li J, Yang S, Sun Y, Zhang K, Xu J, Zheng W, Hu Z, Yang Y, Gao Y, Chen Y, Zhang X, Gao F, Zhang Y (2018) Utilizations of agricultural waste as adsorbent for the removal of contaminants: a review. Chemosphere 211:235–253. https://doi.org/10.1016/j.chemosphere.2018.06.179

Dardouri S, Sghaier J (2017) A comparative study of adsorption and regeneration with different agricultural wastes as adsorbents for the removal of methylene blue from aqueous solution. Chin J Chem Eng 25:1282–1287. https://doi.org/10.1016/j.cjche.2017.01.012

de Andrade JR, Oliveira MF, da Silva MGC, Vieira MGA (2018) Adsorption of pharmaceuticals from water and wastewater using nonconventional low-cost materials: a review. Ind Eng Chem Res 57:3103–3127. https://doi.org/10.1021/acs.iecr.7b05137

De Gisi S, Lofrano G, Grassi M, Notarnicola M (2016) Characteristics and adsorption capacities of low-cost sorbents for wastewater treatment: a review. Sustain Mater Technol 9:10–40. https://doi.org/10.1016/j.susmat.2016.06.002

Demirbas A (2009) Agricultural based activated carbons for the removal of dyes from aqueous solutions: a review. J Hazard Mater 167:1–9. https://doi.org/10.1016/j.jhazmat.2008.12.114

Deng F, Zhong F, Lin DC, Zhao LN, Liu YJ, Huang JH, Luo XB, Luo SL, Dionysiou DD (2017) One-step hydrothermal fabrication of visible-light responsive $AgInS_2/SnIn_4S_8$ heterojunction for highly-efficient photocatalytic treatment of organic pollutants and real pharmaceutical industry wastewater. Appl Catal B Environ 219:163–172. https://doi.org/10.1016/j.apcatb.2017.07.051

Deokar SK, Mandavgane SA, Kulkarni BD (2016) Adsorptive removal of 2,4-dichlorophenoxyacetic acid from aqueous solution using bagasse fly ash as adsorbent in batch and packed-bed techniques. Clean Technol Environ 18:1971–1983. https://doi.org/10.1007/s10098-016-1124-0

Ding D, Ma X, Shi W, Lei Z, Zhang Z (2016) Insights into mechanisms of hexavalent chromium removal from aqueous solution by using rice husk pretreated with hydrothermal carbonization technology. RSC Adv 78:74675–74682. https://doi.org/10.1039/C6RA17707G

Djelloul C, Hamdaoui O (2015) Dynamic adsorption of methylene blue by melon peel in fixed-bed columns. Desalin Water Treat 56:2966–2975. https://doi.org/10.1080/19443994.2014.963158

Du ZL, Zheng T, Wang P, Hao LL, Wang YX (2016) Fast microwave-assisted preparation of a low-cost and recyclable carboxyl modified ligno-cellulose biomass jute fiber for enhanced heavy metal removal from water. Bioresour Technol 201:41–49. https://doi.org/10.1016/j.biortech.2015.11.009

Duan CT, Zhao N, Yu XL, Zhang XY, Xu J (2013) Chemically modified kapok fiber for fast adsorption of Pb^{2+}, Cd^{2+}, Cu^{2+} from aqueous solution. Cellulose 20:849–860. https://doi.org/10.1007/s10570-013-9875-9

El-Deen GES, El-Deen SEAS (2016) Kinetic and isotherm studies for adsorption of Pb(II) from aqueous solution onto coconut shell activated carbon. Desalin Water Treat 57:28910–28931. https://doi.org/10.1080/19443994.2016.1193825

Escudero-Oñate C, Fiol N, Poch J, Villaescusa I (2017) Chapter 17: valorisation of lignocellulosic biomass wastes for the removal of metal ions from aqueous streams: a review. In: Biomass volume estimation and valorization for energy. InTech Open, p 381–407. https://doi.org/10.5772/65958

Etim UJ, Umoren SA, Eduok UM (2016) Coconut coir dust as a low-cost adsorbent for the removal of cationic dye from aqueous solution. J Saudi Chem Soc 20:S67–S76. https://doi.org/10.1016/j.jscs.2012.09.014

Farasati M, Haghighi S, Boroun S (2015) Cd removal from aqueous solution using agricultural wastes. Desalin Water Treat 57:11164–11172. https://doi.org/10.1080/19443994.2015.1043588

Farhan AM, Salem NM, Al-Dujaili AH, Awwad AM (2012) Biosorption studies of Cr(VI) ions from electroplating wastewater by walnut shell powder. Am J Environ Eng 2:188–195. https://doi.org/10.5923/j.ajee.20120206.07

Fathi MR, Asfaram A, Farhangic A (2015) Removal of Direct Red 23 from aqueous solution using corn stalks: isotherms, kinetics and thermodynamic studies. Spectrochim Acta A 135:364–372. https://doi.org/10.1016/j.saa.2014.07.008

Fazal S, Zhang B, Mehmood Q (2015) Biological treatment of combined industrial wastewater. Ecol Eng 84:551–558. https://doi.org/10.1016/j.ecoleng.2015.09.014

Feng J, Qiao K, Pei LY, Lv JP, Xie SL (2015) Using activated carbon prepared from *Typha orientalis* Presl to remove phenol from aqueous solutions. Ecol Eng 84:209–217. https://doi.org/10.1016/j.ecoleng.2015.09.028

Gamiz B, Pignatello JJ, Cox L, Hermosín MC, Celis R (2016) Environmental fate of the fungicide metalaxyl in soil amended with composted olive-mill waste and its biochar: an enantioselective study. Sci Total Environ 541:776–783. https://doi.org/10.1016/j.scitotenv.2015.09.097

Gayathri PV, Yesodharan S, Yesodharan EP (2017) Purification of water contaminated with traces of rhodamine B dye by microwave-assisted, oxidant induced and zinc oxide catalyzed advanced oxidation process. Desalin Water Treat 85:161–174. https://doi.org/10.5004/dwt.2017.21286

Gondhalekar SC, Shukla SR (2015) Biosorption of cadmium metal ions on raw and chemically modified walnut shells. Environ Prog Sustain Energy 34:1613–1619. https://doi.org/10.1002/ep.12161

Guan W, Tian SC (2017) The modified chitosan for dyeing wastewater treatment *via* adsorption and flocculation. Sci Adv Mater 9:1603–1609. https://doi.org/10.1166/sam.2017.3156

Guechi EK, Hamdaoui O (2015a) Biosorption of methylene blue from aqueous solution by potato (*Solanum tuberosum*) peel: equilibrium modelling, kinetic, and thermodynamic studies. Desalin Water Treat 57:10270–10285. https://doi.org/10.1080/19443994.2015.1035338

Guechi EK, Hamdaoui O (2015b) Evaluation of potato peel as a novel adsorbent for the removal of Cu (II) from aqueous solutions: equilibrium, kinetic, and thermodynamic studies. Desalin Water Treat 57:10677–10688. https://doi.org/10.1080/19443994.2015.1038739

Gülen J, Akın B, Ozgür M (2015) Ultrasonic-assisted adsorption of methylene blue on sumac leaves. Desalin Water Treat 57:9286–9295. https://doi.org/10.1080/19443994.2015.1029002

Guo L, Liang LY, Wang YJ, Liu MD (2015) Biosorption of Pb^{2+} from aqueous solution by rice straw modified with citric acid. Environ Prog Sustain 35:359–367. https://doi.org/10.1002/ep.12225

Gupta H, Gupta B (2015) Adsorption of polycyclic aromatic hydrocarbons on banana peel activated carbon. Desalin Water Treat 57:9498–9509. https://doi.org/10.1080/19443994.2015.1029007

Gupta N, Kushwaha AK, Chattopadhyaya MC (2016) Application of potato (*Solanum tuberosum*) plant wastes for the removal of methylene blue and malachite green dye from aqueous solution. Arab J Chem 9:S707–S716. https://doi.org/10.1016/j.arabjc.2011.07.021

Heng KS, Hatti-Kaul R, Adam F, Fukui T, Sudesh K (2017) Conversion of rice husks to polyhydroxyalkanoates (PHA) via a three-step process: optimized alkaline pretreatment, enzymatic hydrolysis, and biosynthesis by *Burkholderia cepacia* USM (JCM 15050). J Chem Technol Biotechnol 92(1):100–108. https://doi.org/10.1002/jctb.4993

Hong GB, Wang YK (2017) Synthesis of low-cost adsorbent from rice bran for the removal of reactive dye based on the response surface methodology. Appl Surf Sci 423:800–809. https://doi.org/10.1016/j.apsusc.2017.06.264

Hosseinzadeh H, Mohammadi S (2016) Biosorption of anionic dyes from aqueous solutions using a novel magnetic nanocomposite adsorbent based on rice husk ash. Sep Sci Technol 51:939–953. https://doi.org/10.1080/01496395.2016.1142564

Ibrahim S, Wang SB, Ang HM (2010) Removal of emulsified oil from oily wastewater using agricultural waste barley straw. Biochem Eng J 49:78–83. https://doi.org/10.1016/j.bej.2009.11.013

Imamoglu M, Şahin H, Aydın Ş, Tosunoglu F, Yılmaz H, Yıldız SZ (2016) Investigation of Pb(II) adsorption on a novel activated carbon prepared from hazelnut husk by K_2CO_3 activation. Desalin Water Treat 57:4587–4596. https://doi.org/10.1080/19443994.2014.995135

Inam EI, Etim UJ, Akpabio EG, Umoren SA (2015) Simultaneous adsorption of lead (II) and 3,7-bis(dimethylamino)-phenothiazin-5-ium chloride from aqueous solution by activated carbon prepared from plantain peels. Desalin Water Treat 57:6540–6553. https://doi.org/10.1080/19443994.2015.1010236

International Agency for Research on Cancer (IARC) (1987) Overall evaluations of carcinogenicity: An updating of IARC monographs volumes, vol 7. WHO, Lyon, pp 1–42. http://monographs.iarc.fr. Accessed 11 Aug 2019

Isa MH, Lang LS, Asaari FAH, Hamidi AA, Ramli NA, Dhas JPA (2007) Low cost removal of disperse dyes from aqueous solution using palm ash. Dyes Pigments 74:446–453. https://doi.org/10.1016/j.dyepig.2006.02.025

Janssens R, Mandal MK, Dubey KK, Luis P (2017) Slurry photocatalytic membrane reactor technology for removal of pharmaceutical compounds from wastewater: towards cytostatic drug elimination. Sci Total Environ 599–600:612–626. https://doi.org/10.1016/j.scitotenv.2017.03.253

Jawad AH, Rashid RA, Mahmuod RMA, Ishak MAM, Kasim NN, Ismail K (2015) Adsorption of methylene blue onto coconut (*Cocos nucifera*) leaf: optimization, isotherm and kinetic studies. Desalin Water Treat 57(19):8839–8853. https://doi.org/10.1080/19443994.2015.1026282

Jung KW, Hwang MJ, Ahn KH, Ok YS (2015) Kinetic study on phosphate removal from aqueous solution by biochar derived from peanut shell as renewable adsorptive media. Intl J Environ Sci Technol 12:3363–3372. https://doi.org/10.1007/s13762-015-0766-5

Karri RR, Jayakumar NS, Sahu JN (2017) Modelling of fluidized bed reactor by differential evolution optimization for phenol removal using coconut shells based activated carbon. J Mol Liq 231:249–262. https://doi.org/10.1016/j.molliq.2017.02.003

Karthick K, Namasivayam C, Pragasan LA (2017) Removal of direct red 12B from aqueous medium by $ZnCl_2$ activated *Jatropha* husk carbon: adsorption dynamics and equilibrium studies. Indian J Chem Technol 24:73–81. http://nopr.niscair.res.in/handle/123456789/39755. Accessed 11 Aug 2019

Kizito S, Wu S, Kirui WK, Lei M, Lu Q, Bah H, Dong R (2015) Evaluation of slow pyrolyzed wood and rice husks biochar for adsorption of ammonium nitrogen from piggery manure anaerobic digestate slurry. Sci Total Environ 505:102–112. 10.1016/j.scitotenv.2014.09.096

Kronenberg M, Trably E, Bernet N, Patureau D (2017) Biodegradation of polycyclic aromatic hydrocarbons: using microbial bioelectrochemical systems to overcome an impasse. Environ Pollut 231:509–523. https://doi.org/10.1016/j.envpol.2017.08.048

Kyzas GZ, Deliyanni EA (2015) Modified activated carbons from potato peels as green environmental-friendly adsorbents for the treatment of pharmaceutical effluents. Chem Eng Res Des 97:135–144. https://doi.org/10.1016/j.cherd.2014.08.020

Lee BG, Rowell RM (2004) Removal of heavy metal ions from aqueous solutions using lignocellulosic fibers. J Nat Fibers 1:97–108. https://doi.org/10.1300/J395v01n01_07

Lee LY, Gan S, Tan MSY, Lim SS, Lee XJ, Lam YF (2015) Effective removal of acid blue 113 dye using overripe *Cucumis sativus* peel as an eco-friendly biosorbent from agricultural residue. J Clean Prod 113:194–203. https://doi.org/10.1016/j.jclepro.2015.11.016

Lin C, Luo W, Luo T, Zhou Q, Li H, Jing L (2018) A study on adsorption of Cr (VI) by modified rice straw: characteristics, performances and mechanism. J Clean Prod 196:626–634. https://doi.org/10.1016/j.jclepro.2018.05.279

Marin-Rangel VM, Cortes-Martines R, Villanueva RAC, Garnica-Romo MG, Martinez-Flores HE (2012) As(V) biosorption in an aqueous solution using chemically treated lemon (*Citrus aurantifolia swingle*) residues. J Food Sci 71:10–14. https://doi.org/10.1111/j.1750-3841.2011.02466.x

Mashhadi S, Javadian H, Ghasemi M, Saleh TA, Gupta VK (2016) Microwave induced H_2SO_4 activation of activated carbon derived from rice agricultural wastes for sorption of methylene blue from aqueous solution. Desalin Water Treat 57:21091–21104. https://doi.org/10.1080/19443994.2015.1119737

Mirctzky P, Cirelli AF (2010) Cr (VI) and Cr (III) removal from aqueous solution by raw and modified lignocellulosic materials: a review. J Hazard Mater 180:1–19. https://doi.org/10.1016/j.jhazmat.2010.04.060

Modenes AN, Espinoza-Quinones FR, Geraldi CAQ, Manenti DR, Trigueros DEG, Oliveira APD, Borba CE, Kroumov AD (2015) Assessment of the banana pseudostem as a low-cost biosorbent for the removal of reactive blue 5G dye. Environ Technol 36:2892–2902. https://doi.org/10.1080/09593330.2015.1051591

Mokkapati RP, Mokkapati J, Ratnakaram VN (2016) Kinetic, isotherm and thermodynamics investigation on adsorption of divalent copper using agro-waste biomaterials, Musa acuminata,

Casuarina equisetifolia L. and Sorghum bicolor. Pol J Chem Technol 18:68–77. https://doi.org/10.1515/pjct-2016-0031

Naushad M, Khan MA, Alothman ZA, Khan MR, Kumar M (2015) Adsorption of methylene blue on chemically modified pine nut shells in single and binary systems: isotherms, kinetics, and thermodynamic studies. Desalin Water Treat 57:15848–15861. https://doi.org/10.1080/19443994.2015.1074121

Nayak AK, Pal A (2017) Green and efficient biosorptive removal of methylene blue by *Abelmoschus esculentus* seed: process optimization and multi-variate modeling. J Environ Manag 200:145–159. https://doi.org/10.1016/j.jenvman.2017.05.045

Nguyen TAH, Ngo HH, Guo WS, Zhang J, Liang S, Yue QY, Li Q, Nguyen TV (2013) Applicability of agricultural waste and by-products for adsorptive removal of heavy metals from wastewater. Bioresour Technol 148:574–585. https://doi.org/10.1016/j.biortech.2013.08.124

Njoku VO, Islam MA, Asif M, Hameed BH (2015) Adsorption of 2,4-dichlorophenoxyacetic acid by mesoporous activated carbon prepared from H_3PO_4-activated langsat empty fruit bunch. J Environ Manag 154:138–144. 10.1016/j.jenvman.2015.02.002

O'Donnel J (2018) Industrial pollution: causes and effects and biggest culprits of global warming. https://www.conservationinstitute.org/industrial-pollution. Accessed 11 Aug 2019

Oei BC, Ibrahim S, Wang SB, Ang HM (2009) Surfactant modified barley straw for removal of acid and reactive dyes from aqueous solution. Bioresour Technol 100:4292–4295. https://doi.org/10.1016/j.biortech.2009.03.063

Ojha AK, Bulasara VK (2015) Adsorption characteristics of jackfruit leaf powder for the removal of Amido Black 10B dye. Environ Prog Sustain 34:461–470. https://doi.org/10.1002/ep.12015

Okoro IA, Okoro SO (2011) Agricultural byproducts as green chemistry absorbents for the removal and recovery of metal ions from wastewater environment. Continental J Water Air Soil Pollut 2:15–22

Omidvar-Hosseini F, Moeinpour F (2016) Removal of Pb(II) from aqueous solutions using *Acacia Nilotica* seed shell ash supported $Ni_{0.5}Zn_{0.5}Fe_2O_4$ magnetic nanoparticles. J Water Reuse Desal 6:562–573. https://doi.org/10.2166/wrd.2016.073

Pachathu A, Ponnusamy K, Srinivasan SKVR (2016) Packed bed column studies on the removal of emulsified oil from water using raw and modified bagasse and corn husk. J Mol Liq 223:1256–1263. https://doi.org/10.1016/j.molliq.2016.09.048

Palma C, Lloret L, Puen A, Tobar M, Contreras E (2016) Production of carbonaceous material from avocado peel for its application as alternative adsorbent for dyes removal. Chin J Chem Eng 24:521–528. https://doi.org/10.1016/j.cjche.2015.11.029

Parlayıcı Ş, Pehlivan E (2017) Removal of metals by Fe_3O_4 loaded activated carbon prepared from plum stone (*Prunus nigra*): kinetics and modelling study. Powder Technol 317:23–30. https://doi.org/10.1016/j.powtec.2017.04.021

Pena D, Lopez-Pineiro A, Albarran A, Rato-Nunes JM, Sanchez-Llerena J, Becerra D, Ramírez M (2016) De-oiled two-phase olive mill waste may reduce water contamination by metribuzin. Sci Total Environ 541:638–645. https://doi.org/10.1016/j.scitotenv.2015.09.019

Portinho R, Zanella O, Feris LA (2017) Grape stalk application for caffeine removal through adsorption. J Environ Manag 202:178–187. https://doi.org/10.1016/j.jenvman.2017.07.033

Quesada HB, Takaoka Alves Baptista A, Cusioli LF, Seibert D, de Oliveira Bezerra C, Bergamasco R (2019) Surface water pollution by pharmaceuticals and an alternative of removal by low-cost adsorbents: a review. Chemosphere 222:766–780. https://doi.org/10.1016/j.chemosphere.2019.02.009

Rattanachueskul N, Saning A, Kaowphong S, Chumha N, Chuenchom L (2016) Magnetic carbon composites with a hierarchical structure for adsorption of tetracycline, prepared from sugarcane bagasse via hydrothermal carbonization coupled with simple heat treatment process. Bioresour Technol 226:164–172. https://doi.org/10.1016/j.biortech.2016.12.024

Rosales E, Meijide J, Pazos M, Sanroman MA (2017) Challenges and recent advances in biochar as low-cost biosorbent: from batch assays to continuous-flow systems. Bioresour Technol 246:176–192. https://doi.org/10.1016/j.biortech.2017.06.084

Safa Y (2016) Utilization of mustard and linseed oil cakes: novel biosorbents for removal of acid dyes. Desalin Water Treat 57:5914–5925. https://doi.org/10.1080/19443994.2015.1007087

Saka C, Sahin O, Kucuk MM (2012) Application on agricultural and forest waste adsorbents for the removal of Pb(II) from contaminated waters. Int J Environ Sci Technol 9:379–394. https://doi.org/10.1007/s13762-012-0041-y

Sandoval A, Hernandez-Ventura C, Klimova TE (2017) Titanate nanotubes for removal of methylene blue dye by combined adsorption and photocatalysis. Fuel 198:22–30. https://doi.org/10.1016/j.fuel.2016.11.007

Sarasidis VC, Plakas KV, Karabelas AJ (2017) Novel water-purification hybrid processes involving *in situ* regenerated activated carbon, membrane separation and advanced oxidation. Chem Eng J 328:1153–1163. https://doi.org/10.1016/j.cej.2017.07.084

Shah BA, Pandya DD, Shah HA (2016) Impounding of ortho-chlorophenol by zeolitic materials adapted from bagasse fly ash: four factor three level Box-Behnken design modelling and optimization. Arab J Sci Eng 42:241–260. https://doi.org/10.1007/s13369-016-2294-0

Shang Y, Zhang JH, Wang X, Zhang RD, Xiao W, Zhang SS, Han RP (2015) Use of polyethyleneimine-modified wheat straw for adsorption of Congo red from solution in batch mode. Desalin Water Treat 57:8872–8883. https://doi.org/10.1080/19443994.2015.1027280

Sharma S, Tiwari DP (2016) Model-fitting approach for methylene blue dye adsorption on *Camelina* and *Sapindus* seeds-derived adsorbents. Adsorpt Sci Technol 34:565–580. https://doi.org/10.1177/0263617416674949

Sherafatmand M, Ng HY (2015) Using sediment microbial fuel cells (SMFCs) for bioremediation of polycyclic aromatic hydrocarbons (PAHs). Bioresour Technol 195:122–130. https://doi.org/10.1016/j.biortech.2015.06.002

Singh S, Shukla SR (2017) Theoretical studies on adsorption of Ni (II) from aqueous solution using *Citrus limetta* peels. Environ Prog Sustain Energy 36:864–872. https://doi.org/10.1002/ep.12526

Singha AS, Guleria A (2014) Utility of chemically modified agricultural waste okra biomass for removal of toxic heavy metal ions from aqueous solution. Eng Agr Environ Food 8:52–60. https://doi.org/10.1016/j.eaef.2014.08.001

Smitha T, Santhi T, Prasad AL, Manonman S (2012) *Cucumis sativus* used as adsorbent for the removal of dyes from aqueous solution. Arab J Chem 10:S244–S251. https://doi.org/10.1016/j.arabjc.2012.07.030

Sohrabi H, Ameri E (2015) Adsorption equilibrium, kinetics, and thermodynamics assessment of the removal of the reactive red 141 dye using sesame waste. Desalin Water Treat 57:18087–18098. https://doi.org/10.1080/19443994.2015.1087345

Song ST, Saman N, Johari K, Mat H (2015) Biosorption of mercury from aqueous solution and oil-field produced water by pristine and sulfur functionalized rice residues. Environ Prog Sustain Energy 34:1298–1310. https://doi.org/10.1002/ep.12116

Subbaiah MV, Kim DS (2016) Adsorption of methyl orange from aqueous solution by aminated pumpkin seed powder: kinetics, isotherms, and thermodynamic studies. Ecotox Environ Safe 128:109–117. https://doi.org/10.1016/j.ecoenv.2016.02.016

Suteu D, Malutan T, Bilba D (2011) Agricultural waste corn cob as a sorbent for removing reactive dye Orange 16: equilibrium and kinetic study. Cellulose Chem Technol 45:413–420. http://www.cellulosechemtechnol.ro. Accessed 11 Aug 2019

Suteu D, Biliuta G, Rusu L, Coseri S, Nacu G (2015) Cellulose cellets as new type of adsorbent for the removal of dyes from aqueous media. Environ Eng Manag J 14:525–532. http://www.eemj.icpm.tuiasi.ro/pdfs/vol14/no3/4_998_Suteu_14.pdf. Accessed 13 Aug 2019

Tang RX, Dai C, Li C, Liu WH, Gao ST, Wang C (2017a) Removal of methylene blue from aqueous solution using agricultural residue walnut shell: equilibrium, kinetic, and thermodynamic studies. J Chem 2017, 10 pages. https://doi.org/10.1155/2017/84049651-10:8404965

Tang CF, Shu Y, Zhang RQ, Li X, Song JF, Li B, Zhang YT, Ou DL (2017b) Comparison of the removal and adsorption mechanisms of cadmium and lead from aqueous solution by activated carbons prepared from *Typha angustifolia* and *Salix matsudana*. RSC Adv 26:16092–16103. https://doi.org/10.1039/C6RA28035H

Tomczak E, Tosik P (2017) Waste plant material as a potential adsorbent of a selected azo dye. Chem Process Eng 38:283–294. https://doi.org/10.1515/cpe-2017-0021

Tonucci MC, Gurgel LVA, Aquino SFD (2015) Activated carbons from agricultural byproducts (pine tree and coconut shell), coal, and carbon nanotubes as adsorbents for removal of sulfamethoxazole from spiked aqueous solutions: kinetic and thermodynamic studies. Ind Crop Prod 74:111–121. https://doi.org/10.1016/j.indcrop.2015.05.003

Toumi LB, Hamdi L, Salem Z, Allia K (2015) Batch adsorption of methylene blue from aqueous solutions by untreated Alfa grass. Desalin Water Treat 53:806–817. https://doi.org/10.1080/19443994.2013.846236

Tran VS, Ngo HH, Guo WS, Zhang J, Liang S, Ton-That C, Zhang XB (2015) Typical low cost biosorbents for adsorptive removal of specific organic pollutants from water. Bioresour Technol 182:353–363. https://doi.org/10.1016/j.biortech.2015.02.003

Trivedi NS, Mandavgane SA, Kulkarn BD (2016) Mustard plant ash: a source of micronutrient and an adsorbent for removal of 2,4-dichlorophenoxyacetic acid. Environ Sci Pollut Res 23(20):20087–20099. https://doi.org/10.1007/s11356-016-6202-7

Tunç O, Tanaci H, Aksu Z (2009) Potential use of cotton plant wastes for the removal of Remazol Black B reactive dye. J Hazard Mater 163:187–198. https://doi.org/10.1016/j.jhazmat.2008.06.078

Wang P, Ma QY, Hu DY, Wang LJ (2015) Adsorption of methylene blue by a low-cost biosorbent: citric acid modified peanut shell. Desalin Water Treat 57:10261–10269. https://doi.org/10.1080/19443994.2015.1033651

Wang H, Chu Y, Fang C, Huang F, Song Y, Xue X (2017) Sorption of tetracycline on biochar derived from rice straw under different temperatures. PLoS One 12:e0182776. https://doi.org/10.1371/journal.pone.0182776

Wang H, Fang C, Wang Q, Chu Y, Song Y, Chen Y, Xue X (2018) Sorption of tetracycline on biochar derived from rice straw and swine manure. RSC Adv 8:16260–16268. https://doi.org/10.1039/C8RA01454J

Weiss M, Haufe J, Carus M, Brandão M, Bringezu S, Hermann B, Patel MK (2012) A review of the environmental impacts of biobased materials. J Ind Ecol 16:S169–S181. https://doi.org/10.1111/j.1530-9290.2012.00468.x

Worch E (2012) Adsorption technology in water treatment: fundamentals, processes, and modeling. Walter de Gruyter, Berlin/Boston, p 6–10. isbn: 978-3-11-024022-1, e-isbn: 978-3-11-024023-8

Yagub MT, Sen TK, Afroze S, Ang HM (2014) Dye and its removal from aqueous solution by adsorption: a review. Adv Colloid Interface Sci 209:172–184. https://doi.org/10.1016/j.cis.2014.04.002

Yin YY, Guo XT, Yang C, Gao LM, Hu YB (2016) An efficient method for tylosin removal from an aqueous solution by goethite modified straw mass. RSC Adv 98:95425–95434. https://doi.org/10.1039/C6RA19172J

Zhang J, Chen H, Chen Z, He J, Shi W, Liu D, Chi H, Fuyi C, Wang W (2016) Microstructured macroporous adsorbent composed of polypyrrole modified natural corn cob-core sponge for Cr (VI) removal. RSC Adv 64:59292–59298. https://doi.org/10.1039/C6RA07687D

Zhang X, Lv L, Qin Y, Xu M, Jia X, Chen Z (2018) Removal of aqueous Cr (VI) by a magnetic biochar derived from *Melia azedarach* wood. Bioresour Technol 256:1–10. https://doi.org/10.1016/j.biortech.2018.01.145

Zheng Y, Wang J, Zhu Y, Wang A (2015) Research and application of kapok fiber as an absorbing material: a mini review. J Environ Sci 27:21–32. https://doi.org/10.1016/j.jes.2014.09.026

Zhou Y, Zhang L, Cheng ZJ (2015) Removal of organic pollutants from aqueous solution using agricultural wastes: a review. J Mol Liq 212:739–762. https://doi.org/10.1016/j.molliq.2015.10.023

Zhou RS, Zhou RW, Zhang XH, Tu S, Yin YW, Yang SZ, Ye LY (2016) An efficient bio-adsorbent for the removal of dye: adsorption studies and cold atmospheric plasma regeneration. J Taiwan Inst Chem Eng 68:372–378. https://doi.org/10.1016/j.jtice.2016.09.030

Zhu C, Duan YF, Wu CY, Zhou Q, She M, Yao T, Zhang J (2016a) Mercury removal and synergistic capture of SO_2/NO by ammonium halides modified rice husk char. Fuel 172:160–169. https://doi.org/10.1016/j.fuel.2015.12.061

Zhu L, Wang Y, He TT, You LJ, Shen XQ (2016b) Assessment of potential capability of water bamboo leaves on the adsorption removal efficiency of cationic dye from aqueous solutions. J Polym Environ 24:148–158. https://doi.org/10.1007/s10924-016-0757-8

Zhu NY, Yan TM, Qiao J, Cao HL (2016c) Adsorption of arsenic, phosphorus and chromium by bismuth impregnated biochar: adsorption mechanism and depleted adsorbent utilization. Chemosphere 164:32–40. https://doi.org/10.1016/j.chemosphere.2016.08.036

Zhu YY, Kolar P, Shah SB, Cheng JJ, Lim PK (2016d) Avocado seed-derived activated carbon for mitigation of aqueous ammonium. Ind Crops Prod 92:34–41. https://doi.org/10.1016/j.indcrop.2016.07.016

Chapter 15
Sonochemistry in Green Processes: Modeling, Experiments, and Technology

Kaouther Kerboua and Oualid Hamdaoui

Contents

K. Kerboua
Department of Second Cycle, Ecole Supérieure de Technologies Industrielles, Annaba, Algeria
e-mail: k.kerboua@esti-annaba.dz

O. Hamdaoui (✉)
Chemical Engineering Department, College of Engineering, King Saud University, Riyadh, Saudi Arabia

Inamuddin, A. M. Asiri (eds.), *Sustainable Green Chemical Processes and their Allied Applications*, Nanotechnology in the Life Sciences,
https://doi.org/10.1007/978-3-030-42284-4_15

Abbreviations

μ	Dynamic viscosity (Pa s)
μ_{th}	Effective thermal viscosity (Pa s)
ρL	Density of the liquid phase (kg/m^3)
ρG	Density of the gas phase (kg/m^3)
σ	Surface tension (N/m)
α	Accommodation coefficient
β	Damping coefficient
η	Efficiency
ξ	Thickness of heat diffusion interface (m)
Γ	Polytropic constant
λ	Thermal conductivity across bubble wall (W/m K)
ω	Angular frequency of the acoustic field (rad)
ω_0	Natural angular frequency of the oscillating bubble (rad)
ΔH_i	Reaction enthalpy of the *i*th reaction (J/mol)
ΔH_f	Enthalpy of formation of fuel gas (J/mol)
ΔH_{rad}	Enthalpy of formation of free radical (J/mol)
AE	Atom economy
c	Sound celerity (m/s)
$\dot{C}$	Flow of energy creation (J/s)
CED	Cumulative energy demand (J)
cv	Isochoric average heat capacity (J/mol K)
c_{v_i}	Isochoric heat capacity of the *i*th species (J/mol K)
E	E factor
Ei	Primary energy used in a process
E_{US}	Acoustic energy of ultrasound
f	Frequency (Hz)
I	Ultrasonic power density (W/m^3)
$\dot{m}$	Evaporation-condensation mass rate (kg/m^2 s)
m	Evaporation-condensation mass acceleration (kg/m^2 s^2)
m_B	Mass of gas phase within a single acoustic cavitation bubble (g)
m_{H_2O}	Mass of process water (g)
m_r	Mass of a reactant (g)
m_s	Mass of a solvent (g)
m_t	Mass of a targeted product (g)
MP	Mass productivity
m_p	Mass of a product (g)
m_u	Mass of an undesirable product (g)
m_w	Mass of waste (g)
M	Molar weight (kg/mol)
MI	Mass intensity
M_p	Molar weight of a product (kg/mol)
M_t	Molar weight of a targeted product (kg/mol)

n	Molar amount (mol)
$\dot{n}$	Time derivate of the molar amount (mol/s)
N	Number of bubbles
$\dot{N}$	Time derivate of the number of bubbles (/s)
N_A	Avogadro number
n_f	Molar yield of a fuel gas (mol)
n_{rad}	Molar yield of a free radical (mol)
P	Pressure in the liquid (Pa)
P_A	Acoustic amplitude (Pa)
P_i	Partial pressure (Pa)
PMI	Process mass intensity
P_{US}	Ultrasonic power (W)
P_v	Saturating pressure (Pa)
p	Pressure of the wave travelling the medium (Pa)
P_∞	Static pressure (Pa)
R	Bubble radius (m)
$\dot{R}$	Bubble wall velocity (m/s)
$\ddot{R}$	Bubble wall acceleration (m/s^2)
r	Reaction rate (mol/s m^3)
ri	Reaction rate of the ith reaction (mol/s m^3)
R_g	Ideal gas constant (J/mol K)
RME	Reaction mass efficiency
S	Section of the bubble wall (m^2)
SI	Solvent intensity
T	Temperature within the bubble (K)
T_∞	Temperature of the liquid (K)
t	Time (s)
V	Volume (m^3)
WWI	Wastewater intensity
X	Dimensionless Laplace tension

15.1 Introduction

Green processing consists in integrating the environmental dimension in the proceeding way. Specifically, a green process is expected to generate less toxic substances, consume lower resources, harness renewable feedstocks, lessen environmental issues, or offer simultaneously some of the previous aspects (Anastas and Warner 1998). Ultrasounds, basically simple mechanical waves oscillating in the frequency range comprised between 20 kHz and 2 MHz, have the phenomenal power to take part in each of the previous cases of green processes (Chatel 2018). Sonochemistry, though being only one of the multiple applications of ultrasound in process engineering, remains the cornerstone of the ultrasound-assisted green processes.

Sonochemistry is defined as the initiation and enhancement of chemical interactions inside a cavitation bubble oscillating under the effect of an ultrasonic wave travelling across a liquid medium (Neppiras 1980). The phenomenon is obviously microscopic but repeated billions of times within a sonochemical reactor. According to the selected medium and operating conditions, the reactor could be the seat of homogeneous sonochemistry of liquids and liquid–gas systems, heterogeneous sonochemistry of liquid–liquid systems, or even heterogeneous sonochemistry involving solid–liquid systems and sonocatalysis (Suslick 2001). A panoply of applications could then legitimately be expected: liquid effluents decontamination, water treatment, organic synthesis, and inorganic synthesis.

Meanwhile, is it enough to irradiate a liquid medium with ultrasound to lead to sonochemistry? Even more, can one deliberately describe as green a process since that it is sonochemical?

Both questions are worth of deep investigations. That is why we suggest in the following sections to have a serious and concise examination of the link between sonochemistry and green processes starting from three facts:

- Green chemistry is based on 12 principles according to Anastas and Werner (1998). Rigorously, only processes respecting these principles can be classified as green.
- Sonochemistry is the result of a physical phenomenon: acoustic cavitation. Evidence of sonochemical activity is well established through experimental observation and quantification of its effects. Nevertheless, experiment is not able neither to figure out the mechanism evolving inside the cavitation volume nor to follow the temporal evolution of complex interacting physical and chemical phenomena.
- Modeling is a key of advance in sonochemistry for more than a reason: It is a convenient and economic tool for describing the complex systems, providing information regarding nonpractically measurable parameters, and understanding and predicting the factors influencing the performance of a system.

15.2 Sonochemistry, a Green Process: To Which Extent?

Among the stations that marked the history of science, the discovery of sonochemistry and the advent of green processing concept are for sure unavoidable ones. Nevertheless, both occurred with a time difference of almost four decades and one-step ahead for sonochemistry. Basically, this latter was first invoked by Richards and Loomis (1927) who reported observations of chemical effects induced by high-frequency acoustic waves, as a part of a series of studies on biological and physical effects of ultrasound. This pioneering research opened perspectives in the field: Sonochemistry was regarded as a miraculous technique to activate and accelerate reactions (Oussaid et al. 1995), homogenize phases through cavitation (Suslick and Price 1987), and even select species (Anto et al. 1991). During this early period, although vision of chemical processing started changing, the classical objectives were still there: Producing more, easier and faster, the environmental dimension was not involved so far.

In the 60s, environmental activists, principally Carson et al. (1962), awoke public's awareness for the first time to the impact of chemicals on the environment. Eco-awareness knew then an ascending span and hazards of several materials and chemicals became better mastered, and hence, processes needed to be revamped to lessen the observed effects. Three decades later, Anastas and Warner published "The Twelve Principles of Green Chemistry" (Anastas and Warner 1998), a revolutionary reference of practical guidelines for scientists and industrials, aiming to reduce the negative environmental impacts of chemical processes. Since then, authors used to present ultrasound-assisted processes as obviously green (Guzen et al. 2007; Banitaba et al. 2013; Nikpassand et al. 2016), and that is exactly where we need to take a break for thinking.

In 2015, Fegade and Trembly (2017) opened a debate on the misuse of the term "green" to qualify some ultrasound-assisted processes. According to these authors, several researches adopt by abuse the attribute green without having enough justifications for it. In other words, some processes qualified of green have no solid reasons to be so. Moreover, the same authors made a rigorous distinction between green and sustainable chemistry. It appeared that some processes should be qualified as "sustainable" rather than "green," which seems to be somehow confusing to the reader. Cintas (2016) took part in the debate. He suggested that it is rather pernicious to differentiate between green and sustainable chemistry, especially that some of the green chemistry principles remain deceitful in practice. In this topic, Ivanković et al. (2017) reviewed the 12 principles of green chemistry in practice. They presented concrete examples of solvents reduction, as in the synthesis of sertraline in ethanol rather than the mixture of methylene chloride, tetrahydrofuran, toluene, and hexane (Cue and Zhang 2009); atom economy, as in the passage from Boots to Hoechst process in the synthesis of ibuprofen (Ahmadi et al. 2014); as well as hazardous material reduction, as in the replacement of toxic carbonyl dichloride COC_2 with carbon dioxide CO_2 in the new Asahi Kasei's polycarbamate synthesis (Fukuoka et al. 2003). They also explored cases of energy-efficient designs, safer solvents and auxiliaries, and biodegradable and renewable catalysts. However, it is noteworthy to mention that Ivanković et al. (2017) affirmed that when designing a green chemistry process, it is impossible to meet simultaneously the requirements of all 12 principles of green chemistry, and one should rather attempt to apply as many principles as possible during certain stages of the process.

The same approach should be applied to sonochemical applications in order to evaluate to which extent sonochemistry can be considered a green process. In fact, very few studies bore a specific interest to this topic. In the following, we will focus on two of them. In an early visionary work, Cintas and Luche (1999) situated sonochemistry in the green chemistry sphere owing to its ability to improve reactional rates, easily form radicals, offer safer reaction conditions, and prepare promising catalysts such as nanostructured ones. Nevertheless, the authors underlined the need to distinguish chemical effects of ultrasound, i.e., sonochemistry, and physical or

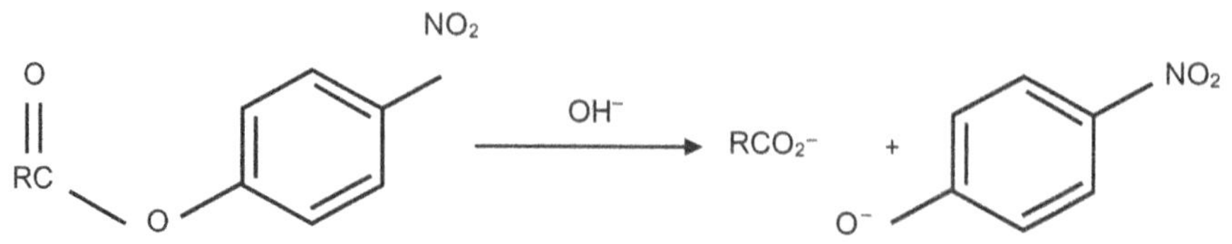

Fig. 15.1 Chemical reaction of ester hydrolysis

mechanical ones, that they qualified as "false sonochemistry." Chatel (2018) reported in a paper that dates from 2017 that a bit more than 33% of articles published by Ultrasonics Sonochemistry in 2016 claimed to be related to green chemistry. Here again, the author invoked precautions to take toward the increasing trend of research on sonochemistry as a green process, as several of them ignore the distinction between "cavitational chemistry," i.e., sonochemistry, and physical effects of ultrasound.

All in all, it seems primarily evident that qualifying a process as "a green sonochemical process" requires two aspects to be valid: the chemical effect of ultrasound and the respect of green chemistry principles.

The first aspect is conditioned by the concise understanding of the process, in order to distinguish mechanical effects of ultrasound from those related to chemistry within acoustic cavitation. Luche et al. (1990) presented a pioneering work to clarify this point by exposing examples from organic synthesis (Kristol et al. 1981; Einhorn and Luche 1986). To illustrate, in the case of ester hydrolysis reported in Fig. 15.1 (Kristol et al. 1981), the reactive ionic intermediate OH^- is not generated by cavitation. Furthermore, the introduction of ultrasound demonstrated significant acceleration in heterogeneous solution, against modest improvement in homogeneous one barely neighboring 15%, which strengthens the interpretation based on physical and mechanical role of the wave rather than that of acoustic cavitation.

Having said that, if in a process the effect of ultrasonic irradiation seems to be rather explained by the mechanical perturbation, such process should be concisely named green ultrasound-assisted process instead of green sonochemical one.

The second aspect, on the other hand, would be rather taken as a compromise where modifications are brought to the classical processes in order to make the design, the materials, and the conditions respecting as much as possible of the principles of green chemistry without completely opposing any of them. This leads to a rational projection of sonochemistry in green chemistry area that can be summarized in being "greener" rather than "green."

15.2.1 *Sonochemistry: A Benign-by-Design Process?*

In 1994, Anastas and Farris introduced the notion of benign-by-design chemistry as the branch of chemistry that accounts for three factors: efficiency, economic viability, and environmental prevention. Similarly, a benign-by-design chemical process needs to respect the productivity condition, to be able to economically survive, and reduce or eliminate the negative impact on environment, not only by delivering a final "clean" outcome but also by monitoring every single stage of the process. In first and last criteria, we easily recognize the concept of green processing; contrariwise, the economic condition does not appear as a principle of green chemistry. A benign-by-design process is then presented as more than a green process but a sustainable one. Sonochemistry is examined for fitting the three criteria of benign-by-design process through the review of examples from literature.

The efficiency of sonochemical processes, namely, sonochemical efficiency, is a nonstandardized parameter, though being of crucial importance in the assessment of the benignity of the process. Several interpretations and formulas are proposed by researchers in the field to more or less approach the idea of an efficient sonochemical process. One commonly used definition, also called "cavitational yield" (Gogate et al. 2001), consists in estimating the ratio of reacted molecules rate, usually per unit volume, to the ultrasonic power density (Gogate et al. 2001; Tuziuti et al. 2008), which is actually equivalent to the ratio of yield of reacted molecules to the ultrasonic energy (Koda et al. 2003) as indicated in Eq. 15.1.

$$\eta = \frac{r}{I} \tag{15.1}$$

Here, the ultrasonic power is measured through calorimetric method, while reacted yield is generally attained through Fricke reaction or KI oxidation within water undergoing sonochemical homolytic cleavage.

The sonochemical efficiency can also be assessed based on atom economy. This concept can be simply interpreted as the maximization of the conversion fraction of reactants into products and has been extensively demonstrated through sonochemical production yields. For instance, Koda et al. (2004) assessed the sonochemical efficiency by measuring the production yields of nitrite and nitrate during the sonolysis of water under an air atmosphere.

Such interpretation of the sonochemical efficiency focuses on the sonochemical kinetics in spite of furnished energy. However, on one hand, input energy acts directly in the formation and dynamics of acoustic cavitation, as energies are converted from acoustical to kinetical, chemical, and works; on the other hand, furnished energy should be reasonably kept inspected, as the process does not deliver only chemical species but also energy in various forms. Clearly, energy and sonochemical rates cannot be dissociated when dealing with efficiency, and consequently, the first definition seems to be more appropriate.

Sonochemical efficiency was as well defined based on the notion of selectivity. In a certain way, the sonochemical process may result in a variety of products with some undesirable ones. Minimizing the quantity of these latter in the benefit of targeted species appears to be a rational criterion to evaluate the efficiency of a sonochemical process. The change in the course of reaction in the presence of ultrasound, namely, sonochemical switching, was explained in some studies by the hot spot conditions that change the mechanism root within the cavitation bubble. An illustrative example is that of sonochemical switching from ionic to radical pathways in the reactions of styrene and trans-β-methylstyrene with lead tetraacetate (Anto et al. 1991). In this case, talking about "sonochemical" switching appears to be well justified. In some other cases, the observed change in the reaction pathway would be rather due to the mechanical effect of ultrasound. One example is that of switching of reaction pathways in solid–liquid two-phase reactions (Ando et al. 1984), where authors concluded that ultrasound assists the contact of potassium cyanide with alumina to decrease the catalytic ability of alumina for the Friedel–Crafts reaction and to enhance nucleophilic attack by the cyanide ion at the alumina surface. Here, the appellation "sonochemical switching" is somehow inappropriate, and the switching is rather ultrasonic than sonochemical. Having said that, if sonochemical switching is proved, one can define a simple metric to evaluate the sonochemical efficiency in this case through a molar yields ratio of undesirable species as compared to targeted ones. Equation 15.2 indicates the formula.

$$\eta = \frac{n_{\mathrm{u}}}{n_{\mathrm{t}}} \tag{15.2}$$

Though the three previous definitions are the most commonly used, visibly, literature does not lack approaches of sonochemical efficiency, and one of the most uncommon are those suggested by Rivas et al. (2010) and Authier et al. (2018). Rivas et al. (2010) perceived sonochemical efficiency as an expression of valuable energy as compared to the involved acoustic energy. Concisely, valuable energy was considered as the enthalpy of formation of radicals, which leads to the following formula:

$$\eta = \frac{\sum \dot{n}_{\mathrm{rad}} \Delta H_{\mathrm{rad}}}{P_{\mathrm{us}}} \tag{15.3}$$

Authier et al. (2018), on the other hand, defined the sonochemical efficiency as the ratio of chemical energy in fuel gases to acoustic energy supplied by ultrasound in water per mass unit of gas phase. The corresponding equation is then:

$$\eta = \frac{\sum n_{\mathrm{f}} \Delta H_{\mathrm{f}}}{E_{\mathrm{us}} / m_{\mathrm{B}} N} \tag{15.4}$$

In spite of the number of interpretations given to sonochemical efficiency, most of them were employed to estimate the relative performance of sonochemical processes which differ in their configurations (reactor geometry, operating conditions, acoustic parameters, medium composition, etc.). Indeed, it is quite rare to find studies that adopted the previous approaches to estimate the efficiency of a sonochemical process as compared to conventional ones.

The few works which estimated the sonochemical efficiency in comparison with other processes founded their comparison on direct confrontation of molar yields, such as in the work of Harada (2001), Rossi et al. (2009), Madhavan et al. (2010a, b, c), Maleki et al. (2010), and Penconi et al. (2015). Their results are summarized and reported in Fig. 15.2. A coarse scanning of Fig. 15.2 demonstrates that assessing the performance of sonochemistry on the basis of molar yields or molar rates of

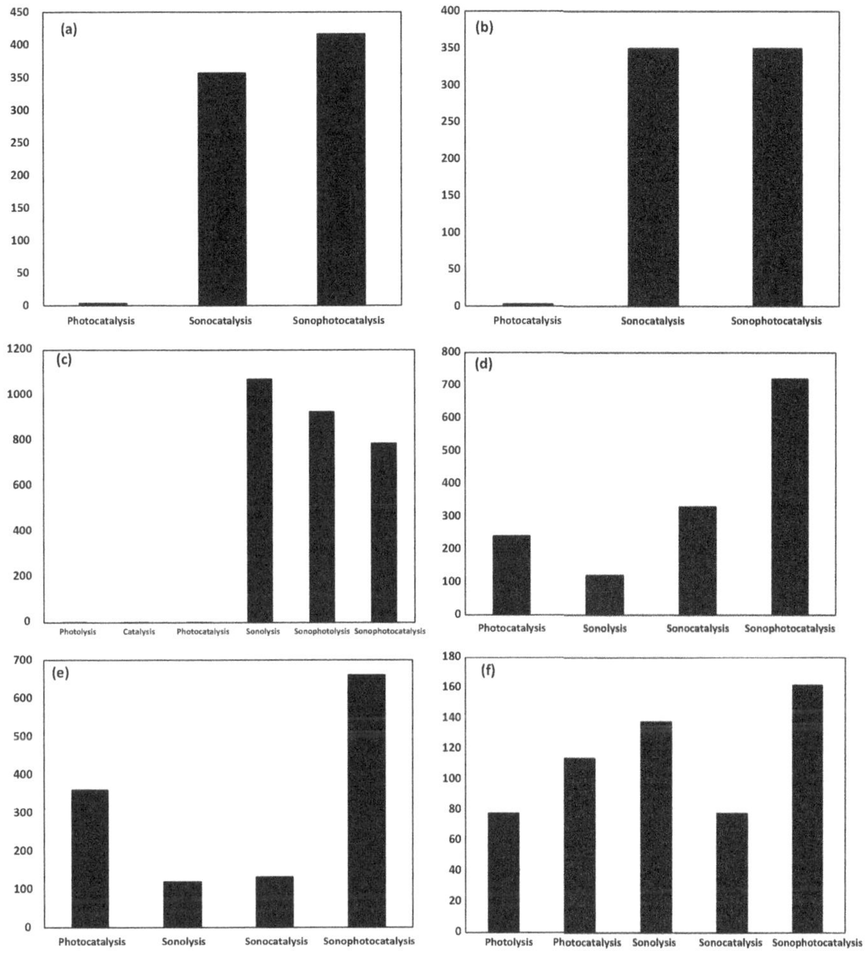

Fig. 15.2 Comparison of the performance of different processes for the production of H_2 (**a**–**c**) and degradation of pollutants (**d**–**f**) expressed in molar rates (mol/L·h) on the vertical axis. (**a**) Adapted from Penconi et al. (2015). 300 mL of ethanol/water solution (20%v), 0.4 g of $Y_{0.8}Ga_{0.2}InO_3$, 38 kHz, 26 W/dm^3, and 1 kW/m^2 of light. (**b**) Adapted from Rossi et al. (2009). 200 mL of ethanol/water (10%v), 0.4 g of $La_{0.8}Ga_{0.2}InO_3$, 38 kHz, 50 W, and 500 W/m^2 of light. (**c**) Adapted from Harada (2001). 40 mL of water, 0.2 g of TiO_2, 200 kHz, 200 W, and 500 W of light. (**d**) Adapted from Madhavan et al. (2010a). Degradation of AR88 in 250 mL of aqueous solution at 0.09 mMol/L, 0.25 g of TiO_2, 213 kHz, 55 W/dm^3, and 450 W of light power. (**e**) Adapted from Madhavan et al. (2010c). Degradation of IBP in 250 mL of aqueous solution at 0.09 mMol/L, 0.25 g of TiO_2, 213 kHz, 55 W/dm^3, and 450 W of light power. (**f**) Adapted from Madhavan et al. (2010b). Degradation of MCP in 250 mL of aqueous solution at 0.12 mMol/L, 0.25 g of TiO_2, 213 kHz, 55 W/dm^3, and 450 W of light power

reacted or produced species (in other words, chemical kinetics) remains strictly applicable to the studied cases. Neither extrapolation nor prediction can be expected from established results.

For instance, according to Madhavan et al. (2010a, c), sonochemical process shows the lowest efficiency in the degradation of AR88 as well as ibuprofen (0.09 mM) among the studied processes (sonocatalysis, photocatalysis, sonophotocatalysis). In the contrary, they observed high relative performance of sonolysis in the degradation of monocrotophos (0.12 mM) (Madhavan et al. 2010a).

To conclude this discussion, one needs to underline three facts. Firstly, no norms exist to homogenize the assessment of sonochemical efficiency as compared to other processes. Secondly, most of bibliographic works tend to suggest concepts of sonochemical efficiency in order to compare different configurations of sonochemical processes rather than comparing these latter to other processes. Finally, no general rule can be established regarding the efficiency of sonochemical processes, whatever the adopted approach. Hence, sonochemistry cannot be qualified as absolutely efficient so far.

The economic viability of sonochemical processes can be simply regarded as technological cost competitiveness and energy efficiency (or in larger scope, functional costs). Both aspects should be considered in a relative point of view. In other words, a sonochemical process can be deemed as economically viable only if serious estimation and comparison are established with other processes delivering similar product, qualitatively and quantitatively.

Starting from this approach, we admit that few research works dealt with the topic in a rigorous way. In most of cases, the economic viability is focused on energy consumption, which leads to the notion of energy efficiency in sonochemical process. At first sight, there is an agreement on the definition of energy efficiency of a sonochemical reactor (Gogate et al. 2001; Son et al. 2010; Toma et al. 2011), as the ratio of energy effectively dissipated into the liquid to the electrical energy supplied to the reactor, according to Eq. 15.5.

$$\eta_{\mathrm{E}} = \frac{E_{\mathrm{US}}}{E_{\mathrm{EL}}} \tag{15.5}$$

Electrical energy can be directly obtained using a multimeter equipped with data logger, while the energy effectively dissipated into the liquid, namely, acoustic or ultrasonic energy, or most often presented as ultrasonic power, is usually estimated through calorimetric measurements:

$$P_{\mathrm{US}} = mC_{\mathrm{p}}\left(\frac{\Delta T}{\Delta t}\right)_{\Delta t \to 0} \tag{15.6}$$

Actually, a part of this energy served first to the generation and oscillation of cavitation. Consequently, what should be intended from the evaluation of energy efficiency of sonochemical process is to maximize the fraction of ultrasonic energy leading to cavitation phenomenon, and all the rest is in fact instantaneously dissipated as heat. As the acoustic bubble implodes, the cavitational energy is recovered as well in the form of heat and accounted in the temperature elevation $\frac{\Delta T}{\Delta t}$ appearing in Eq. 15.6.

Once again, the defined metric was often employed to evaluate the energetic performance of a certain configuration of a sonochemical reactor (González-García et al. 2010). Assessing the performance of a sonochemical process in terms of energy, as compared to other processes, was actually restricted to qualitative descriptions. In addition, energy is not the only economic factor to monitor; economic view of sonochemical processes should be extended to a larger scope, by including, for instance, life cycle assessment for material flows and waste streams reduction or elimination (Saini and Singh 2002). Such elements are crucial to furthering the widespread application of sonochemical processes. Nevertheless, no metrics have been defined yet to evaluate all the aforementioned dimensions. To conclude on this point, bibliography demonstrates a poor attention to the economic aspect of sonochemistry, an aspect without which research and technology axed on sonochemistry will still suffer from insufficient arguments for upscaling and industrial integration. Further still, intuitive qualifications of sonochemical processes as economically profitable owing to their operating conditions or raw material consumption remain contestable as long as no solid quantifications are presented.

Minimizing or eliminating negative environmental impact is, unlike both previous factors, an intrinsic property that characterizes the chemical effect of ultrasonic waves. Prima facie, sonochemistry is per definition a chemistry easily evolving in water, but also naturally occurring under ambient conditions. This is synonym of a process able to efficiently integrate energy in a chemical scheme in such a way that solvent is not necessary (Cintas and Luche 1999) and saving energy classically required to produce harsh operating conditions (high temperatures and pressures).

Moreover, improvement of rates and sonochemical selectivity is a manner for preventing the production of excessive waste and optimizing the exploitation of raw materials.

It is now clear that the three pillars of benign-by-design process are closely dependent in the case of sonochemistry. The question asked in this section should be treated in such a way to lead to the best compromise between these three pillars and suggest metrics to evaluate this compromise. Thus far, qualitative arguments were presented to qualify that sonochemical processes as "benign-by-design," quantitative justifications remain, however, essential to outstrip the theoretical vision of sonochemistry as a green process and move toward implementation of sonochemistry technologies as benign-by-design processes.

15.2.2 Prevention or Remediation?

According to the EPA definition, green chemistry is a chemistry that designs chemical products and processes that are harmless to the environment. Green chemistry should then be thought from a perspective of prevention rather than reparation.

Sustainability, on the other hand, is a broad large sense that gathers concepts and actions aiming to preserve resources for future generations. It can be represented with a three-legged stool having a leg for environmental, social, and financial responsibility (Kummerer and Clark 2016). Remediation, though not involved in the green chemistry vision, remains a branch of chemistry acting for sustainability (Holland 2011). For sure, environmental prevention is an intrinsic property of sonochemical processes, and among their multiple applications (Suslick et al. 1999), environmental remediation constitutes a very interesting one (Sillanpää 2011). It manifests particularly through sonochemical advanced oxidation (Babu et al. 2016). The objective of this section is to emphasize on the fact that remediation itself is not one of the green chemistry principles, and hence, sonochemistry is invoked as a green process owing to its preventing character and not for its environmental remediation capacities.

Environmental remediation is one of the branches of environmental chemistry that benefits of a large, wide techniques having a common goal: reduce installed pollution. Advanced oxidation processes rely on the in situ production of highly reactive oxidants such as $HO^{\bullet}$, a strong and nonselective oxidant capable of degrading a wide range of pollutants in water, principally organic, into small nontoxic molecules or even completely mineralized ones (Babu et al. 2016). Advanced oxidation processes count a broad number of treatment techniques such as photolysis and photocatalysis but also sonolysis (Babu et al. 2016). Sonolysis is an advanced oxidation process that is based on the mechanical effect of cavitational collapse in conjunction with the chemical effect of ultrasound. In 1929, Schmitt et al. (1929) found that cavitation induces oxidation when iodine was liberated during sonication of aqueous potassium iodide. At this stage of knowledge, the oxidation was thought to arise from the formation of hydrogen peroxide. In 1956, Fitzgerald and Sullivan conducted a comprehensive study aiming to figure out how reactions take place in an ultrasonic field, and they concluded that primary reactions occur in gas phase within the cavitation bubble. It was in more recent years by 1994 that the oxidation under ultrasound irradiation was found to be routed in the generation of the highly oxidizing $HO^{\bullet}$, through a research work of Berlan et al. (1994) where the oxidative degradation of phenol in water at 514 kHz was investigated. This report was significant in that it demonstrated not only the complete destruction of phenol in water but also the production of intermediate hydroxylated benzenes such as catechol, which themselves were further oxidized. Since then, successive research showed that under sonication, reactions are expected to occur within the core of cavitation, at the bubble wall interface, and in the bulk of the liquid. This applies as well to the sonochemical advanced oxidation evolving in water. In this case, it is generally accepted to schematize the water molecules cleavage by the following reactions chain:

$$H_2O+))) \rightarrow HO^{\bullet} + H^{\bullet}$$

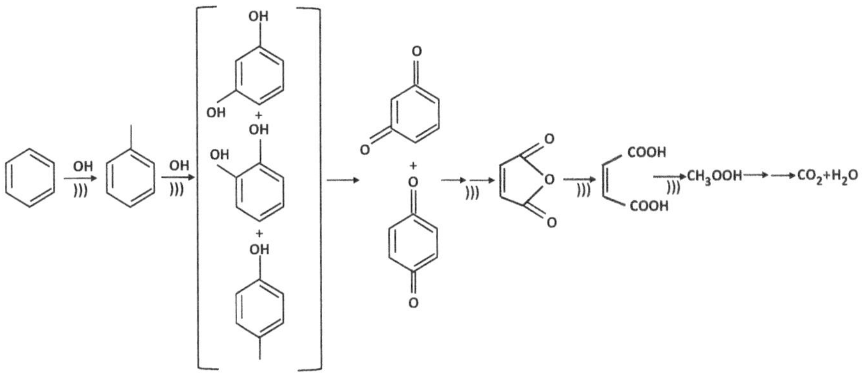

Fig. 15.3 Mechanism of sonochemical degradation of benzene

$$HO^{\bullet} + H^{\bullet} \rightarrow H_2O$$

$$HO^{\bullet} + HO^{\bullet} \rightarrow H_2O_2$$

$$H^{\bullet} + H^{\bullet} \rightarrow H_2$$

In the presence of saturating gases, further free radicals are expected to emerge, mechanisms vary according to the nature and proportions of chemical species in the gas phase, and we exhibit, for illustration, the possible reaction in the case of saturation with pure oxygen.

$$O_2 +))) \rightarrow O + O$$

$$O + H_2O \rightarrow HO^{\bullet} + HO^{\bullet}$$

$$O_2 + H^{\bullet} \rightarrow HO_2^{\bullet}$$

$$HO_2^{\bullet} + HO_2^{\bullet} \rightarrow H_2O_2 + O_2$$

The highly reactive oxidants that emerge from the cleavage of water and saturating gases are capable of cleaving many chemical bonds, especially carbon–carbon, carbon–sulfur, and carbon–nitrogen. For instance, we expose hereafter in Fig. 15.3 the reactions believed to describe the sonochemical degradation of benzene (Pang et al. 2011). In this example, the reaction of $HO^{\bullet}$ with benzene initially produces phenol. Then, electrophilic reaction, which is generally favored at high-electron density position, will cause the $HO^{\bullet}$ to attack on meta-, ortho-, and/or para-positions on the phenol molecule. Phenol will rapidly turn into hydroquinone, resorcinol, and catechol. Then, these compounds will further react with $HO^{\bullet}$ to form *o*- or *p*-benzoquinone. Finally, benzoquinone will be oxidized into fumaric acid, maleic

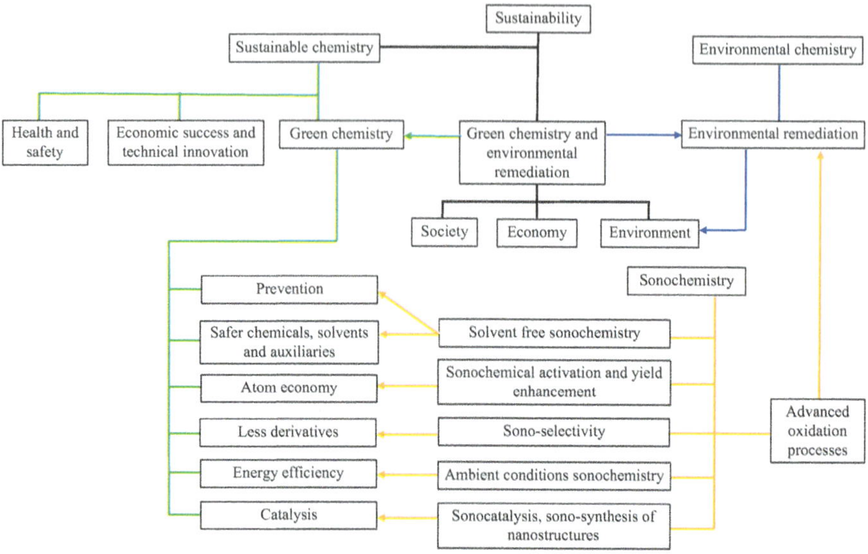

Fig. 15.4 Schematic diagram for the place of sonochemistry, a green process applied in environmental remediation and contributing in the broad sense of sustainability

acid, oxalic acid, and formic acid (through ring opening) which is lastly completely mineralized into CO_2 and H_2O.

Nevertheless, ultrasound can also be associated with other advanced oxidation processes to form hybrid techniques such as sonocatalysis, sonophotolysis, and sonophotocatalysis. Interestingly, these latter showed synergetic effects in pollutants degradation efficiency as compared to isolated techniques, which raised an active interest in the use of sonochemical methods in the presence of UV, catalysts, and chemical additives for treatment of organic pollutants in wastewater (Pang et al. 2011). Consequently, most of emergent organic pollutants can be degraded under sonication or sono-assisted techniques: Pharmaceuticals (Guo et al. 2010), textile industry wastewater (Ferkous et al. 2015), food processing additives (Madhavan et al. 2010a), and more have already been investigated for oxidation under ultrasound (Pétrier et al. 2007).

Sonochemistry is then a green process used for environmental remediation (Adewuyi 2005; Sanghi et al. 2012) through advanced oxidation applications, and this combination leads to a particular classification of a branch of sonochemical processes within "the green processes for environmental remediation." Figure 15.4 exhibits the interconnections between sonochemistry and green processing, on one hand, and sustainability, on the other hand.

15.3 Modeling the Sonochemical Activity: Facts and Challenges to Highlight the Greenness of the Process

15.3.1 The Single Bubble: Inception and Equilibrium

For a long time, bubble inception has been an intriguing problem. Actually, bubble nucleation occurs through two different processes: homogeneous and heterogeneous nucleation. Homogeneous nucleation is the result of apparition of local zones where liquid cohesion is relatively less important as compared to the rest of the liquid volume. Meanwhile, the limit of the cohesion within the liquid is conditioned by the augmentation of the intermolecular distance δ to a sufficient value so as the intermolecular bonds are broken up. The classical nucleation theory brought answers regarding this limit. Nevertheless, we can propose a simplified approach to characterize it. Let's consider an external pressure force F applied on the liquid all along the distance δ to create two distinguished interfaces of surface S. The energy engaged to do so is obviously the interface tension energy E, and hence, the minimum pressure to exert $-P_{\min}$ should verify:

$$-P_{\min}\delta S = 2\sigma S \tag{15.7}$$

Thus, the conclusion is quickly established, and the passage of an ultrasonic wave through a pure liquid can hardly and even impossibly ends up to the creation of cavities. The acoustic cavitation falls then into the second category, i.e., heterogeneous cavitation. This latter occurs provided that dissolved substances, gaseous germs, or impurities are present within the liquid volume. In practice, this is always the case. The formation of cavitation under ultrasound is explained by the existence of dissolved gases. In fact, these latter may disturb the hydrogen bonds and consequently lower the liquid cohesion or in other words reduce the surface tension. However, even at saturation, this modification remains very limited and does not explain the facility of bubble inception in liquids saturated with gases under ultrasound. The dissolved gases form in the liquid gaseous germs, and as soon as the liquid undergoes a depression, the germs tend to expand which brings to the system a metastability provoked by the gases, and not only related to liquid as in the homogeneous nucleation. In order to mathematically describe the phenomenon, we suggest a coarse interpretation of what occurs.

Even in the presence of important depression, until a certain threshold, bubble expansion may be kept very limited due to surface tension effect. In such case, cavitation bubbles tend to oscillate linearly at the same pace as the exciting periodic field. If the exciting field is represented by Eq. 15.8 below,

$$p(t) = -P_{A} \sin(\omega t) \tag{15.8}$$

the linear oscillation is simply modeled by the following formula (Brennen 1995):

$$R(t) = R_0 + \xi \sin(\omega t - \pi) \tag{15.9}$$

Beyond this threshold, surface tension is no longer capable to brake the bubble expansion, and hence, bubbles undergo an explosive phase, during the rarefaction of the wave, followed by an implosive one, during the compression of the wave. The bubbles falling into this category are called inertial cavitation, and the actual limit separating both types of cavitation is known as Blake threshold. Surpassingly, it is the key for cavitation expected to result in physical and chemical effects. To define the Blake threshold in the plane (p, R_0), one can imagine a quasi-static isothermal expansion of a bubble as the pressure is gradually reduced within the liquid from P_∞ to $P_\infty + p$. The bubble radius which is in radial equilibrium should verify (Apfel 1981):

$$\left(P_\infty + \frac{2\sigma}{R_0}\right)\left(\frac{R_0}{R}\right)^3 - \frac{2\sigma}{R} = P_\infty + p \tag{15.10}$$

The term in the left of the above equation passes by a minimum in function of the radius value. Consequently, there exists a pressure value P_{BL}, in reference to Blake pressure, over which the equilibrium is no longer verified. In case the depression p exceeds P_{BL}, a gradient of centripetal pressure appears in the liquid, which explains the passage to inertial cavitation. The Blake threshold is then characterized by the Blake radius and pressure related by the following equations:

$$\frac{P_{BL}}{P_\infty} = \frac{1}{P_\infty} + \sqrt{\frac{4X^3}{27(1+X)}} \tag{15.11}$$

where X represents the dimensionless Laplace tension given by:

$$X = \frac{2\sigma}{P_\infty R_0} \tag{15.12}$$

One needs to keep in mind that Blake threshold has been derived based on quasi-static hypothesis that remains valid under very low-frequency excitation. Dynamics conditions require further considerations and lead to another frontier separating stable oscillation from transient one.

On the other hand, resonance radius is defined as the ambient radius of a bubble oscillating under an exciting frequency which coincides with its natural frequency. The exciting frequency is called in this case resonance frequency. Having said that, it is essential to figure out the notion of natural frequency also known as eigenfrequency, which is the frequency at which a cavitation bubble tends to oscillate in the absence of any driving or damping force. Its expression was determined by several researchers (Minnaert 1933; Briggs et al. 1947). Consequently, the general formula of resonance frequency of a gas bubble is given in function of the dimensionless Laplace tension as:

$$f_{\mathrm{R}} = \frac{1}{2\pi R_0}\sqrt{\frac{3\gamma P_\infty}{\rho}\left(1 + X\left(1 - \frac{1}{3\gamma}\right)\right)} \tag{15.13}$$

In case X is proved to be neglectable, the above forumla is simplified to:

$$f_{\mathrm{R}} = \frac{1}{2\pi R_0}\sqrt{\frac{3\gamma P_\infty}{\rho}} \tag{15.14}$$

For small bubbles, the natural pulsation is deemed to be isothermal, which leads up to a resonance frequency of:

$$f_{\mathrm{R}} = \frac{1}{2\pi R_0}\sqrt{\frac{3\gamma P_\infty}{\rho} + \frac{2\sigma}{\rho R_0}} \tag{15.15}$$

Bigger bubbles are believed to oscillate adiabatically, with the possibility to neglect the surface tension effect. Hence, the resonance frequency is simply expressed as:

$$f_{\mathrm{R}} = \frac{1}{2\pi R_0}\sqrt{\frac{3C_{\mathrm{p}} P_\infty}{C_{\mathrm{v}}\rho}} \tag{15.16}$$

As a result, resonance radius is the ambient radius of a bubble oscillating under an exciting field at its natural frequency. The resonance radius is deduced from the resonance frequency expression with neglecting the effects of surface tension and dissipative phenomena, and it is then given by:

$$R_{\mathrm{R}} = \sqrt{\frac{3\gamma P_\infty}{\rho\omega^2}} \tag{15.17}$$

Usually and for simplification purposes, it is accepted to consider the isothermal resonance radius in all the cases.

If the exiting frequency has a value that exceeds the resonance one, the ambient radius is situated above the resonance size (and vice versa). In this case, and according to the theory of radiation force on bubbles, these latter will be pushed in the open space away from the regions of high acoustic pressure. In other words, they will not survive in the sound field for long. Therefore, this category of cavitation bubbles is not concerned by physical and chemical effects.

When modeling sonochemistry, bubble population is then characterized by a range of ambient sizes comprised between Blake and resonance radii, with a certain distribution and a most probable value within incepted bubbles. In most of numerical studies, the most probable equilibrium radius is adopted as representative of the population, but one must already know what ambient radius is most likely to occur in a population!

When mathematically modeling any phenomenon related to acoustic cavitation bubble, the ambient radius is for sure the starting point. However, we can easily notice that in most of studies, the selection of R_0 value is somehow arbitrary or not deeply investigated and justified. This fact explains in part the disparity in results even when numerical studies are conducted under similar conditions but bring us back into a primordial question: How can modeling predict reliably the sonochemicalactivity and its efficiency if the choice of R_0 may change the whole findings?

15.3.2 Dynamics of Bubble Wall Oscillation: From Stability to Chaos

As an acoustic wave travels a liquid medium, bubbles of ambient radii falling between Blake and resonance thresholds are expected to oscillate nonlinearly as a response to the exciting field. Meanwhile, distinguished rhythms of oscillation are observed. Some bubbles are keeping their shape stable and have accordingly a long lifetime (Church and Carstensen 2001), while some others are rather unstable, namely, transient, and tend, after few cycles, to lose their shape and eventually fragment into daughter bubbles (Zhang et al. 2015), and lastly, extreme instability conducts to chaotic oscillation (Zhang and Zhang 2018). We illustrate in Fig. 15.5 the passage from stable oscillation to chaotic one in the case of an acoustic cavitation bubble evolving in water under an atmosphere of oxygen. The exciting wave is admitted to have a frequency of 300 kHz and an acoustic amplitude of 1.5 atm, while the ambient radius is supposed to vary from 1 to 5 μm, respectively.

On this particular point, we insist on underlining a fact that transient qualification of acoustic cavitation is not always used to refer to shape or dynamics instability; it is as much used to describe acoustic bubbles capable of generating a physical or chemical effect, more concisely active in sonoluminescence or sonochemistry. Such use is somehow confusing knowing that acoustic cavitation bubbles which are both stable and active do exist, and they are called high-energy stable cavitation (Yasui 2011). For this reason, we adopt in this work the most appropriate definition of stable–transient–chaotic bubbles, which is based on either bubble shape is stable or not. The frontier separating stable bubbles from unstable ones is usually called stability or transient threshold. Before dealing with the construction of this threshold, we need to remind briefly the major equations used to describe the bubble oscillation dynamics.

Since the 50s of the last century, researches began to model the observations of the growth and collapse of traveling cavitation bubbles using modifications of Rayleigh's original equation of motion for a spherical bubble. The one developed by Plesset, well known as Rayleigh–Plesset equation (Franc 2007), is applicable in an incompressible medium and given by Eq. 15.18:

$$R\ddot{R}+\frac{3}{2}\dot{R}^2=\frac{1}{\rho}\left(P_B-\frac{2\sigma}{R}-4\mu\frac{\dot{R}}{R}-P_\infty+P_A\sin(\omega t)\right) \tag{15.18}$$

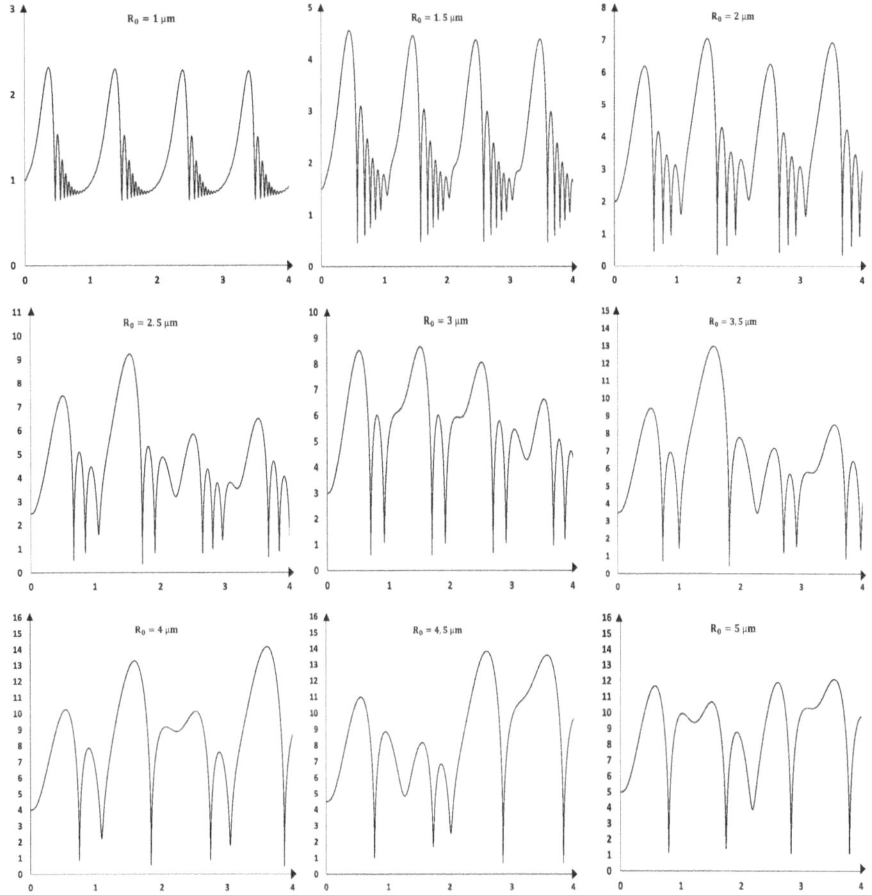

Fig. 15.5 Simulated bubble radius (vertical axis in μm) over four acoustic cycles (horizontal axis) for acoustic bubbles of ambient radii varying in the range of 1–5 μm under an exciting wave of 300 kHz and 1.5 atm. The bubbles are supposed to oscillate in water under oxygen atmosphere

The dynamics of a single spherical gas bubble in an infinite body of compressible viscoelastic material was considered later based on the Keller–Miksis equation (Keller and Miksis 1980) represented below:

$$\left(1-\frac{\dot{R}}{c}\right)R\ddot{R}+\frac{3}{2}\left(1-\frac{1}{3}\frac{\dot{R}}{c}\right)\dot{R}^2=\frac{1}{\rho}\left(1-\frac{\dot{R}}{c}\right)$$
$$\left(P_{\mathrm{B}}-\frac{2\sigma}{R}-4\mu\frac{\dot{R}}{R}-P_{\infty}+\mathrm{P}_{\mathrm{A}}\sin\left(\omega\left(t+\frac{R}{c}\right)\right)\right)+\frac{R}{\rho c}\frac{\mathrm{d}}{\mathrm{d}t}\left(P_{\mathrm{B}}-\frac{2\sigma}{R}-4\mu\frac{\dot{R}}{R}\right) \tag{15.19}$$

In the incompressible limit, the sound speed c tends to infinity, and Eq. 15.19 is reduced to Rayleigh–Plesset equation. The Keller–Miksis equation provides a mechanism for acoustic radiation important in large-amplitude oscillations, which fits with inertial cavitation case. However, it is still based on the idea that the near field is incompressible. Many other versions of both fundamental equations were derived as modified Rayleigh–Plesset or modified Keller–Miksis equations to take into account various aspects such as the nonequilibrium of evaporation and condensation at bubble interface (Yasui 1997).

The passage from stable to transient events, delimited by the stability or transient threshold, is actually assumed to occur if a bubble is susceptible to reach more than twice its original size after few cycles. Noltingk and Neppiras (1950) showed that under incompressible liquid hypothesis, a gas bubble driven by a collapse velocity neighboring the speed of sound, or in other words oscillating at the transient limit, is supposed to grow to a maximum radius of about 2.3 times R_0. This criterion has been adopted by Apfel (1981) in an attempt to explicitly define the stability or transient threshold. We present hereafter in brief the method followed by this last author. According to Apfel's model, the bubble subjected to an ultrasonic irradiation is reasonably expected to expand till the maximum radius R_{max} in two stages. The first stage concerns the expansion while the pressure within the liquid is still positive, and this phase ends up at an intermediate radius R_i given in Eq. 15.20.

$$R_i = \frac{4}{3\omega}\left(\frac{P_A}{P_\infty} - 1\right)\sqrt{\frac{2P_\infty^{\ 2}}{\rho P_A}} \tag{15.20}$$

The second stage describes the expansion of the bubble under a negative pressure in the liquid, and this stage leads to the maximum radius R_{max}, which is simply retrieved through the variation in kinetical energy. R_{max} is finally given by Eq. 15.21.

$$R_{max} = \frac{4}{3\omega}\left(\frac{P_A}{P_\infty} - 1\right)\left(\frac{2P_\infty^{\ 2}}{\rho P_A}\right)^{\frac{1}{2}}\left(1 + \frac{2}{3}\left(\frac{P_A}{P_\infty} - 1\right)\right)^{\frac{1}{3}} \tag{15.21}$$

In case $P_A \gg P_\infty$, the expression of R_{max} is reduced to:

$$R_{max} = \frac{4}{3\omega}\left(\frac{P_A}{P_\infty}\right)^{\frac{5}{6}}\left(\frac{2P_\infty}{\rho}\right)^{\frac{1}{2}}\left(\frac{2}{3}\right)^{\frac{1}{3}} \tag{15.22}$$

Combining both Eqs. 15.21 and 15.22 with the criterion for transient oscillation, the stability or transient threshold would be defined in the plane (P_A, R_0) by the following equations:

$$R_T = \frac{0.82}{\omega}\left(\frac{P_A}{P_\infty} - 1\right)\left(\frac{P_\infty^2}{\rho P_A}\right)^{\frac{1}{2}}\left(1 + \frac{2}{3}\left(\frac{P_A}{P_\infty} - 1\right)\right)^{\frac{1}{3}} \tag{15.23}$$

If $P_A \gg P_\infty$, Eq. 15.23 is reduced to Eq. 15.24.

$$R_T = \frac{0.72}{\omega}\left(\frac{P_A}{P_\infty}\right)^{\frac{5}{6}}\left(\frac{P_\infty}{\rho}\right)^{\frac{1}{2}} \tag{15.24}$$

Once the ambient radius exceeds the transient threshold, instabilities in bubble oscillations dynamics are observed. In fact, the bubble moves gradually from stable to chaotic oscillation as the ambient radius increases. One powerful tool to identify the nature of oscillation is to dress the projection of the trajectory of bubble wall in state space (Zhang and Zhang 2018). This representation consists simply in reporting the bubble velocity $\dot{R}$ in function of the nondimensional bubble radius $\frac{R}{R_0}$. The chaotic oscillation is recognized to the random pattern of bubble wall motion, due to the apparition of subharmonics and ultraharmonics in the response of the bubble to the exciting field. We represent in Fig. 15.6 the projection of bubble wall trajectory in state space corresponding to the bubble dynamics reported previously in Fig. 15.5.

It is worth to note that the previous stability threshold was developed under assumption of mono-frequency excitation. In case of dual or multifrequency exciting fields, the approach is quite different. Zhang et al. (2015) suggested a threshold derivation starting from a hypothetical distortion from spherical shape with harmonics order n as represented in Eq. 15.25.

$$R_s = R + \sum_{n=1}^{\infty} a_n Y_n \tag{15.25}$$

The derivation leads to a stability threshold defined by Eq. 15.26.

$$R_T = \frac{(n-1)(n+1)(n+2)\sigma}{\left(n+\frac{1}{2}\right)\sqrt{\frac{\omega_1^4 P_{A1}^2}{\left|\omega_0^2-\omega_1^2\right|^2}+\frac{\omega_2^4 P_{A2}^2}{\left|\omega_0^2-\omega_2^2\right|^2}}} \tag{15.26}$$

Once the ambient radius exceeds the aforementioned limit, the interface of a bubble oscillating under a dual-frequency field is expected to become instable.

To conclude on this section, according to the character of bubble dynamics in regard to the shape stability, different energetic and sonochemical behaviors are expected. This means that when modeling sonochemical activity, the ambient radius admitted to be representative of the population under the exciting wave would decide of the type of oscillation and consequently orientate the whole acoustic bubble activity. In addition, the nonperiodicity of events constitutes at this stage a real obstacle for extrapolation of numerical simulations results, which are usually limited to the microseconds scale.

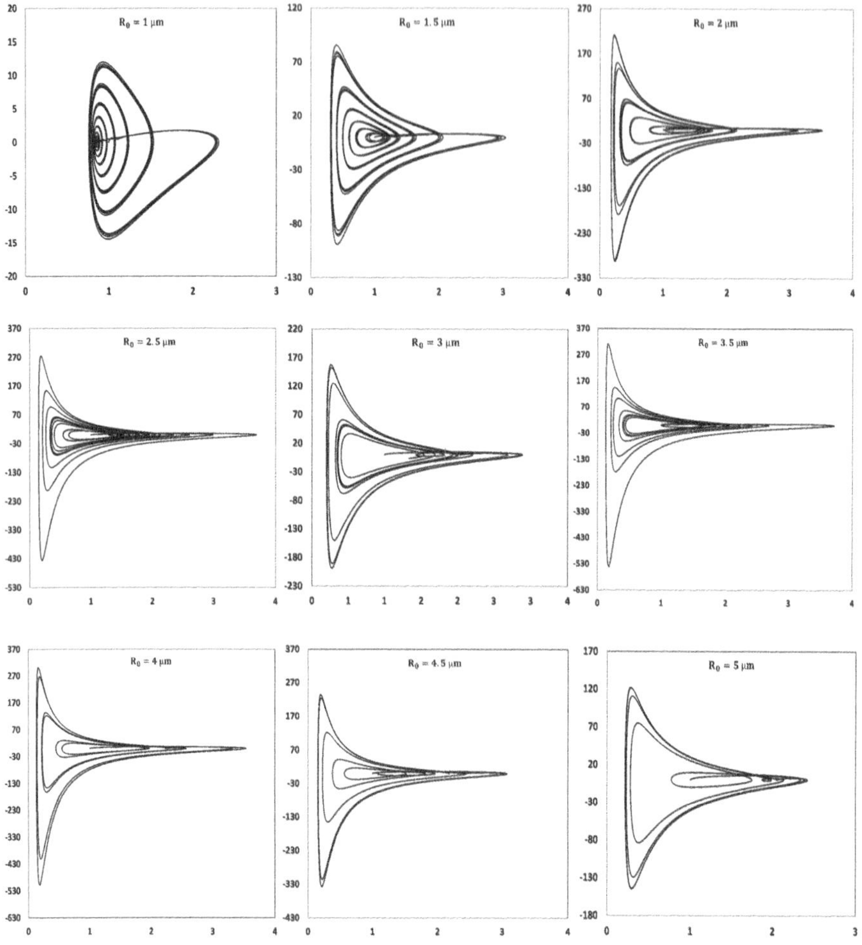

Fig. 15.6 Bubble wall velocity (vertical axis in m/s) in function of reduced radius $\frac{R}{R_0}$ (horizontal axis) for acoustic bubbles of ambient radii varying in the range of 1–5 μm under an exciting wave of 300 kHz and 1.5 atm

15.3.3 Mass and Energy Balances: Assumptions and Findings

The oscillation of bubble wall is necessarily accompanied of mass and energy transformations. The three dimensions of the acoustic cavitation bubble activity (dynamics, energy, and mass evolutions) are in fact closely dependent. In the present section, we will briefly discuss the major characters of energy and mass phenomena related to the acoustic cavitation bubbles starting from its dynamics.

Basically, the bubble oscillation was supposed to be fairly described by an isothermal expansion and an adiabatic compression. Both assumptions were intensively adopted by several authors and justified by the lifetime of bubble oscillation at high frequency which is relatively short with a very rapidly occurring collapse event. This simplified the description of the thermal behavior of the acoustic cavitation bubble to Eqs. 15.27 and 15.28 during expansion and compression, respectively.

$$T = T_{\infty} \tag{15.27}$$

$$T = T_{\infty}\left(\frac{R_{\mathrm{max}}}{R}\right)^{3(\gamma-1)} \tag{15.28}$$

Consequently, it was well accepted to represent the pressure inside the bubble P_B by both Eqs. 15.29 and 15.30 during expansion and compression, respectively, using the ideal gas state equation (Merouani and Hamdaoui 2016).

$$P_{\mathrm{B}} = \frac{R_{\mathrm{g}} T_{\infty} \sum n_i}{\frac{4}{3}\pi R^3} \tag{15.29}$$

$$P_{\mathrm{B}} = \left(P_{\mathrm{v}} + P_{\mathrm{g0}}\left(\frac{R_0}{R_{\mathrm{max}}}\right)^3\right)\left(\frac{R_{\mathrm{max}}}{R}\right)^{3\gamma} \tag{15.30}$$

Nevertheless, previous hypotheses ignore with no doubt some obvious aspects of acoustic cavitation bubble activity. For example, adiabatic compression can be prima facie contested for neglecting the heat exchange due to the nonequilibrium of evaporation and condensation at the bubble interface. The use of ideal gas state equation is as well inappropriate, taking into account the high pressure attained at the strong collapse. It is worth to mention that some other authors suggest a polytropic compression instead of adiabatic (González-García et al. 2010). Actually, this does not change significantly the approach, as the polytropic constant is simplydeemed to be comprised between 1 and $\frac{C_{\mathrm{p}}}{C_{\mathrm{v}}}$. Some others used a real state equation,namely, Van der Waals equation (Kerboua and Hamdaoui 2017, 2018a), but still consider the single bubble as micro reactor closed to heat and mass exchange.

A more accurate view of this question consists in establishing an energy balance considering the various inputs, outputs, creation, and accumulation during the oscillation of the bubble. Here again, numerous works with different assumptions can be found in the literature. We cite, for instance, the energy balance proposed by Yasui (1997) and summarized in Eq. 15.31.

$$\Delta E(t) = -P_{\mathrm{B}}\Delta V(t) + 4\pi R^2 \dot{m}_{\mathrm{H_2O}} c_{\mathrm{v_{H_2O}}} T \Delta t + 4\pi R^2 \lambda \frac{\mathrm{d}T}{\mathrm{d}r}(R)\Delta t + \frac{4}{3}\pi R^3 \left(\sum r_i \Delta H_{\mathrm{i}}\right)\Delta t - \frac{5}{3} m_{\mathrm{H_2O}} \dot{R}\ddot{R}\Delta t \tag{15.31}$$

In this equation, the variation of the thermal energy of a single acoustic cavitation bubble is supposed to come from the respective variations of the work of pressure, the energy carried by evaporating or condensing vapor molecules, the heat of thermal conduction, the heat of chemical reactions, and finally the kinetical energy of gases in the order appearing in the right hand of Eq. 15.31.

Besides, Sivankar and Moholkar (2009) suggested a quite different energy balance based on the physical phenomena within an acoustic cavitation bubble, or more specifically the molecular diffusion at the interface. Hence, this balance neglects the nonequilibrium due to evaporation and condensation. Moreover, it does not consider the heat of reactions. The resulting formula is presented by Eq. 15.32.

$$c_{\mathrm{v}}\sum n_i\frac{\mathrm{d}T}{\mathrm{d}t}=4\pi R^2\lambda\frac{\mathrm{d}T}{\mathrm{d}r}(R)-P_{\mathrm{B}}\frac{\mathrm{d}V}{\mathrm{d}t}+\left(4R_{\mathrm{g}}T_{\infty}-n_{\mathrm{H_2O}}R_{\mathrm{g}}T\left(3+\sum\frac{\frac{\theta_i}{T}}{\mathrm{e}^{\frac{\theta_i}{T}}-1}\right)\right)\frac{\mathrm{d}n_{\mathrm{H_2O}}}{\mathrm{d}t}\tag{15.32}$$

Kerboua and Hamdaoui (2018c), on the other hand, accounted for physical and chemical events occurring at the single bubble level. They assumed that, within the bubble volume, the gas content is static as the ratio ρ_G/ρ_L is very low. The obtained balance is reported in Eq. 15.33.

$$\frac{1}{\sum n_i c_{\mathrm{V}}}\left(-P_{\mathrm{B}}\mathrm{d}V(t)-\sum\Delta H_i r_i V(t)\mathrm{d}t+\sum S(t)\dot{m}_i c_{\mathrm{v}_i}T\right)=\frac{\lambda}{\xi\sum n_i c_{\mathrm{V}}}S(t)(T_{\mathrm{B}}-T_{\infty})\mathrm{d}t+\mathrm{d}T\tag{15.33}$$

Obviously, the mass balance framing each approach is based on the same hypotheses as those governing the energy balance. The variation in mass may emanate from physical phenomena, molecular diffusion and nonequilibrium of phase transformation at bubble interface, as well as chemical activity evolving around the strong collapse. The contribution of molecular diffusion is actually controverted, and according to Yasui (2002, 2011), relative importance of coalescence and rectified diffusion depends on the stage of bubble growth. After acoustic cavitation is fully started, coalescence of bubbles may be the main mechanism of the bubble growth. Contrariwise, at the initial development of acoustic cavitation, rectified diffusion may be the main mechanism as the rate of coalescence is proportional to the square of the number density of bubbles, which is rather small at the initial stage of acoustic cavitation. Consequently, as the time reference is usually taken as the moment when the bubble is fully initiated, the diffusion is neglected in the balances of several models (Yasui 1997; Kerboua and Hamdaoui 2018c). Anyway, the rectified diffusion of gas in the liquid is modeled according to Eller–Flynn theory (Eller and Flynn 1965), and its rate is expressed in Eq. 15.34.

$$\dot{n}_{\text{diff}} = 4\pi R^2 D \frac{A}{B\left(\frac{R_0}{R}\right)^2 R_0}\left(c_i - c_\infty\right) \tag{15.34}$$

On the other hand, the nonequilibrium of evaporation and condensation is simply modeled using the Hertz–Knudsen equation (Kerboua and Hamdaoui 2018b). This equation assumes that among the molecules intercepting the bubble interface, a proportion equal to the accommodation coefficient is expected to adhere to the surface. A molecular flow streamed from or toward the bubble volume is then observed, and it is driven by the pressure difference at the bubble wall and generally taken as positive in the sense of evaporation. Equation 15.35 reports the Hertz–Knudsen law.

$$\dot{m} = \frac{M}{\sqrt{2\pi M R_g}}\alpha \frac{1}{\sqrt{T}}\left(P_v - P_i\right) \tag{15.35}$$

Finally, the chemical kinetics inside an acoustic cavitation bubble is closely linked to the initial composition of the bubble, i.e., the content of both liquid and gas atmosphere, and the selected schema deemed to represent in an acceptable degree the chemical reaction expected to occur inside the bubble under harsh conditions of temperature and pressure. Let's consider a general case where each possible reaction is schematized as shown in Eq. 15.36.

$$\sum_{k=1}^{K} v'_{ki} X_k \leftrightarrow \sum_{k=1}^{K} v''_{ki} X_k \tag{15.36}$$

Accordingly, each species involved in the chemical scheme, i.e., reactants and products, is characterized by a molar rate r_k according to Eq. 15.37.

$$r_k = \sum_{i=1}^{I}\left(v''_{ki} - v'_{ki}\right)\left(k_{fi}\prod_{k=1}^{K}\left[X_k\right]^{v'_{ki}} - k_{ri}\prod_{k=1}^{K}\left[X_k\right]^{v''_{ki}}\right) \tag{15.37}$$

The most challenging point in regard to the kinetics modeling within an acoustic cavitation bubble is not actually the formulation of equations but rather the reading of results. Once the productions at a single acoustic cavitation bubble are obtained through modeling, the crucial question is, How significant are the simulated amounts? Here, we come into the sonochemical activity threshold assumed by Yasui et al. (2008) to be equal to 10^8 molecules or free radicals per second, or in a molar basis, 1.66×10^{-16} mol s^{-1}. A priori, this threshold, though being conventional, can be considered as an absolute criterion for the sonochemical efficiency at single acoustic cavitation bubble. It is noteworthy to remind that most of efficiency criterion seen in Sect. 15.2.1 are defined at the macroscopic scale, which is generally not the scale of sonochemistry modeling.

15.3.4 Bubble Interactions and Population Dimension

In most of numerical studies dealing with sonochemistry, sonochemical activity was assessed at the single acoustic cavitation scale (Gong and Hart 1998; Kanthale et al. 2008; Merouani and Hamdaoui 2016; Kerboua and Hamdaoui 2018a). This approach is for sure an interesting one to a certain extent. It actually provides numerous explanations of macroscopic observation and proposes several possible interpretations on what happened and how inside the cavitation bubble. However, the activity of single acoustic cavitation bubble is only one of the dimensions of macroscopic sonochemistry. All the other dimensions are related to the population of bubbles including many aspects, namely, bubble–bubble interaction, bubble–wave interaction, number density of bubbles, and bubble size and space distribution.

The interactions within bubble population originate from the acoustic radiation forces induced by the interception of the ultrasonic wave with the incepted bubbles (Bjerknes 1906).

Basically, the simplest form of an acoustic radiation force applied by a pressure $p(x, y, z, t)$ on an elementary volume V can be formulated as shown in Eq. 15.38.

$$\vec{F}\left(x,y,z,t\right) = -V\frac{\partial p}{\partial x}, V\frac{\partial p}{\partial y}, V\frac{\partial p}{\partial z} \tag{15.38}$$

According to the source of the pressure p, we can distinguish two cases (Mettin et al. 1997). When the instantaneous local pressure at the position (x, y, z) is the driving ultrasonic wave, the force is called primary Bjerknes force and is expressed as indicated in Eq. 15.39.

$$\vec{F}\left(x,y,z,t\right) = P_{\mathrm{A}}\vec{k}\sin\left(k_x x + k_y y + k_z z\right)V\sin\left(\omega t\right) \tag{15.39}$$

As a result, bubbles having an ambient radius lower than the resonance size and oscillating nonlinearly under a standing ultrasonic wave will be attracted to the pressure antinodes as shown in Fig. 15.7. In a more overall view, under a significant acoustic amplitude, the primary Bjerknes force would push most of the bubbles toward the direction of propagation of the wave, and in their motion, fluid is streamed in the same direction resulting in a phenomenon called acoustic streaming (Leighton et al. 1990).

On the other hand, the radiation force created between two neighbor bubbles is known as the secondary Bjerknes force. It is actually induced by the acoustic pressure radiated by an oscillating bubble and expressed by Eq. 15.40.

$$p = \frac{\rho}{4\pi d^2}\ddot{V} \tag{15.40}$$

We then represent in Eqs. 15.41 and 15.42 the mutual forces resulting between a couple of bubbles 1 and 2, assuming that $\vec{u}_{\mathrm{d}}$ is the unitary vector relating both centers of bubbles, and directed from bubble 1 to bubble 2.

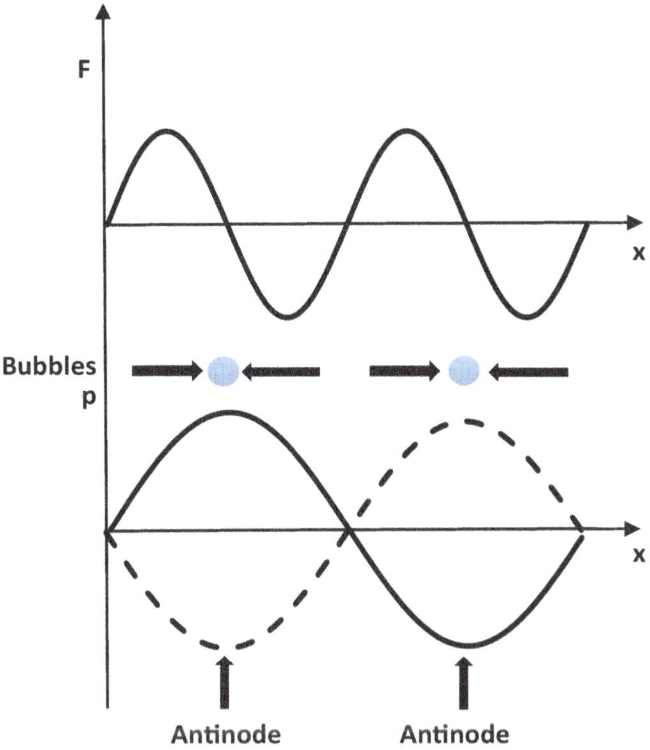

Fig. 15.7 Motion of the bubbles smaller than the resonant size toward pressure antinodes (Adapted from Son 2016)

$$\vec{F}_{1/2} = \frac{\rho}{4\pi d^2} \ddot{V}_1 V_2 \vec{u}_d \tag{15.41}$$

$$\vec{F}_{2/1} = -\frac{\rho}{4\pi d^2} \ddot{V}_2 V_1 \vec{u}_d \tag{15.42}$$

According to the values of ambient radii of both bubbles in the couple, the secondary Bjerknes force may result in either attraction (Leighton 1997) or repulsion (Mettin et al. 1997).

Clearly, the situation becomes more complex when dealing with a bubble population instead of a couple of bubbles as discussed before. If we simply deem that a given bubble of a volume V undergoes the acoustic pressure radiated by a bubble cloud composed of N bubbles, where each bubble i has a volume V_i and is situated at a distance d_i, the resulting radiated pressure can be represented by Eq. 15.43.

$$p = \sum_{i=1}^{N} \frac{\rho}{4\pi d_i^2} \ddot{V}_i \tag{15.43}$$

And consequently, the radiated force $\vec{F}_c$ of the cloud on the given bubble can be expressed as the following:

$$\vec{F}_c = \sum_{i=1}^{N} \frac{\rho}{4\pi d_i^2} \ddot{V}_i V \vec{u}_{d_i} \tag{15.44}$$

where $\vec{u}_{d_i}$ is the unitary vector relating a radiating bubble in the cloud to the radiated bubble and oriented from a bubble i to the radiated bubble.

The only consideration of this first dimension of acoustic bubble population changes systematically the derivation of bubble dynamics equation. Indeed, the inclusion of bubble radiation forces in the derivation of Keller–Miksis model results in a modified equation leading automatically to different results. This was demonstrated by Mettin et al. (1997) and Yasui et al. (2010). Besides, one should keep in mind that the primary Bjerknes force aforementioned was formulated based on a mono-frequency excitation, and once again, if the exciting field is rather dual or multifrequency, the explicit formula will be modified. Servant et al. (2003) deeply investigated the interaction of wave with bubble cloud under mono- and dual-frequency ultrasonic waves. In addition, as the dynamics of bubble is proved to change under multifrequency fields, the secondary Bjerknes force will be implicitly impacted. This aspect was examined by Zhang et al. (2016) in a numerical study.

The second dimension of the bubble population is the number density. In fact, the number density of bubbles was treated in a respectable number of experimental studies using hyperfrequency method (Labouret 1998), laser diffraction (Burdin et al. 1999), phase Doppler (Burdin et al. 1999; Tsochatzidis et al. 2001), electromagnetic reflection (Labouret and Frohly 2004, 2011), sound speed variation (Birkin et al. 2003), and capillary (Iida et al. 2010b) techniques. At the opposite, bibliography is very poor in numerical models dealing with this topic, and we cite, among the rare retrieved works, those of Iida et al. (2010a, b) and Merouani et al. (2015a, b)). Yet one needs to underline that all of them are deductive and not predictive.

Generally, in experimental studies, bubble density is associated to bubble size distribution, as both parameters act on the measurable physical property: the void fraction. In modeling studies, on the other hand, assuming the acoustic bubble population as uniform is so far adopted and largely accepted for simplification purposes. Nevertheless, such hypothesis is far from reflecting the complex and heterogeneous composition of a bubble cloud in terms of sizes, nature of bubbles, lifetimes, and surrounding effects. We will focus in the next brief synthesis on the ambient size aspect. The choice of this axis is motivated actually by the fact that the nature of bubble can be defined according to the ambient radius under a given configuration of operational conditions, as clarified in the previous sections. Hence, the starting line, if assumed the ambient radii, will allow us to assimilate the population to packets of representative bubbles. We choose to present hereafter the approach of Labouret and Frohly (2002) in deducing the ambient size distribution in a bubble cloud.

Labouret and Frohly (2002) founded their approach on the monitoring of dissolution kinetics of acoustic cavitation bubbles starting from the moment of interruption of sonication. On the one hand, each bubble in the cloud is supposed to dissolve from its ambient radius according to a radius evolution law obtained by the simplification of Neppiras (1980) law and formulated as shown in Eq. 15.45.

$$\frac{dR}{dt} = -\left(8\pi\frac{\sigma DC_s}{\rho_G P_\infty}\right)\frac{1}{R^2} \tag{15.45}$$

The integration of this equation between the instant of the interruption of ultrasound and an instant t is easily expressed as:

$$\frac{4}{3}\pi R^3(t) = \frac{4}{3}\pi R_0^3 - \left(8\pi\frac{\sigma DC_s}{\rho_G P_\infty}\right)t \tag{15.46}$$

Hence, the dissolution time for a bubble having an initial radius R_0 within a saturated liquid can be simply written as:

$$t_{dis} = \frac{\frac{4}{3}\pi R_0^3}{\left(8\pi\frac{\sigma DC_s}{\rho_G P_\infty}\right)} \tag{15.47}$$

On the second hand, the bubble volume dissipation speed, or in other words the void rate dissipation speed within the reactor, is proved to be proportional to the number of dissolved bubble since the ultrasound switching off according to the Eq. 15.48.

$$N_{dis}(t) = -\frac{1}{\left(8\pi\frac{\sigma DC_s}{\rho_G P_\infty}\right)}\frac{dV}{dt}(t) \tag{15.48}$$

The void rate dissipation speed is monitored through an electromagnetic resonance frequency technique. From Eq. 15.48, it can be easily demonstrated that the number of bubbles dissolved during a time period comprised between two successive instants t_1 and t_2 is expressed as:

$$N_{dis}(t_2) - N_{dis}(t_1) = -\frac{1}{\left(8\pi\frac{\sigma DC_s}{\rho_G P_\infty}\right)}\left(\frac{dV}{dt}(t_2) - \frac{dV}{dt}(t_1)\right) \tag{15.49}$$

Incremental measurements of void fraction dissipation speed are performed to deduce the corresponding number of dissolved bubbles using Eq. 15.49, and according to the time period elapsed between t_1 and t_2, the dissolution time is calculated and the corresponding ambient radius is retrieved from Eq. 15.47. A distribution size within the cloud of acoustic cavitation bubbles is then constructed.

Several other approaches may be found in the literature (Lee et al. 2005; Brotchie et al. 2009; Iida et al. 2010a, b; Xu et al. 2014). Their common point is that they associate experimental measurement to theoretical notions in order to quantify bubble number density and determine the size distribution.

Finally, it is inevitable when dealing with bubble population to involve the space dimension. Nevertheless, in the sonoreactor, the space dimension itself involves several aspects such as the wave field propagation in space over time, which is closely dependent of the damping phenomenon (viscous, thermal, and acoustic) and the wave speed variation (Zhang et al. 2018a, b, c). The dependency chain does not break at this point; both latter paramount parameters are governed by the bubble population density and the size distribution within it. One can easily notice that this sequence brings us back to the starting point: How to assume a certain size distribution and density of bubbles in a specific location within the reactor volume if the acoustic field is not well defined there?

In fact, the problem of space distribution within a sonochemical reactor is so complex that it can be fairly described as a vicious circle. Obviously, simplifying assumptions are imposed to reach a reasonable modeling of it. We present hereafter an acceptable approach combining the work of Dähnke et al. (1999) and Zhang et al. (2018a, b, c). Let's consider an ultrasonic wave propagating a priori in a homogeneous liquid medium. The Alembert's equation is written for a three-dimensional basis:

$$\frac{\partial^2 p(x,t)}{\partial x^2} = \frac{1}{c^2}\frac{\partial^2 p(x,t)}{\partial t^2} \tag{15.48}$$

$$\frac{\partial^2 p(y,t)}{\partial y^2} = \frac{1}{c^2}\frac{\partial^2 p(y,t)}{\partial t^2} \tag{15.49}$$

$$\frac{\partial^2 p(z,t)}{\partial z^2} = \frac{1}{c^2}\frac{\partial^2 p(z,t)}{\partial t^2} \tag{15.50}$$

The general solution, following the x axis, for example, is known to be written as:

$$p(x,t) = e^{i(kx-\omega t)} \tag{15.51}$$

However, the previous formula for the propagation of an acoustic wave in space over time is admitted provided that the medium is perfectly homogeneous, which is obviously not valid in a bubbly medium. Owing to the existence of bubbles in the liquid, the motion of the wave is significantly affected. More concisely, two major parameters intervene in this case in the wave propagation equation: the wave velocity in the liquid–gas mixture c_{mix} and the damping coefficient β. According to the selected point within the sonoreactor volume, the complex celerity of the wave in the liquid–gas mixture c_{mix} will be defined according to Zhang et al.'s (2018a, b, c) results as:

$$\frac{1}{c_{\text{mix}}^2} = \frac{1}{c^2} + \frac{4\pi n N_{\text{A}} R_0}{\omega_0^2 - \omega^2 + 2i\beta\omega} \tag{15.52}$$

while the damping coefficient β which accounts for thermal, viscous, and acoustic damping mechanisms is derived by Zhang et al. (2018a, b, c) as:

$$\beta = \beta_{\text{vis}} + \beta_{\text{th}} + \beta_{\text{ac}} = \frac{2\mu}{\varepsilon\rho_{\text{L}} R_0^2} + \frac{2\mu_{\text{th}}}{\varepsilon\rho_{\text{L}} R_0^2} + \frac{R_0}{2c}\omega_0^2 \tag{15.53}$$

with:

$$\varepsilon = 1 + \frac{4(\mu + \mu_{\text{th}})R_0}{\rho_{\text{L}} R_0^2 c} \tag{15.54}$$

Hence, considering the bubble cloud effect, the wave propagation equation becomes:

$$p(x,t) = e^{i(kx-\omega t)} e^{i\beta t} \tag{15.55}$$

with:

$$k = \frac{\omega}{c_{\text{mix}}} + i\frac{\beta}{c} \tag{15.56}$$

By calculating the second derivate of Eq. 15.55 in space and time, we retrieve the modified form of Alembert's equation that accounts for the damping phenomena and applies for a bubbly liquid medium:

$$\frac{\beta^2 + \omega^2}{k^2} \frac{\partial^2 p(x,t)}{\partial x^2} = \frac{\partial^2 p(x,t)}{\partial t^2} + 2\beta \frac{\partial p(x,t)}{\partial t} \tag{15.57}$$

$$\frac{\beta^2 + \omega^2}{k^2} \frac{\partial^2 p(y,t)}{\partial y^2} = \frac{\partial^2 p(y,t)}{\partial t^2} + 2\beta \frac{\partial p(y,t)}{\partial t} \tag{15.58}$$

$$\frac{\beta^2 + \omega^2}{k^2} \frac{\partial^2 p(z,t)}{\partial z^2} = \frac{\partial^2 p(z,t)}{\partial t^2} + 2\beta \frac{\partial p(z,t)}{\partial t} \tag{15.59}$$

Other studies available in bibliography coupled the wave propagation equation with a bubble motion model to describe the acoustic cavitation bubble redistribution within the sonoreactor volume. For instance, the hybrid model presented by Mettin et al. (2006) was founded on a linearized bubbly liquid wave equation to describe the wave propagation and a discrete particle approach to characterize the bubble motion. The particle model adopted by these authors accounted for the nonlinear primary and secondary Bjerknes forces which made it more realistic than continuous bubble density models (Servant et al. 2003). Several other experimental works investigated as well the bubble distribution. We cite the work of Ashokkumar et al.

(2010) who used image recording of sonoluminescence, sonophotoluminescence, and sonochemiluminescence to locate the bubble population in the reactor volume, and their observations highlighted the complexities involved in acoustic cavitation.

To conclude the discussion on the bubble population dimensions, it is important to illustrate some direct impacts on modeling the acoustic cavitation bubble. The bubble dynamics is obviously affected by the radiation forces and the resultant of bubble motion and wave propagation in space and time. Besides, the energy associated with the radiation forces between bubbles needs as well to be involved in the energy balance of each single acoustic cavitation bubble. Moreover, with respect to the bubble size distribution, a mass distribution of reactants is automatically defined, which implies a sonochemical production distribution according to bubble size distribution, or in other words in function of space coordinates and time. Clearly, the complexity and interdependency of phenomena impose a certain threshold of simplification according to the target and conditions of the modeling study. A final point to underline is a finding: Sonochemical reaction control requires prediction, whereas several aspects mostly related to bubble population phenomena are still based on deductive approaches from experimental measurements.

15.3.5 Upscaling of Models: The Challenges

In this part, the objective is to highlight the challenges facing the upscaling of models in a green process perspective. This brings us back to both important questions we raised at the beginning of this chapter: the efficiency and the economic viability. The upscaling of numerical models developed so far to describe the activity of acoustic cavitation bubbles is not a simple gathering of fragments to consider the highest number of parameters and phenomena. Upscaling is rather about assigning effective properties to coarse scale from properties of fine scale models, with an acceptable degree of precision. In sonochemistry modeling, upscaling requires not only mastering space and time dimensions but also how to combine them with technological features; the concise definition of initial and boundary conditions is then a critical step (Birkin et al. 2003). In addition, very limited works could be found in bibliography regarding mathematical modeling of sonochemical activity in the presence of continuous flows. We cite, for instance, the works of Hussain et al. (2016, 2017) on continuous transesterification sonoreactors. Unless few exceptions, most of modeling studies focused so far on batch sonochemical reactors. A last point to mention in regard to limitations to overcome is the implementation of energy audit of sonochemical processes. Indeed, though an incredible number of studies dealt and still deal with energetic investigations of sonoreactors (Toma et al. 2011; Son et al. 2012), a homogenized procedure is required to assess, control, and optimize the energy efficiency of sonochemical processes.

15.3.6 *Measuring the Greenness of Sonochemistry: A Modeling Imperative?*

Since the early 90s, green chemistry metrics started gaining attention of scientists and researchers. The interest toward such topic is explained and motivated by several reasons, primarily competitiveness ones. Metrics are indeed powerful tools for measuring the performance of chemical processes and comparing them. Sheldon (2007, 2018) are among the researchers who investigated for long the design of metrics, their applicability, and their improvement. The green process metrics can be summarily classified into four categories. The first category comprises the mass-based metrics, which evaluate various ratios in the flow of mass within a process. This category counts the highest number of metrics defined so far. We cite the atom economy AE (Trost 1991), the E factor (Sheldon 2007), the mass intensity MI (Constable et al. 2002), and the reaction mass efficiency RME (Curzons et al. 2001). A non-exhaustive list of mass-based metrics is presented in Eqs. 15.62–15.69.

$$E = \frac{m_{\mathrm{W}}}{m_{\mathrm{t}}} \tag{15.62}$$

$$\mathrm{AE} = \frac{M_{\mathrm{t}}}{\sum M_{\mathrm{p}}} \tag{15.63}$$

$$\mathrm{MI} = \frac{\sum m_{\mathrm{r}} + \sum m_{\mathrm{p}} - m_{\mathrm{H_2O}}}{m_{\mathrm{t}}} \tag{15.64}$$

$$\mathrm{RME} = \frac{m_{\mathrm{t}}}{\sum m_{\mathrm{r}}} \tag{15.65}$$

$$\mathrm{PMI} = \frac{\sum m_{\mathrm{r}} + \sum m_{\mathrm{p}}}{m_{\mathrm{t}}} \tag{15.66}$$

$$\mathrm{MP} = \frac{m_{\mathrm{t}}}{\sum m_{\mathrm{r}} + \sum m_{\mathrm{p}} + \sum m_{\mathrm{s}}} \tag{15.67}$$

$$\mathrm{WWI} = \frac{m_{\mathrm{H_2O}}}{\sum m_{\mathrm{p}}} \tag{15.68}$$

$$\mathrm{SI} = \frac{\sum m_{\mathrm{s}}}{m_{\mathrm{t}}} \tag{15.69}$$

The second category concerns the metrics which measure the energy efficiency, and the most commonly used metric in this category is the cumulative energy demand CED defined as:

$$\mathrm{CED} = \sum E_{\mathrm{i}} \tag{15.70}$$

The third category is related to the environmental impact of waste, the Q factor, also known as the environmental quotient EQ that is extensively used in this category. This metric is actually a modified E factor that accounts for the threshold limit value of a given waste of a process according to Eq. 15.71.

$$\mathrm{EQ} = Q\frac{m_{\mathrm{W}}}{m_{\mathrm{t}}} \tag{15.71}$$

The fourth category and the least investigated includes the economic viability metrics. In fact, most of works which dealt with the economic metrics in green processes were turned into the circular economy concept (Clark et al. 2016), and several methodologies were adopted according to the treated process in order to assess its economic viability. For instance, Sheldon and Sanders (2015) involved the raw material and capital costs in the process economics. One can deem that functional costs need to be accounted as well for a concise comparison of processes.

The discussion in this section is not about the green processes metrics themselves but rather about their transposition to the sonochemical processes. We exhibited previously in Sect. 15.2.1 several attempts and propositions aiming to measure the efficiency of sonochemical processes. Unfortunately, clear inhomogeneity was noticed which is obviously synonym to inconsistencies in measuring the greenness of sonochemical processes. Hence, the question is, To which extent can the predefined metrics for green processes be adapted and applied to sonochemistry? Further, is it possible to suggest a combination with modeling for a pre-assessment before experiment?

Atom economy, reaction mass efficiency, and process mass intensity can a priori be associated with modeling studies provided that the chemical schema adopted to describe the sonochemical reactions is validated. Meanwhile, most of the other metrics are valid for an industrial-scale process, which constitutes a serious limitation preventing applicability to sonochemistry. Moreover, and for sure, the performance of the process makes sense at the sonoreactor level rather than the single bubble scale. It is then required to reach an upscaled model first before being able to design predictive metrics specific to sonochemistry. This does not exclude the possibility to present differently the experimental studies dealing with sonochemistry at laboratory scale by evaluating mass-based metrics, energy metrics, and environmental metrics (Andraos 2003), especially if the qualifier "green" is attributed to the sonochemical process in question.

Efforts toward the development of green processes will impose a concise harmonized methodology to assess the greenness of any process before claiming it as green. This applies as well to sonochemical processes, which means that in the near future, applying metrics of green sonochemical processes will become an imperative, yet we must first agree on their definition before applying them.

15.4 Experimental Sonochemistry

15.4.1 Sonochemistry and Green Synthesis

Chemical synthesis is about constructing complex chemical compounds from simpler ones. This occurs through a series of physical and chemical transformations. When ultrasound intervene, they can act on both, and if the chemical effect is proven, the process is called sonochemical synthesis.

Organic synthesis is the primary branch of chemical synthesis. Green organic synthesis is consequently concerned by the synthesis of organic compounds using green processes. Green organic synthesis is essentially founded on minimizing or eliminating the common disadvantages of organic processes. We cite the use of solvents that constitute the element of worst impact on environment, cost, safety, and health (Prat et al. 2014). In order to overcome this classical inconvenience of organic synthesis, it is suggested to design processes using reduced quantities of solvents or eliminating them, substituting conventional solvents by those having negligible vapor pressure such as ionic liquid, or simply using water, the most benign solvent (Prat et al. 2015). Sonochemical synthesis can interestingly adhere to the three last alternatives (Lupacchini et al. 2017), which is in a perfect accordance with the fifth green chemistry principle, innocuous solvents, and auxiliaries (Anastas and Warner 1998).

Draye and Kardos (2016) presented in 2016 a well-detailed review of green organic sonochemical processes. Their work covers a wide, large number of synthesis from which we selected relevant cases, not only to illustrate the role of sonochemistry in green organic synthesis but also to underline some misinterpretations of the role of ultrasound in specific organic synthesis examples.

Water represents the ideal solvent in sonochemistry for more than a reason: It is abundant, cheap, and nontoxic, and most importantly, it favors cavitation (Neppiras 1980), the key phenomenon for the existence of sonochemistry. Several studies reported the feasibility of organic synthesis in water. Abaee et al. (2008) used an ultrasound homogenizer of 24 kHz and 400 W to perform aminolysis of epoxides in aqueous media. The reaction conducted at ambient conditions led to the exclusive formation of β-amino alcohol. The process could be rationalized by proposing a favorable hydrogen-bonded association of the epoxide with water molecules under ultrasound, which provides the required activation energy for the reactants to overcome the transition state barrier at room temperature and allow the nucleophilic attack to occur regioselectively at the less hindered side of the epoxide.

In the previous example, beyond the use of water as a solvent, the sonochemical synthesis required no additives, no thermal activation, and no pH control. The translation of these three specificities according to the list of principles of green chemistry demonstrates a clear agreement with the fifth principle, but also with two others: atom and energy efficiencies.

Reaction time and conversion rate constitute as well a couple that acts on the atom economy, the prevention of derivatization, the reduction of waste, and more or less on the energy efficiency. These are, respectively, the second, the eight, the first, and the sixth principles of green chemistry. A telling example is that of Rostamnia and Lamei (2011) who studied the synthesis of derivatives of rhodanine in water submitted to ultrasound using a probe of 20 kHz of frequency and 600 W of maximum power. Under these conditions and only in 3 min of reaction, it was possible to reach 94% of rhodanines, while in the absence of irradiation, 1 h was necessary to lead to barely 10% of product. A similar illustrative example is that of Almeida and Faria (2013) who proposed an experimental protocol in two steps aiming to synthesize penta-substituted pyrroles. Interestingly, the authors retrieved 87% of product after 5 min of exposure to ultrasounds in water, whereas under silent conditions, 1 h was required to yield 56% in water and 80% in methanol. It is noteworthy to underline that both previous processes are moreover catalyst-free, which emphasizes their economic-efficient character.

Sonochemistry offers as well several possibilities for solvent-free synthesis of organic compounds. For instance, Du et al. (2014) implemented a novel solvent-free technique for the Michael addition of carbon-containing nucleophiles and nitrogen to cyclic enones under sonication at 20 kHz and a 675 W and under a fixed temperature of 333 K. This new method allowed a significant increase of four orders of magnitude in the rate of the reaction as compared to when acetonitrile was employed as solvent.

The last case we think worth to be presented in this section is that of ultrasound-assisted organic synthesis in ionic liquid. The first example is that of Deshmukh et al. (2001), who elaborated a Heck reaction at ambient conditions in an ionic liquid under ultrasound irradiation. Reaction rate was considerably enhanced through the formation of Pd–biscarbene complexes and stabilized clusters of zerovalent Pd nanoparticles. The in situ generation of zerovalent Pd species is the result of reduction of the divalent Pd–carbine complex which is accelerated by electron transfer reactions under ultrasound.

As a second example, we present the experimental work reported by Gumel et al. (2012). These authors studied the lipase-catalyzed synthesis of poly-6-hydroxyhexanoate under ultrasound irradiation. They demonstrated ring opening, transesterification, and polymerization of e-caprolactone in the 1-ethyl-3-methylimidazolium tetrafluoroborate, a high viscous ionic liquid. In the absence of ultrasounds, the use of such solvent is expected to limit mass diffusion. In the presence of ultrasound, this was overcoming mostly due to their mechanical effect.

The two preceding examples were obviously chosen in a nonrandom way, but in order to highlight a fact. In both of them, no evidence of any chemical effect of cavitation is established. Hence, they would be called green ultrasound-assisted syntheses rather than green sonochemical syntheses.

On the other hand, sonochemistry has considerable applications in inorganic synthesis, and one of the most illustrative ones is the sonochemical synthesis of metal nanoparticles. Okitsu presented a very interesting reviewing chapter of these processes in theoretical and experimental sonochemistry involving inorganic systems (Pankaj and Ashokkumar 2011). Metal nanoparticles, which are principally used as catalysts, sensors, and adsorbents, have been deeply investigated for sonochemical synthesis (Caruso et al. 2002; Suslick et al. 1996; Yeung et al. 1993). We cite, for instance, the works of Nagata and coworkers on the formation of colloidal silver (Nagata et al. 1992) and gold particles (Nagata et al. 2006) and those of Okitsu and coworkers on the formation of palladium (Okitsu et al. 1996a, b) and noble metals particles (Okitsu et al. 1996a, b). The sonochemical synthesis of metal nanoparticles occurs, thanks to a reduction mechanism (Grieser et al. 1996), with reducing species formed from the sonolysis of organic additives and water. A general chemical schema of the sonochemical reduction of metal ions under an argon atmosphere can be adapted from (Nagata et al. 1992, 2006) and presented as follows (Pankaj and Ashokkumar 2011):

$$H_2O \rightarrow HO^{\bullet} + H^{\bullet}$$

$$\text{Organic additive} + HO^{\bullet} \rightarrow \text{reducing species} + H_2O$$

$$\text{Organic additive} + H^{\bullet} \rightarrow \text{reducing species} + H_2$$

$$\text{Organic additive} + H_2O \rightarrow \text{pyrolysis radicals and unstable products}$$

$$M^{n+} + \text{reductants} \rightarrow M^0$$

$$nM^0 \rightarrow \left(M^0\right)_n$$

$$M^0 + \left(M^0\right)_n \rightarrow \left(M^0\right)_{n+1}$$

As compared to conventional techniques used in the synthesis of metal nanoparticles, such as controlled chemical, metal vaporization, photochemical, and radiation chemical methods, it has been recognized that sonochemistry is one of the new promising techniques for the synthesis of nanostructured materials owing to the unique reactions that can be induced by the irradiation of a liquid even at around ambient conditions. Once again, atom economy and energy efficiency principles are met, which places the sonochemical processes dedicated to inorganic synthesis in the rank of greener processes than conventional ones.

15.4.2 Sonochemistry and Renewable Feedstocks: A Promising Combination?

The seventh principle of green chemistry claims for processes using renewable rather than depleting feedstocks. Sonochemistry can fall into this category if the raw material employed in the sonochemical process is a renewable one. Having said that, the "sonochemical synthesis of biofuels" (Donate 2014) is the perfect example. Hundreds of works which experimented the synthesis of biofuels using ultrasound could be found in bibliography such as those of Stavarache et al. (Stavarache et al. 2003, 2005, 2007a, b) and Gole and Gogate (2012). Jointly, a considerable number of reviews were published to update the state of the art in regard to the evolution in ultrasound-assisted synthesis of biofuels. To illustrate the reviews, we cite Rokhina et al. (2009) and Luo et al. (2014). In fact, ultrasound does not only intervene at the stage of biofuels synthesis but also in the pretreatment of biomass by delignification and acid hydrolysis; in the enzymatic hydrolysis of pretreated biomass; in fermentation of the hydrolysates from acid, enzymatic hydrolysis, and microalgal lipid extraction; and in the biogas digestion. The scope of the present section is neither exposing the works that dealt with the preceding applications nor discussing their configurations; it is rather about highlighting the relation of ultrasound-assisted biofuel synthesis with the green chemistry principles and, more importantly, inspecting the consistence of talking about "sonochemical green biofuels synthesis."

Most of literature published in this area agreed on the beneficial effects of ultrasound on the synthesis of biofuels in terms of yield enhancement, kinetics acceleration, and reduction in the number of process steps. For instance, Stavarache et al. (2005) performed an ultrasound-assisted transesterification of vegetable oil with short-chain alcohols in the presence of base catalyst to obtain biodiesel. They demonstrated that the use of ultrasound at 40 kHz of frequency reduced the reaction time to 10–20 min as compared to mechanical stirring but required as well a quantity of catalyst two or three times lower than in the conventional process. With no doubt, such advantages concur perfectly with the second, the sixth, and the seventh principles of green chemistry. The question is then turned into the existence of the chemical effect of cavitation in the synthesis of biofuels, or in other words, Is the process really sonochemical or just ultrasound-assisted?

In spite of the voluminous literature, very little efforts have been devoted to the mechanistic understanding of the observed enhancement in biofuel synthesis under ultrasound. In the few papers that dealt with the question, the technique of elevated static pressure has been extensively used to determine the relative contribution of physical and chemical effects of ultrasound toward biofuel synthesis. Kalva et al. (2009) presented an interesting comprehensive study coupling experiments to a bubble dynamics model and aiming to discern the physical and chemical effects of ultrasound in regard to biofuel synthesis. The authors proved through simulation that, at the power input used in their experiments, the temperature peak reached in transient collapse of the acoustic cavitation bubble in methanol was too low to produce any radical species. Consequently, they evidenced the noncontribution of sonochemical reactions, and the influence of ultrasound on oil transesterification reaction was then of purely physical nature. The acceleration of reaction was attributed to the formation of fine emulsion between oil and alcohol due to microturbulence generated by cavitation bubbles and resulting in enormous interfacial area. The mechanistic result presented by Kalva et al. (2009) is apparently a point of agreement with several other studies such as those of Parkar et al. (2012) and Choudhury et al. (2013). Having said that, the combination of ultrasound with renewable feedstock to produce biofuels constitutes a process that respects several of the green chemistry principles. However, it is not possible to classify it as a green sonochemical process, since the chemical effect of cavitation is completely rejected in such type of application.

Another association of ultrasound and renewable feedstocks consists in the use of glycerol as solvent is sonochemical processes. Glycerol is one of the main coproducts of biodiesel and oleochemical production. During the last decade, glycerol was proposed as a solvent for sonochemical processes, especially that it provides excellent cavitation (Cintas et al. 2014). In this context, Cravotto et al. (2011) proposed for the first time in 2011 an ultrasound-assisted synthetic protocol using glycerol as solvent. These authors carried out the catalytic transfer hydrogenation of benzaldehyde into benzylic alcohol in glycerol as solvent and hydrogen donor, and catalyzed by Ru(p-cymene)Cl_2 dimer, the mixture was pre-sonicated using a cup horn of 19 kHz of frequency and 100 W of input power and then heated up with a bath of oil. Under those conditions, the duration of the reaction was decreased by factor of two compared to the same reaction without pre-sonication. Neither microwave nor combined ultrasound/microwave could compete with ultrasonic irradiation, which led to 100% yield after 3 h. The crucial role of ultrasound was evidenced in the transfer hydrogenation of benzaldehyde, owing to the capacity of ultrasound to overcome problems of solubility and high viscosity by enhancing heat and mass transfer. Once again, the effect of ultrasound seems to be rather physical than chemical, which accentuates the fact that the use of the attribute "green sonochemical process" should be made and taken with scientific precautions.

15.5 Sonochemical Reactors

15.5.1 Developed Designs

Ultrasonic reactors, also known as sonoreactors, are reactors where the pressure variation induced by the passage of ultrasound results in the generation of cavitation. These latter undergo subsequent growth and violent collapse releasing large magnitudes of energy over a very small volume, resulting in very high energy densities (Gogate and Patil 2016). It is important to bear in mind that ultrasound reactors are usually called sonochemical reactors by abuse. Such appellation does not mean that the reactor houses sonochemical reactions. In most of cases, the effect of ultrasound is rather physical. Meanwhile, in the present section, we will focus on the sonoreactor designs used for sonochemical applications.

Ultrasonic irradiation demonstrated in several studies positive effects on chemical reactions especially in terms of kinetics and selectivity. In environmental area, for example, the sonochemical degradation of toxic organic compounds, such as phenolic compounds, aromatic compounds, esters, and textile dyes, into short-chain organic acids, inorganic ions, or carbon dioxide can be carried out hundreds of times faster than natural oxidation (Babu et al. 2016). Reactors housing such types of sonochemical reactions can vary in their geometry, number of transducers, position of transducers, and even other aspects. In what follows, we attempt to present the main sonoreactor configuration that is suitable for the large-scale operation in respect to some of the green processing principles.

Typically, the sonoreactors are classified according to the contact of ultrasonic irradiation with the reactional medium into direct and indirect. They can also be divided into batch and continuous sonoreactor according to the mass flow. In the following presentation, we will first explore the main batch systems designs, starting from those with indirect irradiation, and we will then investigate the developed designs for continuous flow sonoreactors, which constitute in fact the most suitable category for high-scale or industrial implementation (Feng et al. 2002).

The simplest design of the ultrasonic reactors is with no doubt the bath sonoreactor. In this configuration, the ultrasounds are introduced within a vessel through a transducer generally attached to its bottom; it is generally used in batch mode. The irradiation mode is in this case indirect, and the distribution of cavitation within the reactor volume is closely linked to the number and position of transducers and the geometry of the vessel. Many improvements have then been brought to the basic design of the bath sonoreactor in order to increase the acoustic energy density and enhance the homogeneity of cavitational activity. For instance, Gogate et al. (2011) reported an hexagonal design of batch sonoreactor with transducers attached at the center of each side, and the suggested configuration proved to offer higher energy efficiency. In addition, the acoustic streaming was proved to be insufficient in order to ensure uniform distribution of cavitation in the sonoreactor. Indeed, Sajjadi et al. (2015) performed a comparison of fluid flow in a sonoreactor and classic stirred vessel, and they demonstrated that the axial velocity component was prominent under ultrasound irradiation while the tangential component was more significant in the stirred vessel. This observation justifies the introduction of mechanical stirring in some batch sonoreactor designs in order to improve space uniformity. Indeed, Waghmare et al. (2015), who experimented the ultrasonic synthesis of glycerol carbonate, noticed an enhancement of the conversion percentage from 63.38 to 99.75% when introducing mechanical stirring to the sonoreactor. Other works aiming to define a reference facility for ultrasonic reactors can be found in the literature. In this perspective, Hodnett et al. (2007) proposed a cylindrical cell design for sonochemical processing, and they demonstrated that the suggested configuration offers good reproducibility and acoustic amplitude enhancement.

Systems with direct irradiation mode present obviously the advantage of higher energy density. The basic concept which can provide direct sonication is the horn design, with several possibilities according to the form and the emplacement of the horn in the reacting vessel. We distinguish, for example, cup horn, probe horn, and longitudinal horn sonicators. Nevertheless, not all these configurations are used for direct irradiation, which means that the energy efficiency can significantly vary according to the type of the horn sonicator.

A pioneering review of Berlan and Mason (1992) addressed the problem of scaling up sonochemical reactors in regard to reactions conditions and possible equipment for industrial-scale sonochemistry. Surprisingly, from a technological point of view, the basics of the horn sonicators did not evolve in a significant manner since the early 90s. Typical cup horn and immersed horn (probe horn) exposed in Berlan and Mason's review are still used nowadays in laboratory. These devices are especially adapted for laboratory-scale operations owing to the high cavitational intensity they offer. However, this latter is conditioned by the proper choice of the shape and dimensions of the probe and the liquid height. Moreover, horn sonicators present a major inconvenience of risks of erosion and contamination of the reacting medium especially at longtime use at high power.

The cup horn is a design that combines some advantages of the horn sonicators but is still used for indirect sonication. Gogate et al. (2006) experimented the destruction of formic acid using a cup horn-type high-frequency reactor. They noticed that the extent of degradation strongly depends on the operating parameters. Thus, they optimized the reacting volume, the residence time, and the feeding electrical power. Mixing using mechanical agitation proved as well to be essential to reach uniformity within the treated volume. Titanium crucibles were also exploited as sonoreactors by being adapted to the standard piezoelectric transducers and used as a probe system (Berlan and Mason 1992). The third classic configuration is the probe horn, also known as the immersed horn or commonly called the ultrasonic horn. This design offers shorter micromixing time as compared to the cup horn, for the same energy cost as demonstrated in the comparative study of Monnier et al. (1999).

Some slight modifications of the basic designs were proposed by several authors in order to improve their performance. To illustrate that, we mention Cravotto et al. (2003) whose presented a modified ultrasonic horn with external cooling aiming to maintain the transducer temperature lower than 40 °C. The brought modification permits a continuous use of the sonoreactor, without any modifications in the natural vibrating frequency of the horn. As another example, Bhirud et al. (2004) proposed a new design of sonoreactor with longitudinally vibrating horn. Their comparative experimental study demonstrated that the novel reactor gives about four to five times more cavitational yield as compared to the ultrasonic bath, dual-frequency flow cell, and triple-frequency flow cell and two orders of magnitude higher cavitational yield as compared to ultrasonic horn.

The objective of developing sonoreactors not only for laboratory scale but also for industrial applications imposed a serious investigation of continuous flow designs. Some of the previous bath and horn sonoreactors were actually implemented to work in batch mode but also with continuous flow. For instance, the hexagonal ultrasound sonoreactor reported by Gogate et al. (2011) can easily be adapted to be used in continuous mode. It is also the case of the cup horn, which can be implemented as a cell for continuous flow (Berlan and Mason 1992). In the contrary, it is less obvious to adapt the horn sonicators to large-scale applications because of the decreasing activity observed while going away from the transducers (Kanthale et al. 2003). Many solutions can be found in the literature for middle-scale applications of continuous sonochemistry. For instance, Gondrexon et al. (1998) investigated the hydrodynamic behavior in a continuous-flow sonochemical reactor which consists of a cylindrical PVC body mounted on a stainless steel plate to which a piezoelectric transducer is attached. The authors proved that in their experimental volume range (100, 200, and 300 mL), the continuous-flow sonoreactor behaves as a completely stirred tank reactor. In addition, according to residence time distribution measurements, ultrasonic waves seem to induce a perfect mixing in the whole volume so as no mechanical stirrer has to be added. However, they highlighted the paradox between the ideal hydrodynamic behavior and the low velocities of the acoustic streaming.

Several research works presented designs of large-scale sonoreactors based on the same logic as that for mid-scale sonicators, while some other innovative models were developed from scratch in order to overcome the limitations of volume encountered with the preceding designs. Obviously, it was necessary to use multiple transducers with adjusted arrangements (hexagonal, rectangular, etc.) in order to increase the power distribution and the cavitational active volume. Moreover, interest was turned into flow cell configurations as they give options of recirculating and continuous operation modes. We present hereafter some of these designs.

Suri et al. (2002) suggested to change the locations of two transducers in the design of a cylindrical sonoreactor aiming to ameliorate the mixing properties without any mechanical agitation. One of the transducers was fixed at bottom, while the other was attached on the periphery. With this arrangement, acoustic streaming was generated in both vertical and radial directions resulting in important mixing as compared to the conventional configuration.

Mason (2000) presented a more innovative configuration based on a flexible sheet of embedded piezoelectric pillars. The emitting face in the proposed configuration is a combination of ceramic and plastic, which provides better and uniform acoustic transmission into the liquid. Moreover, the transducer sheet can be molded to fit any shape of reactor. Another design reported by Liu et al. (2008) for the degradation of dimethoate solution constitutes an ultrasonic airlift loop reactor assisted with an advanced oxidation process. The dispersion effects induced by the air loop reactor led to better utilization of ozone, which results in enhanced overall performance of the sonoreactor.

Larger sonoreactors were reported by Hodnett et al. (2007), Asakura et al. (2008), and Son et al. (2009). Hodnett et al. (2007) presented a new design of a cylindrical reactor with operating volume of 25 L equipped with 30 piezoelectric transducers organized in 10 rows on the external periphery. On the other hand, a rectangular sonoreactor with a maximum operating capacity of 112 L was reported by Asakura et al. (2008). Its design comprised of 12 transducers arranged in six per row. Both previous designs admitted maximum power rating of around 600 W. A larger rectangular sonoreactor was invoked by Son et al. (2009), with a working capacity of 250 L. A transducer module consisting of nine transducers operating at 35, 72, 110, and 170 kHz was attached at the center of the sidewall. The maximum power of the transducer module was reported to keep constant at 400 W. The efficiency of this sonoreactor was assessed by estimating the variation in power dissipation, sonochemical efficiency, and damping factor for water and potassium iodide solution, and it was demonstrated that much better results were encountered as compared with the conventional designs.

All in all, it seems that literature contains a considerable number of works suggesting and analyzing more or less innovative designs of sonoreactors. Yet, it is true that a relative limited number of them dealt with the large-scale designs. In addition, divaricated results regarding operating parameters, such as additional mechanical stirring and energy efficiency, indicate the lack of harmonized findings and reference guidelines which are essential to pass from the laboratory scale to the industrial one.

15.5.2 Sonochemical Reactors and Green Engineering Concept

The translation of green chemistry principles when it comes to technological designs constitutes a very complex and challenging task. The design of a new process is the most critical stage to engage in green chemistry. During the design phase of a new process, the scope of possible innovation ranges from incremental or superficial design improvements to completely redesigning the system of production (Mulvihill et al. 2011), and this is exactly the role of "green process engineering."

Green process engineering works on the integration of technical innovation and new environment-friendly chemical routes to achieve green process development. The green process design accounts not only for the economic exploitation but also for the compensation resulting from increased selectivity and savings. With high volume, the problem becomes very complex, since factors such as safety, nonpolluting technologies, reduction of raw materials and energy loss, and product/byproduct recyclability must be considered. Besides, the trend toward global-scale facilities may require a total change of technology, especially that the current technology is no longer capable of being built "just bigger." Hence, the chemical engineers are faced with demands on process intensification, which refers to complex technologies that replace large, expensive, energy-intensive, and polluting process by new process which is ideally smaller, less costly, less polluting, safer, and more efficient and combines multiple operations into a single apparatus or fewer devices (Saroha 2009).

In the case of sonochemical reactors, research works on design improvements should especially consider the reaction kinetics and the operating conditions, namely, the acoustic frequency, the acoustic intensity, the operating pressure and temperature, the sonication time, the liquid physiochemical properties, and the mixing parameters in stirring reactors (Sancheti and Gogate 2017).

Pioneering works in this topic have been presented by Mason (2000, 2003, 2007). In 2000, Mason explored the possibilities for large-scale sonochemical processing, and though the concept of green chemistry was not invoked explicitly in the work in question, the author highlighted the green characters of sonochemical processes (faster reactions, better conversions, improved and even new products) that required further investigations at both fundamental and technological levels. In 2003, Mason presented a very interesting paper (Mason 2003) where several claims are worth to be cited. The initial objective of the author was to demonstrate that future trends of research on power ultrasound are not only concerned by “pure” sonochemistry but also by “false” sonochemistry, referring to the physical effect of ultrasound and cavitation. This was illustrated by Mason through a telling example of ultrasound-assisted extraction, he exhibited the clear link with the principles of green chemistry. Having said that, it was shown that when dealing with upscaling technological designs, physical and chemical effects of ultrasound often intervene simultaneously, and it is then more judicious to talk about ultrasound-assisted green processes rather than “pure” sonochemical green ones. Another work of the same author (Mason 2007) proposed several engineering constructions based on ultrasound application in the domain of environmental protection and remediation. On the other hand, Gogate and coworkers (2016) and Sancheti and Gogate (2017) deeply investigated the effects of operating parameters on the sonochemical activity and proposed guidelines for optimum selection of main operating conditions in accordance with geometric parameters of the sonoreactor, i.e., type of reactor configuration and position and number of transducers. Ebrahimi et al. (2018) recently completed the investigation by exploring the effects of probe size, probe immersion depth, and reactor volume.

This brief review of the state of the arts of ultrasound-assisted technology proves that the main concern in this area cannot be the fundamental understanding of the phenomenon, as research are engaged since many decades in extensively examining the various operational and geometric parameters of the sonoreactor. The question could rather be the economics of scale-up of sonoreactors, which is an essential component for its final exploitation and the less investigated point until this day.

References

Abaee MS, Hamidi V, Mojtahedi MM (2008) Ultrasound promoted aminolysis of epoxides in aqueous media: a rapid procedure with no pH adjustment for additive-free synthesis of β-aminoalcohols. Ultrason Sonochem 15(5):823–827. https://doi.org/10.1016/j.ultsonch.2007.12.006

Adewuyi YG (2005) Sonochemistry in environmental remediation 2. Heterogeneous Sonophotocatalytic oxidation processes for the treatment of pollutants in water. Environ Sci Technol 39(22):8557–8570. https://doi.org/10.1021/es0509127

Ahmadi A et al (2014) Synthesis of ibuprofen with modified and economical process as an NSAID drug. J Appl Chem Res 8(3):91–95. http://jacr.kiau.ac.ir/article_516129_239923d1582447450a36d280e0dd4be6.pdf

Almeida QAR, Faria RB (2013) Synthesis of highly substituted pyrroles using ultrasound in aqueous media. Green Chem Lett Rev 6(2):129–133. https://doi.org/10.1080/17518253.2012.713999

Anastas P, Farris CA (1994) Benign by design chemistry. In: Benign by design, pp 2–22. https://doi.org/10.1021/bk-1994-0577.ch001

Anastas PT, Warner JC (1998) Green chemistry: theory and practice

Ando T et al (1984) Sonochemical switching of reaction pathways in solid-liquid two-phase reactions. J Chem Soc Chem Commun 15(31):439–440. https://doi.org/10.1002/chin.198431129
Andraos J (2003) Unification of reaction metrics for green chemistry: applications to reaction analysis abstract. Org Process Res Dev 9(2):149–163. https://doi.org/10.1021/op049803n
Anto T et al (1991) Sonochemical switching from ionic to radical pathways in the reactions of styrene and trans-β-methylstyrene with lead tetraacetate. Tetrahedron Lett 32(44):6379–6382. https://doi.org/10.1016/0040-4039(91)80174-5
Apfel RE (1981) Acoustic cavitation prediction. J Acoust Soc Am 80(4):1624–1633. https://doi.org/10.1121/1.385939
Asakura Y et al (2008) Development of a large sonochemical reactor at a high frequency. Chem Eng J 139(2):339–343. https://doi.org/10.1016/j.cej.2007.08.007
Ashokkumar M et al (2010) Spatial distribution of acoustic cavitation bubbles at different ultrasound frequencies. ChemPhysChem 11(8):1680–1684. https://doi.org/10.1002/cphc.200901037
Authier O, Ouhabaz H, Bedogni S (2018) 'Modeling of sonochemistry in water in the presence of dissolved carbon dioxide. Ultrason Sonochem 45:17–28. https://doi.org/10.1016/J.ULTSONCH.2018.02.044
Babu SG, Ashokkumar M, Neppolian B (2016) The role of ultrasound on advanced oxidation processes. Top Curr Chem 374(5):1–32. https://doi.org/10.1007/s41061-016-0072-9
Banitaba SH, Safari J, Khalili SD (2013) Ultrasound promoted one-pot synthesis of 2-amino-4,8-dihydropyrano[3,2-b] pyran-3-carbonitrile scaffolds in aqueous media: a complementary "green chemistry" tool to organic synthesis. Ultrason Sonochem 20(1):401–407. https://doi.org/10.1016/j.ultsonch.2012.07.007
Berlan J, Mason TJ (1992) Sonochemistry: from research laboratories to industrial plants. Ultrasonics 30(4):203–212. https://doi.org/10.1016/0041-624X(92)90078-Z
Berlan J et al (1994) Oxidative degradation of phenol in aqueous media using ultrasound. Ultrason Sonochem 1(2):97–102. https://doi.org/10.1016/1350-4177(94)90005-1
Bhirud US et al (2004) Ultrasonic bath with longitudinal vibrations: a novel configuration for efficient wastewater treatment. Ultrason Sonochem 11(3–4):143–147. https://doi.org/10.1016/j.ultsonch.2004.01.010
Birkin PR et al (2003) Experimental and theoretical characterization of sonochemical cells. Part 1. Cylindrical reactors and their use to calculate the speed of sound in aqueous solutions. J Phys Chem A 107(2):306–320. https://doi.org/10.1021/jp014532t
Bjerknes V (1906) Fields of force. The Columbia University Press, New York
Brennen CE (1995) Cavitation and bubble dynamics. Oxford University Press, Oxford
Briggs HB, Johnson JB, Mason WP (1947) Properties of liquids at high sound pressure. J Acoust Soc Am 19(4):664–677. https://doi.org/10.1121/1.1916536
Brotchie A, Grieser F, Ashokkumar M (2009) Effect of power and frequency on bubble-size distributions in acoustic cavitation. Phys Rev Lett 102(February):1–4. https://doi.org/10.1103/PhysRevLett.102.084302
Burdin F et al (1999) Characterisation of the acoustic cavitation cloud by two laser techniques. Ultrason Sonochem 6:43–51. https://doi.org/10.1016/S1350-4177(98)00035-2
Carson R et al (1962) Silent spring. Houghton Mifflin, Boston
Caruso RA, Ashokkumar M, Grieser F (2002) Sonochemical formation of gold sols. Langmuir 18(21):7831–7836. https://doi.org/10.1021/la020276f
Chatel G (2018) How sonochemistry contributes to green chemistry? Ultrason Sonochem 40(B):117–122. https://doi.org/10.1016/j.ultsonch.2017.03.029
Choudhury HA, Malani RS, Moholkar VS (2013) Acid catalyzed biodiesel synthesis from Jatropha oil: mechanistic aspects of ultrasonic intensification. Chem Eng J 231:262–272. https://doi.org/10.1016/j.cej.2013.06.107
Church CC, Carstensen EL (2001) "Stable" inertial cavitation. Ultrasound Med Biol 27(10):1435–1437. https://doi.org/10.1016/S0301-5629(01)00441-0
Cintas P (2016) Ultrasound and green chemistry—further comments. Ultrason Sonochem 28(January):257–258. https://doi.org/10.1016/j.ultsonch.2015.07.024
Cintas P, Luche J (1999) Green chemistry the sonochemical approach. Green Chem 43(1):115–125. https://doi.org/10.1039/A900593E

Cintas P et al (2014) Glycerol: a solvent and a building block of choice for microwave and ultrasound irradiation procedures. Green Chem 16(3):1056–1065. https://doi.org/10.1039/c3gc41955j

Clark JH et al (2016) Circular economy design considerations for research and process development in the chemical sciences. Green Chem 18(14):3914–3934. https://doi.org/10.1039/c6gc00501b

Constable DJC, Curzons AD, Cunningham VL (2002) Metrics to "green" chemistry—which are the best? Green Chem 4(6):521–527. https://doi.org/10.1039/b206169b

Cravotto G et al (2003) The aldol reaction under high-intensity ultrasound: a novel approach to an old reaction. Eur J Org Chem (22):4438–4444. https://doi.org/10.1002/ejoc.200300369

Cravotto G et al (2011) Efficient synthetic protocols in glycerol under heterogeneous catalysis. ChemSusChem 4(8):1130–1134. https://doi.org/10.1002/cssc.201100106

Cue BW, Zhang J (2009) Green process chemistry in the pharmaceutical industry. Green Chem Lett Rev 2(4):193–211. https://doi.org/10.1080/17518250903258150

Curzons AD et al (2001) So you think your process is green, how do you know?—using principles of sustainability to determine what is green—a corporate perspective. Green Chem 3(1):1–6. https://doi.org/10.1039/b007871i

Dähnke S, Swamy KM, Keil FJ (1999) Modeling of three-dimensional pressure fields in sonochemical reactors with an inhomogeneous density distribution of cavitation bubbles. Comparison of theoretical and experimental results. Ultrason Sonochem 6(1–2):31–41. https://doi.org/10.1016/S1350-4177(98)00026-1

Deshmukh RR, Rajagopal R, Srinivasan KV (2001) Ultrasound promoted C-C bond formation: heck reaction at ambient conditions in room temperature ionic liquids. Chem Commun 1(17):1544–1545. https://doi.org/10.1039/b104532f

Donate PM (2014) Green synthesis from biomass. Chem Biol Technol Agric 1(1):1–8. https://doi.org/10.1186/s40538-014-0004-2

Draye M, Kardos N (2016) Advances in green organic sonochemistry. Top Curr Chem 374(5):1–29. https://doi.org/10.1007/s41061-016-0074-7

Du XJ et al (2014) Solvent-free Brønsted acid-catalyzed Michael addition of nitrogen- and carbon-containing nucleophiles by ultrasound activation. Tetrahedron Lett 55(5):1002–1005. https://doi.org/10.1016/j.tetlet.2013.12.059

Ebrahimi SL, Khosravi-Nikou MR, Hashemabadi SH (2018) Sonoreactor optimization for ultrasound assisted oxidative desulfurization of liquid hydrocarbon. Petrol Sci Technol 36(13):959–965. https://doi.org/10.1080/10916466.2018.1458112

Einhorn C, Luche JL (1986) Ready preparation of sugar acetals under ultrasonic irradiation. Carbohydr Res 155(C):258–261. https://doi.org/10.1016/S0008-6215(00)90155-1

Eller A, Flynn H (1965) Rectified diffusion during nonlinear pulsations of cavitation bubbles. J Acoust Soc Am 37(3):493–503. https://doi.org/10.1121/1.1909357

Fegade SL, Trembly JP (2017) Misinterpretation of green chemistry. Ultrason Sonochem 37:686–687. https://doi.org/10.1016/j.ultsonch.2015.04.007

Feng R et al (2002) Enhancement of ultrasonic cavitation yield by multi-frequency sonication. Ultrason Sonochem 9(5):231–236. https://doi.org/10.1016/S1350-4177(02)00083-4

Ferkous H, Hamdaoui O, Merouani S (2015) Sonochemical degradation of naphthol blue black in water: effect of operating parameters. Ultrason Sonochem 26:40–47. https://doi.org/10.1016/j.ultsonch.2015.03.013

Fitzgerald ME, Gripping V, Sullivan J (1956) Chemical effects of ultrasonics—"hot spot" chemistry. J Chem Phys 25(5):926–933. https://doi.org/10.1063/1.1743145

Franc J (2007) The Rayleigh-Plesset equation: a simple and powerful tool to understand various aspects of cavitation. In: Fluid dynamics of cavitation and cavitating turbopumps, pp 1–41. https://doi.org/10.1007/978-3-211-76669-9_1

Fukuoka S et al (2003) A novel non-phosgene polycarbonate production process using by-product CO2 as starting material. Green Chem 5(5):497–507. https://doi.org/10.1039/b304963a

Gogate PR, Patil PN (2016) Sonochemical reactors. Top Curr Chem (Cham) 374(5):61. https://doi.org/10.1007/s41061-016-0064-9

Gogate PR et al (2001) Cavitation reactors: efficiency assessment using a model reaction. AIChE J 47(11):2526–2538. https://doi.org/10.1002/aic.690471115
Gogate PR et al (2006) Destruction of formic acid using high frequency cup horn reactor. Water Res 40(8):1697–1705. https://doi.org/10.1016/j.watres.2006.02.011
Gogate PR, Sutkar VS, Pandit AB (2011) Sonochemical reactors: important design and scale up considerations with a special emphasis on heterogeneous systems. Chem Eng J 166(3):1066–1082. https://doi.org/10.1016/j.cej.2010.11.069
Gole VL, Gogate PR (2012) Intensification of synthesis of biodiesel from nonedible oils using sonochemical reactors. Ind Eng Chem Res 51(37):11866–11874. https://doi.org/10.1021/ie2029442
Gondrexon N et al (1998) Experimental study of the hydrodynamic behaviour of a high frequency ultrasonic reactor. Ultrason Sonochem 5(1):1–6. https://doi.org/10.1016/S1350-4177(97)00043-6
Gong C, Hart DP (1998) Ultrasound induced cavitation and sonochemical yields. J Acoust Soc Am 104(5):2675–2682. https://doi.org/10.1121/1.423851
González-García J et al (2010) Sonochemical treatment of water polluted by chlorinated organocompounds. A review. Water (Switzerland) 2(1):28–74. https://doi.org/10.3390/w2010028
Grieser F et al (1996) Sonochemical reduction processes in aqueous colloidal systems. Ultrasonics 34(2–5):547–550. https://doi.org/10.1016/0041-624X(96)00036-4
Gumel AM et al (2012) Ultrasound assisted lipase catalyzed synthesis of poly-6-hydroxyhexanoate. Ultrason Sonochem 19(3):659–667. https://doi.org/10.1016/j.ultsonch.2011.10.016
Guo W et al (2010) Sonochemical degradation of the antibiotic cephalexin in aqueous solution. Water SA 36(5):651–654. https://doi.org/10.4314/wsa.v36i5.61998
Guzen KP et al (2007) Eco-friendly synthesis of imines by ultrasound irradiation. Tetrahedron Lett 48(10):1845–1848. https://doi.org/10.1016/j.tetlet.2007.01.014
Harada H (2001) Sonophotocatalytic decomposition of water using TiO(2) photocatalyst. Ultrason Sonochem 8(1):55–58. https://doi.org/10.1016/S1350-4177(99)00050-4
Hodnett M, Choi MJ, Zeqiri B (2007) Towards a reference ultrasonic cavitation vessel. Part 1: preliminary investigation of the acoustic field distribution in a 25 kHz cylindrical cell. Ultrason Sonochem 14(1):29–40. https://doi.org/10.1016/j.ultsonch.2006.01.003
Holland KS (2011) A framework for sustainable remediation. Environ Sci Technol 45(17):7116–7117. https://doi.org/10.1021/es202595w
Hussain MN et al (2016) Numerical modelling of sonicated, continuous transesterification and evaluation of reaction kinetics for optimizing biodiesel reactor design. Int J Thermal Environ Eng 11(1):79–86. https://doi.org/10.5383/ijtee.11.01.012
Hussain MN, Al Kaabi S, Janajreh I (2017) Optimizing acoustic energy for better transesterification: a novel sono-chemical reactor design. Energy Procedia 105:544–550. https://doi.org/10.1016/j.egypro.2017.03.354
Iida Y et al (2010a) Bubble population phenomena in sonochemical reactor: I Estimation of bubble size distribution and its number density with pulsed sonication—Laser diffraction method. Ultrason Sonochem 17(2):473–479. https://doi.org/10.1016/j.ultsonch.2009.08.018
Iida Y et al (2010b) Bubble population phenomena in sonochemical reactor: II. Estimation of bubble size distribution and its number density by simple coalescence model calculation. Ultrason Sonochem 17(2):480–486. https://doi.org/10.1016/j.ultsonch.2009.08.017
Ivanković A et al (2017) Review of 12 principles of green chemistry in practice. Int J Sustain Green Energy 6(3):39–48. https://doi.org/10.11648/j.ijrse.20170603.12
Kalva A, Sivasankar T, Moholkar VS (2009) Physical mechanism of ultrasound-assisted synthesis of biodiesel. Ind Eng Chem Res 48(1):534–544. https://doi.org/10.1021/ie800269g
Kanthale PM et al (2003) Mapping of an ultrasonic horn: link primary and secondary effects of ultrasound. Ultrason Sonochem 10(6):331–335. https://doi.org/10.1016/S1350-4177(03)00104-4
Kanthale PM et al (2008) Experimental and theoretical investigations on sonoluminescence under dual frequency conditions. Ultrason Sonochem 15(4):629–635. https://doi.org/10.1016/j.ultsonch.2007.08.006

Kerboua K, Hamdaoui O (2017) Computational study of state equation effect on single acoustic cavitation bubble's phenomenon. Ultrason Sonochem 38:174–188. https://doi.org/10.1016/j.ultsonch.2017.03.005

Kerboua K, Hamdaoui O (2018a) Insights into numerical simulation of controlled ultrasonic waveforms driving single cavitation bubble activity. Ultrason Sonochem 43:237–247. https://doi.org/10.1016/j.ultsonch.2018.01.018

Kerboua K, Hamdaoui O (2018b) Numerical estimation of ultrasonic production of hydrogen: effect of ideal and real gas based models. Ultrason Sonochem 40:194–200. https://doi.org/10.1016/j.ultsonch.2017.07.005

Kerboua K, Hamdaoui O (2018c) Ultrasonic waveform upshot on mass variation within single cavitation bubble: investigation of physical and chemical transformations. Ultrason Sonochem 42:508–516. https://doi.org/10.1016/J.ULTSONCH.2017.12.015

Koda S et al (2003) A standard method to calibrate sonochemical efficiency of an individual reaction system. Ultrason Sonochem 10(3):149–156. https://doi.org/10.1016/S1350-4177(03)00084-1

Koda S et al (2004) Sonochemical efficiency during single-bubble cavitation in water. J Phys Chem A 108(52):11609–11612. https://doi.org/10.1021/JP0461908

Kristol S, Klotz H, Parker RC (1981) The effect of ultrasound on the alkaline hydrolysis of nitrophentl esters. Tetrahedron Lett 22:907–908

Kummerer K, Clark J (2016) Sustainability science. In: Sustainability science, pp 1–606. https://doi.org/10.1007/978-94-017-7242-6_4

Labouret S (1998) Détermination du taux de vide d'un champ de bulles de cavitation ultrasonore par une méthode hyperfréquence: corrélation du taux de vide et de la puissance du bruit de cavitation. University of Valenciennes

Labouret S, Frohly J (2002) Bubble size distribution estimation via void rate dissipation in gas saturated liquid. Application to ultrasonic cavitation bubble fields. Eur Phys J Appl Phys 19:39–54. https://doi.org/10.1051/epjap:2002047

Labouret S, Frohly J (2004) Influence de la fréquence de découpage de l'irradiation ultrasonore sur l'apparition d'un champ de cavitation inertielle. In: CFA/DAGA'04, pp 1209–1210. https://doi.org/10.1037/0033-2909.I26.1.78

Labouret S, Frohly J (2011) Distribution en tailles des bulles d'un champ de cavitation ultrasonore. In: 10ème Congrès Français d'Acoustique. Lyon

Lee J et al (2005) Determination of the size distribution of sonoluminescence bubbles in a pulsed acoustic field. J Appl Chem Soc 127(48):16810–16811. https://doi.org/10.1021/ja0566432

Leighton TG (1997) The acoustic bubble. Academic, Cambridge. https://doi.org/10.1016/B978-0-12-441920-9.50001-8

Leighton T, Walton A, Pickworth MJ (1990) Primary Bjerknes forces. Eur J Phys 11:47–50

Liu YN et al (2008) Study on degradation of dimethoate solution in ultrasonic airlift loop reactor. Ultrason Sonochem 15(5):755–760. https://doi.org/10.1016/j.ultsonch.2007.12.004

Luche JL et al (1990) Organic sonochemistry : a new interpretation and its consequences. Tetrahedron Lett 31(29):4125–4128. https://doi.org/10.1016/S0040-4039(00)97559-2

Luo J, Fang Z, Smith RL (2014) Ultrasound-enhanced conversion of biomass to biofuels. Prog Energy Combust Sci 41(1):56–93. https://doi.org/10.1016/j.pecs.2013.11.001

Lupacchini M et al (2017) Sonochemistry in non-conventional, green solvents or solvent-free reactions. Tetrahedron 73(6):609–653. https://doi.org/10.1016/j.tet.2016.12.014

Madhavan J et al (2010a) Degradation of acid red 88 by the combination of sonolysis and photocatalysis. Sep Purif Technol 74(3):336–341. https://doi.org/10.1016/j.seppur.2010.07.001

Madhavan J et al (2010b) Sonophotocatalytic degradation of monocrotophos using TiO2 and Fe3+. J Hazard Mater 177:944–949. https://doi.org/10.1016/j.jhazmat.2010.01.009

Madhavan J, Grieser F, Ashokkumar M (2010c) Combined advanced oxidation processes for the synergistic degradation of ibuprofen in aqueous environments. J Hazard Mater 178(1–3):202–208. https://doi.org/10.1016/j.jhazmat.2010.01.064

Maleki A et al (2010) Study of photochemical and sonochemical processes efficiency for degradation of dyes in aqueous solution. Korean J Chem Eng 27(6):1805–1810. https://doi.org/10.1007/s11814-010-0261-0

Mason TJ (2000) Large scale sonochemical processing: aspiration and actuality. Ultrason Sonochem 7(4):145–149. https://doi.org/10.1016/S1350-4177(99)00041-3

Mason TJ (2003) Sonochemistry and sonoprocessing: the link, the trends and (probably) the future. Ultrason Sonochem 10:175–179. https://doi.org/10.1016/S1350-4177(03)00086-5

Mason TJ (2007) Sonochemistry and the environment—providing a "green" link between chemistry, physics and engineering. Ultrason Sonochem 14:476–483. https://doi.org/10.1016/j.ultsonch.2006.10.008

Merouani S, Hamdaoui O (2016) The size of active bubbles for the production of hydrogen in sonochemical reaction field. Ultrason Sonochem 32:320–327. https://doi.org/10.1016/j.ultsonch.2016.03.026

Merouani S, Ferkous H et al (2015a) A method for predicting the number of active bubbles in sonochemical reactors. Ultrason Sonochem 22:51–58. https://doi.org/10.1016/j.ultsonch.2014.07.015

Merouani S, Hamdaoui O et al (2015b) Sensitivity of free radicals production in acoustically driven bubble to the ultrasonic frequency and nature of dissolved gases. Ultrason Sonochem 22:41–50. https://doi.org/10.1016/j.ultsonch.2014.07.011

Mettin R et al (1997) Bjerknes forces between small cavitation bubbles in a strong acoustic field. Phys Rev E 56(3):2924–2931. https://doi.org/10.1103/PhysRevE.56.2924

Mettin R et al (2006) Modeling acoustic cavitation with bubble redistribution. In: Sixth International Symposium on Cavitation. Netherland, pp 1–5

Miksis JBKM (1980) Bubble oscillation of large amplitude. J Acoust Soc Am 68(2):628–633

Minnaert M (1933) On musical air-bubbles and the sounds of running water. Philos Mag Series 7 16(104):235–248. https://doi.org/10.1080/14786443309462277

Monnier H, Wilhelm AM, Delmas H (1999) Influence of ultrasound on mixing on the molecular scale for water and viscous liquids. Ultrason Sonochem 6(1–2):67–74. https://doi.org/10.1016/S1350-4177(98)00034-0

Mulvihill MJ et al (2011) Green chemistry and green engineering: a framework for sustainable technology development. Ann Rev Environ Resour 36:271–293. https://doi.org/10.1146/annurev-environ-032009-095500

Nagata Y et al (1992) Formation of colloidal silver in water by ultrasonic irradiation. J Chem Soc Chem Commun 21:1620–1622. https://doi.org/10.1039/C39920001620

Nagata Y et al (2006) 'Sonochemical formation of gold particles in aqueous solution. Radiat Res 146(3):333. https://doi.org/10.2307/3579465

Neppiras EA (1980) Acoustic cavitation. Phys Rep 61(3):159–251. https://doi.org/10.1016/0370-1573(80)90115-5

Nikpassand M, Fekri LZ, Farokhian P (2016) An efficient and green synthesis of novel benzoxazole under ultrasound irradiation. Ultrason Sonochem 28:341–345. https://doi.org/10.1016/j.ultsonch.2015.08.014

Noltingk BE, Neppiras EA (1950) Cavitation produced by ultrasonics. Proc Phys Soc Sect B 63(9):674–685. https://doi.org/10.1088/0370-1301/63/9/305

Okitsu K et al (1996a) Formation of noble metal particles by ultrasonic irradiation. Ultrason Sonochem 3:249–251. https://doi.org/10.1016/S1350-4177(96)00033-8

Okitsu K et al (1996b) Sonochemical preparation of ultrafine palladium particles. Chem Mater 8(2):315–317. https://doi.org/10.1021/cm950285s

Oussaid B, Soufiaoui M, Garrigues B (1995) The Atherton-Todd reactions under Sonochemical activation. Synth Commun 25(6):871–875. https://doi.org/10.1080/00397919508013423

Pang YL, Abdullah AZ, Bhatia S (2011) Review on sonochemical methods in the presence of catalysts and chemical additives for treatment of organic pollutants in wastewater. Desalination 277(1–3):1–14. https://doi.org/10.1016/j.desal.2011.04.049

Pankaj, Ashokkumar M (2011) Theoretical and experimental sonochemistry involving inorganic systems. Springer, Basel

Parkar PA, Choudhary HA, Moholkar VS (2012) Mechanistic and kinetic investigations in ultrasound assisted acid catalyzed biodiesel synthesis. Chem Eng J 187:248–260. https://doi.org/10.1016/j.cej.2012.01.074

Penconi M et al (2015) Hydrogen production from water by photolysis, sonolysis and sonophotolysis with solid solutions of rare earth, gallium and indium oxides as heterogeneous catalysts. Sustainability (Switzerland) 7(7):9310–9325. https://doi.org/10.3390/su7079310
Pétrier C, Combet E, Mason T (2007) Oxygen-induced concurrent ultrasonic degradation of volatile and non-volatile aromatic compounds. Ultrason Sonochem 14(2):117–121. https://doi.org/10.1016/j.ultsonch.2006.04.007
Prat D, Hayler J, Wells A (2014) A survey of solvent selection guides. Green Chem 16(10):4546–4551. https://doi.org/10.1039/c4gc01149j
Prat D et al (2015) CHEM21 selection guide of classical- and less classical-solvents. Green Chem 18(1):288–296. https://doi.org/10.1039/c5gc01008j
Richards WT, Loomis AL (1927) The chemical effects of high frequency sound waves I. A preliminary survey. J Am Chem Soc 49(12):3086–3100
Rivas DF et al (2010) Efficient sonochemistry through microbubbles generated with micromachined surfaces. Angewandte Chem Int Ed 10(49):9699–9701. https://doi.org/10.1002/anie.201005533
Rokhina EV, Lens P, Virkutyte J (2009) Low-frequency ultrasound in biotechnology: state of the art. Trends Biotechnol 27(5):298–306. https://doi.org/10.1016/j.tibtech.2009.02.001
Rossi F et al (2009) Study of catalysts for water photosonolysis to increase the hydrogen production photosonolysis process. In: Hypothesis VIII. Lisbon, pp 1–6
Rostamnia S, Lamei K (2011) A rapid, catalyst-free, three-component synthesis of rhodanines in water using ultrasound. Synthesis 19:3080–3082. https://doi.org/10.1055/s-0030-1260158
Saini RRS, Singh UUR (2002) Green chemistry: environment, economics, and competitiveness. Corp Environ Strateg 9(3):259–266. https://doi.org/10.1016/S1066-7938(02)00068-4
Sajjadi B, Raman AAA, Ibrahim S (2015) A comparative fluid flow characterisation in a low frequency/high power sonoreactor and mechanical stirred vessel. Ultrason Sonochem 27:359–373. https://doi.org/10.1016/j.ultsonch.2015.04.034
Sancheti SV, Gogate PR (2017) A review of engineering aspects of intensification of chemical synthesis using ultrasound. Ultrason Sonochem 36:527–543. https://doi.org/10.1016/j.ultsonch.2016.08.009
Sanghi R, Singh V, Sharma SK (2012) Environment and the role of green chemistry. In: Green chemistry for environmental remediation, pp 1–34. https://doi.org/10.1002/9781118287705.ch1
Saroha AK (2009) Green process engineering. Indian Chem Eng 51(2):v. https://doi.org/10.1080/00194500903361215
Schmitt FO, Johnson CH, Olson AR (1929) Oxidations promoted by ultrasonic radiation. J Am Chem Soc 51(2):370–375
Servant G et al (2003) On the interaction between ultrasound waves and bubble clouds in mono- and dual-frequency sonoreactors. Ultrason Sonochem 10(6):347–355. https://doi.org/10.1016/S1350-4177(03)00105-6
Sheldon RA (2007) The e factor: fifteen years on. Green Chem 9(12):1273–1283. https://doi.org/10.1039/b713736m
Sheldon RA (2018) Metrics of green chemistry and sustainability: past, present, and future. ACS Sustain Chem Eng 6(1):32–48. https://doi.org/10.1021/acssuschemeng.7b03505
Sheldon RA, Sanders JPM (2015) Toward concise metrics for the production of chemicals from renewable biomass. Catal Today 239:3–6. https://doi.org/10.1016/j.cattod.2014.03.032
Sillanpää M (2011) Ultrasound technology in green chemistry. In: Briefs in green chemistry for sustainability, pp 1–21. https://doi.org/10.1007/978-94-007-2409-9
Sivasankar T, Moholkar VS (2009) Physical insights into the sonochemical degradation of recalcitrant organic pollutants with cavitation bubble dynamics. Ultrason Sonochem 16(6):769–781. https://doi.org/10.1016/j.ultsonch.2009.02.009
Son Y (2016) Advanced oxidation processes using ultrasounds technology for water and wastewater treatment. In: Handbook of ultrasonics and sonochemistry, pp 1–22. https://doi.org/10.1007/978-981-287-278-4

Son Y, Lim M, Khim J (2009) Investigation of acoustic cavitation energy in a large-scale sonoreactor. Ultrason Sonochem 16(4):552–556. https://doi.org/10.1016/j.ultsonch.2008.12.004
Son Y et al (2010) 'Estimation of sonochemical reactions under single and dual frequencies based on energy analysis. Jpn J Appl Phys 49(7 PART 2):2–6. https://doi.org/10.1143/JJAP.49.07HE02
Son Y et al (2012) Comparison of calorimetric energy and cavitation energy for the removal of bisphenol-A: the effects of frequency and liquid height. Chem Eng J 183:39–45. https://doi.org/10.1016/j.cej.2011.12.016
Stavarache C et al (2003) Conversion of vegetable oil to biodiesel using ultrasonic irradiation. Chem Lett 32(8):716–717. https://doi.org/10.1246/cl.2003.716
Stavarache C et al (2005) Fatty acids methyl esters from vegetable oil by means of ultrasonic energy. Ultrason Sonochem 12(5):367–372. https://doi.org/10.1016/j.ultsonch.2004.04.001
Stavarache C et al (2007a) Ultrasonically driven continuous process for vegetable oil transesterification. Ultrason Sonochem 14(4):413–417. https://doi.org/10.1016/j.ultsonch.2006.09.014
Stavarache C, Vinatoru M, Maeda Y (2007b) Aspects of ultrasonically assisted transesterification of various vegetable oils with methanol. Ultrason Sonochem 14(3):380–386. https://doi.org/10.1016/j.ultsonch.2006.08.004
Suri C et al (2002) Chaotic mixing generated by acoustic streaming. Ultrasonics 40(1–8):393–396. https://doi.org/10.1016/S0041-624X(02)00150-6
Suslick KS (2001) Sonoluminescence and sonochemistry. In: Encyclopedia of physical science and technology, 3rd edn. Academic, San Diego
Suslick KS, Price GJ (1987) Applications of ultrasound to material chemistry. Ultrasonics 25:35–39. https://doi.org/10.1146/annurev.matsci.29.1.295
Suslick KS, Hyeon T, Fang M (1996) Nanostructured materials generated by high-intensity ultrasound: sonochemical synthesis and catalytic studies. Chem Mater 8(8):2172–2179. https://doi.org/10.1021/cm960056l
Suslick KS et al (1999) Applications of sonochemistry to materials synthesis. In: Sonochemistry and sonoluminescence, pp 291–320
Toma M et al (2011) A calorimetric study of energy conversion efficiency of a sonochemical reactor at 500kHz for organic solvents. Ultrason Sonochem 18(1):197–208. https://doi.org/10.1016/j.ultsonch.2010.05.005
Trost B (1991) The atom economy— a search for synthetic efficiency. Science 254(5032):1471–1477. https://doi.org/10.1126/science.1962206
Tsochatzidis NA et al (2001) Determination of velocity, size and concentration of ultrasonic cavitation bubbles by the phase-Doppler technique. Chem Eng Sci 56:1831–1840. https://doi.org/10.1016/S0009-2509(00)00460-7
Tuziuti T et al (2008) Mechanism of enhancement of sonochemical-reaction efficiency by pulsed ultrasound. J Phys Chem A 112:4875–4878. https://doi.org/10.1021/jp802640x
Waghmare GV, Vetal MD, Rathod VK (2015) Ultrasound assisted enzyme catalyzed synthesis of glycerol carbonate from glycerol and dimethyl carbonate. Ultrason Sonochem 22:311–316. https://doi.org/10.1016/j.ultsonch.2014.06.018
Xu S et al (2014) Bubble size distribution in acoustic droplet vaporization via dissolution using an ultrasound wide-beam method. Ultrason Sonochem 21(3):975–983. https://doi.org/10.1016/j.ultsonch.2013.11.016
Yasui K (1997) Alternative model of single bubble sonoluminescence. Phys Rev E 56(6):6750–6760. https://doi.org/10.1103/PhysRevA.65.054304
Yasui K (2002) Influence of ultrasonic frequency on multibubble sonoluminescence. J Acoust Soc Am 112(4):1405–1413. https://doi.org/10.1121/1.1502898
Yasui K (2011) Fundamentals of acoustic cavitation and sonochemistry. In: Theoretical and experimental sonochemistry involving inorganic systems. National Institute of Advanced Industrial Science and Technology, Anagahora, pp 1–29. https://doi.org/10.1007/978-90-481-3887-6
Yasui K et al (2008) The range of ambient radius for an active bubble in sonoluminescence and sonochemical reactions. J Chem Phys 128:184705. https://doi.org/10.1063/1.2919119

Yasui K et al (2010) Numerical simulations of acoustic cavitation noise with the temporal fluctuation in the number of bubbles. Ultrason Sonochem 17(2):460–472. https://doi.org/10.1016/j.ultsonch.2009.08.014

Yeung SA et al (1993) Formation of gold sols using ultrasound. J Chem Soc Chem Commun (4):378–379. https://doi.org/10.1039/C39930000378

Zhang Y, Zhang Y (2018) Chaotic oscillations of gas bubbles under dual-frequency acoustic excitation. Ultrason Sonochem 40(Pt B):151–157. https://doi.org/10.1016/j.ultsonch.2017.03.058

Zhang Y et al (2015) Instability of interfaces of gas bubbles in liquids under acoustic excitation with dual frequency. Ultrason Sonochem 23:16–20. https://doi.org/10.1016/j.ultsonch.2014.07.021

Zhang Y, Zhang Y, Li S (2016) The secondary Bjerknes force between two gas bubbles under dual-frequency acoustic excitation. Ultrason Sonochem 29:129–145. https://doi.org/10.1016/j.ultsonch.2015.08.022

Zhang Y, Guo Z et al (2018a) Acoustic wave propagation in bubbly flow with gas, vapor or their mixtures. Ultrason Sonochem 40(March):40–45. https://doi.org/10.1016/j.ultsonch.2017.03.048

Zhang Y, Gao Y et al (2018b) Effects of mass transfer on damping mechanisms of vapor bubbles oscillating in liquids. Ultrason Sonochem 40:120–127. https://doi.org/10.1016/j.ultsonch.2017.07.004

Zhang Y, Guo Z, Du X (2018c) Wave propagation in liquids with oscillating vapor-gas bubbles. Appl Thermal Eng 133:483–492. https://doi.org/10.1016/j.applthermaleng.2018.01.056

Chapter 16
Progress on Red Mud-Based Catalysts for the Removal of Environmental Pollutants Through Oxidation and Advanced Oxidation Process

Bikashbindu Das and Kaustubha Mohanty

Contents

Abbreviations

AOP	Advanced oxidation process
ARM	Activated red mud
CNTs	Carbon nanotubes
CO	Carbon monoxide
CVD	Chemical vapor deposition
DBT	Dibenzothiophene
FBCVD	Fluidized bed chemical vapor deposition
MWCNTs	Multi-walled carbon nanotubes
RM	Red mud
VOCs	Volatile organic compounds

B. Das · K. Mohanty (✉)
Department of Chemical Engineering, Indian Institute of Technology Guwahati, Guwahati, India
e-mail: d.bikashbindu@iitg.ac.in; kmohanty@iitg.ac.in

Inamuddin, A. M. Asiri (eds.), *Sustainable Green Chemical Processes and their Allied Applications*, Nanotechnology in the Life Sciences,
https://doi.org/10.1007/978-3-030-42284-4_16

16.1 Introduction

Environmental pollution that includes aquatic, land, and air pollution is becoming a threatening global issue and needs serious attention in the present time. Pollutants in the form of solid, liquid, and gas from various sources are continuously increasing with growing urbanization that in consequence is affecting the whole ecosystems and human health. Wastewater emerging from various industries and regular household use contain varieties of organic and inorganic contaminants that need effective treatment before it is allowed to enter the aquatic environment (Dorgerloh et al. 2008; Tijani et al. 2016; Bagnis et al. 2018). Similarly, the conversion of CO_x, NO_x, and other unburnt hydrocarbons that are emitted due to the incomplete combustion of fuels to less harmful products is considered as a major challenge for the whole research communities (Albonetti et al. 2003; Brijesh and Sreedhara 2013). Catalytic treatment of various organic and inorganic pollutants has become one of the emerging solutions to the ongoing issues related to the environmental pollution (Rezaee et al. 2014; Jain et al. 2018; Krýsa et al. 2018; Vilar et al. 2018; Gerasimov and Pogosbekian 2019).

CO is known as one of the most dangerous pollutants due to its highly toxic nature that adversely affects human and animal life (Soliman 2019). CO is mainly formed in the environment due to the incomplete combustion of carbon substances inside the internal combustion engine (Soliman 2019). It is associated with many human disorders such as sleepiness, impaired vision, and slow reflexes and in certain condition could lead to the death of an individual by replacing oxygen by binding to the iron atoms of hemoglobin (Soliman 2019).

Oxidation in the presence of a suitable catalyst is known as a very favorable technique for the removal of CO and other volatile hydrocarbons (Wang et al. 2011; Sushil and Batra 2012; Kim et al. 2015; de Lima et al. 2018; Shim et al. 2018). Iron catalysts have been significantly used for the oxidation process because of its easy availability and low price. Moreover, iron-based catalysts possess the ability to perform oxidation even when there is no external oxygen supplied during the reaction. When no external oxygen is present, it can release the lattice oxygen adhering to it and can directly oxidize the substrate (Sushil and Batra 2012). In a study, Mondal et al. (2004) [from 70] reported the CO oxidation by using Fe catalyst without the supply of external oxygen. They reported a suitable temperature range of 725–900 °C for the conversion of CO to CO_2. α-Fe_2O_3 prepared from goethite was reported to produce 99% CO conversion that was also without the use of external oxygen (Imai et al. 2001).

Many studies also reported iron oxide as an efficient support for the active metals such as Cu and Au (Liu et al. 2010; Cao et al. 2011; Qiao et al. 2011). These active agents have been found to improve the catalytic efficiency of iron oxides to a significant extent for the CO oxidation that has made iron oxide as a suitable candidate for this process. Similarly, several iron-based catalysts have shown promising activity for the oxidation of volatile organic compounds (Jodaei et al. 2011; Dobosz and Zawadzki 2015; Parvin et al. 2018). Jodaei et al. (2011) reported the Fe-Ag-ZSM-5 catalyst for the removal of ethyl acetate. The highest ethyl acetate removal of 96% was reported at an optimum reaction temperature of 350 °C. Dobosz and Zawadzki (2015) reported the microwave-assisted oxidation of propane by using α-Fe_2O_3 catalyst where they achieved >99% conversion at close to 400 °C reaction temperature.

Advanced oxidation process (AOP) has emerged as an effective solution to treat many toxic and non-biodegradable organic contaminants in the aquatic environment (Jain et al. 2018). Fenton reaction is one such process, which involves the generation of –OH radicals or $-SO^{-}_{4}$ radicals through the interaction of ferrous iron with hydrogen peroxide and persulfate in the reaction system (Poyatos et al. 2010; Feng et al. 2016; Vilar et al. 2018). Fenton process has many advantages over other conventional processes such as high removal efficiency, low-cost process, and requirement of moderate operating conditions and is involved with the non-toxic by-products formation (Poyatos et al. 2010; Jain et al. 2018).

Iron has been extensively used as a catalyst for the AOPs in which the catalytic nature of the redox couple Fe^{+2}/Fe^{+3} produces the –OH or $-SO^{-}_{4}$ radicals necessary for oxidizing the environmental pollutants (Oliveira et al. 2015; Feng et al. 2016). Approximately 90% methylene blue removal was reported in 1 h by using Ti-doped magnetite catalyst (Yang et al. 2009). Incorporation of Ti significantly contributed toward the activity of the iron catalyst. Hu et al. (2011) synthesized magnetic nanoparticles and carbon nanotube composites for the degradation of androgen 17α-methyltestosterone. Carbon nanotubes enhanced the adsorption of trace pollutants due to which a higher degradation efficiency was observed.

Various iron-based catalysts used for the oxidation and advanced oxidation process were synthetic iron catalysts, the high cost of which could make the process expensive. In context to this issue, researchers are now more interested to utilize suitable iron-containing waste materials of different origin that could make the process economical. Red mud is a hugely produced waste from aluminum industries across the world that contains a high amount of Fe. Due to few adverse properties such as huge production (more than 120 million tons), pH of 11–14, and presence of few radioactive elements, RM is considered as a potential threat to the environment (Das and Mohanty 2019b). Due to a large percentage of Fe (40–60%) in RM, it is considered very suitable for the oxidation and AOP for removing various organic and inorganic pollutants.

Several studies have reported RM-based catalysts for the oxidation and AOPs where the aluminum industry waste has shown promising activity. In this article, various RM-based catalysts applied for the oxidation and AOPs have been discussed in details. To the extent of our knowledge, the utilization of RM-derived catalysts for the oxidation and AOPs has not been reported till now. Other reported studies on the catalytic application of RM were mainly focused on processes such as hydrogenation, hydrodechlorination, NO_x reduction, pyrolysis, H_2 production, hydrodeoxygenation, and transesterification (Sushil and Batra 2008; Das and Mohanty 2019c). This article contains the catalytic application of RM for the oxidation and AOPs with a thorough discussion on the process mechanism, catalytic activity, process outputs, and catalytic stability. Moreover, a brief discussion on the physical and chemical characteristics of RM is included in this study.

16.2 Physical and Chemical Characteristics of Red Mud

The physical appearance of RM is red in color, which is due to the presence of iron in it. From the field emission scanning electron microscopy analysis, irregular size and almost circular shape of particles were observed in RM (Fig. 16.1) (Das and Mohanty 2019a). The elemental compositions as obtained from the X-ray fluorescence analysis are provided in Table 16.1 (Das and Mohanty 2019a). The major component of RM was reported as iron present in the form of hematite (Fe_2O_3). Different mineral phases present in RM as observed by the X-ray diffraction technique were Fe_2O_3, $Al(OH)_3$, $CaCO_3$, $Ca_2(SiO)_4$, SiO_2, $NaAlO_2$, $Na_8Al_6(SiO_4)_6Cl_2$, and TiO_2 (Das and Mohanty 2019b). Besides this, iron in the form of FeOOH, $Fe(OH)_3$, and Fe_3O_4 and alumina in the form of AlO(OH) were also found in RM (Sushil and Batra 2008). The surface area of RM was reported to remain between 20 and 30 $m^2\ g^{-1}$ that was greater than the surface area range of hematite (3–5 $m^2\ g^{-1}$) (Sushil and Batra 2008). However, the physiochemical properties of RM could vary with its area of origin, and accordingly its catalytic activity could vary for different processes.

Fig. 16.1 Surface morphology of raw red mud (adapted from Das and Mohanty 2019a)

Table 16.1 Composition of fresh red mud obtained from XRF analysis (adapted from Das and Mohanty 2019a)

Components	Amount (%)
Fe_2O_3	46.51
SiO_2	26.84
Al_2O_3	15.83
TiO_2	6.43
Na_2O	4.02
CaO	3.81
K_2O	0.33
MnO	0.15

16.3 Oxidation and Advanced Oxidation Process by Red Mud-Derived Catalysts

16.3.1 Oxidation by Red Mud-Derived Catalysts

Increasing environmental pollution due to the production of a huge amount of carbon monoxide (CO) from automobiles and industrial exhausts is a matter of concern in recent days. Oxidation of CO to carbon dioxide (CO_2) by using heterogeneous catalyst is considered as an effective way of reducing CO formation. Application of RM-derived catalysts for the oxidation of organic contaminants found in wastewater, sulfur-containing compounds in fuel oil, volatile organic compounds (VOC), and air emission are given in Table 16.2 (Paredes et al. 2004; Sushil and Batra 2012; Kim et al. 2015; Chen et al. 2016; Hu et al. 2016; Oliveira et al. 2017; de Lima et al. 2018; Shim et al. 2018; Tinh et al. 2019).

Catalytic oxidation of carbon monoxide (CO) in the presence of RM, activated RM (ARM), and acid-treated RM was performed by Sushil and Batra (2012). They prepared ARM by treating RM with HCl followed by precipitating the solution by adding aqueous ammonia. The resulting precipitate was washed, dried overnight, and calcined at 500 °C to get the ARM. TRM was prepared by the same method as explained for ARM up to the precipitating step, and then the precipitate was washed and dried. The FeOOH group in the precipitate was preserved by eliminating the calcination process, and the product obtained after simple drying at 120 °C was called TRM. The experimental results for the CO oxidation revealed a higher catalytic performance of TRM as compared to the RM and ARM catalyst. By using RM, ARM, and TRM, complete CO oxidation was reached at 500 °C, 420 °C, and 340 °C, respectively. Munteanu et al. (1997) suggested that the higher activity obtained for TRM compared to RM and ARM was because of the hydroxyl groups that facilitated the reduction of hematite to magnetite during the reaction. CO oxidation was believed to occur in two-step mechanism. In the first step, CO was considered to be oxidized by the oxygen attached to the metal oxide matrix. In the second step, the gaseous dioxygen molecules again reoxidized the reduced metal oxide. Thus, the activity of the oxidation process strongly depends on the ease at which

Table 16.2 Experimental conditions of oxidation process in the presence of RM-based catalysts

Process	Catalyst	Temperature	Pressure	Conversion	Reference
CO oxidation	Acid-treated RM	100–500 °C	–	>90% CO conversion above 400 °C	Sushil and Batra (2012)
CO oxidation	CuO-doped acid-treated RM	170 °C	–	100% CO conversion	Hu et al. (2016)
Methane combustion	Acid-treated RM	300–626 °C	1 atm	100% methane conversion at ~620 °C	Paredes et al. (2004)
Supercritical water oxidation disposal of sewage sludge	RM	400–500 °C	–	TOC removal rate of 99.3% at 400 °C	Chen et al. (2016)
Oxidation of VOCs (toluene)	Pt- and Co-impregnated acid-treated RM	180–300 °C	–	90% toluene conversion at 253 °C	Kim et al. (2015)
p-Xylene oxidation	CuO- and CeO_2-doped RM	275–400 °C	–	90% p-xylene oxidation at 350 °C	Tinh et al. (2019)
Benzene, toluene, and xylene oxidation	TiO_2-treated RM	–	–	80% COD removal in 2 h	de Lima et al. (2018)
Benzene oxidation	Pd-doped acid-treated RM	200–500 °C	–	90% conversion at ~ 280 °C	Shim et al. (2018)
Benzyl alcohol oxidation	Au-based RMC catalyst	130 °C	10 atm O_2	90% conversion in 10 h	Oliveira et al. (2017)

oxygen was released from the metal oxide catalyst. For the case of TRM, the longer bond length between Fe and O in the preserved FeOOH groups resulted in an easy oxygen removal than Fe_2O_3. The calculated activation energy value for RM, ARM, and TRM were found as 13.57, 12.49, and 7.99 kcal mol^{-1}, respectively. RM activated by the above method was found as a suitable catalyst for the CO oxidation. However, the stability of the catalyst for the reaction was not found very promising, and approximately 50 °C rise in temperature to achieve 50% conversion was required in the first reuse of the catalyst compared to the fresh catalytic use.

Hu et al. (2016) extended the work carried out by Sushil and Batra (2012) for increasing the catalytic activity of RM for the CO oxidation. They used CuO as an active agent to modify the catalytic properties of activated red mud (ARM). They prepared ARM by using similar methods as described by Sushil and Batra (2012) and examined the impact of increasing the CuO content on the activity of ARM catalyst for the CO oxidation. It was found that the activity of catalyst increased by increasing the CuO content and Cuo/ARM-20% (20% loading of CuO) produced the highest catalytic effect for CuO oxidation. The activity of CuO catalyst was better than the Cu-doped ARM catalyst. The increase in activity of CuO/ARM as compared to bulk CuO for CO oxidation was due to the improvement in the dispersion of Cu species over the ARM support having a high surface area and rich pore structure (Cao et al. 2008). However, a further increase in CuO content to 30% resulted in decreased catalytic activity. This was because of the discrete crystalline phase of bulk copper oxide formed on the support at a higher percentage of CuO loading (Friedmaan et al. 1978). Another interesting phenomenon observed for the ARM catalyst prepared by Hu et al. (2016) was catalytic activity started at a temperature of 200 °C (with low activity). Whereas, in the earlier work reported by Sushil and Batra (2012), CO and O_2 were found to have very low reactivity even up to a temperature of 600 °C. This might be because of the higher surface area of catalyst with rich pore structure obtained by Hu et al. (2016) as compared to (Sushil and Batra (2008). CuO loading on ARM resulted in a significant increase in the catalytic activity of the reaction. Thus, CuO was considered as an important active agent for oxidation of CO to CO_2. The activity of catalyst with different CuO loading for CO oxidation is shown in Fig. 16.2. By using Cuo/ARM-20% catalyst, complete conversion was obtained at 170 °C as compared to 420 °C reported by Sushil and Batra (2008). Experiments carried out at different calcination temperatures revealed that high-temperature calcination resulted in the collapse of pore structures with a simultaneous change in the crystalline phases in the support that decreased the performance of the catalyst. The catalytic stability of CuO/ARM was found to be very high with almost no change in activity observed for 12 h of reaction.

Paredes et al. (2004) prepared ARM (activated red mud) by treating RM with HCl followed by precipitating the solution with ammonia. Further, the precipitate was dried and calcined at 500 °C and used for the methane oxidation reaction. In

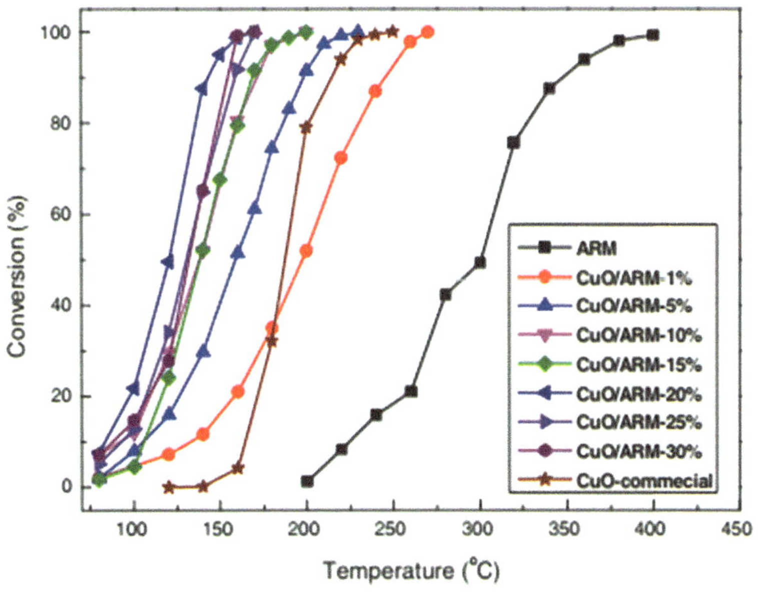

Fig. 16.2 Catalyst activity with different CuO loading calcined at 200 °C (adapted from Hu et al. 2016)

another activation technique to obtain the PARM catalyst, they used a mixture of aqueous HCl and H_3PO_4 and followed a procedure the same as described above. The experimental results revealed that ARM exhibited higher activity than PARM and the activity of ARM was very much comparable to pure hematite on the iron content basis. The use of H_3PO_4 as an additional activating substrate did not improve the catalytic activity and stability of PARM. This suggested that the addition of P to ARM had no significant activity toward methane combustion. The stability of ARM was comparable to that of pure hematite. After 110 h of reaction time, ARM was able to maintain 70% of the initial activity. Temperature-programmed oxidation results confirmed that reduction of hematite (active phase) was not significant during the reaction, thus suggested sintering as the main cause for catalyst deactivation. The mechanism of methane combustion in the presence of RM catalysts was not clear which could be a very much important subject of research for the future. The catalytic performance of Cu-Cr-Ti commercial catalyst was found very lower than the ARM catalyst for the combustion of methane.

Kim et al. (2015) carried out volatile organic compound oxidation by using platinum (Pt)-impregnated acid-treated red mud (Pt/HRM) and compared its activity with Pt/Al catalyst. The calcined HRM (400 °C) produced 90% conversion of toluene at 464 °C. Calcination above 400 °C resulted in the decreased catalytic activity of HRM that clearly indicated the influencing potential of the calcination temperature on the catalytic activity of HRM. Acid treatment decreased the Na and Ca content in HRM, which are generally considered as sintering agents that could affect the catalytic properties when subjected to calcination at different temperature. The phenomenon also increased the surface area of HRM that might have contributed toward the catalytic activity to a reasonable extent. Calcination temperature above 400 °C resulted in sintering of particles that decreased the surface area as well as pore volume. Increasing Pt content from 0.1 wt% up to 1 wt%, the activity of the Pt/HRM catalyst increased which was reflected in the lower conversion temperature than the RM and HRM catalyst. The Pt/HRM catalyst also produced higher activity with increasing Pt loading up to 1 wt% than the Pt/Al catalyst. The toluene conversion with increase in temperature for both Pt/HRM and Pt/Al is shown in Fig. 16.3. The higher activity of Pt/HRM catalyst was attributed to the stronger interaction between platinum oxides and other metal oxides that enhanced the lattice oxygen mobility in the catalytic system. In addition, the better reduction properties of platinum oxide, which resulted in an improved transfer of lattice oxygen on the HRM surface, could also contribute for the higher activity of PT/HRM. The stability of Pt/HRM was found very promising with no decrease in activity observed for 48 h of reaction time. Presence of water vapor is known to affect the oxidation process up to a significant extent by altering the properties of the carrier as well as the active metal on the carrier. For this study, increased water vapor content decreased the activity of RM-based Pt catalysts but had no effect on the temperature for achieving a complete toluene conversion. The mutual effect between water vapor and the support as well as active site is still an undefined subject, which needs further analysis in the future.

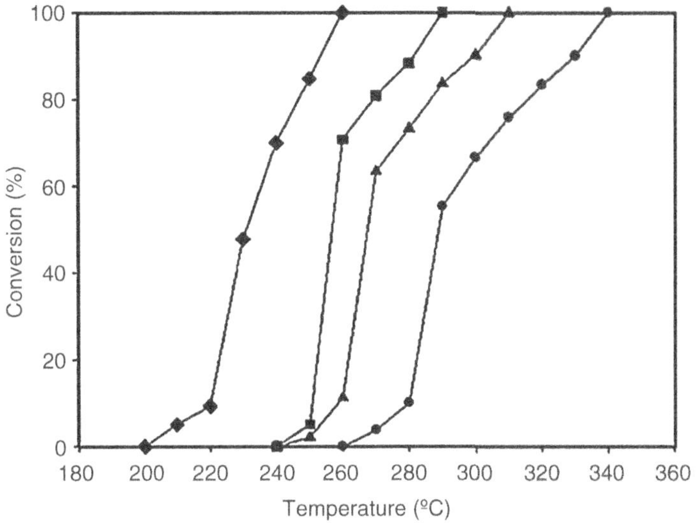

Fig. 16.3 Toluene conversion according to reaction temperature (reaction conditions: toluene concentration = 1000 ppm, GHSV = 75,000 h-1). (●) 0.1 Pt/HRM (400), (▲) 0.3 Pt/HRM (400), (■) 0.5 Pt/HRM (400), (◆) 1 Pt/HRM (400) (adapted from Kim et al. 2015)

16.3.2 Advanced Oxidation Processes by Red Mud-Derived Catalysts

Advanced oxidation process (AOP) applies the Fenton or Fenton-like mechanism for the elimination of various environmental pollutants in wastewater. The mechanism generally demands a catalyst with multiple redox states that can transfer electrons in the catalytic system to activate the oxidant used for the oxidation. Fe as a major constituent of RM has the ability to get oxidized or reduced in the presence of a suitable environment due to which RM has been utilized successfully for the AOPs. Various RM-based catalysts reported till now for the AOPs are shown in Table 16.3 (Costa et al. 2010; Oliveira et al. 2011, 2014, 2015; Saputra et al. 2012; Muhammad et al. 2012; de Resende et al. 2014; Bento et al. 2016; Sahu and Patel 2016; Shao et al. 2016; Xu et al. 2016a, b; Dias et al. 2016; Feng et al. 2016; Rath et al. 2017; Wei et al. 2017; do Prado et al. 2017; Cruceanu et al. 2018; Li et al. 2019).

Saputra et al. (2012) prepared cobalt-impregnated RM (Co/RM) and fly ash (Co/FA) for the oxidation of phenol in wastewater. The process was performed in the presence of oxone and in the temperature range between 25 and 45 °C. The results revealed that only RM and FA were not effective to activate oxone for the sulfate radical production. However, when used with Co, the Co/RM catalyst produced the highest rate for the degradation of phenol (100% conversion in 90 min). The higher activity of Co/RM as compared to Co/FA catalyst was associated with highly dispersed Co_3O_4 on the surface of RM, which provided relatively higher number of

Table 16.3 Experimental conditions for the AOPs in the presence of RM-based catalysts

Process	Catalyst	Temperature	Pressure	Conversion	Reference
Biphasic oxidation of sulfur compound	Gold nanoparticles-impregnated modified RM	–	–	~4.25 mg g_{cat}^{-1} of sulfur removal in 120 min	Oliveira et al. (2015)
Biphasic oxidation of sulfur compounds	RM/CNT composite	–	–	~73% Sudan IV, ~74% thiophene, and ~83% dibenzothiophene removal	Oliveira et al. (2014)
Phenol removal	Co/RM	25 °C	–	100% removal in 90 min	Saputra et al. (2012)
Methylene blue removal	RM/PET composite	–	–	90% removal in 24 h	Bento et al. (2016)
Textile dye degradation	RM/C composite	–	–	100% degradation	Dias et al. (2016)
Sulfadiazine mineralization	RM	20 °C	–	100% degradation	Feng et al. (2016)
Bezafibrate removal	Co-modified RM	–	–	~84% removal in 30 min	Xu et al. (2016b)
Dibenzothiophene removal	RM	25 °C	–	–	de Resende et al. (2014)
Butyl xanthate removal	Acidified/calcined RM	–	–	92% removal	Shao et al. (2016)
Bezafibrate removal	Ce-doped RM	Ambient temperature	–	96% removal	Xu et al. (2016a)
Methylene blue removal	Co-doped neutralized RM	–	–	97.21% removal in 150 min	Sahu and Patel (2016)
Phenol removal	Co/RM catalyst	25 °C	–	100% removal	Muhammad et al. (2012)
Chromium reduction and methylene blue degradation	Reduced RM	–	–	~65% MB removal and ~75% Cr reduction	Costa et al. (2010)
Sulfide removal	Modified RM	–	–	>90% conversion	Cruceanu et al. (2018)
Orange II degradation	Molasses wastewater treated RM	–	–	90% in 6 h	Wei et al. (2017)
Malachite green removal	Mesoporous silica/RM	–	–	97% removal	Rath et al. (2017)
Orange II degradation	RM-based $ZnO-Fe_2O_3$	–	–	~75% removal	Li et al. (2019)
Dibenzothiophene removal	PET/RM	–	–	80% removal	do Prado et al. (2017)
Methylene blue removal	RM/C composite	–	–	>95% removal	Oliveira et al. (2011)

active sites for the activation of oxone. For Co/RM, the optimum oxone concentration to achieve the highest activity was found to be 0.4 g, and beyond that, the catalytic activity started to decrease. The mechanism of oxone interaction with the catalyst was not discussed which could be considered as a subject of interest for future investigation. Increasing the reaction temperature resulted in higher activity of the Co/RM catalyst. A complete phenol conversion was obtained in 30 min at 45 °C as compared to a higher reaction time of 90 min at 25 °C.

Composites based on RM and PET were used by Bento et al. (Bento et al. 2016) for removing methylene blue from wastewater. The reaction was performed by using H_2O_2 as an oxidant in a batch reactor. At low reaction time, the conversion of MB was found very less (<40%), and it significantly increased with reaction time. The highest conversion was found for the RM/PET-15 catalyst (15 wt % PET deposited on RM) after 24 h. The low activity at short reaction time was associated with the presence of less amount of reduced iron Fe^{2+} in the reaction system. Fe^{2+} generally improves the formation of OH radicals from H_2O_2 that in result enhances the rate of oxidation of organic reactants. To confirm that MB removal was not only occurred by adsorption but also oxidation contributed to its removal, they monitored the reaction by ESI-MS. Different peaks obtained from ESI-MS spectra after 24 h reaction time that was related to MB oxidation intermediates confirmed the contribution of oxidation in addition to adsorption for the removal of MB. Catalyst stability study revealed that after two reuse, the catalytic activity slightly increased for the third and fourth reuse. This phenomenon also supported the theory of reduction of Fe^{+3} during reactions to more active Fe^{2+} and Fe^0 ions, which were more active for the OH radical production than Fe^{+3} ions.

16.3.2.1 Carbon Nanotube Production on Red Mud Surface and its Catalytic Applications in Advanced Oxidation Processes

Carbon nanotube (CNT)-based catalysts are known to have advantages of high surface area, increased stability, and features of interparticle mesopores. These properties of CNT help to lower the mass transfer resistance of the reaction system and positively contribute to the overall efficiency of the process. However, the high cost associated with the CNT production limits its wide range of application as catalysts. Research related to the efficient and cost-effective production of CNTs from waste materials is given significant attention in current days. Below paragraphs discuss some of the reported literature for the production of CNT on the surface of RM and its catalytic application for different catalytic processes.

Dunens et al. (2010) reported multi-walled carbon nanotube (MWCNT) production on RM substrate by fluidized bed chemical vapor deposition technique (FBCVD). RM was washed, dried, pulverized, and calcined at 800 °C before subjected to FBCVD reactor where the reaction was carried out at 650 °C under the flow of C_2H_4, H_2, and N_2 (ratio of 1:0.4:1). Carbon yield of 46.8 wt% was confirmed from the TGA analysis of the reaction product, and 75% of MWCNT was formed as a carbon product by this FBCVD process. The wall quality and degree of graphitization were characterized by Raman spectroscopy (Fig. 16.4). The G-band to D-band ratio ($I_G/I_D = 1.65$) was found to be very similar to that of industrial-grade MWCNT. The study clearly discovered a very economical way of CNT production by using a highly produced industrial waste called as red mud.

Dias et al. (2016) used chemical vapor deposition (CVD) technique to synthesize carbon-deposited RM catalyst (RM/C) for the textile dye (RB5) degradation through photo-Fenton process. Different experiments were conducted by varying the H_2O_2 concentration and pH at a fixed catalyst and dye concentration. The characterization results revealed that the reaction of ethanol with RM during the CVD process resulted in the formation of reduced iron phases (formation of Fe^{+2} and Fe^0 from α-Fe_2O_3 and dispersed Fe^{+3}) which were the important active species for the photo-Fenton process. Carbon in the form of nanotubes and nanofilaments was found to be deposited on RM surface (Fig. 16.5). This phenomenon also resulted in an increase in the surface area from 24 $m^2\ g^{-1}$ for pure RM to 68 $m^2\ g^{-1}$ for the RM/C catalyst. The effect of radiation power in the dye degradation process was investigated by keeping the RB5 dye in contact with the catalyst in dark. The amount of dye adsorbed by this technique was found to be 10–22% only. At power radiation of 36 Wt/m^2, 100% decay concentration of the RB5 black occurred. The increased activity of photo-Fenton reaction with the increase in radiation power was associated with the increase in system activity, which favored the radical formation as represented in the below equations:

$$H_2O + h\_ \rightarrow \cdot H + \cdot OH \quad (16.1)$$

$$H_2O_2 + h\_ \rightarrow 2 \cdot OH \quad (16.2)$$

$$R + h\_ \rightarrow R \cdot \quad (16.3)$$

Experiments performed by varying the pH value and keeping the radiation power at 36 ($W\ m^{-2}$) revealed close to 100% degradation of dye at pH 3. At higher pH of 6 and 9, the degradation was very low, i.e., 30% and 40%, respectively. Low pH and the presence of visible solar lights favored the formation of most photoactive Fe^{+3}

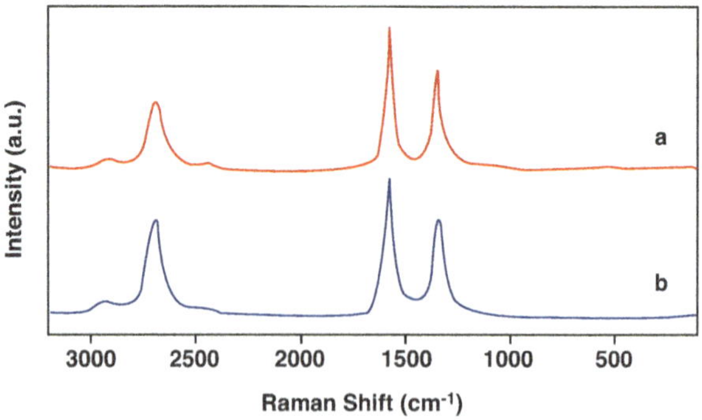

Fig. 16.4 Raman spectra of (**a**) acid-purified MWCNTs and (**b**) as-synthesized MWCNTs produced by fluidized bed chemical vapor deposition using a red mud catalyst and ethylene as the carbon source (adapted from Dunens et al. 2010)

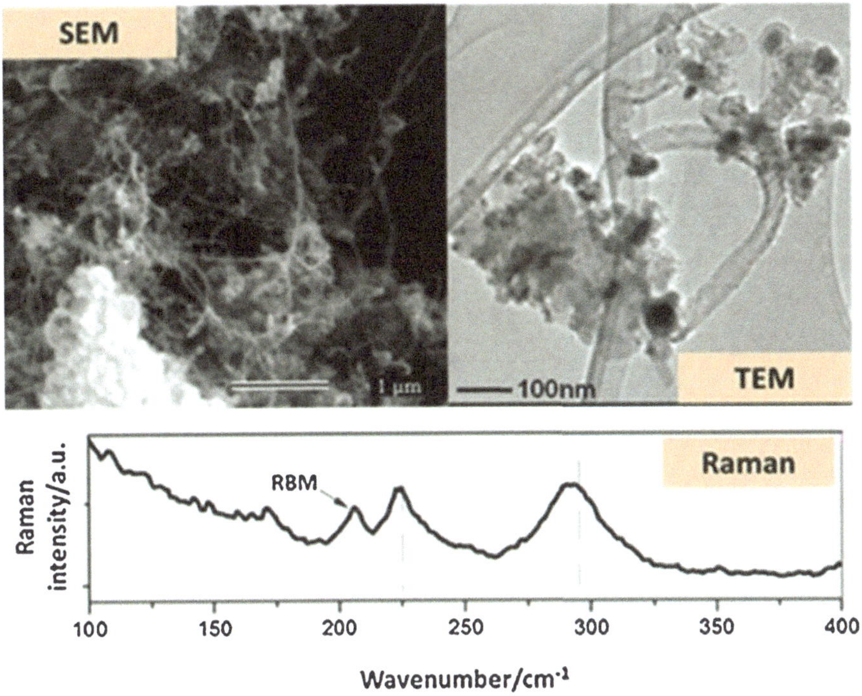

Fig. 16.5 SEM and TEM images and Raman spectrum obtained of the red mud catalyst after CVD process with ethanol (RM/C) (adapted from Dias et al. 2016)

hydroxyl complexes of $Fe(OH)^{2+}$ and more OH radicals, which enhanced the degradation process. Experiments on photo-Fenton reactions with different concentration of H_2O_2 revealed that 11 mM of H_2O_2 at pH 3 produced the best activity with 100% dye degradation. The efficiency decreased when the H_2O_2 concentration increased to 33 and 65 mM. The highest degradation efficiency at 11 mM of H_2O_2 was due to the formation of optimum amounts of –OH radicals at this concentration of H_2O_2. However, at a high H_2O_2 concentration, various undesired parallel reactions between the excess radicals could have occurred that resulted in the formation of H_2O_2, O_2, and hydroperoxy radicals (–OOH). These hydroperoxy radicals provided a lower reduction potential as compared to –OH radicals and decreased the efficiency of the process.

Gold nanoparticles impregnated on different types of RM surface such as raw RM, RM reduced with hydrogen (RmH_2), and partially carbon-coated RM synthesized by chemical vapor deposition (CVD) method (RmEt) were investigated for the oxidation of sulfur compounds by Oliveira et al. (2015). Dibenzothiophene (DBT) was oxidized by using all the above-prepared catalysts, and it was found that efficiency of Au/RmEt (gold nanoparticles supported on RmEt) was twice as compared to other Au-supported catalysts. The high efficiency of RmEt support was attributed to its iron composition, i.e., the presence of magnetite (Fe_3O_4) and metallic iron

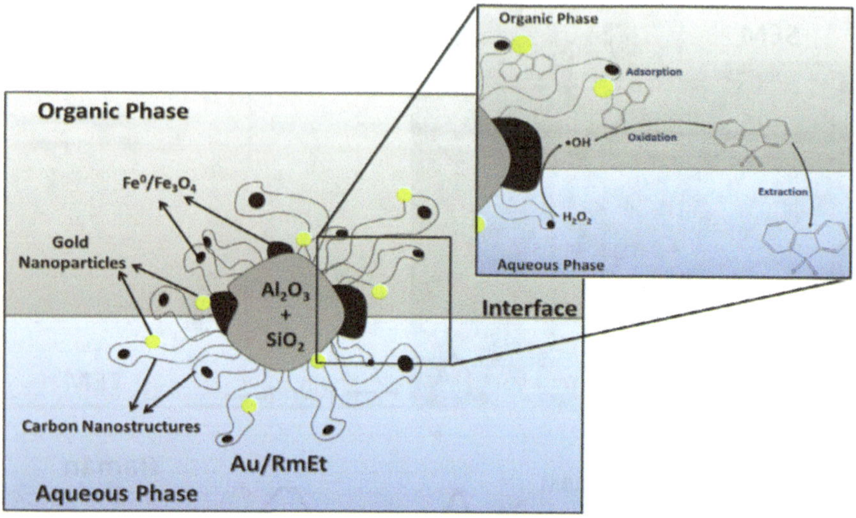

Fig. 16.6 Mechanism of synergistic effect between Au and Fe (adapted from Oliveira et al. 2015)

(Fe^0). Fe^{2+} ions in Fe_3O_4 helped to activate hydrogen peroxide (H_2O_2) that produced more OH radicals for the oxidation of DBT. Moreover, the Fe^0 ions in combination with Fe_3O_4 improved the transfer of electrons from Fe_3O_4 to H_2O_2 to produce OH radicals that initiated the reaction by Haber-Weiss mechanism as given below (Eq. 16.4). The Au/RmEt catalyst was also found amphiphilic in nature, i.e., it was located at the interface of the two immiscible phases, thus simultaneously interacting with the H_2O_2 molecules in the aqueous phase and sulfur contaminants in the organic phase (Fig. 16.6). The synergistic effect between Au nanoparticles and Fe^{2+} (Fig. 16.6) helped in adsorption of DBT on the surface of Au followed by oxidation by the OH radicals. This phenomenon also contributed to the higher activity of Au/RmEt as compared to other two Au-impregnated RM catalysts. The lower efficiency of pure RM was due to the presence of Fe^{3+} oxides such as γ-Fe_2O_3 and α- Fe_2O_3, which exhibited very low activity for H_2O_2 activation. For the case of RmH_2, the lower activity was associated with the presence of only Fe^0 ions, for which the direct transfer of electrons to H_2O_2 at neutral pH condition was a slow process.

$$Fe^{2+}{}_{surf.} + H_2O_2 \rightarrow \cdot OH + OH - + Fe^{3+}{}_{surf.} \qquad (16.4)$$

Oliveira et al. (2011) used RM and ethanol to produce carbon nanotube-coated nanostructured magnetic composite which was used for commercial dye adsorption. They performedchemical vapor deposition (CVD) in the temperature range of 500–950 °C by taking ethanol as carbon precursors to produce carbon nanotubes on the catalyst surface. Mossbauer spectra obtained for the catalyst confirmed that at a temperature of 700 °C, all the hematite of RM converted into metallic iron and iron carbide. The CVD at 700 °C also led to the formation of meso- and macroporous pores that produced a higher catalyst surface area of 79 m^2 g^{-1} as compared to 36 m^2 g^{-1} at 500 °C. Raw RM produced very low adsorption of methylene blue and indigo carmine. Especially for the anionic dye indigo carmine, the negatively charged RM was not suitable for its adsorption at neutral pH. Carbon-deposited RM showed excellent improvement in activity for the adsorption of dyes, which might be because of its higher surface area and hydrophobicity nature.

Carbon nanotube production on RM, its characterization, and application of the RM/CNT composite for lead (Pb) adsorption were reported by Mesgari Abbasi et al. (2016). Temperature range between 600 and 850 °C was selected for the CVD process with the flow of CH_4 and H_2 gas in the ratio of 1:10 for 45 min. During the CVD process, hematite present in RM was converted to magnetite by reacting with methane at 600 °C. By increasing the temperature to 750 °C, Fe_3C was formed due to deposition of carbon on the substrate, which was confirmed from the XRD analysis. FESEM and TEM image revealed that better carbon growth in the form of nanotubes was observed at the temperature of 750 °C with a carbon yield of around 50%. This also led to an increased surface area for the RM/CNT composite. The adsorption capacity of RM/CNT composite increased with time, and the highest value of adsorption was observed with a set of parameters of pH 5, 0.05 g of RM/CNT, and 50 min of adsorption time.

16.4 Conclusion and Future Perspective

Red mud, a hugely produced hazardous waste from aluminum industries, has been applied as a catalyst for the removal of environmental contaminants. The promising activity of RM-based catalysts for the oxidation and AOPs has created an economic pathway of reducing the waste material from the environment. The promising activity of RM-based catalysts for the process of oxidation and AOPs is mostly associated with the presence of iron in it. Besides the higher activity of iron, other metals such as Al, Si, and Ti of RM are also considered to have reasonable activity for these processes. Although the activity of fresh or unmodified RM is not very significant, few modifications as presented in different sections of this study could enhance its activity to a reasonable extent. For example, calcination of the CuO-supported RM at a high temperature resulted in collapsing of the structure of catalyst, thus decreasing the catalytic activity. Other methods of increasing the catalytic activity include impregnation of different active metals and metal oxides on it. For example, impregnation of CuO on RM for the oxidation of CO increased the catalytic activity as compared to bulk CuO catalyst. Pt supported on RM for the VOC oxidation showed higher activity than the Pt/Al catalyst, and the activity was increased with Pt loading. Besides these applications of RM as catalytic agents, it could also act as an excellent support for many catalytic reactions with advantages of high surface area for dispersion of metals, greater physical stability, and ability to generate the lattice O_2 during the oxidation processes. In addition to this, few researchers also reported carbon nanotube- and nanofiber-coated RM for different applications, and this modified method has increased the efficiency of the catalyst up to a significant value. For example, during the oxidation of sulfur compounds, nano-sized gold impregnated on carbon-coated RM resulted in double the efficiency as compared to raw and reduced RM support.

Besides the suitable characteristics of RM for the oxidation and AOPs, it was also found to have certain drawbacks that need to be addressed in the future to make RM applicable for large-scale purposes.

1. Advanced oxidation method (AOP) which involves the Fenton mechanism is considered as a promising technology for organic pollutant degradation. Many studies have been reported in this area by using different types of homogeneous and heterogeneous catalysts, out of which iron has emerged as an economic and efficient catalyst for this process. Researchers are more interested in using heterogeneous catalysts because of its various advantages over homogeneous catalysts. The use of artificial UV for the AOP is a very expensive process, and the amount of UV contained in the sunlight reaching earth is only 4%. Thus, the development of heterogeneous photo-Fenton catalysts for organic pollutant removal by using sunlight irradiation is a very much desirable process, for which RM can be considered as a promising catalyst. Few works have reported the photo-Fenton catalytic reaction for pollutant degradation by using RM-based catalysts, but further in-depth research is required to increase its efficiency toward the process in the future.
2. Preserving the FeOOH group in the activated RM was observed to be very productive for the oxidation of CO. However, the stability of the catalyst when used for the second cycle was not found as good as the fresh catalyst. CuO/ARM catalyst produced very promising activity for the conversion of CO to CO_2 with reasonable stability. Therefore, further studies regarding the behavior of other active metal oxides on RM surface for the oxidation process need to be done. In addition, the use of this kind of catalyst development methods, which showed favorable activity, requires further investigations toward other oxidation reactions that would make RM as an oxidation catalyst for large-scale purposes in future. Presence of water vapor during the oxidation process could play an important role, but the effect varied in a random manner for the reported RM catalyst, which needs further study.
3. Few recent works have successfully produced carbon nanotubes on the surface of RM by the CVD process, which has significantly contributed toward the catalytic activity for the AOPs. The RM/CNT-based catalysts showed higher magnetic properties, higher surface area, and low leaching characteristics. Nevertheless, sufficient work has not been performed to widen this method of catalytic reaction toward AOPs as well as other industrial processes. In-depth research on this catalyst preparation technique and its application for other chemical processes could make the RM/CNT catalyst suitable to replace other conventional catalysts.

References

Albonetti S, Blasioli S, Bugani M et al (2003) Effect of silica additive on the thermal stability of catalysts for NOx abatement. Environ Chem Lett 1:197–200. https://doi.org/10.1007/s10311-003-0036-5

Bagnis S, Fitzsimons MF, Snape J et al (2018) Processes of distribution of pharmaceuticals in surface freshwaters: implications for risk assessment. Environ Chem Lett 16:1193–1216. https://doi.org/10.1007/s10311-018-0742-7

Bento NI, Santos PSC, de Souza TE et al (2016) Composites based on PET and red mud residues as catalyst for organic removal from water. J Hazard Mater 314:304–311. https://doi.org/10.1016/j.jhazmat.2016.04.066

Brijesh P, Sreedhara S (2013) Exhaust emissions and its control methods in compression ignition engines: a review. Int J Automot Technol 14:195–206. https://doi.org/10.1007/s12239

Cao JL, Shao GS, Wang Y et al (2008) CuO catalysts supported on attapulgite clay for low-temperature CO oxidation. Catal Commun 9:2555–2559. https://doi.org/10.1016/j.catcom.2008.07.016

Cao J, Wang Y, Ma T et al (2011) Synthesis of porous hematite nanorods loaded with CuO nanocrystals as catalysts for CO oxidation. J Nat Gas Chem 20:669–676. https://doi.org/10.1016/S1003-9953(10)60238-1

Chen H, Wang G, Xu Y et al (2016) Application of red mud as both neutralizer and catalyst in supercritical water oxidation (SCWO) disposal of sewage sludge. RSC Adv 6:54202–54214. https://doi.org/10.1039/c6ra07458h

Costa RCC, Moura FCC, Oliveira PEF et al (2010) Controlled reduction of red mud waste to produce active systems for environmental applications: heterogeneous Fenton reaction and reduction of Cr(VI). Chemosphere 78:1116–1120. https://doi.org/10.1016/j.chemosphere.2009.12.032

Cruceanu A, Zăvoianu R, Pavel OD et al (2018) Alternative valorization of red mud waste as functional materials with catalytic activity for sulfide oxidation in wastewater. Int J Environ Sci Technol 15:895–908. https://doi.org/10.1007/s13762-017-1449-1

Das B, Mohanty K (2019a) A green and facile production of catalysts from waste red mud for the one-pot synthesis of glycerol carbonate from glycerol. J Environ Chem Eng 7(1):102888. https://doi.org/10.1016/j.jece.2019.102888

Das B, Mohanty K (2019b) Exploring the promotional effects of K, Sr, and Mg on the catalytic stability of red mud for the synthesis of glycerol carbonate from renewable glycerol. Ind Eng Chem Res 58(35):15803–15817. https://doi.org/10.1021/acs.iecr.9b00420

Das B, Mohanty K (2019c) A review on advances in sustainable energy production through various catalytic processes by using catalysts derived from waste red mud. Renew Energy 143:1791–1811. https://doi.org/10.1016/j.renene.2019.05.114

de Lima BA, de Castro PPR, França AB et al (2018) Evaluation of the effectiveness of red mud-supported catalysts in combination with ozone and TiO2 in the treatment of solution containing benzene, toluene, and xylene. Environ Monit Assess 190(9):560. https://doi.org/10.1007/s10661-018-6924-8

de Resende EC, Carvalho IDRG, Schlaf M, Guerreiro MC (2014) Red mud waste from the Bayer process as a catalyst for the desulfurization of hydrocarbon fuels. RSC Adv 4:47287–47296. https://doi.org/10.1039/c4ra07635d

Dias FF, Oliveira AAS, Arcanjo AP et al (2016) Residue-based iron catalyst for the degradation of textile dye via heterogeneous photo-Fenton. Appl Catal B Environ 186:136–142. https://doi.org/10.1016/j.apcatb.2015.12.049

do Prado NT, Heitmann AP, Mansur HS et al (2017) PET-modified red mud as catalysts for oxidative desulfurization reactions. J Environ Sci (China) 57:312–320. https://doi.org/10.1016/j.jes.2017.01.011

Dobosz J, Zawadzki M (2015) Total oxidation of lean propane over α-Fe_2O_3 using microwaves as an energy source. React Kinet Mech Catal 114:157–172. https://doi.org/10.1007/s11144-014-0776-1

Dorgerloh U, Becker R, Nehls I (2008) Determination of volatile organic sulfur compounds in contaminated groundwater. Environ Chem Lett 6:101–106. https://doi.org/10.1007/s10311-007-0117-y

Dunens OM, MacKenzie KJ, Harris AT (2010) Synthesis of multi-walled carbon nanotubes on "red mud" catalysts. Carbon 48:2375–2377. https://doi.org/10.1016/j.carbon.2010.02.039

Feng Y, Wu D, Liao C et al (2016) Red mud powders as low-cost and efficient catalysts for persulfate activation: pathways and reusability of mineralizing sulfadiazine. Sep Purif Technol 167:136–145. https://doi.org/10.1016/j.seppur.2016.04.051

Friedmaan RM, Freeman JJ, Farrel LW (1978) Characterization of Cu/Al2O3 catalysts. J Catal 55:10–28. https://doi.org/10.1016/0021-9517(78)90181-1

Gerasimov G, Pogosbekian M (2019) Nanostructured catalysts in vehicle exhaust control systems. Handb Ecomater 3:1679–1700. https://doi.org/10.1007/978-3-319-68255-6_120

Hu X, Liu B, Deng Y et al (2011) Adsorption and heterogeneous Fenton degradation of 17α-methyltestosterone on nano Fe3O4/MWCNTs in aqueous solution. Appl Catal B Environ 107:274–283. https://doi.org/10.1016/j.apcatb.2011.07.025

Hu ZP, Zhu YP, Gao ZM et al (2016) CuO catalysts supported on activated red mud for efficient catalytic carbon monoxide oxidation. Chem Eng J 302:23–32. https://doi.org/10.1016/j.cej.2016.05.008

Imai T, Matsui T, Fujii Y et al (2001) Oxidation catalyst of iron oxide suppressing dioxin formation in polyethylene combustion. J Mater Cycles Waste Manage 3:103–109. https://doi.org/10.1007/s10163-000-0064-3

Jain B, Singh AK, Kim H et al (2018) Treatment of organic pollutants by homogeneous and heterogeneous Fenton reaction processes. Environ Chem Lett 16:947–967. https://doi.org/10.1007/s10311-018-0738-3

Jodaei A, Niaei A, Salari D (2011) Performance of nanostructure Fe-Ag-ZSM-5 catalysts for the catalytic oxidation of volatile organic compounds: process optimization using response surface methodology. Korean J Chem Eng 28:1665–1671. https://doi.org/10.1007/s11814-011-0052-2

Kim SC, Nahm SW, Park YK (2015) Property and performance of red mud-based catalysts for the complete oxidation of volatile organic compounds. J Hazard Mater 300:104–113. https://doi.org/10.1016/j.jhazmat.2015.06.059

Krýsa J, Mantzavinos D, Pichat P, Poulios I (2018) Advanced oxidation processes for water and wastewater treatment. Environ Sci Pollut Res 25:34799–34800. https://doi.org/10.2166/9781780403076

Li Y, Wei G, Shao L et al (2019) Green synthesis of red mud based ZnO–Fe2O3 composite used for photo-Fenton reaction under visible light. J Clean Prod 207:717–727. https://doi.org/10.1016/j.jclepro.2018.10.051

Liu Y, Jia CN, Yamasaki J et al (2010) Highly active iron oxide supported gold catalysts for CO oxidation: how small must the gold nanoparticles be? Angew Chem Int Ed Engl 49:5771–5775. https://doi.org/10.1002/anie.201000452

Mesgari Abbasi S, Rashidi A, Ghorbani A, Khalaj G (2016) Synthesis, processing, characterization, and applications of red mud/carbon nanotube composites. Ceram Int 42:16738–16743. https://doi.org/10.1016/j.ceramint.2016.07.146

Mondal K, Lorethova H, Hippo E et al (2004) Reduction of iron oxide in carbon monoxide atmosphere—reaction controlled kinetics. Fuel Process Technol 86:33–47. https://doi.org/10.1016/j.fuproc.2003.12.009

Muhammad S, Saputra E, Sun H et al (2012) Heterogeneous catalytic oxidation of aqueous phenol on red mud-supported cobalt catalysts. Ind Eng Chem Res 51:15351–15359. https://doi.org/10.1021/ie301639t

Munteanu G, Ilieva L, Andreeva D (1997) Kinetic parameters obtained from TPR data for α-Fe2O3 and systems. Thermochim Acta 291:171–177. https://doi.org/10.1016/S0040-6031(96)03097-3

Oliveira AAS, Tristão JC, Ardisson JD et al (2011) Production of nanostructured magnetic composites based on FeO nuclei coated with carbon nanofibers and nanotubes from red mud waste and ethanol. Appl Catal B Environ 105:163–170. https://doi.org/10.1016/j.apcatb.2011.04.007

Oliveira AAS, Teixeira IF, Christofani T et al (2014) Biphasic oxidation reactions promoted by amphiphilic catalysts based on red mud residue. Appl Catal B Environ 144:144–151. https://doi.org/10.1016/j.apcatb.2013.07.015
Oliveira AAS, Costa DAS, Teixeira IF, Moura FCC (2015) Gold nanoparticles supported on modified red mud for biphasic oxidation of sulfur compounds: a synergistic effect. Appl Catal B Environ 162:475–482. https://doi.org/10.1016/j.apcatb.2014.07.003
Oliveira AAS, Costa DS, Teixeira IF et al (2017) Red mud based gold catalysts in the oxidation of benzyl alcohol with molecular oxygen. Catal Today 289:89–95. https://doi.org/10.1016/j.cattod.2016.10.028
Paredes JR, Ordóñez S, Vega A, Díez FV (2004) Catalytic combustion of methane over red mud-based catalysts. Appl Catal B Environ 47:37–45. https://doi.org/10.1016/S0926-3373(03)00325-4
Parvin F, Nayna OK, Tareq SM et al (2018) Facile synthesis of iron oxide nanoparticle and synergistic effect of iron nanoparticle in the presence of sunlight for the degradation of DOM from textile wastewater. Appl Water Sci 8:1–11. https://doi.org/10.1007/s13201-018-0719-5
Poyatos JM, Muñio MM, Almecija MC et al (2010) Advanced oxidation processes for wastewater treatment: state of the art. Water Air Soil Pollut 205:187–204. https://doi.org/10.1007/s11270-009-0065-1
Qiao B, Wang A, Lin J et al (2011) Highly effective CuO/Fe(OH)x catalysts for selective oxidation of CO in H2-rich stream. Appl Catal B Environ 105:103–110. https://doi.org/10.1016/j.apcatb.2011.03.040
Rath D, Nanda B, Parida K (2017) Sustainable nano composite of mesoporous silica supported red mud for solar powered degradation of aquatic pollutants. J Environ Chem Eng 5:6137–6147. https://doi.org/10.1016/j.jece.2017.11.037
Rezaee A, Rangkooy H, Khavanin A, Jafari AJ (2014) High photocatalytic decomposition of the air pollutant formaldehyde using nano-ZnO on bone char. Environ Chem Lett 12:353–357. https://doi.org/10.1007/s10311-014-0453-7
Sahu MK, Patel RK (2016) Novel visible-light-driven cobalt loaded neutralized red mud (Co/NRM) composite with photocatalytic activity toward methylene blue dye degradation. J Ind Eng Chem 40:72–82. https://doi.org/10.1016/j.jiec.2016.06.008
Saputra E, Muhammad S, Sun H et al (2012) Red mud and fly ash supported Co catalysts for phenol oxidation. Catal Today 190:68–72. https://doi.org/10.1016/j.cattod.2011.10.025
Shao L, Wei G, Wang Y et al (2016) Preparation and application of acidified/calcined red mud catalyst for catalytic degradation of butyl xanthate in Fenton-like process. Environ Sci Pollut Res 23:15202–15207. https://doi.org/10.1007/s11356-016-6691-4
Shim WG, Nah JW, Jung HY et al (2018) Recycling of red mud as a catalyst for complete oxidation of benzene. J Ind Eng Chem 60:259–267. https://doi.org/10.1016/j.jiec.2017.11.012
Soliman NK (2019) Factors affecting CO oxidation reaction over nanosized materials: a review. J Mater Res Technol 8(2):2395–2407. https://doi.org/10.1016/j.jmrt.2018.12.012
Sushil S, Batra VS (2008) Catalytic applications of red mud, an aluminium industry waste : a review. Appl Catal B Environ 81:64–77. https://doi.org/10.1016/j.apcatb.2007.12.002
Sushil S, Batra VS (2012) Modification of red mud by acid treatment and its application for CO removal. J Hazard Mater 203–204:264–273. https://doi.org/10.1016/j.jhazmat.2011.12.007
Tijani JO, Fatoba OO, Babajide OO, Petrik LF (2016) Pharmaceuticals, endocrine disruptors, personal care products, nanomaterials and perfluorinated pollutants: a review. Environ Chem Lett 14:27–49. https://doi.org/10.1007/s10311-015-0537-z
Tinh NT, Van NTT, Anh NP et al (2019) CuO and CeO2 -doped catalytic material synthesized from red mud and rice husk ash for p-xylene deep oxidation. J Environ Sci Health A Tox Hazard Subst Environ Eng 54(4):352–358. https://doi.org/10.1080/10934529.2018.1551649
Vilar VJP, Silva AMT, Rizzo L (2018) New challenges in the application of advanced oxidation processes. Environ Sci Pollut Res 25:27673–27675. https://doi.org/10.1007/s11356-018-2653-3
Wang X, Lu G, Guo Y et al (2011) Role of Rh promoter on increasing stability of Au/Al2O3 catalyst for CO oxidation at low temperature. Environ Chem Lett 9:185–189. https://doi.org/10.1007/s10311-009-0261-7

Wei G, Shao L, Mo J et al (2017) Preparation of a new Fenton-like catalyst from red mud using molasses wastewater as partial acidifying agent. Environ Sci Pollut Res 24:15067–15077. https://doi.org/10.1007/s11356-017-9126-y

Xu B, Qi F, Sun D et al (2016a) Cerium doped red mud catalytic ozonation for bezafibrate degradation in wastewater: efficiency, intermediates, and toxicity. Chemosphere 146:22–31. https://doi.org/10.1016/j.chemosphere.2015.12.016

Xu B, Qi F, Zhang J et al (2016b) Cobalt modified red mud catalytic ozonation for the degradation of bezafibrate in water: catalyst surface properties characterization and reaction mechanism. Chem Eng J 284:942–952. https://doi.org/10.1016/j.cej.2015.09.032

Yang S, He H, Wu D et al (2009) Decolorization of methylene blue by heterogeneous Fenton reaction using Fe3-xTixO4 ($0 \leq x \leq 0.78$) at neutral pH values. Appl Catal B Environ 89:527–535. https://doi.org/10.1016/j.apcatb.2009.01.012

Chapter 17
Recent Trends in the Utilization of Biosurfactant for the Treatment of Textile Waste and Industrial Effluents

Charles Oluwaseun Adetunji

Contents

C. O. Adetunji (✉)
Applied Microbiology, Biotechnology and Nanotechnology Laboratory,
Department of Microbiology, Edo University, Iyamho, Edo State, Nigeria
e-mail: adetunji.charles@edouniversity.edu.ng

Inamuddin, A. M. Asiri (eds.), *Sustainable Green Chemical Processes and their Allied Applications*, Nanotechnology in the Life Sciences,
https://doi.org/10.1007/978-3-030-42284-4_17

Abbreviations

^{13}CNMR	Carbon Nuclear Magnetic Resonance
16S rRNA	16S ribosomal ribonucleic acid
18S rRNA	18S ribosomal ribonucleic acid
^{1}HNMR	Proton Nuclear Magnetic Resonance
CH4	Methane gas
COD	Chemical oxygen demand
CTAB	Cetyltrimethylammonium bromide
E24	Emulsification index
FT-IR	Fourier Transform Infrared
HPLC	High-Pressure Liquid Chromatography
LC-MS	Liquid Chromatography-Mass Spectrometry
$NaNO_3$	Sodium nitrate
rDNA	Recombinant DNA
STP	Standard Temperature and Pressure
TLC	Thin Layer Chromatography

17.1 Introduction

The rate of pollution in the environment has been highlighted as a main point of concern affecting mankind globally which might be linked to the high-level upsurge in the amount of anthropogenic activities and increase in the number of human population worldwide. Some of these anthropogenic activities include the release of textile waste and industrial effluents into the body of water, thereby increasing the level of pollution in these water bodies and making them unfit for some other industrial and domestic purposes. Moreover, it has been discovered that the most significant and common sources of xenobiotic and recalcitrant compounds in the form of wastewater being released into the water body are dye and wastewaters from food and agricultural processing industries like abattoir and textile industry (Li et al. 2015). Some of the adverse effects of the discharged pollutants include changes in the color of the water making it unfit for consumption, high level of toxicity to every living organism, high chemical oxygen demand, and bioaccumulation of heavy metals (Joshi et al. 2010). Furthermore, textile industries require large volume of water because they demand high level of quality water but discharge wastewater possessing different characteristics which are complex in nature (Spagni et al. 2012).

The application of natural technique-sourced microorganisms or their metabolites or biological products for effective treatment of heavily contaminated environment either water or soil has been given considerable attention in contemporary time (Cheung and Gu 2007; Drogui et al. 2005). Microorganisms have been recognized as natural machinery capable of recycling toxic substances into less nontoxic products. Some of these beneficial microorganisms include fungus, bacteria, and actinomycetes that possess numerous mechanisms that could convert toxic substances available in polluted soil into nontoxic substances (Parales et al. 2002). This includes their application in the decolorization of dyestuffs, treatment of heavy metal, and bioremediation of heavily polluted water body with textile effluent (Arunprasad and Rao 2010; Oturkar et al. 2013; Balapure et al. 2015).

Textile industry has been recognized as one of the biggest industries that utilizes several liters of water globally, and the end product from their industries discharges several contaminants such as heavy metals, dyes, inorganic salts, degradable organics, stabilizing agents, and detergents. Also, most of these untreated wastewaters from industry are normally discharged into the water bodies without applying any form of treatment. The main processes normally applied during the process of production in textile industry include printing, pretreatment, dyeing, and finishing. All these processes generate large quantities of wastewater such as organic compounds, chemicals, biological oxygen demand, and reactive dyes. However, most of these techniques are much very costly and time-consuming. Most of the conventional techniques used in the treatment of water include combined treatment processes and physicochemical and biological techniques such as utilization of efficient microorganisms and their biological compounds. Moreover, the enforcement of ecological standards is currently being given the highest priority most especially in the textile unit. The high level of strict implementation of environmental standards has sensitized most textile industries to search for an eco-friendly, cost-effective, and sustainable solution that will substitute the utilization of chemicals which possess several numerous hazards and harmful substances. This will go a long way in resolving all the flaws encountered in the handling of toxic end product from the textile industry (Siddique et al. 2017).

Moreover, synthetic dyes are manufactured to be chemically stable, but they continue to remain viable in the environments for a long period of time (Saratale et al. 2011; Ramalho et al. 2004; Wang et al. 2009). About 10–15% of the dye used during the process of dyeing is normally released into the water bodies which eventually constitutes wastewater that inferred serious health and environmental threats (Arora et al. 2007). The discharge of these dyes into the aquatic ecosystem decreases photosynthetic actions by preventing light dispersion into deeper layers (Saratale et al. 2011; Asad et al. 2007; Daneshvar et al. 2006) which normally results into diminution of dissolved oxygen and causes several uncountable damages on the biodiversity in the aquatic environment. Also, most of these wastewaters obtained from dye industry possess some anaerobic metabolites of dyes which are mutagenic in nature, toxic, and carcinogenic to human, microorganisms, and other aquatic life (Hsueh and Chen 2008). The process of removing the dye from these wastewaters imposes more challenges compared to the elimination of the soluble colorless organic substances (Banat et al. 1996).

Chemical, physical, and biological techniques have been exploited for the management of industrial wastes and textile dyes (Saratale et al. 2011; dos Santos et al. 2007). Moreover, the application of chemical and physical techniques has been documented to be effective for treatment of color of wastewaters but has several challenges such as development of dangerous by-products, exhaustive energy consumption, and high operative cost (Arora et al. 2007; Daneshvar et al. 2006; Ramya et al. 2008).

The application of biological techniques such as the utilization of microorganisms (Sreelatha et al. 2015; Dos Santos et al. 2007) and biologically active molecules obtained from them could be an alternative to the synthetic methods because of several exceptional qualities such as lower sludge production, high effectiveness, and environmentally friendly nature (Chen et al. 2003). Examples of such beneficial microorganisms that could decolorize dye wastewaters include bacteria (Ramalho et al. 2004; Daneshvar et al. 2006; Asad et al. 2007; Jadhav et al. 2008; Banat et al. 1996) and algae (Daneshvar et al. 2006).

The utilization of such microorganisms most especially dye-decolorizing bacterial strain isolated from textile effluent discharging sites (Asad et al. 2007; Arora et al. 2007; Jadhav et al. 2008; Kalyani et al. 2008; Bhatt et al. 2005; Chand et al. 2014; Nigam et al. 1996) will be a sustainable biotechnological technique for the management of dye wastewaters in developing and developed countries (Bhatt et al. 2005). However, only limited researchers have utilized biological active molecules from microorganisms such as biosurfactant as well as beneficial microorganism for the treatment of wastewaters from dye-processing industry (Guo et al. 2008; Leena and Raj 2008; Olukanni et al. 2006).

Therefore, in view of the aforementioned, there is great need to search for a sustainable technology that could be utilized for the handling of textile waste and industrial effluents. Biosurfactant has been highlighted as a green and natural biological molecule that could help in the ecorestoration of contaminated water most especially for the management of textile waste and industrial effluents (Ramani et al. 2012; Bachmann et al. 2014).

Biosurfactant has been discovered to vary in terms of their structure which includes two moieties: hydrophobic and hydrophilic. The first moiety entails saccharides which might be mono-, di-, or polys, amino acids, anions or cations, peptides, and saturated and unsaturated fatty acids (Basak and Das 2014). Furthermore, biosurfactant has been categorized into different groups which include fatty acids, neutral lipids, glycolipids, phospholipids, lipopolysaccharides, and lipopeptides, respectively. Some of the various benefits that portend them as a suitable candidate for the bioremediation of heavily polluted environments compared to synthetic surfactant include their lower toxicity, enhanced environmental compatibility, economical, improved biodegradability, maximum selectivity, capability to exhibit numerous action at extreme pH, salinity, temperatures, and their capability to utilize agro-industrial wastes for their mass production (Lazarkevich et al. 2015; Adamu et al. 2015).

Therefore, this chapter discusses extensively current progresses in the fabrication, extraction, and purification of biosurfactant utilized for the management of textile waste and industrial effluents, various sources of biosurfactant utilized in this regard, optimization/parameters that could influence the mass production of biosurfactant, physicochemical properties of biosurfactant, and characterization of biosurfactant which has been utilized for treatment of textile waste and industrial effluents using NMR, HPLC, GCMS, FTIR, and TLC. This chapter also suggests some relevant conclusions and future recommendations that could increase the application of biosurfactant toward the treatment of heavily contaminated environment with textile waste and industrial effluents.

17.2 Molecular Techniques Used for the Identification and Characterization of Biosurfactant-Producing Genes

Several biotechnological techniques have been highlighted for the selection of beneficial isolates, bacterial and fungal isolates, that portend the capability to produce high level of biosurfactant using 16s rRNA and 18s rRNA, respectively. Satpute et al. (2008) provide details of molecular techniques used for the identification of biosurfactant isolated from bacteria dwelling in a marine environment. Hsieh et al. (2004) utilized *sfp* locus for polymerase chain reaction for the screening of biosurfactant fabricating attributes of *B. circulans* and *B. amyloliquefaciens*, respectively. Furthermore, Morita et al. (2006) utilized rDNA sequence for the identification of mannosylerythritol lipid from strain *P. rugulosa* NBRC 10877. Whiteley et al. (1999) stated a technique that could identify genes that regulate quorum-sensing signal and modulators.

17.2.1 Recent Methodologies that Are Used for the Screening of Biosurfactant

Biosurfactants are natural biomolecules that possess numerous variations in their structure. The various types of biosurfactants are lipopeptides or phospholipids, lipopolysaccharides, and glycolipids. The simplest methodology is to access the surface activity potential and the interface of producing strains with hydrophobic membranes from the beneficial strains that possess the capability to produce biosurfactant. These methods gave a better quantitative and qualitative analysis after the biosurfactant has been produced during the biological process. Examples of these methodologies include surface/interfacial activity, du Nouy ring method, emulsification activity, oil spreading method, surface tension measurements, oil displacement assay, and drop collapse test.

17.2.2 Surface/Interfacial Activity

This technique is normally applied based on the interfacial activity or their surface reaction with the surfactant produced by microorganisms that portend the capability to produce biosurfactant before grouping them into various categories. This methodology is normally used in the preliminary grouping of microorganism that portends the capability to produce biosurfactant. Most of the time, this is established on the determination of surface tension of the liquid.

17.2.3 du Nouy Ring Method

This methodology is used for preliminary classification of biosurfactant-producing microorganism. There is a need for a force to separate wire loop or ring from the superficial of an agent which forms the basis for the classification of the various types of biosurfactant. This is normally determined with the aids of tensiometer which works on the principle based on the amount of force that is proportionate to interfacial surface tension using Wilhelmy Plate Methods techniques. The wire loop attached to the machine is normally sterilized to eradicate the presence of any impurity.

17.2.4 Emulsification Activity

This technique works based on the formula that was developed by Goldenberg and Cooper for the determination and classification of microorganisms that portend the capability to produce biosurfactant. E24 is evaluated.

E24 = height of emulsion layer × 100/total amount of solution (Cooper and Goldenberg 1997).

17.2.5 Oil Spreading Method

This methodology was established by Morikawa, which forms a methodology for the classification of surfactant which works on the principles of oil distribution. The length or diameter of the clear loop of oil is determined in petri plate after the addition of 10 μL of sample agents on 100 μL of crude oil base layered together with 50 mL of distilled water. The diameter of oil bubble depicts the rate of dispersion action of oil compared to negative control.

17.2.6 Surface Tension Determination

This technique is established on the uninterrupted determination of surface tension produced by the biosurfactant from microorganism with the aids of tensiometer. This is normally performed at temperature of 24 °C force which is evaluated to remove platinum loop from the liquid-air boundary present in the supernatant or the tested liquid obtained from the biosurfactant-producing microorganisms (Morikawa et al. 2000; Cha et al. 2008; Batista et al. 2006; Volchenko et al. 2007; Plaza et al. 2006).

17.2.7 Oil Displacement Assay

This is normally performed in a petri plate when 30 mL of purified water is added to 1 mL of crude oil, after which 20 μL of biosurfactant was added to the total reaction combination. The length of the diameter of the oil is used in the determination of the activity of the microorganism that could produce biosurfactant (Plaza et al. 2006).

17.2.8 Drop Collapse Test

This drop collapse technique is the easiest method used in the determination of biosurfactant-producing capability of different microorganism. It doesn't require any special equipment but require only small volume. This is carried out by placing small amount of crude oil on a clean microscopic glass slide while 10 μL of biosurfactant solution was dropped with the aids of micropipette without having adverse effect on the dome shape of the oil. The drop collapsed within the period of 1 min was shown to be considered as drop collapse test (Plaza et al. 2006).

17.2.9 Bacterial Adhesion to Hydrocarbons (BATH) Assay

This technique is normally applied to evaluate the hydrophobicity of the cell surface using the protocol developed by Rosenberg et al. (1980). They normally use this technique by putting 2 mL of cell suspension together with 100 μL of hydrocarbon (crude oil, xylene, and hexane) followed by vortexing, and then reaction mixture needs to undergo shaking for 3 min in test tubes. The aqueous phase and hydrocarbon will then be allowed to separate after shaking the reaction mixture for a period of 1 h. The level of decrease in the turbidity of the aqueous segment will be expressed as percentage of adherence to hydrocarbon. This can be expressed by the formula as designated by Van der Vegt et al. (1991): 1 − (optical density of the aqueous phase/optical density of initial cell suspension) × 100.

17.2.10 Hydrocarbon Overlay Agar Method

This technique involved the application of crude oil-coated Luria-Bertani agar plates after inoculating with overnight-grown culture of biosurfactant-producing microorganism. The reaction mixture needs to be incubated at 30 °C for 48–72 h. The colony surrounded by emulsified halos will then be measured for biosurfactant fabrication (Satpute et al. 2008).

17.2.11 CTAB Agar Plate Technique

This is normally used for preliminary screening for the presence of biosurfactant on blue agar plates containing 0.2 mg/mL of cetyltrimethylammonium bromide (CTAB) in addition to 5 mg/mL of methylene blue. This is shown as the manifestation of dark blue halos around the colonies (Siegmund and Wagner 1991).

17.2.12 Retrieval and Purification of Biosurfactants

The reclamation and partial or total refinement of products of biotechnological interests have been discovered to account up to 60% of the total manufacturing cost used for the total production which alludes to the extremely high price of producing biosurfactant. Therefore, the application of techniques that will minimize cost by the utilization of renewable substrate using agro-industrial wastes which are inexpensive will go a long way in increasing the profit derived during the bioprocessing (Desai and Banat 1997; Banat et al. 2000; Makkar and Cameotra 1997). Moreover, it has been discovered that huge amount of money is necessary during the refinement process (Rodrigues et al. 2006). Also, there are several challenges that are encountered during the fermentation process which include presence of contaminant, and some other factors like solubility, ionic charge (chromatography), and position whether inside, outside, or cell bound will all determine the types of purification process that will be utilized for the processing of biosurfactant. Generally, ammonium sulfate is utilized in the precipitation and refinement of high-molecular-weight biosurfactant, the process which is followed by the process of dialysis which ensures the final removal of small molecules. Some other techniques involve the application of solvent system which includes chloroform/methanol, trichloroacetic acid, ethanol, and acetone precipitation. The other techniques which have been described over the years were included in Table 17.1.

Table 17.1 The various techniques utilized for the recovery and purification of biosurfactant

Conventional methods	Techniques	Microorganisms	References
Acetone precipitation	Biosurfactant is extracted from the cell-free supernatant mixed with ice-cold acetone to precipitate at 4 °C for 15–20 h	*Arthrobacter* spp.	Rosenberg et al. (1979), Patil and Chopade (2001a, b, 2003)
Ethanol precipitation	Biosurfactant is obtained from the culture broth after centrifugation at 11,000 × *g*/20 min/4 °C. The biosurfactant is precipitated from the supernatant by using cold ethanol	*Acinetobacter* spp., *Pseudomonas* spp., *Bacillus* spp., *Cyanobacterium* spp., yeast species; *A. calcoaceticus* subsp. *anitratus* SM7	Phetrong et al. (2008)
Ammonium sulfate precipitation	This is normally used for high-molecular-weight biosurfactant such as emulsan, biodispersion (protein-rich compounds) which is precipitated using $(NH_4)_2SO_4$	*A. calcoaceticus* BD413, *A. venetianus* RAG-1	Bach et al. (2003), Toren et al. (2001, 2002), Youssef et al. (2005)
Acid precipitation	This techniques is normally used for the recovery of biosurfactant such as surfactin, lipopeptides, glycolipids, etc.	*Bacillus* spp.	Al-Bahrya et al. (2013)
Surfactin using acid hydrolysis	Biosurfactant is purified from cell-free supernatant using acid hydrolysis involving concentrated HCl to bring down the pH 2.0. This led to the precipitation of proteins and lipid-containing biosurfactant when left at 4 °C overnight centrifugation. The obtained pellet is further extracted by using various solvents	*Bacillus* spp.	Arima et al. (1968), Mukherjee et al. (2006), Nitschke and Pastore (2006), Thaniyavarn et al. (2003), Jennings and Tanner (2004)
Precipitation of rhamnolipid by the addition of HCl to precipitate glycolipid	The rhamnolipid present in the cultural medium is extracted with HCl through the process of acidification, centrifugation, and extraction procedure which are similar for other glycolipids	*Pseudomonas aeruginosa*	Haba et al. (2000), Smyth et al. (2009), Nunez et al. (2001), Rapp et al. (1979)
Adsorption-desorption	This works on the principle that enhances biosurfactant to adsorb and desorb from Amberlite XAD 2 or 16 polystyrene resins, and the interaction is normally utilized for the purification of biosurfactant	*P. aeruginosa*	Reiling et al. (1986), Abalos et al. (2001), Dubey et al. (2005)
Ion exchange chromatography	This is used for the recovery of charged biosurfactant such as rhamnolipids which had negative charge at higher pH. The biosurfactant is normally eluted with buffer containing 10% (v/v) ethanol	*Pseudomonas* spp.	Matsufuji et al. (1997)

(continued)

Table 17.1 (continued)

Conventional methods	Techniques	Microorganisms	References
Centrifugation	This technique is used after acid precipitation to obtain biosurfactant present in the broth which can be derived using the process of centrifugation at 12,000 rpm for 15 min at 4 °C so as to obtain the crude product and the pellet obtained is then dried under N2 and extracted with solvents		Nitschke and Pastore (2006)
Crystallization	Biosurfactant is obtained through precipitation and by redissolving an organic solvent. Most of the time glycolipids such as rhamnolipids are concentrated, i.e., obtain pure crystals of Rha-Rha-C10-C10	*Pseudomonas aeruginosa*	Manso Pajarron et al. (1993)
Filtration and precipitation	Biosurfactant is obtained through precipitation using ethanol, acetone, and ethanol acetic acid (1%)/5 N HCl in an equal volume of culture liquid. The process of extraction is normally carried out twice to enhance the yield of biosurfactant	*P. aeruginosa*	Turkovskaya et al. (2001)
Foam fractionation	The foam is obtained using fractionation column and acidified with HCl down to pH 1.0–2.0 to precipitate the presence of biosurfactant, which can be extracted with solvents	*Alcanivorax borkumensis*, *Bacillus* spp.	Abraham et al. (1998), Neu and Poralla (1990), Noah et al. (2002)
Isoelectric focusing	This is a novel technique applied for purification of biosurfactant. The crude biosurfactant is utilized in IEF and pH, and emulsification activity needs to be confirmed	Marine yeast *Yarrowia lipolytica* NCIM	Zinjarde et al. (1997)
Solvent extraction	This technique is based on the fact that hydrophobic moieties of biosurfactant are soluble in some solvents which help in extraction and separation of crude product. Examples of solvents that have been reported includes isopropanol, chloroform, di ethyl ether, methanol, butanol, ethyl acetate, di-chloromethane, pentane, and chloromethane. It has been used for the purification of rhamnolipid, trehalose lipids, sophorolipids, cellobiolipids, and liposan. The organic phases are evaporated to dryness and analyzed by TLC technique	*Pseudomonas aeruginosa*	Desai and Banat (1997), Smyth et al. (2009), Tuleva et al. (2002)

(continued)

Table 17.1 (continued)

Conventional methods	Techniques	Microorganisms	References
Ultrafiltration	It is used for the purification of biosurfactant such as surfactin and rhamnolipids. Examples include Amicon filter paper or other types of filters of 0.22 μ and 0.45 μ pore sizes. Biosurfactant are obtained from the fermentation broth using ultra-filtration with 30 kDa molecular weight cutoffs (MWCO)	Yeast *Yarrowia lipolytica* NCIM, *B. licheniformis* mutant JF2, *Bacillus* spp., *P. putida* ML2	Mulligan and Gibbs (1990), Lin and Jiang (1997), Zinjarde et al. (1997), Lin et al. (1998a), Cooper and Goldenberg (1987), Lin et al. (1998b), Bonilla et al. (2005)
Preparative thin layer chromatography (TLC)	This technique allows purification of biosurfactant using preparative TLC on a silica-coated glass plate with variable thickness applied with BS sample and allowed to run in a solvent system	*Pseudomonas aeruginosa*	Adetunji et al. (2017)
Dialysis and lyophilization	Seamless cellulose tubing dialysis bags are utilized for the purification of biosurfactant. This techniques has been utilized for the resolution of sophorolipids	*A. calcoaceticus* BD4 and BD413	Kaplan and Rosenberg (1982), Shah and Prabhune (2007)

17.2.13 Application of Biosurfactant for the Management of Textile Waste and Industrial Effluents

Damasceno et al. (2018) tested the synergetic effect of crude enzyme containing 0.5% weight per volume (w/v) secreted by *Penicillium brevicompactum* during solid-state fermentation together with rhamnolipid biosurfactant containing 27 milligrams per liter (mg/L) generated during submerged fermentation. The rhamnolipid was used in the handling of wastewater obtained from a poultry slaughterhouse using a bioreactor containing bench upflow anaerobic sludge blanket. The bioreactor consists of pretreated wastewater, while the other was fed with unprocessed water serving as control. It was observed that the bioreactor containing pretreatment exhibits more enhanced methane production with greater chemical oxygen demand (COD) (91.2%), enhanced specific methane fabrication (67.8 mL CH4 (STP)/g), and oil and grease removal (95.8%) compared to the control that generated 36.1–16.9 mL CH4 (STP)/g of chemical oxygen demand removed with 72.5% of chemical oxygen demand elimination and 48% of oil and grease elimination under 11 kg COD/m^3·day. Moreover, it was discovered that the bioreactor without any treatment showed five clogging episodes which might be linked to the buildup of fat on the superficial, while the bioreactor containing the treatment did not exhibit any operational problem. The level of froth generated in the treated bioreactor indicated ten times amounts of fat compared to the one without any treatment. The cost-effectiveness evaluation showed that this substitute technology is more economically viable when associated to the unadventurous technology.

Makkar et al. (2011) wrote a comprehensive review on the application of biosurfactant in several sectors like oil- and food-processing industries, health care, bioremediation, and environmental, including treatment of industrial wastes. Some of the unique properties that enable and enhance their wide application include dispersing, emulsifying, detergency, foaming, eco-friendly, and low toxicity. The authors highlighted several advances in the utilization of several wastes derived from agriculture and industry for the fabrication of biosurfactant in order to decrease the cost of invention.

It has been discovered that some strains of microorganism and their biological compounds of interest have been documented as a biotechnological tool for the bioremediation of textile and industrial wastes. Rajendran et al. (2015) isolated 30 local bacterial strains from denim textile effluent of which four isolates were discovered to be effective in the degradation of crude indigo carmine when assayed using decolorization technique. The bacterial isolates were *Staphylococcus* sp. (PK28), *Actinomyces* sp. (PK07), *Stenotrophomonas* sp. (PK23), and *Pseudomonas* sp. (PK18) based on the microscopic and biochemical test. The immobilized bacterial cell showed that strain PK23 exhibited the maximum cells of 1×10^9 CFU/mL with the highest effectiveness of 92% on polyurethane foam (PU foam).

Rajendran et al. (2015) utilized the PU foam immobilized bacterial cells for the ecorestoration of denim industry wastewaters. The best condition that enhanced the activity of the immobilized bacteria which was attained at pH 6, 27 °C, and 100% of substrate concentration led to the development of biofilm for 1 day (1.5% w/v). The result obtained from the scanning electron microscopy showed that the immobilized cell using polyurethane foam enhanced that stability of the various extracellular enzymes at various higher levels compared to the suspended forms. The structural elucidation using NMR and TLC showed the presence of anthranilic acid and isatin as the product of degradation. Moreover, it was discovered that the mathematical model that designates bacterial transport that regulates the process of biodegradation which entails several reactions with nonstandard mathematical techniques centered on the time step structure.

Guadie et al. (2017) isolated bacteria that is capable of decolorizing azo dye from Chamo and Abaya using Reactive Red 239 (RR239) dye. The best active strain CH12 was further used to validate the influence of nutrient supplementation, temperature, dye concentration, types of decolorization, dissolved oxygen, and pH. The molecular characterization reveals that CH12 is a *Bacillus* sp. The result revealed that the decolorization efficiency of the bacterial isolate was enhanced with organic nitrogen (~100%) supplement and carbon (≥98%). Moreover, total decolorization was also exhibited under anaerobic and aerobic conditions, pH of 10, and temperature of 30 °C, but the decolorization efficiency was decreased when under aerobic culturing condition (≤6%), after the basal media was supplemented with $NaNO_3$ (1–8%) which shows that RR239 was chosen as electron acceptor. Their study showed that strain CH12 could be utilized for decolorization uses which might be linked to its capability to adequately decolorize concentration that varies from 50 to

250 mg/L of higher RR239 and six additional dyes. Generally, Bacillus spp CH12 can be an auspicious candidate for decolorization usages owing to its impending to successfully decolorize higher RR239 concentrations other six additional dyes used during their study.

Jamaly et al. (2015) wrote a widespread review on the current and recently developed methods for oily wastewater treatment for effective removal of inorganics, grease, fats and oil. The authors highlighted several recent techniques that could be used in the bioremediation such as the utilization of biosurfactants, biological treatment, and treatment through vacuum ultraviolet radiation, destabilization of emulsions using natural minerals and of zeolites, membrane filtration, and electrochemical treatment. Their study highlighted novel and innovative technology used for the management of oily wastewater that led to decrease in the adverse influence of the release of wastewater containing oil into the environment. They also highlighted several challenges that could affect the best performance of this technology during the treatment of oily wastewater. They also presented several future research areas that need to be addressed that could lead to adequate provision of sustainable technologies for the management of wastewater.

Santos et al. (2016) wrote a comprehensive review on the multifunctional application of biosurfactant most especially their practical application as a biotechnological tool that could mitigate several challenges facing mankind in the twenty-first century. They also highlighted several unique properties that make their utilization more accepted in several sectors. Some of these special properties include detergency, emulsification, biodegradability, dispersal, low toxicity, and foam formation. Some of the several sectors include cosmetic, petroleum, agricultural, health, and chemical industries.

Tripathy and Mishra (2016) wrote a comprehensive review on the application of biosurfactant obtained from microorganisms. They also listed specific examples of biosurfactants from some microorganisms such as sophorolipids manufactured from *Candida bombicola,* surfactin manufactured from *Bacillus subtilis,* emulsan produced from *Acinetobacter calcoaceticus*, and rhamnolipids produced from *Pseudomonas aeruginosa.* They also highlighted some current application of biosurfactant as a biotechnology tool in the bioremediation that could help in resolving several environmental challenges

Perez-Ameneiro et al. (2015) utilized lipopeptide biosurfactant derived from corn steep liquor amended in the lignocellulosic biocomposite which is a formulation utilized for remediation of wastewater. The result acquired shows that the dye sorption ability of the hydrogel having hydrolyzed vineyard trimming waste can be enhanced through modification of surfactant using synthetic surfactant from detergent. It was observed that the removal of the dye compound and sulfates was enhanced with the concentration of 10% and 62%, correspondingly, when biosurfactant was improved with biocomposite. The mode of action utilized by this lipopeptide biosurfactant for the bioremediation involved bioadsorption together with pseudo-second-order kinetic model, liquid film diffusion, and intraparticle diffusion. Their study showed that biosurfactant could be utilized in the invention of environmentally friendly adsorbents for the management of heavily polluted environment most especially textile waste and industrial effluent.

Akbari et al. (2018) wrote a comprehensive review on the application of biosurfactant for several industrial and domestic applications. They are biotechnological tools that could be utilized for the bioremediation solution to mitigate all the challenges facing mankind such as climate change for the next generations, environmental pollutions, agricultural problems, and decontamination and treatment of textile wastewater. The authors also highlighted the various applications of biosurfactant in diverse sectors including textile, cosmetic, petroleum, pharmaceutical, and food and in the treatment of wastewater. They also stated several advantages and demerit of biosurfactant over synthetic surfactant.

The discharge of wastewater into the environment has been highlighted as a major issue that constitutes several hazards most especially to the water body. Several sources of pollution includes fish canning, sewage sludge, oil and gas industry, dairy, and domestic wastewater. The authors also highlighted the potentials of local bacteria isolates of *Bacillus salmalaya* 139SI which improve the sorption capability with great reclamation of Cr(VI) from aqueous solution and water solubility of crude oil waste (Dadrasnia et al. 2015; Ismail and Dadrasnia 2015).

17.2.14 Modes of Action of Biosurfactants

The application of biologically active compound like biosurfactant has been highlighted as a biotechnological tool for the bioremediation of heavily polluted environment. Biosurfactants are obtained from the yeast or cell membrane. They showed minimum amount of noxiousness, and they are decomposable in nature because they possessed the capability to reduce the interfacial and surface tension and the critical micelle concentration. Moreover, they could withstand extreme conditions such as pH and temperature while they also possess the capability to influence the interface of molecules. They have a wider range of application in the environment and in the petrochemical industries because of the presence of some macromolecules such as lipoproteins, fatty acids, and glycolipids in biosurfactants (Al-Bahrya et al. 2013; Rufino et al. 2014).

17.2.15 Structural Elucidation of Biosurfactant

Several structural elucidation techniques have been utilized to validate the various structural multiplicity of biosurfactant manufactured by different microorganisms. Examples of these techniques include Thin Layer Chromatography (TLC), High-Pressure Liquid Chromatography (HPLC), Fourier Transform Infrared (FT-IR), Liquid Chromatography-Mass Spectrometry (LC-MS), Proton Nuclear Magnetic

Resonance (^{1}HNMR), and Carbon Nuclear Magnetic Resonance (^{13}CNMR) (Qiao and Shao 2010; Adetunji et al. 2017, 2018).

17.3 Conclusion and Future Direction

This chapter has provided a comprehensive detail about the production of biosurfactant from beneficial microorganism and their effectual utilization for the ecorestoration of heavily polluted environment with textile waste and industrial effluents. Moreover, there is a need to focus more attention on the production of cost-effective and high-yielding strain that will maximize profit during the downstream processing. The utilization of agro-industrial wastes will be a sustainable solution that could serve as cheap carbon and nitrogen sources during the production and solving of environmental challenges by recycling the agricultural waste products which would further reduce the cost of production during the downstream processing. This also will go a long way by making biosurfactant a permanent replacement to synthetic surfactant. Furthermore, several analytical techniques (NMR, HPLC, GCMS, FTIR, TLC) used for the structural elucidation of novel surface active biosurfactant have been discussed in details. Therefore, this study will be a baseline study for several researchers and policy makers to enable them to implement strategies on the application of biosurfactant as a typical example of innovative technology that could help in the bioremediation of heavily polluted environment with textile waste and industrial effluents.

References

Abalos A, Pinazo A, Infante MR, Casals M, García F, Manresa A (2001) Physicochemical and antimicrobial properties of new rhamnolipids produced by *Pseudomonas aeruginosa* AT10 from soybean oil refinery wastes. Langmuir 17:1367–1371

Abraham WR, Meyer H, Yakimov M (1998) Novel glycine containing glucolipids from the alkane using bacterium *Alcanivorax borkumensis*. Biochim Biophys Acta 1393:57–62

Adamu A, Ijah UJ, Riskuwa ML, Ismail HY, Ibrahim UB (2015) Study on biosurfactant production by two *Bacillus* species. Int J Sci Res 3(1):30–20

Adetunji CO, Oloke JK, Pradeep M, Jolly RS, Anil KS, Swaranjit SC, Bello OM (2017) Characterization and optimization of a rhamnolipid from *Pseudomonas aeruginosa* C1501 with novel biosurfactant activities. Sustain Chem Pharm 6:26–36. https://doi.org/10.1016/j.scp.2017.07.001

Adetunji CO, Adejumo IO, Afolabi IS, Adetunji JB, Ajisejiri ES (2018) Prolonging the shelf-life of 'Agege Sweet' orange with chitosan-rhamnolipid coating. Hortic Environ Biotechnol 59(5):687–697. https://doi.org/10.1007/s13580-018-0083-2

Akbari S, Abdurahman NH, Yunus RM, Fayaz F, Alara OR (2018) Biosurfactants a new frontier for social and environmental safety: a mini review. Biotechnol Res Innov 2(1):81–90. https://doi.org/10.1016/j.biori.2018.09.001

Al-Bahrya SN, Al-Wahaibib YM, Elshafiea AE, Al-Bemanib AS, Joshia SJ, Al-Makhmaria HS, Al-Sulaimani HS (2013) Biosurfactant production by *Bacillus subtilis* B20 using date molasses and its possible application in enhanced oil recovery. Int Biodeterior Biodegrad 81:141–146

Arima K, Kakinuma A, Tamura G (1968) Surfactin, a crystalline peptidelipid surfactant produced by *Bacillus subtilis*: isolation, characterization and its inhibition of fibrin clot formation. Biochem Biophys Res Commun 31:488–494

Arora S, Sain HS, Singh K (2007) Decolorization optimization of a mono azo disperse dye with *Bacillus firmus*: identification of a degradation product. Color Technol 123:184–190

Arunprasad AS, Rao B (2010) Physio-chemical characterization of textile effluent and screening for dye decolorizing bacteria. Glob J Biotechnol Biochem 5(2):80–86

Asad S, Amoozegar MA, Pourbabaee AA, Sarbolouki MN, Dastghei SMM (2007) Decolorization of textile azo dyes by newly isolated halophilic and halotolerant bacteria. Bioresour Technol 98:2082–2088

Bach H, Berdichevsky Y, Gutnick D (2003) An exocellular protein from the oil-degrading microbe *Acinetobacter venetianus* RAG-1 enhances the emulsifying activity of the polymeric bioemulsifier emulsan. Appl Environ Microbiol 69:2608–2615

Bachmann RT, Johnson AC, Edyvean RG (2014) Biotechnology in the petroleum industry: an overview. Int Biodeterior Biodegrad 86:225–273

Balapure K, Bhatt N, Dutta M (2015) Mineralization of reactive azo dyes present in simulated textile waste water using down flow microaerophilic fixed film bioreactor. Bioresour Technol 175:1–7

Banat IM, Nigam P, Singh D, Marchant R (1996) Microbial decolorization of textile dye containing effluents: a review. Bioresour Technol 58:217–227

Banat IM, Makkar RS, Cameotra SS (2000) Potential commercial applications of microbial surfactants. Appl Microbiol Biotechnol 53:495–508

Basak G, Das N (2014) Characterization of sophorolipid biosurfactant produced by *Cryptococcus* sp. VITGBN2 and its application on Zn (II) removal from electroplating wastewater. J Environ Biol 35:1087–1094

Batista S, Mounteer A, Amorim F, Tótola MR (2006) Isolation and characterization of biosurfactant/bioemulsifier-producing bacteria from petroleum contaminated sites. Bioresour Technol 97(6):868–875

Bhatt N, Patel K, Haresh C, Madmwar D (2005) Decolorization of diazo-dye reactive blue 172 by *Pseudomonas aeruginosa* NBAR12J. J Basic Microbiol 45:407–418

Bonilla M, Olivaro C, Corona M, Vazquez A, Soubes M (2005) Production and characterization of a new bioemulsifier from *Pseudomonas putida* ML2. J Appl Microbiol 98:456–463

Cha M, Lee N, Kim M, Kim M, Lee S (2008) Heterologous production of *Pseudomonas aeruginosa* EMS1 biosurfactant in *Pseudomonas putida*. Bioresour Technol 99:2192–2199, PMID: 17611103.

Chand N, Sajedi RH, Nateri AS, Khajeh K, Rassa M (2014) Fermentative desizing of cotton fabric using alpha-amylase-producing *Bacillus* strain: optimization of simultaneous enzyme production and desizing. Process Biochem 49:1884–1888

Chen K, Wu J, Liou D, Hwang S (2003) Decolorization of the textile dyes by newly isolated bacterial strains. J Biotechnol 101:57–68

Cheung KH, Gu JD (2007) Mechanisms of hexavalent chromium detoxification by microorganisms and bioremediation application potential: a review. Int Biodeterior Biodegrad 59:8–15

Cooper DG, Goldenberg BG (1987) Surface-active agents from two *Bacillus* species. Appl Environ Microbiol 53:224–229

Cooper D, Goldenberg B (1997) Surface-active agents from two Bacillus species. Appl Environ Microbiol 53(2):224–229

Dadrasnia A, Chuan Wei KS, Shahsavari N, Azirun MS, Ismail S (2015) Biosorption potential of *Bacillus salmalaya* strain 139SI for removal of Cr(VI) from aqueous solution. Int J Environ Res Public Health 12(12):15321–15338

Damasceno FRC, Cavalcanti-Oliveira ED, Kookos IK, Koutinas AA, Cammarota MC, Freire DMG (2018) Treatment of wastewater with high fat content employing an enzyme pool and biosurfactant: technical and economic feasibility. Braz J Chem Eng 35(2):531–542. https://doi.org/10.1590/0104-6632.20180352s20160711

Daneshvar N, Ayazloo M, Khataee A, Pourhassan M (2006) Biological decolorization of dye solution containing malachite green by microalgae *Cosmarium* sp. Bioresour Technol 12:121–128
Desai JD, Banat IM (1997) Microbial production of surfactants and their commercial potential. Microbiol Mol Biol Rev 61:47–64
dos Santos AB, Cervantes FJ, van Lier JB (2007) Review paper on current technologies for decolorization of textile wastewaters: perspectives for anaerobic biotechnology. Bioresour Technol 98:2369–2385
Drogui P, Blais JF, Mercier G (2005) Hybrid process for heavy metal removal from wastewater sludge. Water Environ Res 77:372–380
Dubey KV, Juwarkar AA, Singh SK (2005) Adsorption-desorption process using wood-based activated carbon for recovery of biosurfactant from fermented distillery wastewater. Biotechnol Prog 21:860–867
Guadie A, Tizazu S, Melese M, Guo W, Ngo HH, Xia S (2017) Biodecolorization of textile azo dye using *Bacillus* sp. strain CH12 isolated from alkaline lake. Biotechnol Rep (Amst) 15:92–100. https://doi.org/10.1016/j.btre.2017.06.007
Guo J, Zhou J, Wang D, Tamura K, Wang P, Uddin MS (2008) A novel moderately halophilic bacterium for decolorization azo dye under high salt condition. Biodegradation 19:15–19
Haba E, Espuny MJ, Busquets M, Manresa A (2000) Screening and production of rhamnolipids by *Pseudomonas aeruginosa* 47T2 NCIB 40044 from waste frying oils. J Appl Microbiol 88:379–387
Hsieh FC, Li MC, Lin TC, Kao SS (2004) Rapid detection and characterization of surfactin-producing *Bacillus subtilis* and closely related species based on PCR. Curr Microbiol 49:186–191
Hsueh CC, Chen BY (2008) Exploring effects of chemical structure on azo dye decolorization characteristics by *Pseudomonas luteola*. J Hazard Mater 154:703–710
Ismail S, Dadrasnia A (2015) Biotechnological potential of *Bacillus salmalaya* 139SI: a novel strain for remediating water polluted with crude oil waste. PLoS One 10(4):e0120931
Jadhav UU, Dawkar VV, Ghodake GS, Govindwar SP (2008) Biodegradation of direct red 5B, a textile dye by newly isolated *Comamonas* sp. UVS. J Hazard Mater 158:507–516
Jamaly S, Giwa A, Hasan SW (2015) Recent improvements in oily wastewater treatment: progress, challenges, and future opportunities. J Environ Sci 37:15–30
Jennings EM, Tanner RS (2004) The effect of a *Bacillus* biosurfactant on methanogenic hexadecane degradation. Biorem J 8:79–86
Joshi SM, Inamdar SA, Telke AA, Tamboli DP, Govindwar SP (2010) Exploring the potential of natural bacterial consortium to degrade mixture of dyes and textile effluent. Int Biodeterior Biodegradation 64:622–628
Kalyani DC, Patil PS, Jadhav JP, Govindwar SP (2008) Biodegradation of reactive textile dye red BLI by an isolated bacterium *Pseudomonas* sp. SUK1. Bioresour Technol 99:4635–4641
Kaplan N, Rosenberg E (1982) Exopolysaccharide distribution of and bioemulsifier production by *Acinetobacter calcoaceticus* BD4 and BD413. Appl Environ Microbiol 44:1335–1341
Lazarkevich I, Sotirova A, Avramova T, Stoitsova S, Paunova-Krasteva T, Galabova D (2015) Antibacterial activity of methyltiosulfonate and its complexes with rhamnolipid and trehalose lipid against *Pseudomonas aeruginosa* NBIMCC 1390. Res J Pharm Biol Chem Sci 6:282–290
Leena R, Raj DS (2008) Bio-decolourization of textile effluent containing reactive black-B by effluent-adapted and non-adapted bacteria. Afr J Biotechnol 7:3309–3313
Li C, Zhang Z, Li Y, Cao J (2015) Study on dyeing wastewater treatment at high temperature by MBBR and the thermotolerant mechanism based on its microbial analysis. Process Biochem 50:1934–1941
Lin SC, Jiang HJ (1997) Recovery and purification of the lipopeptide biosurfactant of *Bacillus subtilis* by ultrafiltration. Biotechnol Tech 11:413–416
Lin SC, Lin KG, Lo CC, Lin YM (1998a) Enhanced biosurfactant production by a *Bacillus licheniformis* mutant. Enzyme Microbiol Technol 23:267–272
Lin S, Carswell KS, Sharma MM, Georgiou G (1998b) Continuous production of the lipopeptide biosurfactant ob *Bacillus licheniformis* JF-2. Appl Microbiol Biotechnol 41:281–285

Makkar RM, Cameotra SS (1997) Biosurfactant production by a thermophilic *Bacillus subtilis* strain. J Ind Microbiol Bioeng 18:37–42
Makkar RS, Cameotra SS, Banat IM (2011) Advances in utilization of renewable substrates for biosurfactant production. AMB Express 1:5
Manso Pajarron A, De Koster CG, Heerma W, Schmidt M, Haverkamp J (1993) Structure identification of natural rhamnolipid mixtures by fast atom bombardment tandem mass spectrometry. Glycoconj J 10:219–226
Matsufuji M, Nakata K, Yoshimoto A (1997) High production of rhamnolipids by *Pseudomonas aeruginosa* growing on ethanol. Biotechnol Lett 19(12):1213–1215
Morikawa M, Hirata Y, Imanaka TA (2000) A study on the structure function relationship of lipopeptide bio-surfactants. Biochim Biophys Acta 1488(3):211–218
Morita T, Konishi M, Fukuoka T, Imura T, Kitamoto D (2006) Discovery of *Pseudozyma rugulosa* NBRC 10877 as a novel producer of the glycolipid biosurfactants, mannosylerythritol lipids, based on rDNA sequence. Appl Microbiol Biotechnol 73:305–313
Mukherjee S, Das P, Sen R (2006) Towards commercial production of microbial surfactants. Trends Biotechnol 24:509–515
Mulligan CN, Gibbs BF (1990) Recovery of biosurfactants by ultrafiltration. J Chem Technol Biotechnol 47:23–29
Neu TR, Poralla K (1990) Emulsifying agent from bacteria isolated during screening for cells with hydrophobic surfaces. Appl Microbiol Biotechnol 32:521–525
Nigam P, Banat IM, Singh D, Marchant R (1996) Microbial process for the decolorization of textile effluent containing azo, diazo and reactive dyes. Process Biochem 31:435–442
Nitschke M, Pastore GM (2006) Production and properties of a surfactant obtained from *Bacillus subtilis* grown on cassava wastewater. Bioresour Technol 97:336–341
Noah KS, Fox SL, Bruhn DF, Thompson DN, Bala GA (2002) Development of continuous surfactin production from potato process effluent by *Bacillus subtilis* in an airlift reactor. Appl Biochem Biotechnol 98–100:803–813
Nunez A, Ashby R, Foglia TA, Solaiman DKY (2001) Analysis and characterization of sophorolipids by liquid chromatography with atmospheric pressure chemical ionization. Chromatographia 53:673–677
Olukanni OD, Osuntoki AA, Gbenle GO (2006) Textile effluent biodegradation potentials of textile effluent-adapted and non-adapted bacteria. Afr J Biotechnol 5:1980–1984
Oturkar C, Patole MS, Gawai KR, Madmwar D (2013) Enzyme based cleavage strategy of *Bacillus lentus* BI377 in response to metabolism of azoic recalcitrant. Bioresour Technol 130:360–365
Parales RE, Bruce NC, Schmid A, Wackett LP (2002) Biodegradation, biotransformation, and biocatalysis (b3). Appl Environ Microbiol 68:4699–4709
Patil JR, Chopade BA (2001a) Distribution and *in vitro* antimicrobial susceptibility of *Acinetobacter* species on the skin of healthy humans. Natl Med J India 14:204–208
Patil JR, Chopade BA (2001b) Studies on bioemulsifier production by *Acinetobacter* strains isolated from healthy human skin. J Appl Microbiol 91:290–298
Patil JR, Chopade BA (2003) Bioemulsifier production by *Acinetobacter* strains isolated from healthy human skin. US Patent no. US 2004/0138429 A1
Perez-Ameneiro M, Vecino X, Freire JMC, Moldes AB (2015) Wastewater treatment enhancement by applying a lipopeptide biosurfactant to a lignocellulosic biocomposite. Carbohydr Polym 131:186–196
Phetrong K, H-Kittikun A, Maneerat S (2008) Production and characterization of bioemulsifier from a marine bacterium, *Acinetobacter calcoaceticus* subsp. *anitratus* SM7. Songklanakarin J Sci Technol 30(3):297–305
Plaza G, Zjawiony I, Banat I (2006) Use of different methods for detection of thermophilic biosurfactants producing bacteria from hydrocarbon contaminated and bio remediated soils. J Petrol Sci Eng 50(1):71–77
Qiao N, Shao Z (2010) Isolation and characterization of a novel biosurfactant produced by hydrocarbon-degrading bacterium *Alcanivorax dieselolei* B-5. J Appl Microbiol 108(4):1207–1216. https://doi.org/10.1111/j.1365-2672.2009.04513.x

Rajendran P, Prabhavathi P, Karthiksundaram S, Pattabi S, Kumar SD, Santhanam P (2015) Biodecolorization and bioremediation of denim industrial wastewater by adapted bacterial consortium immobilized on inert polyurethane foam (PUF) matrix: a first approach with biobarrier model. Pol J Microbiol 64(4):329–338

Ramalho PA, Cardoso MH, Cavaco-Paulo A, Ramalho MT (2004) Characterization of azo reduction activity in a novel ascomycete yeast strain. Appl Environ Microbiol 70:2279–2288

Ramani K, Jain SC, Mandal AB, Sekaran G (2012) Microbial induced lipoprotein biosurfactant from slaughterhouse lipid waste and its application to the removal of metal ion from aqueous solution. Colloids Surf B Biointerfaces 97:254–263

Ramya M, Anusha B, Kalavathy S (2008) Decolorization and biodegradation of indigo carmine by a textile soil isolate *Paenibacillus larvae*. Biodegradation 19:283–291

Rapp P, Bock H, Wray V, Wagner F (1979) Formation, isolation and characterization of trehalose dimycolates from *Rhodococcus erythropolis* grown on n-alkanes. J Gen Microbiol 115:491–503

Reiling HE, Thanei-Wyss U, Guerra-Santos LH, Hirt R, Käppeli O, Fiechter A (1986) Pilot plant production of rhamnolipid biosurfactant by *Pseudomonas aeruginosa*. Appl Environ Microbiol 51:985–989

Rodrigues L, Banat IM, Teixeira J, Oliveira R (2006) Biosurfactants: potential applications in medicine. J Antimicrob Chemother 57:609–618

Rosenberg E, Zuckerberg A, Rubinovitz C, Gutnick DL (1979) Emulsifier of *Arthrobacter* RAG-1: isolation and emulsifying properties. Appl Environ Microbiol 37:402–408

Rosenberg M, Gutnick D, Rosenberg E (1980) Adherence of bacteria to hydrocarbons a simple method for measuring cell-surface hydrophobicity. FEMS Microbiol Lett 9:29–33

Rufino RD, Luna JM, Takaki GMC, Sarubbo LA (2014) Characterization and properties of the biosurfactant produced by Candida lipolytica UCP 0988. Electron J Biotechnol 17(1):34–38

Santos DKF, Rufino RD, Luna JM, Santos VA, Sarubbo LA (2016) Review biosurfactants: multifunctional biomolecules of the 21st century. Int J Mol Sci 17(3):401. https://doi.org/10.3390/ijms17030401

Saratale RG, Saratale GD, Chang JS, Govindwar SP (2011) Bacterial decolorization and degradation of azo dyes: a review. J Taiwan Inst Chem Eng 42:138–157

Satpute SK, Bhawsar BD, Dhakephalkar PK, Chopade BA (2008) Assessment of different screening methods for selecting biosurfactant producing marine bacteria. Indian J Geo Mar Sci 37:243–250

Shah S, Prabhune A (2007) Purification by silica gel chromatography using dialysis tubing and characterization of sophorolipids produced from *Candida bombicola* grown on glucose and arachidonic acid. Biotechnol Lett 29:267–272

Siddique K, Rizwan M, Shahid MJ, Ali S, Ahmad R, Rizvi H (2017) Textile wastewater treatment options: a critical review. In: Anjum NA et al (eds) Enhancing cleanup of environmental pollutants. Springer International Publishing AG, Basel, pp 183–207. https://doi.org/10.1007/978-3-319-55423-5_6

Siegmund I, Wagner F (1991) New method for detecting rhamnolipids excreted by *Pseudomonas* species during growth on mineral agar. Biotechnol Tech 5:265–268

Smyth TJP, Perfumo A, Marchant R, Banat IM (2009) Isolation and analysis of low molecular weight microbial glycolipids: microbiology of hydrocarbons, oils, lipids, and derived compounds. In: Timmis KN (ed) Handbook of hydrocarbon and lipid microbiology. Springer, Berlin

Spagni A, Casu S, Grilli S (2012) Decolourisation of textile wastewater in a submerged anaerobic membrane bioreactor. Bioresour Technol 117:180–185

Sreelatha S, Reddy CN, Velvizhi G, Mohan SV (2015) Reductive behaviour of acid azo dye based wastewater: biocatalyst activity in conjunction with enzymatic and bio-electro catalytic evaluation. Bioresour Technol 188:2–8

Thaniyavarn J, Roongsawang N, Kameyama T, Haruki M, Imanaka T, Morikawa M, Kanaya S (2003) Production and characterization of biosurfactants from *Bacillus licheniformis* F2.2. Biosci Biotechnol Biochem 67:1239–1244

Toren A, Navon-Venezia S, Ron EZ, Rosenberg E (2001) Emulsifying activities of purified Alasan proteins from *Acinetobacter radioresistens* KA53. Appl Environ Microbiol 67:1102–1106

Toren A, Orr E, Paitan Y, Ron EZ, Rosenberg E (2002) The active component of the bioemulsifier alasan from *Acinetobacter radioresistens* KA53 is an OmpA-like protein. J Bacteriol 184:165–170

Tripathy DB, Mishra A (2016) Sustainable biosurfactants. https://doi.org/10.1002/9781119951438.eibc2433

Tuleva BK, Ivanov GR, Christova NE (2002) Biosurfactant production by a new *Pseudomonas putida* strain. Z Naturforsch C J Biosci 57:356–360

Turkovskaya OV, Dmitrieva TV, Muratova AY (2001) A biosurfactant-Producing *Pseudomonas aeruginosa* strain. Appl Biochem Microbiol 37(1):71–75

Van der Vegt W, Van der Mei H, Noordmans J (1991) Assessment of bacterial biosurfactant production through axisymmetrical drop shape-analysis by profile. Appl. Microbiol Biotechnol 35:766–770

Volchenko NN, Karasev SG, Nimchenko DV, Karaseva EV (2007) Cell hydrophobicity as a criterion of selection of bacterial producers of biosurfactants. Microbiology 76:112–114. https://doi.org/10.1134/S0026261707010158

Wang H, Su JQ, Zheng XW, Tian Y, Xiong XJ, Zheng TL (2009) Bacterial decolorization and degradation of the reactive dye reactive red 180 by *Citrobacter* sp. CK3. Int Biodeterior Biodegrad 63:395–399

Whiteley M, Lee KM, Greenberg EP (1999) Identification of genes controlled by quorum sensing in *Pseudomonas aeruginosa*. Proc Natl Acad Sci U S A 96:13904–13909

Youssef NH, Duncan KE, McInerney MJ (2005) Importance of 3-hydroxy fatty acid composition of lipopeptides for biosurfactant activity. Appl Environ Microbiol 71:7690–7695

Zinjarde S, Chinnathambi S, Lachke AH, Pant A (1997) Isolation of an emulsifier from *Yarrowia lipolytica* NCIM 3589 using a modified mini isoelectric focusing unit. Lett Appl Microbiol 24:117–121

Chapter 18
Water Depollution by Advanced Oxidation Technologies

Vittorio Loddo, Marianna Bellardita, Giovanni Camera Roda, Leonardo Palmisano, and Francesco Parrino

Contents

V. Loddo (✉) · M. Bellardita · L. Palmisano
Dipartimento di Ingegneria, University of Palermo, Palermo, Italy
e-mail: vittorio.loddo@unipa.it; marianna.bellardita@unipa.it; leonardo.palmisano@unipa.it

G. C. Roda
Dipartimento di Ingegneria Civile, Chimica, Ambientale e dei Materiali, University of Bologna, Bologna, Italy
e-mail: giovanni.cameraroda@unibo.it

F. Parrino
Dipartimento di Ingegneria Industriale, University of Trento, Trento, Italy
e-mail: francesco.parrino@unitn.it

Inamuddin, A. M. Asiri (eds.), *Sustainable Green Chemical Processes and their Allied Applications*, Nanotechnology in the Life Sciences,
https://doi.org/10.1007/978-3-030-42284-4_18

Abbreviations

2,6-DCP	2,6 dichlorophenol
4-CP	4-chlorophenol
AOPs	Advanced oxidation processes
CBZ	Carbamazepine
CT	Charge transfer
D	Generic substrate
D*	Generic substrate in its excited state
DBS	Dibutyl sulphide
DMSO	Dimethyl sulphoxide
DNA	Deoxyribonucleic acid
DOM	Dissolved organic matter
DTZ	Diatrizoate
EDCs	Endocrine disrupting chemicals
FXT	Fluoxetine
LED	Light-emitting diode
NDMA	N-nitrosodimethylamine
NOM	Natural organic matter
OET	Optical electron transfer
PET	Photo-induced electron transfer
PFOA	Perfluorooctanoic acid
PFOS	Perfluorooctane sulphonate
POPs	Persistent organic pollutants
RNA	Ribonucleic acid
ROS	Reactive oxygen species
TCEP	Tris-2-chloroethyl phosphate
TOrC	Trace of organic compounds
TPA	Terephthalic acid
UV	Ultraviolet
VUV	Vacuum ultraviolet

18.1 Introduction

Human activity has always required significant consumption of water taken from groundwater, lakes and rivers, while the use of water from other sources such as seawater desalination is more limited. Nowadays, many water resources are polluted by anthropogenic sources, including household, agricultural waste and industrial processes. Public concern over the environmental impact of wastewater pollution has increased.

The sustainable exploitation of water may be considered one of the great challenges of the current century (Bouwer 2000). Generally, both natural and anthropogenic contaminants may affect water quality (Berner and Berner 1989; Nigro et al. 2018).

In these last years, the growth of water demand has made wastewater recycling almost an essential process. Thus, it is necessary to develop methods suitable to decontaminate waters efficiently and ecologically. Several conventional chemical and physical wastewater treatment techniques have been applied to remove contaminants, but some limitations especially linked to the economic aspects should be faced.

Moreover, in recent times, traces of organic compounds (TOrCs) from pharmaceutical products and their metabolites, industrial chemicals and consumer products have been found in the environment (Huerta-Fontela et al. 2010). Among the most significant TOrC emitters are wastewater treatment plant effluents and urban and agricultural run-offs (Lim 2008; Gros et al. 2010; Luo et al. 2014). Indeed, generally, TOrCs remain in effluents from wastewater treatment plants as conventional biological and physical methods only partially transform these pollutants and the remaining fraction is consequently discharged into surface waters (Lim 2008; Luo et al. 2014; Zhang et al. 2008).

Advanced oxidation processes (AOPs) are efficient and ecological methods suitable for destroying and/or transforming persistent organic pollutants (POPs) present in waters and wastewater. According to Glaze et al. (1987), 'Advanced oxidation processes are defined as those which involve the generation of hydroxyl radicals in sufficient quantity to affect water purification'.

Indeed, most of the AOPs produce in situ powerful hydroxyl radicals ($^{\bullet}OH$) oxidising agents at a such concentration to successfully decontaminate waters. However, Glaze's definition does not contemplate important processes involving other types of radicals as chlorine ($Cl^{\bullet}$) and sulphate radicals ($SO_4^{\bullet -}$) (Miklos et al. 2018). Thus, a more general definition for AOPs could be 'a process that generates in situ highly reactive chemical radicals for the degradation of water pollutants'. AOPs are recognised as highly effective in delivering safe drinking water free of organics, inorganics and microbes (Abdel-Raouf et al. 2012; Bin and Sobera-Madej 2012). The significant advantage of using AOPs with respect to other existing chemical and biological processes is that they are totally 'environmental friendly' as they neither transfer pollutants from one phase to another (as in chemical precipitation and adsorption) nor produce massive amounts of hazardous sludge (Ayoub et al. 2010; Bebelis et al. 2013; Bethi et al. 2016; Fernández-Castro et al. 2015).

All AOPs consist of (1) the in situ generation of primary oxidant species and (2) the subsequent interaction of these species with pollutants. The mechanism of generation of the reactive species (generally radicals) depends on specific parameters of the process and can be influenced by reactor design and contaminant composition. In addition to radical scavenging, also other parameters such as radical mass transfer in surface-based AOPs and hydrodynamics are important in order to efficiently destroy the contaminants.

Among the different AOPs, treatments that use ozone and irradiation by UV light are well experienced and are applied at full scale for the purification of drinking water and for the abatement of noxious species in water reuse facilities. However, new results on innovative AOPs for water treatment are continuously reported by researchers (i.e. ultrasound- or microwave-based AOPs, electrochemical AOP, plasma, electron beam etc.) (Stefan 2018). The rising number of new technologies and the combinations of different processes call for a critical evaluation of AOPs with respect to cost-benefit analyses (i.e. general feasibility, chemical input, energy consumption and sustainability) to compare their efficiency with that of other alternative processes.

This chapter presents a description of the foremost types of AOPs based on chemical and photochemical reactions. Principles, advantages, drawbacks and performances of AOPs, coupled with other methodologies and their applications to waters and wastewater treatment, have been analysed.

18.2 UV-Based AOPs

UV-based AOPs take place in systems that use UV irradiation alone or in combination with radical promoters. Then there are UV AOPs that are radical-based processes in which primary oxidant species are produced by UV irradiation only, and there are others that also use other species such as ozone, H_2O_2, persulphate (to form sulphate radicals) and Cl_2 (to form radical chlorine species and $^{\bullet}OH$ radicals).

UV radiation is emitted in the interval of wavelengths ranging from 1 to 380 nm. The radiations can be classified by considering the emission wavelengths as UV-A, which is emitted in the range between 380 and 315 nm; UV-B, between 315 and 280 nm; UV-C, between 280 and 200 nm; vacuum UV (VUV), between 200 and 100 nm; and extreme UV, between 100 and 1 nm (Zoschke et al. 2014).

The most used UV radiation sources for UV-based AOPs are low-pressure Hg vapour lamps with partial pressure of Hg of about 1 Pa at the lamp wall. These lamps show efficiencies between 25 and 45% in the emission wavelength range. The emission spectrum of low-pressure Hg lamps exhibit well-defined lines at ca. 254 and 185 nm. Table 18.1 shows the relative intensity of the emission of low-pressure Hg lamps related to the wavelength of 254 (quoted as 100%) (Masschelein 2002).

The 254-nm emission line is directly useful for disinfection. Indeed, UV light can inactivate microorganisms causing DNA or RNA molecule damages, preventing their reproduction (Blatchley et al. 2008; Bolton and Cotton 2008).

In germicidal lamps, the emission line at 185 nm is generally cut off by using suitable materials such as quartz doped with TiO_2 or optical glass (Masschelein 2002). The materials that allow the passage of 185-nm radiation are high-purity synthetic quartz (Suprasil) (Zoschke et al. 2014).

Other lamps that emit in the VUV range are Xe-excimer radiators, which emit at 172 nm and are mainly applied for advanced oxidation technologies (Zoschke et al. 2014).

Among the irradiation technologies, UV light-emitting diodes (LED) have been investigated for disinfection purposes (Song et al. 2016).

Table 18.1 Emission intensity of low-pressure mercury VUV lamps relative to emission at 254 nm (Masschelein 2002)

Wavelength (nm)	Emitted intensity (I_0, rel)
184.9	8
296.7	0.2
248.2	0.01
253.7	100
265.2–265.5	0.05
275.3	0.03
280.4	0.02
289.4	0.04
405.5–407.8	0.39
302.2–302.8	0.06
312.6–313.2	0.6
334.1	0.03
365.0–366.3	0.54

UV LEDs allow to use, for disinfection purposes, the optimum wavelength instead of the 254 nm emitted by low-pressure mercury lamps. Other benefits of UV LEDs with respect to classical medium- and low-pressure mercury lamps are the following (Würtele et al. 2011):

- No mercury content
- Compact size and robust design: easy handling and more duration (no glass or filaments)
- No warm-up time required
- Low-power requirements, lower voltages
- Potential for higher energy efficiency
- Longer lifetimes

However, the irradiation efficiencies of current diodes are less than 10%, and the predicted UV-LED efficiencies is about 75% in 2020.

18.2.1 UV Irradiation

AOPs that use UV light irradiation alone exploit different mechanisms.

One pathway is the reaction with singlet oxygen deriving from dissolved organic matter (DOM), which is excited by a radiative source of suitable energy. However, the steady-state concentration of singlet oxygen in solution is low because it is rapidly quenched to the ground state in the presence of water.

The main steps in a process involving singlet oxygen are the following (Haag and Hoigné 1986):

$$S \xrightarrow{h\nu} S^* \xrightarrow{O_2} {}^1O_2 + S \quad (18.1)$$

$$^1O_2 \xrightarrow{k_d} O_2\,(\text{physical quenching}) \tag{18.2}$$

$$^1O_2 + A \xrightarrow{k_q} O_2 + A\,(\text{physical quenching by A}) \tag{18.3}$$

$$^1O_2 + A \xrightarrow{k_r} \text{Products}\,(\text{chemical reaction with A}) \tag{18.4}$$

in which S is the species that can be excited by radiation and A represents a species present in the reaction environment.

The absorption of VUV radiations can generate also homolysis and photochemical ionisation of water:

$$H_2O \xrightarrow{h\nu\,(185\,\text{nm})} {}^{\cdot}OH + {}^{\cdot}H\,(\text{Homolysis}) \tag{18.5}$$

$$H_2O \xrightarrow{h\nu\,(185\,\text{nm})} {}^{\cdot}OH + H^+ + e^-_{aq}\,(\text{Photochemical ionization}) \tag{18.6}$$

Therefore, VUV photolysis of water represents an efficient method for the generation of hydroxyl radicals, and their formation rate is comparable to other advanced oxidation processes (Legrini et al. 1993).

Moreover, UV light can also afford homolytic cleavage of chemical bonds of the pollutants, resulting in the formation of two radicals. Wavelengths less than 190 nm can effectively break highly stable C-F bond, whereas wavelengths in the 210–230-nm range can break C–Cl bond (Mishra et al. 2017):

$$RX \xrightarrow{h\nu} {}^{\cdot}R + {}^{\cdot}X \tag{18.7}$$

$${}^{\cdot}R + O_2 \rightarrow RO_2 \tag{18.8}$$

Radicals can react with oxygen (Eq. 18.8) or may initiate further redox reactions with other molecules in solution.

The absorption of VUV radiation by oxygen (in the gas phase) could cause its photolysis.

The subsequent reaction of photo-generated oxygen atom with another oxygen molecule generates ozone. This reaction takes place in the presence of a further molecule that absorbs the excessive kinetic energy (Bolton and Denkewicz 2007).

The major advantage of this method is that it requires shorter treatment time, and it does not require the addition of any chemical so that no residual products are formed during the treatment.

This process has been used for the degradation of different industrial solvents and organic pollutants, including endocrine disrupting chemicals (EDCs), with diverse applications, such as flame retardants, pesticides, cosmetics and plastic additives, which may contaminate food and other products. Therefore, EDCs may be released as residue from the products that contain them.

It has been proven that the method is efficient in removing N-nitrosodimethylamine (NDMA), a well-known potent carcinogen found in trace amounts in wastewater (Mishra et al. 2017).

Low-pressure UV lamps have been used to degrade pesticides as alachlor, atrazine, diuron, pentachlorophenol, chlorphenvinphos and isoproturon, and most of them, except isoproturon were degraded completely (Sanches et al. 2010). Also, pharmaceuticals (Sakai et al. 2012; Kim et al. 2009; Avisar et al. 2010; Real et al. 2009) and anabolic-androgenic steroids (Błedzka et al. 2010; Błedzka et al. 2012) present in trace amounts were successfully degraded.

18.2.2 UV-H_2O_2

The most used combination of UV-based AOP is with H_2O_2. The primary oxidant species are generated by the photolytic breakage of the O-O bond of the H_2O_2 molecule with the formation of two hydroxyl radicals:

$$H_2O_2 \xrightarrow{h\nu} 2\,{}^{\cdot}OH \tag{18.9}$$

Moreover, as far as the pH is concerned:

$$H_2O_2 \rightleftarrows HO_2^- + H^+ \,(\text{acidic medium}) \tag{18.10}$$

$$H_2O_2 + OH^- \rightleftarrows HO_2^- + H_2O\,(\text{alkaline medium}) \tag{18.11}$$

However, H_2O_2 has a relatively low molar absorption coefficient ($\varepsilon = 18.6\,M^{-1}\,cm^{-1}$ at $\lambda = 254$ nm), and consequently the turnover of its decomposition is less than 10%.

Equation 18.9 represents the initiation step, followed by the propagation ones (Eqs. 18.12–18.14):

$${}^{\cdot}OH + H_2O_2 \rightarrow H_2O + HO_2^{\cdot} \tag{18.12}$$

$${}^{\cdot}OH + HO_2^- \rightarrow HO_2^{\cdot} + OH^- \tag{18.13}$$

$$HO_2^{\cdot} + H_2O_2 \rightarrow {}^{\cdot}OH + H_2O + O_2 \tag{18.14}$$

and the termination steps (Eqs. 18.15–18.17).

$$2HO_2^{\cdot} \rightarrow H_2O_2 + O_2 \tag{18.15}$$

$${}^{\cdot}OH + HO_2^{\cdot} \rightarrow H_2O + O_2 \tag{18.16}$$

$$2\,{}^{\cdot}OH \rightarrow H_2O_2 \tag{18.17}$$

Notably, the rate of production of free radicals depends on different factors, such as type of UV lamp and physicochemical properties of the reacting solution (turbidity, pH, UV radiation distribution etc.) (Oturan and Aaron 2014).

As far as pH is concerned (Eqs. 18.10 and 18.11), due to the fact that HO_2^- anion can absorb UV light to generate radicals such as $HO_2^{\cdot}$ and $^{\cdot}OH$, the reaction rate should be higher in alkaline medium (pH > 10). Nevertheless, a conflicting effect exists as bicarbonate and carbonate ions could interact with $^{\cdot}OH$ species at very high pH values, thus acting as scavengers (Boczkaj and Fernandes 2017):

$$^{\cdot}OH + HCO_3^- \rightarrow CO_3^{\cdot-} + H_2O \tag{18.18}$$

$$^{\cdot}OH + CO_3^{2-} \rightarrow CO_3^{\cdot-} + OH^- \tag{18.19}$$

Moreover, absorption of CO_2 from the air increases at alkaline pHs, and the reaction must be carried out in a closed vessel to avoid this effect.

The concentration of carbonate and bicarbonate ions decreases as the pH decreases, thus increasing the amount of free hydroxyl radicals in solution, improving the efficiency of the process (Ameta et al. 2013).

However, a drawback of this process, as mentioned before, is the low absorption coefficient of H_2O_2 in the UV region (when using low-pressure lamps), and for an efficient degradation of organic pollutants to occur, very high amounts of hydrogen peroxide are needed, making necessary the removal of residual H_2O_2 in a subsequent step. The doses of H_2O_2 to be used are determined mainly by considering the economic aspects. However, H_2O_2 in excess concentrations can act also as a scavenger of OH radicals affecting the radical yield (Wang and Xu 2012):

$$^{\cdot}OH + H_2O_2 \rightarrow H_2O + HO_2^{\cdot} \tag{18.20}$$

The H_2O_2-UV process has been applied to the treatment of model and real wastewater containing inorganic and organic compounds. Species like 4chloro-2nitrophenol (Saritha et al. 2007), model dye (Alaton et al. 2002), naphthenic acids (Liang et al. 2011), bisphenol A (Sharma et al. 2015a) and salicylic acid (Saien et al. 2015) were recalcitrant to oxidation, needing more than 3 h to be degraded at high extent. In contrast, with other compounds such as cyanide (Malhotra et al. 2005), dimethyl sulphoxide (DMSO), phenol, dibutyl sulphide (DBS), diatrizoate (DTZ), terephthalic acid (TPA), metoprolol, carbamazepine (CBZ), nonylphenol and norfloxacin, high degradation efficiencies (>73%) have been obtained in less than 3 h in the 9–12 pH range (Boczkaj and Fernandes 2017).

The results of experiments carried out by using real wastewater showed that this process is not effective for the treatment of refinery wastewater at basic pH due to the behaviour of hydrogen peroxide in alkaline solution. In general, H_2O_2-UV AOP showed low efficiency at basic pHs, with the exception of some model species such as norfloxacin, metoprolol and nonylphenol. However, the efficiency of the treatment increased at neutral or acidic values (Boczkaj and Fernandes 2017).

Full-scale applications of this technology have been developed for both disinfection and advanced oxidation process. An example is the full-scale UV-H_2O_2 installation for pre-treated IJssel Lake water (Martijn et al. 2007). The maximum capacity of the plant is 4500 m^3/h, and it has been operational since November 2004.

18.2.3 UV-O_3

In the combined UV and O_3 process, irradiation with wavelengths less than 300 nm allows the decomposition of ozone in O_2 and oxygen excited atoms in singlet state $O(^1D)$, which can react with water to produce thermally excited hydrogen peroxide (Garoma and Gurol 2004; Song et al. 2008; Reisz et al. 2003):

$$O_3 \xrightarrow{h\nu} O_2 + O(^1D) \tag{18.21}$$

$$O(^1D) + H_2O \rightarrow H_2O_2 \tag{18.22}$$

Hydrogen peroxide can be excited by light and can decompose to hydroxyl radicals (Eq. 18.9).

However, only small proportions of H_2O_2 decomposes due to recombination, as confirmed by $^{\bullet}OH$ quantum yield of only 10% (Reisz et al. 2003).

A net enhancement of organic degradation rate occurs in the presence of DOM due to direct and indirect production of $^{\bullet}OH$.

The molar extinction coefficient of ozone is $\varepsilon = 3300\ M^{-1}\ cm^{-1}$ at $\lambda = 254$ nm, higher than that of hydrogen peroxide at the same wavelength.

Ozone can be decomposed by $^{\bullet}OH$ in a short reaction chain, described in the following reactions (Garoma and Gurol 2004; Song et al. 2008; Reisz et al. 2003):

$$O_3 + {}^{\cdot}OH \rightarrow HO^{\cdot}_2 + O_2 \tag{18.23}$$

$$HO^{\cdot}_2 \rightleftarrows H^+ + {}^{\cdot}O_2^- \tag{18.24}$$

$$O_3 + {}^{\cdot}O_2^- \rightarrow {}^{\cdot}O_3^- + O_2 \tag{18.25}$$

$$O_3 + HO^{\cdot}_2 \rightarrow {}^{\cdot}OH + 2O_2 \tag{18.26}$$

The presence of OH^-, as an initiator, could generate HO_2^- by reaction with ozone (Eq. 18.27) (Wang and Xu 2012):

$$O_3 + OH^- \rightarrow HO_2^- + O_2 \tag{18.27}$$

Ozone can react with organics both directly by electrophilic interaction or indirectly by radical-type reaction. It can be noticed that the main mechanism of the UV-O_3 process is the reaction of hydroxyl radical with the organic species because the reaction rate of direct oxidation with molecular ozone is slower than that of $^{\bullet}OH$ (Garoma and Gurol 2004). The main advantage of the process is that it allows to degrade a wide range of organic compounds with some exceptions. Indeed, as examples, dimethyl sulphoxide (DMSO) and fluoxetine (FXT) were successfully degraded (~90%) in less than 90 min in a 9–11 pH range, whereas model dyes like perfluorooctane sulphonate (PFOS) and perfluorooctanoic acid (PFOA) were degraded with higher difficulty (over 120 min to reach high degrees of degradation) [37]. However, the process needs a high electrical energy demand (for UV lamps and ozone generators), denoting low-energy efficiencies for radical generation. This fact could explain the absence of data on full-scale UV-O_3 applications.

The addition of H_2O_2 to the UV-O_3 process has been investigated to accelerate the degradation of organic pollutants. Indeed, an increased rate of $^{\bullet}OH$ radical generation was found (Kusic et al. 2006). Comparative studies of UV, UV-H_2O_2, UV-O_3 and UV-H_2O_2-O_3 processes carried out in the presence of organic pollutants indicated that the last process showed the highest degree of mineralisation (Kusic et al. 2006; Peternel et al. 2006). After 1 h of reaction, the following order was found: UV < UV-H_2O_2 < UV-O_3 < UV-H_2O_2-O_3 for the mineralisation degree of the organic dye C.I. Reactive Red 45 (Peternel et al. 2006).

18.2.4 UV-Cl_2

Another UV-assisted AOP is the UV-Cl_2 process, in which UV-activated chlorine species such as $Cl^{\bullet}$ are formed. The photolysis of aqueous chlorine was studied both under sunlight and ultraviolet wavelengths (Nowell and Hoigné 1992a, b; Watts and Linden 2007).

In aqueous solution with pH in the 4–8 range and in the absence of amines and ammonia, chlorine is present as hypochlorous acid (HOCl) and hypochlorite (ClO^-) in equilibrium (Nowell and Hoigné 1992a). If the solution containing chlorine is irradiated by sunlight or UV light, the direct formation of hydroxyl radicals occurs predominantly (Watts and Linden 2007; Kishimoto 2019) (Eq. 18.28):

$$HOCl \xrightarrow{h\nu} {}^{\cdot}OH + Cl^{\cdot} \quad (18.28)$$

Chlorine radicals can also be indirectly formed by protonation of the species O^- according to Eqs. 18.29 and 18.30 (Nowell and Hoigné 1992a; Buxton and Subhani 1972; Vogt and Schindler 1992):

$$ClO^- \xrightarrow{hn} Cl^{\cdot} + O^- \quad (18.29)$$

$$ClO^- \xrightarrow{hn} Cl^{\cdot} + O(^1D) \quad (18.30)$$

Apart from the primary photolysis of HOCl, in the presence of organic species (R), other chain reactions could be possible in which HOCl works as chain promoters and Cl• or •OH are the chain carriers (Eqs. 18.31–18.34) in such a way as to maintain a steady-state concentration of the radicals produced (Nowell and Hoigné 1992a):

$$^{\cdot}OH + RH \rightarrow R^{\cdot} + H_2O \quad (18.31)$$

$$R^{\cdot} + HOCl \rightarrow RCl + {}^{\cdot}OH \quad (18.32)$$

$$Cl^{\cdot} + RH \rightarrow R^{\cdot} + HCl \quad (18.33)$$

$$R^{\cdot} + HOCl \rightarrow ROH + Cl^{\cdot} \quad (18.34)$$

Chlorine radical is an oxidant more selective than hydroxyl radical as it can react preferably with electron-rich species (Fang et al. 2014). The molar absorption coefficients can be influenced by pH as the ratio between HOCl and ClO^- may considerably change. The UV-Cl_2 process is favoured at low pH values as in water originated from reverse osmosis permeate (Watts and Linden 2007). Oxidative chlorine species like ClO^- and ClO• formed during the process could be transformed by •OH to perchlorate, chlorate and halogenated by-products, resulting in a drawback for the process. A full-scale plant of this process was developed for potable reuse, and it is placed at Los Angeles Terminal Island Water Reclamation Plant (Xylem 2015).

18.2.5 UV-$SO_4^{\cdot-}$

The UV-$SO_4^{\cdot-}$ process generates mainly sulphate radicals ($SO_4^{\cdot-}$), which are a good alternative to AOPs •OH based, for the degradation of organic water contaminants (Miklos et al. 2018). These radicals have the same oxidising power than the •OH ones, but they are more selective (Lutze et al. 2015).

Moreover, the sulphate radicals have a longer half-life than the •OH ones as they mainly act through electron transfer reactions, while the oxidation mechanism of •OH provides the less dominant hydrogen-atom abstraction (Rastogi et al. 2009).

Sulphate radical generation can occur by two currently used processes, which use peroxydisulphate (PS) and peroxymonosulphate (PMS). They can be formed from PS through its homolitic cleavage by UV-C radiations or its activation by the use of a catalyst or by heat:

$$S_2O_8^{2-} \xrightarrow{hv} 2SO_4^{\cdot-} \quad (18.35)$$

$$S_2O_8^{2-} \xrightarrow{\Delta} 2SO_4^{\cdot -} \quad E_a = 33.5\,\text{kcal/mol} \tag{18.36}$$

$$S_2O_8^{2-} + e^- \rightarrow SO_4^{2-} + SO_4^{\cdot -} \tag{18.37}$$

In alkaline medium, in the presence of water, sulphate radicals can be decomposed to form hydroxyl radicals:

$$SO_4^{\cdot -} + H_2O \rightarrow {}^{\cdot}OH + HSO_4^- \tag{18.38}$$

$$SO_4^{\cdot -} + OH^- \rightarrow {}^{\cdot}OH + SO_4^{2-} \tag{18.39}$$

In recent years, the activation of PS by NaOH according to the following mechanism has been proposed (Furman et al. 2010):

$$2S_2O_8^{2-} + 2H_2O \xrightarrow{OH^-} 3SO_4^{2-} + SO_4^{\cdot -} + O_2^{\cdot -} + 4H^+ \tag{18.40}$$

At acidic pHs, H^+ could interact with PS to generate both sulphate and radical sulphate according to the following equation:

$$S_2O_8^{2-} + e^- \xrightarrow{H^+} SO_4^{2-} + SO_4^{\cdot -} \tag{18.41}$$

The mechanisms of oxidation in aqueous solution are electron transfer, hydrogen atom transfer and addition (Long et al. 2014). The advantages of using sulphate radicals are their high oxidation power, high stability and high solubility in aqueous medium. The main drawback is the production of ions as sulphate, potassium and sodium (Huang et al. 2005).

Also, the PMS anion (HSO_5^-) can be activated by light of suitable energy, by heat or by using a catalyst for the production of sulphate and hydroxyl radicals:

$$HSO_5^- \xrightarrow{h\nu} SO_4^{\cdot -} + {}^{\cdot}OH \tag{18.42}$$

$$HSO_5^- \xrightarrow{\Delta} SO_4^{\cdot -} + {}^{\cdot}OH \quad E_a = 33.5\,\text{kcal/mol} \tag{18.43}$$

Moreover, the activation of PMS can be also achieved by using an alkaline compound (NaOH) (Qi et al. 2016), and the suggested mechanism consists of a first step in which sulphate anion and H_2O_2 are generated in alkaline medium, as described in the following equations:

$$HSO_5^- \xrightarrow{NaOH} H^+ + SO_5^{2-} \tag{18.44}$$

$$SO_5^{2-} + H_2O \rightarrow H_2O_2 + SO_4^{2-} \tag{18.45}$$

Afterwards, H_2O_2 can generate hydroxyl radicals according to the mechanism described in the previous section.

As far as the acidic medium is concerned, PMS reacts with H^+ forming sulphate ions and water (Oh et al. 2016). In this case, the activation of PMS is inhibited and the efficiency of the process decreases dramatically with respect to the case of alkaline pH:

$$HSO_5^- + H^+ + 2e^- \rightarrow H_2O + SO_4^{2-} \quad (18.46)$$

HSO_5^- can be obtained from potassium or sodium salts or as a triple salt ($2KHSO_5 \cdot KHSO_4 \cdot K_2SO_4$), called Oxone (Sharma et al. 2015b). HSO_5^- can decompose and react at alkaline pHs (Sharma et al. 2015b; Qi et al. 2016; Zhang et al. 2013). Moreover, as previously described (Eq. 18.38), $SO_4^{\bullet -}$ generates $^\bullet OH$ in the presence of water.

In the literature, several studies on the application of UV/PMS are present (Antoniou et al. 2010; Khan et al. 2014), either utilising radical-driven processes or direct electron transfer, and they are very powerful tools for the treatment of a broad range of organics. However, the reactivity of persulphates is largely dependent on the composition of water matrix and on activation methods. Furthermore, direct reactions of peroxymonosulphate (PMS) are rather slow and mostly unsuitable for pollutant degradation (Wacławek et al. 2017).

18.3 Ozone-Based AOPs

Ozone is a strong oxidant able to degrade organic compounds into smaller molecules or in CO_2, H_2O and mineral acids (Chandrasekara et al. 2009; Cañizares et al. 2009; Abu Amr et al. 2013; Puspita et al. 2015). In aqueous media, it has a low solubility and a short lifetime.

Different techniques have been studied to overcome the low solubility in water, such as line mixers, stirring, fixed beds, contact towers, diffusers or large bubble columns to increase the retention time (Litter 2005; Glaze and Kangt 1989; Shahidi et al. 2015). Ozone is directly produced in situ by suitable equipment in which an oxygen or air stream is subjected to an electric discharge.

Ozone has been used for a long time as oxidant or disinfectant in water treatment. It is very selective and reacts mainly with species having double bonds and electron-rich functional groups like activated aromatic rings (e.g. phenols) and amines. Ozonation is considered as an AOP because its reactions often involve the formation of $^\bullet OH$ in actual aqueous solutions. Indeed, $^\bullet OH$ can be formed from the reaction of ozone with hydroxide ions (Merényi et al. 2010). Moreover, ozonation is effective in water treatment for the degradation of microbes (Von Gunten 2003), micro-pollutant removal (Gottschalk et al. 2009; Bourgin et al. 2017), decolorisation (Konsowa 2003), non-protonated amines (Camel and Bermond 1998) and odour and taste removal (Von Gunten 2003; Camel and Bermond 1998).

18.3.1 Ozonation

Processes using O_3 are defined as AOPs in alkaline solutions in combination with transition metal cations or UV/sunlight, as previously described.

In acidic medium, ozone is decomposed by reaction with protons preventing a direct reaction of O_3 with pollutants (Shahidi et al. 2015):

$$O_3 + 2H^+ + 2e^- \rightarrow O_2 + H_2O \tag{18.47}$$

In alkaline solution, the presence of OH^- is beneficial as ozone reacts directly with it producing $^{\bullet}OH$, according to the following equations:

$$O_3 + OH^- \rightarrow {}^{\bullet}O_3^- + {}^{\bullet}OH \tag{18.48}$$

$${}^{\bullet}O_3^- \rightarrow {}^{\bullet}O^- + O_2 \tag{18.49}$$

$${}^{\bullet}O^- + H^+ \rightarrow {}^{\bullet}OH \tag{18.50}$$

Other mechanisms of hydroxyl radical production are described elsewhere (Litter 2005; Glaze and Kangt 1989; Shahidi et al. 2015).

The amount of OH^- is linked to $^{\bullet}OH$ production and thus to indirect ozonation. This process might be a promising process, particularly at pHs higher than 8.

18.3.2 O_3-H_2O_2 (Peroxone) Process

The combination of ozone and hydrogen peroxide is called the peroxone process. Organic compounds are oxidised both indirectly and directly (Ameta et al. 2013). The combination of O_3 and H_2O_2 causes a greater increase of oxidation efficiency by enhancing the conversion of O_3 to $^{\bullet}OH$ and improving O_3 transfer from the gas to the liquid phase (Ameta et al. 2013).

Alternatively, ozone can activate H_2O_2 with the generation of radicals such as $^{\bullet}OH$ and $HO_2^{\bullet}$:

$$H_2O_2 + O_3 \rightarrow {}^{\bullet}OH + HO_2^{\bullet} + O_2 \tag{18.51}$$

O_3 can interact also with $^{\bullet}OH$ producing less reactive radicals like $HO_2^{\bullet}$, which generates O_2 and $^{\bullet}OH$ by reacting with ozone:

$${}^{\bullet}OH + O_3 \rightarrow O_2 + HO_2^{\bullet} \tag{18.52}$$

$$O_3 + HO_2^{\bullet} \rightarrow 2O_2 + {}^{\bullet}OH \tag{18.53}$$

At acidic pHs, the decomposition of ozone to form hydroxyl radicals is inhibited (see Eq. 18.47), whereas it is favoured in alkaline medium (Eqs. 18.48–18.50).

Before discharging in the environment, the water treated with the peroxone process, the residual hydrogen peroxide, must be eliminated. Typical ozone concentrations range between 1 and 20 mg/L, whereas the optimum molar ratio between H_2O_2 and O_3 is equal to 0.5 mol/mol (Katsoyiannis et al. 2011; Pisarenko et al. 2012).

Hydrogen peroxide can be also formed from the reaction of ozone with water, but its contribution to the subsequent formation of $^{\bullet}OH$ is negligible.

Although peroxone is a well-established process for the treatment of drinking water and water recycle, recent researches indicate that the high competition between the reactions limits the advantages of this process. However, a benefit for its application is the minimisation of bromate formation during ozone processes (Miklos et al. 2018). It can be noticed that the process, by using ozone alone, is more efficient for radical formation than the peroxone one (Hübner et al. 2015).

18.3.3 O_3 Catalysts

The addition of a catalyst to a system that uses ozone can increase both the efficiency of the process and the adsorption of ozone in solution and consequently the production of hydroxyl radicals. This treatment is called catalytic ozonation. Catalytic ozonation can be a homogeneous or a heterogeneous process, depending on the solubility of the catalyst in water. The catalysts studied for homogeneous catalytic ozonation are generally transition metals such as Fe(II), Mn(II), Ni(II), Co(II), Cd(II), Cu(II), Ag(I), Cr(III) and Zn(II). The type of metal affects reaction rate, selectivity and ozone consumption (Kasprzyk-Hordern et al. 2003).

The mechanism of the reaction consists of an initiation step of ozone decomposition favoured by the metallic ions present in solution and followed by the formation of $^{\bullet}OH$ radicals (Kasprzyk-Hordern et al. 2003).

Typical examples of the mechanisms of the possible pathways of production are reported below.

As an example, by using Mn(II) as the catalyst (Wu et al. 2008), the following species are formed (Eq. 18.54):

$$Mn^{2+} + O_3 + H^+ \rightarrow Mn^{3+} + {}^{\bullet}OH + O_2 \tag{18.54}$$

The reactions of ozone with Fe(II) to produce $^{\bullet}OH$ are described by Eqs. 18.55 and 18.59 (Wang and Xu 2012):

$$Fe^{2+} + O_3 \rightarrow Fe^{3+} + O_3^{\bullet -} \tag{18.55}$$

$$O_3^{\bullet -} + H^+ \rightarrow O_2 + {}^{\bullet}OH \tag{18.56}$$

$$Fe^{2+} + O_3 \rightarrow FeO^{2+} + O_2 \tag{18.57}$$

$$FeO^{2+} + H_2O \rightarrow Fe^{3+} + OH^- + {}^{\cdot}OH \tag{18.58}$$

$$Fe^{3+} + O_3 + H_2O \rightarrow FeO^{2+} + H^+ + {}^{\cdot}OH + O_2 \tag{18.59}$$

A rule exists for the mechanism of homogeneous catalytic ozonation by using transition metals like Fe(II) and Mn(II), and it can be described by the following reactions:

$$M^{n+} + O_3 + H^+ \rightarrow M^{(n+1)+} + O_2 + {}^{\cdot}OH \tag{18.60}$$

$$M^{n+} + {}^{\cdot}OH \rightarrow M^{(n+1)+} + OH^- \tag{18.61}$$

where *n* represents the oxidation number of the metal before the reaction.

It can be noticed that if the catalyst amount is higher than the optimum one, hydroxyl radicals can be scavenged (Eq. 18.61), with the consequent decrease of the degradation rate (Wu et al. 2008).

As far as the system O_3 and Co(II) is concerned, the mechanism proposed is the following (Wang and Xu 2012):

$$Co^{2+} + O_3 + H_2O \rightarrow Co(OH)^{2+} + {}^{\cdot}OH + O_2 \tag{18.62}$$

$${}^{\cdot}OH + O_3 \rightarrow HO_2^{\cdot} + O_2 \tag{18.63}$$

$$HO_2^{\cdot} + Co(OH)^{2+} \rightarrow Co^{2+} + H_2O + O_2 \tag{18.64}$$

The same mechanism is valid for Ni^{2+}.

The application of catalytic ozonation by transition metal ions greatly improves the efficiency of degradation with respect to ozonation alone. Humic substances have been successfully removed from wastewater, and the best results were obtained for Mn(II) (62% TOC) and Ag(I) (61% TOC).

Moreover, the catalytic ozonation allows a low ozone consumption (Gracia et al. 1996).

Catalytic ozonation by Mn(II) and Fe(II) ions was found effective in the degradation of chlorinated benzene derivatives. The main by-products were formaldehyde and methylglyoxal in Fe(II)/O_3 systems and mainly glycoxal in Mn(II)/O_3 ones (Cortés et al. 2000). These catalysts were found to be effective also in the treatment of wastewater from the production of organochloride pesticides (dicofol and tetradifon), which contain chlorinated benzene derivatives (Pines and Reckhow 2002). Oxalic acid does not react directly with ozone, and it can be easily degraded by using trace amounts of cobalt(II) (Pines and Reckhow 2002).

The efficiency of catalytic ozonation can be further increased by combining it with UV irradiation. Indeed, 2,4-dichlorophenoxyacetic acid (2,4-D) was significantly degraded by using Fe^{2+}-O_3 and Fe^{2+}-UV systems (Piera et al. 2000). However, the complete degradation of 2,4-D in aqueous media occurred only with the O_3-Fe^{2+}-UV system, which was also found to be efficient for aniline and 2,4-chlorophenol removal.

The performance of the homogeneous catalytic ozonation depends on different parameters, such as ozone concentration, pH and catalyst amount. Generally, the degradation of organics proceeds through $^{\bullet}OH$ radical attack or the formation of complexes between the organic species and the metallic ions, followed by an ozone molecule attack (Wang and Xu 2012; Gracia et al. 1996; Wu et al. 2008).

The use of a solid support in catalytic processes allows to avoid the leakage of the catalyst by precipitation or dissolution and contemporaneously increases the surface area and enhances adsorption. In the heterogeneous catalytic ozonation process, metal oxides such as TiO_2, Al_2O_3 and MnO_2 and supported metal oxides like Cu-TiO_2, Fe_2O_3-Al_2O_3, Cu-Al_2O_3 and Ru-CeO_2 are used as catalysts. In heterogeneous systems, it is possible to hyphotesise three different mechanisms (Beltrán et al. 2002):

1. The chemisorption of ozone generating active species that react with organic species in solution
2. The chemisorption of organic molecules, both associative or dissociative, and their subsequent reaction with gaseous or aqueous ozone
3. The chemisorption of both organic molecules and ozone and their further interaction

Adsorption of ozone followed by its decomposition generates surface-bound O radicals and hydroxyl radicals on the catalysts' surface (*S*) (Kasprzyk-Hordern et al. 2003; Gracia et al. 1996).

The mechanisms that describe this process depend on the pH range. For pHs ranging from 2 to 6, the mechanism is described by Eqs. 18.65–18.67, and surface-bound oxygen species are generated instead of free radicals, thus indicating a lower catalytic decomposition of ozone:

$$O_3 + S \rightleftarrows S \cdot O_3 \tag{18.65}$$

$$S \cdot O_3 \rightleftarrows S \cdot O + O_2 \tag{18.66}$$

$$O_3 + S \cdot O \rightleftarrows 2O_2 + S \tag{18.67}$$

For pHs higher than 6, the following mechanism is proposed:

$$OH^- + S \rightleftarrows S \cdot OH \tag{18.68}$$

$$O_3 + S \cdot OH \rightleftarrows S \cdot O_3^{\cdot} + {}^{\cdot}OH \tag{18.69}$$

$$S \cdot O_3^{\cdot} \rightleftarrows S \cdot O^{\cdot} + O_2 \tag{18.70}$$

$$O_3 + S \cdot O^{\cdot} \rightleftarrows O_2^{\cdot -} + O_2 + S \tag{18.71}$$

The hydroxyl radical formation in the above reactions indicates a faster mineralisation of organics (Gracia et al. 1996).

18.4 Fenton-like Like Reactions

A considerable amount of publications is present in the pertinent literature about Fenton and Fenton-like reactions, supporting the efficiency of these processes in water and soil treatments (Neyens and Baeyens 2003; Pignatello et al. 2006), the complex mechanisms of the reactions (Pignatello et al. 2006; Masarwa et al. 2005) and the application of the Fenton method for the treatment of industrial wastewater (Bautista et al. 2008).

Fenton-like reactions are economical and efficient techniques that use peroxides (generally hydrogen peroxide) or dissolved oxygen and iron ions to produce radicals for the abatement of organic and inorganic species (Masarwa et al. 2005). Generally, the oxidising species is the hydroxyl radical, and its production is the main step in the radical Fenton chain reaction.

18.4.1 Fenton Process

The homogeneous Fenton process consists of a series of cyclic reactions initiated by the production of $^{\cdot}OH$ radicals at acidic pH in the presence of ferric or ferrous ions and hydrogen peroxide (Eqs. 18.72 and 18.73) (Pignatello et al. 2006; Masarwa et al. 2005; Bautista et al. 2008):

$$Fe^{2+} + H_2O_2 \rightarrow Fe^{3+} + {}^{\cdot}OH + OH^- \tag{18.72}$$

$$Fe^{3+} + H_2O_2 \rightarrow Fe^{2+} + HO_2^{\cdot} + H^+ \tag{18.73}$$

$$Fe^{3+} + H_2O_2 \rightleftarrows Fe(OOH)^{2+} + H^+ \tag{18.74}$$

$$Fe(OOH)^{2+} \rightarrow Fe^{2+} + HO_2^{\cdot} \tag{18.75}$$

The reaction is propagated by Fe^{3+} reduction with H_2O_2 to regenerate Fe^{2+} (Eqs. 18.73–18.75). In particular, Eq. 18.73 is coupled with a series of reactions in which peroxo complexes of Fe(III) are produced in an equilibrium reaction (Eq. 18.74) and subsequently decomposed into Fe^{2+} and $HO_2^{\cdot}$ (Eq. 18.75) (Wang and Xu 2012).

The decomposition of a generic organic pollutant (RH) by hydroxyl radicals occurs mainly via hydrogen abstraction, oxygen abstraction or nitrogen abstraction, and the addition of $HO^{\bullet}$ to aromatic rings or to carbon-carbon bonds depends on the redox potential of the organic molecule:

$$RH + {}^{\cdot}OH \rightarrow R^{\cdot} + H_2O \tag{18.76}$$

$$R^{\cdot} + Fe^{3+} \rightarrow Fe^{2+} + R^{+} \tag{18.77}$$

$$R^{+} + OH^{-} \rightarrow ROH \tag{18.78}$$

$$R^{\cdot} + H_2O_2 \rightarrow ROH + {}^{\cdot}OH \tag{18.79}$$

$$R^{\cdot} + O_2 \rightarrow ROO^{\cdot} \tag{18.80}$$

$${}^{\cdot}OH + Fe^{2+} \rightarrow OH^{-} + Fe^{3+} \tag{18.81}$$

$${}^{\cdot}OH + H_2O_2 \rightarrow H_2O + HO_2^{\cdot} \tag{18.82}$$

$${}^{\cdot}OH + HO_2^{\cdot} \rightarrow H_2O + O_2 \tag{18.83}$$

$${}^{\cdot}OH + {}^{\cdot}OH \rightarrow H_2O_2 \tag{18.84}$$

The attack of hydroxyl radical to the organic species produces radicals ($R^{\bullet}$) (Eq. 18.76) that can react both with Fe^{3+} to regenerate Fe^{2+} and to form R^{+} (Eq. 18.77) and with H_2O_2 (Eq. 18.79) to form ROH, which can be further oxidised (Pignatello et al. 2006; Masarwa et al. 2005; Bautista et al. 2008).

Moreover, the organic radical may react with oxygen to form a peroxyl species ($ROO^{\bullet}$) (Eq. 18.80), which can be finally oxidised to CO_2, H_2O and organic acids (Wang and Xu 2012).

It can be noticed that a series of competitive reactions that can scavenge the primary oxidant species can occur (Eqs. 18.81–18.84).

During the Fenton process, under specified operative conditions, Fe(IV) hypervalent iron complexes that are able to oxidise organic molecules by electron transfer have been found:

$$Fe^{2+} + H_2O_2 \rightarrow \left[Fe(OH)_2\right]^{2+} \rightarrow Fe^{3+} + {}^{\cdot}OH + OH^{-} \tag{18.85}$$

It is worth to be noticed that both ferrous and ferric ions are simultaneously present in the system, as shown in the previous reactions, then it is possible to use Fe^{3+} instead of Fe^{2+} to decompose hydrogen peroxide and it is possible to consider both as homogeneous Fenton processes (Siedlecka et al. 2008).

The Fenton-like processes have some disadvantages due to the high cost of hydrogen peroxide and the risk related to its storage and transportation, together with the removal of iron sludge. Moreover, also the acidification of the effluent at a pH range between 2 and 4 before the treatment and the further neutralisation of the treated solution must be taken into account (Brillas et al. 2000).

To overcome some disadvantages of this process, heterogeneous Fenton processes using solid iron-containing catalysts have been used (Deng et al. 2008). The catalysts mainly used were zero-valent iron and iron oxides, such as Fe_3O_4, Fe_2O_3 and Fe_0/Fe_3O_4.

In general, the Fe-containing solid catalysts, in the presence of H_2O_2 or O_2, generate both Fe^{2+} and Fe^{3+} ions, and subsequently the oxidising species are produced through the homogeneous Fenton reaction. The oxidation of organic species can occur by dissolved iron ions or by reactions taking place between species in solution and surface-bound ones.

The disadvantages of the Fenton process can be overcome also by using hybrid processes such as photo-Fenton, which allows to enhance degradation efficiency.

18.4.2 Photo-Fenton Process

The photo-Fenton process is a Fenton process supported by light irradiation (ultraviolet or visible), which increases degradation efficiency (Deng et al. 2008). In addition to the reactions described above, the following processes contemporaneously occur, i.e. photolysis of H_2O_2 if the UV light used has a wavelength less than 285 nm (maximum absorbance of hydrogen peroxide at 210–230 nm) (Elmolla and Chaudhuri 2009) and photoreduction of aqueous ferric ions in ferrous ones, which can take place with wavelength less than 365 nm, enhancing the production of hydroxyl radicals:

$$H_2O_2 \xrightarrow{h\nu} 2\ ^{\cdot}OH \tag{18.86}$$

$$Fe^{3+} + e^- \overset{h\nu}{\rightleftarrows} Fe^{2+} \tag{18.87}$$

In aqueous solution, iron aquo complexes, which absorb in the UV-Vis range, can be formed, and they can both efficiently regenerate Fe^{2+} and produce hydroxyl radicals, as shown in the following reaction (Lapertot et al. 2006):

$$\left[Fe^{3+}\left(OH^-\right)_x\left(H_2O\right)_y\right] \xrightarrow{h\nu} Fe^{2+} + (x-1)OH^- + yH_2O + {}^{\cdot}OH \tag{18.88}$$

The photo-Fenton process is very effective for the degradation of various dyes under visible light irradiation due to a transfer of electrons from the excited dyes to Fe^{3+}, which promotes the catalytic transition from Fe^{3+} to Fe^{2+} (Ma et al. 2005):

$$\text{Dye} \xrightarrow{hv} \text{Dye}* \tag{18.89}$$

$$\text{Dye}* + \text{Fe}^{3+} \rightarrow \text{Fe}^{2+} + \text{Dye}^{+} \tag{18.90}$$

Heterogeneous photo-Fenton processes have been carried out by using solid iron-containing photocatalysts, and they were studied by considering the mechanism of both photocatalysis and the Fenton process (Wang and Xu 2012).

Indeed, TiO_2 and iron oxide on a polymer film showed a synergistic effect in the degradation of hydroquinone (Mazille et al. 2009).

18.5 Heterogeneous Photocatalysis

Heterogeneous photocatalysis is an advanced oxidation process based on the use of irradiated semiconductors in order to promote light-induced electron transfer reactions. Generally, (organic or inorganic) semiconductors can be viewed as 'very big' molecules in which the energy difference between some molecular orbitals can be neglected, giving rise to an energetic continuum called energy band. The opto-electronic properties of a semiconductor can be roughly associated with the energetic features of two bands: the conduction and the valence bands, which correspond to the lowest unoccupied molecular orbitals and to the highest unoccupied molecular orbitals, respectively. The energetic difference between these two bands is named band gap. In semiconductors, this value is typically ranging between 1.0 and 4.0 eV, so it is possible to increase conductivity through thermal- or photo-excitation. In particular, upon excitation by suitable energy, electrons are promoted from the valence to the conduction band so that, in the excited state, charge separation is achieved. The potential of the photo-generated holes and electrons determines their oxidation and reduction ability, respectively. The energy level with a 50% of probability of occupation by an electron is called Fermi level. Fermi level is almost equally distant from the valence and conduction bands in intrinsic semiconductors, while it is close to the conduction or to the valence band in n-type and p-type semiconductors, respectively. TiO_2, the most used and studied photocatalyst, is an n-type semiconductor because of the presence of oxygen vacancies, which make electrons the majority charge carriers. The conduction band of TiO_2 presents mainly a Ti3d character, while the valence-band edge an O2p character (Kisch 2015).

This electronic structure is useful to drive photocatalytic reactions. In fact, the semiconductor is capable to absorb light with an energy equal to or higher than its band gap, thus generating electron-hole pairs. These charges can undergo recombination, releasing radiative or non-radiative energy, or can migrate to the surface of the semiconductor being stabilised by trapping on surface active sites. The enhanced lifetime enables efficient interfacial electron transfer with adsorbed species possessing a suitable redox potential. Species whose redox potential lies within the potentials of the photo-generated charges, which roughly correspond to the energy levels of the conduction and valence bands, can undergo interfacial charge transfer (Braslavsky et al. 2011). Notably, the electronic structure discussed above is useful for practical applications, but it is oversimplified as it does not take into account the heterogeneity of the semiconductor surface. In fact, the localised interfacial electron transfer occurring at the surface active sites depends on the electronic nature of the different active sites, which can dramatically change upon the adsorption of chemical compounds or under irradiation. For this reason, unlike in thermal catalysis, the electronic density of the active sites rather than their number is a useful parameter. This can be macroscopically (and indirectly) taken into account by considering the quantum efficiency of the observed process, which can be expressed as the amount of molecules transformed upon the considered amount of absorbed photons of a certain energy (i.e. wavelength). On the other hand, the electronic density of states possessing similar energy can be experimentally measured by means of photoelectrospectroscopic measurements (Kobielusz et al. 2018) or by photoacoustic spectroscopy (Nitta et al. 2016). A precise electronic map reporting the density of state distribution can be obtained, which univocally identifies a photocatalyst. However, for practical purposes, simplified schemes, as that shown in Fig. 18.1, showing the energetic levels of conduction and valence bands, along with the band gap of the semiconductor, are useful to determine if the desired reaction is thermodynamically allowed or not.

It is worth to note that even if a photocatalytic reaction is thermodynamically allowed, it can be limited by kinetic factors depending primarily on the surface properties of the photocatalyst and on reaction conditions.

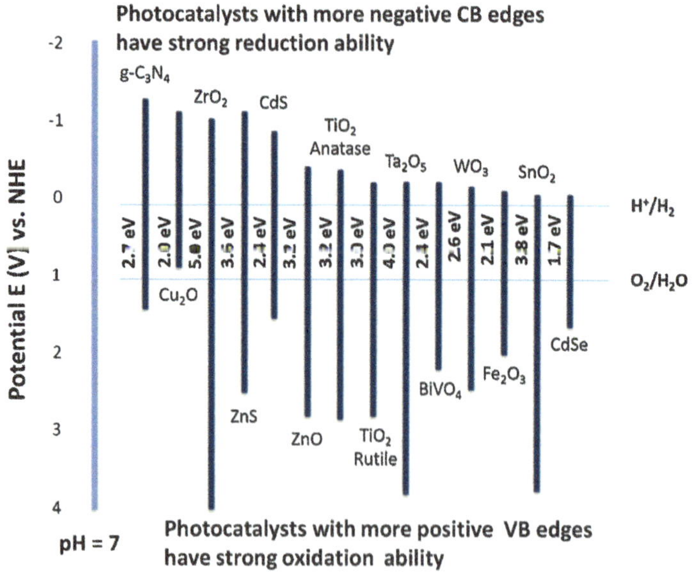

Fig. 18.1 Band gaps and band edge positions of some semiconductors

Unlike in a photoelectrochemical cell, where photo-induced oxidation and reduction reactions occur at the anode and cathode, respectively, in a single particle of an irradiated semiconductor (SC), they are spatially less separated and recombination phenomena can be significant and can limit the interfacial electron transfer. These steps are reported in Eqs. 18.91–18.93, which describe the generation of electrons (e^-) and holes (h^+) (Eq. 18.91), the radiative or non-radiative charge recombination (Eq. 18.92) and the primary electron transfer from the excited semiconductor to electron donors (D) or acceptors (A) (Augugliaro et al. 2011):

$$SC \xrightarrow{h\nu} SC\left(e^-, h^+\right) \tag{18.91}$$

$$SC\left(e^-, h^+\right) \rightarrow SC + h\nu \, / \, \text{heat} \tag{18.92}$$

$$SC\left(e^-, h^+\right) + A + D \rightarrow SC + A^{\cdot -} + D^{\cdot +} \tag{18.93}$$

It is evident that the photo-generated holes can directly oxidise adsorbed species, as shown in Eq. 18.93 ($D \rightarrow D^{\cdot +}$). However, reactions involving oxygen and water at the surface of the photocatalyst produce highly reactive oxygen species (ROS) such as hydroxyl radicals ($^{\cdot}OH$), peroxides and superoxide radical anions (protonated $HO_2^{\cdot}$ or non-protonated $O_2^{\cdot -}$), which are mainly responsible for photocatalytic oxidation reactions. In particular, oxygen acts as an electron acceptor (Eq. 18.94), and further reactions, such as those expressed by Eqs. 18.95–18.100, produce ROS species:

$$O_2 + e^- \rightarrow O_2^{\cdot -} \tag{18.94}$$

$$O_2^{\cdot -} + H^+ \rightarrow HO_2^{\cdot} \tag{18.95}$$

$$HO_2^{\cdot} + HO_2^{\cdot} \rightarrow H_2O_2 + O_2 \tag{18.96}$$

$$O_2^{\cdot -} + HO_2^{\cdot} \rightarrow O_2 + HO_2^- \tag{18.97}$$

$$HO_2^- + H^+ \rightarrow H_2O_2 \tag{18.98}$$

$$H_2O_2 + O_2^{\cdot -} \rightarrow {}^{\cdot}OH + OH^- + O_2 \tag{18.99}$$

$$H_2O_2 + e^- \rightarrow {}^{\cdot}OH + OH^- \tag{18.100}$$

Water can also act as an electron donor producing hydroxyl radicals (Eq. 18.101):

$$H_2O + h^+ \rightarrow {}^{\cdot}OH + H^+ \tag{18.101}$$

ROS species are able to oxidise almost all the organic compounds. For this reason, photocatalysis has been traditionally considered as a tool for environmental remediation (Pichat 2013). In fact, it not only allows degradation of organic and inorganic pollutants but also presents antimicrobial capacity.

The photocatalytic mechanisms discussed above can be referred to as 'direct photocatalysis'. In fact, in this case, the semiconductor itself is the light-absorbing species. However, it is possible that other species such as dyes or semiconductor-substrate charge transfer complexes absorb the impinging radiation inducing photocatalytic processes, which in this case can be named as 'indirect' (Parrino and Palmisano 2018).

When the substrate (D) is able to absorb the impinging radiation, its excited state (D∗) can inject an electron into the conduction band of the semiconductor, so the electron can reduce an electron acceptor (A). This mechanism, also referred to as photo-induced electron transfer (PET), is expressed in Eqs. 18.102–18.104 (Sakata et al. 1990):

$$D \xrightarrow{h\nu} D* \tag{18.102}$$

$$D* + SC \rightarrow SC\left(e^-\right) + D^+ \tag{18.103}$$

$$SC\left(e^-\right) + A \rightarrow SC + A^{\cdot -} \tag{18.104}$$

By taking into account this mechanism, it is evident that the use of dyes as model pollutants to evaluate the photocatalytic activity of novel semiconductors must be avoided, especially under visible light irradiation. In fact, the efficiency of degradation of the dye only reflects PET efficiency rather than the activity of the semiconductor (Rochkind et al. 2015).

Other compounds such as sulphur dioxide (Parrino et al. 2012), or aromatic compounds such as 1,2 diols (Macyk et al. 2010), present a very strong electronic interaction with the semiconductor, locally modifying the electronic structure of the charge transfer (CT) complex thereby formed (Eq. 18.105). Therefore, although neither the semiconductor nor the substrate absorb visible light, their CT complex does (Eq. 18.106), and the separated charges can induce redox reactions (Eq. 18.107):

$$SC + D \rightarrow \left[SC^{\delta -} \cdots D^{\delta +}\right] \tag{18.105}$$

$$\left[SC^{\delta -} \cdots D^{\delta +}\right] \xrightarrow{h\nu} \left[SC\left(e^-\right) \cdots D^+\right] \tag{18.106}$$

$$\left[SC\left(e^-\right) \cdots D^+\right] A \rightarrow SC + D^+ + A^{\cdot -} \tag{18.107}$$

In this case, one speaks of optical electron transfer (OET). Both by optical and photo-induced electron transfer, the oxidised form of the substrate ($D^{\bullet +}$) plays the role of a photo-generated hole.

Notably, the last mechanisms presuppose a direct interaction between the substrate and the semiconductor. However, the ROS species generated in photocatalytic suspensions live enough to oxidise also species that are not adsorbed but only close to the surface. This is the case of the photocatalytic degradation of cyanide ions, which occurs efficiently at basic pH values when the repulsive Coulombic interaction with the surface of TiO_2 hinders their adsorption.

Basic photocatalytic mechanisms have been lately very well understood and unravelled. However, it is always difficult to predict features of photocatalytic reactions in terms of conversion and selectivity. In fact, the complex interplay of various parameters differently contributing to the whole process is difficult to foresee. These parameters are related to the interaction between the solvent, substrates and semiconductor; to the semiconductor itself and to its physico-chemical features; to the operation conditions; and, finally, to light-matter interactions, by considering photons as immaterial reactants (Balzani et al. 2015).

If some of the products of a photocatalytic reactions accumulate selectively in the system, the reaction is of synthetic interest and can be considered as a 'green' option to traditional methods of syntheses due to the mild reaction conditions, the use of non-toxic materials and solvents and the possibility to exploit sunlight as the driving energy and oxygen as the oxidant.

However, photocatalysis has been traditionally applied for the degradation of organic and inorganic pollutants, which are dangerous for humans and the environment. It is worth to note that the degradation of pollutants does not imply that the effluent (gaseous or liquid) is purified. In fact, if the degradation affording only CO_2, H_2O is not complete, and possibly other species in their highest oxidation state such as phosphate, nitrate, sulphate and various reaction intermediates can still exist, their toxicity could be even higher than the compounds from which they are derived. For this reason, for practical applications, studying the degradation path of the considered pollutant and the toxicity of the intermediates is important.

Natural water resources often present contamination of various organic and inorganic compounds and of microorganisms. The organic compounds can mainly derive from pesticides and other products used in agriculture, from personal care products and from pharmaceuticals (Fagan et al. 2015). The latter class of compounds is of great concern due to the serious and often unpredictable changes in the metabolisms of living organisms (Petrie et al. 2015) and because their interaction with microbes can develop antibiotic-resistant bacterial strains that are a direct threat to humans (Marti et al. 2014). These compounds are not mineralised in traditional wastewater treatment plants (WWTPs), which are their primary source (Nakada et al. 2017). Moreover, the sludge derived from WWTPs are often used as soil fertilisers, with a consequent easy access to these compounds in the food chain. This problem could be faced by introducing photocatalytic treatments downstream to WWTPs. More than 200 different pharmaceutical compounds are classified as emerging contaminants and have been detected worldwide in various natural water streams (Hughes et al. 2012). For this reason, while up to some decades ago researchers focused on photocatalytic degradation of single contaminants, recently much efforts have been devoted to simultaneous degradation of various contaminants, highlighting issues closer to the real application. It is worth to mention that for this kind of application, robust, efficient, cheap and reusable semiconductors are desired more than the elegant and complex photocatalysts often reported in literature. TiO_2 meets this criteria, it being cheap, stable, efficient and abundant. Some recent examples will be reported here. Five contaminants (1,4 dioxane; n-nitrosodimethylamine (NDMA); tris-2-chloroethyl phosphate (TCEP); gemfibrozil; and 17 β estradiol) have been simultaneously degraded under UV light irradiation in the presence of TiO_2 (Alvarez-Corena et al. 2016). The fastest degradation was observed for gemfibrozil (95%) and the slowest for TCEP (45%) after 30-min irradiation. The simultaneous degradation of two phenolic compounds (4-chlorophenol 4-CP and 2,6 dichlorophenol, 2,6-DCP) was investigated, together with the competition between the phenolic compounds. It has been reported that operative parameters (pH, concentration, light irradiation) and the physico-chemical properties of the photocatalyst surface are relevant parameters (Pino and Encinas 2012). In another study, the selective degradation of nonylphenol from an aqueous mixture of different phenols was achieved by modifying the surface of TiO_2 with alkyl groups, which preferentially addressed adsorption of the less polar nonylphenol with respect to the other phenols (Inumaru et al. 2011).

Immobilising TiO_2 on opportune substrates such as glass allows avoiding costly post-process separation steps, thus making the process economically sustainable, although the photocatalytic efficiency is generally reduced. Fifteen emerging contaminants (acetaminophen, antipyrine, atrazine, carbamazepine, diclofenac, flumequine, hydroxybiphenyl, ibuprofen, isoproturon, ketorolac, ofloxacin, progesterone, sulphamethoxazole and triclosan) have been degraded in the presence of TiO_2 deposited on a glass substrate in a pilot plant equipped with compound parabolic collectors. The removal reached almost 90% for all of the pollutants within 2 h of UV irradiation (Miranda-García et al. 2011). Similar results have been obtained in the presence of TiO_2 deposited onto quartz fibres (Arlos et al. 2016) or onto a stainless steel mesh (Murgolo et al. 2016). In the latter case, the simultaneous degradation of ten emerging contaminants was more efficient than that obtained in the presence of commercial TiO_2 P25 (Evonik) dispersed in suspension, and accurate biotoxicity analyses have been performed. To simulate natural conditions, photocatalytic degradation of contaminants is often carried out in the presence of natural organic matter (NOM) (Peng et al. 2017; Gora and Andrews 2017). NOM often agglomerates onto the TiO_2 particles and effectively scavenge photo-generated holes and ROS radicals, thus reducing the efficiency of the process. Recently, Long et al. reported that these detrimental effects of NOM can be minimised by using phosphate-modified TiO_2 (Long et al. 2016).

It is worth to mention that the toxicity level of some compounds is associated with their concentration. Recently, photocatalytic degradation of micropollutants such as drugs or antibiotics has been studied downstream of other oxidation processes. In this case, the low concentration of pollutants requires a careful analysis of the amount of oxidising species photocatalytically generated in order to perform purification as efficiently as possible. Worth to be mentioned is the approach of performing photocatalytic reactions in raceway pond reactors instead of on reactors equipped with parabolic collectors for the degradation of micro-pollutants (Soriano-Molina et al. 2019).

Photocatalysis allows also to get rid of inorganic contaminants such as arsenite (Wang et al. 2016) and hexavalent chromium (George et al. 2016). For instance, toxic As(III) can be oxidised to innocuous As(V) directly by the photo-generated holes (Yoon and Lee 2005), while Cr(VI) can be used as electron trap, giving rise to innocuous Cr(III) and simultaneously oxidising organic endocrine disruptors (Kim et al. 2015).

18.5.1 *TiO_2-O_3 Photocatalysis*

Ozone can oxidise species in solution directly by addition to double bonds or nucleophilic attack (Oyama 2000) or indirectly by inducing formation of ROS at opportune pH conditions, as seen in the previous sections.

In fact, O_3 molecules react with hydroxyl ions at alkaline pH values giving rise to hydroxyl radicals, according to Eqs. 18.108–18.111:

$$O_3 + OH^- \rightarrow O_2^{\cdot-} + HO_2^{\cdot} \quad (18.108)$$

$$O_3 + HO_2^{\cdot} \rightarrow 2O_2 + {}^{\cdot}OH \quad (18.109)$$

or alternatively:

$$2HO_2^{\cdot} \rightarrow O_2 + H_2O_2 \quad (18.110)$$

$$H_2O_2 + O_2^{\cdot-} \rightarrow {}^{\cdot}OH + OH^- + O_2 \quad (18.111)$$

The simultaneous action of photocatalysis and ozonation generally provides process intensification. In fact, under certain experimental conditions, the integrated process affords efficiencies higher than the sum of the two single processes acting separately. In fact, in a photocatalytic suspension, three photons are required to generate one $^{\cdot}OH$ radical in the presence of oxygen, while only one photon is able to generate one $^{\cdot}OH$ radical in the presence of ozone. This is evident by considering the following Eqs. 18.112 and 18.113:

$$O_{3(ads)} + e^- \rightarrow O_3^{\cdot-} \quad (18.112)$$

$$O_3^{\cdot-} + H_2O \rightarrow {}^{\cdot}OH + OH^- + O_2 \quad (18.113)$$

However, the synergy of photocatalysis and ozonation was still under debate until a few years ago as it could not be observed by some authors in certain experimental conditions. Recently, Parrino et al. (2015) proposed a rational approach to evidence the process intensification by defining an intensification factor (E_i) as in Eq. 18.114:

$$E_i = \frac{\text{oxidation rate in integrated process}}{\text{oxidation rate in PC} + \text{oxidation rate in AOP}} \qquad (18.114)$$

E_i higher than 1 expresses the presence of synergy, while $E_i = 1$ expresses the absence of intensification. The intensification factor has been experimentally obtained as a function of the ratio between the rates of the component processes, and the optimum intensification between photocatalysis and ozonation (E_i equal to ca. 60%) was obtained when the rate of photocatalysis was ca. 0.3 times the rate of ozonation. This result is general and allows rationalising the intensification problem in the different experimental conditions often reported in literature.

Various organic pollutants and model molecules have been degraded by using photocatalytic ozonation (Piera et al. 2000) due to the higher efficiency of this integrated process. Recently, coupling photocatalysis and ozonation has been proposed as a promising tool to control the formation of toxic ozonation by-products. The by-product of main concern is bromate, which is quantitatively generated under the ozonation of bromide-containing water effluents and whose maximum allowed concentration in potable water is 10 μg L^{-1} (World Health Organization 1993). This stable, toxic and carcinogenic compound, however, can be successfully reduced to innocuous bromide by means of photocatalysis (Parrino et al. 2014). A thorough kinetic study that was recently published reports an assessment of the photocatalytic ozonation process capable of maximising treatment efficiency and avoiding the formation of bromate ions. Moreover, these results have been recently applied in a long-term test for the purification of a real reef coral aquarium (Camera-Roda et al. 2019).

18.6 Impact of AOPs on Oxidation By-Products of Real Wastewater

Generally, in the production of drinking water from raw water or wastewater generated by urban centres, industries and farms, a great amount of pathogenic microbial organisms must be deactivated. Moreover, wastewater may contain halogens, nitrogen and dissolved organic matter (DOM) that allow the formation of organic halogenated and inorganic by-products during the disinfection treatment by AOPs. Moreover, in order to prevent re-infection of potable water, post-treatments of chlorination (with chlorine) or chloramination (with chloramine, NH_2Cl) are performed in order to maintain residual chlorine or chloramine in the treated water for secondary disinfection (National Research Council 1987). During the chemical disinfection, highly genotoxic and cytotoxic inorganic and organic species called disinfection by-products can be formed. Indeed, considering the most commonly advanced oxidation processes, under realistic water treatment conditions, an increase of the possibility to form disinfection by-products after chloramination exists (Metz et al. 2011).

The presence of halogens, nitrogen and DOM may contribute to the formation of organic halogenated by-products (e.g. trihalomethanes, haloacetonitriles, haloacetic acids) and inorganic species such as bromate, chlorate and perchlorate. The production of different oxidation by-products depends on radical types, water composition (presence of radical scavenger) and the presence of chemical oxidants (ozone or chlorine).

Therefore, during water treatments performed with different AOPs, we can have different distributions of disinfection by-products.

Indeed, by using UV irradiation, in which DOM can be directly excited with the significant formation of low-weight species containing carbon with a low mineralisation, high amounts of disinfection by-products could be generated (Thomson et al. 2004). By using disinfection UV doses less than or equal to 300 mJ/cm^2, after chloramination, an increase of hyrophilicity and of phenol content in the treated DOM, with the formation of haloacetic acids and a reduction of trihalomethanes, was found (Wang et al. 2015). In the presence of nitrate, by using medium-pressure UV lamps, its photolysis may lead to DOM nitration with the formation of chloropicrin, dichloronitromethane and chloronitromethane (Shah et al. 2011). In contrast, no increase of halomethane formation has been found by low-pressure UV irradiation, whereas a significant increase of halomethane concentration was observed by irradiating with either low-pressure or medium-pressure lamps in the treatment of nitrate-containing water (Lyon et al. 2012). During the treatment with low-pressure UV irradiation of water containing both nitrate and bromide ions, after chloramination, bromopicrin was produced (Lyon et al. 2012). These results can be explained by considering that the use of low-pressure UV irradiation at high doses may favour the photolysis of nitrate with formation of nitrate DOM. Bromine, produced in the presence of bromide ions, can convert nitrated DOM to halomethanes more effectively than in the presence of chlorine due to its superior halogenating capacity (Westerhoff et al. 2004).

In UV-O_3 treatment, a better reduction of formation of trihalomethane was found with respect to UV and O_3 treatments applied separately, whereas during ozone-based AOPs in the presence of bromide, bromate can be generated by reaction with ozone. This reaction, in AOPs, is limited by the fast decomposition of ozone (Von Gunten 2003), but a high bromate amount was observed by using high ozone concentrations in the peroxone process (Hübner et al. 2015). On the other hand, O_3-H_2O_2 was found to substantially increase the formation potential of disinfection by-products with respect to UV-O_3. The results show the general insight of the non-selective oxidation of dissolved organic matter by reactive radicals in contrast to the selective reaction of ozone.

As far as UV-H_2O_2 is concerned, it was found that initially oxidation increases the reactivity of DOM to chlorine, with the consequent higher production of disinfection by-products, which decreases by continuing the oxidation process due mainly to mineralisation (Chu et al. 2014). Only a few studies on the post-chloramination of Fenton and Fenton-like processes are present in the pertinent literature. The studies generally show a significant reduction of the formation of trihalomethanes almost always associated with a reduction of dissolved organic carbon, indicating that the formation of disinfection by-products is limited by mineralisation (Moncayo-Lassoa et al. 2008).

Results on the formation of post-chloramination disinfection by-products after sulphate radical AOPs showed that a significant reduction of disinfection by-products was found by using UV-peroxydisulphate with respect to UV-H_2O_2. This result was attributed to the lower formation of low-molecular-weight species containing carbon and important precursors in the halogenated disinfection by-product production during UV/peroxydisulphate treatment (Deborde and von Gunten 2008).

In TiO_2-based photocatalytic systems, a substantial decrease of trihalomethane formation was reported under different irradiation conditions (i.e. simulated solar irradiation, monochromatic UVA or UVC irradiations). The reduced amount of residual disinfection by-products was mainly due to the transformation of residual dissolved organic matter in less potent disinfection by-product precursors. Moreover, in the presence of bromide ions, no formation of bromate was found during photocatalytic and photocatalytic ozonation processes (Parrino et al. 2014).

References

Abdel-Raouf N, Al-Homaidan AA, Ibraheem IBM (2012) Microalgae and wastewater treatment. Saudi J Biol Sci 19:257–275

Abu Amr SS, Aziz HA, Adlan MN (2013) Optimization of stabilized leachate treatment using ozone/persulfate in the advanced oxidation process. Waste Manag 33:1434–1441

Alaton IA, Balcioglu IA, Bahnemann DW (2002) Advanced oxidation of a reactive dyebath effluent: comparison of O_3, H_2O_2/UV-C and TiO_2/UV-A processes. Water Res 36:1143–1154

Alvarez-Corena JR, Bergendahl JA, Hart FL (2016) Advanced oxidation of five contaminants in water by UV/TiO_2: reaction kinetics and byproducts identification. J Environ Manage 181:544–551

Ameta R, Kumar A, Punjabi PB, Ameta SC (2013) Advanced oxidation processes: basics and principles. In: Rao DG, Senthilkumar R, Byrne JA, Feroz S (eds) Wastewater treatment: advanced processes and technologies, 2013th edn. CRC Press, Boca Raton, pp 61–107

Antoniou MG, de La Cruz AA, Dionysiou DD (2010) Degradation of microcystin-LR using 779 sulfate radicals generated through photolysis, thermolysis and e-transfer mechanisms. Appl Catal B Environ 96:290–298

Arlos MJ, Hatat-Fraile MM, Liang R, Bragg LM, Zhou NY, Andrews SA, Servos MR (2016) Photocatalytic decomposition of organic micropollutants using immobilized TiO_2 having different isoelectric points. Water Res 101:351–361

Augugliaro V, Loddo V, Pagliaro M, Palmisano G, Palmisano L (2011) Clean by light irradiation: practical applications of supported TiO_2. RSC Publishing, Cambridge

Avisar D, Lester Y, Mamane H (2010) pH induced polychromatic UV treatment for the removal of a mixture of SMX, OTC and CIP from water. Hazard Mater 175:1068–1074

Ayoub K, van Hullebusch ED, Cassir M, Bermond A (2010) Application of advanced oxidation processes for TNT removal: a review. J Hazard Mater 178:10–28

Balzani V, Bergamini G, Ceroni P (2015) Light: a very peculiar reactant and product. Angewandte Chemie Int Ed 54:11320–11337

Bautista P, Mohedano AF, Casas JA, Zazo JA, Rodriguez JJ (2008) An overview of the application of Fenton oxidation to industrial wastewaters treatment. J Chem Technol Biotechnol 83:1323–1338

Bebelis S, Bouzek K, Cornell A, Ferreira MGS, Kelsall GH, Lapicque F, Walsh FC (2013) Highlights during the development of electrochemical engineering. Chem Eng Res Des 91:1998–2020

Beltrán FJ, Rivas J, Álvarez P, Montero-de-Espinosa R (2002) Kinetics of heterogeneous catalytic ozone decomposition in water on an activated carbon. Ozone Sci Eng 24:227–237

Berner EK, Berner RA (1989) The global water cycle: geochemistry and environment, environmental conservation. Prentice-Hall Inc., Englewood Cliffs

Bethi B, Sonawane SH, Bhanvase BA, Gumfekar SP (2016) Nanomaterials-based advanced oxidation processes for wastewater treatment: a review. Chem Eng Process 109:178–189

Bin AK, Sobera-Madej S (2012) Comparison of the advanced oxidation processes (UV, UV/H_2O_2 and O_3) for the removal of antibiotic substances during wastewater treatment. Ozone Sci Eng 34:136–139

Blatchley ER, Shen C, Scheible OK, Robinson JP, Ragheb K, Bergstrom DE, Rokjer D (2008) Validation of large-scale, monochromatic UV disinfection systems for drinking water using dyed microspheres. Water Res 42:677–688

Błedzka D, Gmurek M, Gryglik M, Olak M, Miller JS, Ledakowicz S (2010) Photodegradation and advanced oxidation of endocrine disruptors in aqueous solutions. Catal Today 151:125–130

Błedzka D, Gmurek M, Gryglik M, Olak M, Miller JS, Ledakowicz S (2012) Photodegradation of n-butylparaben in natural water from Sulejow reservoir. Ecol Chem Eng S 19:517–525

Boczkaj G, Fernandes A (2017) Wastewater treatment by means of advanced oxidation processes at basic pH conditions: a review. Chem Eng J 320:608–633

Bolton JR, Cotton CA (2008) The ultraviolet disinfection handbook. American Water Works Association, Denver

Bolton JR, Denkewicz R (2007) Synergistic disinfection of drinking water using ultraviolet and ozone co-generated from the same UV lamp. In: World Water Congress on Ozone and Ultraviolet Technologies, Los Angeles, CA

Bourgin M, Borowska E, Helbing J, Hollender J, Kaiser HP, Kienle C, von Gunten U (2017) Effect of operational and water quality parameters on conventional ozonation and the advanced oxidation process O_3/H_2O_2: kinetics of micropollutant abatement, transformation product and bromate formation in a surface water. Water Res 122:234–245

Bouwer H (2000) Integrated water management: emerging issues and challenges. Agric Water Manag 45:217–228

Braslavsky SE, Braun AM, Cassano AE, Emeline AV, Litter MI, Palmisano L, Parmon VN, Serpone N (2011) Glossary of terms used in photocatalysis and radiation catalysis (IUPAC recommendations 2011). Pure Appl Chem 83:931–1014

Brillas E, Calpe JC, Casado J (2000) Mineralization of 2,4-D by advanced electrochemical oxidation processes. Water Res 34:2253–2262

Buxton GV, Subhani MS (1972) Radiation chemistry and photochemistry of oxychlorine ions. 2. Photodecomposition of aqueous solutions of hypochlorite ions. J Chem Soc, Faraday Trans 1 68:958–969

Camel V, Bermond A (1998) The use of ozone and associated oxidation processes in drinking water treatment. Water Res 32:3208–3222

Camera-Roda G, Loddo V, Palmisano L, Parrino F (2019) Photocatalytic ozonation for a sustainable aquaculture: a long-term test in a seawater aquarium. Appl Catal B Environ 253:69–76

Cañizares P, Paz R, Sáez C, Rodrigo MA (2009) Costs of the electrochemical oxidation of wastewaters: a comparison with ozonation and Fenton oxidation processes. J Environ Manage 90:410–420

Chandrasekara PK, Kwon TO, Moon IS (2009) Degradation of wastewater from terephthalic acid manufacturing process by ozonation catalyzed with Fe^{2+}, H_2O_2 and UV light: direct versus indirect ozonation reactions. Appl Catal B Environ 91:319–328

Chu W, Gao N, Yin D, Krasner SW, Mitch WA (2014) Impact of UV/H_2O_2 pre-oxidation on the formation of haloacetamides and other nitrogenous disinfection byproducts during chlorination. Environ Sci Technol 48:12190–12198

Cortés S, Sarasa J, Ormad P, Gracia R, Ovelleiro JL (2000) Comparative efficiency of the systems O_3/high pH and O_3/catalyst for the oxidation of chlorobenzenes in water. Ozone Sci Eng 22:415–426

Deborde M, von Gunten U (2008) Reaction of chlorine with inorganic and organic compounds during water treatment-kinetics and mechanisms: a critical review. Water Res 42:13–51

Deng JH, Jiang JY, Zhang YY, Lin XP, Du CM, Xiong Y (2008) $FeVO_4$ as a highly active heterogeneous Fenton-like catalyst towards the degradation of orange II. Appl Catal B Environ 84:468–473

Elmolla ES, Chaudhuri M (2009) Degradation of the antibiotics amoxicillin, ampicillin and cloxacillin in aqueous solution by the photo-Fenton process. J Hazard Mater 172:1476–1481

Fagan R, McCormack DE, Dionysiou DD, Pillai SC (2015) A review of solar and visible light active TiO_2 photocatalysis for treating bacteria, cyanotoxins and contaminants of emerging concern. Mater Sci Semicond Process 42:2–14

Fang J, Fu Y, Shang C (2014) The roles of reactive species in micropollutant degradation in the UV/free chlorine system. Environ Sci Technol 48:1859–1868

Fernández-Castro P, Vallejo M, San Román M, Ortiz I (2015) Insight on the fundamentals of advanced oxidation processes. Role and review of the determination methods of reactive oxygen species. J Chem Technol Biotechnol 90:796–820

Furman OS, Teel AL, Watts RJ (2010) Mechanism of base activation of persulfate. Environ Sci Technol 44:6423–6428

Garoma T, Gurol MD (2004) Degradation of tert-butyl alcohol in dilute aqueous solution by an O_3/UV process. Environ Sci Technol 38:5246–5252

George R, Bahadur N, Singh N, Singh R, Verma A, Shukla A (2016) Environmentally benign TiO_2 nanomaterials for removal of heavy metal ions with interfering ions present in tap water. Mater Today Proc 3:162–166

Glaze WH, Kangt J (1989) Advanced oxidation processes. Test of a kinetic model for the oxidation of organic compounds with ozone and hydrogen peroxide in a semibatch reactor. Ind Eng Chem Res 28:1580–1587

Glaze WH, Kang J-W, Chapin DH (1987) The chemistry of water treatment processes involving ozone, hydrogen peroxide and ultraviolet radiation. Ozone Sci Eng 9:335–352

Gora SL, Andrews SA (2017) Adsorption of natural organic matter and disinfection byproduct precursors from surface water on TiO_2 nanoparticles: pH effects, isotherm modelling and implications for using TiO_2 for drinking water treatment. Chemosphere 174:363–370

Gottschalk C, Libra JA, Saupe A (2009) Ozonation of water and waste water: a practical guide to understanding ozone and its applications. Wiley, Hoboken

Gracia R, Aragües JL, Ovelleiro JL (1996) Study of the catalytic ozonation of humic substances in water and their ozonation byproducts. Ozone Sci Eng 18:195–208

Gros M, Petrović M, Ginebreda A, Barceló D (2010) Removal of pharmaceuticals during wastewater treatment and environmental risk assessment using hazard indexes. Environ Int 36:15–26

Haag WR, Hoigné J (1986) Singlet oxygen in surface water. 3. Photochemical formation and steady-state concentrations in various type of waters. Environ Sci Technol 20:341–348

Huang KC, Zhao Z, Hoag GE, Dahmani A, Block PA (2005) Degradation of volatile organic compounds with thermally activated persulfate oxidation. Chemosphere 61:551–560

Hübner U, Zucker I, Jekel M (2015) Options and limitations of hydrogen peroxide addition to enhance radical formation during ozonation of secondary effluents. J Water Reuse Desal 5:8–16
Huerta-Fontela M, Galceran MT, Ventura F (2010) Fast liquid chromatography quadrupole-linear ion trap mass spectrometry for the analysis of pharmaceuticals, and hormones in water resources. J Chromatogr A 1217:4212–4222
Hughes SR, Kay P, Brown LE (2012) Global synthesis and critical evaluation of pharmaceutical data sets collected from river systems. Environ Sci Technol 47:661–677
Inumaru K, Yasui M, Kasahara T, Yamaguchi K, Yasuda A, Yamanaka S (2011) Nanocomposites of crystalline TiO_2 particles and mesoporous silica: molecular selective photocatalysis tuned by controlling pore size and structure. J Mater Chem 21:12117–12125
Kasprzyk-Hordern B, Ziółek M, Nawrocki J (2003) Catalytic ozonation and methods of enhancing molecular ozone reactions in water treatment. Appl Catal B Environ 46:639–669
Katsoyiannis IA, Canonica S, von Gunten U (2011) Efficiency and energy requirements for the transformation of organic micropollutants by ozone, O_3/H_2O_2 and UV/H_2O_2. Water Res 45:3811–3822
Khan JA, He X, Shah NS, Khan HM, Hapeshi E, Fatta-Kassinos D, Dionysiou DD (2014) Kinetic and mechanism investigation on the photochemical degradation of atrazine with activated H_2O_2, $S_2O_8^{2-}$ and HSO_5^-. Chem Eng J 252:393–403
Kim I, Yamashita N, Tanaka H (2009) Performance of UV and UV/H_2O_2 processes for the removal of pharmaceuticals detected in secondary effluent of a sewage treatment plant in Japan. J Hazard Mater 166:1134–1140
Kim Y, Joo H, Her N, Yoon Y, Sohn J, Kim S, Yoon J (2015) Simultaneously photocatalytic treatment of hexavalent chromium (Cr(VI)) and endocrine disrupting compounds (EDCs) using rotating reactor under solar irradiation. J Hazard Mater 288:124–133
Kisch H (2015) Semiconductor photocatalysis: principles and applications. Wiley-VCH Verlag GmbH, Weinheim
Kishimoto N (2019) State of the art of UV/Chlorine advanced oxidation processes: their mechanism, byproducts formation, process variation, and applications. J. of Water and Environment Technology 17:302–335
Kobielusz M, Pilarczyk K, Świętek E, Szaciłowski K, Macyk W (2018) Spectroelectrochemical analysis of TiO_2 electronic states—implications for the photocatalytic activity of anatase and rutile. Catal Today 39:35–42
Konsowa AH (2003) Decolorization of wastewater containing direct dye by ozonation in a batch bubble column reactor. Desalination 158:233–240
Kusic H, Koprivanac N, Bozic AL (2006) Minimization of organic pollutant content in aqueous solution by means of AOPs: UV- and ozone-based technologies. Chem Eng J 123:127–137
Lapertot M, Pulgarín C, Fernández-Ibáñez P, Maldonado MI, Pérez-Estrada L, Oller I, Gernjak W, Malato S (2006) Enhancing biodegradability of priority substances (pesticides) by solar photo-Fenton. Water Res 40:1086–1094
Legrini O, Oliveros E, Braun AM (1993) Photochemical processes for water treatment. Chem Rev 93:671–698
Liang X, Zhu X, Butler EC (2011) Comparison of four advanced oxidation processes for the removal of naphthenic acids from model oil sands process water. J Hazard Mater 190:168–176
Lim MH (2008) Fate of wastewater-derived contaminants in surface waters. Dissertation, University of California, Berkeley
Litter M (2005) Introduction to photochemical advanced oxidation processes for water treatment. In: Boule P, Bahnemann DW, Robertson PKJ (eds) Environmental photochemistry part II, Springer, Berlin, pp 325–366
Long A, Lei Y, Zhang H (2014) Degradation of toluene by a selective ferrous ion activated persulfate oxidation process. Ind Eng Chem Res 53:1033–1039
Long M, Brame J, Qin F, Bao J, Li Q, Alvarez PJ (2016) Phosphate changes effect of humic acids on TiO_2 photocatalysis: from inhibition to mitigation of electron–hole recombination. Environ Sci Technol 51:514–521

Luo Y, Guo W, Ngo HH, Nghiem LD, Hai FI, Zhang J, Liang S, Wang XC (2014) A review on the occurrence of micropollutants in the aquatic environment and their fate and removal during wastewater treatment. Sci Total Environ 473-474:619–641

Lutze HV, Kerlin N, Schmidt TC (2015) Sulfate radical-based water treatment in presence of chloride: formation of chlorate, inter-conversion of sulfate radicals into hydroxyl radicals and influence of bicarbonate. Water Res 72:349–360

Lyon BA, Dotson AD, Linden KG, Weinberg HS (2012) The effect of inorganic precursors on disinfection byproducts formation during UV- chlorine/chloramine drinking water treatment. Water Res 46:4653–4664

Ma JH, Song WJ, Chen CC, Ma WH, Zhao JC, Tang YL (2005) Fenton degradation of organic compounds promoted by dyes under visible irradiation. Environ Sci Technol 39:5810–5815

Macyk W, Szaciłowski K, Stochel G, Buchalska M, Kuncewicz J, Łabuz P (2010) Titanium(IV) complexes as direct TiO_2 photosensitizers. Coord Chem Rev 254:2687–2701

Malhotra S, Pandit M, Kapoor JC, Tyagi DK (2005) Photo-oxidation of cyanide in aqueous solution by the UV/H_2O_2 process. J Chem Technol Biotechnol 80:13–19

Marti E, Variatza E, Balcazar JL (2014) The role of aquatic ecosystems as reservoirs of antibiotic resistance. Trends Microbiol 22:36–41

Martijn BJ, Kruithof JC, Rosenthal LPM (2007) Design and implementation of UV/H_2O_2 treatment in a full scale drinking water treatment plant. Proc Water Environ Feder 1:134–153

Masarwa A, Rachmilovich-Calis S, Meyerstein N, Meyerstein D (2005) Oxidation of organic substrates in aerated aqueous solutions by the Fenton reagent. Coord Chem Rev 249:1937–1943

Masschelein WJ (2002) Ultraviolet light in water and wastewater sanitation. Lewis Publishers, Boca Raton

Mazille F, Schoettl T, Pulgarin C (2009) Synergistic effect of TiO_2 and iron oxide supported on fluorocarbon films. Part 1: effect of preparation parameters on photocatalytic degradation of organic pollutant at neutral pH. Appl Catal B Environ 89:635–644

Merényi G, Lind J, Naumov S, von Sonntag C (2010) The reaction of ozone with the hydroxide ion: mechanistic considerations based on thermokinetic and quantum chemical calculations and the role of HO_4^- in superoxide dismutation. Chemistry 16:1372–1377

Metz DH, Meyer M, Dotson A, Beerendonk E, Dyonisiou DD (2011) The effect of UV/H_2O_2 treatment on disinfection by-products formation potential under simulated distribution system conditions. Water Res 45:3969–3980

Miklos DB, Remy C, Jekel M, Linden KG, Drewes JE, Hübner U (2018) Evaluation of advanced oxidation processes for water and wastewater treatment—a critical review. Water Res 139:118–131

Miranda-García N, Suárez S, Sánchez B, Coronado JM, Malato S, Maldonado MI (2011) Photocatalytic degradation of emerging contaminants in municipal wastewater treatment plant effluents using immobilized TiO_2 in a solar pilot plant. Appl Catal B Environ 103:294–301

Mishra NS, Reddy R, Kuila A, Rani A, Mukherjee P, Nawaz A, Pichiah S (2017) A review on advanced oxidation processes for effective water treatment. Curr World Environ 12:470–490

Moncayo-Lassoa A, Pulgarin C, Bentez N (2008) Degradation of DBPs' precursors in river water before and after low sand filtration by photo-Fenton process. Water Res 42:4125–4132

Murgolo S, Yargeau V, Gerbasi R, Visentin F, El Habra N, Ricco G, Lacchetti I, Carere M, Curri ML, Mascolo G (2016) A new supported TiO_2 film deposited on stainless steel for the photocatalytic degradation of contaminants of emerging concern. Chem Eng J 318:103–111

Nakada N, Hanamoto S, Jürgens MD, Johnson AC, Bowes MJ, Tanaka H (2017) Assessing the population equivalent and performance of wastewater treatment through the ratios of pharmaceuticals and personal care products present in a river basin: application to the river Thames basin, UK. Sci Total Environ 575:1100–1108

National Research Council (1987) Drinking water and health: disinfectants and disinfectant by-products. National Academic Press, Washington

Neyens E, Baeyens JA (2003) Review of classic Fenton's peroxidation as an advanced oxidation technique. J Hazard Mater 98:33–50

Nigro A, Sappa G, Barbieri M (2018) Boron isotopes in groundwater: evidence from contamination and interaction with terrigenous–evaporitic sequence, east-central Italy. Geochem Explor Environ Anal 18:343–350

Nitta A, Takase M, Takashima M, Murakami N, Ohtani B (2016) A fingerprint of metal-oxide powders: energy-resolved distribution of electron traps. Chem Commun 52:12096–12099

Nowell LH, Hoigné J (1992a) Photolysis of aqueous chlorine at sunlight and ultraviolet wavelengths. I. Degradation rates. Water Res 26:593–598

Nowell LH, Hoigné J (1992b) Photolysis of aqueous chlorine at sunlight and ultraviolet wavelengths. II. Hydroxyl radical production. Water Res 26:599–605

Oh W-D, Dong Z, Lim T-T (2016) Generation of sulfate radical through heterogeneous catalysis for organic contaminants removal: current development, challenges and prospects. Appl Catal B Environ 194:169–201

Oturan MA, Aaron J-J (2014) Advanced oxidation processes in water/wastewater treatment: principles and applications. A review. Crit Rev Environ Sci Technol 44:2577–2641

Oyama ST (2000) Chemical and catalytic of ozone. Catal Rev Sci Eng 42:279–322

Parrino F, Palmisano L (2018) Reactions in the presence of irradiated semiconductors: are they simply photocatalytic? Mini-Rev Org Chem 15:157–164

Parrino F, Ramakrishnan A, Damm C, Kisch H (2012) Visible-light induced sulfoxidation of alkanes in the presence of titania. ChemPlusChem 77:713–720

Parrino F, Camera-Roda G, Loddo V, Palmisano G, Augugliaro V (2014) Combination of ozonation and photocatalysis for purification of aqueous effluents containing formic acid as probe pollutant and bromide ion. Water Res 50:189–199

Parrino F, Camera-Roda G, Loddo V, Augugliaro V, Palmisano L (2015) Photocatalytic ozonation: maximization of the reaction rate and control of undesired by-products. Appl Catal B Environ 178:37–43

Peng H, Chen Y, Mao L, Zhang X (2017) Significant changes in the photo-reactivity of TiO_2 in the presence of a capped natural dissolved organic matter layer. Water Res 110:233–240

Peternel I, Koprivanac N, Kusic H (2006) UV-based processes for reactive azo dye mineralization. Water Res 40:525–532

Petrie B, Barden R, Kasprzyk-Hordern B (2015) A review on emerging contaminants in wastewaters and the environment: current knowledge, understudied areas and recommendations for future monitoring. Water Res 72:3–27

Pichat P (2013) Photocatalysis and water purification: from fundamentals to recent applications. Wiley-VCH Verlag GmbH, Weinheim

Piera E, Calpe JC, Brillas E, Domènech X, Peral J (2000) 2,4-Dichlorophenoxyacetic acid degradation by catalysed ozonation: TiO_2/UVA/O_3 and Fe(II)/UVA/O_3 systems. Appl Catal B Environ 27:169–177

Pignatello JJ, Oliveros E, MacKay A (2006) Advanced oxidation processes for organic contaminant destruction based on the Fenton reaction and related chemistry. Crit Rev Environ Sci Technol 36:1–84

Pines DS, Reckhow DA (2002) Effect of dissolved cobalt(II) on the ozonation of oxalic acid. Environ Sci Technol 36:4046–4051

Pino E, Encinas MV (2012) Photocatalytic degradation of chlorophenols on TiO_2-325 mesh and TiO_2-P25. An extended kinetic study of photodegradation under competitive conditions. J Photochem Photobiol A Chem 242:20–27

Pisarenko AN, Stanford BD, Yan D, Gerrity D, Snyder SA (2012) Effects of ozone and ozone/peroxide on trace organic contaminants and NDMA in drinking water and water reuse applications. Water Res 46:316–326

Puspita P, Roddick F, Porter N (2015) Efficiency of sequential ozone and UV-based treatments for the treatment of secondary effluent. Chem Eng J 268:337–347

Qi C, Liu X, Ma J, Lin C, Li X, Zhang H (2016) Activation of peroxymonosulfate by base: implications for the degradation of organic pollutants. Chemosphere 151:280–288

Rastogi A, Al-Abed SR, Dionysiou DD (2009) Sulfate radical-based ferrous–peroxymonosulfate oxidative system for PCBs degradation in aqueous and sediment systems. Appl Catal B Environ 85:171–179

Real FJ, Benitez FJ, Acero JL, Sagasti JJ, Casas F (2009) Kinetics of the chemical oxidation of the pharmaceuticals primidone, ketoprofen, and diatrizoate in ultrapure and natural waters. Ind Eng Chem Res 48:3380–3388

Reisz E, Schmidt W, Schuchmann HP, von Sonntag C (2003) Photolysis of ozone in aqueous solutions in the presence of tertiary butanol. Environ Sci Technol 37:1941–1948

Rochkind M, Pasternak S, Paz Y (2015) Using dyes for evaluating photocatalytic properties: a critical review. Molecules 20:88–110

Saien J, Osali M, Soleymani AR (2015) UV/persulfate and UV/hydrogen peroxide processes for the treatment of salicylic acid: effect of operating parameters, kinetic, and energy consumption. Desalin Water Treat 56:3087–3095

Sakai H, Takamatsu T, Kosaka K, Kamiko N, Takizawa S (2012) Effects of wavelength and water quality on photodegradation of N-nitrosodimethylamine (NDMA). Chemosphere 89:702–707

Sakata T, Hashimoto K, Hiramoto M (1990) New aspects of electron transfer on semiconductor surface: dye-sensitization system. J Phys Chem 94:3040–3045

Sanches S, Crespo MTB, Pereira VJ (2010) Drinking water treatment of priority pesticides using low pressure UV photolysis and advanced oxidation processes. Water Res 44:1809–1818

Saritha P, Aparna C, Himabindu V, Anjaneyulu Y (2007) Comparison of various advanced oxidation processes for the degradation of 4-chloro-2 nitrophenol. J Hazard Mater 149:609–614

Shah AD, Dotson AD, Linden KG, Mitch WA (2011) Impact of UV disinfection combined with chrorination/chloramination on the formation of halonitromethanes and haloacetonitriles in drinking water. Environ Sci Technol 45:3657–3664

Shahidi D, Roy R, Azzouz A (2015) Advances in catalytic oxidation of organic pollutants—prospects for thorough mineralization by natural clay catalysts. Appl Catal B Environ 174–175:277–292

Sharma J, Mishra IM, Kumar V (2015a) Degradation and mineralization of Bisphenol A (BPA) in aqueous solution using advanced oxidation processes: UV/H_2O_2 and UV/$S_2O_8^{2-}$ oxidation systems. J Environ Manag 156:266–275

Sharma J, Mishra IM, Dionysiou DD, Kumar V (2015b) Oxidative removal of Bisphenol A by UV-C/peroxymonosulfate (PMS): kinetics, influence of coexisting chemicals and degradation pathway. Chem Eng J 276:193–204

Siedlecka EM, Mrozik W, Kaczyński Z, Stepnowski P (2008) Degradation of 1-butyl-3-methylimidazolium chloride ionic liquid in a Fenton-like system. J Hazard Mater 154:893–900

Song S, Xu X, Xu LJ, He ZQ, Ying HP, Chen JM, Yan B (2008) Mineralization of CI reactive yellow 145 in aqueous solution by ultraviolet enhanced ozonation. Ind Eng Chem Res 47:1386–1391

Song K, Mohseni M, Taghipour F (2016) Application of ultraviolet light-emitting diodes (UV-LEDs) for water disinfection: a review. Water Res 94:341–349

Soriano-Molina P, Plaza-Bolaños P, Lorenzo A, Agüera A, García Sánchez JL, Malato S, Sánchez Pérez JA (2019) Assessment of solar raceway pond reactors for removal of contaminants of emerging concern by photo-Fenton at circumneutral pH from very different municipal wastewater effluents. Chem Eng J 366:141–149

Stefan MI (ed) (2018) Advanced oxidation processes for water treatment: fundamentals and applications. IWA Publishing, London

Thomson J, Roddick FA, Drikas M (2004) Vacuum ultraviolet irradiation for natural organic matter removal. J Water Supply Res Technol AQUA 53:193–206

Vogt R, Schindler RN (1992) Product channels in the photolysis of HOCl. J Photochem Photobiol A Chem 66:133–140

Von Gunten U (2003) Ozonation of drinking water: part I. Oxidation kinetics and product formation. Water Res 37:1443–1467

Wacławek S, Lutze HV, Grübel K, Padil VVT, Černík M, Dionysiou DD (2017) Chemistry of persulfates in water and wastewater treatment: a review. Chem Eng J 330:44–62

Wang JL, Xu LJ (2012) Advanced oxidation processes for wastewater treatment: formation of hydroxyl radical and application. Crit Rev Environ Sci Technol 42:251–325

Wang H, Zhu Y, Hu C, Hu X (2015) Treatment of NOM fraction of reservoir sediments: effect of UV and chlorination on formation of DBPs. Sep Purif Technol 154:228–235

Wang Y, Duan J, Li W, Beecham S, Mulcahy D (2016) Aqueous arsenite removal by simultaneous ultraviolet photocatalytic oxidation-coagulation of titanium sulfate. J Hazard Mater 303:162–170

Watts MJ, Linden KG (2007) Chlorine photolysis and subsequent OH radical production during UV treatment of chlorinated water. Water Res 41:2871–2878

Westerhoff P, Chao P, Mash H (2004) Reactivity of natural organic matter with aqueous chlorine and bromine. Water Res 38:1502–1513

World Health Organization (1993) Guidelines for drinking water quality. WHO, Geneva

Wu CH, Kuo CY, Chang CL (2008) Homogeneous catalytic ozonation of C.I. Reactive red 2 by metallic ions in a bubble column reactor. J Hazard Mater 154:748–755

Würtele MA, Kolbe T, Lipsz M, Külberg A, Weyers M, Kneissl M, Jekel M (2011) Application of GaN-based ultraviolet-C light emitting diodes–UV LEDs—for water disinfection. Water Res 45:1481–1489

Xylem, City of Los Angeles Terminal Island Water Reclamation Plant selects innovative water reuse solution to address need for water security, Press release from June 3, 2015

Yoon S-H, Lee JH (2005) Oxidation mechanism of As(III) in the UV/TiO_2 system: evidence for a direct hole oxidation mechanism. Environ Sci Technol 39:9695–9701

Zhang Y, Geissen S-U, Gal C (2008) Carbamazepine and diclofenac: removal in wastewater treatment plants and occurrence in water bodies. Chemosphere 73:1151–1161

Zhang T, Zhu H, Croué JP (2013) Production of sulfate radical from peroxymonosulfate induced by a magnetically separable $CuFe_2O_4$ spinel in water: efficiency, stability, and mechanism. Environ Sci Technol 47:2784–2791

Zoschke K, Börnick H, Worch E (2014) Vacuum-UV radiation at 185 nm in water treatment—a review. Water Res 52:131–145

Chapter 19
Review of Progress in Microalgal Biotechnology Applied to Wastewater Treatment

Erfan Sadatshojaei, Dariush Mowla, and David A. Wood

Contents

E. Sadatshojaei (✉) · D. Mowla
Department of Chemical Engineering, Shiraz University, Shiraz, Iran

Environmental Research Center, Department of Chemical Engineering, Shiraz University, Shiraz, Iran
e-mail: dmowla@shirazu.ac.ir

D. A. Wood
DWA Energy Limited, Lincoln, UK
e-mail: dw@dwasolutions.com

Inamuddin, A. M. Asiri (eds.), *Sustainable Green Chemical Processes and their Allied Applications*, Nanotechnology in the Life Sciences,
https://doi.org/10.1007/978-3-030-42284-4_19

Abbreviations

OPM Outside-cellular polymeric materials
WWT Wastewater treatment

19.1 Introduction

Distributed around the world are new technologies that are critical for sustaining human life and modern lifestyles, such as enhancing oil recovery (Sadatshojaei et al. 2016, 2018, 2019; Choubineh et al. 2017; Wood 2019) and wastewater treatment processing. These technologies are the focus of intense research studies aimed at improving their handling, performance, and quality such that they can be exploited by humankind in safer, more efficient, and sustainable way. Algae have characteristics that can be exploited for these purposes by wastewater treatment industries. Algae contain different cellular types of chlorophyll including single and multicellular varieties not associated with plant leaves, stems, and roots. Biofilms, constructed by algae, develop as intricate aggregations of microorganisms on solid substrates surrounded by matrices made up of extracellular polymeric substances (EPS) (Leadbeater and Callow 1992). Algal biofilm aggregations tend to occupy those areas where nutrients and moisture are present (Leadbeater and Callow 1992; Jarvie et al. 2002). A key attribute of algal biofilms is that they possess the capability of (1) adapting to changing environment conditions, (2) separating existence as specific masses or isolated colony, and (3) maintaining colonies over larger areas (Menicucci Jr 2010).

Algal biofilms are exploited as a nutrient elimination option for wastewater treatment and also the source for providing algal biomass, demonstrating the potential to evolve as succeeding generations of algal biomass with improving performance (Christenson and Sims 2011; Kesaano and Sims 2014). Nevertheless, expanding the effective implementation of algal biofilm wastewater treatment technologies faces some limitations. These include nutrient elimination capacities, limited knowledge on algal growth, and biofilm surface requirements to make them sustainable (Olguín 2012; Boelee et al. 2012). The optimum proficiency and procedures associated with algal biofilm wastewater treatment technologies are insufficiently understood to confidently facilitate the scale-up of the nutrient removal systems they drive (Kesaano and Sims 2014). Some useful knowledge does exist; however, their optimum growth conditions do not require them to form the heterotrophic systems that support biofilms utilized in wastewater treatment (Nicolella et al. 2000; Gullicks et al. 2011).

19.2 Stable Film Technologies for Wastewater Treatment

Wastewater treatment utilizing stable film systems is achieved by stimulating and/or enhancing conditions that favor microorganism consortia which possess the capabilities to metabolize the specific pollutants. For instance, some aerobic microorganisms connected to plastic or rock surfaces are able to reduce the organic substances present in wastewater by executing a tricking filter procedure (Hoffmann 1998). Algal biofilm systems are able to eliminate pollutants through plant adsorption processes, which help to sustain biomass growth and development (Hoffmann 1998). Some wastewater treatment equipment enhances microalgae's capability to use phosphorous and nitrogen to achieve reduction of effluent nutrients. The outcome of such process is to produce bio-products.

The National Pollutant Discharge Elimination System (NPDES) under the direction of the United States Environmental Protection Agency (USEPA) set the nutrient restrictions needed for the effluent discharged from wastewater treatment plants, utilizing the accessible technology (Kang et al. 2008). Based on origin, crude wastewater typically possesses adequate nutrients for algal growing (Pittman et al. 2011). However, sometimes it is necessary to optimize the molar stoichiometric ratios of nitrogen, phosphorus, and carbon (C:N:P) in order to stimulate growth by adding nutrients. This is typically required when algal biofilms are employed in tertiary wastewater treatment processes (i.e., effluent polishing) (Olguín 2012).

For algal biofilm water treatment systems, the maximum percentage removal of N and P is associated with the organism's metabolic uptake (Pizarro et al. 2002; Su et al. 2011; De Godos et al. 2009). Subsequently, the organisms release N and P compounds as chemical sediments, phosphorus in the presence of magnesium and calcium ions, and nitrogen in the form of ammonia releasing. This is stimulated by increasing pH values of the fluids as a consequence of algal photosynthesis (Wei et al. 2008). The phosphorus and nitrogen concentrations in algal cells in biofilms used in water treatment systems, in dry weight terms, fluctuate from 0.3% to 2% and 2.9% to 7.5%, respectively (Boelee et al. 2011; Kebede-Westhead et al. 2003; Wilkie and Mulbry 2002).

An increase in nutrient concentrations in the wastewater has generally resulted in the enhancement of phosphorus and nitrogen concentrations in the algae in most experiments conducted (Boelee et al. 2011, Kebede-Westhead et al. 2003; Wilkie and Mulbry 2002; Kebede-Westhead et al. 2006). However, Posadas et al. (2013) reported no change in phosphorus and nitrogen content as nutrient loading increased. Changes in nutrient concentrations in the vicinity of algal biomass growths are likely due to their high selective absorption of N and P. The algal cells tend to store phosphorus and nitrogen in surplus quantities to those required to sustain their growth (Boelee et al. 2012).

The biofilm growth cycle tends to be coordinated with the nutrient elimination process. While there is only slight absorption potential observed at the beginning of the algal growth phase, absorption rates increase as algal growth accelerates. On the other hand, during the growth rate decline and ultimate death of the algae, a decrease in nutrient absorption is observed (Boelee et al. 2011). Low nutrient elimination at the beginning of the algal growth cycle is thought to be a consequence of inadequate coordination and compatibility at the community level of the algae within the biofilm (De Godos et al. 2009). During the decline phase, reduced nutrient elimination is due to lack of biofilm integration due to sloughing (Boelee et al. 2011). Enhancing the rate of nutrient loading resulted in average algal growth and higher phosphorus and nitrogen adsorption rates (Kebede-Westhead et al. 2003; Boelee et al. 2012; De Godos et al. 2009). However, once the highest P and N adsorption potential is reached, algal growth tends to pause (Boelee et al. 2011).

Thirty percent to 50% ammonium elimination was achieved in tests conducted by Posadas et al. (2013) through nitrification in algal biofilm treatment. The substantial increase in nitrification rates observed in fixation ponds was due to the existence of algal biofilms in investigations conducted by Babu (2011). That study concluded that the algal biofilms established a defensive barrier that prevented the washout of the nitrifying bacteria, thereby enhancing the impacts of the nitrifying bacteria (Choi et al. 2010).

High nitrification rates related to algal biofilms were observed to enhance the concentrations of discharged nitrate above influent levels in treated dairy wastewater (Mulbry and Wilkie 2001). The limited nitrate ion (NO^{3-}) adsorption is the result of small amounts of soluble carbon in the wastewater and the adsorption of ammonium ions. Specifically, in dairy wastewater, control of nitrification was attempted by enhancing the stability time and lowering the nitrogen-loading rate. However, the reduction in nitrification rates achieved were not sufficient (Wilkie and Mulbry 2002).

Carbon dioxide is exploited for various applications in different industries including catalyst regeneration (Sadatshojaei et al. 2018) and wastewater treatment. Kebede-Westhead et al. (2003) illustrated that nitrification became limited by high irradiance and bubbling of carbon dioxide through wastewater as these conditions stimulate increased algal photosynthesis. On the other hand, nitrifying bacteria growth was likely influenced by rivalry for solved inorganic carbon, daily variation of oxygen concentrations, and rivalry with heterotrophic bacteria. Such competitive actions could be influenced by algae changing the biofilm chemical microenvironment through breathing and photosynthesis (Wolf et al. 2007; Nils 2003).

The function of connected and hanging algal systems in wastewater treatment was investigated by Hoffmann (1998). That study demonstrated that nutrient elimination rates of the two systems were different. Phosphorus elimination rates of 74–95% resulted for high-throughput rate algae ponds with algal biofilms achieving a 95% rate.

19.3 Establishment and Growth of a Biofilm Consortium

Figure 19.1a shows microalgae species tend to combine and build communities by aggregating onto solid substrates by generating mucilage/holdfast (Sekar et al. 2004; Liu et al. 2017). Outside-cellular polymeric materials (OPM), a polymer

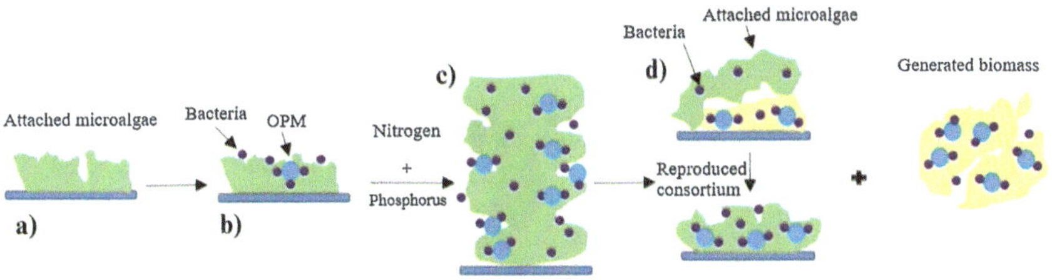

Fig. 19.1 Diagram showing the formation and growth of microalgae/bacteria consortia. (**a**) Microalgae aggregations onto solid substrates. (**b**) Establishment of connected microalgae/bacteria consortia improved by outside-cellular polymeric materials (OPM). (**c**) Microalgae/bacteria consortia relying on the presence of phosphorus and nitrogen nutrients. (**d**) Mature microalgae/bacteria consortia with components subject to substrate detachment (modified after Liu et al. 2017)

blend excreted by algae/bacteria, bind the microorganism community together (Nadell et al. 2015). This helps to establish the broad netlike form of biofilms (Fig. 19.1b). The algae or bacteria consortia adsorb nutritious substances from wastewater by performing various conversion process on the nutrients. These include denitrification, nitrification, and ammonification that occur during the algal growth phases (Su et al. 2017). As the biofilm reaches its maximum movement capability relative to its solid substrate, biomass separation tends to occur. The fragmented biofilm is then substituted by fresh biofilm growth on the substrate (Fig. 19.1d).

19.4 Biotic Interplays of Microalgae/Bacteria Consortia

There is substantial biodiversity in the makeup of such consortia due to a spectrum of potential interactions between bacteria and microalgal components. These include collaborative interplays to advance simultaneous growth of both bacterial and microalgal components and competitive interplays for nutrients and space (Fig. 19.2) (Gonçalves et al. 2017). Therefore, the combined interplays within a consortium act to create a consistent aggregation of the biofilm structure and help to stabilize its ecological functionality (Lindemann et al. 2016). Fundamentally, microalgae possess the ability to maintain bacterial growth by supplying organic carbon materials such as proteins and carbohydrates. On the other hand, carbon dioxide liberation by the bacteria and their consumption of oxygen tend to decrease the photosynthetic oxygen concentration and increase the concentration of carbon dioxide, respectively. Such changes stimulate additional algal growth (Fig. 19.2a) (Gonçalves et al. 2017). Also, bacteria parse organic matter into the mineral structure and release outside-cellular metabolites (Ramanan et al. 2016). These include vitamin B_{12} and auxins which tend to promote microalgal growth (Fig. 19.2a).

Substrate precolonization investigations by Roeselers et al. (2007) demonstrated that they promote growth through exploiting Betaproteobacteria. Also, they pro-

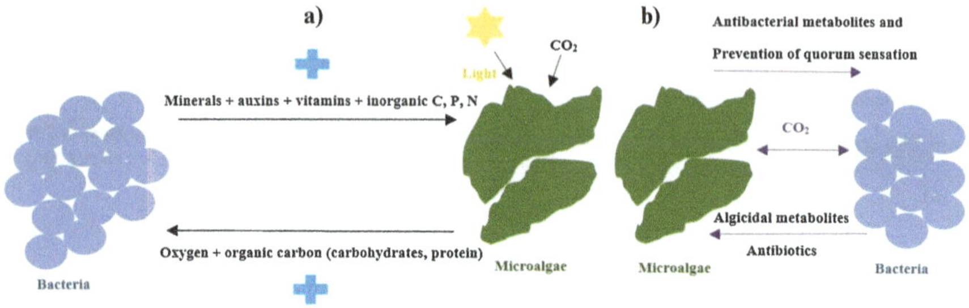

Fig. 19.2 (**a**) Cooperative interplays between bacteria and microalgae in a consortium attached to a substrate. Bacteria supply carbon dioxide, nitrogen, phosphorus, minerals, vitamins, and auxins to microalgae; in exchange, microalgae supply oxygen and organic carbon to bacteria. (**b**) Competitive interplays between bacteria and microalgae in a consortium attached to a substrate. Microalgae release antibacterial materials to prevent bacterial growth and quorum sensing. In response, bacteria generate antibiotics that disrupt algal photosynthesis. Algicidal metabolites are also released by some bacteria that tend to impede microalgae growth (modified after Liu et al. 2017)

mote adhesion of *Microcoleus vaginatus* to polycarbonate areas. Combinations of certain bacteria species can have substantial impacts on algal growth rates. For instance, the combination of *Brevundimonas sp.* and *Chlorella ellipsoidea* was found to lengthen the exponential growth stage and results in a 50-fold increase in biomass efficiency (Park et al. 2008). Moreover, *Brevundimonas sp.* entered a second exponential growth phase following a four-day coculture with *Chlorella ellipsoidea*. This resulted in five times more cell density compared to that obtained in a *Brevundimonas sp.* monoculture.

Competitive and opposing interplays occurring between bacteria and microalgae in a consortium have been actually observed and recorded (Fig. 19.2b). Some studies suggest that certain microalgal metabolites have a bactericidal impact. For instance, Gonçalves et al. (2017) reported that chlorellin induces bactericidal impacts on Gram bacteria. Quorum-sensing bacteria release chemical signal molecules, so-called autoinducers, that typically increase in concentration as a function of cell density (Miller and Bassler 2001). However, specific metabolites, such as malyngolide produced by *Lyngbya spp.*, can act to impede or suppress the quorum-sensing abilities of certain bacteria (Borowitzka 2016).

19.5 Components of Microalgae Cultivation

Attached microalgae cultivations refer to microalgae that are grown on solid substrates to initiate biofilm development (Zhuang et al. 2018). Figure 19.3a illustrates attached microalgae cultivation and the biofilm development process. Connections to the support media are established primarily through the hydrophobic interplays and distinctive acid-base interplays between the support media and the microalgae

(Gross and Wen 2014; Ozkan and Berberoglu 2013). Prerequisites for microalgae growth are nutrients, carbon dioxide, water, and light; these are basic requirements for both suspended and attached microalgae. Clearly, the attached microalgal networks also depend upon robust supporting components (Fig. 19.3c). The biofilm surfaces supply culture and light. The biofilm areas closest to the supporting components are referred to as the "internal section." The biofilm areas most distal from the supporting components are referred to as the "external section." Also, the culture broth used to grow biofilms in laboratory conditions and light are usually absorbed by the exterior sections of microalgae biofilm. If the support media is luminous or porous, e.g., glass or filter paper, light or culture broth, respectively, can be supplied in laboratory conditions from the internal section.

As algal growth reactor structures vary, attached microalgae cultures can exist as immersed, semi-immersed, or non-immersed with respect to location of the support media compared to the culture broth. They may also be categorized as vertical, horizontal, or rotational regarding the directional orientation of the support media

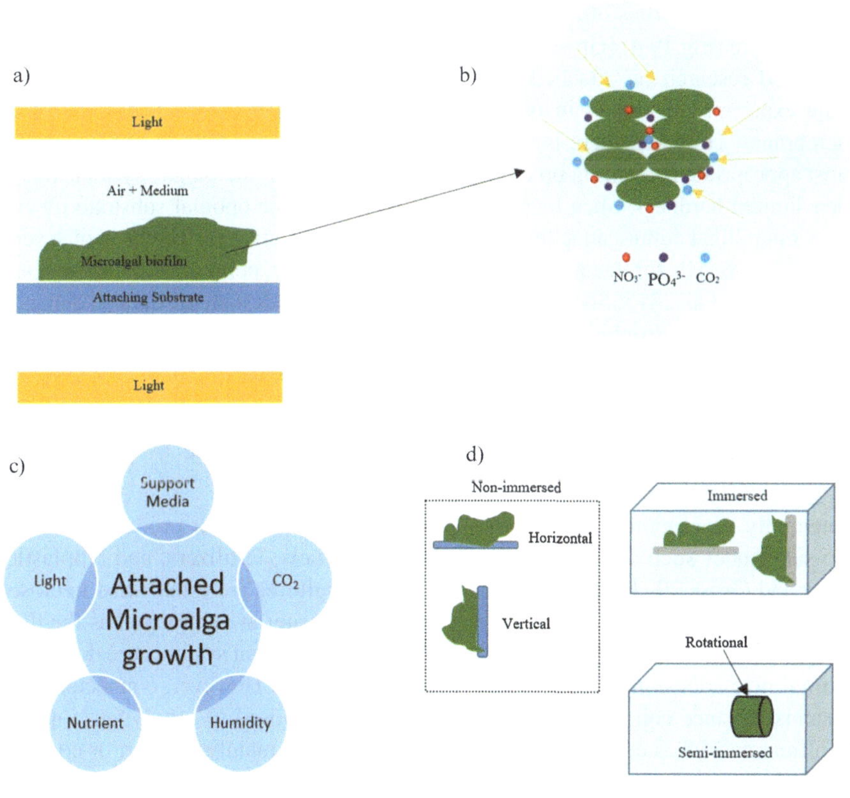

Fig. 19.3 Schematic illustration of cultivation of microalgae and its component parts (modified after Zhuang et al. 2018)

(Fig. 19.3d). Most semi-immersed microalgae/bacteria culture systems are also rotational systems. All microalgae consortia, in suspended and attached conditions, are constrained and impacted by the outside-cellular organic materials (OOMs), nutrients soluble in the OOM, and water. This is illustrated microscopically in Fig. 19.3b. The concentrations of these outside-cellular materials vary significantly in cultivation-reactor configurations, particularly the OOMs (Shen et al. 2016). Microalgae density in suspended cultures is much lower than in attached systems. As attached microalgae exist in constant fixed positions, the light and mass transfer conditions tend to be different than for suspended cultures. Such differences result in distinct growth patterns for attached microalgae versus suspended microalgae. This is particularly evident for their EOM liberation and growth rates.

19.6 Microalgae Consortia Support Media Distribution

The most commonly used materials for attached microalgae culture system are cotton and various membranes consisting of cellulose acetate/nitrate and polycarbonates. These are the materials utilized for attached microalgae culture systems reported in the literature (Fig.19.4) (Gross et al. 2016; Mulbry and Wilkie 2001). More than 90% published research on attached microalgae cultivation are stand-alone laboratory-scale experiments, mostly involving polycarbonate and/or cellulose acetate/nitrate membranes. However, there is no evidence to suggest that these two substances are most appropriate for scaling up the cultivation of attached microalgae systems due to their limited hardness. Shen recognized cotton cord as the optimal substrate for primary microalgal culture attachment by comparing the relative performance of several potential materials for that purpose, including acrylic, jute, polyester, cotton cord, and cotton (Kim et al. 2015; Shen et al. 2016). Cotton possesses some distinct advantages, i.e., coarseness, hydrophilicity, non-biotoxicity, low cost, and porosity.

19.7 Algal Biomass Production on a Larger Scale

Potentially, algal biomass possesses the ability to be utilized for producing various bio-substances such as nutrients, animal feed, biofuels, fertilizers, and bioplastics (Pulz and Gross 2004; Spolaore et al. 2006). The cultures are established in closed photobioreactors or open ponds. Seawater or fresh water are employed as the fluid media to cultivate algae in bulk (Chisti 2007). Hanging algal systems are developed to minimize adhesion of cells to the bioreactor surfaces. However, connected algal systems advance cell adhesion onto any surface composed of a polymeric matrix (Hoffmann 1998), i.e., a composite material composed mainly of short or continuous fibers composed of organic polymers. Surface colonization by microalgae has been observed within 24 h (Irving and Allen 2011; Menicucci Jr 2010). Moreover, long-term durability and stability of algal colonies generated have been observed, contrary to the generically dynamic nature of algal biofilms.

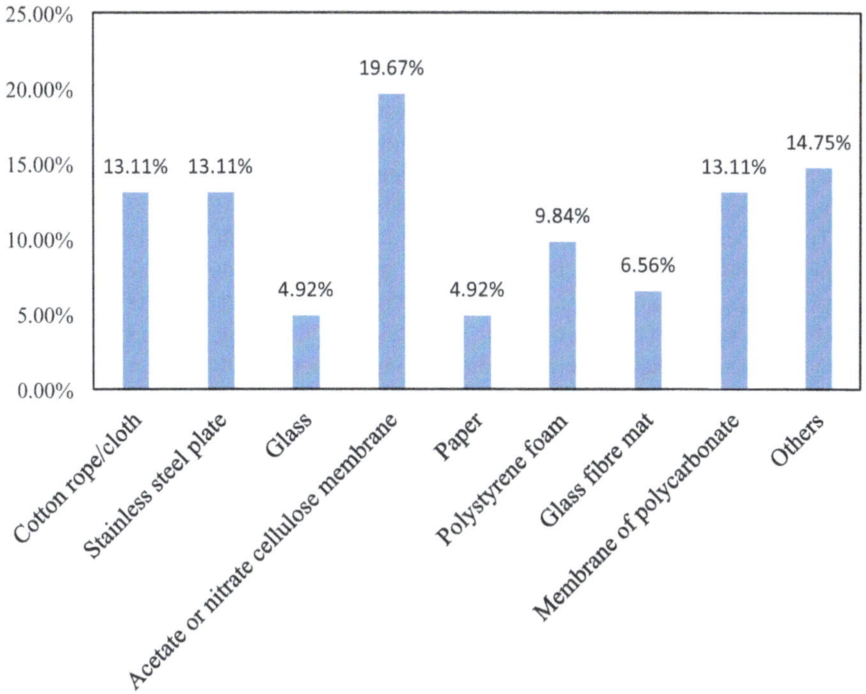

Fig. 19.4 Support media considered for attached microalgae cultivation systems (data from Zhuang et al. 2018)

There are some factors inhibiting the bulk cultivation of biomass. These include public understanding and confidence, the absence of a long-term proven trials and a track record, the potential contamination of water systems, the potential inability to sustain homogeneous nutrient supply, and the potential for pathogenic organisms to develop and breed in the biomass. These concerns and issues mean that large-scale wastewater utilization of cultivated microalgae to generate customized fit-for-purpose biofilms remains limited (Greenwell et al. 2009; Olguín 2012). Some regulatory initiatives are helping to simplify the scale-up process and promote the industrialization of bio-substances generated from algal biomass (Gellenbeck 2012; Doe 2010).

19.8 Influences on Algal Biofilm Development

19.8.1 Outside-Cellular Polymeric Materials (OPM) Impacts

OPM consists of lipids, nucleic acids, proteins, humic acids, and polysaccharides (Romaní et al. 2008; Di Pippo et al. 2009), the relative concentrations of which affect its chemical and physical characteristics. The OPM impacts algal biofilm aggregations in several ways. These include helping cell motion (Smith and

Underwood 1998), avoiding cell dehydration (Häubner et al. 2006), isolating cells from potentially toxic materials (García-Meza et al. 2005), and supplying stability through its adhesive properties (Sutherland 2001). There are various factors that influence OPM generation. These include the age of the biofilm, nutrient accessibility, species content, and reaction to stress (Sutherland 2001; Barranguet et al. 2005; Di Pippo et al. 2009). They are indirectly related to the light and temperature conditions in which algal photosynthesis and growth are sustained (Di Pippo et al. 2012).

The highest OPM to biomass ratios tend to occur in fresh and strongly grazed algal biofilms. This feature is likely driven by OPM generation functioning as a perpetuity mechanism (Barranguet et al. 2005). Wolfstein and Stal (2002) recorded the highest rate of OPM generation at 15 and 25 °C during the early stationary growing phase. In that temperature range, aging impacts on OPM generation tended to decrease. Additionally, the presence of diatom biomass and cyanobacteria has positive impacts on OPM generated in wastewater algal biofilms (Di Pippo et al. 2012). The lack of light does not completely prevent OPM synthesis. *Nitzschia sigma*, *Navicula perminuta*, and axenic monocultures of *Cylindrotheca closterium* have been shown to possess the ability to secrete considerable quantities of OPM by consuming formerly accumulated glucan as a carbon source (Smith and Underwood 1998). Such actions are believed to assist algal cells' departures from formerly dark conditions. Wastewater algal biofilms exist in environments that are susceptible to grazers and repetitious alterations in cultivation conditions. This often leads to periodic and seasonal instabilities in algal growth rates. It is therefore essential to screen algal species for their OPM generation capacity and minimize such carbon restrictions that might inhibit or prevent biofilm growth.

19.8.2 Substratum Impacts

Most algal adherence studies have focused on characterizing the impact of area specifications and substance content on biofilm establishment to determine what conditions are ideal to maintain biofilm growth and advance cell attachment (Sekar et al. 2002; Cao et al. 2009). Surface impacts on microalgae attachment vary according to the type of substrata. Porous and rough surfaces are generally associated with enhanced cell attachment. This is due to their enhanced surface area and better resistance to hydraulic shear forces generated by the fluid medium (Babu 2011). Low-porosity materials, such as nylon sponge, loofah, and polyurethane, possess sufficient hardness to sustain algal biomass cultivation in their pores and cavities (Johnson and Wen 2010). The aggregation of periphyton, freshwater organisms attached or clinging to biomass and other objects projecting above a basal substrate, has been studied in this regard. Tuchman and Blinn (1979) showed that periphyton aggregations developed more extensively on aluminum areas than on glass surfaces. The aluminum surfaces included microgrooves and cuts making them more analogous to the microtopography of natural vegetation and substrates than the smooth surface of glass. Cao et al. (2009) reported that adherence of *Scenedesmus dimor-*

phus cells to stainless steel sheets, treated with a microscale laser to provide it with a textured surface, was better than to regular (smoother) steel surfaces. Testing material compositions and textures to establish their suitability as algal biofilm substrate is an area of intense ongoing research. Although hydrophobic surfaces typically expedite the growth of biofilms with greater cell densities, area hydrophobicity was found to only have a very slight impact on the adherence capabilities of algal biofilms (Ozkan and Berberoglu 2011). On the other hand, Irving and Allen (2011) reported that there was no relationship between surface hydrophobicity and cell adherence density of *Chlorella vulgaris* and *Scenedesmus obliquus* biofilms cultivated in wastewater. Their studies suggested that the substance characteristics of a substratum had only limited influence on algal biofilm development compared to other factors.

19.8.3 Temperature Impacts

Temperature variations often do influence algal biofilm systems as they do for any biological systems applied to wastewater treatment (Werker et al. 2002). Schroepfer et al. (1952) demonstrated that temporary microbial aggregations associated with trickling filters could be easily adapted to temperature changes. The quantity of biofilm on filters typically changes with temperature, mostly caused by changes in grazing behaviors and microbial interactions (Honda and Matsumoto 1983). Other studies have also shown that temperature changes variably impact the development rates of different algal species and the grazing interactions within algal biofilm aggregations (Rao 2010).

An enhancement in algal cell densities can be achieved by temperature variations, selecting a temperature that establishes optimum cell densities for a specific system. However, algae tend to develop greater cell densities on surfaces with periphyton aggregations existing in a warm pond environment. Tuchman and Blinn (1979) reported that cell densities were greater in warmer sections of an algal pond (i.e., at 30.5 °C) compared to its intake area (at 23.1 °C) or at a weir within the system (at 27.5 °C).

Arrhenius identified the temperature impacts of algal growth rates during optimal nutrient conditions and stable light intensity (Goldman and Carpenter 1974). The responses of algae species to temperature fluctuations are changeable in natural biofilm aggregations, commonly depending on the associated nutrient conditions and light intensities in laboratory tests (Defew et al. 2004). De Nicola describes the commonly reported temperature trend in natural ecosystems in which certain microorganisms cease to thrive. For green algae, that is at 15–30 °C; for cyanobacteria, that is greater than 30 °C; and for diatoms, that is 5–15 °C (DeNicola 1996).

Photobioreactors used to cultivate biofilms are typically defenseless in respect of excessive evaporation rates and temperature alterations. The reason for this is their high surface area to volume ratios. Algal biofilm reactors, therefore, have distinct water requirements compared to open-pond systems with most of the water lost to vaporization in typical biofilm reactors (Ozkan et al. 2012). This leads to concerns over vaporized casualties caused by the needs for water conservation, sustaining a constant chemical content in the reactor water, and avoiding algal cell dehydration. An extended thermal model for an algal biofilm photobioreactor in the absence of active cooling calculated vaporized water shortages of about 1.0 L/m^2/day, 3.4 L/m^2/day, 7.3 L/m^2/day, and 6.0 L/m^2/day corresponding to prevailing seasonal conditions for winter, fall, summer, and spring months, respectively (Murphy 2012). Algal biofilm photobioreactor vaporization rates reported from laboratory experiments vary from 1 to 5 L/m^2/day (Ozkan et al. 2012). That is consistent with vaporization rates in open ponds equipped with hanging algal cultivation systems (Doucha and Lívanský 2009).

19.8.4 Nutrient Impacts

Nutrients accessibility to a biofilm aggregation impacts algal development and the type of biofilm ultimately created in terms of substitution and species content (Sekar et al. 2002). Phototrophic biofilms, those that capture photons to provide their energy supply, prosper in the presence of nutrients and light (Hillebrand et al. 2002). On the other hand, media with substantial quantities of biodegradable organic substances favors heterotrophic biofilms (Olapade and Leff 2006), i.e., a form of nutrition involving organisms depending upon other organisms for their food supply. Nutrients that are present in the bulk solution diffuse into algal cells through a so-called concentration boundary layer (Liehr et al. 1989). The Monod equation defines a mathematical model for the kinetic growth of microorganisms and has been extensively applied in algal biofilm studies to express algal growth in terms of nutrient concentrations (Hill et al. 2009) and to determine the rate of layer usage (Flora et al. 1993). For stream biofilms, the Monod equation was used to determine a growth saturation threshold of 25 μg/L for dissolved reactive phosphorus related to algal growth. Biofilm development is also linked to consecutive alterations of the microalgae types in response to changes in nutrients in the growth space (Sekar et al. 2002). Initially, green algae types tend to dominate biofilm aggregation, followed by diatom species and finally cyanobacteria in the nitrogen-deficient regions.

19.8.5 Species Interplays

There are various microbial cells existing in natural biofilm aggregations. These include bacteria, fungi, protozoa, and flagellates (Di Pippo et al. 2009; Leadbeater and Callow 1992). Factors influencing which species are dominant and which are displaced from dominance are the growth-confining variables, a biofilm's age, the presence and type of grazers, and the various micro-medias present in and around the biofilm (Barranguet et al. 2005). There are interplays within biofilm aggregations between the autotrophic organisms, those producing organic materials from light energy, and the heterotrophic organisms (Roeselers et al. 2007; Eschar and Characlis 1982), whereas only limited details are typically available regarding their relative contributions to the growth and structure of evolving biofilm aggregations. Indeed, the mechanisms involved and influences on those mechanisms that drive the interplays between algae and bacteria in biofilm aggregations remain poorly understood.

A significant quantity of *Chlorella vulgaris* and *Scenedesmus obliquus* species prefers to remain in suspension once mature, exploiting the sterile axenic conditions, i.e., away from the influence of other microorganisms, compared to rapidly growing biofilms in non-sterile wastewater (Irving and Allen 2011). On the other hand, symbiotic interplays of bacteria supplying inorganic carbon and vitamins for the benefit of the algae, coupled with algae providing oxygen and organic carbon for the benefit of the bacteria, are well documented (Romani and Sabater 2000). Some surfaces primarily colonized by bacteria have been observed to accelerate algal biofilm development (Ács et al. 2007). This cannot be considered as a ubiquitous outcome as there are some bacterial strains that inhibit algal growth due to cell-to-cell collisions and generation of substances that are toxic to algae (Holmström and Kjelleberg 2000). For instance, Rivas et al. (2010) reported, at a temperature of 20 °C, improved growth of *Botryococcus braunii* biofilms in the presence of *Rhizobium sp.* but inhibited growth of those biofilms in the presence of *Acinetobacter* sp.

19.8.6 Light Impacts

Light is an essential requirement for microalgal cultivation, being the energy source used in its photosynthesis. However, algal growth is inhibited by exposure to too much light, e.g., in the upper layers of a biofilm, and by exposure to too little light, e.g., in highly shaded sections of the biofilm (Wolf et al. 2007). Fortunately for algae, they are able to regulate their rate of photosynthesis according to the incident light intensity which is essential for avoiding more extensive light damage (Defew et al. 2004).

The degree to which light restrictions impact algal development depends on the type of biofilm aggregation and its cultivation status. Algal biofilm growth rates in a range of 12–88 μmol $m^{-2}s^{-1}$ were considered to be light limited in some tests (Hill and Fanta 2008). On the other hand, others have reported light-saturation impacts on a range of algal biofilms at growth rates of 60 μmol m^{-2} s^{-1} (Guzzon et al. 2008), 100 μmol m^{-2} s^{-1} (Hill et al. 2009), and 150 μmol m^{-2} s^{-1} (Liu et al. 2013), respectively.

Kinetic equations linking the growth rates measured for particular algae to light intensity, together with models fitted using empirical data, are broadly in agreement with each other for hanging algal cultures in the estimates they provide for algal productivity (Litchman 2000). On the other hand, kinetic models have only been applied to algal biofilms to a limited extent due to the incongruity and intricacy of biofilm aggregations (Wolf et al. 2007). Illuminated biofilm aggregations tend to have enhanced algal densities compared to biofilms cultivated in dark environments (Sekar et al. 2002). Green algae were observed to be the predominant heterotrophic bacteria in restricted light conditions, whereas early area colonizers dominated the consortia in high light intensities (Roeselers et al. 2007).

Diatoms were found to adapt better to limited light conditions than green algae within light-limited biofilms (Guzzon et al. 2008). Biofilms with enhanced cellular phosphorus quantities were recorded for the range of light intensities of 15, 30, 60, and 120 μmol m^{-2} s^{-1} with phosphorus storage capacity in the cells of 120 μmol m^{-2} s^{-1} (Guzzon et al. 2008).

Algal biofilm cells aging under similar light conditions displayed higher cellular nutrient contents, i.e., $N = 7.1\%$ and $P = 1.47\%$ (Mulbry and Wilkie 2001). In contrast, algal biofilm cells tested at two higher light intensities, i.e., 270 and 390 μmol $m^{-2}s^{-1}$, displayed $N = 3.6\%$ and $P = 0.65\%$ (Kebede-Westhead et al. 2003). Improved nutrient storage obtained at lower incident light intensity was also associated with cell growth at lower rates (Kebede-Westhead et al. 2003).

19.9 Conclusion

Algal biofilm technology has substantial scope to become more widely exploited in wastewater treatment. Its advantages include cost-effectiveness, potential for high biomass productivity, flexibility, adaptability, efficient interplays of certain species of algae with bacteria, and the ability to be deployed on industrial scales. By critically assessing the capabilities of microalgal cultivation methods in terms of their functionality at the cell level, it is possible to identify key factors that influence their ability to remove pollutants from wastewater and grow at sustainable rates. Recent studies have advanced our understanding considerably regarding how outside-cellular polymeric materials, substrata, temperature, nutrient supply, species interplays, and light both independently and collectively interact to influence nutrient storage capacities and biomass growth rates. Our recently improved ability to understand and control the subtle interactions between these critical factors is expected to lead to substantial improvements in the near future of the efficiency of microalgal wastewater technologies and plant designs.

References

Ács É, Borsodi AK, Kröpfl K, Vladár P, Záray G (2007) Changes in the algal composition, bacterial metabolic activity and element content of biofilms developed on artificial substrata in the early phase of colonization. Acta Bot Croat 66(2):89–100

Babu M (2011) Effect of algal biofilm and operational conditions on nitrogen removal. In: Waste stabilization ponds. UNESCO-IHE PhD thesis, CRC Press

Barranguet C, Veuger B, Van Beusekom SA, Marvan P, Sinke JJ, Admiraal W (2005) Divergent composition of algal-bacterial biofilms developing under various external factors. Eur J Phycol 40(1):1–8. https://doi.org/10.1080/09670260400009882

Boelee NC, Temmink H, Janssen M, Buisman CJN, Wijffels RH (2011) Nitrogen and phosphorus removal from municipal wastewater effluent using microalgal biofilms. Water Res 45(18):5925–5933. https://doi.org/10.1016/j.watres.2011.08.044

Boelee NC, Temmink H, Janssen M, Buisman CJ, Wijffels RH (2012) Scenario analysis of nutrient removal from municipal wastewater by microalgal biofilms. Water 4(2):460–473. https://doi.org/10.3390/w4020460

Borowitzka MA (2016) Chemically-mediated interactions in microalgae. In: The physiology of microalgae. Springer, Cham, pp 321–357. https://doi.org/10.1007/978-3-319-24945-2_15

Cao J, Yuan W, Pei ZJ, Davis T, Cui Y, Beltran M (2009) A preliminary study of the effect of surface texture on algae cell attachment for a mechanical-biological energy manufacturing system. J Manuf Sci Eng 131(6):064505. https://doi.org/10.1115/1.4000562

Chisti Y (2007) Biodiesel from microalgae. Biotechnol Adv 25(3):294–306. https://doi.org/10.1016/j.biotechadv.2007.02.001

Choi O, Das A, Yu CP, Hu Z (2010) Nitrifying bacterial growth inhibition in the presence of algae and cyanobacteria. Biotechnol Bioeng 107(6):1004–1011. https://doi.org/10.1002/bit.22860

Choubineh A, Ghorbani H, Wood DA, Moosavi SR, Khalafi E, Sadatshojaei E (2017) Improved predictions of wellhead choke liquid critical-flow rates: modelling based on hybrid neural network training learning based optimization. Fuel 207:547–560. https://doi.org/10.1016/j.fuel.2017.06.131

Christenson L, Sims R (2011) Production and harvesting of microalgae for wastewater treatment, biofuels, and bioproducts. Biotechnol Adv 29(6):686–702. https://doi.org/10.1016/j.biotechadv.2011.05.015

De Godos I, González C, Becares E, García-Encina PA, Muñoz R (2009) Simultaneous nutrients and carbon removal during pretreated swine slurry degradation in a tubular biofilm photobioreactor. Appl Microbiol Biotechnol 82(1):187–194. https://doi.org/10.1007/s00253-008-1825-3

Defew EC, Perkins RG, Paterson DM (2004) The influence of light and temperature interactions on a natural estuarine microphytobenthic assemblage. Biofilms 1(1):21–30. https://doi.org/10.1017/S1479050503001054

DeNicola DM (1996) Periphyton responses to temperature at different ecological levels. In: Algal ecology: freshwater benthic ecosystems, p 149–181

Di Pippo F, Bohn A, Congestri R, De Philippis R, Albertano P (2009) Capsular polysaccharides of cultured phototrophic biofilms. Biofouling 25(6):495–504. https://doi.org/10.1080/08927010902914037

Di Pippo F, Ellwood NTW, Guzzon A, Siliato L, Micheletti E, De Philippis R, Albertano PB (2012) Effect of light and temperature on biomass, photosynthesis and capsular polysaccharides in cultured phototrophic biofilms. J Appl Phycol 24(2):211–220. https://doi.org/10.1007/s10811-011-9669-0

Doe US (2010) National algal biofuels technology roadmap. US Department of Energy, Office of Energy Efficiency and Renewable Energy, Biomass Program, Washington

Doucha J, Lívanský K (2009) Outdoor open thin-layer microalgal photobioreactor: potential productivity. J Appl Phycol 21(1):111–117. https://doi.org/10.1007/s10811-008-9336-2

Eschar A, Characlis WG (1982) Algal-bacterial interactions within aggregates. Biotechnol Bioeng 24(10):2283–2290. https://doi.org/10.1002/bit.260241017

Flora JR, Suidan MT, Biswas P, Sayles GD (1993) Modeling substrate transport into biofilms: role of multiple ions and pH effects. J Environ Eng 119(5):908–930. https://doi.org/10.1061/(ASCE)0733-9372(1993)119:5(908)

García-Meza JV, Barrangue C, Admiraal W (2005) Biofilm formation by algae as a mechanism for surviving on mine tailings. Environ Toxicol Chem 24(3):573–581. https://doi.org/10.1897/04-064R.1

Gellenbeck KW (2012) Utilization of algal materials for nutraceutical and cosmeceutical applications—what do manufacturers need to know? J Appl Phycol 24(3):309–313. https://doi.org/10.1007/s10811-011-9722-z

Goldman JC, Carpenter EJ (1974) A kinetic approach to the effect of temperature on algal growth 1. Limnol Oceanogr 19(5):756–766. https://doi.org/10.4319/lo.1974.19.5.0756
Gonçalves, AZ, Srivastava, DS, Oliveira PS, Romero GQ (2017) Effects of predatory ants within and across ecosystems in bromeliad food webs. J Anim Ecol 86(4):790–799
Greenwell HC, Laurens LML, Shields RJ, Lovitt RW, Flynn KJ (2009) Placing microalgae on the biofuels priority list: a review of the technological challenges. J R Soc Interface 7(46):703–726. https://doi.org/10.1098/rsif.2009.0322
Gross M, Wen Z (2014) Yearlong evaluation of performance and durability of a pilot-scale revolving algal biofilm (RAB) cultivation system. Bioresour Technol 171:50–58. https://doi.org/10.1016/j.biortech.2014.08.052
Gross M, Zhao X, Mascarenhas V, Wen Z (2016) Effects of the surface physico-chemical properties and the surface textures on the initial colonization and the attached growth in algal biofilm. Biotechnol Biofuels 9(1):38. https://doi.org/10.1186/s13068-016-0451-z
Gullicks H, Hasan H, Das D, Moretti C, Hung YT (2011) Biofilm fixed film systems. Water 3(3):843–868. https://doi.org/10.3390/w3030843
Guzzon A, Bohn A, Diociaiuti M, Albertano P (2008) Cultured phototrophic biofilms for phosphorus removal in wastewater treatment. Water Res 42(16):4357–4367
Häubner N, Schumann R, Karsten U (2006) Aeroterrestrial microalgae growing in biofilms on facades—response to temperature and water stress. Microb Ecol 51(3):285–293. https://doi.org/10.1007/s00248-006-9016-1
Hill WR, Fanta SE (2008) Phosphorus and light colimit periphyton growth at subsaturating irradiances. Freshw Biol 53(2):215–225. https://doi.org/10.1111/j.1365-2427.2007.01885.x
Hill WR, Fanta SE, Roberts BJ (2009) Quantifying phosphorus and light effects in stream algae. Limnol Oceanogr 54(1):368–380. https://doi.org/10.4319/lo.2009.54.1.0368
Hillebrand H, Kahlert M, Haglund AL, Berninger UG, Nagel S, Wickham S (2002) Control of microbenthic communities by grazing and nutrient supply. Ecology 83(8):2205–2219. https://doi.org/10.1890/0012-9658(2002)083[2205:COMCBG]2.0.CO;2
Hoffmann JP (1998) Wastewater treatment with suspended and nonsuspended algae. J Phycol 34(5):757–763. https://doi.org/10.1046/j.1529-8817.1998.340757.x
Holmström C, Kjelleberg S (2000) Bacterial interactions with marine fouling organisms. In: Biofilms: recent advances in their study and control, p 101–115
Honda Y, Matsumoto J (1983) The effect of temperature on the growth of microbial film in a model trickling filter. Water Res 17(4):375–382. https://doi.org/10.1016/0043-1354(83)90132-X
Irving TE, Allen DG (2011) Species and material considerations in the formation and development of microalgal biofilms. Appl Microbiol Biotechnol 92(2):283–294. https://doi.org/10.1007/s00253-011-3341-0
Jarvie HP, Neal C, Warwick A, White J, Neal M, Wickham HD, Andrews MC (2002) Phosphorus uptake into algal biofilms in a lowland chalk river. Sci Total Environ 282:353–373. https://doi.org/10.1016/S0048-9697(01)00924-X
Johnson MB, Wen Z (2010) Development of an attached microalgal growth system for biofuel production. Appl Microbiol Biotechnol 85(3):525–534. https://doi.org/10.1007/s00253-009-2133-2
Kang SJ, Olmstead K, Takacs K, Collins J (2008) Municipal nutrient removal technologies reference document. US Environmental Protection Agency, Washington
Kebede-Westhead E, Pizarro C, Mulbry WW, Wilkie AC (2003) Production and nutrient removal by periphyton grown under different loading rates of anaerobically digested flushed dairy manure. J Phycol 39(6):1275–1282
Kebede-Westhead E, Pizarro C, Mulbry WW (2006) Treatment of swine manure effluent using freshwater algae: production, nutrient recovery, and elemental composition of algal biomass at four effluent loading rates. J Appl Phycol 18(1):41–46. https://doi.org/10.1007/s10811-005-9012-8
Kesaano M, Sims RC (2014) Algal biofilm based technology for wastewater treatment. Algal Res 5:231–240. https://doi.org/10.1016/j.algal.2014.02.003

Kim BH, Kim DH, Choi JW, Kang Z, Cho DH, Kim JY et al (2015) Polypropylene bundle attached multilayered stigeoclonium biofilms cultivated in untreated sewage generate high biomass and lipid productivity. J Microbiol Biotechnol 25(9):1547–1554

Leadbeater BSC, Callow ME (1992) Formation, composition and physiology of algal biofilms. In: Biofilms—science and technology. Springer, Dordrecht, pp 149–162

Liehr SK, Suidan MT, Eheart JW (1989) Effect of concentration boundary layer on carbon limited algal biofilms. J Environ Eng 115(2):320–335. https://doi.org/10.1061/(ASCE)0733-9372(1989)115:2(320)

Lindemann SR, Bernstein HC, Song HS, Fredrickson JK, Fields MW, Shou W et al (2016) Engineering microbial consortia for controllable outputs. ISME J 10(9):2077. https://doi.org/10.1038/ismej.2016.26

Litchman E (2000) Growth rates of phytoplankton under fluctuating light. Freshw Biol 44(2):223–235. https://doi.org/10.1046/j.1365-2427.2000.00559.x

Liu T, Wang J, Hu Q, Cheng P, Ji B, Liu J et al (2013) Attached cultivation technology of microalgae for efficient biomass feedstock production. Bioresour Technol 127:216–222. https://doi.org/10.1016/j.biortech.2012.09.100

Liu J, Wu Y, Wu C, Muylaert K, Vyverman W, Yu HQ et al (2017) Advanced nutrient removal from surface water by a consortium of attached microalgae and bacteria: a review. Bioresour Technol 241:1127–1137. https://doi.org/10.1016/j.biortech.2017.06.054

Menicucci JA, Jr (2010) Algal biofilms, microbial fuel cells, and implementation of state-of-the-art research into chemical and biological engineering laboratories. Doctoral dissertation, Montana State University-Bozeman, College of Engineering

Miller MB, Bassler BL (2001) Quorum sensing in bacteria. Ann Rev Microbiol 55(1):165–199. https://doi.org/10.1146/annurev.micro.55.1.165

Mulbry WW, Wilkie AC (2001) Growth of benthic freshwater algae on dairy manures. J Appl Phycol 13(4):301–306. https://doi.org/10.1023/A:1017545116317

Murphy TE (2012) Temperature fluctuation and evaporative loss rate in an algae biofilm photobioreactor. J Solar Energy Eng 134(1):011002. https://doi.org/10.1115/1.4005088

Nadell CD, Drescher K, Wingreen NS, Bassler BL (2015) Extracellular matrix structure governs invasion resistance in bacterial biofilms. ISME J 9(8):1700. https://doi.org/10.1038/ismej.2014.246

Nicolella C, Van Loosdrecht MCM, Heijnen JJ (2000) Wastewater treatment with particulate biofilm reactors. J Biotechnol 80(1):1–33. https://doi.org/10.1016/S0168-1656(00)00229-7

Nils RP (2003) Coupled nitrification-denitrification in autotrophic and heterotrophic estuarine sediments: on the influence of benthic microalgae. Limnol Oceanogr 48(1):93–105. https://doi.org/10.4319/lo.2003.48.1.0093

Olapade OA, Leff LG (2006) Influence of dissolved organic matter and inorganic nutrients on the biofilm bacterial community on artificial substrates in a northeastern Ohio, USA, stream. Can J Microbiol 52(6):540–549. https://doi.org/10.1139/w06-003

Olguín EJ (2012) Dual purpose microalgae–bacteria-based systems that treat wastewater and produce biodiesel and chemical products within a biorefinery. Biotechnol Adv 30(5):1031–1046. https://doi.org/10.1016/j.biotechadv.2012.05.001

Ozkan A, Berberoglu H (2011) Adhesion of Chlorella vulgaris on hydrophilic and hydrophobic surfaces. In: ASME 2011 international mechanical engineering congress and exposition. American Society of Mechanical Engineers Digital Collection, p 169–178. https://doi.org/10.1115/IMECE2011-64133

Ozkan A, Berberoglu H (2013) Cell to substratum and cell to cell interactions of microalgae. Colloids Surf B Biointerfaces 112:302–309. https://doi.org/10.1016/j.colsurfb.2013.08.007

Ozkan A, Kinney K, Katz L, Berberoglu H (2012) Reduction of water and energy requirement of algae cultivation using an algae biofilm photobioreactor. Bioresour Technol 114:542–548. https://doi.org/10.1016/j.biortech.2012.03.055

Park Y, Je KW, Lee K, Jung SE, Choi TJ (2008) Growth promotion of Chlorella ellipsoidea by co-inoculation with Brevundimonas sp. isolated from the microalga. Hydrobiologia 598(1):219–228. https://doi.org/10.1007/s10750-007-9152-8

Pittman JK, Dean AP, Osundeko O (2011) The potential of sustainable algal biofuel production using wastewater resources. Bioresour Technol 102(1):17–25. https://doi.org/10.1016/j.biortech.2010.06.035
Pizarro C, Kebede-Westhead E, Mulbry W (2002) Nitrogen and phosphorus removal rates using small algal turfs grown with dairy manure. J Appl Phycol 14(6):469–473. https://doi.org/10.1023/A:1022338722952
Posadas E, García-Encina PA, Soltau A, Domínguez A, Díaz I, Muñoz R (2013) Carbon and nutrient removal from centrates and domestic wastewater using algal–bacterial biofilm bioreactors. Bioresour Technol 139:50–58. https://doi.org/10.1016/j.biortech.2013.04.008
Pulz O, Gross W (2004) Valuable products from biotechnology of microalgae. Appl Microbiol Biotechnol 65(6):635–648. https://doi.org/10.1007/s00253-004-1647-x
Ramanan R, Kim BH, Cho DH, Oh HM, Kim HS (2016) Algae–bacteria interactions: evolution, ecology and emerging applications. Biotechnol Adv 34(1):14–29. https://doi.org/10.1016/j.biotechadv.2015.12.003
Rao TS (2010) Comparative effect of temperature on biofilm formation in natural and modified marine environment. Aquat Ecol 44(2):463–478. https://doi.org/10.1007/s10452-009-9304-1
Rivas MO, Vargas P, Riquelme CE (2010) Interactions of Botryococcus braunii cultures with bacterial biofilms. Microb Ecol 60(3):628–635. https://doi.org/10.1007/s00248-010-9686-6
Roeselers G, Van Loosdrecht MCM, Muyzer G (2007) Heterotrophic pioneers facilitate phototrophic biofilm development. Microb Ecol 54(3):578–585. https://doi.org/10.1007/s00248-007-9238-x
Romani AM, Sabater S (2000) Influence of algal biomass on extracellular enzyme activity in river biofilms. Microb Ecol 40(1):16–24. https://doi.org/10.1007/s002480000041
Romaní AM, Fund K, Artigas J, Schwartz T, Sabater S, Obst U (2008) Relevance of polymeric matrix enzymes during biofilm formation. Microb Ecol 56(3):427–436. https://doi.org/10.1007/s00248-007-9361-8
Sadatshojaei E, Jamialahmadi M, Esmaeilzadeh F, Ghazanfari MH (2016) Effects of low-salinity water coupled with silica nanoparticles on wettability alteration of dolomite at reservoir temperature. Pet Sci Technol 34(15):1345–1351. https://doi.org/10.1080/10916466.2016.1204316
Sadatshojaei E, Esmaeilzadeh F, Fathikaljahi J, Barzi SEH, Wood DA (2018) Regeneration of the midrex reformer catalysts using supercritical carbon dioxide. Chem Eng J 343:748–758. https://doi.org/10.1016/j.cej.2018.02.038
Sadatshojaei E, Jamialahmadi M, Esmaeilzadeh F, Wood DA, Ghazanfari MH (2019) The impacts of silica nanoparticles coupled with low-salinity water on wettability and interfacial tension: experiments on a carbonate core. J Dispers Sci Technol 1–15. https://doi.org/10.1080/01932691.2019.1614943
Schroepfer GJ, Al-Hakim MB, Seidel HF, Ziemke NR (1952) Temperature effects on trickling filters. Sewage Ind Waste 24:705–722
Sekar R, Nair KVK, Rao VNR, Venugopalan VP (2002) Nutrient dynamics and successional changes in a lentic freshwater biofilm. Freshw Biol 47(10):1893–1907. https://doi.org/10.1046/j.1365-2427.2002.00936.x
Sekar R, Venugopalan VP, Satpathy KK, Nair KVK, Rao VNR (2004) Laboratory studies on adhesion of microalgae to hard substrates. In: Asian Pacific phycology in the 21st century: prospects and challenges. Springer, Dordrecht, pp 109–116. https://doi.org/10.1007/978-94-007-0944-7_14
Shen Y, Zhu W, Chen C, Nie Y, Lin X (2016) Biofilm formation in attached microalgal reactors. Bioprocess Biosyst Eng 39(8):1281–1288. https://doi.org/10.1007/s00449-016-1606-9
Smith DJ, Underwood GJ (1998) Exopolymer production by intertidal epipelic diatoms. Limnol Oceanogr 43(7):1578–1591
Spolaore P, Joannis-Cassan C, Duran E, Isambert A (2006) Commercial applications of microalgae. J Biosci Bioeng 101(2):87–96. https://doi.org/10.1263/jbb.101.87
Su Y, Mennerich A, Urban B (2011) Municipal wastewater treatment and biomass accumulation with a wastewater-born and settleable algal-bacterial culture. Water Res 45(11):3351–3358. https://doi.org/10.1016/j.watres.2011.03.046

Su J, Kang D, Xiang W, Wu C (2017) Periphyton biofilm development and its role in nutrient cycling in paddy microcosms. J Soils Sediments 17(3):810–819. https://doi.org/10.1007/s11368-016-1575-2

Sutherland IW (2001) Biofilm exopolysaccharides: a strong and sticky framework. Microbiology 147(1):3–9. https://doi.org/10.1099/00221287-147-1-3

Tuchman M, Blinn DW (1979) Comparison of attached algal communities on natural and artificial substrata along a thermal gradient. Br Phycol J 14(3):243–254. https://doi.org/10.1080/00071617900650281

Wei Q, Hu Z, Li G, Xiao B, Sun H, Tao M (2008) Removing nitrogen and phosphorus from simulated wastewater using algal biofilm technique. Front Environ Sci Eng China 2(4):446–451. https://doi.org/10.1007/s11783-008-0064-2

Werker AG, Dougherty JM, McHenry JL, Van Loon WA (2002) Treatment variability for wetland wastewater treatment design in cold climates. Ecol Eng 19(1):1–11. https://doi.org/10.1016/S0925-8574(02)00016-2

Wilkie AC, Mulbry WW (2002) Recovery of dairy manure nutrients by benthic freshwater algae. Bioresour Technol 84(1):81–91. https://doi.org/10.1016/S0960-8524(02)00003-2

Wolf G, Picioreanu C, van Loosdrecht MC (2007) Kinetic modeling of phototrophic biofilms: the PHOBIA model. Biotechnol Bioeng 97(5):1064–1079. https://doi.org/10.1002/bit.21306

Wolfstein K, Stal LJ (2002) Production of extracellular polymeric substances (EPS) by benthic diatoms: effect of irradiance and temperature. Mar Ecol Prog Ser 236:13–22

Wood DA (2019) Microbial improved and enhanced oil recovery (MIEOR): review of a set of technologies diversifying their applications. Adv Geoenergy Res 3(2):122–140. https://doi.org/10.26804/ager.2019.02.02

Zhuang LL, Yu D, Zhang J, Liu FF, Wu YH, Zhang TY et al (2018) The characteristics and influencing factors of the attached microalgae cultivation: a review. Renew Sust Energ Rev 94:1110–1119. https://doi.org/10.1016/j.rser.2018.06.006

Chapter 20
Sustainable Development in Textile Processing

S. Basak, T. Senthilkumar, G. Krishnaprasad, and P. Jagajanantha

Contents

Abbreviations

BOD	Biological oxygen demand
CF	Fluorocarbon
COD	Chemical oxygen demand
DMDHEU	Dimethyl dihydroxy ethylene urea
DNA	Deoxyribonucleic acid
LOI	Limiting oxygen index
nm	Nanometer
UV	Ultraviolet

S. Basak (✉) · T. Senthilkumar · G. Krishnaprasad · P. Jagajanantha
ICAR—Central Institute for Research on Cotton Technology, Mumbai, India

Inamuddin, A. M. Asiri (eds.), *Sustainable Green Chemical Processes and their Allied Applications*, Nanotechnology in the Life Sciences,
https://doi.org/10.1007/978-3-030-42284-4_20

20.1 Introduction

Due to global warming and environmental pollution, for the last 30–40 years, various researches focusing on sustainable textile processing have been conducted. Among the researches, lesser usage of water and chemicals, foam finishing, enzyme-based treatment of textiles, and different irradiation techniques are mostly the predominant topics. For the last 10–20 years, more new technologies have emerged in the research field of sustainable textile processing. Among the developed technologies, nanoparticle-formulation-based textile finishing, plasma, UV irradiation for different value additions of textiles, and use of natural biomolecule for functionalization are important (Samanta et al. 2019). In most cases, industrial textile processing requires large quantities of alkalis, acids, chlorinated bleaching agents, toxic synthetic dyes, finishing chemicals. etc. These mentioned ingredients are very toxic and are quite capable of increasing the total dissolved solids, BOD, and COD of effluent water. Some of the mostly explored synthetic dyes, especially sulfur dye and azo-based dyes, are very much toxic and have strong carcinogenic effect on any living being. Polyester textile dyeing is very costly and requires high-temperature high-pressure machine for the accomplishment of the dyeing process. Therefore, researchers are trying to replace polyester dyeing with waterless supercritical carbon-dioxide-based dyeing process. Concerning antimicrobial treatment of textile materials, chlorinated compounds, positive-charge-containing quaternary ammonium compounds, and polyhexamethylene biguanide are generally explored by textile industries (Samanta et al. 2017a). For UV-protective finishing, different dark-color high-molecular-weight compounds like phenyl salicylate, benzophenone, and benzotriazole-based chemicals are continuously and commonly used by textile industries. However, they are carcinogenic and have some irritant smell and odor. With regard to flame-retardant finishing, formaldehyde-based chemicals are more popular in the commercial sector for its outstanding durability and effectivity for flame retardancy. The phosphorus-, chlorine-, and nitrogen-based Pyrovatex, Proban, and formaldehyde resin cross-linkers are used most popularly in flame-retardant finishing of cotton textiles (Samanta et al. 2014a). Due to the difficulty in the process control and the toxicity of the treatment, very few Indian industries are running these processes. For fire retardancy of synthetic materials (polyester, nylon, acrylic. etc.), toxic halogen-based chemicals are commonly used by industries. However, use of halogens has been slowly abolished because of dioxins and furan release during processing. On the whole, most of the auxiliaries used by textile industries in textile processing have adverse effects on the environment. In the past, various steps have been taken by scientists and researchers in making eco-friendly synthetic chemicals for textile processing. Di- and tricarboxylic acids and dimethylol-dihydroxy-ethylene-urea-based cross-linkers have already been used by earlier researchers. In addition, use of larger quantity of salt during reactive dyeing, use of high alkali and hydrose in vat dyeing, sulfur dyeing, reduction clearing of polyester (hydrose and soda ash required), and color stripping from dyed fabric (hydrose required) are not eco-friendly processes. However, these are the running process in most of the textile industries in India and other Asian countries (Lal et al. 2018). All these mentioned processes adopted by textile industries discharge larger quantity of unfixed dyes, alkalis, acids, hydrose, oxidative bleaching agents, cyanide, etc. in the effluent water, and they are some of the major challenges in the sustainable processing of textiles.

A number of research groups, industries, and scientists around the world are involved in sustainable activities connected with textile processing. They are continuously trying to develop newer methodologies and process conditions for the value addition and processing of textile materials. Different groups of scientists are working on different ways to achieve the same target: sustainability. Some of them have also been adopted by textile industries. Application of various plant-based extracts (molecules) for textile coloration and functionalization; water-saving processes like plasma, UV, and electron beam irradiation; and nanomaterial-based value addition of textiles are also becoming popular and are discussed below in details.

20.2 Eco-Friendly Approaches for Making Textile Material

In order to come up with an eco-friendly textile processing, researches are being conducted in different directions. Most of the researches are based on the usage of nanomaterials and plant-based extracts. These materials are capable of replacing various synthetic materials for the value addition of textile. As minimization of the exploration of water is another important challenge, different energy- and water-saving techniques like irradiation technologies, zero-liquor processes, padding techniques, foam finishing, etc. have been developed and commercialized in some of the application areas.

20.2.1 Sustainable Textile Processing by Using Biomacromolecule

20.2.1.1 Textile Dyeing by Natural Dye

Different plant-based and animal-based biomolecules have been explored for making natural dye and also value-added textile materials. Concerning the use of natural dye, during the Vedic time period, India explored different herbal and medicinal plants for dyeing textile materials. Indeed, during that time, they have made "Ayurvastra" type textile products by using their own indigenous technologies. From the literature, it has been found that the fabric made by them has various well-being properties. However, natural dyeing was costly, and the availability of the source material was difficult. Lots of other issues like shade variations, shade limitations, and lack of wash durability were also present. In those days, only the elite class of people could wear dyed dresses. Therefore, after the emergence of synthetic dyes, natural dye processing became a big setback, and the world started to move in the synthetic direction. It was a synthetic revolution, and dyes like vat, reactive, acid, and basic became very popular in the textile industries as these dyes are capable of providing nice shades and color, with more amount of fabric in less time but with more durability. However, for the last 30 years, due to environmental challenges and government legislation, the world is trying to move towards the eco-friendly direction in every sector. Therefore, natural-dyed textiles are again in demand. Now, in this area, a major challenge for researchers is to maintain repeatability of shades, durability, etc. (Samanta et al. 2016).

Chemically, natural dyes have different alternating double bond and single bond groups. Most of them contain long chain of quinines, anthraquinones, napthaquinones, and indigoid-based chromophores. Natural dye can be extracted from various herbal sources by water or alcoholic extraction. However, in most of the cases, prior mordanting is required before application of the natural dye on the natural textile material. Generally tannic acid, alum used as mordants for coloration of textile materials with different natural dyes. Apart from it, other salts of high-valance metals like copper, chromium, iron, etc. have been used by researchers for mordanting purposes. These mordanting materials assist to hold the dye molecule and the primary –OH group of cotton cellulose by suitable chemical bridge formation. In connection to dyeing, madder, turmeric, onions, beetroot (dye from root sources), Sappan, khair, sandalwood (dye from bark sources), indigo, lemongrass (dye from leaf sources), marigold, dahlia, Tesu, Kusum (dye from flower sources), myrobalan, pomegranate rind, and Latkan (dye from fruit sources) have been explored popularly for the coloration of cotton textile materials. However, in most cases, washing durability and wet rubbing fastness are major challenging areas. Another major challenge in the natural dye sector is the limitation of shades. Only the colors yellow, blue, brown, orange, and red are mostly popular in this area. By changing the metal of the mordant (iron, copper, chromium, aluminum, etc.), the color of the dyed material can also be changed. It may be because of the fact that metal presents in the mordant, reacts with the chromophore group of natural dye, and change its electron stability (Samanta et al. 2017b).

Researchers have reported that the effectivity of natural dye on proteinous wool and jute textiles was found suitable because of the more amorphous zone and the presence of lignin, carboxylic group, etc. The amphoteric nature of wool and the presence of lignin in the jute textile assist them in taking in more amount of natural dye at neutral pH condition. As per the report of the researchers concerning jute textiles, harda and alum are an efficient combination for mordanting, and a mordanted fabric can hold natural dye when it has been dyed at elevated temperature at an alkaline pH condition. As sustainability is one of the major concerns today, even the coloration of synthetic fiber has been attempted by using natural dye extracts. One recent research study reported on the dyeing of polyester and nylon fabric using the extract of the bark of a babool tree. They have reported that a wider range of shades is possible for natural-dyed synthetic textile materials.

20.2.1.2 Value-Added Finishing of Textile by Biomolecule

Different plant biomolecules have been explored for making antimicrobial, UV-protective, fire-retardant textile materials. In addition to this, recently various well-being textiles, vitamin E finishing, and mosquito-repellent textiles have also been made available in the market with an eco-friendly caption.

Antimicrobial Textile

The natural dyes with antimicrobial functionalities are useful in textiles as they not only impart colors but also provide antimicrobial finishing, which can help to increase the life of the fabric and is useful for medicinal garments, sanitary napkins, socks, etc. Cellulosic fabrics like cotton and jute are more prone to be attacked by bacteria, microorganisms, etc. The main source of natural dyes is either plants or microorganisms. Researchers have reported on the use of the bark extract of *Acacia arabica* (babool), *Acacia catechu* (catechu), *Terminalia chebula* (harda) and *Acacia arabica* wild (babool) for making antimicrobial cotton fabric. They have reported that the dye molecule extracted from the mentioned sources is rich in tannin, which may help in the formation of a strong bond with the fiber and thereby assist in the proper fixation of the dye on the fibrous material. Researchers have studied the antibacterial finishing of cotton textile using a neem (*Azadirachta indica*) extract. As per report, researchers have prepared the neem extract from the bark and seed part of the neem tree. The main active ingredient of the neem leaf is limonoid, which is consist of Azadirachtin A, nimbin, and salanin. As per structure, all three components of the neem extract contain polyphenolic group with the presence of anthraquinonoid ring (C=O) in the structure. It also contains tannin group in its structure. Bark extract exhibits antimicrobial activity against both the positive and negative bacteria, respectively, when it has been applied on cotton fabric (Samanta et al. 2015a). One research group has studied the antimicrobial efficacy of sal (*Shorea robusta gaertn F*) bark dye on a mordanted silk fiber. The mordants added on the textile had a significant impact on the color intensity, variation, and fastness properties of the dyed fabric. A maximum change in color has been observed in 3% of the copper-sulfate-mordant-treated samples. They reported that the sal dye also exhibited antibacterial properties, but it has not been studied on dyed fabrics (Samanta et al. 2015b). Likewise, another research group, in their study of the effect of the bark extract of a votiyar tree (*Odina wodier)* on cotton fabric, reported that the said bark extract has good antibacterial and antifungal properties, but the effect has not been studied on the dyed fabric (Chattopadhyay et al. 2016). Apart from it, extracts of various other natural sources like tulasi, pomegranate, turmeric, clove, karanga, henna, cashew shell, cumin, etc. have also been explored in detail by the researchers for making sustainable antimicrobial textile materials. Improving the durability of the antimicrobial treatment is one of the major concerns nowadays, and some of the researchers have used the cross-linking method (by using polycarboxylic acid) and the microencapsulation method (by using guar gum, alginate, etc.). It has been found that the microencapsulation method is more efficient for providing long-lasting effectivity of the finishing after repeated washing. Very recently, researchers are trying to make such kind of textile, which indigenously (does not require any chemical) has antimicrobial effectivity, depending on the structure of the fibrous material. This kind of research concept has been inspired from the various nano-structure surface of water-living animals.

UV-Protective Textile

UV-protective textiles guard living beings from harmful UV rays, which may potentially damage the skin or cause cancer due to overexposure. UV rays have been classified into UVA (315–400 nm), UVB, and UVC (280–315 nm) regions. UVC has a wavelength less than 280 nm and, having a very high intensity, is very much harmful to human beings. It may cause cancer and deadly diseases. Generally, the ozone layer of the atmosphere prevents UVC rays of the sun from entering the earth. With the depletion of some of the areas of the ozone layer, UVC rays are slowly entering the earth. Therefore, people need UV-protective materials like textile to protect their life. The UV protection performance of textiles is measured by the UV protection factor (UPF), with highest protection at values greater than 50 and no protection at values less than 10 (Samanta et al. 2017b). As per scientific report, UPF value of 5–10 may cause skin cancer, 10–20 may tan or occasionally burn the skin, and 20–35 may cause sufficient level of melanin pigment secretion and prevent the skin from burning (Samanta et al. 2017a).

Textile cotton, jute fabric, etc. can be made UV protective by using the extract of different plant products that provide a dark color and contain polyphenolic compounds, tannin, etc. Mongkholrattanasit et al. investigated that dye extracted from the bark of Kurz (*Garcinia dulcis*) not only was useful as a dye but also has UV protection property when applied to silk fabrics. As per their report, the crude application of the extracted dye on the silk fabric gives a light pale yellow color. Mongkholrattanasit R. et.al studied the UV-protective ability of proteinous silk fabric dyed with the bark extract of Magnifera (*Rhizophora apiculata Blume*). The study confirmed that the bark extract provided a UV-protective finishing to the silk fabric. The extent of UV protection also varied from good to excellent, not only with an increase in the concentrations of the dye but also with an increase in the concentration of the mordant. Another research group has investigated the dyeing of cotton fabrics with the extracts of a mangrove plant (*Xylocarpus granatum*), which is commonly used in textile dyeing. The dye is rich in tannin content and is reddish brown in color. The color strength of the dye improves with the application of mordant, which is due to the formation of bridges between the dye, the mordant, and the fabric (Samanta et al. 2015c, d; Alongi et al. 2014a).

Fire-Retardant Textile by Using Biomolecule

Most of the flame retardants used by the textile industries are not eco-friendly and release high ppm of formaldehyde during finishing, which is also released by the finished fabric during storage and use. Therefore, there is a research gap, and researchers are trying to find a kind of chemical that would be easy for textile application, would be less expensive and sustainable, and would have less negative effect on the mechanical properties of the textile. Focusing on this concept, for the last 10 years, different research works have been accomplished. One research group in Italy has reported that DNA extracted from herring sperm and salmon fish consists of phosphate, sugar glucose units of polysaccharide, and some amino acids. The DNA contains phosphorous- and nitrogen-containing groups, and the cotton fabric treated by DNA shows a limiting oxygen index (LOI) of 28, whereas the control fabric shows an LOI value of 18 and burnt easily within 1 min. They have also reported that DNA has more carbonaceous char formation capability and ammonia release property (Alongi et al. 2014b). Efforts were also directed to make cotton fabric flame retardant with other natural products (wastage of paneer industries, chicken slaughter waste, fungus sources, etc.), which are rich in phosphate, disulfide, and protein that can influence the pyrolysis mechanism and reduces flammable gas liberation (Basak and Ali 2018). Apart from this, natural chitosan, maleic acid, phytic acid, tannic acid, vegetable starch, fish bone powder, rice husk silicate, etc. have also been explored by the researchers. Some plant-based materials like banana stem waste, coconut shell extract, pomegranate rind extract, spinach juice, etc. have been applied and reported by the researchers as well (Shukla et al. 2016; Basak and Ali 2019a, b; Basak and Smanta 2018; Kambli et al. 2018).

20.3 Irradiation-Based Technology for Sustainable Textile Processing

As mentioned in the above section, various groups of researchers have engaged in sustainable approaches by using biomolecule-based processing (Basak et al. 2015a, b, c, 2016; Basak and Ali 2016), nanomaterials for textile processing (Seshama et al. 2017; Sharma et al. 2018; Gupta and Basak 2010), irradiation-based processing (Samanta et al. 2014b, 2017a; Chattopadhyay et al. 2016), etc. In this section of the chapter, we have tried to enlist some of the basic irradiation approaches for the modification or functionalization of textile materials.

Various chemical methods such as chlorination, alkali treatment, and synthetic polymer coating have been reported for modifying the fiber surface of textiles. However, these processes are often found to be very harsh and non-eco-friendly, modifying the bulk properties of textile materials, and also generating effluents (Maclaren and Milligan 1981; Jocic et al. 1993). With the increasing environmental concern in recent years and the various government legislation on eco-friendliness and stringent norms on effluent discharge in many countries, alternate environmentally friendly methods with lower pollution load are being explored. One such method is the treatment of textile fabric by irradiation techniques. Different types of irradiation techniques are being utilized as alternatives to the chemical processing of natural and synthetic textiles. Of these, plasma and UV methods are considered to be a clean, cheaper, and multipurpose option. These processes are environmentally friendly, dry, and an energy-saving textile treatment compared to many other traditional methods. Plasma irradiation has been used by researchers for modifying the surface of textiles by physico-chemical reaction. As a result of the physico-chemical changes, the resultant textile material can easily take in dye and other finishing chemicals at lower temperature and in less time. The UV irradiation techniques are normally used to change the nanoscale surface of textiles and also to change the rate of the dyeing of silk, wool, and polyester (Gupta and Basak 2010; Periyasamy et al. 2007a; Gulrajani et al. 2008; Basak and Gupta 2013). Researchers are also using this high-energized VUV irradiation technology for imparting functional finishing and improved serviceability to both natural and synthetic fabrics. Most of the applications reported in the literature are on fabrics made of protein because of their amphoteric nature. It has been reported that irradiation technique can be effectively used to shrink resist wool fabric. The 254-nm UV treatment can modify the wool fabric, lessen its shrinkage, and enhance its dyeing rate and pill-resist properties (Xin et al. 2002). Laser-based surface modification is normally used in aerospace, optoelectronics, automotive industries, the medical sector, etc. However, in some areas, laser rays have also been explored for modifying textile surface and designing textile materials (Morgan et al. 2018; Kan 2008). Very recently, electron beam technology has emerged in the market, which can be used for eco-friendly and clean biodegradable effluent treatment (Deogaonkar et al. 2019; Henniges et al. 2013).

20.3.1 Plasma Irradiation

Plasma is a fourth state of matter and consists of highly ionized, electrically conductive gaseous substances. It emits highly energized rays, which have significant capability to change the physico-chemical properties of the textile surface in a nano scale. However, irradiation treatment cannot alter the major inherent bulk properties of textiles. In general, plasma has been used by various researchers in the material science field for surface modification, grafting, cross-linking, etc. When this highly energized ray falls upon the textile surface, it causes some physico-chemical changes in it. Researchers have reported that cotton fabric can be made hydrophilic and hydrophobic by plasma irradiation in different atmospheres. Plasma irradiation in air and oxygen atmosphere assists to generate carboxylic (–COOH) and hydroxyl (–OH) groups on the fabric surface, and as a result, its hydrophilicity increases. Increased level of hydrophilicity of textile materials help to absorb more amount of chemicals (like dye and finishing chemicals) from the treatment bath in less time.

Indeed, the rate of dyeing or finishing has increased after irradiation. The effect of helium/oxygen plasma treatment on the cotton fabric has been explained by the researchers (Samanta et al. 2010a). As per report, the effect of hydrophilicity is more prominent for the wool fiber because of its amphoteric nature and because its structural composition consists of nanolevel fatty layer. Plasma irradiation helps to break the fatty surface layer of wool and makes it more accessible to the dye or finish chemicals. The irradiation process is sustainable as it minimizes the use of various chlorinated compounds, which has the potential to break the fatty layer of wool. As per report, plasma ray can also be effective on the fabric surface in inert nitrogen atmosphere. This atmosphere helps to generate more positively charged amine groups on the wool surface, which help to pick up more anionic dye from the treatment bath. As a result, dye wastage can be minimized; the temperature required for the treatment is also less compared to the untreated wool textile. On the other hand, plasma can also be applied on the textile surface in helium-flurocarbon gas atmosphere. As per report of the researcher, after treatment, hydrophilic cotton turned into a highly hydrophobic one, and 37 μL was not absorbed by the fabric's surface even after 60 min. In contrast, the control cotton fabric absorbed the same amount of water within 0.05 min (Basak et al. 2015c). After performing various characterizations, the author has confirmed that CF_X molecules deposited on the fabric surface after treatment assist to hold the water droplet on the cotton surface. Apart from this, different phosphorous- and halogen-containing gas plasmas have also been used by the researchers to make fire-resistant cotton fabric. The CF_4 plasma gas was responsible for imparting hydrophobicity and thermal stability of the cotton textile. Plasma irradiation also shows on curing, cross-linking of polymer, grafting, polymerization of monomer on the textile surface, etc. (Samanta et al. 2009, 2010b).

20.3.2 UV Irradiation

UV irradiation lower than 200 nm is also effective for modifying fabric surface with functional groups and physical etching. Mainly it emits highly energized photons that react with atmospheric air and form an ozone gas in the ground state, which again dissociates into oxygen and nascent oxygen (excited state). It is a continuous association and dissociation process. When the excited species falls upon the textile material surface, it generates some free radicals on the surface, and physically the surface of the material is etched. Like plasma, UV rays are also more effective on wool and silk textiles because of their amphoteric nature. Most of the literature reported that UV irradiation helps to break the scales of the wool and also etches the nanolayer thick fatty scales on its surface and reduces the directional frictional effect. As a result, the wool can be made shrink proof without using any chemicals. Another research group has reported that wool can be made easily accessible to anionic acid dye when it has been irradiated by UV rays at inert nitrogen atmosphere. It may be due to the creation of more amount of amine groups on the fabric's surface, which help to catch more acid dye molecules from the dye bath. On the

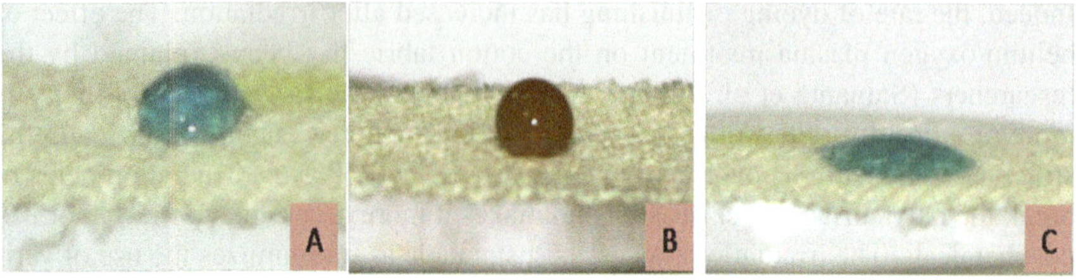

Fig. 20.1 Contact angle measurement of water droplet control wool fabric (**a**), F/C treated nonirradiated side (**b**), and F/C treated 30-min irradiated side (**c**) (Basak et al. 2015d; Periyasamy et al. 2007b)

other hand, UV irradiation in air or oxygen atmosphere creates negatively charged carboxylic and hydroxyl groups on the fabric and helps to catch more basic dye from the dye bath. One of the challenges in this process is that there are major chances of uneven dyeing and ring dyeing after UV treatment. In addition, active species generated on the fabric surface has the possibility of getting aged with time, and so its effectivity could gradually lessen. Researchers have reported the details of the effect of dyeing on UV-irradiated wool fabric in their published papers. The same research group has also made double hydrophilic/hydrophobic wool fabric by using 172-nm UV treatment. According to their report, a multifunctional wool fabric with hydrophobic property on one face and hydrophilicity on the opposite face was prepared by first padding it with fluorocarbon-based composition and then irradiating one side of the fabric with the use of 172-nm UV excimer lamps. The treated samples were then characterized by measurements of contact angle and vertical wicking behavior (Sharma et al. 2018).

It can be seen from Fig. 20.1 that on the irradiated side, the contact angle dropped to 10° after UV treatment. It might be because of irradiation using UV excimer lamp (172 nm); the F/C finish, which has high C–F bond breaking energy, has dissociated and undergone a photo oxidation involving defluorination of the surface (Shukla et al. 2016; Basak and Ali 2019a). Further, some extra polar groups like C=O and –COOH might have also been generated on the 30-min irradiated surface of the treated fabric. On the opposite side, the wool fabric surface shows a lotus effect behavior (Basak et al. 2015d).

UV rays also have a significant effect on the silk fabric surface. Periyasamy et al. (Periyasamy et al. 2007b) have reported surface modification of silk fabric by using an excimer lamp. It has been reported in their manuscript that UV treatment has significantly reduced the wetting time of the irradiated silk fabric compared to the untreated silk fabric. In addition, the rate of wicking of the treated fabric also has been improved significantly. He also elaborates the technical process for developing double hydrophilic/hydrophobic mulberry silk fabric by also using a 172-nm excimer lamp and reported on the detail mechanism behind it in their published paper (Periyasamy et al. 2007b).

20.3.3 Laser Irradiation

Laser emits electromagnetic radiation, depending on optical amplification. Researchers have reported on the effect of laser irradiation on the properties (fabric weight, fiber diameter, abrasion, bending, luster, wetting, etc.) of polyester fabric. They indicated that the laser ray only affects the surface properties of the polyester fabric. However, the performance and comfort properties of the laser-irradiated polyester fabric changed after laser ablation (Kan 2008). Recently, it also has been established the dyeing behavior of CO_2-laser-treated wool fabric. They have reported that laser irradiation approach has a promising effect of reducing energy and water consumption. It has also been reported that laser irradiation improves the dye uptake of the wool textile. However, some tonal variations have been observed on the wool fabric. The same research group has also created various designs on the wool fabric surface through color fading with the help of laser rays (Morgan et al. 2018; Deogaonkar et al. 2019). Chow et al. (2012) have treated cotton fabric with CO_2 laser. They have varied the pixel time and the resolution of the laser for application purposes and have reported that fabric weight and strength decreased with the increase of resolution and pixel time [47, 48]. However, as per report, the treated fabric turned yellowish after treatment. Very recently, another research group has reported on the effect of CO_2 laser treatment on the fabric hand of cotton and cotton/polyester blend fabric. They have explained the effect of laser treatment on the hand properties of the cotton and polyester/cotton blend fabric. In connection to hand properties, they have measured stiffness, smoothness, softness, wrinkle recovery, drapability, etc. and concluded that laser treatment has sufficient capability to change the mentioned hand properties of the fabric.

20.3.4 Electron Beam Irradiation

Electron irradiation is composed of highly energized electrons. Normally, the energy of electrons varies from the KeV to MeV range, depending on the depth of the penetration required. The irradiation dose of the electron beam radiation is generally measured in grays. For the last 5 years, electron beam technology has been used popularly for textile effluent treatment. A researcher has reported on the treatment of textile effluent and nonionic ethoxylated surfactant by electron beam for the removal of color and toxicity (Samanta et al. 2010a). In the same line of work, Deogaonkar et al. (2019) have also reported on the use of electron beam irradiation treatment for the degradation of nonbiodegradable contaminants in textile wastewater. As per this report, electron beam radiation (80 kGy) helps in the biodegradation of effluent, generated from textile desizing, scouring, bleaching, dyeing, and finishing. They have also reported on the biological oxygen demand (BOD) and chemical oxygen demand (COD) of the waste effluent before and after the electron beam treatment (Samanta et al. 2010b).

20.4 Nanotechnology and Other Sustainable Textile Processing

In the past decade, nanotechnology had been widely used by researchers for making value-added textile products. Earlier, different chlorinated products and large molecular weight cationic compounds, like polyhexamethylene biguanide and others, have been widely used by researchers for making antimicrobial textile. However, add-ons, chemical consumption, and toxicity were the biggest challenges. Therefore, nanosilver-based chemicals emerged on the market, and now they are widely used by researchers for making antimicrobial cotton textiles. For wash and wear and durable press finishing of garnets, formaldehyde-based resin, along with sodium hypophosphite, is frequently used by researchers and textile industries. These chemicals have very good cross-linking property and assist in achieving a wrinkle recovery angle of more than 150. However, the treatment affects the strength properties of the fabric, and formaldehyde release is another major concern. Therefore, researchers have tried and developed nano- and polycarboxylic-acid-based, environmentally friendly chemicals for making crease-resistant cotton fabric. As per report, a few textile industries are using poly-carboxylic acids, such as 1,2,3,4-butanetetracarboxylic acid (BTCA), and catalysts, like sodium hypophosphite, for making durable press fabric. However, it is also costly for end users. Self-cleaning technology and lotus effect are more predominant in recent years because of their smart application in textile materials. Superhydrophobic surfaces have also been prepared by using nano titanium dioxide, aluminum oxide, zinc oxide (ZnO) nanoparticles, etc. Recently, nanotechnology has also been explored for making sustainable fire-retardant textile materials. Nano zinc oxide, zinc carbonate, etc. have been explored successfully by researchers for making fire-retardant textiles (Seshama et al. 2017; Sharma et al. 2018). However, as per report, the durability and uniformity of the treatment are the major challenges for its application. Antimony oxide nanoparticle in combination with zinc salt assists to improve the thermal stability of the treated cotton fabric, and the treated fabric shows an LOI value of 24, compared to the LOI value of 18 of the control cotton (Alongi et al. 2014b). Fire-resistant property has been further improved by increasing the add-on percentage on the fabric's surface. Recently, researchers have used ZnO nanoparticles for improving the thermal stability and harmful ultraviolet ray protection functionalities in cotton and blended textiles. The presence of nanodispersed montmorillonite (MMT) clay showed a significant thermal stable property after application on natural polymer substrates (Shukla et al. 2016).

Most of the industries (especially textile industries) generated effluent, which is very much harmful for living beings, as well as for ecosystems. Effluent water has lost its physical, chemical, and biological properties due to mixing with some other contaminants like dyes, heavy metals, pathogens, other inorganic and organic materials, etc. Nanomaterials like metal oxide, metal nanoparticles, zeolite, etc. have already been explored effectively in the field of wastewater purification due to their lower size, high surface area, and size-dependent properties.

20.5 Conclusion

Sustainable development in the textile field is one of the biggest concerns all over the world. Different researches, workshops, and industrial trials are ongoing for the implementation of the different sustainable processes in the textile sector. Among all the technologies developed by researchers, biomolecule-based technologies, water-free processing, and nanoprocessing of textile materials are important. Different technologies like foam finishing, enzymatic treatment, and nano-based processing are commercializing slowly in the industries. However, the cost of treatment, batch-to-batch variation, and repeatability of the treatment are important concerns in most of these areas. In addition, washing and storage durability also need to be taken into consideration. More research and hard work are still required to achieve the sustainability goal in the textile processing sector.

References

Alongi J, Blasio AD, Cuttica F, Carosio F, Malucelli G (2014a) Bulk or surface treatments of ethylene vinyl acetate copolymers with DNA: investigation on the flame retardant properties. Eur Polym J 51:112–119

Alongi J, Cuttica F, Blasio AD, Carosio F, Malucelli G (2014b) Intumescent features of nucleic acids and proteins. Thermochim Acta 591:31–39

Basak S, Ali SW (2016) Sustainable fire retardancy of textiles using bio-macromolecules. Polym Degrad Stab 133:47–64

Basak S, Ali SW (2018) Fire resistant behavior of cellulosic textile functionalized with wastage plant bio-molecules: a comparative scientific report. Int J Biol Macromol 114:169–180

Basak S, Ali SW (2019a) Wastage pomegranate rind extracts (PRE): a one step green solution for bioactive and naturally dyed cotton substrate with special emphasis on its fire protection efficacy. Cellulose 26(5):3601–3623

Basak S, Ali SW (2019b) Sodium tri polyphosphate in combination with pomegranate rind extracts as a novel fire retardant composition for cellulose polymer. J Therm Anal Calorim 137:1233–1247

Basak S, Gupta D (2013) Advanced processing of woollen textile pretreated with UV excimer radiation. Manmade Textile in India 41(6):204–209

Basak S, Smanta KK (2018) Thermal behavior and the cone calorimetric analysis of the jute fabric treated in different pH condition. J Therm Anal Calorim 135(6):3095–3105

Basak S, Samanta KK, Chattopadhyay SK, Narkar R (2015a) Banana pseudostem sap: wastage byproduct for making fire retardant cellulosic paper. Cellulose 22(4):2767–2776

Basak S, Samanta KK, Saxena S, Chattopadhyay SK, Narkar R, Mahangade R (2015b) Fire retardant cellulosic textile using banana pseudostem sap. Int J Cloth Sci Technol 27:247–261

Basak S, Samanta KK, Chattopadhyay SK, Narkar R (2015c) Self-extinguishable ligno-cellulosic fabric made by banana pseudostem sap. Curr Sci 108:372–383

Basak S, Samanta KK, Chattopadhyay SK, Narkar R (2015d) Development of dual hydrophilic/hydrophobic wool fabric by 172nm VUV irradiation. J Sci Ind Res 75:439–443

Basak S, Patil PG, Shaikh AJ, Samanta AK (2016) Green coconut shell extract and boric acid: new formulation for making thermally stable cellulosic paper. J Chem Technol Biotechnol 91:2871–2881

Chattopadhyay SK, Samanta KK, Basak S (2016) Potential of ligno-cellulosic and protein fibres in sustainable fashion. In: Muthu S (ed) Sustainable fibre for fashion industry, vol 2. Springer, Singapore, pp 61–109

Chow YLF, Chan A, Kan CW (2012) Effect of CO_2 laser irradiation on the properties of cotton fabric. Text Res J 82(12):245–253

Deogaonkar SC, Wakode P, Rawat KP (2019) Electron beam irradiation post treatment for degradation of non-biodegradable contaminants in textile waste water. Radiat Phys Chem 165:108377

Gulrajani ML, Gupta D, Periyasamy S, Muthu SG (2008) Preparation and application of silver nanoparticles on silk for antimicrobial properties. J Appl Polym Sci 108(1):614–623

Gupta D, Basak S (2010) Surface functionalization of wool using 172nm UV excimer lamp. J Appl Polym Sci 117:3448–3453

Henniges U, Hasani M, Potthast A, Westmass G, Rosenau T (2013) Electron beam irradiation of cellulosic materials—opportunities and limitations. Materials 6:1584–1598

Jocic D, Jovancic P, Trajkovic R, Seles G (1993) Influence of a chlorination treatment on wool dyeing. Text Res J 63(11):619–626

Kambli ND, Samanta KK, Basak S, Chattopadhyay SK, Patil PG, Deshmukh RR (2018) Characterization of the corn husk fibre and improvement in its thermal stability by banana pseudostem sap. Cellulose 25:5241–5257

Kan CW (2008) Effects of laser irradiation on polyester textile properties. J Appl Polym Sci 107(3):1584–1589

Lal S, Arora S, Kumar V, Rani S, Sharma C, Kumar P (2018) Thermal and biological studies of Schiff bases of chitosan derived from heteroaryl aldehydes. J Therm Anal Calorim 132:1707–1716

Maclaren JA, Milligan B (1981) Wool science: the chemical reactivity of wool fibre. Science Press, Marricksville

Morgan L, Tyrer J, Kane F (2018) The effect of CO2 laser irradiation on surface and dyeing properties of wool for textile design. J Laser Appl 2014:1

Periyasamy S, Gupta D, Gulrajani ML (2007a) Modification of one side of mulberry silk fabric by monochromatic UV excimer lamp. Eur Polym J 43(10):4573–4581

Periyasamy S, Gulrajani ML, Gupta D (2007b) Preparation of a multifunctional mulberry silk fabric having hydrophobic and hydrophilic surface using VUV excimer lamp. Surf Coat Technol 201(16–17):7286–7291

Samanta KK, Jassal M, Agrawal AK (2009) Improvement in water and oil absorbancy of the textile substrate by atmospheric pressure cold plasma treatment. Surf Coat Technol 203:1336–1342

Samanta KK, Jassal M, Agrawal AK (2010a) Antistatic effect of atmospheric pressure glow discharge cold plasma treatment on textile substrates. Fibers Polym 11(3):431–437

Samanta KK, Jassal M, Agrawal AK (2010b) Atmospheric pressure plasma polymerization of 1,3-butadiene for hydrophobic finishing of textile substrates. J Phys Conf Ser 208:012098

Samanta KK, Basak S, Chattopadhyay SK (2014a) Eco-friendly coloration and functionalization of textile using plant extracts. In: Muthu S (ed) Roadmap to sustainable textiles and clothing. Springer, Singapore, pp 263–287

Samanta KK, Basak S, Chattopadhyay SK (2014b) Environment friendly processing using plasma and UV treatment. In: Muthu S (ed) Roadmap to sustainable textiles and clothing. Springer, Singapore, pp 161–201

Samanta KK, Basak S, Chattopadhyay SK (2015a) Sustainable UV-protective apparel textile. In: Muthu S (ed) Handbook of sustainable apparel production. Routledge association with GSE Research, Abingdon, pp 113–137

Samanta KK, Basak S, Chattopadhyay SK (2015b) Speciality chemical finishes for sustainable luxurious textiles. In: Muthu S (ed) Handbook of sustainable luxury textiles and fashion. Springer, Singapore, pp 145–184

Samanta KK, Basak S, Chattopadhyay SK (2015c) Water free plasma processing and finishing of apparel textiles. In: Muthu S (ed) Handbook of sustainable apparel production, vol 3. Routledge association with GSE Research, Abingdon, pp 3–37

Samanta KK, Basak S, Chattopadhyay SK (2015d) Recycled fibrous and non fibrous biomass for value added textile and non textile applications. In: Muthu S (ed) Environmental implications of recycling and recycled products. Springer, Singapore, pp 167–212

Samanta KK, Basak S, Chattopadhyay SK (2016) Potentials of fibrous and nonfibrous materials. In: Muthu S (ed) Biodegradable packaging. Springer, Singapore, pp 75–113

Samanta SK, Basak S, Chattopadhyay SK (2017a) Environmental friendly denim processing using water free technologies. In: Muthu S (ed) Sustainability in Denim. Springer, Singapore, pp 319–348

Samanta KK, Basak S, Chattopadhyay SK (2017b) Sustainable dyeing and finishing of textiles using natural ingredients and water free technologies. In: Muthu S (ed) Textiles and clothing sustainability. Springer, Singapore, pp 99–131

Samanta KK, Pandit P, Samanta P, Basak S (2019) Water consumption in textile processing and sustainable approaches for its conservation. In: Muthu S (ed) Water in textiles and fashion. Woodhead Publishing, Cambridge, pp 41–59

Seshama M, Khatri H, Suther M, Basak S, Ali SW (2017) Bulk vs nano ZnO: influence of fire retardant behaviour on sisal fibre yarn. Carbohydr Polym 175:257–264

Sharma V, Basak S, Rishabh K, Umaria H, Ali SW (2018) Synthesis of zinc carbonate nanoneedles, a potential flame retardant for cotton textiles. Cellulose 25(10):6191–6205

Shukla S, Basak S, Ali SW, Chattopadhyay R (2016) Development of fire retardant sisal yarn. Cellulose 24:423–434

Xin JH, Zhu R, Hua J, Shen J (2002) Surface modification and low temperature dyeing properties of wool treated by UV radiation. Color Technol 118(4):169–173

Chapter 21
Third Generation of Biofuels Exploiting Microalgae

Erfan Sadatshojaei, David A. Wood, and Dariush Mowla

Contents

Abbreviations

DOE US Department of Energy
GMO Genetically modified organisms
Mg Megagram
NREL National Renewable Energy Laboratory
ORP Open raceway pond
PBR Photobioreactor

E. Sadatshojaei (✉) · D. Mowla
Environmental Research Center, Department of Chemical Engineering, Shiraz University, Shiraz, Iran
e-mail: dmowla@shirazu.ac.ir

D. A. Wood
DWA Energy Limited, Lincoln, UK
e-mail: dw@dwasolutions.com

Inamuddin, A. M. Asiri (eds.), *Sustainable Green Chemical Processes and their Allied Applications*, Nanotechnology in the Life Sciences,
https://doi.org/10.1007/978-3-030-42284-4_21

21.1 Introduction

New technologies applied to improve efficiency to better exploit existing resources and develop more useful products from them play key roles in improving quality of life. Various innovative technologies have emerged in recent years that target efficiency and sustainability associated with improving the production of energy fuels. For example, relating to enhanced oil recovery (Sadatshojaei et al. 2016, 2019; Choubineh et al. 2017), various applications of supercritical fluid (Sadatshojaei et al. 2018) and exploiting algae (Wood 2019) and, in particular, their future potential role in the generation of biofuel (Chowdhury et al. 2019) have the ability to make fuel supply more efficient and sustainable. Biofuels offer a more environmentally friendly alternative to crude-oil-derived gasoline and diesel, which are currently the dominant vehicle fuel source worldwide. Ongoing adjustments to the quality and type of feedstocks used to produce biofuels enable them to further improve the environmental benefits they bring and improve public confidence in using them more extensively as a viable and sustainable fuel source.

Emerging biofuel production techniques are promising to provide essentially nontoxic fuel supplies that will meet the toughening environmentally driven quality standards expected to be introduced over the coming decades. Various existing conversion technologies enable biomass from diverse sources to be relatively easily converted into various forms of biofuel, particularly gasoline and diesel. The term "third-generation biofuel" specifically refers to biofuels that involve algal biomass feedstock in their production.

Third-generation biofuels include those that involve various algal species such as feedstock and offer several advantages with respect to biofuel production methods that rely on various crops and organic waste materials. These advantages are as follows: (1) algal feedstocks remove competition between crops supplying food and crops grown to provide biofuels, (2) algae-based biofuel can be produced by simple cultivation techniques with a small land footprint, (3) growth of certain algal species can be sustained at high rates, and (4) a vast diversity of growth media can be used to support and sustain algal growth, including wastewater, seawater, and various carbon-dioxide-rich effluents. However, algal biofuel production does involve some potential disadvantages. These include (1) low efficiency in terms of extracting the lipids generated by the algal cells from the algal biomass, (2) the possibility of that algal biomass grown in open ponds to become contaminated or to escape into the environment to cause contamination itself, and (3) high energy consumption over the entire algal biofuel production cycle due to the processes involved at each stage, which include mixing, filtration, and centrifugation.

Microalgae play a crucial role in the production processes being developed for third-generation biofuels. This is because microalgae are able to overcome some of the disadvantages that have adversely affected the uptake of first-generation and second-generation biofuels. Consequently, there has been a significant focus on developing techniques that could lead to the exploitation of microalgae for biofuel production over the past 15 years (Li et al. 2008; Chisti 2007; Dragone et al. 2010; Nigam and Singh 2011). Microalgae are capable of yielding a wide range of biofuel sources, such as biodiesel, hydrogen, and methane (Spolaore et al. 2006; Gavrilescu and Chisti 2005; Kapdan and Kargi 2006). A key advantage in producing biofuel from algae is that, depending on the species and production method applied, it has the potential to produce between 15 and 300 times more biodiesel than biofuel production processes that rely upon traditional crop-based feedstock (Dragone et al. 2010). Moreover, there is no requirement to dedicate large areas of high-quality agricultural land to cultivate microalgae for the purpose of generating biofuels on commercially viable scales. Further positive features are that the cultivation cycle of microalgae is very short, and their growth rates are very high, both of which lead to improved production efficiency (Dragone et al. 2010; Scott et al. 2010; Schenk et al. 2008).

Biofuels consist of products produced by converting biologically derived and cultivated materials or organic substances, naturally occurring or waste materials, into substances that can be used as fuels. Those fuels may be solid, fluid, or gaseous in terms of their physical states (Lü et al. 2011; Mobin and Alam 2018). Figure 21.1 illustrates the categorization of biomass into four distinct generations of feedstock used to produce sustainable biofuels, plus various residues also used as fuels of

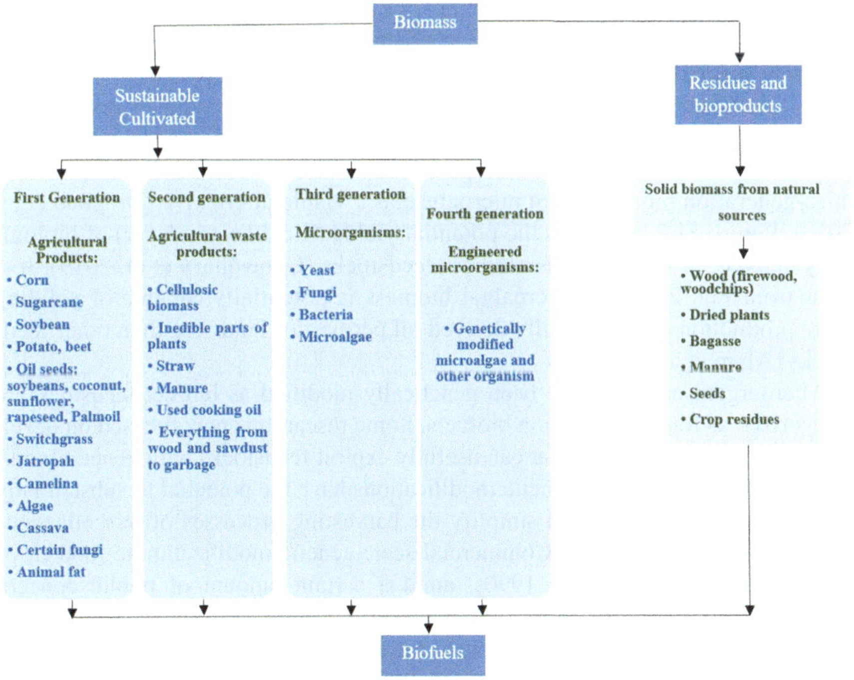

Fig. 21.1 A categorization for biomass feedstock to produce biofuel (modified after Chowdhury et al. 2019)

organic origin. Biomass can be usefully divided into two distinct groups in terms of source type: "sustainably cultivated" and "residues and biproducts." Most biofuels are generated as the primary products of commercial cultivation projects and can be categorized by generation methods. On the other hand, biomass constituting residues and biproducts are derived from natural sources, which are often used as subsistence fuels and may or may not be renewable or sustainable in their collection methods. This latter group includes wood, dried plants, seeds, and manure.

First-generation biomass from cultivated crops is associated with inherent economic and sustainability problems inhibiting the scaling up of their production volumes. They require vast dedicated agricultural land resources that encroach upon and diminish land available for crops grown for human and animal food consumption and animal grazing. Furthermore, the procedures used to provide first-generation biofuels often involve negative environmental impact, including deforestation in the case of some sugar cane and palm oil feedstocks, and significant fuel consumption in their production processes. Consequently, the motivation and public acceptance for consuming first-generation biomass to produce biofuels of questionable negligible overall benefits compared to fossil fuels have decreased.

Second-generation biofuels are generated from miscellaneous nonfood sources that overcome at least some of the issues associated with large-scale first-generation biofuel production. However, it is typically not industrially or commercially viable to generate at scale and supply second-generation biomass to produce biofuel (Brennan and Owende 2010). The main reasons for this are that the feedstocks are in limited supply and their production procedures are expensive and inefficient, requiring specialist and novel technologies (Dragone et al. 2010).

As a consequence of the shortcomings of first-generation and second-generation biofuels (Chisti 2007), industry and academia have focused more concertedly in recent years (Nigam and Singh 2011) on the research and development of third-generation biomass from microalgae as a viable alternative (Dragone et al. 2010). Figure 21.2 compares the potential yields (liters/hectare/year) of biologically derived oil from various biomass feedstocks (Chowdhury et al. 2019). It is clear from Fig. 21.2 that microalgal biomass is potentially capable of yielding more production of biologically derived oil per unit of land area than other feedstocks (Alam et al. 2015).

Microorganisms that have been genetically modified as biofuel feedstock are categorized as fourth-generation biofuels. Some research is now focused on developing biofuel technologies that can usefully exploit feedstocks with genetic modifications (Lü et al. 2011). Genetic modifications have the potential to substantially improve the productivity and simplify the harvesting processes of miscellaneous potential biofuel feedstocks. Commercial-scale genetic modification to food crops began with tomatoes in the 1990s, amid a certain amount of public concern regarding their long-term safety and unknown impact on ecosystems. Those initial modifications to tomatoes enabled them to remain edible for a longer period post harvesting, which was a significant step forward for commercial markets at that time (Stave 1999). Following on from the initial successes in the 1990s, many genetically modified crops are now cultivated. These include corn, cotton, papaya,

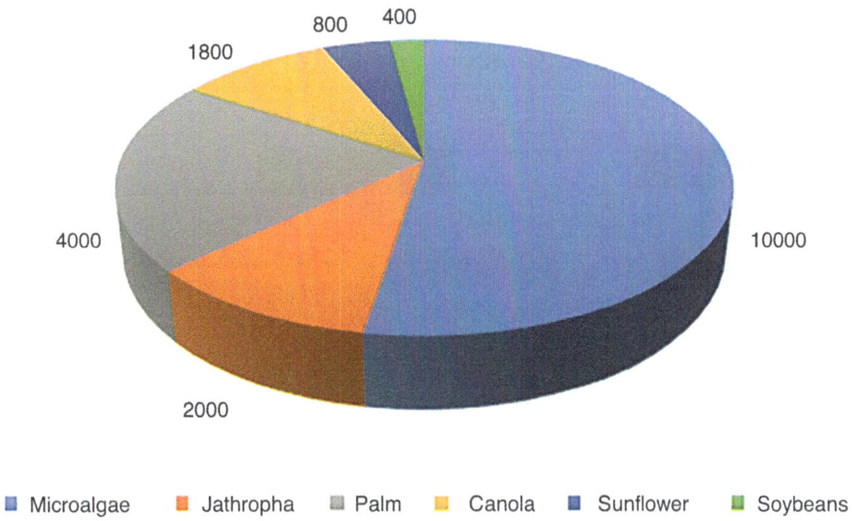

Fig. 21.2 Potential yield of crude biologically derived oil from miscellaneous biomass (liter/hectare/year) (data from: Chowdhury et al. 2019)

soya beans, and sugar beet. It is these successes that have led researchers to exploit genetic modification technologies in attempts to enhance the productivity of lipid-like liquids by microalgae. It has been shown that some microalgal species can be beneficially modified genetically in terms of their potential biofuel production capabilities (Lü et al. 2011).

21.2 Microalgae Cultivation Strategies

There are several distinctive industrial-scale cultivation methods for microalgae (Fig. 21.3). Fundamentally, there are two distinct approaches: (1) immobile—growing microalgae as static/fixed cultures such as biofilms attached to a rigid grid and (2) suspended—growing microalgae as supported but free-hanging cultures in open or closed systems. Almost universally, the algal raceway ponds have been utilized to date for scaled-up cultivation systems. These consist of relatively shallow but open-to-the-atmosphere water tanks that are mounted on wheels. The wheels enable mobility to enable carbon dioxide and nutrients to better circulate within these systems and thereby enhance algal growth. Raceways are typically associated with capital and operating costs and are relatively easy to operate and maneuver. On the other hand, raceways are vulnerable and exposed to a range of potentially negative external impacts, which typically leads to them achieving relatively low productivity (Mata et al. 2010). A particular issue for open pond microalgal cultivation systems is their exposure to contamination by unfavorable

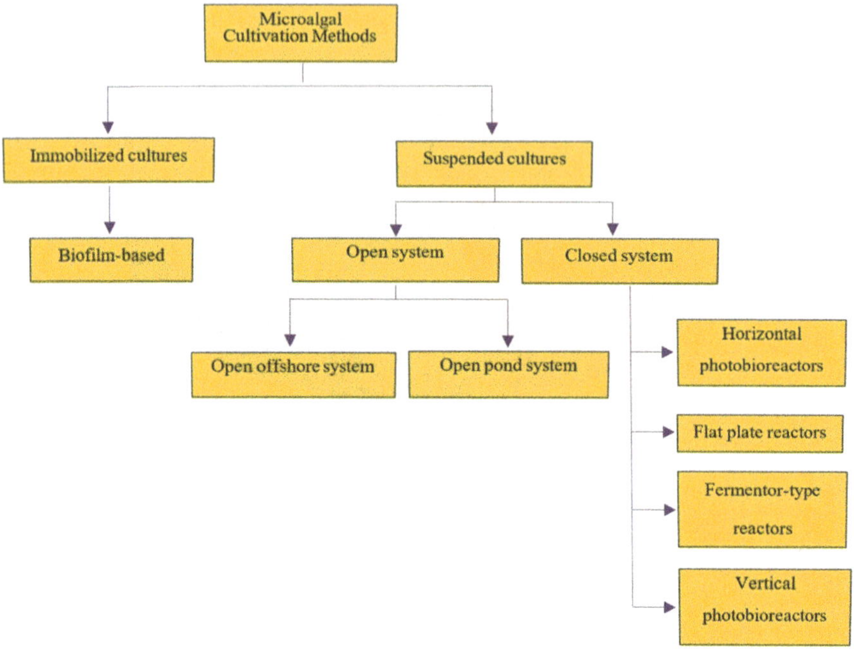

Fig. 21.3 Microalgae cultivation methods (after Chowdhury et al. 2019)

photosynthetic microorganisms through air or rainwater entry. For this issue, one solution to illustrate is to utilize closed systems such as a photobioreactors, which are more controllable and provide a stable environment to cultivate algae.

Tubular photobioreactors have several valuable advantages compared to open-pond systems. These include (a) improved temperature control, (b) maintaining more desirable acidity-alkalinity balance (pH), (c) facilitating high-quality blending, (d) achieving higher cell densities, (e) limiting vaporization, and (f) providing more favorable protection against contamination (Mata et al. 2010). However, the main drawback of closed photobioreactors is that they are more costly in terms of capital to build and operating requirements.

Tubular photobioreactors dominate the closed systems used to cultivate microalgae at commercial scales (Chisti 2007). Five distinct types of photobioreactors are employed. These are (1) fermenter type, (2) helical, (3) vertical, (4) flat plate, and (5) horizontal. There are good and bad aspects associated with each of these photobioreactor designs. For instance, fermenter-type reactors tend not to have favorable surface area/volume ratios (Koh and Ghazoul 2008). This characteristic lessens their efficiency in terms of the ability of external light from the sun to gain entry (Lardon et al. 2009). Consequently, there are not a large number of fermenter-type reactors used for large-scale microalgae cultivation.

The basic challenge for closed systems is to access and utilize light efficiently such that productivity is optimized. Helical photobioreactors have proved to be the easiest and cheapest way to scale up for commercial production for that reason. A key challenge for raceways, which also applies to tubular photobioreactors, is that harvesting the biomass also involves dewatering. Dewatering requirements are much reduced in the case of attached or immobilized algal cultivation systems (Hoffmann 1998), which favor biofilm-type cultures. However, such systems require larger surface areas (Scott et al. 2010). A wide range of innovative methods for algae production are now in use and under development. These include raceway ponds, closed-system photobioreactors, fermentation tanks, and combinations of these components. As yet, there is no standardized design adopted for industrial-scale algal cultivation. Certain microalgae species develop and survive more reliably in closed systems rather than in open systems.

The primary strategy for microalgal cultivation is to focus on maximizing algal growth in order to maximize biofuel production. However, there are several variables that influence algal biofuel productivity. These include the algal species involved, the desired final biofuel product, local preferences for fuel types and their specifications, weather conditions, geographic characteristics of location, accessibility, and costs of resources to supply specific sites.

21.3 Determining Appropriate Sites and Support Systems

Microalgae will grow in a wide range of water-based fluids. These include seawater, freshwater, urban (utility-derived) wastewaters, and industrial effluents. Certain nutrients are essential for them to sustain growth, i.e., phosphate (P), carbon (C), and nitrogen (N), plus various trace constituents (Zhou et al. 2014). Seawater can be enhanced by the addition of fertilizers, specifically nitrates and phosphates, to promote the growth of marine microalgae (Grima et al. 2003). Freshwater also possesses insufficient nutrients to sustain the rapid growth of microalgal cultures, and supplemental nutrients are required. On the other hand, wastewaters typically include abundant nutrients that promote microalgae growth.

The challenge with many wastewaters is that their composition varies over the short term, and this makes nutrient concentrations difficult to regulate (Mobin and Alam 2014). This issue leads to various difficulties in the large-scale application of wastewater-based algae systems. Wastewaters are characterized by diversity in terms of origin, composition, temperatures, weather-related influences, and pre-treatment requirements. Their varying C/N and N/P ratios cause fluctuations in nutrient availability, disrupting the rate of algal growth. Variations in the coloration of wastewaters due to the presence of pigments and suspended solid particles cause fluctuations in the rate of light transmission. This again can disrupt algal growth rates, as can the presence of microflora caused by biofouling and traces of other toxins, which tend to be exacerbated in recycled fluids (Zhou et al. 2014).

For low operating costs, microalgal cultivation sites need to located close to their source of fluid medium. They also need to benefit from bright, high sunlight conditions, which will rule out some geographic and highly shaded locations. The sites also need to benefit from low-cost utilities, specifically electricity and freshwater, ideally with competition among suppliers. Supplemental supplies of nutrients (CO_2, N, and P) need to be available nearby, in bulk, and at low delivered cost. By paying attention to such requirements, the operating costs of the best sites can be about half the cost of the less favorable sites (Slade and Bauen 2013).

21.4 Commercial Viability of Biofuels Derived from Microalgae

Figure 21.4 describes the sequential stages of the microalgal feedstock to biofuel supply chain. These stages are (1) feedstock growth; (2) feedstock harvesting, including dewatering; (3) feedstock extraction; (4) conversion into specification fuel products, including blending; and (5) marketing and distribution and, in some cases, additional blending. There is an operating cost component associated with each stage, adding incrementally to influence the overall cost of supply. Various components contribute to the cost of supply in providing biofuels to their ultimate end users. Feedstock cultivation has semi-fixed and semi-variable operating costs. Semi-fixed costs are land rentals, site maintenance and security expense, algal species development, research expense for modified algal species, and cultivation sys-

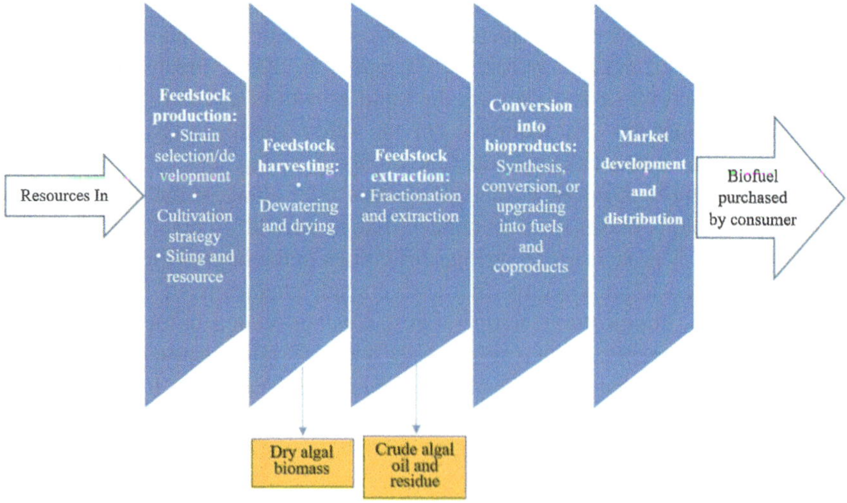

Fig. 21.4 Microalgal feedstock to biofuel supply chain (modified after Chowdhury et al. 2019)

tem expense. Semi-variable costs are for utilities, electricity, water, fertilizer, other supplementary nutrients and chemicals, and employees.

There are significant costs to harvest and extract the crude lipids from the feedstock. These include costs to dewater and, in most cases, to dry the mature biomass cultures. There are plant and machinery capital costs and operating and maintenance costs associated with these activities. They can be both energy and labor intensive, resulting in significant electricity and/or other fuel costs and staff costs. Similar cost components are also associated with conversion of the extracted lipids into specification biofuels, typically biodiesel and bioethanol. The processing costs are particularly costs associated with process equipment, power, and staff costs.

21.5 Factors that Control Microalgal Growth Rates

The justifications to exploit microalgae to produce biomass become more compelling year by year as more knowledge of microalgae and new designs for production plants materialize. Key justifications are that microalgal-biofuel systems are relatively inexpensive and are more readily scaled up to commercially viable projects than those based on food crops. The fundamental elements required to grow microalgae are a growth or culture medium, light, carbon dioxide, and nutrients. Sugar can be an alternative for light in systems that utilize fermentation-type production processes to produce bioethanol. The interaction of all these elements needs to be optimized and stabilized for each specific microalgae culturing system in order for them to achieve the maximum productivity of lipid components per unit of biomass grown.

The specifics associated with each of these influencing elements are described in points (1) to (4):

1. ***Growing medium***: water plays a primary role in the growth of algae. Consequently, water content, salinity, pH, and temperature collectively have significant impact on the efficiency of algal growth. For most species, the optimum range of temperature for the growth medium is 16–27 °C. However, there are certain microalgal species that prefer lower or higher temperatures.
2. ***Light***: through the photosynthesis process, algae consume light, and that assists their growth. Consequently, light exposure time and light intensity are the controlling factors that influence algal growth. Mixing of algal cultures prevents and/or disturbs algal settlement by gravitational forces. Mixing or stirring the cultures therefore improves and regularizes the quantity of light reaching all algal cells on average. Utilizing freely available sunlight to the maximum helps to reduce costs and speed up algal growth. However, diurnal and seasonal changes in light intensity impact most locations, and these need to be carefully accounted for and managed to ensure stable algal systems (Chisti 2007; Chaumont 1993; Pulz 2001).

3. ***Carbon dioxide***: the input of carbon dioxide into cultures is essential to sustain algal growth. Consequently, it is necessary to monitor and regulate the quantity of carbon dioxide throughout the growing algal cultures. The flow of carbon dioxide into the cultures needs to be sufficiently aerated with oxygen and nitrogen to optimize the algal growth rate.
4. ***Nutrients***: the essential nutrients required to continuously feed algal cells are mainly silicon (Si), nitrogen (N), iron (Fe), and phosphorus (P). These nutrients do not only need to be introduced to the culture systems; they also need to be continuously circulated within the culture to effectively reach all of the algal cells.

21.6 Algal Biofuel Costs of Supply

Figure 21.5 shows the components of the supply-chain-cost-associated algal-derived biofuels. In 2016, the U.S. Department of Energy (DOE) calculated the feedstock cost of algal-derived biofuel at about $3.67/L (Barry et al. 2016). Almost 80% of the total cost of supply is associated with feedstock. The delivered price of produced biofuel in the U.S.A. is offset to a small degree (~2%) by additional revenue streams associated with the sale of recycled/recovered heat, energy, and coproducts (Barry et al. 2016). The biofuels produced and sold in the U.S.A. in recent years are mainly sourced from soybean, cellulosic biomass, and corn feedstocks.

Figure 21.6 compares the costs of biomass production for biofuel feedstock in lentic algal systems with that of terrestrial crop-based biofuel feedstocks. Lentic systems are clearly more expensive than terrestrial crop-based systems. Lentic projects involve high capital costs and higher operating costs related to the workforce and direct energy consumption to cool, dry, and centrifuge the biomass. The photobioreactor-algal system (Arizona) is tenfold costlier than first-generation, terrestrial crop-based biofuel feedstocks. It is also twofold more expensive than the other algal systems considered in Fig. 21.6. Switchgrass is associated with the low-

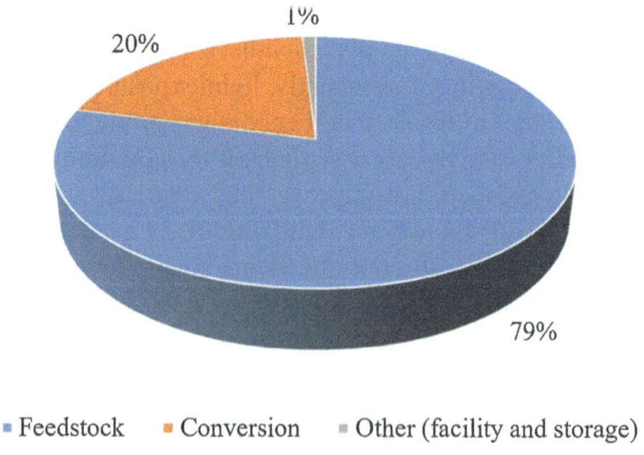

Fig. 21.5 Biofuel cost of supply components associated with production from algal biomass feedstock (after Chowdhury, 2019)

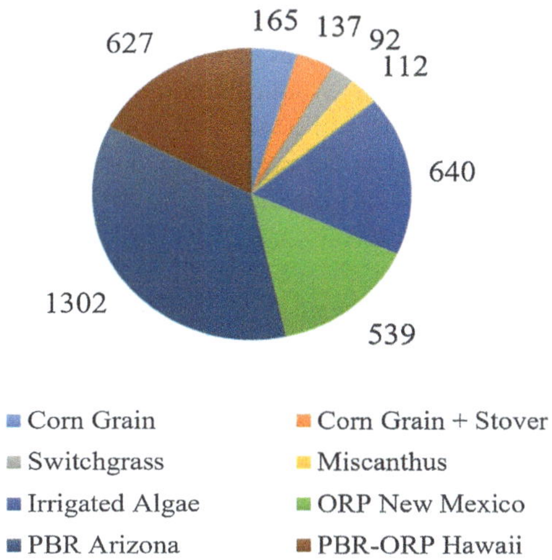

Fig. 21.6 Biomass generation expenses for biofuel feedstocks based on two scenarios: terrestrial crop-based and algal-based production (data from Christiansen et al. 2018) (unit in figure is $ per Mg (megagram); ORP is open raceway pond; PBR is photobioreactors)

est cost of production of the systems considered in Fig. 21.6. This is because switchgrass (a second-generation biofuel feedstock) can be grown on cheaper-to-rent low-grade land and has low cultivation costs because of its perennial nature. Miscanthus grass has similar attributes to switchgrass, and although it is slightly more expensive to grow than switchgrass, it has substantially higher biofuel yields (Scagline-Mellor et al. 2018).

Bioethanol produced from cellulosic feedstock (i.e., second-generation biofuel) costs between 0.62 and 1.00 $ per liter. That cost range was comparable with the cost of biodiesel produced from soya bean, which was about $0.94/L in 2018 (Center for Agricultural and Rural Development, Iowa State University, U.S.A., CARD 2018). However, the cost of bioethanol production from corn was much lower, between 0.2 and 0.32 $ per liter (CARD 2018). A 2017 U.S. Government National Renewable Energy Laboratory (NREL) study (Theis 2017) suggested that second-generation biofuels could be produced on a commercial basis at about $0.57/L. On the other hand, biodiesel produced from algal biomass in recent years has required much higher costs of supply (Langholtz et al. 2016) of about $4.67–4.91 per liter (Barry et al. 2016). This makes algal biofuel currently more expensive to produce than other biofuels. These cost of supply comparisons suggest that algal biofuel technologies require about a tenfold reduction in their costs of supply to become commercially competitive with first- and second-generation biofuels. Much more research and development effort are required to achieve that.

21.7 Conclusions

Biofuels offer the opportunity to provide sustainable supplies of fuels for decades to come and mitigate the negative environmental impact of fossil fuels. However, first-generation biofuels from food crops and second-generation biofuels from cellulosic materials and waste come with their own adverse impact. Excessive land use, competition with food crops, deforestation impact of crop-derived biofuels, and production inefficiencies from cellulosic materials limit their long-term sustainability. On the other hand, third-generation biofuels produced from microalgae offer the potential to overcome these problems. The high lipid yields and growth rates of some microalgae mean that they can be developed with small land footprints. They also have the potential to use cheap sources of growth media fluids, such as industrial wastewater, and cheap sources of carbon dioxide, plus nitrogen-rich and phosphorus-rich nutrients, contained in effluents and emissions from some industrial plants. The use of such waste materials has the potential to reduce costs but also brings some challenges in terms of maintaining stable and sustainable growth in microalgal cultures. Microalgae can be grown at commercial scales in open ponds or closed systems, such as tubular photobioreactors. The former are cheaper to build and operate at larger scales, but the latter are more controllable and lead to more reliable and sustainable microalgal growth and biofuel production. The major hurdle for microalgal biofuel technologies is overcoming scale-up challenges related to their overall cost of supply, which currently is about ten times that of food crop-derived bioethanol and about six times that of crop-derived biodiesel. In order for algal biofuels to achieve their full potential and substitute for fossil fuels and first-generation biofuels, much future investment, research, and development are required. New, more resilient high-yield microalgae strains need to be developed. Fourth-generation biofuel technologies, involving genetically engineered modifications to certain algal strains, are likely to deliver more efficient and sustainable biofuel production. Careful selection of sites for large-scale algal biofuel production is also essential. Such sites need to be located close to their key input resources, minimizing transport and other supply chain costs, as well as providing favorable light, temperature, and air-circulation conditions. Further economies of scale also need to be achieved in the harvesting, drying, extraction, conversion, and blending stages of the algal biofuel production cycle. Such improvements would help it on its way to becoming commercially viable from large-scale production sites, and ultimately become competitive with first-generation biofuels.

References

Alam F, Mobin S, Chowdhury H (2015) Third generation biofuel from algae. Procedia Eng 105:763–768. https://doi.org/10.1016/j.proeng.2015.05.068

Barry A, Wolfe A, English C, Ruddick C, Lambert D (2016) 2016 National Algal Biofuels Technology Review (no. DOE/EE-1409). U.S. DOE Office of Energy Efficiency and Renewable Energy (EERE), Bioenergy Technologies Office (EE-3B). https://doi.org/10.2172/1259407

Brennan L, Owende P (2010) Biofuels from microalgae—a review of technologies for production, processing, and extractions of biofuels and co-products. Renew Sust Energ Rev 14(2):557–577. https://doi.org/10.1016/j.rser.2009.10.009

CARD (2018) C.f.A.a.R.D. Prices for ethanol, corn, and natural gas. [cited 31.05.18]. www.card.iastate.edu

Chaumont D (1993) Biotechnology of algal biomass production: a review of systems for outdoor mass culture. J Appl Phycol 5(6):593–604. https://doi.org/10.1007/BF02184638

Chisti Y (2007) Biodiesel from microalgae. Biotechnol Adv 25(3):294–306. https://doi.org/10.1016/j.biotechadv.2007.02.001

Choubineh A, Ghorbani H, Wood DA, Moosavi SR, Khalafi E, Sadatshojaei E (2017) Improved predictions of wellhead choke liquid critical-flow rates: modelling based on hybrid neural network training learning-based optimization. Fuel 207:547–560. https://doi.org/10.1016/j.fuel.2017.06.131

Chowdhury H, Loganathan B, Mustary I, Alam F, Mobin SM (2019) Algae for biofuels: the third generation of feedstock. In: Second and third generation of feedstocks. Elsevier, p 323–344. https://doi.org/10.1016/B978-0-12-815162-4.00012-4

Christiansen K, Raman DR, Hu G, Anex R (2018) First-order estimates of the costs, input-output energy analysis, and energy returns on investment of conventional and emerging biofuels feedstocks. Biofuel Res J 5(4):894. https://doi.org/10.18331/BRJ2018.5.4.4

Dragone G, Fernandes BD, Vicente AA, Teixeira JA (2010) Third generation biofuels from microalgae. In: Méndez-Vilas A (ed) Current research, technology and education topics in applied microbiology and microbial biotechnology. Formatex Research Center, Spain

Gavrilescu M, Chisti Y (2005) Biotechnology—a sustainable alternative for chemical industry. Biotechnol Adv 23(7–8):471–499. https://doi.org/10.1016/j.biotechadv.2005.03.004

Grima EM, Belarbi EH, Fernández FA, Medina AR, Chisti Y (2003) Recovery of microalgal biomass and metabolites: process options and economics. Biotechnol Adv 20(7–8):491–515. https://doi.org/10.1016/S0734-9750(02)00050-2

Hoffmann JP (1998) Wastewater treatment with suspended and nonsuspended algae. J Phycol 34(5):757–763. https://doi.org/10.1046/j.1529-8817.1998.340757.x

Kapdan IK, Kargi F (2006) Bio-hydrogen production from waste materials. Enzyme Microb Tech 38(5):569–582. https://doi.org/10.1016/j.enzmictec.2005.09.015

Koh LP, Ghazoul J (2008) Biofuels, biodiversity, and people: understanding the conflicts and finding opportunities. Biol Conserv 141(10):2450–2460. https://doi.org/10.1016/j.biocon.2008.08.005

Langholtz MH, Stokes BJ, Eaton LM (2016) Billion-ton report: advancing domestic resources for a thriving bioeconomy, Volume 1: Economic availability of feedstock. Oak Ridge National Laboratory, Oak Ridge, Tennessee, managed by UT-Battelle, LLC for the US Department of Energy, pp 1–411

Lardon L, Hélias A, Sialve B, Steyer JP, Bernard O (2009) Life-cycle assessment of biodiesel production from microalgae. Environ Sci Technol 43:6475–6481. https://doi.org/10.1021/es900705j

Li Y, Horsman M, Wu N, Lan CQ, Dubois-Calero N (2008) Biofuels from microalgae. Biotechnol Prog 24(4):815–820. https://doi.org/10.1021/bp070371k

Lü J, Sheahan C, Fu P (2011) Metabolic engineering of algae for fourth generation biofuels production. Energy Environ Sci 4(7):2451–2466. https://doi.org/10.1039/C0EE00593B

Mata TM, Martins AA, Caetano NS (2010) Microalgae for biodiesel production and other applications: a review. Renew Sust Energ Rev 14(1):217–232. https://doi.org/10.1016/j.rser.2009.07.020

Mobin S, Alam F (2014) Biofuel production from algae utilizing wastewater. In: 19th Australasian fluid mechanics conference, p 1–7

Mobin SM, Alam F (2018) A review of microalgal biofuels, challenges and future directions. In: Application of thermo-fluid processes in energy systems. Springer, Singapore, pp 83–108. https://doi.org/10.1007/978-981-10-0697-5_4

Nigam PS, Singh A (2011) Production of liquid biofuels from renewable resources. Prog Energy Combust Sci 37(1):52–68. https://doi.org/10.1016/j.pecs.2010.01.003

Pulz O (2001) Photobioreactors: production systems for phototrophic microorganisms. Appl Microbiol Biotechnol 57(3):287–293. https://doi.org/10.1007/s002530100702

Sadatshojaei E, Jamialahmadi M, Esmaeilzadeh F, Ghazanfari MH (2016) Effects of low-salinity water coupled with silica nanoparticles on wettability alteration of dolomite at reservoir temperature. Pet Sci Technol 34(15):1345–1351. https://doi.org/10.1080/10916466.2016.1204316

Sadatshojaei E, Esmaeilzadeh F, Fathikaljahi J, Barzi SEH, Wood DA (2018) Regeneration of the midrex reformer catalysts using supercritical carbon dioxide. Chem Eng J 343:748–758. https://doi.org/10.1016/j.cej.2018.02.038

Sadatshojaei E, Jamialahmadi M, Esmaeilzadeh F, Wood DA, Ghazanfari MH (2019) The impacts of silica nanoparticles coupled with low-salinity water on wettability and interfacial tension: experiments on a carbonate core. J Dispers Sci Technol 1–15. https://doi.org/10.1080/01932691.2019.1614943

Scagline-Mellor S, Griggs T, Skousen J, Wolfrum E, Holaskova I (2018) Switchgrass and giant miscanthus biomass and theoretical ethanol production from reclaimed mine lands. Bioenergy Res 11(3):562–573. https://doi.org/10.1007/s12155-018-9915-2

Schenk PM, Thomas-Hall SR, Stephens E, Marx UC, Mussgnug JH, Posten C et al (2008) Second generation biofuels: high-efficiency microalgae for biodiesel production. Bioenergy Res 1(1):20–43. https://doi.org/10.1007/s12155-008-9008-8

Scott SA, Davey MP, Dennis JS, Horst I, Howe CJ, Lea-Smith DJ, Smith AG (2010) Biodiesel from algae: challenges and prospects. Curr Opin Biotechnol 21(3):277–286. https://doi.org/10.1016/j.copbio.2010.03.005

Slade R, Bauen A (2013) Micro-algae cultivation for biofuels: cost, energy balance, environmental impacts and future prospects. Biomass Bioenergy 53:29–38. https://doi.org/10.1016/j.biombioe.2012.12.019

Spolaore P, Joannis-Cassan C, Duran E, Isambert A (2006) Commercial applications of microalgae. J Biosci Bioeng 101(2):87–96. https://doi.org/10.1263/jbb.101.87

Stave JW (1999) Detection of new or modified proteins in novel foods derived from GMO–future needs. Food Control 10(6):367–374. https://doi.org/10.1016/S0956-7135(99)00077-8

Theis K (2017) At $2.15 a gallon, cellulosic ethanol could be cost competitive. [cited 31.05.18]. https://www.nrel.gov/continuum/sustainable_transportation/cellulosic_ethanol.html

Wood DA (2019) Microbial improved and enhanced oil recovery (MIEOR): review of a set of technologies diversifying their applications. Adv Geoenergy Res 3(2):122–140. https://doi.org/10.26804/ager.2019.02.02

Zhou W, Chen P, Min M, Ma X, Wang J, Griffith R et al (2014) Environment-enhancing algal biofuel production using wastewaters. Renew Sust Energ Rev 36:256–269. https://doi.org/10.1016/j.rser.2014.04.073

Correction to: Sustainable Green Chemical Processes and their Allied Applications

Inamuddin and Abdullah M. Asiri

Correction to:
Inamuddin, A. M. Asiri (eds.), *Sustainable Green Chemical Processes and their Allied Applications*, Nanotechnology in the Life Sciences, https://doi.org/10.1007/978-3-030-42284-4

This book was inadvertently published without the middle initial of the volume editor Abdullah M. Asiri and the second affiliation of Inamuddin. This has now been corrected throughout the book.

The updated online version of this chapter can be found at
https://doi.org/10.1007/978-3-030-42284-4

Inamuddin, A. M. Asiri (eds.), *Sustainable Green Chemical Processes and their Allied Applications*, Nanotechnology in the Life Sciences,
https://doi.org/10.1007/978-3-030-42284-4_22

Index

Inamuddin, A. M. Asiri (eds.), *Sustainable Green Chemical Processes and their Allied Applications*, Nanotechnology in the Life Sciences,
https://doi.org/10.1007/978-3-030-42284-4

D

H

The manufacturer's authorised representative in the EU is Springer Nature Customer Service Centre GmbH, Europaplatz 3, 69115 Heidelberg, Germany. If you have any concerns regarding our products, please contact ProductSafety@springernature.com

Printed and bound by CPI Group (UK) Ltd, Croydon, CR0 4YY
30/09/2026
02225693-0008